MW00581135

DECISION NEUROSCIENCE

DECISION NEUROSCIENCE

AN INTEGRATIVE PERSPECTIVE

Edited by

JEAN-CLAUDE DREHER

LÉON TREMBLAY

Institute of Cognitive Science (CNRS), Lyon, France

AMSTERDAM • BOSTON • HEIDELBERG • LONDON
NEW YORK • OXFORD • PARIS • SAN DIEGO
SAN FRANCISCO • SINGAPORE • SYDNEY • TOKYO
Academic Press is an imprint of Elsevier

Academic Press is an imprint of Elsevier
125 London Wall, London EC2Y 5AS, United Kingdom
525 B Street, Suite 1800, San Diego, CA 92101-4495, United States
50 Hampshire Street, 5th Floor, Cambridge, MA 02139, United States
The Boulevard, Langford Lane, Kidlington, Oxford OX5 1GB, United Kingdom

Library of Congress Cataloging-in-Publication Data
A catalog record for this book is available from the Library of Congress

British Library Cataloguing-in-Publication Data
A catalogue record for this book is available from the British Library

ISBN: 978-0-12-805308-9

For information on all Academic Press publications
visit our website at https://www.elsevier.com/

www.elsevier.com • www.bookaid.org

Publisher: Mara Conner
Acquisition Editor: April Farr
Editorial Project Manager: Timothy Bennett
Production Project Manager: Edward Taylor
Designer: Matthew Limbert

Typeset by TNQ Books and Journals

Contents

I

ANIMAL STUDIES ON REWARDS, PUNISHMENTS, AND DECISION-MAKING

1. Anatomy and Connectivity of the Reward Circuit

S.N. HABER

2. Electrophysiological Correlates of Reward Processing in Dopamine Neurons

W. SCHULTZ

3. Appetitive and Aversive Systems in the Amygdala

S. BERNARDI AND D. SALZMAN

4. Ventral Striatopallidal Pathways Involved in Appetitive and Aversive Motivational Processes

Y. SAGA AND L. TREMBLAY

5. Reward and Decision Encoding in Basal Ganglia: Insights From Optogenetics and Viral Tracing Studies in Rodents

J. TIAN, N. UCHIDA AND N. ESHEL

II

HUMAN STUDIES ON MOTIVATION, PERCEPTUAL, AND VALUE-BASED DECISION-MAKING

III

SOCIAL DECISION NEUROSCIENCE

V

GENETIC AND HORMONAL INFLUENCES ON MOTIVATION AND SOCIAL BEHAVIOR

List of Contributors

B. Ahmed   University of Oxford, Oxford, United Kingdom

B.W. Balleine   University of Sydney, Camperdown, NSW, Australia

S. Ballesta   Centre National de la Recherche Scientifique, Bron, France; Université Lyon 1, Villeurbanne, France; The University of Arizona, Tucson, AZ, United States

S. Bernardi   Columbia University, New York, New York, United States

A. Blangero   University of Oxford, Oxford, United Kingdom

L.A. Bradfield   University of Sydney, Camperdown, NSW, Australia

W. Chaisangmongkon   King Mongkut's University of Technology Thonburi, Bangkok, Thailand; New York University, New York, NY, United States

G. Chierchia   Max Planck Institute for Human Cognitive and Brain Sciences, Leipzig, Germany

L. Clark   University of British Columbia, Vancouver, BC, Canada

A. Dagher   McGill University, Montreal, QC, Canada

J.W. Dalley   University of Cambridge, Cambridge, United Kingdom

J.A. Diaz   University of Glasgow, Glasgow, United Kingdom

J.-C. Dreher   Institute of Cognitive Science (CNRS), Lyon, France

J.-R. Duhamel   Centre National de la Recherche Scientifique, Bron, France; Université Lyon 1, Villeurbanne, France

N. Eshel   Harvard University, Cambridge, MA, United States

L.K. Fellows   McGill University, Montreal, QC, Canada

R.D. Fernald   Stanford University, Stanford, CA, United States

G. Fernández   Radboud University Medical Centre, Nijmegen, The Netherlands

P.C. Fletcher   University of Cambridge, Cambridge, United Kingdom

J.M. Fuster   University of California Los Angeles, Los Angeles, CA, United States

S. Gherman   University of Glasgow, Glasgow, United Kingdom

P.W. Glimcher   New York University, New York, NY, United States

D.W. Grupe   University of Wisconsin–Madison, Madison, WI, United States

S.N. Haber   University of Rochester School of Medicine, Rochester, NY, United States

J.-E. Han   McGill University, Montreal, QC, Canada

M.J.A.G. Henckens   Radboud University Medical Centre, Nijmegen, The Netherlands

E.J. Hermans   Radboud University Medical Centre, Nijmegen, The Netherlands

K. Izuma   University of York, York, United Kingdom

M. Jazayari   Centre National de la Recherche Scientifique, Bron, France; Université Lyon 1, Villeurbanne, France

M. Joëls   University Medical Center Utrecht, Utrecht, The Netherlands

T. Kahnt   Northwestern University Feinberg School of Medicine, Chicago, IL, United States

K. Krug   University of Oxford, Oxford, United Kingdom

D. Lee   Yale University, New Haven, CT, United States

A. Lefevre   Institut des Sciences Cognitives Marc Jeannerod, UMR 5229, CNRS, Bron, France; Université Claude Bernard Lyon 1, Lyon, France

R. Ligneul   Institute of Cognitive Science (CNRS), Lyon, France

K. Louie   New York University, New York, NY, United States

R.B. Mars   University of Oxford, Oxford, United Kingdom

G.K. Murray   University of Cambridge, Cambridge, United Kingdom

S. Neseliler   McGill University, Montreal, QC, Canada

F.X. Neubert   University of Oxford, Oxford, United Kingdom

M.P. Noonan   University of Oxford, Oxford, United Kingdom

N. Ortner   Medical University of Vienna, Vienna, Austria

S. Palminteri   University College London, London, United Kingdom; Ecole Normale Supérieure, Paris, France

M. Pessiglione   Institut du Cerveau et de la Moelle (ICM), Inserm U1127, Paris, France; Université Pierre et Marie Curie (UPMC-Paris 6), Paris, France

L. Pezawas   Medical University of Vienna, Vienna, Austria

M.G. Philiastides   University of Glasgow, Glasgow, United Kingdom

U. Rabl   Medical University of Vienna, Vienna, Austria

T.W. Robbins   University of Cambridge, Cambridge, United Kingdom

C.C. Ruff   University of Zurich, Zurich, Switzerland

Y. Saga   Institute of Cognitive Sciences (CNRS), Lyon, France

J. Sallet   University of Oxford, Oxford, United Kingdom

D. Salzman   Columbia University, New York, New York, United States; New York State Psychiatric Institute, New York, New York, United States

W. Schultz   University of Cambridge, Cambridge, United Kingdom

H. Seo   Yale University, New Haven, CT, United States

T. Singer   Max Planck Institute for Human Cognitive and Brain Sciences, Leipzig, Germany

A. Sirigu   Institut des Sciences Cognitives Marc Jeannerod, UMR 5229, CNRS, Bron, France; Université Claude Bernard Lyon 1, Lyon, France

J. Smith   University of Oxford, Oxford, United Kingdom

A. Soltani   Dartmouth College, Hanover, NH, United States

C. Summerfield   University of Oxford, Oxford, United Kingdom

J. Tian   Harvard University, Cambridge, MA, United States

P.N. Tobler   University of Zurich, Zurich, Switzerland

L. Tremblay   Institute of Cognitive Science (CNRS), Lyon, France

C. Tudor-Sfetea   University of Cambridge, Cambridge, United Kingdom

N. Uchida   Harvard University, Cambridge, MA, United States

G. Ugazio   University of Zurich, Zurich, Switzerland

A.R. Vaidya   McGill University, Montreal, QC, Canada

V. Voon   University of Cambridge, Cambridge, United Kingdom; Cambridgeshire and Peterborough NHS Foundation Trust, Cambridge, United Kingdom

X.-J. Wang   New York University, New York, NY, United States; NYU Shanghai, Shanghai, China

K. Witt   Christian Albrecht University, Kiel, Germany

Preface

Decision Neuroscience: an Integrative Perspective addresses fundamental questions about how the brain makes perceptual, value-based, and more complex decisions in nonsocial and social contexts. This book presents recent and compelling neuroimaging, electrophysiological, lesional, and neurocomputational studies, in combination with hormonal and genetic studies, that have led to a clearer understanding of the neural mechanisms behind how the brain makes decisions. The neural mechanisms underlying decision-making processes are of critical interest to scientists because of the fundamental role that reward plays in a number of cognitive processes (such as motivation, action selection, and learning) and because they have theoretical and clinical implications for understanding dysfunctions of major neurological and psychiatric disorders.

The idea for this book grew up from our edition of the *Handbook of Reward and Decision Making* (Academic Press, 2009). We originally thought to revise and reedit this book, addressing one fundamental question about the nature of behavior: how does the brain process reward and makes decisions when facing multiple options? However, given the developments in this active area of research, we decided to feature an entirely different book with new contents, covering results on the neural substrates of rewards and punishments; perceptual, value-based, and social decision-making; clinical aspects such as behavioral addictions; and the roles of genes and hormones in these various aspects. For example, an exciting topic from the field of social neuroscience is to know whether the neural structures engaged with various forms of social interactions are cause or consequence of these interactions (Fernald, Chapter 28).

A mechanistic understanding of the neural encoding underlying decision-making processes is of great interest to a broad readership because of their theoretical and clinical implications. Findings in this research field are also important to basic neuroscientists interested in how the brain reaches decisions, cognitive psychologists working on decision-making, as well as computational neuroscientists studying probabilistic models of brain functions. Decision-making covers a wide range of topics and levels of analysis, from molecular mechanisms to neural systems dynamics, neurocomputational models, and social system levels. The contributions to this book

are forward-looking assessments of the current and future issues faced by researchers. We were fortunate to assemble an outstanding collection of experts who addressed various aspects of decision-making processes. The book is divided into five parts that address distinct but interrelated topics.

STRUCTURE OF THE BOOK

A decision neuroscience perspective requires multiple levels of analyses spanning neuroimaging, electrophysiological, behavioral, and pharmacological techniques, in combination with molecular and genetic tools. These approaches have begun to build a mechanistic understanding of individual and social decision-making. This book highlights some of these advancements that have led to the current understanding of the neuronal mechanisms underlying motivational and decision-making processes.

Part I is devoted to animal studies (anatomical, neurophysiological, pharmacological, and optogenetics) on rewards/punishments and decision-making. In their natural environment, animals face a multitude of stimuli, very few of which are likely to be useful as predictors of reward or punishment. It is thus crucial that the brain learns to predict rewards, providing a critical evolutionary advantage for survival. This first part of the book offers a comprehensive view of the specific contributions of various brain structures as the dopaminergic midbrain neurons, the amygdala, the ventral striatum, and the prefrontal cortex, including the lateral prefrontal cortex and the orbitofrontal cortex, to the component processes underlying reinforcement-guided decision-making, such as the representation of instructions, expectations, and outcomes; the updating of action values; and the evaluation process guiding choices between prospective rewards. Special emphasis is made on the neuroanatomy of the reward system and the fundamental roles of dopaminergic neurons and the basal ganglia in learning stimulus–reward associations.

Chapter 1 (Haber SN) describes the anatomy and connectivity of the reward circuit in nonhuman primates. It describes how cortical–basal ganglia loops are topographically organized and the key areas of convergence between functional regions.

Chapter 2 describes three novel electrophysiological properties of the classical dopamine reward-prediction error (RPE) signal (Schultz W). Studies have identified three novel properties of the dopamine RPE signal. In particular, concerning its roles in making choices, the dopamine RPE signal may not only reflect subjective reward value and formal economic utility but could also fit into formal competitive decision models. The RPE signal may code the chosen value suitable for updating or immediately influencing object and action values. Thus, the dopamine utility prediction error signal bridges the gap between animal learning theory and economic decision theory.

Chapter 3 focuses on the electrophysiological properties of another important component of the reward system in primates, namely the amygdala (Bernardi S and Salzman D). The amygdala contains distinct appetitive and aversive networks of neurons. Processing in these two amygdalar networks can both regulate and be regulated by diverse cognitive operations.

Chapter 4 extends the concept of appetitive and aversive motivational processes to the striatum (Saga Y and Tremblay L). This chapter describes how the ventral striatum and the ventral pallidum, two parts of the limbic circuit in the basal ganglia, are involved not only in appetitive rewarding behavior, as classically believed, but also in negative motivational behavior. These results can be linked with the control of approach/avoidance behavior in a normal context and with the expression of anxiety-related disorders. The disturbance of this pathway may induce not only psychiatric symptoms, but also abnormal value-based decision-making.

Chapter 5 (Tian J, Uchida N, and Eshel N) highlights new advances in the physiology, function, and circuit mechanism of decision-making, focusing especially on the involvement of dopamine and striatal neurons. Using optogenetics in rodents, molecular techniques, and genetic techniques, this chapter shows how these tools have been used to dissect the circuits underlying decision-making. It describes exciting new avenues to understand a circuit, by recording from neurons with knowledge of their cell type and patterns of connectivity. Furthermore, the ability to manipulate the activity of specific neural types provides an important means to test hypotheses of circuit function.

Chapter 6 (Bradfield L and Balleine B) describes the neural bases of the learning and motivational processes controlling goal-directed action. By definition, the performance of such action respects both the current value of its outcome and the extant contingency between that action and its outcome. This chapter identifies the neural circuits mediating distinct processes, including the acquisition of action-outcome contingencies, the

encoding and retrieval or incentive value, the matching of that value to specific outcome representations, and finally the integration of this information for action selection. It also shows how each of these individual processes are integrated within the striatum for successful goal-directed action selection.

Chapter 7 (Robbins TW and Dalley JW) describes animal models (mostly in rodents) of impulsivity and risky choices. It reviews the neural and neurochemical basis of various forms of impulsive behavior by distinguishing three main forms of impulsivity: waiting impulsivity, risky choice impulsivity, and stopping impulsivity. It shows that dopamine- and serotonin-dependent functions of the nucleus accumbens are implicated in waiting impulsivity and risky choice impulsivity, as well as cortical structures projecting to the nucleus accumbens. For stopping impulsivity, dopamine-dependent functions of the dorsal striatum are implicated, as well as circuitry including the orbitofrontal cortex and dorsal prelimbic cortex. Differences and commonalities between the forms of impulsive responding are highlighted. Importantly, various applications to human neuropsychiatric disorders such as drug addiction and attention deficit hyperactivity disorder are also discussed.

Chapter 8 (Fuster JM) proposes that the neural mechanisms of decision-making are understandable only in the structural and dynamic context of the perception—action cycle, defined as the biocybernetic processing of information that adapts the organism to its environment. It presents a general view of the role of the prefrontal cortex in decision-making, in the general framework of the perception—action cycle, including prediction, preparation toward decision, execution, and feedback from decision.

Part II covers the topic of the neural representation of motivation, perceptual decision-making, and value-based decision-making in humans, mostly combining neurocomputational models and brain imaging studies.

Chapter 9 (Tobler P and Kahnt T) reviews several definitions of value and salience, and describes human neuroimaging studies that dissociate these variables. Value increases with the magnitude and probability of reward but decreases with the magnitude and probability of punishment, whereas salience increases with the magnitude and probability of both reward and punishment. At the neural level, value signals arise in striatum, orbitofrontal and ventromedial prefrontal cortex, and superior parietal areas, whereas magnitude-based salience signals arise in the anterior cingulate cortex and the inferior parietal cortex. By contrast, probability-based salience signals have been found in the ventromedial prefrontal cortex.

Chapter 10 (Louie K and Glimcher PW) reviews an approach centered on basic computations underlying neural value coding. It proposes that neural information processing in valuation and choice relies on computational principles such as contextual modulation and divisive normalization. Divisive normalization is a nonlinear gain control algorithm widely observed in multiple sensory modalities and brain regions. Identification of these computations sheds light on how the underlying neural circuits are organized, and neural activity dynamics provides a link between biological mechanism and computations.

Chapter 11 (Philiastides M, Diaz J, and Gherman S) introduces the general principles guiding perceptual decision-making. Perceptual decisions occur when perceptual inputs are integrated and converted to form a categorical choice. It reviews the influence of a number of factors that interact and contribute to the decision process, such as prestimulus state, reward and punishment, speed—accuracy trade-off, learning and training, confidence, and neuromodulation. It shows how these decision modulators can exert their influence at various stages of processing, in line with predictions derived from sequential-sampling models of decision-making.

Chapter 12 (Summerfield C) reviews the neural and computational mechanisms of perceptual decisions. It addresses current controversial questions, such as how we decide when to draw our decisions to a conclusion, and how perceptual decisions are biased by prior information.

Chapter 13 (Soltani A, Chaisangmongkon W, and Wang XJ) presents possible biophysical and circuit mechanisms of valuation and reward-dependent plasticity underlying adaptive choice behavior. It reviews mathematical models of reward-dependent adaptive choice behavior, and proposes a biologically plausible, reward-modulated Hebbian synaptic plasticity rule. It shows that a decision-making neural circuit endowed with this learning rule is capable of accounting for behavioral and neurophysiological observations in a variety of decision-making tasks.

Part III of the book focuses on the rapidly developing field of social neuroscience, integrating neuroscience data from both nonhuman primates and humans. Primates are fundamentally social animals, and they may share common neural mechanisms in diverse forms of social behavior. Examples of such behavior include tracking intentions and beliefs from others, being observed by others during prosocial decisions, or learning the social hierarchy in a group of individuals. It is also likely that at the macroscopic level, important differences exist concerning social brain structures and connectivity, and there is a need to directly compare between species to answer this fundamental question. Indeed, studies in both humans and monkeys report not only an increase in gray matter density of specific brain

structures relative to the size of our social network, but also species differences in prefrontal—temporal brain connectivity. Furthermore, this part of the book presents neurocomputational approaches starting to provide a mechanistic understanding of social decisions. For example, reinforcement learning models and strategic reasoning models can be used when learning social hierarchies or during social interactions.

A social neuroscience understanding requires multiple approaches, such as electrophysiology and neuroimaging in both monkeys (Chapters 14, 15, 19) and humans (Chapters 16, 18, 20), as well as causal (Chapter 21), neurocomputational (Chapters 17—19), endocrinological, genetics, and clinical approaches (Part V).

Chapter 14 (Duhamel JR and colleagues) presents monkey electrophysiological data revealing that the orbitofrontal cortex is tuned to social information. For example, in one experiment, macaque monkeys worked to collect rewards for themselves and two monkey partners. Single neurons encoded the meaning of visual cues that predicted the magnitude of future rewards, the motivational value of rewards obtained in a social context, and the tracking of social preferences and partner's identity and social rank. The orbitofrontal cortex thus contains key neuronal mechanisms for the evaluation of social information. Moreover, macaque monkeys take into account the welfare of their peers when making behavioral choices bringing about pleasant or unpleasant outcomes to a monkey partner. Thus, this chapter reveals that prosocial decision-making is sustained by an intrinsic motivation for social affiliation and controlled through positive and negative vicarious reinforcements.

Chapter 15 (Sallet J and colleagues) reviews the similarities between monkeys and humans in the organization of the social brain. Using MRI-based connectivity methods, they compare human and macaque social areas, such as the organization of the medial prefrontal cortex. They revealed that the connectivity fingerprint of macaque area 10 best matched that of the human frontal pole, suggesting that even high-level areas share features between species. They also showed that animals housed in large social groups had more gray matter volume in bilateral mid-superior temporal sulcus and rostral prefrontal cortex. Beyond species similarities, there are also distinct differences between human and macaque prefrontal—temporal brain connectivity. For example, functional connections between the temporal cortex and the lateral prefrontal cortex are stronger in humans compared to connections with the medial prefrontal cortex in humans, but the opposite pattern is observed in macaques.

Chapter 16 (Izuma K) focuses on two forms of social influence, the audience effect, which is an increased prosocial tendency in front of other people, and social conformity, which consists in adjusting one's attitude or behavior to those of a group. This chapter discusses fMRI findings in healthy humans in these two types of social influence and also shows how reputation processing is impaired in individuals with autism. It also links social conformity and reward-based learning (reinforcement learning).

Chapter 17 (Ligneul R and Dreher JC) examines how the brain learns social dominance hierarchies. Social dominance refers to relationships wherein the goals of one individual prevail over the goals of another individual in a systematic manner. Dominance hierarchies have emerged as a major evolutionary force to drive dyadic asymmetries in a social group. This chapter proposes that the emergence of dominance relationships are learned incrementally, by accumulating positive and negative competitive feedbacks associated with specific individuals and other members of the social group. It considers such emergence of social dominance as a reinforcement learning problem inspired by neurocomputational approaches traditionally applied to nonsocial cognition. This chapter also reports how dominance hierarchies induce changes in specific brain systems, and it reviews the literature on interindividual differences in the appraisal of social hierarchies, as well as the underlying modulations of cortisol, testosterone, and serotonin/dopamine systems, which mediate these phenomena.

Chapter 18 (Seo H and Lee D) describes reinforcement learning models and strategic reasoning during social decision-making. It shows that dynamic changes in choices and decision-making strategies can be accounted for by reinforcement learning in a variety of contexts. This framework has also been successfully adopted in a large number of neurobiological studies to characterize the functions of multiple cortical areas and basal ganglia. For complex decision-making, including social interactions, this chapter shows that multiple learning algorithms may operate in parallel.

Chapter 19 (Ugazio G and Ruff C) reports brain stimulation studies on social decision-making, which test the causal relationship between neural activity and different types of processes underlying these decisions, including social emotions, social cognition, and social behavioral control.

Chapter 20 (Chierchia G and Singer T) shows that two important social emotions, empathy and compassion, engage distinct neurobiological mechanisms, as well as different affective and motivational states. Empathy for pain engages a network including the anterior insula and anterior midcingulate cortex, areas associated with negative affect; compassionate states engage the medial orbitofrontal cortex and ventral striatum and are associated with feelings of warmth, concern, and positive affect.

Part IV of the book focuses on clinical aspects involving disorders of decision-making and of the reward system that link together basic research areas, including systems, cognitive, and clinical neuroscience. Dysfunction of the reward system and decision-making is present in a number of neurological and psychiatric disorders, such as Parkinson's disease, schizophrenia, drug addiction, and focal brain lesions. The study of pathological gambling, for example, and other motivated states associated with, and leading to, compulsive behavior provides an opportunity to learn about the dysfunctions of reward system activity, independent of direct pharmacological activation of brain reward circuits. On the other hand, because drugs of abuse directly activate brain systems, they provide a unique challenge in understanding how pharmacological activation influences reward mechanisms leading to persistent compulsive behavior.

Chapter 21 (Murray GK, Tudor-Sfetea C, and Fletcher PC) shows that principles of reinforcement learning are useful to understand the neural mechanisms underlying impaired learning, reward, and motivational processes in schizophrenia. Two symptoms characteristic of this disease is considered in this framework, namely delusions and anhedonia.

Chapter 22 (Vaidya AR and Fellows LK) takes a neuropsychological approach to review focal frontal lobe damage effects on value-based decisions. It reveals the necessary contributions of specific subregions (ventromedial, lateral, and dorsomedial prefrontal cortex) to decision-making, and provides evidence as to the dissociability of component processes. It argues that the ventromedial frontal lobe is required for optimal learning from reward under dynamic conditions and contributes to specific aspects of value-based decision-making. It also shows a necessary contribution of the dorsomedial frontal lobe in representing action-value expectations.

Chapter 23 (Palminteri S and Pessiglione M) reviews reinforcement learning models applied to reward and punishment learning. These studies include fMRI and neural perturbation following drug administration and/or pathological conditions. They propose that distinct brain systems are engaged, one in reward learning (midbrain dopaminergic nuclei and ventral prefrontostriatal circuits) and another in punishment learning, revolving around the anterior insula.

Chapter 24 (Voon V) discusses decision-making impairments and impulse control disorders in Parkinson's disease. The author reports enhancement of the gain associated with levodopa, reinforcing properties of dopaminergic medications, and enhancement of delay discounting in these patients. Lower striatal dopamine transporter levels preceding medication exposure, and decreased midbrain D2 autoreceptor sensitivity, may underlie enhanced ventral striatal dopamine release and activity in response to salient reward cues, anticipated and unexpected rewards, and gambling tasks.

Impairments in decisional impulsivity (delay discounting, reflection impulsivity, and risk taking) implicate the ventral striatum, orbitofrontal cortex, anterior insula, and dorsal cingulate. These findings provide insight into the role of dopamine in decision-making processes in addiction and suggest potential therapeutic targets.

Chapter 25 (Witt K) reports that motor control is the result of a balance between activation and inhibition of movement patterns. It points to a central role of the subthalamic nucleus within the indirect basal ganglia pathway, acting as a brake on the motor system. This subthalamic nucleus function occurs when an automatic response must be suppressed to have more time to choose between alternative responses.

Chapter 26 (Grupe DW) discusses value-based decision-making as one of a key behavioral symptoms present in anxiety disorders. This chapter highlights alterations to specific processes: decision representation, valuation, action selection, outcome evaluation, and learning. Distinct anxious phenotypes may be characterized by differential alterations to these processes and their associated neurobiological mechanisms.

Chapter 27 (Clark L) presents a conceptualization of disordered gambling as a behavioral addiction driven by an exaggeration of multiple psychological distortions that are characteristic of human decision-making, and underpinned by neural circuitry subserving appetitive behavior, reinforcement learning, and choice selection. The chapter discusses the neurobiological basis of pathological gambling behavior in loss aversion, probability weighting, perceptions of randomness, and the illusion of control.

Part V focuses on the roles of hormones and genes involved in motivation and social decision-making processes. The combination of molecular genetic, endocrinology, and neuroimaging has provided a considerable amount of data that help in the understanding of the biological mechanisms influencing decision processes. These studies have demonstrated that genetic and hormonal variations have an impact on the physiological response of the decision-making system. These variations may account for some of the inter- and intra-individual behavioral differences observed in social cognition.

Chapter 28 (Fernald RD) presents an original approach for cognitive neuroscientists by focusing on the difficult question of how an animal's behavior or perception of its social and physical surroundings shapes its brain. Using a fish model system that depends on complex social interactions, this chapter reports how the social context influences the brain and, in turn, alters the behavior and neural circuitry of animals as they interact. Gathering of social information vicariously produces rapid changes in gene expression in key brain nuclei and

these genomic responses prepare the individual to modify its behavior to move into a different social niche. Both social success and failure produce changes in neuronal cell size and connectivity in key brain nuclei. This approach bridges the gap between social information gathering from the environment and the levels of cellular and molecular responses.

Chapter 29 (Rabl U, Ortner N, and Pezawas L) examines the use of imaging genetics to explore the relationships between major depressive disorder and decision-making.

Chapters 30–32 report neuroendocrinological findings in social decision-making, likening variations in the levels of different types of hormones (cortisol, oxytocin, ghrelin/leptin) to brain systems engaged in social decisions and food choices. Chapter 30 (Hermans EJ and colleagues) integrates knowledge of the effects of stress at the neuroendocrine, cellular, brain systems, and behavioral levels to quantify how stress-related neuromodulators trigger time-dependent shifts in the balance between two brain systems: a "salience" network, which supports rapid but rigid decisions, and an "executive control" network, which supports flexible, elaborate decisions. This simple model elucidates paradoxical findings reported in human studies on stress and cognition.

Chapter 31 (Lefevre A and Sirigu A) reviews evidence for a role for oxytocin in individual and social decision-making. It discusses animal and human studies to link the behavioral effects of oxytocin to its underlying neurophysiological mechanisms.

Chapter 32 (Dagher A, Neseliler S, and Han JE) examines the neurobehavioral factors that determine food choices and food intake. It reviews findings on the interactions between brain systems that mediate feeding behavior and the gut and adipose peptides that signal the current state of energy balance.

Chapter 33 (Dreher, Tremblay, and Schultz) concludes this decision neuroscience book by integrating perspectives from all contributors.

We anticipate that while some readers may read the volume from the first to the last chapter, other readers may read only one or more chapters at a time, and not necessarily in the order presented in the book. This is why we encouraged an organization of this volume whereby each chapter can stand alone, while making references to others and minimizing redundancies across the volume. Given the consistent acceleration of advances in the various approaches described in this book on decision neuroscience, you are about to be dazzled by a first look at the new stages of an exciting era in brain research. Enjoy!

Jean-Claude Dreher
Léon Tremblay

ANIMAL STUDIES ON REWARDS, PUNISHMENTS, AND DECISION-MAKING

1

Anatomy and Connectivity of the Reward Circuit

S.N. Haber

University of Rochester School of Medicine, Rochester, NY, United States

Abstract

While cells in many brain regions are responsive to reward, the cortical–basal ganglia circuit is at the heart of the reward system. The key structures in this network are the anterior cingulate cortex, the orbital prefrontal cortex, the ventral striatum, the ventral pallidum, and the midbrain dopamine neurons. In addition, other structures, including the dorsal prefrontal cortex, amygdala, thalamus, and lateral habenular nucleus, are key components in regulating the reward circuit. Connectivity between these areas forms a complex neural network that is topographically organized, thus maintaining functional continuity through the corticobasal ganglia pathway. However, the reward circuit does not work in isolation. The network also contains specific regions in which convergent pathways provide an anatomical substrate for integration across functional domains.

INTRODUCTION

The reward circuit is a complex neural network that underlies the ability to effectively assess the likely outcomes of different choices. A key component to good decision-making and appropriate goal-directed behaviors is the ability to accurately evaluate reward value, predictability, and risk. While the hypothalamus is central for processing information about basic, or primary, rewards higher cortical and subcortical forebrain structures are engaged when complex choices about these fundamental needs are required. Moreover, choices often involve secondary rewards, such as money, power, challenge, etc., that are more abstract (compared to primary needs), and not as dependent on direct sensory stimulation. Although cells that respond to various aspects of reward such as anticipation, value, etc., are found throughout the brain, at the center of this neural network is the ventral corticobasal ganglia circuit. The basal ganglia (BG) are traditionally considered to process information in parallel and segregated functional streams consisting of reward processing, cognition, and motor control areas [1]. Moreover, within the ventral BG, there are microcircuits thought to be associated with various aspects of reward processing. However, a key component for learning and adaptation of goal-directed behaviors is the ability not only to evaluate various aspects of reward but also develop appropriate action plans and inhibit maladaptive choices on the basis of previous experience. This requires integration between various aspects of reward processing as well as interaction between reward circuits and brain regions involved in cognition. Thus, while parallel processing provides throughput channels by which specific actions can be expressed while others are inhibited, the BG also plays a central role in learning new procedures and associations, implying the necessity for integrative processing across circuits. Indeed, we now know that the network contains multiple regions in which integration across circuits occurs [2–8]. Therefore, while the ventral BG network is at the heart of reward processing, it does not work in isolation. This chapter addresses not only the connectivities within this circuit, but also how this circuit anatomically interfaces with other BG circuits. Reward and aversive processes work together in learning and decision-making. Aversive processing associated with punishment and negative outcomes is addressed by other authors in this book.

The frontal–BG network, in general, mediates all aspects of action planning, including reward and motivation, cognition, and motor control. However, specific regions within this network play a unique role in various aspects of reward processing and evaluation of outcomes, including reward value, anticipation, predictability, and risk. The key structures are prefrontal areas [anterior cingulate cortex (ACC) and orbital prefrontal cortex (OFC)], the ventral striatum (VS), the ventral pallidum (VP), and the midbrain dopamine (DA) neurons. The ACC and OFC prefrontal areas mediate different aspects of reward-based behaviors, error prediction, value, and the choice between short- and long-term gains. Cells in the VS and VP respond to anticipation of reward and reward detection (see Chapter 4). Reward-prediction and error-detection signals are generated, in part, from the midbrain DA cells (see Schultz in this volume). While the VS and the ventral tegmental area (VTA) DA neurons are the BG areas most commonly associated with reward, reward-responsive activation is not restricted to these, but found throughout the striatum and substantia nigra pars compacta (SNc). In addition, other structures including the dorsal prefrontal cortex (DPFC), amygdala, hippocampus, thalamus, subthalamic nucleus (STN), and lateral habenula (LHb) are part of the reward circuit (Fig. 1.1).

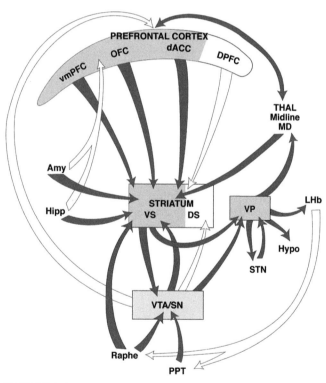

FIGURE 1.1 Schematic illustrating key structures and pathways of the reward circuit. *Shaded areas* and *gray arrows* represent the basic ventral cortical—basal ganglia structures and connections. *Amy*, amygdala; *dACC*, dorsal anterior cingulate cortex; *DPFC*, dorsal prefrontal cortex; *DS*, dorsal striatum; *Hipp*, hippocampus; *Hypo*, hypothalamus; *LHb*, lateral habenula; *MD*, mediodorsal nucleus of the thalamus; *OFC*, orbital frontal cortex; *PPT*, pedunculopontine nucleus; *SN*, substantia nigra pars compacta; *STN*, subthalamic nucleus; *THAL*, thalamus; *vmPFC*, ventral medial prefrontal cortex; *VP*, ventral pallidum; *VS*, ventral striatum; *VTA*, ventral tegmental area.

PREFRONTAL CORTEX

Although cells throughout the cortex fire in response to various aspects of reward processing, the main components of evaluating reward value and outcome are the orbital (OFC) and anterior cingulate (ACC) prefrontal cortices. These regions comprise several specific cortical areas: the orbital cortex is divided into areas 11, 12, 13, 14, and, often, caudal regions referred to as either parts of the insular cortex or periallo- and proisocortical areas; the ACC is divided into dorsal and subgenual regions, areas 24, 25, and 32 [9,10]. Based on specific roles for mediating different aspects of reward processing and emotional regulation, these regions can be functionally grouped into: (1) the OFC; (2) the ventral medial prefrontal cortex (VMPFC), which includes medial OFC and subgenual ACC; and (3) the dorsal ACC (DACC). In addition to the DACC, OFC, and VMPFC, the DPFC, in particular, areas 9 and 46, are engaged when working memory is required for monitoring incentive-based behavioral responses. The DPFC also encodes reward amount and becomes active when anticipated rewards signal future outcomes [11,12].

A key function of the OFC is to link sensory representations of stimuli to outcomes, which is consistent with its connections to both sensory and reward-related regions [12—15]. The OFC can be generally parceled into somewhat functionally different regions based on a caudal—rostral axis, with more caudal regions receiving stronger inputs from primary sensory areas and rostral regions connected to highly processed sensory areas. The OFC's unique access to both primary and highly processed sensory information, coupled with connections to the amygdala and cingulate, explains many of the functional properties of the region. Indeed, the two cardinal tests of OFC function are reward devaluation paradigms and stimulus-outcome reversal learning [16—18], both of which have been demonstrated with OFC lesions across species. Consistent with connectional differences between caudal and rostral OFC, there is an apparent gradient between primary reward representations in more caudal OFC/insular cortex and representations of secondary rewards such as money in more rostral OFC regions [19]. Such dissociations may rely on the differential inputs from early versus higher sensory representations to caudal versus rostral and OFC, respectively, or amygdala connections to caudal OFC and dorsolateral prefrontal cortex and a frontal pole to rostral OFC.

In contrast to the OFC, the ACC has a relative absence of sensory connections, but contains within in it a representation of many diverse frontal lobe functions, including motivation cognition and motor control, which is reflected in its widespread connections with other limbic, cognitive, and motor cortical areas. This is a complex area, but the overall role of the ACC appears to be involved in monitoring these functions for action selection [20−22]. Overall, the ACC can be divided functionally along its dorsal−ventral axis. The ventral ACC (areas 25 and ventral parts of 32) is closely associated with visceral and emotional functions and has strong connections to the hypothalamus, amygdala, and hippocampus. From a functional perspective, imaging and lesion studies have identified an area referred to as the VMPFC. Depending on the specific study, this region may include different combinations of these regions, but overall involves area 25, parts of 32, medial OFC (area 14 and 11), and ventromedial area 10. VMPFC cells track values in the context of internally generated states such as satiety [23]. Moreover, this area plays a role, not only in the valuation of stimuli, but also in selecting between these values. However, perhaps the most remarkable feature of the VMPFC valuation signal is its flexibility. Whereas several other brain regions rely on experience to estimate values, the VMPFC can encode values that must be computed quickly often just prior to, or during, an action [24].

The DACC, primarily area 24, is tightly linked with many PFC areas, including lateral regions associated with cognitive control, and more caudally, with motor control areas. Thus, the DACC sits at the connectional intersection of the brain's reward and action networks. Unlike OFC lesions, DACC lesions have little effect on learning reward reversals based on sensory cues, but they do if the rewards are tied to two different actions (such as "turn" or "push") [25,26]. Imaging studies show ACC activation in a wide variety of tasks, many of which can be explained in the context of selecting between different actions [27].

Anatomical relationships both within and between these PFC regions are complex. As such, several organizational schemes have been proposed based on combinations of cortical architecture and connectivity [6,9,10,28]. In general, cortical areas within each prefrontal group are highly interconnected. However, these connections are quite specific in that a circumscribed region within each area projects to specific regions of other areas, but not throughout. For example, a given part of area 25 of the VMPFC projects to only specific parts of area 14. Overall the VMPFC is primarily interconnected to other medial subgenual regions and to the DACC, with few connections to areas 9 and 46. In contrast, the DACC is tightly linked to area 9 and pre- and supplementary motor cortices. Different OFC regions are highly interconnected, but as with the VMPFC, these are specific connections. For example, not all of area 11 projects throughout area 13. Lateral OFC is connected to parts of the ventrolateral PFC. DPFC regions are interconnected and also project to the DACC, lateral OFC, area 8, and rostral premotor regions.

Finally, the VMPFC, OFC, and DACC are linked to the amygdala [29−31]. These projections arise primarily from the basal and lateral nuclear complex and terminate in specific regions along the ventral and medial cortical surface (VMPFC, OFC, and DACC). Few projections reach the lateral surfaces, although there are some dense terminals in the ventrolateral PFC. The amygdalocortical projections are bidirectional. Here, the OFC−amygdalar projections target the intercalated masses, whereas terminals from the VMPFC and DACC are more diffuse. An important additional connection primarily to the VMPFC, but also to parts of the OFC, is derived from the hippocampus. In contrast, there are few hippocampal projections to DACC or the dorsal and lateral PFC areas.

THE VENTRAL STRIATUM

The link between the nucleus accumbens and reward was first demonstrated as part of the self-stimulation circuit originally described by Olds and Milner [32]. Since then, the nucleus accumbens (and the VS in general) has been a central site for studying reward and drug reinforcement and for the transition between drug use as a reward and habit [33,34]. The term VS, coined by Heimer, includes the nucleus accumbens and the broad continuity between the caudate nucleus and putamen ventral to the rostral internal capsule, the olfactory tubercle, and the rostrolateral portion of the anterior perforated space adjacent to the lateral olfactory tract in primates [35]. From a connectional perspective, it also includes the medial caudate nucleus, rostral to the anterior commissure [36] (see later).

Human imaging studies demonstrate the involvement of the VS in reward prediction and reward prediction errors [37,38] and, consistent with physiological results in nonhuman primate studies, the region is activated during reward anticipation [39] (also see Fig. 4.1). Interestingly, the relative size of the cortical input to the nucleus accumbens compared against the size of hippocampal/amygdala nucleus accumbens projections predicts subjects' relative propensity for reward-seeking versus novelty-seeking behavior [40]. The VS blood oxygen level-dependent signal is regularly found to code for a reward prediction error [41,42]: the reward minus the expectation of that reward. It has been suggested that this signal depends on dopaminergic input, as DA cells are known for a similar pattern of reward coding [43], but other VS-projecting regions discussed earlier also code for rewards. It is therefore likely that the VS signal contains several combined reward representations. Such an idea is supported by data in which connectional and functional data were acquired in the same subjects. The extent to which the VS signal looks like a reward prediction error, rather than a simple coding of reward, depends on the strength of connection (measured by diffusion MRI tractography) between the VS and the dopaminergic midbrain [44]. Collectively, these studies demonstrate its key role in the acquisition and development of reward-based behaviors and its involvement in drug addiction and drug-seeking behaviors. However, it has been recognized for some time now that cells in the primate dorsal striatum as well as the VS respond to the anticipation, magnitude, and delivery of reward [45–47]. The striatum, as a whole, is the main input structure of the BG and receives a massive and topographic input from cerebral cortex (and thalamus). These afferent projections to the striatum terminate in a general topographic manner, such that the ventromedial striatum receives input from the VMPFC, OFC, and DACC; the central striatum receives input from the DPFC, including areas 9 and 46. Indeed, the different prefrontal cortical areas have corresponding striatal regions that are involved in various aspects of reward evaluation and incentive-based learning [38,48,49] and are associated with pathological risk-taking and addictive behaviors [50,51].

Special Features of the Ventral Striatum

While the VS is similar to the dorsal striatum in most respects, there are also some unique features. The VS contains a subterritory, called the shell, which plays a particularly important role in the circuitry underlying goal-directed behaviors, behavioral sensitization, and changes in affective states [52,53]. Whereas several transmitter and receptor distribution patterns distinguish the shell/core subterritories, calbindin is the most consistent marker for the shell across species [54]. Although a calbindin-poor region marks the subterritory of the shell, staining intensity of other histochemical markers in the shell distinguish it from the rest of the VS. In general, GluR1, GAP-43, acetylcholinesterase, serotonin, and substance P are particularly rich in the shell, while the μ receptor is relatively low in the shell compared to the rest of the striatum [55–58]. (Fig. 1.2A). Finally, the shell has some unique connectivities compared to the rest of the VS. These are indicated in the following.

In addition to the shell compartment, several other characteristics are unique to the VS. The DA transporter is relatively low throughout the VS including the core. This pattern is consistent with the fact that the dorsal tier DA neurons express relatively low levels of mRNA for the DA transporter compared to the ventral tier [59] (see later regarding the substantia nigra). In addition, unlike the dorsal striatum, the VS contains numerous cell islands, including the islands of Calleja, which are thought to contain quiescent immature cells that remain in the adult brain [60–62]. The VS also contains many pallidal elements that invade this ventral

territory (see later regarding the VP). Of particular importance is the fact that, whereas both the dorsal and the ventral striatum receive input from the cortex, thalamus, and brain stem, the VS alone also receives a dense projection from the amygdala and hippocampus (Fig. 1.1). While collectively these are important distinguishing features of the VS, its dorsal and lateral border is continuous with the rest of the striatum and neither cytoarchitectonic nor histochemical distinctions mark a clear boundary between it and the dorsal striatum. Moreover, because reward-related cells are found throughout a large region of the rostral striatum, the best way to outline where this function lies is by its afferent projections from cortical areas that mediate different aspects of reward processing, the VMPFC, OFC, and DACC. Thus, in the following sections, we refer to the VS as that area of the striatum that receives inputs from these cortical regions. Note that this expands the traditional boundaries of the VS into the dorsomedial striatum.

Connections of the Ventral Striatum (See Fig. 1.3)

The VS is the main input structure of the ventral BG. Like the dorsal striatum, afferent projections to the VS are derived from three major sources: a massive, generally topographic input from the cerebral cortex; a large input from the thalamus; and a smaller, but critical input from the brain stem, primarily from the midbrain dopaminergic cells. The traditional way of viewing the overall organization of corticostriatal projections was to divide the striatum into the VS, input from limbic areas; the central striatum, input from associative cortical areas; and the dorsolateral striatum, input from sensorimotor areas [36,63]. However, as indicated above, the striatal region receiving input from reward-related areas expands outside of the traditional boundaries of the VS.

Cortical Projections to the Ventral Striatum

Corticostriatal terminals are organized in densely distributed patches [6,64,65] (Fig. 1.2B), which can be visualized at relatively low magnification. The overall general distribution of terminals from different cortical regions, viewed separately or in selective groupings, is the foundation for the concept of parallel and segregated cortical–BG circuits. However, as described in more detail later, the collective distribution of cortical inputs from different cortical areas demonstrates extensive converge of cortical terminals throughout the striatum, creating a complex interface between inputs from functionally distinct cortical regions. Indeed, each cortical region, while projecting topographically to the striatum, does not maintain the narrow funneling that

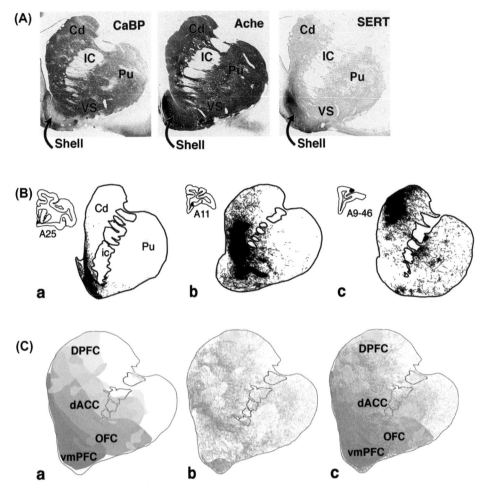

FIGURE 1.2 Histochemistry and connections of the rostral striatum. (A) Photomicrographs of the rostral striatum stained with three different histochemical markers to illustrate the shell: calbindin (*CaBP*), acetylcholinesterase (*Ache*), and serotonin (*SERT*). *Cd*, caudate nucleus; *IC*, internal capsule; *Pu*, putamen; *VS*, ventral striatum. (B) Schematic chartings of labeled fibers following injections into various prefrontal regions. (a) VMPFC injection site (area 25); (b) OFC injection site (area 11); (c) DPFC injection site (area 9/46). The dense projection fields are indicated in *solid black*. Note the diffuse projection fibers outside of the dense projection fields. (C) Schematics demonstrating convergence of cortical projections from various reward-related regions and dorsal prefrontal areas. (a) Convergence between dense projections from various prefrontal regions. (b) Distribution of diffuse fibers from various prefrontal regions. (c) Combination of dense and diffuse fibers. *dACC*, dorsal anterior cingulate cortex; *DPFC*, dorsal lateral prefrontal cortex; *OFC*, orbital prefrontal cortex; *vmPFC*, ventral medial prefrontal cortex. *Red*, inputs from vmPFC; *dark orange*, inputs from OFC; *light orange*, inputs from dACC; *yellow*, inputs from DPFC.

would be required to "fit" a large cortex into a far smaller striatal volume. For example, the volume occupied by the collective dense terminal fields from the VMPFC, DACC, and OFC is approximately 22% of the striatum, a larger cortical input than would be predicted by the relative cortical volume of these areas. Together, projections from these cortical areas terminate primarily in the rostral, medial, and ventral parts of the striatum and define the ventral striatal territory [6,66,67]. The large extent of this region is consistent with the findings that diverse striatal areas are activated following reward-related behavioral paradigms [51,68,69]. The dense projection field from the VMPFC is the most limited (particularly from area 25) and is concentrated within and just lateral to the shell (Fig. 1.2B(a)). The innervation of the shell receives the

densest input from area 25, although fibers from areas 14 and 32 and from agranular insular cortex also terminate here. The VMPFC also projects to the medial wall of the caudate nucleus, adjacent to the ventricle. In contrast, the central and lateral parts of the VS (including the ventral caudate nucleus and putamen) receive inputs from the OFC (Fig. 1.2B(b)). These terminals also extend dorsally, along the medial caudate nucleus, but lateral to those from the VMPFC. There is some medial-to-lateral and rostral-to-caudal topographic organization of the OFC terminal fields. For example, projections from area 11 terminate rostral to those from area 13 and those from area 12 terminate laterally. Despite this general topography, overlap between the OFC terminals is significant. Projections from the DACC extend from the rostral pole of the

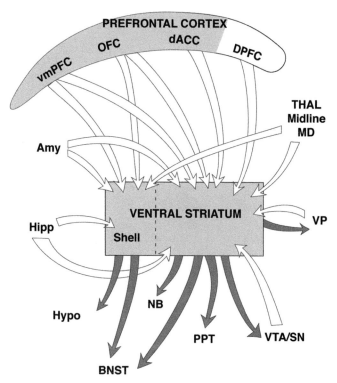

FIGURE 1.3 Schematic illustrating the connections of the ventral striatum. *White arrows*, inputs; *gray arrows*, outputs. *Amy*, amygdala; *BNST*, bed nucleus stria terminalis; *dACC*, dorsal anterior cingulate cortex; *DPFC*, dorsal lateral prefrontal cortex; *Hipp*, hippocampus; *Hypo*, hypothalamus; *MD*, mediodorsal nucleus of the thalamus; *OFC*, orbital frontal cortex; *PPT*, pedunculopontine nucleus; *SN*, substantia nigra pars compacta; *THAL*, thalamus; *vmPFC*, ventral medial prefrontal cortex; *VP*, ventral pallidum; *VTA*, ventral tegmental area.

the dense terminal fields of the VMPFC, OFC, and DACC provide an anatomical substrate for integration between reward-processing circuits within specific striatal areas and may represent critical "hubs" for integrating reward value, predictability, and salience.

In addition to convergence between VMPFC, DACC, and OFC dense terminals, projections from DACC and OFC also converge with inputs from the DPFC, demonstrating that functionally diverse PFC projections also converge in the striatum. At rostral levels, DPFC terminals converge with those from both the DACC and OFC, although each cortical projection also occupies its own territory. Here, projections from all PFC areas occupy a central region, the different cortical projection extending into nonoverlapping zones. Convergence is less prominent caudally, with almost complete separation of the dense terminals from the DPFC and DACC/OFC just rostral to the anterior commissure. Importantly, convergence does not take place only at the boundaries or edges of different functional domains. Rather, there are dense clusters of invading terminals from, for example, the DPFC, embedded deep within the dense projection field of the OFC. Because medium spiny neurons have a dendritic field spanning 0.5 mm [70], the area of convergence is likely to be larger than estimated based solely on the relationship between the afferent projections.

Using a computational model to characterize the consistency of cortical projection zones into the striatum predicts striatal locations that receive specific cortical inputs. These computations demonstrate that there is an exponential decay in overlap as a function of distance [7]. Thus, convergence of corticostriatal terminals from cortical areas that are separated by 5 mm is 50% in nonhuman primates. This overlap decreased to <20% from regions separated by 30 mm. This distance between cortical regions can span several functional domains. For example, there is a region within the rostral, anterior striatum that receives projections from across the prefrontal cortical regions, including the VMPFC, OFC, DACC, and dorsal and lateral PFC.

There has been growing interest in the idea of broad associative cortical networks, which feature nodes (or connection hubs) that integrate and distribute information across multiple systems [71,72]. These studies indicate that there are specific regions that receive inputs from multiple cortical areas that cross-link between functional systems. The aforementioned data suggest that these hubs exist in the striatum as well as the cortex. In particular, this rostral striatal region may serve as a hub for VMPFC, OFC, and DACC to connect with dorsal and lateral PFC regions that integrates motivational, reward, and cognitive control information. It is possible that this convergence of cortical areas on specific locations in the striatum may thus facilitate value

striatum to the anterior commissure and are located in both the central caudate nucleus and the putamen. They primarily avoid the shell region. These fibers terminate somewhat lateral to those from the OFC. Thus, the OFC terminal fields are positioned between the VMPFC and the DACC. In contrast, the DPFC projects primarily in the head of the caudate and part of the rostral putamen (Fig. 1.2B(c)). Terminals from areas 9 and 46 are somewhat topographically organized and extend from the rostral pole, through the rostral, central putamen and much of the length of the caudate nucleus [6,65].

Despite the general topography described above, dense terminal fields from the VMPFC, OFC, and DACC show a complex interweaving and convergence, providing an anatomical substrate for modulation between circuits [6,7] (Fig. 1.2C). Dense projections from the DACC and OFC regions do not occupy completely separate territories in any part of the striatum, but converge most extensively at rostral levels. By contrast, there is greater separation of projections between terminals from the VMPFC and the DACC/OFC, particularly at caudal levels. These regions of convergence between computations across diverse domains into a common currency. Thus, a coordinated activation of DPFC, DACC, and/or OFC terminals in specific striatal regions would produce a unique combinatorial activation at the specific sites suggesting specific subregions for reward-based incentive drive to influence long-term strategic planning and habit formation. In contrast to this, some striatal regions, particularly posterior and lateral portions, receive inputs from only a few prefrontal regions, and therefore they may serve more specialized computational roles.

The Amygdala and Hippocampal Projections to the Ventral Striatum

The amygdala is a prominent limbic structure that plays a key role in the emotional coding of environmental stimuli and provides contextual information used for adjusting motivational level that is based on immediate goals. It comprises the basolateral nuclear group (BLNG), the corticomedial region, and the central amygdaloid complex. The BLNG, the primary source of amygdaloid input to the VS outside the shell, processes higher order sensory inputs in all modalities except olfactory, and responds to highly palatable foods and "multimodal" cues. Overall, the basal nucleus and the magnocellular division of the accessory basal nucleus are the main sources of inputs to the VS [73,74]. The lateral nucleus has a relatively minor input to the VS. The amygdala has few inputs to the dorsal striatum in primates. The basal and accessory basal nuclei innervate both the shell and the ventromedial striatum outside the shell. Moreover, there is a topographic organization to that innervation, such that inputs from basal nucleus subdivisions target different regions of the VS. The shell is set apart from the rest of the VS by a specific set of connections derived from the medial part of the central nucleus, periamygdaloid cortex, and medial nucleus of the amygdala. Thus, much like the topography of prefrontal inputs to the VS, nuclei of the amygdala that mediate various functions target subterritories of the ventromedial striatum. In contrast to the amygdala, the hippocampal formation projects to a more limited region of the VS. The main terminal field is located in the most medial and ventral parts of the VS, essentially confined to the shell region. However, additional fibers do extend outside the boundary of the shell. The subiculum appears to provide the main source of this input. Moreover, some afferent fibers are also derived from the parasubiculum and part of CA1 [75]. Inputs from both the hippocampus and the amygdala terminate in overlapping regions of the shell [75,76].

Thalamic Projections to the Ventral Striatum

The midline and medial intralaminar thalamic nuclei project to medial prefrontal areas, the amygdala, and hippocampus and, as such, are considered the limbic-related thalamic nuclear groups. The VS receives dense projections from the midline thalamic nuclei and from the medial parafascicular nucleus [77,78]. The midline nuclei include the anterior and posterior paraventricular, paratenial, rhomboid, and reuniens thalamic nuclei. The shell of the nucleus accumbens receives the most limited projection. The medial shell is innervated almost exclusively by the anterior and posterior paraventricular nuclei and the medial parafascicular nucleus. The ventral shell receives input from these nuclei as well as from the paratenial, rhomboid, and reuniens midline groups. The medial wall of the caudate nucleus receives projections, not only from the midline and the medial intralaminar nuclei, but also from the central superior lateral nucleus. In contrast, the central part of the VS receives a limited projection from the midline thalamic nuclei, predominantly from the rhomboid nucleus. It also receives input from the parafascicular nucleus and the central superior lateral nucleus. In addition to the midline and intralaminar thalamostriatal projections, in primates there is a large input from the "specific" thalamic BG relay nuclei, the medial dorsalis nucleus (MD) and ventral anterior and ventral lateral nuclei [79,80]. The VS also receives these direct afferent projections, which are derived primarily from the MD and a limited input from the magnocellular subdivision of the ventral anterior nucleus.

Efferent Projections From the Ventral Striatum (See Fig. 1.4)

Efferent projections from the ventral striatum, like those from the dorsal striatum, project primarily to the pallidum and substantia nigra/VTA [81,82]. Specifically, they terminate topographically in the subcommissural part of the globus pallidus (classically defined as the VP), the rostral pole of the external segment, and the rostromedial portion of the internal segment. The more central and caudal portions of the globus pallidus do not receive this input. Fibers from the VS projecting to the substantia nigra are not as confined to a specific region as those projecting to the globus pallidus. Although the densest terminal fields occur in the medial portion, numerous fibers also extend laterally to innervate the dorsal tier of the midbrain dopaminergic neurons (see section on the substantia nigra). This projection extends throughout the rostral–caudal extent of the substantia nigra. Projections from the medial part of the VS also project more caudally, terminating in the pedunculopontine nucleus and, to some extent, in the medial central gray. In addition to projections to the typical BG output structures, the VS also projects to non-BG regions. The shell sends fibers caudally and medially into the lateral hypothalamus. Axons from the medial VS (including the shell) travel to and terminate in the bed nucleus of the stria terminalis and parts of the ventral regions of the VS terminate in the nucleus basalis [81]. A projection to the nucleus basalis in the basal forebrain is of particular interest, because this is the main source of cholinergic fibers to the cerebral cortex and the amygdala. These data indicate that the VS may influence the cortex directly, without going through the pallidal, thalamic circuit. A potential connection between ventral striatal fibers and cholinergic neurons in the basal forebrain has been demonstrated at the light microscopic level in several species, including primates, and verified at the electron microscope level and, more recently, using optogenetic methods in rodents [83–88]. Likewise, the projection to the bed nucleus of the stria terminalis indicates direct striatal influence on the extended amygdala.

VENTRAL PALLIDUM (FIG. 1.4)

The VP is an important component of the reward circuit. These cells respond specifically during the learning and performance of reward-incentive behaviors and are an area of focus in the study of addictive behaviors [89,90]. The VP, like the dorsal pallidum, has indirect (via the medial STN and adjacent hypothalamus) and direct projections to the MD thalamus, thus completing the loop back to the cortex [91]. The STN connection with the VP, along with a hyperdirect pathway from the VMPFC, OFC, and ACC to the medial STN, highlights the role of the STN in the reward pathway [8,91]. This will be discussed in more detail later. Pallidal cells, including the VP, send direct input to the LHb [91,92]. The LHb, in turn, projects indirectly to the DA cells. Thus, the VP, STN, and LHb are important

components of the reward circuit. VP cells respond specifically during the learning and performance of reward-incentive behaviors. The complexity of the VP circuitry coupled with its central position in the reward circuit indicates that this structure is likely to be activated during imaging studies. Neuroimaging studies that document ventral striatal activation often document overlapping ventral pallidal activation. However, the lack of sufficient spatial resolution makes it difficult to distinguish the VP from the VS. LHb cells are inhibited by a reward-predicting stimulus, but fire following a nonreward signal. This stimulation of the LHb directly or following a nonreward signal inhibits DA cells [93,94]. Interestingly, few fibers from the LHb directly reach the SNc in primates; rather these cells project primarily to the rostromedial tegmental nucleus (RMTg), which, in turns projects to DA cells. An event-related fMRI study featuring adequate spatial and temporal resolution to visualize habenular activity indicated that negative but not positive feedback activates the habenular complex, consistent with findings from nonhuman primate electrophysiology [95].

The term VP was first used to describe in rats the region below the anterior commissure, extending into the anterior perforated space, based both on histological criteria and on its input from the then defined VS [96]. However, in primates, it is best defined by its input from the entire input from reward-related VS. As indicated earlier, that would include not only the subcommissural regions, but also the rostral pole of the external segment and the medial rostral internal segment of the globus pallidus. Like the dorsal pallidum, the VP contains two parts: a substance-P-positive and enkephalin-positive component, which project to the thalamus and STN, respectively [91,97–99]. Pallidal neurons have a distinct morphology that is outlined with immunoreactivity for these peptides, making these stains particularly useful for determining the boundaries and extent of the VP [99–103]. These fibers, which appear as tubular-like structures, referred to as "wooly fibers," demonstrate the extent of the VP and its continuity with the dorsal pallidum [102]. The VP reaches not only ventrally, but also rostrally to invade the rostral and ventral portions of the VS, sending finger-like extensions into the anterior perforated space. In the human brain the VP extends far into the anterior perforated space, where the structure appears to be rapidly broken up into an interconnected lacework of pallidal areas, interdigitating with striatal cell islands and microcellular islands, including possibly the deep islands of Calleja. The identification of the VP and VS simplified the structural analysis of the ventral forebrain, demonstrating that a large part of the area referred to as "substantia innominata" is actually   an   extension   of   the   reward-related

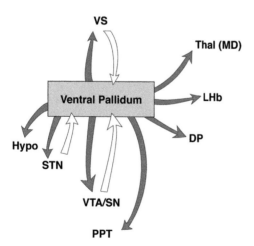

FIGURE 1.4   Schematic illustrating the connections of the ventral pallidum. *DP*, dorsal pallidum; *Hypo*, hypothalamus; *LHb*, lateral habenula; *MD*, mediodorsal nucleus of the thalamus; *PPT*, pedunculopontine nucleus; *SN*, substantia nigra; *STN*, subthalamic nucleus; *Thal*, thalamus; *VS*, ventral striatum; *VTA*, ventral tegmental area.

striatopallidal complex [55]. In addition to the GABAergic ventral striatal input, there is a glutamatergic input from the STN nucleus and a dopaminergic input from the midbrain [104,105].

Descending efferent projection from the VP terminates primarily in the medial STN, extending into the adjacent lateral hypothalamus [91,106–108]. Axons continue into the substantia nigra, terminating medially in the SNc, substantia nigra pars reticulata (SNr), and VTA. Projections from the VP to the STN and the lateral hypothalamus are topographically arranged. This arrangement demonstrates that pathways from distinct pallidal regions that receive specific striatal input terminate in specific subthalamic/hypothalamic regions, thus maintaining a topographic arrangement. In contrast, terminating fibers from the VP in the substantia nigra overlap extensively, suggesting convergence of terminals from different ventral pallidal regions. Fibers continue caudally to innervate the pedunculopontine nucleus. As with the dorsal pallidum, components of the VP project to the thalamus, terminating in the midline nuclei and medial MD nucleus. Pallidal fibers entering the thalamus give off several collaterals that form branches that terminate primarily onto the soma and proximal dendrites of thalamic projection cells. In addition, some synaptic contact is also made with local circuit neurons. This terminal organization indicates that, while pallidal projections to the thalamus are primarily inhibitory on thalamic relay neuron cells, they may also function to disinhibit projection cells via the local circuit neurons [109,110]. In addition, the VP also projects to both the internal and the external segments of the dorsal pallidum. This is a unique projection, in that the dorsal pallidum does not seem to project ventrally.

Parts of the VP (along with the dorsal pallidum) project to the LHb nucleus. Most cells that project here are located in the region that circumvents the internal segment and, in particular, are embedded within the accessory medullary lamina that divides the lateral and medial portions of the dorsal internal segment project [106,111]. Finally, part of the ventral pallidal area [as with the external segment of the globus pallidus (GPe)] also projects to the striatum [112]. This pallidostriatal pathway is extensive in the monkey and is organized in a topographic manner preserving a general, but not strict, medial-to-lateral and ventral-to-dorsal organization. The terminal field is widespread and nonadjacent pallidal regions send fibers to the striatum that overlap considerably indicating convergence of terminals from different pallidal regions. Thus, the pallidostriatal pathway contains both a reciprocal and a nonreciprocal pathway. The nonreciprocal component suggests that this feedback projection may play an additional role in integrating information between ventral striatal subcircuits.

THE MIDBRAIN DOPAMINE NEURONS

DA neurons play a central role in the reward circuit [43,113]. While behavioral and pharmacological studies of DA pathways have led to the association of the mesolimbic pathway and nigrostriatal pathway with reward and motor activity, respectively, more recently both of these cell groups have been associated with reward. The midbrain DA neurons project widely throughout the brain. However, studies of the rapid signaling that is associated with incentive learning and habit formation focus on the DA striatal pathways.

The midbrain DA neurons are divided into the VTA and the SNc. Based on projection and chemical signatures, these cells are also referred to as the dorsal and ventral tier neurons [59] (Fig. 1.5A). The dorsal tier includes the VTA and the dorsal part of the SNc (also referred to as the retrorubral cell group). The cells of the dorsal tier are calbindin positive and contain relatively low levels of mRNA for the DA transporter and D2 receptor subtype. They project to the VS, cortex, hypothalamus, and amygdala. The ventral tier of DA cells are calbindin negative, have relatively high levels of mRNA for the DA transporter and D2 receptor, and project primarily to the dorsal striatum. Ventral tier cells (calbindin poor, DA transporter and D2 receptor rich) are

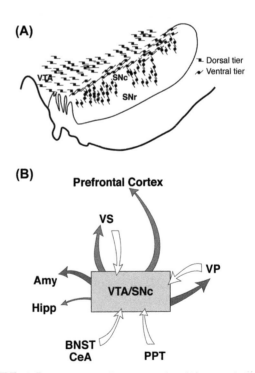

FIGURE 1.5 Schematic illustrating the (A) organization and (B) connections of the midbrain dopamine cells. *Amy,* amygdala; *BNST,* bed nucleus stria terminalis; *CeA,* central amygdala nucleus; *Hipp,* hippocampus; *PPT,* pedunculopontine nucleus; *SNc,* substantia nigra pars compacta; *SNr,* substantia nigra pars reticulata; *VP,* ventral pallidum; *VS,* ventral striatum; *VTA,* ventral tegmental area.

more vulnerable to degeneration in Parkinson's disease and to N-methyl-4-phenyl-1,2,3,6-tetrahydropyridine-induced toxicity, whereas the dorsal tier cells are selectively spared [114]. As mentioned before, despite these distinctions, both cell groups respond to unexpected rewards.

Afferent Projections (Fig. 1.5B)

Input to the midbrain DA neurons is primarily from the striatum, from both the GPe and the VP, and from the brain stem. In addition, there are projections to the dorsal tier from the bed nucleus of the stria terminalis, the sublenticular substantia innominata, and the extended amygdala (the bed nucleus of the stria terminalis and the central amygdala nucleus).

As described earlier, the striatonigral projection is the most massive projection to the SN and terminates in both the VTA/SNc and the SNr [115–118]. The ventral (like the dorsal) striatonigral connection terminates throughout the rostrocaudal extent of the substantia nigra. There is an inverse ventral/dorsal topography to the striatonigral projections. The dorsolateral striatonigral inputs are concentrated in the ventrolateral SN. These striatal cells project primarily to the SNr, but also terminate on the cell columns of DA neurons that penetrate deep into the SNr. In contrast, the ventral striatonigral inputs terminate in the VTA, the dorsal part of the ventral tier, and in the medial and dorsal SNr. Thus the VS not only projects throughout the rostrocaudal extent of the substantia nigra, but also covers a wide mediolateral range [5,118].

Descending projections from the central nucleus of the amygdala also terminate in a wide mediolateral region, but are limited primarily to the dorsal tier cells. In addition, there are projections to the dorsal tier from the bed nucleus of the stria terminalis and from the sublenticular substantia innominata that travel together with those from the amygdala [119,120]. Both the GPe and the VP project to the substantia nigra. The pallidal projection follows an inverse dorsal/ventral organization similar to that of the striatonigral projection. Thus, the VP projects dorsally, primarily to the dorsal tier and dorsal SNc. The pedunculopontine nucleus sends a major glutamatergic input to the dopaminergic cell bodies. In addition, there is a serotonergic innervation from the dorsal raphe nucleus, though there is disagreement regarding whether fibers terminate primarily in the pars compacta or pars reticulata. Other brain-stem inputs to the DA neurons include those from the superior colliculus, the parabrachial nucleus, and the locus coeruleus. These inputs raise the interesting possibility that DA cells receive a direct sensory input. The collicular input, in particular, has been suggested to be responsible for the short-latency burst-firing activity of the DA cells in response to a salient or rewarding stimuli [67]. Finally, in primates, there is a small and limited projection from the PFC to the midbrain DA neurons, to both the VTA and the SNc. While considerable attention has been given to this projection, relative to the density of its other inputs, it is weak in primates [121].

Efferent Projections (Fig. 1.5B)

The largest midbrain DA neuron efferent projection is to the striatum [116,117,122]. As with the descending striatonigral pathway, there is a mediolateral and an inverse dorsoventral topography arrangement to the projection. The ventral pars compacta neurons project to the dorsal striatum and the dorsally located DA neurons project to the VS. The shell region receives the most limited input, primarily derived from the medial VTA [122]. The rest of the VS receives input from the entire dorsal tier. In addition, there are some afferent projections from the medial region and dorsal part of the SNc. In contrast to the VS, the central striatal area (the region innervated by the DPFC) receives input from a wide region of the SNc. The dorsolateral (motor-related) striatum receives the largest midbrain projection from cells throughout the ventral tier. In contrast to the dorsolateral region of the striatum, the VS receives the most limited DA cell input. Thus, in addition to an inverse topography, there is also a differential ratio of DA projections to the various striatal areas [5].

The dorsal tier cells also project widely throughout the primate cortex and are found not only in granular areas but also in agranular frontal regions, parietal cortex, temporal cortex, and, albeit sparsely, in occipital cortex [123,124]. The majority of DA cortical projections are from the parabrachial pigmented nucleus of the VTA and the dorsal part of the SNc. The VTA also projects to the hippocampus, albeit to a lesser extent than in neocortex. The DA cells that project to functionally different cortical regions are intermingled with one another, in that individual neurons send collateral axons to different cortical regions. Thus the nigrocortical projection is a more diffuse system compared to the nigrostriatal system, and can modulate cortical activity at several levels. DA fibers are located in superficial layers, including a prominent projection throughout layer I. This input provides a general modulation of cortical cells at the distal apical dendrites. DA fibers are also found in the deep layers in specific cortical areas [125,126]. Projections to the amygdala arise primarily from the dorsal tier. These terminals form symmetric synapses primarily with spiny cells of specific subpopulations in the amygdala [127]. As indicated earlier, DA fibers also project to the VP.

The Striatonigrostriatal Network

While the role of DA in reward is well established, the latency between the presentation of the reward stimuli and the activity of the DA cells is too short to reflect higher cortical processing necessary for linking a stimulus with its rewarding properties. The fast, burst-firing activity is likely, therefore, to be generated from other inputs such as in brain-stem glutamatergic nuclei [67]. An interesting issue, then, is how do the DA cells receive information concerning reward value? The largest forebrain input to the DA neurons is from the striatum. However, this is a relatively slow GABAergic inhibitory projection, unlikely to result in the immediate, fast burst-firing activity. Nonetheless, the collective complex network of PFC, amygdala, and hippocampal inputs to the VS integrates information related to reward processing and memory to modulate striatal activity. These striatal cells then have a direct impact on a subset of medial DA neurons, which, through a series of connections described later, influences the dorsal striatum.

As mentioned before, projections from the striatum to the midbrain are arranged in an inverse dorsal—ventral topography and there is also an inverse dorsal—ventral topographic organization to the midbrain striatal projection. Considered separately, each limb of the system creates a loose topographic organization, the VTA and medial SN being associated with the limbic system, and the central and ventral SN with the associative and motor striatal regions, respectively. However, each functional region differs in their proportional projections, which significantly alter their relationship to one another. The VS receives a limited midbrain input, but projects to a large region. In contrast, the dorsolateral striatum receives a wide input, but projects to a limited region. In other words, the VS influences a wide range of DA neurons, but is itself influenced by a relatively limited group of DA cells. On the other hand, the dorsolateral striatum influences a limited midbrain region, but is affected by a relatively large midbrain region.

With this arrangement, while the VS receives input from the VMPFC, OFC, DACC, and amygdala, its efferent projection to the midbrain extends beyond the tight VS/dorsal tier DA/VS circuit. It terminates also in the ventral tier, to influence the dorsal striatum. Moreover, this part of the ventral tier is reciprocally connected to the central (or associative) striatum. The central striatum also projects to a more ventral region than it receives input from. This region, in turn, projects to the dorsolateral (or motor) striatum. Taken together, the interface between different striatal regions via the midbrain DA cells is organized in an ascending spiral interconnecting different functional regions of the striatum and creating a feed-forward organization, from reward-related regions of the striatum, to cognitive and motor areas (see Fig. 1.6). Thus, although the short-latency burst-firing activity of DA that signals immediate reinforcement is likely to be triggered from brain-stem nuclei, the corticostriato-midbrain pathway is in the position to influence DA cells to distinguish rewards and modify responses to incoming salient stimuli over time. This pathway is further reinforced via the nigrostriatal pathway, placing the striatonigrostriatal pathway in a pivotal position for transferring from the VS to the dorsal striatum during learning and habit formation. Indeed, cells in the dorsal striatum are progressively recruited during various types of learning, from simple motor tasks to drug self-administration [128—132]. Moreover, when the striatonigrostriatal circuit is interrupted, information transfer from Pavlovian to instrumental learning does not take place [133].

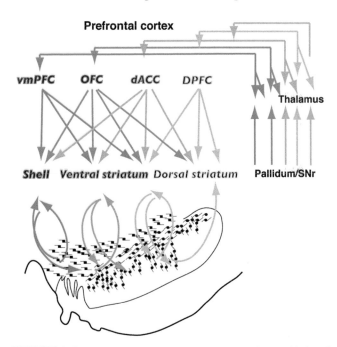

FIGURE 1.6 Three networks of integration through cortical—basal ganglia pathways. (1) Fibers from different prefrontal areas converge within subregions of the striatum. (2) Through the organization of striatonigrostriatal projections, the ventral striatum can influence the dorsal striatum [5]. Midbrain projections from the shell target both the VTA and the ventromedial SNc. Projections from the VTA to the shell form a "closed," reciprocal loop, but also project more laterally to impact on DA cells projecting to the rest of the ventral striatum forming the first part of a feed-forward loop (or spiral). The spiral continues through the striatonigrostriatal projections through which the ventral striatum impacts on cognitive and motor striatal areas via the midbrain DA cells. (3) The nonreciprocal corticothalamic projection carries information from reward-related regions, through cognitive and motor controls. *dACC*, dorsal anterior cingulate cortex; *DPFC*, dorsal prefrontal cortex; *OFC*, orbital prefrontal cortex; *SNr*, substantia nigra pars reticulata; *vmPFC*, ventral medial prefrontal cortex. *Red*, vmPFC pathways; *dark orange*, OFC pathways; *light orange*, dACC pathways; *yellow*, DPFC pathways; *green*, output to motor control areas.

COMPLETING THE CORTICAL–BASAL GANGLIA REWARD CIRCUIT

In addition to the PFC, VS, VP, amygdala, and hippocampus, other key components of the circuit include the STN, the LHb, and the thalamus. Each of these structures has complex connectivities with multiple brain regions that we briefly review here.

Subthalamic Nucleus, the Hyperdirect Cortical Pathway

Like the corticostriatal connection, frontal areas project to or within the immediate region of the STN in a patchy manner and follow a general functional topography. Motor and premotor projections are concentrated in the lateral, dorsal, and caudal parts of the nucleus. In contrast, prefrontal projections are concentrated in the anterior, ventral, and medial half of the STN. Moreover, regions within the motor control areas and prefrontal cortex terminate in a topographic manner. For example, M1 projects to the dorsolateral STN and area 6 projects ventromedially to M1 projections [134–136]. Overall area 8 and caudal 9 and 46 terminals are located ventral and medial to premotor and motor projections [135]. Rostral areas 10, 9, and 46 terminate in the medial half of the STN [8]. DACC terminal fields are located medial to these projections in the anterior, medial tip of the STN. The location of the terminals from VMPFC/OFC and DACC are of particular interest here. Whereas DPFC dense projections are contained within the conventional medial border of the STN, DACC projections straddle this border and VMPFC/OFC terminals are located outside of it in the adjacent lateral hypothalamus (LH). Thus, the cone-shaped region that surrounds the ventromedial tip of the STN is occupied by VMPFC/OFC and part of the DACC projections. Indeed, the cortical terminal fields in this area are in the topographic continuity of the other PFC projections. Interestingly, terminals from and cells to the VP, a known STN input/output, are in the same location surrounding the medial tip of the STN [91]. Thus, the hyperdirect pathway defines a topographic anatomic connection, composed of a rostral limbic component, concentrated in the medial tip of the STN and the adjacent LH; a cognitive component in the medial half; and a more lateral and caudal motor component centered in the lateral half of the STN. This is consistent with the topography of the pallido-STN interconnection [91,137,138]. However, importantly, as seen with the corticostriatal projection, overlap between the various prefrontal projections is extensive, suggesting complex integration between these [8]. Thus, OFC and VMPFC terminals overlap with DACC dense terminals in the medial tip of the STN and the LH. DACC converges with DPFC. DPFC terminals also overlap with those from area 6.

Similar to the pallidum, the morphology of STN dendrites indicates that convergence between STN cortical inputs from different functional areas may be greater than it appears based on projection patterns. STN dendrites are oriented along the long axis of the nucleus and occupy approximately two-thirds of its volume [139]. Each dendrite stretches across multiple functional regions and receives inputs along this entire length. Indeed, inputs from the VP and globus pallidus have been shown to converge onto a single STN neuron [140]. Thus, STN neurons at the interface of functional territories are likely to receive convergent inputs onto their proximal dendrites.

The Lateral Habenula, a Negative-Reward Signal

The LHb also plays a central role in the reward circuit through its interaction with DA cells [93,95,141]. LHb cells fire following a nonreward signal and are inhibited by reward-predicting stimuli [93,142]. Stimulation of the LHb nuclei results in a negative-reward-related signal in the SNc [93]. Indeed, LHb firing is strongest when there is 0% probability of a reward or a 100% probability of an aversive outcome [143]. These cells project to and excite the GABAergic cells in the RMTg that, in turn, inhibits DA burst firing in response to reward, thus driving a negative reward signal [94,144–146]. The inputs to the LHb are primarily from the LH and globus pallidus, including, but not limited to the VP [147,148]. In primates, cells projecting to the LHb arise primarily from forebrain areas: the internal globus pallidus segment (GPi) and the limbic-related VP, the adjacent LH, and the perifornical area [91,92,148]. The GPi and VP receive indirect topographic input from prefrontal regions associated with reward value and reward learning via the striatum.

The Thalamus, the Final Link Back to Cortex

The MD and midline thalamic nucleus are a critical piece of the frontal–BG circuit that mediates reward, motivation, and emotional drive. As noted earlier, each PFC region of the frontal cortex is connected with specific areas of striatum, pallidum, and thalamus. The thalamus represents the final BG link and is often treated as a simple "one-way relay" back to cortex. The medial MD nucleus projects to the PFC, and is the final link in the reward circuit [91,149,150]. However, these connections are bidirectional. Moreover, while corticothalamic projections of the specific thalamic relay nuclei follow a general rule of reciprocity, the corticothalamic projections of the MD (as seen in other thalamocortical systems) are more extensive than its thalamocortical projections [150–152]. Importantly, they are, in part, derived from areas not innervated by the same thalamic region, indicating a nonreciprocal corticothalamic component. In particular, within the reward circuit, whereas the MD completes the reward cortical–BG circuit, its nonreciprocal input is derived from functionally distinct frontal cortical areas. For example, within the reward circuit, the central MD has reciprocal projections with the OFC, but also a nonreciprocal input from the VMPFC. More lateral MD areas are reciprocally connected to the DPFC, but also have a nonreciprocal connection to and from the OFC [150].

The Lateral Habenula, Pedunculopontine Tegmental Nucleus, and the Raphe Serotonergic Systems

Studies have emphasized the potential importance of the LHb in regulating the DA reward signal. In particular, experiments show that stimulation of the LHb nuclei in primates results in a negative-reward-related signal in the SNc, by inhibiting DA activity when an expected reward does not occur [93]. These LHb cells are inhibited by a reward-predicting stimulus, but fire following a non-reward signal. Stimulation of the LHb inhibits DA cells, suggesting that it plays a key role in regulating the DA reward (or rather negative reward) signal. Interestingly, few fibers from the LHb directly reach the SNc in primates, indicating an indirect regulation of the DA signal. There are several possible routes by which the LHb might impact on the DA cell. Connections described in rats, in addition to those to BG structures, include the basal forebrain, preoptic area of the hypothalamus, interpeduncular nucleus, pedunculopontine nucleus, raphe nucleus, superior colliculus, pretectal area, central gray, VTA, and reticular formation [91,145,147]. The pedunculopontine tegmental nucleus is connected to multiple BG structures and provides one of the strongest excitatory inputs to the midbrain DA cells [153,154]. Moreover, the cells in this brain-stem area receive input from the LHb. Anatomical and physiological studies, coupled with the central role of DA for reward prediction error, led to studies that support the hypothesis that the pedunculopontine nucleus may play a role in this reward signal [155]. The brain stem serotonergic system plays an important role in reinforcement behaviors by encoding expected and received rewards [156]. This reward signal could arise from a number of brain regions, but perhaps the strongest might arise from inputs derived from the OFC and VMPFC, the amygdala, the substantia nigra, and the LHb [157].

Complex Network Features of the Reward Circuit

The reward circuit comprises several cortical and subcortical regions that form a complex network to mediate different aspects of incentive learning, leading to adaptive behaviors and good decision-making. The cortical–BG network is at the center of this circuit. This complex network involves a system in which the PFC exploits the BG for additional processing of reward to effectively modulate learning that leads to the development of goal-directed behaviors and action plans. To develop an appropriate behavioral response to external environmental stimuli, information about motivation and reward needs to be combined with a strategy and an action plan for obtaining the goal. In other words, it is not sufficient to desire or want to, for example, to win at a card game. One has to understand the rules of the game, remember the cards played, etc., before executing the play. In addition, there is a complex interaction between the desire to put cards in play and the inhibition of impulse to play them too early. Thus action plans developed toward obtaining a goal require a combination of reward processing, cognition, and motor control. Yet, theories related to cortical–BG processing have emphasized the separation between functions (including different reward circuits), highlighting separate and parallel pathways [158–160]. The pathways and connections reviewed in this chapter clearly show that there are dual cortical–BG systems permitting both parallel and integrative processing.

Thus the reward circuit does not work in isolation. As such, this complex circuitry interfaces with pathways that mediate cognitive function and motor planning (Fig. 1.6). Within each of the cortical–BG structures, there are regions of convergence linking up areas that are associated with different functional domains. Convergence between terminals derived from different cortical areas in the striatum and STN permits cortical information to be disseminated across multiple functional regions. In addition, there are several interconnections in the system that contain reciprocal–nonreciprocal networks. The two major ones are the striatonigrostriatal pathway and the corticothalamocortical network. In addition, the VP–VS–VP also has a nonreciprocal component. Through these networks, the reward circuit influences cognition and motor control through several different interactive routes allowing information about reward to be channeled through cognitive and motor control circuits to mediate the development of appropriate action plans.

References

[1] Alexander GE, Crutcher MD, DeLong MR. Basal ganglia-thalamocortical circuits: parallel substrates for motor, oculomotor, "prefrontal" and "limbic" functions. Prog Brain Res 1990; 85:119–46.

[2] Bevan MD, Smith AD, Bolam JP. The substantia nigra as a site of synaptic integration of functionally diverse information arising from the ventral pallidum and the globus pallidus in the rat. Neuroscience 1996;75:5–12.

[3] Brown LL, Smith DM, Goldbloom LM. Organizing principles of cortical integration in the rat neostriatum: corticostriate map of the body surface is an ordered lattice of curved laminae and radial points. J Comp Neurol 1998;392:468–88.

[4] Kasanetz F, Riquelme LA, Della-Maggiore V, O'Donnell P, Murer MG. Functional integration across a gradient of corticostriatal channels controls UP state transitions in the dorsal striatum. Proc Natl Acad Sci USA 2008;105:8124–9.

[5] Haber SN, Fudge JL, McFarland NR. Striatonigrostriatal pathways in primates form an ascending spiral from the shell to the dorsolateral striatum. J Neurosci 2000;20:2369–82.

[6] Haber SN, Kim KS, Mailly P, Calzavara R. Reward-related cortical inputs define a large striatal region in primates that interface with associative cortical connections, providing a substrate for incentive-based learning. J Neurosci 2006;26:8368–76.

[7] Averbeck BB, Lehman J, Jacobson M, Haber SN. Estimates of projection overlap and zones of convergence within frontal-striatal circuits. J Neurosci 2014;34:9497–505.

[8] Haynes WI, Haber SN. The organization of prefrontal-subthalamic inputs in primates provides an anatomical substrate for both functional specificity and integration: implications for basal ganglia models and deep brain stimulation. J Neurosci 2013;33:4804–14.

[9] Carmichael ST, Price JL. Architectonic subdivision of the orbital and medial prefrontal cortex in the macaque monkey. J Comp Neurol 1994;346:366–402.

[10] Barbas H. Architecture and cortical connections of the prefrontal cortex in the rhesus monkey. In: Chauvel P, Delgado-Escueta AV, editors. Advances in neurology. New York: Raven Press, Ltd; 1992. p. 91–115.

[11] Goldman-Rakic PS, Funahashi S, Bruce CJ. Neocortical memory circuits. Cold Spring Harb Symp Quant Biol 1990;55:1025–38.

[12] Wallis JD, Miller EK. Neuronal activity in primate dorsolateral and orbital prefrontal cortex during performance of a reward preference task. Eur J Neurosci 2003;18:2069–81.

[13] Padoa-Schioppa C, Assad JA. Neurons in the orbitofrontal cortex encode economic value. Nature 2006;441:223–6.

[14] Tremblay L, Schultz W. Reward-related neuronal activity during go-nogo task performance in primate orbitofrontal cortex. Jour Neurophysiol 2000;83:1864–76.

[15] Roesch MR, Olson CR. Neuronal activity related to reward value and motivation in primate frontal cortex. Science 2004;304:307–10.

[16] McEnaney KW, Butter CM. Perseveration of responding and nonresponding in monkeys with orbital frontal ablations. J Comp Physiol Psychol 1969;68:558–61.

[17] Fellows LK, Farah MJ. Ventromedial frontal cortex mediates affective shifting in humans: evidence from a reversal learning paradigm. Brain 2003;126:1830–7.

[18] Izquierdo A, Suda RK, Murray EA. Bilateral orbital prefrontal cortex lesions in rhesus monkeys disrupt choices guided by both reward value and reward contingency. J Neurosci 2004;24:7540–8.

[19] Sescousse G, Redoute J, Dreher JC. The architecture of reward value coding in the human orbitofrontal cortex. J Neurosci 2010;30:13095–104.

[20] Walton ME, Bannerman DM, Alterescu K, Rushworth MF. Functional specialization within medial frontal cortex of the anterior cingulate for evaluating effort-related decisions. J Neurosci 2003;23:6475–9.

[21] Paus T. Primate anterior cingulate cortex: where motor control, drive and cognition interface. Nat Rev Neurosci 2001;2:417–24.

[22] Vogt BA, Vogt L, Farber NB, Bush G. Architecture and neurocytology of monkey cingulate gyrus. J Comp Neurol 2005;485:218–39.

[23] Bouret S, Richmond BJ. Ventromedial and orbital prefrontal neurons differentially encode internally and externally driven motivational values in monkeys. J Neurosci 2010;30:8591–601.

[24] Behrens TE, Hunt LT, Woolrich MW, Rushworth MF. Associative learning of social value. Nature 2008;456:245–9.

[25] Rudebeck PH, Behrens TE, Kennerley SW, Baxter MG, Buckley MJ, Walton ME, et al. Frontal cortex subregions play distinct roles in choices between actions and stimuli. J Neurosci 2008;28:13775–85.

[26] Camille N, Tsuchida A, Fellows LK. Double dissociation of stimulus-value and action-value learning in humans with orbitofrontal or anterior cingulate cortex damage. J Neurosci 2011;31:15048–52.

[27] Alexander WH, Brown JW. Medial prefrontal cortex as an action-outcome predictor. Nat Neurosci 2011;14:1338–44.

[28] Carmichael ST, Price JL. Connectional networks within the orbital and medial prefrontal cortex of macaque monkeys. J Comp Neurol 1996;371:179–207.

[29] Carmichael ST, Price JL. Limbic connections of the orbital and medial prefrontal cortex in macaque monkeys. J Comp Neurol 1995;363:615–41.

[30] Barbas H, Blatt GJ. Topographically specific hippocampal projections target functionally distinct prefrontal areas in the rhesus monkey. Hippocampus 1995;5:511–33.

[31] Ghashghaei HT, Barbas H. Pathways for emotion: interactions of prefrontal and anterior temporal pathways in the amygdala of the rhesus monkey. Neuroscience 2002;115:1261–79.

[32] Olds J, Milner P. Positive reinforcement produced by electrical stimulation of septal area and other regions of rat brain. J Comp Physiol Psychol 1954;47:419–27.

[33] Taha SA, Fields HL. Inhibitions of nucleus accumbens neurons encode a gating signal for reward-directed behavior. J Neurosci 2006;26:217–22.

[34] Kalivas PW, Volkow N, Seamans J. Unmanageable motivation in addiction: a pathology in prefrontal-accumbens glutamate transmission. Neuron 2005;45:647–50.

[35] Heimer L, De Olmos JS, Alheid GF, Person J, Sakamoto N, Shinoda K, et al. The human basal forebrain. Part II. In handbook of chemical neuroanatomy. In: Bloom FE, Bjorkland A, Hokfelt T, editors. The primate nervous system, Part III, vol. 15. Amsterdam: Elsevier; 1999. p. 57–226.

[36] Haber SN, McFarland NR. The concept of the ventral striatum in nonhuman primates. Ann NY Acad Sci 1999;877:33–48.

[37] Pagnoni G, Zink CF, Montague PR, Berns GS. Activity in human ventral striatum locked to errors of reward prediction. Nat Neurosci 2002;5:97–8.

[38] Knutson B, Adams CM, Fong GW, Hommer D. Anticipation of increasing monetary reward selectively recruits nucleus accumbens. J Neurosci 2001;21:RC159.

[39] Schultz W. Multiple reward signals in the brain. Nat Rev Neurosci 2000;1:199–207.

[40] Cohen MX, Schoene-Bake JC, Elger CE, Weber B. Connectivity-based segregation of the human striatum predicts personality characteristics. Nat Neurosci 2009;12:32–4.

[41] O'Doherty J, Dayan P, Schultz J, Deichmann R, Friston K, Dolan RJ. Dissociable roles of ventral and dorsal striatum in instrumental conditioning. Science 2004;304:452–4.

[42] Rutledge RB, Dean M, Caplin A, Glimcher PW. Testing the reward prediction error hypothesis with an axiomatic model. J Neurosci 2010;30:13525–36.

[43] Schultz W. Getting formal with dopamine and reward. Neuron 2002;36:241–63.

[44] Chowdhury R, Guitart-Masip M, Lambert C, Dayan P, Huys Q, Duzel E, et al. Dopamine restores reward prediction errors in old age. Nat Neurosci 2013;16:648–53.

[45] Cromwell HC, Schultz W. Effects of expectations for different reward magnitudes on neuronal activity in primate striatum. J Neurophysiol 2003;89:2823–38.

[46] Watanabe K, Hikosaka O. Immediate changes in anticipatory activity of caudate neurons associated with reversal of position-reward contingency. J Neurophysiol 2005;94:1879–87.

[47] Tremblay L, Hollerman JR, Schultz W. Modifications of reward expectation-related neuronal activity during learning in primate striatum. J Neurophysiol 1998;80:964–77.

[48] Elliott R, Newman JL, Longe OA, Deakin JF. Differential response patterns in the striatum and orbitofrontal cortex to financial reward in humans: a parametric functional magnetic resonance imaging study. J Neurosci 2003;23:303–7.

[49] Schultz W, Tremblay L, Hollerman JR. Reward processing in primate orbitofrontal cortex and basal ganglia. Cereb Cortex 2000; 10:272–84.

[50] Volkow N, Li TK. The neuroscience of addiction. Nat Neurosci 2005;8:1429–30.

[51] Kuhnen CM, Knutson B. The neural basis of financial risk taking. Neuron 2005;47:763–70.

[52] Ito R, Robbins TW, Everitt BJ. Differential control over cocaine-seeking behavior by nucleus accumbens core and shell. Nat Neurosci 2004;7:389–97.

[53] Carlezon WA, Wise RA. Rewarding actions of phencyclidine and related drugs in nucleus accumbens shell and frontal cortex. J Neurosci 1996;16:3112–22.

[54] Meredith GE, Pattiselanno A, Groenewegen HJ, Haber SN. Shell and core in monkey and human nucleus accumbens identified with antibodies to calbindin-D$_{28k}$. J Comp Neurol 1996; 365:628–39.

[55] Alheid GF, Heimer L. New perspectives in basal forebrain organization of special relevance for neuropsychiatric disorders: the striatopallidal, amygdaloid, and corticopetal components of substantia innominata. Neuroscience 1988;27:1–39.

[56] Ikemoto K, Satoh K, Maeda T, Fibiger HC. Neurochemical heterogeneity of the primate nucleus accumbens. Exp Brain Res 1995;104:177–90.

[57] Sato K, Kiyama H, Tohyama M. The differential expression patterns of messenger RNAs encoding non-N-methyl-D-aspartate glutamate receptor subunits (GluR1–4) in the rat brain. Neuroscience 1993;52:515–39.

[58] Martin LJ, Blackstone CD, Levey AI, Huganir RL, Price DL. AMPA glutamate receptor subunits are differentially distributed in rat brain. Neuroscience 1993;53(2):327–38.

[59] Haber SN, Ryoo H, Cox C, Lu W. Subsets of midbrain dopaminergic neurons in monkeys are distinguished by different levels of mRNA for the dopamine transporter: comparison with the mRNA for the D2 receptor, tyrosine hydroxylase and calbindin immunoreactivity. J Comp Neurol 1995;362:400–10.

[60] Chronister RB, Sikes RW, Trow TW, DeFrance JF. The organization of the nucleus accumbens. In: Chronister RB, DeFrance JF, editors. The neurobiology of the nucleus accumbens. Brunswick (ME): Haer Institute; 1981. p. 97–146.

[61] Meyer G, Gonzalez-Hernandez T, Carrillo-Padilla F, Ferres-Torres R. Aggregations of granule cells in the basal forebrain (islands of Calleja): Golgi and cytoarchitectonic study in different mammals, including man. J Comp Neurol 1989;284:405–28.

[62] Bayer SA. Neurogenesis in the olfactory tubercle and islands of Calleja in the rat. Int J Devl Neurosci 1985;3:135–47.

[63] Parent A. Comparative neurobiology of the basal ganglia. New York: John Wiley and Sons; 1986.

[64] Calzavara R, Mailly P, Haber SN. Relationship between the corticostriatal terminals from areas 9 and 46, and those from area 8A, dorsal and rostral premotor cortex and area 24c: an anatomical substrate for cognition to action. Eur J Neurosci 2007;26: 2005–24.

[65] Selemon LD, Goldman-Rakic PS. Longitudinal topography and interdigitation of corticostriatal projections in the rhesus monkey. J Neurosci 1985;5:776–94.

[66] Haber SN, Kunishio K, Mizobuchi M, Lynd-Balta E. The orbital and medial prefrontal circuit through the primate basal ganglia. J Neurosci 1995;15:4851–67.

[67] Dommett E, Coizet V, Blaha CD, Martindale J, Lefebvre V, Walton N, et al. How visual stimuli activate dopaminergic neurons at short latency. Science 2005;307:1476–9.

[68] Tanaka SC, Doya K, Okada G, Ueda K, Okamoto Y, Yamawaki S. Prediction of immediate and future rewards differentially recruits cortico-basal ganglia loops. Nat Neurosci 2004;7:887–93.

[69] Corlett PR, Aitken MR, Dickinson A, Shanks DR, Honey GD, Honey RA, et al. Prediction error during retrospective revaluation of causal associations in humans: fMRI evidence in favor of an associative model of learning. Neuron 2004;44: 877–88.

[70] Wilson CJ. The basal ganglia. In: Shepherd GM, editor. The synaptic organization of the brain. 5th ed. New York (NY): Oxford University Press; 2004. p. 361–413.

[71] Buckner RL, Andrews-Hanna JR, Schacter DL. The brain's default network: anatomy, function, and relevance to disease. Ann NY Acad Sci 2008;1124:1–38.

[72] Power JD, Schlaggar BL, Lessov-Schlaggar CN, Petersen SE. Evidence for hubs in human functional brain networks. Neuron 2013;79:798–813.

[73] Russchen FT, Bakst I, Amaral DG, Price JL. The amygdalostriatal projections in the monkey. an anterograde tracing study. Brain Res 1985;329:241–57.

[74] Fudge JL, Kunishio K, Walsh P, Richard C, Haber SN. Amygdaloid projections to ventromedial striatal subterritories in the primate. Neuroscience 2002;110:257–75.

[75] Friedman DP, Aggleton JP, Saunders RC. Comparison of hippocampal, amygdala, and perirhinal projections to the nucleus accumbens: combined anterograde and retrograde tracing study in the Macaque brain. J Comp Neurol 2002;450:345–65.

[76] Berendse HW, Groenewegen HJ, Lohman AHM. Compartmental distribution of ventral striatal neurons projecting to the mesencephalon in the rat. J Neurosci 1992;12(6):2079–103.

[77] Gimenez-Amaya JM, McFarland NR, de las Heras S, Haber SN. Organization of thalamic projections to the ventral striatum in the primate. J Comp Neurol 1995;354:127–49.

[78] Berendse HW, Groenewegen HJ. The organization of the thalamostriatal projections in the rat, with special emphasis on the ventral striatum. J Comp Neurol 1990;299:187–228.

[79] McFarland NR, Haber SN. Convergent inputs from thalamic motor nuclei and frontal cortical areas to the dorsal striatum in the primate. J Neurosci 2000;20:3798–813.

[80] McFarland NR, Haber SN. Organization of thalamostriatal terminals from the ventral motor nuclei in the macaque. J Comp Neurol 2001;429:321–36.

[81] Haber SN, Lynd E, Klein C, Groenewegen HJ. Topographic organization of the ventral striatal efferent projections in the rhesus monkey: an anterograde tracing study. J Comp Neurol 1990; 293:282–98.

[82] Zahm DS, Heimer L. Two transpallidal pathways originating in the rat nucleus accumbens. J Comp Neurol 1990;302:437–46.

[83] Martinez-Murillo R, Blasco I, Alvarez FJ, Villalba R, Solano ML, Montero-Caballero MI, et al. Distribution of enkephalin-immunoreactive nerve fibers and terminals in the region of the nucleus basalis magnocellularis of the rat: a light and electron microscopic study. J Neurocytol 1988;17:361–76.

[84] Chang HT, Penny GR, Kitai ST. Enkephalinergic-cholinergic interaction in the rat globus pallidus: a pre-embedding double-labeling immunocytochemistry study. Brain Res 1987;426:197–203.

[85] Beach TG, Tago H, McGeer EG. Light microscopic evidence for a substance P-containing innervation of the human nucleus basalis of Meynert. Brain Res 1987;408:251−7.

[86] Haber S. Anatomical relationship between the basal ganglia and the basal nucleus of Meynert in human and monkey forebrain. Proc Natl Acad Sci USA 1987;84:1408−12.

[87] Zaborszky L, Cullinan WE. Projections from the nucleus accumbens to cholinergic neurons of the ventral pallidum: a correlated light and electron microscopic double-immunolabeling study in rat. Brain Res 1992;570:92−101.

[88] Saunders A, Oldenburg IA, Berezovskii VK, Johnson CA, Kingery ND, Elliott HL, et al. A direct GABAergic output from the basal ganglia to frontal cortex. Nature 2015;521:85−9.

[89] Smith KS, Berridge KC. Opioid limbic circuit for reward: interaction between hedonic hotspots of nucleus accumbens and ventral pallidum. J Neurosci 2007;27:1594−605.

[90] Tindell AJ, Smith KS, Pecina S, Berridge KC, Aldridge JW. Ventral pallidum firing codes hedonic reward: when a bad taste turns good. J Neurophysiol 2006;96:2399−409.

[91] Haber SN, Lynd-Balta E, Mitchell SJ. The organization of the descending ventral pallidal projections in the monkey. J Comp Neurol 1993;329:111−28.

[92] Parent A, Gravel S, Boucher R. The origin of forebrain afferents to the habenula in rat, cat and monkey. Brain Res Bull 1981;6:23−38.

[93] Matsumoto M, Hikosaka O. Lateral habenula as a source of negative reward signals in dopamine neurons. Nature 2007; 447:1111−5.

[94] Ji H, Shepard PD. Lateral habenula stimulation inhibits rat midbrain dopamine neurons through a GABA(A) receptor-mediated mechanism. J Neurosci 2007;27:6923−30.

[95] Ullsperger M, von Cramon DY. Error monitoring using external feedback: specific roles of the habenular complex, the reward system, and the cingulate motor area revealed by functional magnetic resonance imaging. J Neurosci 2003;23:4308−14.

[96] Heimer L. The olfactory cortex and the ventral striatum. In: Livingston KE, Hornykiewicz O, editors. Limbic mechanisms. New York: Plenum Press; 1978. p. 95−187.

[97] Haber SN, Wolfe DP, Groenewegen HJ. The relationship between ventral striatal efferent fibers and the distribution of peptide-positive woolly fibers in the forebrain of the rhesus-monkey. Neuroscience 1990;39:323−38.

[98] Russchen FT, Amaral DG, Price JL. The afferent input to the magnocellular division of the mediodorsal thalamic nucleus in the monkey, *Macaca fascicularis*. J Comp Neurol 1987;256:175−210.

[99] Mai JK, Stephens PH, Hopf A, Cuello AC. Substance P in the human brain. Neuroscience 1986;17(3):709−39.

[100] DiFiglia M, Aronin N, Martin JB. Light and electron microscopic localization of immunoreactive leu-enkephalin in the monkey basal ganglia. J Neurosci 1982;2(3):303−20.

[101] Haber SN, Watson SJ. The comparative distribution of enkephalin, dynorphin and substance P in the human globus pallidus and basal forebrain. Neuroscience 1985;14:1011−24.

[102] Haber SN, Nauta WJ. Ramifications of the globus pallidus in the rat as indicated by patterns of immunohistochemistry. Neuroscience 1983;9:245−60.

[103] Fox CH, Andrade HN, Du Qui IJ, Rafols JA. The primate globus pallidus. A Golgi and electron microscope study. J Hirnforsch 1974;15:75−93.

[104] Turner MS, Lavin A, Grace AA, Napier TC. Regulation of limbic information outflow by the subthalamic nucleus: excitatory amino acid projections to the ventral pallidum. J Neurosci 2001;21:2820−32.

[105] Klitenick MA, Deutch AY, Churchill L, Kalivas PW. Topography and functional role of dopaminergic projections from the ventral mesencephalic tegmentum to the ventral pallidum. Neuroscience 1992;50(2):371−86.

[106] Haber SN, Groenewegen HJ, Grove EA, Nauta WJ. Efferent connections of the ventral pallidum: evidence of a dual striato pallidofugal pathway. J Comp Neurol 1985;235:322−35.

[107] Zahm DS. The ventral striatopallidal parts of the basal ganglia in the rat. II. Compartmentation of ventral pallidal efferents. Neuroscience 1989;30:33−50.

[108] Francois C, Grabli D, McCairn K, Jan C, Karachi C, Hirsch EC, et al. Behavioural disorders induced by external globus pallidus dysfunction in primates II. Anatomical study. Brain 2004;127: 2055−70.

[109] Ilinsky IA, Yi H, Kultas-Ilinsky K. Mode of termination of pallidal afferents to the thalamus: a light and electron microscopic study with anterograde tracers and immunocytochemistry in *Macaca mulatta*. J Comp Neurol 1997;386:601−12.

[110] Arecchi-Bouchhioua P, Yelnik J, Francois C, Percheron G, Tande D. Three-dimensional morphology and distribution of pallidal axons projecting to both the lateral region of the thalamus and the central complex in primates. Brain Res 1997;754: 311−4.

[111] Parent A, De Bellefeuille L. Organization of efferent projections from the internal segment of the globus pallidus in the primate as revealed by fluorescence retrograde labeling method. Brain Res 1982;245:201−13.

[112] Spooren WP, Lynd-Balta E, Mitchell S, Haber SN. Ventral pallid-ostriatal pathway in the monkey: evidence for modulation of basal ganglia circuits. J Comp Neurol 1996;370:295−312.

[113] Wise RA. Brain reward circuitry: insights from unsensed incentives. Neuron 2002;36:229−40.

[114] Lavoie B, Parent A. Dopaminergic neurons expressing calbindin in normal and parkinsonian monkeys. Neuroreport 1991;2(10): 601−4.

[115] Lynd-Balta E, Haber SN. Primate striatonigral projections: a comparison of the sensorimotor-related striatum and the ventral striatum. J Comp Neurol 1994;345:562−78.

[116] Szabo J. Strionigral and nigrostriatal connections. Anatomical studies. Appl Neurophysiol 1979;42:9−12.

[117] Hedreen JC, DeLong MR. Organization of striatopallidal, striatonigral, and nigrostriatal projections in the Macaque. J Comp Neurol 1991;304:569−95.

[118] Sgambato-Faure V, Worbe Y, Epinat J, Feger J, Tremblay L. Cortico-basal ganglia circuits involved in different motivation disorders in non-human primates. Brain Struct Funct 2014;221: 345−64.

[119] Fudge JL, Haber SN. The central nucleus of the amygdala projection to dopamine subpopulations in primates. Neuroscience 2000;97:479−94.

[120] Fudge JL, Haber SN. Bed nucleus of the stria terminalis and extended amygdala inputs to dopamine subpopulations in primates. Neuroscience 2001;104:807−27.

[121] Frankle WG, Laruelle M, Haber SN. Prefrontal cortical projections to the midbrain in primates: evidence for a sparse connection. Neuropsychopharmacology 2006;31:1627−36.

[122] Lynd-Balta E, Haber SN. The organization of midbrain projections to the striatum in the primate: sensorimotor-related striatum versus ventral striatum. Neuroscience 1994;59: 625−40.

[123] Gaspar P, Stepneiwska I, Kaas JH. Topography and collateralization of the dopaminergic projections to motor and lateral prefrontal cortex in owl monkeys. J Comp Neurol 1992;325:1−21.

[124] Lidow MS, Goldman-Rakic PS, Gallager DW, Rakic P. Distribution of dopaminergic receptors in the primate cerebral cortex: quantitative autoradiographic analysis using [^{3}H] raclopride, [^{3}H] spiperone and [^{3}H]SCH23390. Neuroscience 1991;40(3): 657−71.

[125] Lewis DA. The catecholaminergic innervation of primate prefrontal cortex. J Neural Transm Suppl 1992;36:179−200.

[126] Goldman-Rakic PS, Bergson C, Krimer LS, Lidow MS, Williams SM, Williams GV. Chapter V the primate mesocortical dopamine system. In: Bloom FE, Bjorklund A, Hokfelt T, editors. Handbook of chemical neuroanatomy, vol. 15. Amsterdam: Elsevier Science; 1999. p. 403–28.

[127] Brinley-Reed M, McDonald AJ. Evidence that dopaminergic axons provide a dense innervation of specific neuronal subpopulations in the rat basolateral amygdala. Brain Res 1999;850:127–35.

[128] Volkow ND, Wang GJ, Telang F, Fowler JS, Logan J, Childress AR, et al. Cocaine cues and dopamine in dorsal striatum: mechanism of craving in cocaine addiction. J Neurosci 2006;26:6583–8.

[129] Porrino LJ, Lyons D, Smith HR, Daunais JB, Nader MA. Cocaine self-administration produces a progressive involvement of limbic, association, and sensorimotor striatal domains. J Neurosci 2004;24:3554–62.

[130] Lehericy S, Benali H, Van de Moortele PF, Pelegrini-Issac M, Waechter T, Ugurbil K, et al. Distinct basal ganglia territories are engaged in early and advanced motor sequence learning. Proc Natl Acad Sci USA 2005;102:12566–71.

[131] Pasupathy A, Miller EK. Different time courses of learning-related activity in the prefrontal cortex and striatum. Nature 2005;433:873–6.

[132] Everitt BJ, Robbins TW. Neural systems of reinforcement for drug addiction: from actions to habits to compulsion. Nat Neurosci 2005;8:1481–9.

[133] Belin D, Everitt BJ. Cocaine seeking habits depend upon dopamine-dependent serial connectivity linking the ventral with the dorsal striatum. Neuron 2008;57:432–41.

[134] Kunzle H, Akert K. Efferent connections of cortical, area 8 (frontal eye field) in *Macaca fascicularis*. A reinvestigation using the autoradiographic technique. J Comp Neurol 1977;173:147–64.

[135] Hartmann-von Monakow K, Akert K, Künzle H. Projections of precentral and premotor cortex to the red nucleus and other midbrain areas in *Macaca fascicularis*. Exp Brain Res 1979;34: 91–105.

[136] Nambu A, Tokuno H, Inase M, Takada M. Corticosubthalamic input zones from forelimb representations of the dorsal and ventral divisions of the premotor cortex in the macaque monkey: comparison with the input zones from the primary motor cortex and the supplementary motor area. Neurosci Lett 1997; 239:13–6.

[137] Shink E, Bevan MD, Bolam JP, Smith Y. The subthalamic nucleus and the external pallidum: two tightly interconnected structures that control the output of the basal ganglia in the monkey. Neuroscience 1996;73:335–57.

[138] Karachi C, Yelnik J, Tande D, Tremblay L, Hirsch EC, Francois C. The pallidosubthalamic projection: an anatomical substrate for nonmotor functions of the subthalamic nucleus in primates. Mov Disord 2005;20:172–80.

[139] Yelnik J, Percheron G. Subthalamic neurons in primates: a quantitative and comparative analysis. Neuroscience 1979;4:1717–43.

[140] Bevan MD, Clarke NP, Bolam JP. Synaptic integration of functionally diverse pallidal information in the entopeduncular nucleus and subthalamic nucleus in the rat. J Neurosci 1997;17: 308–24.

[141] Hong S, Hikosaka O. The globus pallidus sends reward-related signals to the lateral habenula. Neuron 2008;60:720–9.

[142] Salas R, Baldwin P, de Biasi M, Montague PR. BOLD responses to negative reward prediction errors in human habenula. Front Hum Neurosci 2010;4:36.

[143] Hikosaka O. The habenula: from stress evasion to value-based decision-making. Nat Rev Neurosci 2010;11:503–13.

[144] Christoph GR, Leonzio RJ, Wilcox KS. Stimulation of the lateral habenula inhibits dopamine-containing neurons in the substantia nigra and ventral tegmental area of the rat. J Neurosci 1986; 6:613–9.

[145] Araki M, McGeer PL, Kimura H. The efferent projections of the rat lateral habenular nucleus revealed by the PHA-L anterograde tracing method. Brain Res 1988;441:319–30.

[146] Herkenham M, Nauta WJH. Efferent connections of the habenular nuclei in the rat. J Comp Neurol 1979;187:19–48.

[147] Herkenham M, Nauta WJH. Afferent connections of the habenular nuclei in the rat. A horseradish peroxidase study, with a note on the fiber-of-passage problem. J Comp Neurol 1977;173: 123–46.

[148] Bromberg-Martin ES, Matsumoto M, Hong S, Hikosaka O. A pallidus-habenula-dopamine pathway signals inferred stimulus values. J Neurophysiol 2010;104:1068–76.

[149] Ray JP, Price JL. The organization of projections from the mediodorsal nucleus of the thalamus to orbital and medial prefrontal cortex in macaque monkeys. J Comp Neurol 1993; 337:1–31.

[150] McFarland NR, Haber SN. Thalamic relay nuclei of the basal ganglia form both reciprocal and nonreciprocal cortical connections, linking multiple frontal cortical areas. J Neurosci 2002;22: 8117–32.

[151] Darian-Smith C, Tan A, Edwards S. Comparing thalamocortical and corticothalamic microstructure and spatial reciprocity in the macaque ventral posterolateral nucleus (VPLc) and medial pulvinar. J Comp Neurol 1999;410:211–34.

[152] Sherman SM, Guillery RW. Functional organization of thalamocortical relays. J Neurophysiol 1996;76:1367–95.

[153] Lavoie B, Parent A. Pedunculopontine nucleus in the squirrel monkey: projections to the basal ganglia as revealed by anterograde tract-tracing methods. J Comp Neurol 1994;344: 210–31.

[154] Blaha CD, Allen LF, Das S, Inglis WL, Latimer MP, Vincent SR, et al. Modulation of dopamine efflux in the nucleus accumbens after cholinergic stimulation of the ventral tegmental area in intact, pedunculopontine tegmental nucleus-lesioned, and laterodorsal tegmental nucleus-lesioned rats. J Neurosci 1996;16: 714–22.

[155] Kobayashi Y, Okada K. Reward prediction error computation in the pedunculopontine tegmental nucleus neurons. Ann NY Acad Sci 2007;1104:310–23.

[156] Nakamura K, Matsumoto M, Hikosaka O. Reward-dependent modulation of neuronal activity in the primate dorsal raphe nucleus. J Neurosci 2008;28:5331–43.

[157] Peyron C, Petit JM, Rampon C, Jouvet M, Luppi PH. Forebrain afferents to the rat dorsal raphe nucleus demonstrated by retrograde and anterograde tracing methods. Neuroscience 1998;82: 443–68.

[158] Alexander GE, Crutcher MD. Functional architecture of basal ganglia circuits: neural substrates of parallel processing. Trends Neurosci 1990;13:266–71.

[159] Middleton FA, Strick PL. Basal-ganglia 'projections' to the prefrontal cortex of the primate. Cereb Cortex 2002;12:926–35.

[160] Price JL, Carmichael ST, Drevets WC. Networks related to the orbital and medial prefrontal cortex; a substrate for emotional behavior? Prog Brain Res 1996;107:523–36.

2

Electrophysiological Correlates of Reward Processing in Dopamine Neurons

W. Schultz

University of Cambridge, Cambridge, United Kingdom

Abstract

Studies have identified three novel properties of the dopamine reward-prediction error signal. First, the dopamine response reports initially and unselectively many salient, potentially rewarding events and subsequently processes more specifically the reward-prediction error. This two-component structure restricts the earlier claimed salience coding to the initial component and explains aversive activations by physical impact rather than punishment. Second, the dopamine prediction error signal reflects subjective reward value and, more stringently, formal economic utility. A dopamine utility prediction error signal would be particularly useful for economic choices that maximize utility. Third, the dopamine signal fits well into formal competitive decision models, whereby it codes the output variable (chosen value) suitable for updating or immediately influencing main input variables (object value and action value). With these properties, the dopamine utility prediction error signal bridges the gap between animal learning theory (prediction error) and economic decision theory (utility).

INTRODUCTION

Dopamine has several functions in the brain. The neurologist knows dopamine from the movement disorders in Parkinsonian patients, who lack dopamine; the psychiatrist knows dopamine from psychotics, who may suffer from excessive dopamine processes; the cognitive psychopharmacologist knows dopamine from the many deficits caused by underfunctioning or overstimulated dopamine receptors in humans and animals; and the neurophysiologist knows dopamine from the reward responses of dopamine neurons. There seems to be no such thing with dopamine as "one neurotransmitter—one function." Research advances relate these diversities to different time courses of dopamine function, from tonic via slow to rapid, and to the various brain structures innervated by dopamine neurons [1].

This chapter will consider the most rapid dopamine function, which concerns primarily the signaling of reward in relation to its prediction. This signal is very similar in latency, duration, and sensitivity across dopamine neurons and contrasts with the widely diverse neuronal categories in other brain structures. I will describe advances in characterizing the dopamine reward-prediction error signal using behavioral tools of experimental economics, which are based on the well-developed concepts of economic choice theory. I will then formulate some hypotheses about the way this dopamine signal might function to update economic decision variables and to directly influence economic decisions. More details are found in a review from which this text is distilled [2].

BASICS OF DOPAMINE REWARD-PREDICTION ERROR CODING

Reward-Prediction Error

One of the simplest forms of learning involves error and error correction. Beyond the basic notion of incorrect behavior, "error" refers to the difference between an outcome and a prediction. In reinforcement learning, prediction errors drive learning according to the equation

$$V(t + 1) = V(t) + \alpha \times PE(t) \qquad (2.1)$$

with

$$PE(t) = \lambda(t) - V(t) \qquad (2.2)$$

where PE is the prediction error, V indicates reward prediction, α the learning coefficient, t the trial, and λ the reward. A reward that is better than predicted will increase the associative strength, or prediction, carried by a stimulus; a reward that is worse than predicted with reduce the associative strength, or prediction; and a reward that occurs exactly as predicted will generate no change in associative strength or prediction [3]. The term of associative strength in animal learning theory can be translated into the notion of value in machine learning and economic decision theory. Thus, a stimulus acquires motivational value through prediction-error learning; it becomes a higher order reinforcer that itself can induce learning. The suitability of conditioned stimuli, along with unconditioned rewards, for learning is captured in temporal difference (TD) learning [4] by the equation

$$V(t+1) = V(t) + \alpha \times TDPE(t) \qquad (2.3)$$

with

$$TDPE(t) = \left[\lambda(t) + \gamma \sum V(t)\right] - V(t-1) \qquad (2.4)$$

where γ indicates the value decay associated with later rewards (temporal discounting) and t the time step during a trial. Actions do not have value themselves but can lead to the acquisition of valuable objects for the animal, which motivates the shorthand term of action value.

The majority of dopamine neurons (60–95%) in the pars compacta of the substantia nigra (SNc) and in the ventral tegmental area (VTA) report reward-prediction errors in humans, monkeys, rats, and mice (Fig. 2.1A); the remaining dopamine neurons fail to respond at all [5–10]. With food or liquid rewards that are better than predicted, or completely unpredicted, the response consists of a short, phasic increase in activity at latencies of <100 ms and durations of <200 ms ("activation"). With rewards that are worse than predicted, or completely omitted, the response consists of a depression of activity with similar latencies but somewhat longer durations. Rewards that occur exactly as predicted draw no dopamine response. Thus, the response reflects fully the bidirectional nature of prediction errors as conceptualized in Rescorla–Wagner learning [11,12]. Importantly, the response is sensitive to the temporal occurrence of reward and thus reflects a prediction error in time [13]. The reward response fulfills two formal prediction error tests. In blocking, a stimulus is not learned when the reward is fully predicted by another stimulus, and the dopamine response reflects the absence of prediction by the unlearned stimulus at the time of reward [14]. In conditioned inhibition, a stimulus explicitly predicts reward absence, and a reward occurring after that stimulus elicits an exaggerated positive-prediction error and correspondingly strong dopamine activation [15].

Unpredicted reward-predicting stimuli induce phasic activations in most dopamine neurons (Fig. 2.1A) [6,16] and induce striatal dopamine release [17]. These responses appear to concur with prediction V of the Rescorla–Wagner models but, on closer inspection, depend on the prediction of the reward-predicting stimuli and thus reflect higher order reward-prediction errors that are compatible with TD models (Fig. 2.1B and C) [5,7,18–20].

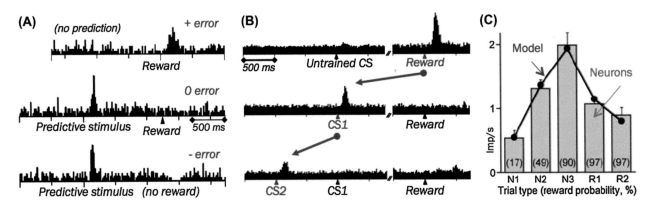

FIGURE 2.1 Dopamine reward-prediction error responses in monkeys. (A) Reward-prediction error responses at time of reward (right) and reward-predicting visual stimuli (left in bottom two graphs). The dopamine neuron is activated by the unpredicted reward, eliciting a positive reward-prediction error (blue +error, top); shows no response to the fully predicted reward, eliciting no prediction error (0 error, middle); and is depressed by the omission of predicted reward, eliciting a negative prediction error (−error, bottom). *(From Schultz W, Dayan P, Montague RR. A neural substrate of prediction and reward. Science 1997;275:1593–9.)* (B) Propagation of dopamine response from reward to first reward-predicting stimulus. CS2 and CS1, reward-predicting stimuli in delayed response task. *(From Schultz W. Predictive reward signal of dopamine neurons. J Neurophysiol 1998;80:1–27.)* (C) Reward-prediction error responses in serial reward task matching formal temporal difference prediction errors (averaged from 26 dopamine neurons). *Gray bars* indicate dopamine response amplitude at different sequence steps, *numbers* indicate reward probabilities in percentage. *Black line* shows time course of modeled prediction errors. *(From Enomoto K, Matsumoto N, Nakai S, Satoh T, Sato TK, Ueda Y, et al. Dopamine neurons learn to encode the long-term value of multiple future rewards. Proc Natl Acad Sci USA 2011;108:15462–7.)*

Human neuroimaging studies report blood oxygen level-dependent neuromagnetic (fMRI) signals reflecting reward-prediction errors in the ventral midbrain [21]. Reward-prediction error signals are usually stronger in the ventral striatum [22,23], where they presumably reflect dopamine inputs to the area [24].

Response Components

The dopamine reward prediction error response develops over 50–200 ms from an initial, brief, and unselective activation component that detects any noticeable event, including "neutral" stimuli, conditioned inhibitors, and punishers [25–31]. Other cognitive and reward neurons show similar gradual developments from initial unselective reports to subsequent specific stimulus identification and discrimination [32,33], although inferotemporal feature detectors capture stimulus specificity immediately [34]. The components of dopamine responses are often not discriminable with easily identifiable stimuli but become distinctive with more demanding stimuli [19]. The unselective response part is boosted by physical stimulus intensity [29–31], similarity to rewarded stimuli (generalization) [26], rewarded context [35], and novelty [25]. These factors reflect physical salience, positive motivational salience, and novelty salience, respectively. In this way, dopamine neurons detect a wide spectrum of stimuli unselectively and thus are unlikely to miss major rewards. The short latency of the initial response component allows subjects to prepare actions very early, and the different forms of salience are likely to sharpen neuronal processing in the subsequent reward component. As the main dopamine response component recognizes the absent or negative reward value of neutral stimuli and punishers, the neurons process the value of all stimuli accurately after the initial unselective response and thus will not mislead the animal.

The initial lack of selectivity of the dopamine response may explain why responses to unrewarded stimuli and punishers were interpreted as salience and aversive signals. Testing neutral stimuli without assessing reward-prediction error reveals only the unselective response, which could give rise to the interpretation of a prime salience function of dopamine responses [27], whereas salience concerns only the initial response. Testing punishers without controlling for physical impact, reward generalization, and reward context may encourage interpretations toward aversive activations and general motivational salience processing [36,37], whereas adequate controls reveal that the activations to punishers reflect their physical impact rather than their aversiveness [29]. Thus, recognition of the gradual development of the two-component dopamine response may help to identify its true character.

Reward Learning

Reflecting reward-prediction errors, dopamine responses should follow the development of reward predictions during learning. The changes should become apparent when reward delivery elicits prediction errors. While learning cognitive tasks via intermediate, more simple subtasks, dopamine neurons show repeatedly transient responses to rewards in each subtask when predictions are not yet fully established; these responses disappear after acquisition in every subtask (Fig. 2.2A and B) [5]. With the disappearing reward responses, responses to reward-predicting stimuli appear gradually during learning [5,14] and then disappear again with extinction [28]. Voltammetrically measured dopamine concentrations in the rat ventral striatum show comparable reductions in reward responses and increases in stimulus responses during learning [17]. Thus, dopamine responses develop systematically in parallel with reward-prediction errors during learning.

Comparisons with the Rescorla–Wagner and TD learning rules suggest that the bidirectional dopamine reward-prediction error response would be an ideal teaching signal for learning and updating of decision variables. Indeed, electrical and optogenetic stimulation of dopamine neurons induces operant learning of lever pressing and nose poking in rodents [38–41]. The learning function may involve, in its most simple way, the action of dopamine release on postsynaptic neurons. In a hypothetical, anatomically based model [5], the signal from SNc and VTA dopamine neurons would impact on postsynaptic neurons in dorsal and ventral striatum (caudate nucleus, putamen, nucleus accumbens), frontal cortex, and amygdala through diverging, widespread projections (Fig. 2.2C and D). Dopamine release would affect neuronal transmission at a triad arrangement composed of dendritic spines, presynaptic afferents, and dopamine inputs [42], such that the dopamine signal enhances synaptic transmission from recently active inputs while leaving inactive synapses unchanged. Experimental evidence demonstrates dopamine-dependent plasticity at striatal neurons carrying D1 dopamine receptors [43,44]. In correspondence to these dopamine effects on behavioral learning and neuronal plasticity, lesion and psychopharmacologic studies suggest the necessary role of dopamine impulse responses and neurotransmission in simple forms of associative learning [45,46]. Thus, the dopamine reward-prediction error signal may exert a behavioral function on basic forms of reward learning.

In addition to reward learning [38–41], stimulation of dopamine neurons elicits also immediate behavioral reactions [40], suggesting that dopamine activation induces motivational value (and satisfaction if we accept emotional terms). This activation mimics the neuronal

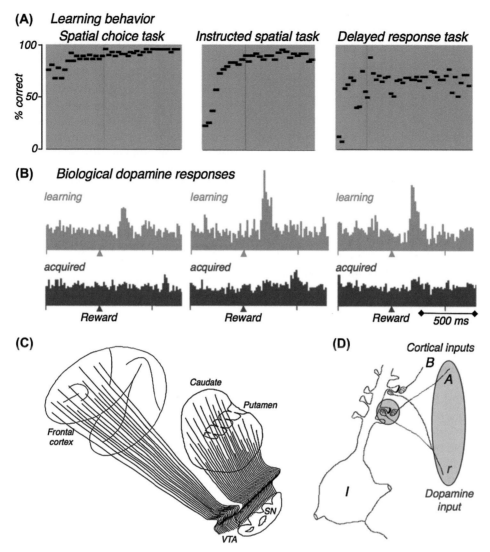

FIGURE 2.2 Dopamine responses and their hypothetical influences during learning. (A) Stepwise learning of spatial delayed-response task via intermediate spatial subtasks. Each dot shows mean percentage of correct behavioral performance in 20–30 trials during learning (red) and asymptotic performance (blue). (B) Positive dopamine reward-prediction error responses during learning of each subtask, and their disappearance during acquired performance. Each histogram shows averaged responses from 10–35 monkey dopamine neurons during the subtasks shown in (A). ((A) and (B) are from Schultz W, Apicella P, Ljungberg T. Responses of monkey dopamine neurons to reward and conditioned stimuli during successive steps of learning a delayed response task. J Neurosci 1993;13:900–13.) (C) Global dopamine reward signal advancing to striatum and cortex. The population response of dopamine neurons in substantia nigra (SN) pars compacta and ventral tegmental area (VTA) can be schematized as synchronous, parallel activity advancing from the midbrain to the striatum (caudate and putamen) and cortex. (From Schultz W. Predictive reward signal of dopamine neurons. J Neurophysiol 1998;80:1–27.) (D) Differential influence of dopamine signal on selectively active corticostriatal transmission. The dopamine reinforcement signal (r) modifies conjointly active synapses (A) at the striatal neuron (I) but leaves inactive synapses (B) unchanged. Gray circle and ellipse indicate simultaneously active elements. (From Smith AD, Bolam JP. The neural network of the basal ganglia as revealed by the study of synaptic connections of identified neurones. Trends Neurosci 1990;13:259–65; Schultz W. Potential vulnerabilities of neuronal reward, risk, and decision mechanisms to addictive drugs. Neuron 2011;69:603–17.)

activation evoked by a positive reward-prediction error (Figs. 2.1A,B, 2.2B, and 2.3D,F). It is reasonable to assume that a natural dopamine activation induced by positive reward-prediction errors might have similar effects on behavior. The effects may be more biasing or modulatory, as a real reward induces also nondopamine responses that may mitigate the dopamine effects. Nevertheless, only positive reward-prediction errors induce dopamine activations and thus biasing effects on behavior, whereas simple reward occurrence without positive-prediction error does not stimulate dopamine neurons. Thus, the biasing effects of dopamine neurons occur only when a reward is better than predicted. However, the positive reward-prediction error increases at the same time as the reward prediction, in correspondence with Rescorla–Wagner and TD learning [3,4]. The next

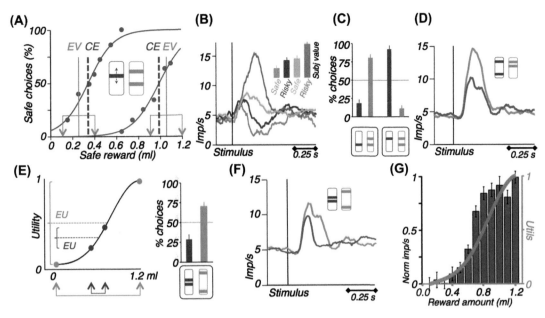

FIGURE 2.3 Risk and utility coding in monkey dopamine neurons. (A) Psychophysical assessment of risk attitude in oculomotor choices. The certainty equivalent (CE) indicates the subjective value of the risky gamble as the magnitude of the adjustable safe outcome (blue) at choice indifference against the gamble with the same expected value (EV, red). A CE > EV indicates risk seeking (left), a CE < EV indicates risk avoidance (right). *Red arrows* indicate the two outcomes of each gamble. *Insets* show choice set with adjustable safe outcome (blue) and fixed gamble (red). The *higher bar* indicates higher reward amount, *two bars* indicate a binary equiprobable gamble ($p = .5$ for each outcome). *(From Stauffer WR, Lak A, Schultz W. Dopamine reward prediction error responses reflect marginal utility. Curr Biol 2014;24:2491–500.)* (B) Risk increases subjective value responses of dopamine neurons. The subjective value for safe black currant juice (green) is higher than for safe orange juice (blue) and higher for binary gambles than safe outcome of the two juices (risk seeking), as assessed psychophysically by choice indifference against a common reference reward *(inset)*. Amounts are 0.45 mL for safe juices and 0.3 and 0.6 mL for the two gamble outcomes. Dopamine neurons show analogously ranked neuronal responses to the reward-predicting stimuli. *(From Lak A, Stauffer WR, Schultz W. Dopamine prediction error responses integrate subjective value from different reward dimensions. Proc Natl Acad Sci USA 2014;111:2343–8.)* (C) Choices between safe and risky outcomes. Monkeys avoid the low certain reward *(single blue bar, bottom)*, which is equal to the low gamble outcome *(double red bars, left)* and prefer the high safe reward (equal to high gamble outcome) (right) in choices against an equiprobable gamble. These choices satisfy first-order stochastic dominance, indicating rational preferences. (D) Dopamine responses to upper gamble outcomes (positive-prediction error) satisfy first-order stochastic dominance (averages from 52 neurons). The red gamble dominates the blue gamble, as the *top bars* indicate equal reward amounts and the *lower bar* is higher for the red compared to the blue gamble. (E) Risk-seeking behavior in gambles with mean-preserving spread. Left: Plotting gamble values onto the individually assessed utility function demonstrates higher expected utility (EU) for the more widely spread gamble (red, compared to blue). Right: In the tested range, monkeys prefer equal-chance gambles with higher mean-preserving spread (red over blue gamble), thus satisfying second-order stochastic dominance for risk seeking and indicating meaningful incorporation of risk into choices. (F) Dopamine responses to upper gamble outcomes satisfy second-order stochastic dominance (averages from 52 neurons). (G) Utility-prediction error signal in dopamine neurons. *Red*, Utility function; *Black*, Corresponding, nonlinear increase in population response to unpredicted juice rewards (positive-prediction error; $n = 14$ dopamine neurons). *((C)–(G) are from Stauffer WR, Lak A, Schultz W. Dopamine reward prediction error responses reflect marginal utility. Curr Biol 2014; 24:2491–500.)*

time the same reward occurs, the prediction has already increased, and the reward elicits only a lower or no positive-prediction error and dopamine activation, and the behavioral stimulating effect is lower. Only a higher reward will now have the same influence on behavior. To bring down the prediction again requires a negative reward-prediction error from a smaller reward evoking a dopamine depression, but smaller-than-predicted rewards are typically avoided. In this way, for rewards to have similarly motivating influences on behavior via the dopamine signal, they need to become higher and higher. This circle of stimulating behavior and at the same time increasing the prediction may explain why we strive for ever higher rewards. If we argue in terms of satisfaction that is induced by dopamine activation,

only higher-than-expected rewards produce satisfaction, whereas the same old rewards leave us unimpressed. But as higher rewards induce higher predictions, to obtain the same satisfaction from a positive-prediction error-signaling dopamine response will require continuously increasing rewards.

SUBJECTIVE VALUE AND FORMAL ECONOMIC UTILITY CODING

Subjective Value

The function of rewards reflects biological needs for nutritional and other substances. The value of rewards depends on the organism's momentary requirements and attitudes and thus is subjective. Subjective value cannot be derived directly from physical or sensory properties of rewards but requires behavioral measures. Particularly useful methods involve choices between different rewards and their psychophysical assessment against a common reference reward. Subjective value estimated in these ways can be expressed in physical units of the reference reward at points of equal preference (indifference points, measured in milliliters of juice, numbers of pellets, pounds, euros, or dollars). The reference reward defines a common currency, which is called a numeraire in economics; all other rewards can be quantitatively stated in real-number multiples of the numeraire.

Although subjective value can be measured in external, physical units, we do not know how much a milliliter of juice or a pound is worth to the individual decision-maker; we do not know how these physical units are scaled internally. All we know is that the internal value usually relates in a positive monotonic fashion to the externally scaled subjective reward value, as long as satiation is avoided. Higher externally scaled subjective value means higher internal value. However, this relationship is not necessarily proportional along the whole value range; thus it provides only rank-ordered, ordinal internal measures. We will see below how to obtain internal value measures on a continuous, numeric, cardinal scale by estimating utility with specific choice protocols.

Subjective Value Derived From Risk

One of the major determinants of subjective value is risk, which reflects the general notion that all environmental events, including rewards, are inherently uncertain. The simplest and most controlled form of risk can be tested with binary, equiprobable gambles ($p = .5$ for each outcome), in which a large and a small reward occur with equal chance. Here risk is defined entirely by the statistical variance and can be varied by changing the outcomes by equal amounts in opposite directions without changing the statistical expected value (EV; "mean-preserving spread") [47]; another form of risk, skewness risk, is absent, and subjective probability distortion [48] is minimal and symmetric. The influence of risk on subjective value is conveyed by the notions of risk avoidance and seeking, which indicate reduced or enhanced preferences, and thus subjective value, for risky compared to safe rewards of the same average magnitude.

Risk attitudes are conveniently assessed by the certainty equivalent (CE) in psychophysical choices against an adjustable safe reward; the physical amount of safe reward at choice indifference indicates the subjective

value of the gamble. In low reward ranges, rhesus monkeys show CEs higher than EVs, indicating risk seeking, whereas in higher ranges, CEs are lower than EVs, indicating risk avoidance (Fig. 2.3A) [49]. The employed test stimuli contain bars whose vertical height indicates reward amount; two bars predict an equal chance of obtaining the upper or lower juice amount (Fig. 2.3A, inset). Dopamine neurons show stronger responses to more preferred liquid reward of the same amount (eg, black currant over orange juice) and, corresponding to the enhancing influence of risk on subjective value in the low reward range, the responses are increased with risky rewards of identical EV (Fig. 2.3B) [50]. Corresponding to the electrophysiological responses, voltammetrically assessed dopamine release to risky cues in the ventral striatum is reduced in risk-avoiding rats and enhanced in risk seekers [51]. Monkeys' choices of risky rewards are meaningful by satisfying first-order stochastic dominance; a binary gamble whose lower outcome equals that of a safe reward is preferred over that reward, because the subject may win and cannot lose anything by choosing the gamble; by contrast, a gamble whose upper outcome is equal to a safe reward is dispreferred over that reward because the subject may get the low gamble outcome and thus lose compared to the safe reward (Fig. 2.3C) [49]. Dopamine neurons show correspondingly stronger responses to first-order stochastically dominating gambles (Fig. 2.3D). Thus, in this and other experiments, monkeys are risk seekers in small reward ranges, risk affects subjective reward value in a corresponding manner, and dopamine neurons reflect the influence of risk on value and thus code subjective, rather than objective, reward value.

Utility

Whereas subjective value estimated from choices follows an objective, physical scale (eg, milliliters of juice at choice indifference), formal economic utility goes one step further and reflects a private, internal measure of subjective value [52] (often called utils). Utility is a mathematical representation of the preference structure of the individual decision-maker that can be described by a continuous function of physical value, such as u(x) [53]. Neuronal response functions vary in a near-continuous way and have cardinal properties; they are valid irrespective of offset and gain (positive affine transformation). Therefore, the represented utility should also be continuous, quantitative, numeric, and cardinal. Such properties can be achieved in choices involving gambles with known probabilities [54], such as the binary gambles already described. The iterative, chaining procedure provides a convenient method for estimating cardinal utility from such gambles [55,56]. The utility functions estimated in monkeys are convex in low reward ranges and concave at higher ranges (Fig. 2.3E) [49], replicating the pattern of risk seeking and risk avoidance seen in the direct risk tests (Fig. 2.3A). When calculations on the utility function suggest that a mean-preserving riskier gamble has higher expected utility than a less risky gamble, which indicates risk seeking, monkeys show corresponding preferences for the riskier gamble (Fig. 2.3E); these choices satisfy second-order stochastic dominance for risk seeking [49,57], indicating meaningful reflection of risk in the utility function. Dopamine responses show correspondingly higher responses to the riskier gamble (Fig. 2.3F) [49] and thus follow second-order stochastic dominance, suggesting adequate incorporation of risk into neuronal utility.

The systematic variation in risk attitude and the behavioral and neuronal satisfaction of first- and second-order stochastic dominance suggest that the behavioral tools for assessing utility are adequate for monkeys. Already the simplest reward test suggests that dopamine neurons code utility; delivery of free, unpredicted reward juice, which constitutes a positive reward-prediction error, induces a nonlinear increase in dopamine responses that follows the behavioral utility function (Fig. 2.3G) [49]. Economic theory requires gambles with known probabilities for estimating numeric utility; testing with gambles satisfying this requirement, such as the binary gambles described above, results in very similar nonlinear changes in dopamine responses [49]. With gambles of identical variance, the dopamine prediction error response to the upper gamble outcome is small in lower reward ranges in which the utility slope is flat, larger in higher reward ranges with steeper utility slope, and smaller again at highest reward ranges with flatter utility slope. Thus, the dopamine response reflects a prediction error in utility. These nonmonotonic responses reflect, but do not necessarily code, the nonmonotonic marginal utility (first derivative of utility) that underlies the inflected utility function. Although neuronal coding is indistinguishable between marginal utility and utility prediction error with these equal-risk gambles, the dopamine responses to free reward increase with reward amount rather than marginal utility, which identifies coding of utility prediction error rather than marginal utility. The two terms relate well to each other, as marginal utility and prediction error conceptualize deviations from a reference (current wealth and prediction, respectively), but dopamine neurons distinguish between the two by coding utility prediction error and not marginal utility.

Taken together, three lines of argument suggest that dopamine neurons meaningfully incorporate risk into subjective value and utility coding. Dopamine reward-prediction error responses increase when risk enhances subjective value (Fig. 2.3B), they follow second-order stochastic dominance (Fig. 2.3F), and they increase nonlinearly with free reward (Fig. 2.3G) and gambles in close correspondence to formal economic utility [49]. In coding utility, dopamine neurons process reward information in the most formal and constrained way defined by economic choice theory.

DOPAMINE AND DECISION-MAKING

Updating of Decision Variables

A popular minimal generic model of decision-making [58,59] involves the competition between two options. The hypothetical model involves forward excitation of independent input activity, mutual lateral inhibition, and a threshold mechanism, such that only the stronger input leads to an output, whereas the weaker input is lost (Fig. 2.4). This winner-take-all (WTA) mechanism turns the graded difference between the inputs into an all-or-none output that reflects only the value of the stronger option. The basic input variables to this model are object value, action value, and their derivatives and combinations. Object value denotes the value of a stimulus or reward object and is appropriate for decisions about goods, as conceptualized in economics. Action value indicates the value obtained by an action and is appropriate for decisions about actions, as modeled in machine learning. To constitute a neuronal decision mechanism, separate pools of neurons would code the value of each input object or action separately and irrespective of the behavioral choice. Neurons coding object value or action value are found in the orbitofrontal cortex [60] and striatum [61], respectively; these structures receive dense inputs from dopamine neurons.

The reward-prediction error reflected in dopamine neurons results from the subtraction of predicted reward value from experienced value. The experienced value is the received reward for Rescorla—Wagner models (Eq. (2.2)) and may include higher order rewarded stimuli for TD models (Eq. (2.4)). The experienced value reflects the outcome of the decision process and is called the chosen value. Chosen value reflects the value of the option that won the competitive decision process; it is the result of the decision rather than being its input and thus is usually considered to be a postdecision variable. The predicted reward value is the current object value or action value of the chosen option. Predicted values derive exclusively from recent outcomes in model-free reinforcement processes but may incorporate predictions from model-based reinforcement learning. Although being particularly suitable for model-free learning, the dopamine signal incorporates predictions from models of the world, as shown with different probability distributions, sequential

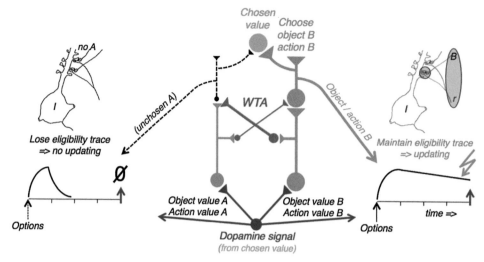

FIGURE 2.4 Hypothetical model for updating of decision variables by the dopamine signal. In a competitive decision process, the global dopamine reward-prediction error signal acts on postsynaptic neurons via eligibility traces. Eligibility traces are selectively stabilized by neurons coding the chosen object or chosen action (right) but decay for neurons coding the unchosen option (left). Thus, the dopamine signal updates only object values or action values of options that were chosen. The dopamine prediction error derives from chosen value (experienced minus predicted object value or action value of chosen option). *Gray zones at top right* indicate common activations. The weight of *dots* and *lines* indicates levels of neuronal activity. *Dotted lines* indicate inputs from unchanged neuronal activities. *Crossed green connections* are inhibitory; *WTA,* winner-take-all mechanism. The weights of dots and lines in the circuit model indicate level of neuronal activity. *I,* striatal neuron; *r,* dopamine reinforcement signal. Scheme developed together with Fabian Grabenhorst. *Modified from Schultz W. Neuronal reward and decision signals: from theories to data. Physiol Rev 2015;95:853–951.*

movements, and reversal sets [15,62,63]. A prediction error suitable for updating object value and action value would result from the subtraction of object value or action value of the chosen option from the experienced chosen value. This term constitutes a reward prediction error in chosen value. Indeed, dopamine neurons and voltammetric striatal dopamine release code such a chosen value prediction error during choices [51,64]. Although we use the generic term of value, the best-defined and adequate metric for value that should be coded in all these signals is economic utility. The dopamine signal codes formal economic utility [49] and integrates reward delay, reward risk, and various liquid and food rewards [50,65].

With these characteristics, the dopamine prediction error signal is suitable for updating the input variables to the competitive decision process for the next trial, namely object value and action value, following (Eqs. (2.1)–(2.4)). The hypothetical updating model shown in Fig. 2.4 represents a possible mechanism. Despite the indiscriminate propagation of the dopamine signal to postsynaptic neurons (Fig. 2.2C), specificity in updating may result from the eligibility traces that make previously active neurons and synapses selectively sensitive to the reinforcing effects of the reward resulting from the chosen option [4]. Eligibility traces may decay rapidly unless they are stabilized by neuronal activity associated with the chosen object or action. In this way, the chosen value signal of dopamine neurons would selectively affect only object values or action

values in neurons that contributed to the decision, as marked by maintained eligibility traces but leave the values of the unchosen option unchanged. For example, choice of object B would lead to a dopamine reward-prediction error signal derived from the chosen value minus object value B that would serve to selectively update object value B. The selectivity in updating is due to the eligibility trace from object B being maintained by activity from the chosen object B, whereas the eligibility trace from object A decays before the reinforcement signal can act on it. The dopamine updating influence would involve known synaptic plasticity [43,44]. Taken together, the updating of decision variables by the dopamine prediction error signal in chosen value can be conceptualized on the basis of a simple, biologically plausible WTA decision mechanism.

Direct Dopamine Influences on Choices

In addition to learning and updating, electrical or optogenetic stimulation of dopamine neurons or striatal neurons carrying specific dopamine receptors induces immediate behavioral actions and differential choice biases [40,66]. These effects may involve direct dopamine actions on striatal synapses [67]. Correspondingly, reduction of D1 receptor stimulation impairs spatial working memory and memory-related neuronal activity in monkey frontal cortex [68,69]; reduced dopamine bursting activity in mice with knocked out NMDA receptors on dopamine neurons is associated with

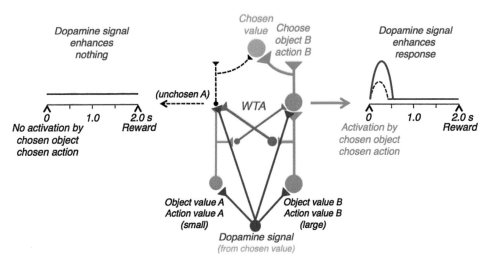

FIGURE 2.5   Hypothetical model for immediate influence of the dopamine signal on neuronal decision mechanisms. The dopamine prediction error signal of chosen value may have two distinguishable influences. First, it may boost differences between neuronal signals coding object value or action value at the input to the competitive decision mechanism (bottom). Second, it may enhance activities selected by the winner-take-all (WTA) mechanism that represent chosen object or chosen action (top right), without affecting activity that was lost in the WTA competition (top left). *Modified from Schultz W. Neuronal reward and decision signals: from theories to data. Physiol Rev 2015;95:853–951.*

prolonged reaction times [45]. Thus, dopamine signals may have immediate effects on decision-making, distinct from their reinforcement function.

The hypothetical competitive decision model may be modified to capture these immediate dopamine effects (Fig. 2.5). The dopamine prediction error signal derived from chosen value may exaggerate differences in neuronal activity in two possible ways. First, the signal may affect object value or action value signals at the input of the decision process. Local optogenetic stimulation of striatal neurons carrying specific dopamine receptors affects action values for contralateral actions [66]. Although this experimental effect may rely on locally specific dopamine influences, which may not exist in natural choices, it may also function by enhancing the frequently occurring differences in postsynaptic value signals in many behavioral tasks [66]. Second, the dopamine signal may boost exclusively neuronal activity reflecting the chosen object or chosen action that survived the WTA competition, without affecting signals that were lost by the WTA mechanism (Fig. 2.5, right); only the neuronal signal for the chosen option, and not for the unchosen option, would benefit from the immediate effects of the dopamine prediction error signal, thus sharpening the WTA selection. Taken together, the immediate influence of the dopamine prediction error signal on behavioral choices can be captured by a simple WTA decision mechanism. The possible implementation of this hypothesis needs to be worked out by considering the particular neuronal activities and circuits in the postsynaptic structures such as striatum, frontal cortex, and amygdala.

Acknowledgments

Our work is supported by the Wellcome Trust (Principal Research Fellowship, Programme and Project Grants 058365, 093270, 095495), European Research Council (ERC Advanced Grant 293549), and NIH Caltech Conte Center (P50MH094258).

References

[1] Schultz W. Multiple dopamine functions at different time courses. Ann Rev Neurosci 2007;30:259–88.

[2] Schultz W. Neuronal reward and decision signals: from theories to data. Physiol Rev 2015;95:853–951.

[3] Rescorla RA, Wagner AR. A theory of Pavlovian conditioning: variations in the effectiveness of reinforcement and nonreinforcement. In: Black AH, Prokasy WF, editors. Classical conditioning II: current research and theory. New York: Appleton Century Crofts; 1972. p. 64–99.

[4] Sutton RS, Barto AG. Toward a modern theory of adaptive networks: expectation and prediction. Psychol Rev 1981;88: 135–70.

[5] Schultz W, Apicella P, Ljungberg T. Responses of monkey dopamine neurons to reward and conditioned stimuli during successive steps of learning a delayed response task. J Neurosci 1993; 13:900–13.

[6] Schultz W, Dayan P, Montague RR. A neural substrate of prediction and reward. Science 1997;275:1593–9.

[7] Schultz W. Predictive reward signal of dopamine neurons. J Neurophysiol 1998;80:1–27.

[8] Pan W-X, Schmidt R, Wickens JR, Hyland BI. Dopamine cells respond to predicted events during classical conditioning: evidence for eligibility traces in the reward-learning network. J Neurosci 2005;25:6235–42.

[9] Zaghloul KA, Blanco JA, Weidemann CT, McGill K, Jaggi JL, Baltuch GH, et al. Human substantia nigra neurons encode unexpected financial rewards. Science 2009;323:1496–9.

[10] Cohen JY, Haesler S, Vong L, Lowell BB, Uchida N. Neuron-type-specific signals for reward and punishment in the ventral tegmental area. Nature 2012;482:85–8.

[11] Ljungberg T, Apicella P, Schultz W. Responses of monkey midbrain dopamine neurons during delayed alternation performance. Brain Res 1991;586:337–41.

[12] Mirenowicz J, Schultz W. Importance of unpredictability for reward responses in primate dopamine neurons. J Neurophysiol 1994;72:1024–7.

[13] Hollerman JR, Schultz W. Dopamine neurons report an error in the temporal prediction of reward during learning. Nat Neurosci 1998;1:304–9.

[14] Waelti P, Dickinson A, Schultz W. Dopamine responses comply with basic assumptions of formal learning theory. Nature 2001; 412:43–8.

[15] Tobler PN, Fiorillo CD, Schultz W. Adaptive coding of reward value by dopamine neurons. Science 2005;307:1642–5.

[16] Satoh T, Nakai S, Sato T, Kimura M. Correlated coding of motivation and outcome of decision by dopamine neurons. J Neurosci 2003;23:9913–23.

[17] Day JJ, Roitman MF, Wightman RM, Carelli RM. Associative learning mediates dynamic shifts in dopamine signaling in the nucleus accumbens. Nat Neurosci 2007;10:1020–8.

[18] Montague PR, Dayan P, Sejnowski TJ. A framework for mesencephalic dopamine systems based on predictive Hebbian learning. J Neurosci 1996;16:1936–47.

[19] Nomoto K, Schultz W, Watanabe T, Sakagami M. Temporally extended dopamine responses to perceptually demanding reward-predictive stimuli. J Neurosci 2010;30:10692–702.

[20] Enomoto K, Matsumoto N, Nakai S, Satoh T, Sato TK, Ueda Y, et al. Dopamine neurons learn to encode the long-term value of multiple future rewards. Proc Natl Acad Sci USA 2011;108: 15462–7.

[21] D'Ardenne K, McClure SM, Nystrom LE, Cohen JD. BOLD Responses reflecting dopaminergic signals in the human ventral tegmental area. Science 2008;319:1264–7.

[22] O'Doherty J, Dayan P, Friston K, Critchley H, Dolan RJ. Temporal difference models and reward-related learning in the human brain. Neuron 2003;28:329–37.

[23] McClure SM, Berns GS, Montague PR. Temporal prediction errors in a passive learning task activate human striatum. Neuron 2003; 38:339–46.

[24] Pessiglione M, Seymour B, Flandin G, Dolan RJ, Frith CD. Dopamine-dependent prediction errors underpin reward-seeking behaviour in humans. Nature 2006;442:1042–5.

[25] Ljungberg T, Apicella P, Schultz W. Responses of monkey dopamine neurons during learning of behavioral reactions. J Neurophysiol 1992;67:145–63.

[26] Mirenowicz J, Schultz W. Preferential activation of midbrain dopamine neurons by appetitive rather than aversive stimuli. Nature 1996;379:449–51.

[27] Horvitz JC, Stewart T, Jacobs BL. Burst activity of ventral tegmental dopamine neurons is elicited by sensory stimuli in the awake cat. Brain Res 1997;759:251–8.

[28] Tobler PN, Dickinson A, Schultz W. Coding of predicted reward omission by dopamine neurons in a conditioned inhibition paradigm. J Neurosci 2003;23:10402–10.

[29] Fiorillo CD, Song MR, Yun SR. Multiphasic temporal dynamics in responses of midbrain dopamine neurons to appetitive and aversive stimuli. J Neurosci 2013;33:4710–25.

[30] Fiorillo CD, Yun SR, Song MR. Diversity and homogeneity in responses of midbrain dopamine neurons. J Neurosci 2013;33: 4693–709.

[31] Fiorillo CD. Two dimensions of value: dopamine neurons represent reward but not aversiveness. Science 2013;341:546–9.

[32] Hedgé J. Time course of visual perception: coarse-to-fine processing and beyond. Prog Neurobiol 2008;84:405–39.

[33] Bisley JW, Goldberg ME. Attention, intention, and priority in the parietal lobe. Ann Rev Neurosci 2010;33:1–21.

[34] Hung CP, Kreiman G, Poggio T, DiCarlo JJ. Fast readout of object identity from macaque inferior temporal cortex. Science 2005;310: 863–6.

[35] Kobayashi S, Schultz W. Reward contexts extend dopamine signals to unrewarded stimuli. Curr Biol 2014;24:56–62.

[36] Joshua M, Adler A, Mitelman R, Vaadia E, Bergman H. Midbrain dopaminergic neurons and striatal cholinergic interneurons encode the difference between reward and aversive events at different epochs of probabilistic classical conditioning trials. J Neurosci 2008;28:11673–84.

[37] Matsumoto M, Hikosaka O. Two types of dopamine neuron distinctively convey positive and negative motivational signals. Nature 2009;459:837–41.

[38] Corbett D, Wise RA. Intracranial self-stimulation in relation to the ascending dopaminergic systems of the midbrain: a moveable microelectrode study. Brain Res 1980;185:1–15.

[39] Tsai H-C, Zhang F, Adamantidis A, Stuber GD, Bonci A, de Lecea L, et al. Phasic firing in dopaminergic neurons is sufficient for behavioral conditioning. Science 2009;324:1080–4.

[40] Kim KM, Baratta MV, Yang A, Lee D, Boyden ES, Fiorillo CD. Optogenetic mimicry of the transient activation of dopamine neurons by natural reward is sufficient for operant reinforcement. PLoS One 2012;7:e33612.

[41] Steinberg EE, Keiflin R, Boivin JR, Witten IB, Deisseroth K, Janak PH. A causal link between prediction errors, dopamine neurons and learning. Nat Neurosci 2013;16:966–73.

[42] Freund TF, Powell JF, Smith AD. Tyrosine hydroxylase-immunoreactive boutons in synaptic contact with identified striatonigral neurons, with particular reference to dendritic spines. Neuroscience 1984;13:1189–215.

[43] Reynolds JNJ, Hyland BI, Wickens JR. A cellular mechanism of reward-related learning. Nature 2001;413:67–70.

[44] Pawlak V, Kerr JND. Dopamine receptor activation is required for corticostriatal spike-timing-dependent plasticity. J Neurosci 2008; 28:2435–46.

[45] Zweifel LS, Parker JG, Lobb CJ, Rainwater A, Wall VZ, Fadok JP, et al. Disruption of NMDAR-dependent burst firing by dopamine neurons provides selective assessment of phasic dopamine-dependent behavior. Proc Natl Acad Sci USA 2009;106:7281–8.

[46] Puig MV, Miller EK. The role of prefrontal dopamine D1 receptors in the neural mechanisms of associative learning. Neuron 2012;74: 874–86.

[47] Rothschild M, Stiglitz JE. Increasing risk: I. A definition. J Econ Theory 1970;2:225–43.

[48] Stauffer WR, Lak A, Bossaerts P, Schultz W. Economic choices reveal probability distortion in monkeys. J Neurosci 2015;35: 3146–54.

[49] Stauffer WR, Lak A, Schultz W. Dopamine reward prediction error responses reflect marginal utility. Curr Biol 2014;24:2491–500.

[50] Lak A, Stauffer WR, Schultz W. Dopamine prediction error responses integrate subjective value from different reward dimensions. Proc Natl Acad Sci USA 2014;111:2343–8.

[51] Sugam JA, Day JJ, Wightman RM, Carelli RM. Phasic nucleus accumbens dopamine encodes risk-based decision-making behavior. Biol Psychiatry 2012;71:199–205.

[52] Luce RD. Individual choice behavior: a theoretical analysis. New York: Wiley; 1959.

[53] Kagel JH, Battalio RC, Green L. Economic choice theory: an experimental analysis of animal behavior. Cambridge: Cambridge University Press; 1995.

[54] von Neumann J, Morgenstern O. The theory of games and economic behavior. Princeton: Princeton University Press; 1944.

[55] Caraco T, Martindale S, Whitham TS. An empirical demonstration of risk-sensitive foraging preferences. Anim Behav 1980;28:820–30.

[56] Machina MJ. Choice under uncertainty: problems solved and unsolved. J Econ Perspect 1987;1:121–54.

[57] O'Neill M, Schultz W. Coding of reward risk distinct from reward value by orbitofrontal neurons. Neuron 2010;68:789—800.

[58] Bogacz R, Brown E, Moehlis J, Holmes P, Cohen JD. The physics of optimal decision making: a formal analysis of models of performance in two-alternative forced-choice tasks. Psychol Rev 2006; 113:700—65.

[59] Wang X-J. Decision making in recurrent neuronal circuits. Neuron 2008;60:215—34.

[60] Padoa-Schioppa C, Assad JA. Neurons in the orbitofrontal cortex encode economic value. Nature 2006;441:223—6.

[61] Samejima K, Ueda Y, Doya K, Kimura M. Representation of action-specific reward values in the striatum. Science 2005;310: 1337—40.

[62] Nakahara H, Itoh H, Kawagoe R, Takikawa Y, Hikosaka O. Dopamine neurons can represent context-dependent prediction error. Neuron 2004;41:269—80.

[63] Bromberg-Martin ES, Matsumoto M, Hon S, Hikosaka O. A pallidus-habenula-dopamine pathway signals inferred stimulus values. J Neurophysiol 2010;104:1068—76.

[64] Morris G, Nevet A, Arkadir D, Vaadia E, Bergman H. Midbrain dopamine neurons encode decisions for future action. Nat Neurosci 2006;9:1057—63.

[65] Kobayashi S, Schultz W. Influence of reward delays on responses of dopamine neurons. J Neurosci 2008;28:7837—46.

[66] Tai LH, Lee AM, Benavidez N, Bonci A, Wilbrecht L. Transient stimulation of distinct subpopulations of striatal neurons mimics changes in action value. Nat Neurosci 2012;15:1281—9.

[67] Hernández-López S, Bargas J, Surmeier DJ, Reyes A, Galarraga E. D1 receptor activation enhances evoked discharge in neostriatal medium spiny neurons by modulating an L-type Ca^{2+} conductance. J Neurosci 1997;17:3334—42.

[68] Sawaguchi T, Goldman-Rakic PS. D1 dopamine receptors in prefrontal cortex: involvement in working memory. Science 1991; 251:947—50.

[69] Vijayraghavan S, Wang M, Birnbaum SG, Williams GV, Arnsten AFT. Inverted-U dopamine D1 receptor actions on prefrontal neurons engaged in working memory. Nat Neurosci 2007;10:376—84.

[70] Smith AD, Bolam JP. The neural network of the basal ganglia as revealed by the study of synaptic connections of identified neurones. Trends Neurosci 1990;13:259—65.

[71] Schultz W. Potential vulnerabilities of neuronal reward, risk, and decision mechanisms to addictive drugs. Neuron 2011;69:603—17.

3

Appetitive and Aversive Systems in the Amygdala

S. Bernardi[1], D. Salzman[1,2]

[1]Columbia University, New York, New York, United States; [2]New York State Psychiatric Institute, New York, New York, United States

Abstract

The amygdala has been traditionally considered to process information about aversive stimuli, but it is now recognized to mediate emotional behaviors elicited by stimuli of both valences. Electrophysiological studies have shown that the amygdala contains distinct appetitive and aversive networks of neurons. The development of new molecular tools for interrogating neural circuit function has rapidly advanced our understanding of these networks. Selective activation of appetitive or aversive amygdalar networks can elicit innate valence-specific autonomic responses, and can induce Pavlovian and operant learning. An important goal for scientific inquiry is to understand the functional role of the bidirectional interactions between appetitive and aversive systems in the amygdala and networks in the cerebral cortex that implement cognitive functions; the pathophysiology of psychiatric disorders probably involves dysfunctional interactions between these cognitive and emotional systems. Studies have begun to describe how activity in appetitive and aversive systems is updated by cognitive operations, and how such activity may also influence cognitive functions like attention.

INTRODUCTION

Anticipatory emotional responses to sensory stimuli require an agent to predict accurately the positive or negative emotional outcome associated with a stimulus presentation. This prediction relies on more than knowing the link between a sensory stimulus and a reinforcement, because such associations may differ depending upon the circumstances, as may the emotional meaning of the associations. For example, a particular stimulus, such as an unopened bottle of Romanée—Conti, may elicit a positive anticipatory emotional feeling only when it appears in circumstances in which opening the bottle is a possibility. Even then, its appearance may produce a positive emotional response only if one is not tired of drinking wine from this vineyard. The link between a sensory stimulus and an emotional response thereby relies on knowing how a stimulus, potential actions, and current circumstances— or context—together predict a particular outcome. Of note, context here is defined broadly as the set of circumstances in which the subject operates, including internal (abstract cognitive and homeostatic) and external (social and environmental) states [1–3]. As a result, a neural representation of the emotional significance of a stimulus must take into account contextual information, which includes information about internal and external variables as well as operant contingencies.

A neural representation of emotional significance bears similarity to neural representations posited to play a critical role in economic decision-making. In economic decision-making, a neural representation of stimulus value for two or more stimulus options must be computed online, taking into account context-dependent stimulus—outcome relations as well as internal and external contextual information that can adjust the value of a particular reinforcement outcome at a given moment [4–7]. Moreover, the values of stimuli, the values of actions, and "state value"—the value of the overall situation of an organism at a given moment— are essential variables in theoretical accounts of learning [8,9], and these theoretical accounts remain influential in understanding how previously neutral stimuli develop the capacity to generate emotional responses through such learning. Therefore, the acquisition of neural representations of the value of stimuli is also critical for economic decision-making. The coordination of anticipatory emotions, value-based decision-making, and reinforcement learning all probably share computational mechanisms that allow an agent to predict the relationship between sensory stimulus, actions, context, and meaningful outcomes.

Many of the brain areas that have been implicated in emotional processing—such as the amygdala—also are thought to play an important role in value-based decision-making [2,10—14]. Traditionally studied as a center essential for the acquisition and expression of defensive behaviors elicited by an aversive unconditioned stimulus (US) or by a conditioned stimulus (CS) paired with an aversive US (reviewed in [13]), the amygdala was confirmed by a 2014 meta-analysis to be an area of overlapping activity in value processing for both aversion and appetitive information in human functional imaging studies [15]. Substantial evidence indicates that the amygdala is involved in learning the relationship between a previously neutral CS and both appetitive or aversive reinforcement (US), as well as in coordinating responses to these stimuli [10,13,14,16—19].

The focus of this chapter is to review studies that have begun to describe in more detail the role of the amygdala in both appetitive and aversive processes in animal model systems. This work spans rodents and monkeys. Work in rodents has been particularly successful at characterizing detailed circuit-level mechanisms within the amygdala that mediate various forms of valence-specific emotional behavior. Studies in nonhuman primates have been particularly helpful for examining the relationship between the amygdala and cortical processing mediating various cognitive functions. Interactions between the amygdala and the prefrontal cortex is likely to be critical for updating or regulating valence-specific neuronal representations in the amygdala in a context-dependent manner. At the same time, neural representations of the emotional or motivational significance of stimuli in the amygdala are likely to influence cognitive functions, including processes such as decision-making and attention. Overall, the investigation of how amygdalar and cortical processing interact is likely to prove critical for understanding the pathophysiology of many psychiatric disorders.

Valence in the Amygdala

A long tradition of work has implicated the amygdala in processes related to emotional valence [10,13,14,16,20—25]. Classic studies linked the amygdala to the acquisition and expression of behavioral and physiological responses to aversive CSs [26,27]. It remained unclear, however, whether the same or different neurons participated in emotional learning and behavior for both positive and negative valences because prior studies typically did not examine the physiological response properties of neurons during both appetitive and aversive classical conditioning procedures.

Paton et al. [22] established that in fact the amygdala appears to contain separate appetitive and aversive neural systems. They applied a trace-conditioning task (a version of classical conditioning in which a brief temporal gap is inserted between the CS and the US). Novel, abstract visual CSs were followed by a US: a liquid reward, nothing, or an aversive air puff directed at a monkey's face. During conditioning, single-neuron activity in the amygdala was recorded. Neural signals related to sensory properties of the visual CSs themselves were disentangled from signals related to reinforcement contingencies by reversing the contingencies during each experiment. Individual amygdala neurons were often sensitive to this reversal: neurons would respond preferentially to either positively or negatively conditioned CSs both before and after the reversal. These neurons updated their response to a CS with a time course fast enough to account for behavioral learning of reversed reinforced contingencies.

Subsequent studies demonstrated that the neuronal populations encoding the valence of a CS provide a representation of the positive or negative overall value of the animal's state throughout the various events occurring during an entire trial (as opposed to being limited to the CS presentation) [28]. For example, the response to a fixation point, which monkeys voluntarily foveate to commence a trial, reflects the mildly positive value that this stimulus possesses for motivated subjects. Neurons belonging to the appetitive system tend to increase their firing rate in response to fixation-point appearance, but neurons of the aversive system tend to decrease their firing rate to the same stimulus. The responses to USs also reflected the valence-specific encoding properties of these neuronal populations. Other studies have reported neural signals in the amygdala that reflect expected reward value over long timescales during a task that involves planning [11], consistent with the encoding of state value across a series of planned actions.

The existence of cells encoding valence raises a fundamental question about the organization of circuits within the amygdala: do cells that respond preferentially to either rewarding or aversive events form distinct circuits? Reconstruction of the anatomical location of valence-encoding cells had failed to reveal clustering of neurons encoding a particular valence [22,28], suggesting that neurons within the appetitive and aversive systems in the amygdala were intermingled. Zhang et al. [29] applied a cross-correlogram (CCG) analysis to examine the functional connectivity between neurons simultaneously recorded during trace conditioning. Functional interactions between pairs of amygdala cells were related to their response properties, and these interactions also were observed to be task modulated [29].

Peaks in CCGs at short latencies were significantly more likely to be observed if two neurons shared the same valence preference than if they did not. Thus valence-encoding neurons in the amygdala appear to form distinct circuits with other neurons encoding the same valence; these circuits could potentially guide relevant valence-specific behaviors, such as approach or avoidance [29]. However, because traditional methods for manipulating neural activity typically rely on anatomical clustering of neurons sharing physiological response properties in order to activate or inactivate neurons encoding a similar signal, the apparent lack of anatomical organization in the amygdala presented an experimental challenge both for understanding the causal role of neurons in mediating behavior and for understanding the circuit-level mechanisms that mediate valence-specific functions.

A link between neural activity in valence-specific circuits in the basolateral amygdala (BLA) and appropriate behavioral output has been established by exploiting new molecular techniques for identifying, marking, and manipulating neural activity based upon the response properties of targeted neurons. These approaches utilize activity-dependent and inducible systems to label only cells involved in valence-specific processes. For example, Gore et al. [24] used a lentivirus construct in which the expression of the immediate early gene c-fos, a molecular marker of neural activity, also drove the expression of channelrhodopsin (ChR2) linked to enhanced yellow fluorescent protein and a nuclear-localized mCherry [14,24]. The temporal expression pattern of this viral construct provided a two-time-point molecular marker of neuronal activity because c-fos is only transiently expressed after neural activity is elicited. During this early time window, expression of mCherry cannot be detected; by contrast, the detection of mCherry occurs with a longer latency after neural activity is elicited. As a result, c-fos marks neurons activated by a recent stimulus (e.g., 1 h earlier), and mCherry marks neuron activity in response to a stimulus that occurred further back in time (e.g., 18 h previous). Consistent with electrophysiological studies, using this approach Gore et al. [24] identified distinct, but anatomically intermingled, neuronal ensembles in the BLA that responded differentially to an innately appetitive US (nicotine, at a dose that elicited conditioned place preference) and an innately aversive US (footshock, which can be used to drive fear conditioning). Moreover, if the authors presented only a single stimulus to activate neurons, ChR2 could be expressed in either the appetitive or the aversive population of neurons selectively. Thus, despite the fact that neurons belonging to appetitive and aversive systems appear to be anatomically intermingled, a means for selective activation of these populations had been achieved.

Gore et al. used this approach to demonstrate that optical activation of nicotine-responsive and footshock-responsive neuronal ensembles in BLA elicited distinct, valence-specific, physiological responses (opposite changes in heart and respiratory rate) [24]. In addition, activation of footshock-sensitive neurons elicited freezing, a defensive response. Thus representations of USs in the BLA have the capacity to elicit valence-specific innate emotional responses.

Activity within the representation of an appetitive or aversive US in the amygdala was also demonstrated to be sufficient to drive Pavlovian conditioning if US activation was temporally paired with a CS. Photoactivation of neurons responsive to footshock could induce freezing if paired with a previously neutral auditory tone, or, if paired with a previously neutral odor, could induce avoidance of that odor. Photoactivation of neurons responsive to nicotine paired with a previously neutral odor could induce approach to that odorant [24]. Moreover, photoactivation of nicotine-responsive calls could also induce instrumental learning, as mice would learn to nosepoke preferentially to a portal paired with optical activation even in the absence of any other reinforcement. Thus appetitive and aversive neuronal ensembles in the BLA were linked causally to a variety of valence-specific behaviors: innate physiological responses, both innate and learned freezing, approach and avoidance induced by Pavlovian conditioning, and learned approach induced by operant conditioning [24].

The capacity of neural representations of USs in BLA to drive both innate and learned behaviors raises the possibility that these representations provide a link between earlier stage sensory representations of CSs and USs and the neuronal circuits that must direct valence-specific behavior itself. If this notion is correct, then sensory information about CSs must engage neural representations of USs in the BLA to direct learned emotional behavior. Gore et al. addressed this hypothesis by first labeling cells responsive to an aversively conditioned CS in the BLA. Optical activation of the cell population responsive to an aversively conditioned CS indeed elicited the expression of aversively conditioned behavior. Thus the BLA contains a representation of a CS after learning that is sufficient to direct appropriate behavioral responses [24].

In a subsequent experiment, Gore et al. showed that the expression of behavior indicative of aversive learning requires activation of a US representation in the BLA. In these experiments, the c-fos promoter drove the expression of halorhodopsin, allowing the authors to inhibit cells based on their physiological response properties [24]. Photoinactivation of shock-sensitive neurons in BLA prevented the expression of freezing in response to a footshock-conditioned tone, and it prevented the expression of avoidance in response to a footshock-conditioned odor. The representation of a CS must therefore normally activate a US representation in the BLA to generate appropriate behavioral responses for aversively conditioned stimuli. The US representation in the BLA can therefore be conceptualized as providing a critical link between earlier stage sensory representations of stimuli and the output pathways that mediate appropriate behavior [24]. Optogenetic interrogation of anatomical pathways originating from the BLA also indicates that neurons within the BLA mediate valence-specific behavioral responses [30–33].

Valence-specific networks of neurons in the BLA have also been identified by Redondo et al. using a transgenic approach and again exploiting activity-dependent expression of ChR2 [34]. In this study, complex stimuli comprising both a CS and a US, or exposure to a mating experience, were used to label neurons in the aversive and appetitive systems. By subsequently reactivating those cells with photostimulation of the BLA, valence-specific behavioral responses were observed in the absence of the CS, as mice avoided the context associated with aversive conditioning and approached the context paired with an appetitive experience. Moreover, consistent with electrophysiological data demonstrating that amygdala neurons preserve their valence encoding even after switches in reinforcement contingencies, reversing the contexts associated with each valence did not affect the valence specificity observed upon photoactivation of a particular amygdala population of neurons.

The new molecular approaches for identifying, marking, and manipulating neural activity in anatomically intermingled yet distinct neural circuits have propelled the field forward by providing causal evidence linking the physiological response properties of neurons to predictable and appropriate behavior. Although much work remains to understand amygdalar appetitive and aversive circuitry in greater detail, and for a broader range of stimuli, the new methodology probably corresponds to a watershed event for characterizing neural circuit function, at least in relation to relatively simple appetitively and aversively motivated behaviors.

Updating the Valence Representation in the Amygdala: Learning, Extinction, and Cognitive Regulation

Physiological data indicate that the representation of valence in the amygdala can be rapidly updated. How does this occur so quickly and flexibly? Associative learning, such as that which links a sensory stimulus with punishment or reward, is frequently assumed to arise gradually via changing synaptic weights; computational models of reinforcement learning have often

worked from this assumption. The seminal Rescorla–Wagner model posits that the value representation should be updated on a trial-by-trial basis by "error signals"—that is, signals reflecting the difference between expected and received reinforcement—and recent theories [e.g., temporal difference (TD) models] have extended this model so that reinforcement learning may be described quantitatively in real time [35,36]. TD models require a neural representation of value as a function of time, with error signals being computed continuously by taking the difference in the value of situations or "states" at successive time steps. Reward prediction errors guiding value update have been shown in various areas of the brain, starting with the seminal studies of the dopamine signal in the midbrain [9,37] (see Chapter 2 for a detailed description of prediction error and dopamine signal) and extending to the medial prefrontal cortex [38–40], striatum [41–43], globus pallidus [44], and lateral habenula [45]. Neurons in this last nucleus in particular encode reward prediction error in the opposite direction compared to dopaminergic neurons [45].

Electrophysiological evidence indicates that prediction error signals probably influence processing within the amygdala. Belova et al. [23] showed that responses to reinforcement were often stronger when rewards or aversive stimuli occur unexpectedly, such as immediately after a reversal, or if the reinforcement occurs at a random time [23]. These responses differed from classical prediction error signals in that the response profile of amygdala neurons did not match the prototypical phasic, short latency response profile of neurons encoding TD errors. Moreover, amygdala neurons generally did not exhibit response modulation in relation to omitted reinforcement, a signature of neurons computing an error signal.

The modulation of neural responses to reinforcement by expectation did reveal the presence of neural processing related to both valence-specific and valence-nonspecific functions in the amygdala. Some amygdala neurons modulated their responses to an unexpected reward or to an aversive stimulus, but not to both. By contrast, other neurons modulated responses to both valences of USs when they were delivered unexpectedly. These data indicate that there exist valence-nonspecific networks in the amygdala, an observation supported by other studies [46]. Consistent with this notion, a series of studies has now linked amygdala neural activity—and targets of the amygdala—to various aspects of processing relevant for modulating spatial attention to emotionally significant stimuli (see later and [47–51]). The specific mechanisms by which the effects of expectation—and the potential processing of TD error signals—modulate neural circuit function during learning in the amygdala still await characterization at a more detailed level. However, at a cellular level, substantial progress has been made in characterizing synaptic plasticity during learning in the amygdala (for example, [17]).

Although our understanding of how TD errors impinge on amygdala processing during learning remains somewhat limited, much progress has been made in understanding neural circuitry related to extinction. Extinction provides a powerful means for changing behavioral responses to previously conditioned appetitive or aversive stimuli. A detailed analysis of circuitry within the central amygdala, as well as the contribution of the prefrontal cortex to extinction and its relation to amygdala function, has been advanced [52–54]. These topics have been reviewed, and will not be summarized in detail here [55,56]. Instead, we turn our attention to issues related to the cognitive regulation of emotion, a topic of considerable interest for understanding the normal flexible regulation of emotion, which does not require a process of extinction for adjusting responses to a particular stimulus.

The regulation of emotions can be extremely flexible. Responses to a sensory cue can be adjusted just by invoking a cognitive process, like the knowledge of the rules of a game. For example, in blackjack, the same sensory stimulus can be rewarding in one hand, because a player makes "21," but punishing in the next hand, because the player goes bust. The player has no difficulty adjusting emotional responses to the same card, so long as the rules are understood. The mechanisms underlying this flexibility are not well understood.

Historically, the flexible regulation of reinforcement expectation has often been studied using reversal-learning tasks. In these tasks, two stimuli and/or actions reverse reinforcement contingencies, and animals must learn these new associations. In principle, flexible adjustment of reinforcement expectation during reversal learning can involve two different sorts of mechanisms, a learning process or a cognitive control mechanism. In a learning process, each CS–US association is learned independently upon the reversal. Although behavioral training can lead to faster learning rates, fundamentally the learning about each CS–US pair is a distinct process. In a cognitive control mechanism, subjects can exploit a rule embedded in a reversal-learning task, namely that two CS–US contingencies switch at the same time. If a subject grasps this rule, then upon experiencing one CS as having switched its reinforcement contingency, a subject can infer that the other CS has switched. Two recent studies have explored neurophysiological processing in the amygdala and the prefrontal cortex during the performance of reversal-learning tasks in which subjects invoke either learning or cognitive control strategies.

Morrison et al. [57] targeted simultaneous physiological recording in the amygdala and orbitofrontal cortex (OFC), an area that has been implicated in mediating reversal learning in some (but not all) studies [58–65]. Neurons in the OFC encode positive and negative anticipated reinforcement outcomes [65–71] and have often been studied during decision-making tasks. During classical conditioning tasks, OFC neural responses are correlated with monkeys' behavioral use of information about both rewarding and aversive CSs, and positive and negative neurons in the OFC—like in the amygdala—track value in a consistent manner across the various sensory events in a trial—including the fixation point and CS and US presentations—even though those stimuli differ in sensory modality [66,67].

Morrison et al. [57] examined the OFC and amygdala neurophysiological responses while monkeys performed a task in which one CS predicted a liquid reward and another CS predicted an aversive air puff; the two CSs switched contingencies without warning during recording experiments. Behavioral evidence indicated that monkeys updated their reinforcement expectation for each CS independently upon reversal (a learning mechanism) [57]; there was no evidence of monkeys employing a cognitive strategy. Although neuroscientists have long believed that the prefrontal cortex drives reversal learning, this assertion has often been made without distinguishing between the types of strategies that might be employed and how neural processing may differ depending upon the strategy [58–62,64,65,72].

Morrison et al. [57] were able to examine the dynamics of learning in positive- and negative-valenced neurons in both amygdala and OFC when monkeys were not employing a cognitive strategy, but instead relied on independent learning about each changed CS–US association. Appetitive and aversive networks in the OFC and amygdala exhibited different learning rates, and—surprisingly—the observed differences depended on the valence preference of the cell populations in question. For positive cells, changes in OFC neural activity after reversal were largely complete many trials earlier than in positive cells in the amygdala; for negative cells, the opposite was true, as the aversive network in the amygdala learned more rapidly than the aversive network in the OFC. In each case, the faster-changing area was completing its transition around the time of the onset of changes in behavior; meanwhile the other, more slowly changing, area did not complete the shift in firing pattern until many trials after the behavioral responses began to change (Fig. 3.1).

Thus, signals appropriate for driving behavioral learning are present in both brain structures, with the putative aversive system in the amygdala and the appetitive system in the OFC being particularly rapid in

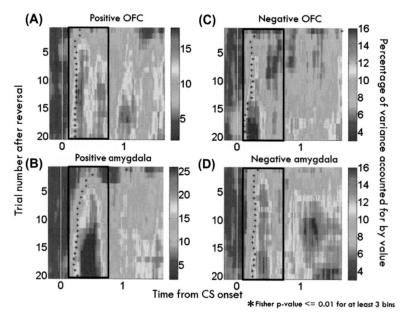

FIGURE 3.1 (A–D) Time course of changes in value-related signals in the amygdala and orbitofrontal cortex (OFC) as a function of reversal learning. For each bin, an index of the contribution-to-image value was computed by performing a two-way analysis of variance and calculating the proportion of variance accounted for by image value divided by the total variance (see Ref. [57]), and then averaged across cells. These index values are plotted as a function of trials after reversal and time from image onset (*white line*). Average contribution-of-image value to neural activity from positive value-coding neurons in (A) OFC and (B) amygdala, and from negative value-coding cells in (C) OFC and (D) amygdala. *Asterisks*, time when the contribution-of-value index becomes significant (for asterisks placed in the center of the first of at least three consecutive significant bins, Fisher $p < .01$). Bin size, 200 ms. Bin step, 20 ms. Notice the different learning rates in appetitive and aversive networks across the two brain areas. *Adapted with permission from Morrison SE, Saez A, Lau B, Salzman CD. Different time courses for learning-related changes in amygdala and orbitofrontal cortex. Neuron [Internet] 2011;6:1127–40 [cited August 23, 2014]. Available from: http://www.sciencedirect.com/science/article/pii/S089662731100643X.*

reflecting changes in reinforcement contingencies. These findings suggest that distinct sequences of neural processing lead to the updating of representations within the appetitive and aversive networks. Perhaps the faster learning response of the aversive network in the amygdala reflects the preservation across evolution of an aversive system that learns very quickly in order to avoid threats. Strikingly, however, after learning occurs, and the new CS–US contingencies are established, both populations of OFC cells—positively and negatively valenced—predict reinforcement earlier in the trial than their counterparts in the amygdala. In addition, analyses of local field potentials suggested that the directional flow of information between amygdala and OFC evolves during learning, with amygdala-to-OFC directionality being stronger earlier during the learning process. Overall, these data indicate that dynamic interactions between amygdala and OFC during reversal learning may be more complex than previously appreciated, a finding supported by fMRI data [61].

An alternative mechanism underlying flexible responses to changes in CS–US contingencies involves the application of cognitive control. Knowledge of the rules governing the environment (the rule that both CS and US contingencies switch at the same time) can allow a subject to infer the new contingency associated with one CS by first experiencing the other switched CS. Saez et al. [73] investigated the use of this type of cognitive control in the process of stimulus-value updating in changing environments. In this study, monkeys acquired and retained information about different sets of contingencies associated with the same stimulus and regulated behavior accordingly. A serial reversal-learning task was performed, in which two CS–US pairs switched contingencies many times in every experiment. Every day the monkeys would be presented with new CS–US pairs, with each set of CS–US pairs comprising a "task set." In one task set, one CS was associated with reward and the other CS was associated with nothing; in the other task set, the first CS was no longer associated with reward and the second CS was. The monkeys switched back and forth between task sets. Sometimes an additional visual cue marked the task set within a trial, but on the majority of trials no cue was present so that the monkey had to rely on memory of the task set to guide reinforcement expectation upon CS presentation.

Unlike in the study by Morrison et al. [57], in this study, probably by virtue of the serial reversals in the task design, monkeys demonstrated that they employed a cognitive strategy to update expectations of reinforcement expectation. Saez et al. [73] measured anticipatory licking to assay expected reinforcement. Monkeys adjusted their anticipatory licking in a switch-like manner after recognizing that reinforcement contingencies had changed. As soon as they experienced the first presentation of a CS with new contingencies, they could infer that the second CS had also changed contingencies and they adjusted their anticipatory licking. In other words, after the monkey had experienced a task set in which CS A was rewarded and CS B was not, the monkey could experience one trial in which CS A was not rewarded and then instantaneously switch its anticipatory licking for CS B. Monkeys were therefore able to generalize their knowledge of task structure (i.e., the task sets that defined two abstract contexts) immediately to new CS—US combinations and infer reinforcement accurately. This occurred even when CS—US pairings had never been experienced previously, such as on the first reversal of an experiment (CSs were novel in every experiment). Monkeys' performance on this task therefore did not derive from having extensive experience viewing the two CSs in each context. These results indicate that the monkeys had acquired and retained a "mental map," an abstract cognitive representation of the rules of the task (stimulus contingencies) and did not need to rely on relearning new associations with each reversal.

Saez et al. [73] recorded single units from the amygdala, the anterior cingulate cortex (ACC; another part of the prefrontal cortex bidirectionally connected with the amygdala), and the OFC. They used a linear decoder to analyze task-relevant signals as a function of time by considering populations of neurons collectively. The chief advantage of using a decoder is that it avoids having to select cells based on their response properties; this was important because neurons often exhibit different types of selectivity in different time windows, such as selectivity to the task set (or abstract context) during the fixation interval and reinforcement expectation selectivity during the trace interval. In addition, neurons were often observed that exhibited mixed selectivity (selectivity to multiple parameters), and these neurons could also be included in the analysis. A 2013 study identifies information provided by mixed-selectivity neurons as important for computations that may endow cognitive flexibility [74].

The linear decoder was trained to read out trial features from the spike counts of the populations of neurons recorded in each brain area. The decoder accuracy provided a single summary statistic that quantified how accurately the whole neuronal population represented information about any particular variable. Information about abstract context (task set) was encoded not only in the ACC and OFC but also in the amygdala [73].

The context signal reflected a process of abstraction, as it indicated knowledge of the task sets that defined context in the study (Fig. 3.2C). Strikingly, errors in behavior (licking when no reward was going to be delivered, or not licking when reward was going to occur) were correlated with the failure to maintain a representation of abstract context in the amygdala.

The prefrontal cortex had previously been proposed to provide a cognitive map of task space [75]. The data from Saez et al. [73] indicate that this role is not limited to prefrontal cortex; moreover, the amygdala does not merely receive gated input from the prefrontal cortex, but instead it actually holds abstract contextual information relevant to forming predictions about anticipated reinforcement and applies it to compute outcome expectations that will lead to behavioral output. This may correspond to a neurophysiological signature of a cognitive control mechanism.

The Influence of Valenced Representations in the Amygdala on Cognitive Processes

Experiments have made clear that the amygdala processes information relevant for cognitive functions beyond the regulation of valenced representations. Most prominently, one cognitive process that interacts with valence attribution is attention (see Chapter 10 on the interaction between salience and valence). Cognitive resources (attention) are typically directed toward stimuli that promote or threaten survival [76]. However, a threatening or rewarding stimulus must be identified and located to guide attention. A traditional view of amygdala function has held that amygdala neurons do not encode spatial information. Instead, the amygdala had been thought to induce arousal, a spatially nonspecific means of augmenting vigilance [77], which can then influence attention.

Peck et al. [50] explored how neural activity in the amygdala is linked to spatial attention and found that amygdala neurons often combine information about both the spatial configuration of visual stimuli and the rewards predicted by stimuli. Monkeys performed a task in which the motivational significance of visual cues modulates visuospatial attention, as measured by reaction time and performance. Amygdala neurons encoded both the spatial configuration and the reward value of motivational cues. Moreover, neural responses were also correlated with measures of spatial attention. A strong positive correlation between value and spatial selectivity was observed, demonstrating that individual amygdala neurons systematically and selectively combine information about space and reward value to signal the location of reward-predictive stimuli with both negative and positive excursions in firing rate

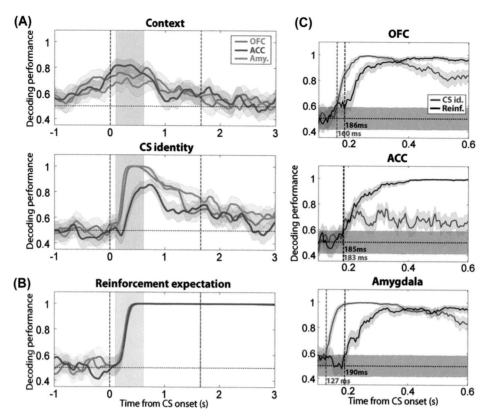

FIGURE 3.2 Population-level encoding of context and conditioned stimulus identity occurs fast enough to account for the correct anticipation of reinforcement. (A) Linear decoder performance at decoding context, and CS identity during "correct" trials. The decoding accuracy was computed on a 250-ms sliding window stepped every 50 ms across the trial for three brain areas separately: *blue*, OFC; *purple*, ACC; *green*, amygdala. The number of cells from each population used in the decoder was equalized for comparisons across areas. *Shaded areas* indicate 95% confidence intervals (bootstrap). *Vertical dashed lines* represent CS onset and earliest possible US onset. *Gray-shaded area* corresponds to 500-ms window from 100 to 600 ms after CS onset used in subsequent analyses. (B) Performance of the linear decoder at decoding reinforcement expectation. (C) Relative timing between the CS identity signal (*blue*) and the reinforcement expectation signal (*black*) in OFC, ACC, and amygdala. The performance of the linear decoder was computed on a 50-ms sliding window stepped every 5 ms across the 500-ms time window shown in (A) and (B) (*shaded area*). *Vertical dashed lines* and corresponding labels indicate the first time bin in which the decoding performance is significantly above chance level and remains above it for the 10 subsequent time bins. *Shaded areas* around chance level indicate 95% confidence intervals for the decoding performances when the two trial types (CS1 and CS2, or reward and no reward) were randomly labeled. *ACC*, anterior cingulate cortex; *CS*, conditioned stimulus; *OFC*, orbitofrontal cortex; *US*, unconditioned stimulus. *Adapted with permission from Saez A, Rigotti M, Ostojic S, Fusi S, Salzman CD. Abstract context representations in primate amygdala and prefrontal cortex. Neuron [Internet] 2015;4:869—81 [cited August 19, 2015]. Available from: http://www.ncbi.nlm.nih.gov/pubmed/26291167.*

[50] (Fig. 3.3). Moreover, the strength of the spatial signal in the amygdala is modulated by task demands, suggesting that it is related to attentional engagement [47]. In a subsequent study, Peck & Salzman. [48] extended these observations to consider how the processing of aversive stimuli is related to the allocation of spatial attention. Stimuli associated with an aversive stimulus (an air puff) attracted spatial attention. Amygdala neurons represented the spatial location of both reward- and punishment-predicting stimuli, and modulation occurred in the same direction for both types of stimuli.

In principle, neural representations within the amygdala that combine spatial information with motivational (appetitive or aversive) information might influence the degree of attention devoted to a stimulus through multiple anatomical pathways. The amygdala

projects densely to the basal forebrain (BF) [78], a collection of subcortical nuclei that are the main source of acetylcholine for the brain [79]; this pathway may be a route through which the amygdala could influence dorsal stream processing, including parietal areas known to be important in attention [80]. Much like amygdala neurons, BF neurons respond selectively depending upon the spatial location and associated reward of stimuli [49].

The amygdala may also influence ventral stream processing through direct projections to visual areas [81]. This would provide a means for motivation to modulate attention and thereby improve perception at one location at the expensive of another. Indeed, the relative value of stimuli at different locations controls preferential processing to one location at the expense of others.

(A)

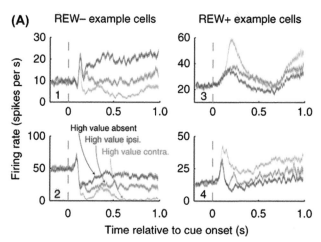

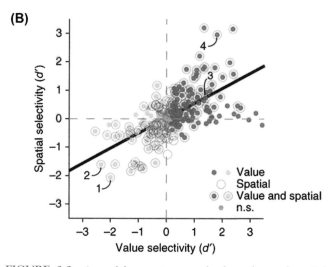

(B)

FIGURE 3.3 Amygdala neurons encode the value and spatial configuration of cues. (A) Peristimulus time histograms showing average firing rate plotted as a function of time relative to cue onset for four amygdala neurons (30-ms bins shifted by 2 ms; shading, SEM). REW− neurons map onto the aversive network, as they respond more strongly to conditioned stimuli paired with no reward than to those paired with reward. REW+ neurons by definition belong to the appetitive work, as they increase firing rate as reward expectation increases. The activity of each example neuron was significantly (bootstrap, $p < .05$) dependent on the presence and location of a cue associated with a large reward (high-value cue, which could appear ipsi- or contralateral to the recorded cell; a low-value cue appeared either in the opposite visual hemifield when a high-value cue was present or in both hemispheres). Expected reward and the spatial configuration of cues frequently had strong effects on neural responses. (B) A spatial-selectivity index was calculated comparing trials in which the high-value cue appeared contralaterally to those in which it appeared ipsilaterally. A value-selectivity index was also calculated in the same time window as for the spatial-selectivity index but comparing high-value-present trials to high-value-absent trials. Neurons that signaled the presence of a high-value cue with an increase in activity also tended to respond more when the high-value cue was contralateral. Scatter plot displays spatial-versus value-selectivity indices for each individual neuron ($n = 326$ neurons). Value-selectivity indices >0 indicate higher activity when a high-value cue was present, and indices <0 indicate higher activity when the high-value cue was absent. Spatial-selectivity indices >0 indicate higher activity when the high-value cue was contralateral; values <0 indicate higher activity when the high-value cue was ipsilateral. Symbol style

Baruni et al. [51] have characterized how reward expectation can differentially modulate neural representations in one ventral stream visual area, V4. V4 has often been the focus of physiological studies investigating the neural correlates of spatial attention, and directing attention into the receptive fields of V4 neurons enhances firing rate [82] as well as gamma-band power and synchronization [83], while diminishing trial-to-trial variability and interneuronal noise correlations [84]. Baruni et al. showed that in contrast with behavior, which was controlled by the relative value of reward expectation, V4 neurons were modulated primarily by the absolute value of the stimulus associated with a receptive field stimulus. Modulation of V4 activity therefore does not necessarily predict improved perceptual performance at that location, and the modulation can be decoupled from the perceptual benefits of attention. Instead, selection mechanisms and not changes in V4 activity that reflect enhanced signal-to-noise at an attended stimulus location probably confer the perceptual benefits of attention. A ripe area for investigation concerns whether the modulation of V4 activity by absolute reward is driven at least in part by the amygdala-to-V4 connection.

CONCLUSION

Neurophysiological and molecular markers of neuronal activity have now clearly established the presence of appetitive and aversive networks within the amygdala. Moreover, optogenetic approaches have established a causal role for activity in these networks in mediating valence-specific emotional behavior. The combination of these approaches, along with appropriately designed behavioral tasks, promises to reveal in greater detail the circuit-level mechanisms within the amygdala that mediate different forms of emotional behavior. Electrophysiological studies have also begun to provide insight into how processing in appetitive and aversive amygdalar networks can be regulated by cognitive operations and how these representations may also influence cognitive functions. Continued

indicates the significance of selectivity for each neuron (*green*, REW+; *yellow*, REW−); *black line* represents the weighted least-squares regression fit ($\beta = 0.53$, $p < 10^{-6}$). Individual amygdala neurons selectively combine information about space and value to signal the location of reward-predictive stimuli with both negative and positive excursions in firing rate. Numbers indicate data points corresponding to the example neurons in (A). *N.S.*, not significant. *Adapted with permission from Peck CJ, Lau B, Salzman CD. The primate amygdala combines information about space and value. Nat Neurosci [Internet] Nat Publishing Group 2013;3:340−8 [cited September 23, 2014]. Available from: http://www.pubmedcentral.nih.gov/articlerender.fcgi?artid=3596258&tool= pmcentrez&rendertype=abstract.*

efforts along these lines, potentially incorporating new tools for delineating neural circuit function, is likely to be critical for unraveling how complex neural circuit dysfunction leads to psychiatric pathophysiology.

In humans, a disruption in the neural mechanisms underlying the attribution of value, and therefore the modulation of anticipatory emotions and decision-making, can lead to maladaptive patterns of behavior. For example, addiction is commonly understood as resulting from a disruption in circuitry that processes reward-predictive stimuli and that must regulate behavioral responses to such stimuli [85]. Deficits in contextual processing can lead to interpersonal sensitivity with failures in emotional regulation such as that observed in social anxiety and borderline personality disorders [86] and schizophrenia [87]. Moreover, the inability to contextualize how stimuli and actions are related to aversive events may lead to posttraumatic stress disorder [88]. The development of transformative treatment strategies for a range of psychiatric disorders therefore probably relies on our acquiring a much more detailed understanding of neural circuitry dedicated to processing rewarding and aversive information, and an understanding of how these circuitries interact with circuitries responsible for cognitive functions [89].

References

[1] Maren S, Phan KL, Liberzon I. The contextual brain: implications for fear conditioning, extinction and psychopathology. Nat Rev Neurosci [Internet] 2013;6:417−28 [cited July 9, 2014]. Available from: http://www.ncbi.nlm.nih.gov/pubmed/23635870.

[2] Salzman CD, Fusi S. Emotion, cognition, and mental state representation in amygdala and prefrontal cortex. Annu Rev Neurosci [Internet] 2010:173−202 [cited July 10, 2014]. Available from: http://www.pubmedcentral.nih.gov/articlerender.fcgi?artid=3108339&tool=pmcentrez&rendertype=abstract.

[3] Wallis JD, Anderson KC, Miller EK. Single neurons in prefrontal cortex encode abstract rules. Nature [Internet] 2001;6840:953−6 [cited October 2, 2014]. Available from: http://www.ncbi.nlm.nih.gov/pubmed/11418860.

[4] Rangel A, Camerer C, Montague PR. A framework for studying the neurobiology of value-based decision making. Nat Rev Neurosci [Internet] 2008;7:545−56 [cited July 9, 2014]. Available from: http://www.pubmedcentral.nih.gov/articlerender.fcgi?artid=4332708&tool=pmcentrez&rendertype=abstract.

[5] Montague PR, Berns GS. Neural economics and the biological substrates of valuation. Neuron [Internet] 2002;2:265−84 [cited November 11, 2015]. Available from: http://www.ncbi.nlm.nih.gov/pubmed/12383781.

[6] Sugrue LP, Corrado GS, Newsome WT. Matching behavior and the representation of value in the parietal cortex. Science [Internet] 2004;5678:1782−7 [cited December 7, 2014]. Available from: http://www.ncbi.nlm.nih.gov/pubmed/15205529.

[7] Padoa-Schioppa C. Neurobiology of economic choice: a good-based model. Annu Rev Neurosci [Internet] 2011:333−59 [cited May 26, 2015]. Available from: http://www.pubmedcentral.nih.gov/articlerender.fcgi?artid=3273993&tool=pmcentrez&rendertype=abstract.

[8] Sutton RS. Learning to predict by the methods of temporal differences. Mach Learn [Internet] 1988;1:9−44 [cited June 25, 2015]. Available from: http://link.springer.com/10.1007/BF00115009.

[9] Schultz W, Dayan P, Montague PR. A neural substrate of prediction and reward. Science [Internet] 1997;5306:1593−9 [cited January 5, 2015]. Available from: http://www.ncbi.nlm.nih.gov/pubmed/9054347.

[10] Morrison SE, Salzman CD. Re-valuing the amygdala. Curr Opin Neurobiol [Internet] 2010;2:221−30 [cited March 22, 2015]. Available from: http://www.sciencedirect.com/science/article/pii/S0959438810000292.

[11] Hernádi I, Grabenhorst F, Schultz W. Planning activity for internally generated reward goals in monkey amygdala neurons. Nat Neurosci [Internet] 2015;3:461−9 [cited June 25, 2015]. Available from: http://www.ncbi.nlm.nih.gov/pubmed/25622146.

[12] Grabenhorst F, Hernadi I, Schultz W. Prediction of economic choice by primate amygdala neurons. Proc Natl Acad Sci [Internet] 2012;46:18950−5 [cited June 25, 2015]. Available from: http://www.pubmedcentral.nih.gov/articlerender.fcgi?artid=3503170&tool=pmcentrez&rendertype=abstract.

[13] LeDoux JE. Emotion circuits in the brain. Annu Rev Neurosci [Internet] 2000:155−84 [cited December 5, 2014]. Available from: http://www.ncbi.nlm.nih.gov/pubmed/10845062.

[14] Gore F, Schwartz EC, Salzman CD. Manipulating neural activity in physiologically classified neurons: triumphs and challenges. Philos Trans R Soc Lond B Biol Sci [Internet] 2015;1677. http://dx.doi.org/10.1098/rstb.2014.0216 [cited August 5, 2015]. Available from: http://www.ncbi.nlm.nih.gov/pubmed/26240431.

[15] Hayes DJ, Duncan NW, Xu J, Northoff G. A comparison of neural responses to appetitive and aversive stimuli in humans and other mammals. Neurosci Biobehav Rev [Internet] 2014:350−68 [cited April 20, 2015]. Available from: http://www.sciencedirect.com/science/article/pii/S0149763414001560.

[16] Baxter MG, Murray EA. The amygdala and reward. Nat Rev Neurosci [Internet] 2002;7:563−73 [cited March 27, 2015]. Available from: http://www.ncbi.nlm.nih.gov/pubmed/12094212.

[17] Tye KM, Stuber GD, de Ridder B, Bonci A, Janak PH. Rapid strengthening of thalamo-amygdala synapses mediates cue−reward learning. Nature [Internet] 2008;7199:1253−7 [cited October 10, 2015]. Available from: http://www.pubmedcentral.nih.gov/articlerender.fcgi?artid=2759353&tool=pmcentrez&rendertype=abstract.

[18] Nabavi S, Fox R, Proulx CD, Lin JY, Tsien RY, Malinow R. Engineering a memory with LTD and LTP. Nature [Internet] 2014;7509:348−52 [cited July 10, 2014]. Available from: http://www.pubmedcentral.nih.gov/articlerender.fcgi?artid=4210354&tool=pmcentrez&rendertype=abstract.

[19] Resnik J, Paz R. Fear generalization in the primate amygdala. Nat Neurosci [Internet] 2015;2:188−90 [cited December 3, 2015]. Available from: http://www.ncbi.nlm.nih.gov/pubmed/25531573.

[20] Janak PH, Tye KM. From circuits to behaviour in the amygdala. Nature [Internet] 2015;7534:284−92 [cited January 14, 2015]. Available from: http://www.pubmedcentral.nih.gov/articlerender.fcgi?artid=4565157&tool=pmcentrez&rendertype=abstract.

[21] Salzman CD, Paton JJ, Belova MA, Morrison SE. Flexible neural representations of value in the primate brain. Ann NY Acad Sci [Internet] Blackwell Publishing Inc. 2007;1:336−54 [cited June 5, 2015]. Available from: http://www.pubmedcentral.nih.gov/articlerender.fcgi?artid=2376754&tool=pmcentrez&rendertype=abstract.

[22] Paton JJ, Belova MA, Morrison SE, Salzman CD. The primate amygdala represents the positive and negative value of visual stimuli during learning. Nature [Internet] 2006;7078:865−70 [cited March 20, 2015]. Available from: http://dx.doi.org/10.1038/nature04490.

[23] Belova MA, Paton JJ, Morrison SE, Salzman CD. Expectation modulates neural responses to pleasant and aversive stimuli in primate amygdala. Neuron [Internet] 2007;6:970−84 [cited August 6, 2014]. Available from: http://www.sciencedirect.com/science/article/pii/S0896627307006150.

[24] Gore F, Schwartz EC, Brangers BC, Aladi S, Stujenske JM, Likhtik E, et al. Neural representations of unconditioned stimuli in basolateral amygdala mediate innate and learned responses. Cell [Internet] 2015;1:134−45 [cited July 4, 2015]. Available from: http://www.ncbi.nlm.nih.gov/pubmed/26140594.

[25] Gründemann J, Lüthi A. Ensemble coding in amygdala circuits for associative learning. Curr Opin Neurobiol [Internet] 2015:200−6 [cited November 14, 2015]. Available from: http://www.ncbi.nlm.nih.gov/pubmed/26531780.

[26] Quirk GJ, Repa C, LeDoux JE. Fear conditioning enhances short-latency auditory responses of lateral amygdala neurons: parallel recordings in the freely behaving rat. Neuron [Internet] 1995;5:1029−39 [cited November 5, 2015]. Available from: http://www.ncbi.nlm.nih.gov/pubmed/7576647.

[27] Repa JC, Muller J, Apergis J, Desrochers TM, Zhou Y, LeDoux JE. Two different lateral amygdala cell populations contribute to the initiation and storage of memory. Nat Neurosci [Internet] 2001;7:724−31 [cited June 6, 2015]. Available from: http://www.ncbi.nlm.nih.gov/pubmed/11426229.

[28] Belova MA, Paton JJ, Salzman CD. Moment-to-moment tracking of state value in the amygdala. J Neurosci [Internet] 2008;40:10023−30 [cited June 7, 2015]. Available from: http://www.pubmedcentral.nih.gov/articlerender.fcgi?artid=2610542&tool=pmcentrez&rendertype=abstract.

[29] Zhang W, Schneider DM, Belova MA, Morrison SE, Paton JJ, Salzman CD. Functional circuits and anatomical distribution of response properties in the primate amygdala. J Neurosci [Internet] 2013;2:722−33 [cited June 8, 2015]. Available from: http://www.jneurosci.org/content/33/2/722.long.

[30] Felix-Ortiz AC, Beyeler A, Seo C, Leppla CA, Wildes CP, Tye KM. BLA to vHPC inputs modulate anxiety-related behaviors. Neuron [Internet] 2013;4:658−64 [cited July 16, 2015]. Available from: http://www.pubmedcentral.nih.gov/articlerender.fcgi?artid=4205569&tool=pmcentrez&rendertype=abstract.

[31] Nieh EH, Kim S-Y, Namburi P, Tye KM. Optogenetic dissection of neural circuits underlying emotional valence and motivated behaviors. Brain Res [Internet] 2013:73−92 [cited February 24, 2015]. Available from: http://www.sciencedirect.com/science/article/pii/S0006899312017763.

[32] Stuber GD, Sparta DR, Stamatakis AM, van Leeuwen WA, Hardjoprajitno JE, Cho S, et al. Excitatory transmission from the amygdala to nucleus accumbens facilitates reward seeking. Nature [Internet] 2011;7356:377−80 [cited October 7, 2015]. Available from: http://www.pubmedcentral.nih.gov/articlerender.fcgi?artid=3775282&tool=pmcentrez&rendertype=abstract.

[33] Tye KM, Prakash R, Kim S-Y, Fenno LE, Grosenick L, Zarabi H, et al. Amygdala circuitry mediating reversible and bidirectional control of anxiety. Nature [Internet] 2011;7338:358−62 [cited September 7, 2015]. Available from: http://www.pubmedcentral.nih.gov/articlerender.fcgi?artid=3154022&tool=pmcentrez&rendertype=abstract.

[34] Redondo RL, Kim J, Arons AL, Ramirez S, Liu X, Tonegawa S. Bidirectional switch of the valence associated with a hippocampal contextual memory engram. Nature [Internet] 2014;7518:426−30 [cited August 27, 2014]. Available from: http://www.pubmedcentral.nih.gov/articlerender.fcgi?artid=4169316&tool=pmcentrez&rendertype=abstract.

[35] Rescorla RA, W AR. A theory of Pavlovian conditioning: variations in the effectiveness of reinforcement and nonreinforcement. In: Prokasy WF, B AH, editors. Classical conditioning II: current research and theory. New York: Appleton Century Crofts; 1972. p. 64−99.

[36] Sutton RS, Barto AG. Reinforcement learning: an introduction. Cambridge, Massachusetts: MIT Press; 1998.

[37] Hollerman JR, Schultz W. Dopamine neurons report an error in the temporal prediction of reward during learning. Nat Neurosci [Internet] 1998;4:304−9 [cited April 24, 2015]. Available from: http://www.ncbi.nlm.nih.gov/pubmed/10195164.

[38] Seo H, Lee D. Temporal filtering of reward signals in the dorsal anterior cingulate cortex during a mixed-strategy game. J Neurosci [Internet] 2007;31:8366−77 [cited June 9, 2015]. Available from: http://www.pubmedcentral.nih.gov/articlerender.fcgi?artid=2413179&tool=pmcentrez&rendertype=abstract.

[39] Matsumoto M, Matsumoto K, Abe H, Tanaka K. Medial prefrontal cell activity signaling prediction errors of action values. Nat Neurosci [Internet] 2007;5:647−56 [cited April 9, 2015]. Available from: http://www.ncbi.nlm.nih.gov/pubmed/17450137.

[40] Kennerley SW, Behrens TEJ, Wallis JD. Double dissociation of value computations in orbitofrontal and anterior cingulate neurons. Nat Neurosci [Internet] Nature Publishing Group 2011;12:1581−9 [cited January 31, 2015]. Available from: http://www.nature.com.ezproxy.cul.columbia.edu/neuro/journal/v14/n12/full/nn.2961.html.

[41] Oyama K, Hernádi I, Iijima T, Tsutsui K-I. Reward prediction error coding in dorsal striatal neurons. J Neurosci [Internet] 2010;34:11447−57 [cited June 26, 2015]. Available from: http://www.ncbi.nlm.nih.gov/pubmed/20739566.

[42] Kim H, Sul JH, Huh N, Lee D, Jung MW. Role of striatum in updating values of chosen actions. J Neurosci [Internet] 2009;47:14701−12 [cited June 26, 2015]. Available from: http://www.ncbi.nlm.nih.gov/pubmed/19940165.

[43] Apicella P. Leading tonically active neurons of the striatum from reward detection to context recognition. Trends Neurosci [Internet] 2007;6:299−306 [cited April 28, 2015]. Available from: http://www.ncbi.nlm.nih.gov/pubmed/17420057.

[44] Hong S, Hikosaka O. The globus pallidus sends reward-related signals to the lateral habenula. Neuron [Internet] 2008;4:720−9 [cited June 1, 2015]. Available from: http://www.pubmedcentral.nih.gov/articlerender.fcgi?artid=2638585&tool=pmcentrez&rendertype=abstract.

[45] Matsumoto M, Hikosaka O. Lateral habenula as a source of negative reward signals in dopamine neurons. Nature [Internet] 2007;7148:1111−5 [cited September 15, 2014]. Available from: http://www.ncbi.nlm.nih.gov/pubmed/17522629.

[46] Shabel SJ, Janak PH. Substantial similarity in amygdala neuronal activity during conditioned appetitive and aversive emotional arousal. Proc Natl Acad Sci [Internet] 2009;35:15031−6 [cited September 28, 2015]. Available from: http://www.pubmedcentral.nih.gov/articlerender.fcgi?artid=2736461&tool=pmcentrez&rendertype=abstract.

[47] Peck EL, Peck CJ, Salzman CD. Task-dependent spatial selectivity in the primate amygdala. J Neurosci [Internet] 2014;49:16220−33 [cited January 13, 2015]. Available from: http://www.ncbi.nlm.nih.gov/pubmed/25471563.

[48] Peck CJ, Salzman CD. Amygdala neural activity reflects spatial attention towards stimuli promising reward or threatening punishment. Elife [Internet] Elife Sci Publications Limited 2014:e04478 [cited June 27, 2015]. Available from: http://elifesciences.org/content/3/e04478.abstract.

[49] Peck CJ, Salzman CD. The amygdala and basal forebrain as a pathway for motivationally guided attention. J Neurosci [Internet] 2014;41:13757−67 [cited June 27, 2015]. Available from: http://www.jneurosci.org.ezproxy.cul.columbia.edu/content/34/41/13757.long#sec-16.

[50] Peck CJ, Lau B, Salzman CD. The primate amygdala combines information about space and value. Nat Neurosci [Internet] Nat Publishing Group 2013;3:340–8 [cited September 23, 2014]. Available from: http://www.pubmedcentral.nih.gov/articlerender.fcgi?artid=3596258&tool=pmcentrez&rendertype=abstract.

[51] Baruni JK, Lau B, Salzman CD. Reward expectation differentially modulates attentional behavior and activity in visual area V4. Nat Neurosci [Internet] Nat Publishing Group 2015;11:1656–63 [cited October 27, 2015]. Available from: http://www.nature.com.ezproxy.cul.columbia.edu/neuro/journal/v18/n11/full/nn.4141.html.

[52] Cho J-H, Deisseroth K, Bolshakov VY. Synaptic encoding of fear extinction in mPFC-amygdala circuits. Neuron [Internet] 2013;6:1491–507 [cited October 30, 2015]. Available from: http://www.pubmedcentral.nih.gov/articlerender.fcgi?artid=3872173&tool=pmcentrez&rendertype=abstract.

[53] Sotres-Bayon F, Sierra-Mercado D, Pardilla-Delgado E, Quirk GJ. Gating of fear in prelimbic cortex by hippocampal and amygdala inputs. Neuron [Internet] 2012;4:804–12 [cited September 3, 2015]. Available from: http://www.pubmedcentral.nih.gov/articlerender.fcgi?artid=3508462&tool=pmcentrez&rendertype=abstract.

[54] Asede D, Bosch D, Lüthi A, Ferraguti F, Ehrlich I. Sensory inputs to intercalated cells provide fear-learning modulated inhibition to the basolateral amygdala. Neuron [Internet] 2015;2:541–54 [cited November 30, 2015]. Available from: http://www.ncbi.nlm.nih.gov/pubmed/25843406.

[55] Maren S. Out with the old and in with the new: synaptic mechanisms of extinction in the amygdala. Brain Res [Internet] 2015:231–8 [cited November 29, 2015]. Available from: http://www.sciencedirect.com/science/article/pii/S0006899314013626.

[56] Likhtik E, Paz R. Amygdala-prefrontal interactions in (mal)adaptive learning. Trends Neurosci [Internet] 2015;3:158–66 [cited August 25, 2015]. Available from: http://www.sciencedirect.com/science/article/pii/S0166223614002355.

[57] Morrison SE, Saez A, Lau B, Salzman CD. Different time courses for learning-related changes in amygdala and orbitofrontal cortex. Neuron [Internet] 2011;6:1127–40 [cited August 23, 2014]. Available from: http://www.sciencedirect.com/science/article/pii/S089662731100643X.

[58] Rudebeck PH, Murray EA. The orbitofrontal oracle: cortical mechanisms for the prediction and evaluation of specific behavioral outcomes. Neuron [Internet] 2014;6:1143–56 [cited November 7, 2015]. Available from: http://www.ncbi.nlm.nih.gov/pubmed/25521376 [cited December 17, 2014]. Available from: http://www.cell.com/article/S0896627314009969/fulltext.

[59] Fellows LK, Farah MJ. Ventromedial frontal cortex mediates affective shifting in humans: evidence from a reversal learning paradigm. Brain [Internet] 2003;(Pt 8):1830–7 [cited December 3, 2015]. Available from: http://www.ncbi.nlm.nih.gov/pubmed/12821528.

[60] Rushworth MFS, Noonan MP, Boorman ED, Walton ME, Behrens TE. Frontal cortex and reward-guided learning and decision-making. Neuron [Internet] 2011;6:1054–69 [cited March 6, 2015]. Available from: http://www.ncbi.nlm.nih.gov/pubmed/21689594.

[61] Chau BKH, Sallet J, Papageorgiou GK, Noonan MP, Bell AH, Walton ME, et al. Contrasting roles for orbitofrontal cortex and amygdala in credit assignment and learning in *Macaques*. Neuron [Internet] 2015;5:1106–18 [cited November 10, 2015]. Available from: http://www.pubmedcentral.nih.gov/articlerender.fcgi?artid=4562909&tool=pmcentrez&rendertype=abstract.

[62] Rudebeck PH, Behrens TE, Kennerley SW, Baxter MG, Buckley MJ, Walton ME, et al. Frontal cortex subregions play distinct roles in choices between actions and stimuli. J Neurosci [Internet] 2008;51:13775–85 [cited May 16, 2015]. Available from: http://www.ncbi.nlm.nih.gov/pubmed/19091968.

[63] Izquierdo A, Suda RK, Murray EA. Bilateral orbital prefrontal cortex lesions in rhesus monkeys disrupt choices guided by both reward value and reward contingency. J Neurosci [Internet] 2004;34:7540–8 [cited June 8, 2015]. Available from: http://www.ncbi.nlm.nih.gov/pubmed/15329401.

[64] Rudebeck PH, Saunders RC, Prescott AT, Chau LS, Murray EA. Prefrontal mechanisms of behavioral flexibility, emotion regulation and value updating. Nat Neurosci [Internet] Nat Publishing Group 2013;8:1140–5 [cited July 23, 2015]. Available from: http://www.nature.com/neuro/journal/v16/n8/full/nn.3440.html#ref1.

[65] Schoenbaum G, Setlow B, Saddoris MP, Gallagher M. Encoding predicted outcome and acquired value in orbitofrontal cortex during cue sampling depends upon input from basolateral amygdala. Neuron [Internet] 2003;5:855–67 [cited May 26, 2015]. Available from: http://www.ncbi.nlm.nih.gov/pubmed/12948451.

[66] Morrison SE, Salzman CD. Representations of appetitive and aversive information in the primate orbitofrontal cortex. Ann NY Acad Sci [Internet] 2011:59–70 [cited December 3, 2015]. Available from: http://www.pubmedcentral.nih.gov/articlerender.fcgi?artid=3683838&tool=pmcentrez&rendertype=abstract.

[67] Tremblay L, Schultz W. Relative reward preference in primate orbitofrontal cortex. Nature [Internet] 1999;6729:704–8 [cited May 8, 2015]. Available from: http://www.ncbi.nlm.nih.gov/pubmed/10227292.

[68] Murray EA, Izquierdo A. Orbitofrontal cortex and amygdala contributions to affect and action in primates. Ann NY Acad Sci [Internet] 2007:273–96 [cited June 8, 2015]. Available from: http://www.ncbi.nlm.nih.gov/pubmed/17846154.

[69] Wallis JD, Miller EK. Neuronal activity in primate dorsolateral and orbital prefrontal cortex during performance of a reward preference task. Eur J Neurosci [Internet] 2003;7:2069–81 [cited April 27, 2015]. Available from: http://www.ncbi.nlm.nih.gov/pubmed/14622240.

[70] Padoa-Schioppa C, Assad JA. Neurons in the orbitofrontal cortex encode economic value. Nature [Internet] 2006;7090:223–6 [cited March 18, 2015]. Available from: http://www.pubmedcentral.nih.gov/articlerender.fcgi?artid=2630027&tool=pmcentrez&rendertype=abstract.

[71] Roesch MR, Olson CR. Neuronal activity related to reward value and motivation in primate frontal cortex. Science [Internet] 2004;5668:307–10 [cited Mar 21, 2015]. Available from: http://www.ncbi.nlm.nih.gov/pubmed/15073380.

[72] Rolls ET. The orbitofrontal cortex. Philos Trans R Soc Lond B Biol Sci [Internet] 1996;1346:1433–43 [cited June 8, 2015]. discussion 1443–4. Available from: http://www.ncbi.nlm.nih.gov/pubmed/8941955.

[73] Saez A, Rigotti M, Ostojic S, Fusi S, Salzman CD. Abstract context representations in primate amygdala and prefrontal cortex. Neuron [Internet] 2015;4:869–81 [cited August 19, 2015]. Available from: http://www.ncbi.nlm.nih.gov/pubmed/26291167.

[74] Rigotti M, Barak O, Warden MR, Wang X-J, Daw ND, Miller EK, et al. The importance of mixed selectivity in complex cognitive tasks. Nature [Internet] 2013;7451:585–90 [cited July 14, 2014]. Available from: http://www.ncbi.nlm.nih.gov/pubmed/23685452.

[75] Wilson RC, Takahashi YK, Schoenbaum G, Niv Y. Orbitofrontal cortex as a cognitive map of task space. Neuron [Internet] 2014;2:267–79 [cited October 23, 2014]. Available from: http://www.pubmedcentral.nih.gov/articlerender.fcgi?artid=4001869&tool=pmcentrez&rendertype=abstract.

[76] Lang PJ, Davis M. Emotion, motivation, and the brain: reflex foundations in animal and human research. Prog Brain Res [Internet] 2006:3–29 [cited June 27, 2015]. Available from: http://www.ncbi.nlm.nih.gov/pubmed/17015072.

[77] Davis M, Whalen PJ. The amygdala: vigilance and emotion. Mol Psychiatry [Internet] 2001;1:13−34 [cited September 27, 2014]. Available from: http://www.ncbi.nlm.nih.gov/pubmed/11244481.

[78] Russchen FT, Amaral DG, Price JL. The afferent connections of the substantia innominata in the monkey, Macaca fascicularis. J Comp Neurol [Internet] 1985;1:1−27 [cited June 27, 2015]. Available from: http://www.ncbi.nlm.nih.gov/pubmed/3841131.

[79] Mesulam MM, Mufson EJ, Levey AI, Wainer BH. Cholinergic innervation of cortex by the basal forebrain: cytochemistry and cortical connections of the septal area, diagonal band nuclei, nucleus basalis (substantia innominata), and hypothalamus in the rhesus monkey. J Comp Neurol [Internet] 1983;2:170−97 [cited June 27, 2015]. Available from: http://doi.wiley.com/10.1002/cne.902140206.

[80] Corbetta M, Shulman GL. Control of goal-directed and stimulus-driven attention in the brain. Nat Rev Neurosci [Internet] 2002;3:201−15 [cited July 10, 2014]. Available from: http://www.ncbi.nlm.nih.gov/pubmed/11994752.

[81] Freese J, Amaral D. In: Whalen P, Phelps E, editors. The human amygdala. Guilford Press; 2009. p. 3−42.

[82] Moran J, Desimone R. Selective attention gates visual processing in the extrastriate cortex. Science [Internet] 1985;4715:782−4 [cited July 17, 2015]. Available from: http://www.ncbi.nlm.nih.gov/pubmed/4023713.

[83] Fries P, Reynolds JH, Rorie AE, Desimone R. Modulation of oscillatory neuronal synchronization by selective visual attention. Science [Internet] 2001;5508:1560−3 [cited September 24, 2015]. Available from: http://www.ncbi.nlm.nih.gov/pubmed/11222864.

[84] Cohen MR, Maunsell JHR. Attention improves performance primarily by reducing interneuronal correlations. Nat Neurosci [Internet] 2009;12:1594−600 [cited October 28, 2015]. Available from: http://www.pubmedcentral.nih.gov/articlerender.fcgi?artid=2820564&tool=pmcentrez&rendertype=abstract.

[85] Volkow ND, Wang G-J, Fowler JS, Tomasi D, Telang F. Addiction: beyond dopamine reward circuitry. Proc Natl Acad Sci USA [Internet] 2011;37:15037−42 [cited November 30, 2015]. Available from: http://www.pubmedcentral.nih.gov/articlerender.fcgi?artid=3174598&tool=pmcentrez&rendertype=abstract.

[86] Schaffer Y, Barak O, Rassovsky Y. Social perception in borderline personality disorder: the role of context. J Pers Disord [Internet] 2013;29(2) [cited October 1, 2014]. Available from: http://www.ncbi.nlm.nih.gov/pubmed/23445472.

[87] MacDonald AW, Pogue-Geile MF, Johnson MK, Carter CS. A specific deficit in context processing in the unaffected siblings of patients with schizophrenia. Arch Gen Psychiatry [Internet] 2003;1:57−65 [cited September 30, 2014]. Available from: http://www.ncbi.nlm.nih.gov/pubmed/12511173.

[88] Milad MR, Pitman RK, Ellis CB, Gold AL, Shin LM, Lasko NB, et al. Neurobiological basis of failure to recall extinction memory in posttraumatic stress disorder. Biol Psychiatry [Internet] 2009;12:1075−82 [cited September 30, 2014]. Available from: http://www.pubmedcentral.nih.gov/articlerender.fcgi?artid=2787650&tool=pmcentrez&rendertype=abstract.

[89] Ramirez S, Liu X, MacDonald CJ, Moffa A, Zhou J, Redondo RL, et al. Activating positive memory engrams suppresses depression-like behaviour. Nature [Internet] 2015;7556:335−9 [cited June 17, 2015]. Available from: http://www.ncbi.nlm.nih.gov/pubmed/26085274.

4

Ventral Striatopallidal Pathways Involved in Appetitive and Aversive Motivational Processes

Y. Saga, L. Tremblay

Institute of Cognitive Sciences (CNRS), Lyon, France

Abstract

The basal ganglia are widely known to play a role in appetitive rewarding behavior. Imaging and psychiatric clinical studies have provided evidence that the basal ganglia, especially the limbic territory, also play a role in aversive information processing. Inappropriate aversive information processing could induce an anxious state, which is widely observed in a variety of psychiatric diseases. Here, we describe how the ventral striatum and the ventral pallidum, two parts of the limbic circuit in the basal ganglia, are involved in negative motivational behavior, which can be linked with the control of approach/avoidance behavior in a normal context and with the expression of anxiety-related disorders. The disturbance of this pathway would induce not only psychiatric symptoms, but also abnormal value-based decision-making.

INTRODUCTION

Every day, humans and animals must make decisions in situations or contexts that lead them to direct their actions toward events that they want to occur (approach toward appetitive stimuli) or away from events that they want to avoid (avoidance of aversive stimuli). A broad range of results from studies in humans and animals show that common and separate neuronal networks control these opposite contexts [1–7]. Although the limbic territory of the basal ganglia such as the ventral striatum (VS) and the ventral pallidum (VP) has been investigated in positive motivational behavior (approaching an appetitive goal), studies have shown that these areas are also involved in negative motivational behavior (avoidance of aversive stimuli). Neuronal representation of positive motivational behavior in these regions is obviously useful for appropriate decision-making, but that of negative motivational behavior is also useful to avoid potentially dangerous and harmful future events. For that, a process is required to discriminate conditioned stimuli (CS), which enables one to predict an occurrence of a positive or negative event in the near future, leading to anticipation of the consequences. If these processes are disturbed, this can lead to addiction (reinforced appetitive conditioning) or anxiety (excessive anticipation of negative consequence) [8,9]. The manifestation of anxiety can be observed in a variety of psychiatric disorders such as obsessive–compulsive disorders (OCDs), panic disorders, anorexia, and depression, and there is an increased demand for better understanding of the underlying neuronal mechanisms. Thus, an investigation into the role of the limbic territories, including the VS, the VP, and other regions, in aversive processes will lay the foundation for a better understanding of the physiopathology of such psychiatric disorders.

The role of the limbic territory of the basal ganglia in the processes of reward, motivation, and decision-making is now generally accepted based on a broad range of results coming from neuroimaging studies in humans and local pharmacological disturbance studies in animals, mostly rats and monkeys. In the rat, it was shown that the disturbance of the dopaminergic and opioid transmission in the nucleus accumbens induced compulsive behaviors such that decision-making and motivation were strictly directed toward food or drug-taking, two stimuli with strong reward properties [10–12]. This type of result has led someone to suggest

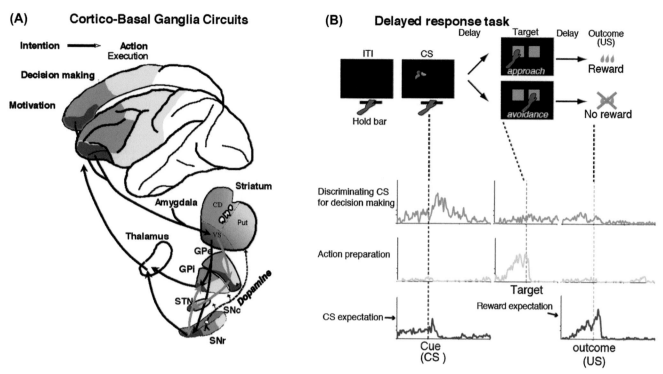

FIGURE 4.1 (A) Schematic representation of the anatomical and functional circuits of the corticobasal ganglia and (B) a schematic overview of the main forms of activity that could be observed in the striatum when monkeys performed a delayed response task. (A) The colors *purple, green,* and *yellow* delineate territories and activities that reflect respectively motivation, cognitive, and motor processes. Inside the striatum the territories have been characterized based on anatomical and electrophysiological investigations, whereas the territories inside the two segments of the globus pallidus (internal, *GPi;* and external, *GPe*) as well as the substantia nigra pars reticulata (*SNr*) are based only on anatomical investigation. Return projections to the cortex are illustrated only for the limbic circuit, which is presumed to be dedicated to motivation processes. The direct and indirect pathways from the ventral striatum (*VS*) to the output structure of the basal ganglia are indicated in *black* and *red arrows*, respectively. In order not to overload the picture, some basal ganglia projections are not illustrated. (B) Behavioral task in appetitive context. Subjects hold a bar to start a trial in which an appetitive conditioned stimulus (*CS*) is presented. Trials are separated by intertrial interval (*ITI*). Different examples of neuronal activities illustrated were recorded in the striatum of monkeys performing this task to obtain drops of juice as reward. *CD,* caudate; *Put,* putamen; *SNc,* substantia nigra pars compacta; *STN,* subthalamic nucleus; *US,* unconditioned stimulus.

that the VS, also called the "accumbens," functions largely as a hedonic structure. For others, it is a structure dedicated to processing the reinforcing aspect of reward, underlying the establishment of stimulus–response or context–habit associations via procedural or instrumental learning. Finally, others regard the VS as being a structure primarily implicated in the determination of the goal of an action, the "thing" that we want to obtain (e.g., food, water, sex: "positive value") or to avoid (e.g., disgusting food, pain, loss of money: "negative value"). It is also important when selecting among alternative goals and actions. In this last conception, the VS and the VP seem to function as an interface between motivation and action [13]. Over the past several years, a large number of neuroimaging studies have appeared, further highlighting these concepts and expanding the range of domains of reward and motivation processed by the VS and the VP from variables such as food [14] and sex [15,16] to financial [17] and social [18] domains. However, while results such as these may provide additional insight into the

types of variables treated by the VS and the VP, they do not show how the neurons of the VS and the VP contribute to these functions. How is the basic information represented in the neuronal activity in the VS and the VP? How is this modified in a specific context of choice or decision-making? What could be the relative contribution of the VS and the VP in comparison with other cerebral structures implicated in the same network, and how do the basal ganglia interact with these cortical areas?

In this chapter, we will primarily focus on results from studies carried out on not only the appetitive aspect, but also the aversive aspect, by introducing nonhuman primate studies and human imaging studies (studies on rodents will be discussed in the next chapters). Related to this, we briefly review anatomical connections including the corticobasal ganglia, but also intrinsic connections in the basal ganglia (see also Haber's Chapter 1). Last, we will focus on the implications of the striatopallidal pathway in anxiety, which is a pathological aspect of anticipation of negative future events and is enhanced in uncertain states, as will be described in a later chapter by Grupe (see Chapter 30).

THE CORTICOBASAL GANGLIA FUNCTIONAL CIRCUITS: THE RELATION BETWEEN THE CORTEX AND THE BASAL GANGLIA

Within the basal ganglia and in particular within the striatum, the processing of motor and nonmotor information is carried out by the massive, topographically and functionally organized cortical projections [19–21] that provide anatomical evidence for the functional subdivision into sensorimotor, associative, and limbic territories within the basal ganglia. The basal ganglia were largely considered to comprise a motor complex in which different inputs converged to produce an output specific to the control of motor execution. At that time, the anatomical approach with classic tracers had shown that cortical regions projected into the striatum with a topographical organization in which regions could be identified based on different functional properties, i.e., the VS (limbic territory), the caudate nucleus (associative territory), and the posterior putamen (sensorimotor territory) (Fig. 4.1A). However, some cortical areas project into a similar territory of the striatum (convergence) and the reduction in the number of neurons from the striatum to the output structures of the basal ganglia, the internal segment of the globus pallidus (GPi), and the substantia nigra pars reticulata (SNr), and the large dendrites of the neurons in these output nuclei [22], strongly suggested that a large functional convergence could occur at these levels of the basal ganglia circuit before the information returned to the cortex via the thalamic relay [27,28]. In addition, only the basal ganglia return projections to the motor and the higher motor cortical areas were well described at this time, thus strengthening the view that the basal ganglia exclusively influenced motor function. It was from this viewpoint that Mogenson and collaborators [13] first suggested that the VS and the VP were perfect candidates to be the interface between motivation and action. The specific afferents of VS come from limbic structures such as the amygdala, the anterior insula, and the medial prefrontal and the orbitofrontal cortex, areas that are well known to be involved in motivational and emotional processing [23] (see also Chapter 1). Combined with the prevailing view that the principal if not exclusive output of the basal ganglia was through the motor cortex, the idea of motivation and action being integrated in the basal ganglia via the VS was compelling, and this concept continues to be expressed frequently in the literature.

An alternative view of basal ganglia circuitry can be seen to have arisen from the specific investigation of the basal ganglia output pathway(s) using rabies virus (a retrograde transsynaptic tracer) injections inside various frontal areas. This work, done by the team of Peter Strick, provided two major new pieces of information: (1) the output of the basal ganglia could project not only to the motor cortex but also to more anterior areas inside the prefrontal cortex dedicated to nonmotor functions [24,25] and (2) the output pathways to the cortex are organized in different parallel circuits where different functional domains could be processed independently [24,26]. Although Strick's team did not study the basal ganglia projections to the orbitofrontal (OFC) or the anterior cingulate cortex (ACC), their conclusions from data on other prefrontal territories strongly suggested the existence of similar closed loops in which information relative to motivation originating from the OFC and/or ACC would return to the same cortical areas after passing through the basal ganglia. In this way, as illustrated in Fig. 4.1A, the VS and the return pathways to these structures could specifically process motivational variables without necessarily exerting a direct influence on the selection or execution of action. As will be developed in the following section, this circuit could process only the goal aspect of the action (what one wants) without determining the appropriate action (which action one has to perform to obtain what one wants). As illustrated in Fig. 4.1A, a second circuit (in *green* in Fig. 4.1A) involving the dorsolateral prefrontal cortex and the caudate nucleus could be involved in cognitive processes concerning decision-making, and a third separate circuit (in *yellow* in Fig. 4.1A) involving the premotor and motor cortices together with the putamen appears to be more specifically dedicated to the preparation and execution of action. In this view, at least three corticobasal ganglia loops are implicated in the various functional processes (motivation, decision-making, and execution of action) involved in translating an intention into an action. The interaction between these different loops could happen inside basal ganglia or at the thalamic level and cortical level. Indeed, the hierarchical projection from the anterior prefrontal areas to the premotor and the motor cortex is well known [27] and these corticocortical projections could provide the link between the parallel corticobasal ganglia circuits.

THE DIRECT AND INDIRECT PATHWAYS: FROM INHIBITION OF COMPETITIVE MOVEMENT TO AVERSIVE BEHAVIORS

The striatum receives massive inputs from a variety of cortical areas; subsequently they project to the external segment of the globus pallidus (GPe; indirect pathway) and the GPi/SNr (direct pathway). It is well known that the neurons of the output structures, the GPi and the SNr, have high spontaneous activity (i.e., around 60 spikes/s) and thereby exert a sustained inhibitory (GABAergic) effect on their ascending targets (thalamic neurons projecting into the cortex) or their descending targets (pedunculopontine nucleus for the GPi and superior colliculus for the SNr). Hikosaka and Wurtz [28] have well demonstrated, in a series of elegant experiments using microinjections of GABA agonists and antagonists in monkeys trained to perform memory-guided saccadic eye movement tasks, that the inhibitory tone of the SNr has a direct suppressive effect on the production of saccadic eye movements. Complementary results indicate that when this inhibitory tone is suppressed or inhibited, the production of saccade is facilitated. In another series of experiments performed in rats using single-unit recording, Deniau and Chevalier [29,30] demonstrated that neurons in the SNr are directly inhibited by activation of striatal neurons. The striatal neurons can directly inhibit neurons in the SNr and thereby disinhibit the thalamic neurons under the tonic inhibitory influence of these SNr neurons. This disinhibition mechanism, whereby a series of two inhibitions has an activating effect on the target, is particularly prominent in processing within the basal ganglia. In the case of the striatal neurons involved in the direct pathway, their activation results in a "permissive" effect on their thalamic targets by this disinhibition mechanism. In contrast, activation of the striatal neurons that project to the GPe (i.e., the indirect pathway) would lead to disinhibition of the basal ganglia output structures, producing a suppressive effect on the target neurons in the thalamus. Several authors have suggested that these dual and opposing pathways could be the substrates for the role of the basal ganglia in a selection of movements [31,32]. In the motor domain attributed to the posterior part of the striatum, it has been suggested that activation of the direct pathway permits (via disinhibition) the execution of a desired motor program, whereas activation of the indirect pathway has a suppressing effect on other competing motor programs. If the same mechanism could be attributed to the VS in the domain of motivation or reward processing, the direct pathway could be engaged in the selection of a specific motivated goal, and the indirect pathway, such as the VP, would suppress other competing goals (e.g., those that are not appropriate to the current need, situation, or context). This hypothesis, based on our knowledge about the physiology and physiopathology of the dorsal striatum, is supported by studies involving local perturbations of function in the VS or the VP of behaving monkeys. Manipulations of the two pathways have revealed opposite roles in not only motor control, but also motivated behaviors [51–54].

SINGLE-UNIT RECORDING IN AWAKE ANIMALS TO INVESTIGATE THE NEURAL BASES

Most of our decisions and actions are driven by the combination of immediate or delayed expectation of reward and corresponding internal motivational factors. Numerous neuroimaging studies in humans have contributed in recent years to identifying brain areas and neural networks that could be involved in specific functional processes that drive decisions and actions in healthy subjects as well as in pathological contexts. Parts 2 and 4 of this book are dedicated specifically to reviewing these studies and their implications. One aspect of neuronal processing not accessible in neuroimaging studies, however, involves how information is represented at the neuronal level in the various brain areas identified by imaging, and thus what role each area plays in its functional networks. Knowledge about processing at the neuronal level comes primarily from animal studies using electrophysiological single-unit neuronal recording or local pharmacological modulation of cerebral circuits. This part of the chapter is dedicated to a description of the electrophysiological studies performed in the VS and the VP of monkeys in appetitive and aversive behavior.

Recording the activity of single neurons in identified brain structures of awake monkeys using inserted microelectrodes has been a classic approach for acquiring knowledge of diverse functions. Despite the tremendous advancements in the field of functional neuroimaging, electrophysiological neuronal recording retains the advantage of being a direct measure of the neuronal activity, with a notably better temporal resolution (milliseconds compared to seconds for the most advanced fMRI), which is a crucial factor when attempting to determine relationships to transient sensory, motor, and cognitive events. However, recording single neurons is a time-consuming approach, as it takes time to accumulate data from a sufficient number of neurons to have confidence that a representative sample of the structure being studied has been obtained. Even at that point, it can be argued that the sample is restricted to a relatively small proportion of neurons in that structure and that proportion could be biased by the experimental conditions. In contrast, with fMRI the entire brain can be measured virtually simultaneously in a single session; however, because of their small size (especially the VP) and relative depth, structures such as the basal ganglia have been difficult to analyze precisely in earlier fMRI studies.

TASKS TO INVESTIGATE APPETITIVE APPROACH BEHAVIOR AND POSITIVE MOTIVATION

Approach behavior and reward-seeking, which frequently occur to achieve a goal or to obtain a reward, are fundamental positive motivational behaviors [33]. These behaviors are crucial in intake behavior such as eating and drinking [34]. To examine these behaviors, researchers use two different types of behavioral tasks, namely Pavlovian conditioning (classical conditioning) and operant conditioning (instrumental conditioning). In the former, subjects learn a CS, which is associated with an unconditioned stimulus (US). Through performing trials, subjects can predict a US (e.g., a drop of juice, one euro, etc.) from a presented CS. Subjects are required to see a CS, but not generate action. In operant conditioning, subjects learn a CS associated with a US and they have to make a decision and a response to a target presented after the CS to obtain the US (i.e., reaching a target, releasing a bar, etc.). Two delay periods are interleaved between CS presentation and the target to be reached, and between the subject's response and the US (Fig. 4.1B). If they make an incorrect response, they cannot obtain any US and another trial will start. By using this task, we can investigate the involvement of neurons in specific functions associated with behavioral events (see Fig. 4.1B).

A crucial difference between the protocols is that, whereas Pavlovian conditioning involves passive engagement in tasks, subjects have to make choices and responses voluntarily (actively engage) in operant conditioning. This difference is crucially important regarding the decision-making processes present in the operant task and not in the Pavlovian task. Fig. 4.1B shows an example of a delayed response task. To start a trial, a monkey has to hold a response bar. Then CSs associated with reward (or aversive outcome) are presented (CS presentation). After several seconds of delay, the monkey has to choose a target to obtain a reward (or avoid it if preferred). He can obtain a CS-associated outcome (US) according to his response followed by the delay period (reward delay). While monkeys perform this task, experimenters can investigate if neurons respond to behavioral events (Fig. 4.1B, lower panels). For example, before CS presentation, the monkeys expect a CS (motivational process: purple activity) and look at the CS image on the screen (attentional process: green activity). But dissociation of two possibilities would be difficult in this task because subjects are required to keep their attention on the screen and they normally have motivation to perform the tasks. To dissociate these two aspects (i.e., attention vs motivation) in neuronal activity, several different domains of stimuli

that have different motivational value (i.e., positive, negative, and neutral) are used. For example, if neurons encode motivational value, neurons should show different responses between positive and negative motivational contexts. On the other hand, if neurons encode salience (potentially related to attention), neurons should show responses to both motivational stimuli [35,36]. The example of the neuron also shows CS response with phasic activity. Then a neuron could show increased activity during the first delay period while the target was being chosen, considered as preparation for action or movement-related activity (yellow). After the response, the monkeys wait for US delivery (anticipation period). The activity during this period would reflect anticipation of the US outcome or reward expectation (purple, right).

This task structure allows one to examine the characteristics of neurons in the specific territories of the basal ganglia. In the striatum and the pallidum, previous studies have shown different types of neuronal activities involved in motivated goal-directed behaviors for appetitive contexts [37–43]. A summary of the main recorded types inside the striatum is illustrated in Fig. 4.1B.

VENTRAL STRIATUM AND VENTRAL PALLIDUM ARE ALSO INVOLVED IN AVERSIVE BEHAVIORS: FIRST EVIDENCE OF LOCAL INHIBITORY DYSFUNCTION

Nevertheless, few studies in monkeys have ever taken neuronal recordings with aversive stimuli, although the aversive network has been identified in humans and animals [44]. The primary goal of previous research has been to determine the role of the VS in processing the motivational value of visual stimuli that predict primary aversive stimuli. We know from anatomical studies that the VS receives projections from cerebral structures such as the amygdala, the lateral OFC, and the insula, all of which process aversive stimuli [45–51]. Moreover, selective activation inside the central part of the VS, which produces stereotyped behaviors (an anxiety manifestation; Fig. 4.2B), induces specific activation of these three cerebral structures, which are involved in processing aversive events. Berridge has already proposed the existence of a subregion inside the VS implied in the aversive processes, based on behavioral reactions induced by pharmacological disturbances in rats [52]. MRI studies have also identified a similar subterritory in humans inside the VS that is involved in the anticipation and avoidance of aversive stimuli or unpleasant events [53,54]. We suspect that a specific territory inside the VS is dedicated to the processing of aversive stimuli. The level of heterogeneity

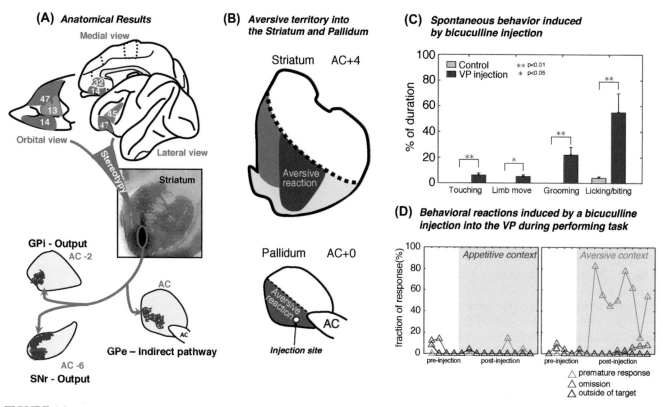

FIGURE 4.2 (A) Corticobasal ganglia circuit involved in stereotyped behaviors induced by bicuculline injection into the ventral striatum (VS) and identified by injection of biotinylated dextran amine, a retrograde and anterograde anatomical tracer (*Modified from Worbe Y, Sgambato-Faure V, Epinat J, Chaigneau M, Tande D, Francois C, et al. Towards a primate model of Gilles de la Tourette syndrome: anatomo-behavioural correlation of disorders induced by striatal dysfunction. Cortex 2013;49:1126—40, with permission.*). (B) Hypothesized aversive territory inside the VS and the ventral pallidum (*VP*). The VS is illustrated at the level of 4 mm anterior to the anterior commissure (AC + 4), which shows the territory where negative behavioral effects were induced by bicuculline injections [63]. The VP is drawn at the level of AC 0, where the AC is still observed. The small white circle represents the bicuculline injection site. Results are shown in (C) and (D). (C) Spontaneous behavioral effect induced by bicuculline injection. The vertical axis represents the percentage of the tested duration of the behavioral patterns. Behavioral patterns are classified into touching, limb movement, grooming, and licking and biting fingers. Compulsive grooming and licking and biting fingers are the main aversive reactions induced by bicuculline injection. The *gray bars* represent control sessions (injection saline), *blue bars* indicate bicuculline injections. (D) During the task performance, bicuculline induced escape behaviors (premature responses) only in an aversive context (right). The *white zone* indicates behavioral results before injection and the *gray zone* indicates postinjection. The colored marks show each type of behavioral reaction. *p < 0.05, **p < 0.01. *GPe*, external segment of the globus pallidus; *GPi*, internal segment of the globus pallidus; *SNr*, substantia nigra pars reticulata.

within the VS is an important question. The heterogeneity of the VS implies that neuronal disturbances in specific territories could be responsible for various psychiatric disorders that affect the following: food intake (bulimia and anorexia), money attraction (gambling), sex activity (hypersexuality disorder), or social motivation (social phobia).

To examine the heterogeneity of the VS function, we initiated bicuculline injections into various subregions of the VS. Inhibitory dysfunction by injection of bicuculline into the VS induces stereotyped behaviors (repetitive licking and biting of the fingers and compulsive grooming), hypoactivity with loss of food motivation (reduced food intake), and sexual manifestations, depending on the injection site [55]. Subsequently, electrical stimulation of the VS induces similar effects except for sexual manifestations [56]. Stereotypy may reflect an

increased anxious state because humans with anxiety frequently show similar behaviors [57—60]. Hypoactivity may reflect a reduced general motivational state or a specific loss of food motivation, which could be linked to the apathy observed in depression or loss of food motivation in anorexia [61,62]. These different aberrant activations induced from different VS territories all suggest aversive effects. Interestingly, we never observed positive effects such as enhancement of intake behaviors or motivation for performing behavioral tasks.

As for the striatum, we also manipulated pallidum activity by injection of bicuculline. The bicuculline injections induced different behavioral effects in the dorsal and ventral parts [63]. As seen in a striatal injection study, whereas injection into the dorsal part induced hyperactivity and attention deficit, injections of bicuculline into the VP induced stereotyped behaviors (Fig. 4.2B;

increase of grooming and licking/biting fingers). Thus, the VP, like the VS, is one of the components that can produce aversive behaviors and seems to exert control on behavioral reactions to aversive events.

THE VENTRAL STRIATUM AND THE VENTRAL PALLIDUM ENCODE AVERSIVE FUTURE EVENTS FOR PREPARATION OF AVOIDANCE AND FOR CONTROLLING ANXIETY LEVEL

Unlike appetitive approach behavior and positive motivation, negative motivation and behavior to avoid harmful, fearful, or dangerous stimuli are critically important to survive in a natural environment [64]. In addition to this, aberrant processing of aversive information could lead to an anxious state. Therefore, control of these processes by neuronal activity is crucial for understanding anxiety and related disorders. Indeed, patients with anxiety-related disorders, such as posttraumatic stress disorder (PTSD) and OCD, show excessive avoidance behavior and loss of anxiety regulation [9,65,66]. Moreover, some cognitive processes are disturbed in patients with anxiety, resulting in bias to attending to potential threat information and a tendency to interpret uncertain events negatively, and these abnormal processes make them more aversive [67,68]. Thus, people with anxious traits and related disorders have been shown that their aversive information processes are abnormal and excessive.

Over the last decade, researchers have begun experiments focused on aversive information processing and negative motivation. Human imaging studies have identified the regions that process aversive motivational information, such as the ACC, the anterior insula, and the amygdala, but also the VS [54,69,70]. These areas play roles in detection of negative emotional stimuli and anticipation of negative outcomes. Human studies have identified components of the basal ganglia, such as the VS, as critical contributors to aversive as well as appetitive learning [47,53,54,69,70]. One of the pioneer works in human study by Jensen et al. [54] showed anticipation of an aversive stimulus (electrical shock on a finger), and when the subjects avoided this stimulus, the ACC, the anterior insula, and the VS were activated. More recently, common and distinct approach and avoidance networks in humans have been shown [71]. Thus, imaging studies have already shown that the brain has networks for appetitive and aversive motivational behavior, particularly in the limbic circuit.

As already mentioned, Berridge and his colleagues proposed the existence of a subregion inside the VS implied in the aversive processes, based on behavioral reactions induced by pharmacological disturbances in rats [52,72]. In addition, after the rats learned to avoid aversive outcomes (e.g., electrical shock to the feet), $GABA_A$ agonist (muscimol) injection into the VS, amygdala, and prelimbic cortex (homolog of the dorsal ACC in primates), inducing inactivation of the target region, impaired conditioned avoidance [73–76]. These results suggest that the VS plays a role in processing aversive information and learning avoidance behavior in collaboration with other cortical regions.

Although several studies have been conducted on rats, a few studies have investigated monkeys. The ACC, which projects to the VS, induces negative decision-making by applying electrical stimulation [77]. In this study, monkeys were offered outcomes and selected to approach them or avoid them by manipulating a lever. Monkeys could obtain a reward by choosing the "approach" response, but they had to receive an air puff (i.e., an aversive stimulus) before receiving the reward. The strength of the air puff and the volume of the reward were indicated by conditioned cues; therefore the parameters of the air puff and the reward varied trial by trial. Depending on the offered stimulus, monkeys approached (received an air puff to obtain a reward) or avoided (no outcome, but actually given a small reward to maintain their motivation). The stimulation of the ACC allowed the monkeys to make negative-biased decisions (more avoidance). This finding suggests that the ACC controls the decision point regarding the cost–benefit balance and regulates positive- and negative-motivated behavior.

The VP, the recipient of the VS, was also investigated in a similar way, but mainly with regard to the positive motivational context. A study of healthy human subjects showed that the VP is involved in transforming positive motivation to action [78]. In this study, the subjects were presented with an instruction that indicated a potential available outcome (money) and the handgrip power necessary to obtain it. The VP was more activated when subjects used a handgrip in higher motivational trials to obtain a greater money reward than in lower motivational trials, suggesting the involvement of positive motivational behavior. Patients with bilateral pallidum lesions showed an apathetic state, which is a state of reduced motivation [62,79,80]. On the other hand, when healthy human subjects look at disgusting food pictures, the VP and anteroventral insula are activated [81]. In rats, a VP lesion or inactivation caused an excessive disgust reaction [82]. Thus, the VP may play a role in negative or aversive information processing. However, few studies show aversive information processing in the VP of monkeys.

We have developed behavioral tasks in which monkeys were required to choose adapted behavior (approach or avoidance) faced to the value (appetitive or aversive) of the expected outcome (Fig. 4.1B). The monkeys were trained to perform different delayed-response tasks with appetitive and aversive outcomes (i.e., a drop of juice and an air puff, respectively). To start a trial, the monkey had to hold a lever and then a CS associated with either appetitive or aversive outcome (US) was presented on a screen in front of the monkey. After a delay period without CS, the monkey had to select one of the two targets presented on the screen. When the monkey selected the target, there were optional actions, approach or avoidance. If the monkey chose the target in the same position in which the CS was presented, he could obtain the US associated with the CS (i.e., approach). Contrary to this, if the monkey chose the opposite target, he could avoid the US and get nothing as the outcome. That meant the monkey showed active avoidance. Depending on the context of a trial (i.e., appetitive or aversive), the nature of the actions was differ (neuron 2) and a VP neuron (neuron 4) responded with transient responses to aversive CS selectively. In addition to this, we found that a number of neurons anticipated specific outcomes (see examples of appetitive and aversive anticipation in Fig. 4.3B and C, right). Interestingly, most of these anticipatory activities did not reflect the chosen action (i.e., approach or avoidance), suggesting encoding in contextual information positive and negative value rather than behavioral consequence. As in previous studies, appetitive coding

neurons were found (neurons 1 and 3 in Fig. 4.3B and C). The numbers of these appetitive and aversive neurons were even and distributed equally. Although two different types of population were distributed in an intermixed manner, it is still an open question whether a specific population projects to a direct or indirect pathway. Because the VP is a nucleus of the indirect pathway, theoretically, it is possible to receive GABAergic inhibition from the VS, which is subsequently projected to the GPi or SNr (Fig. 4.3A, neurons 2 and 3). It is crucial to identify which of the two different populations projects to the GPi/SNr (direct pathway: neuron 1) and the VP (indirect pathway: neuron 2), and to what extent. Therefore, suppression activity observed in appetitive-preferring neurons in the VP might receive inputs preferentially from aversive-preferring neurons in the VS. On the other hand, neurons showing excitation responses to appetitive and aversive stimuli receive inputs from the subthalamic nucleus, whose major neurotransmitter is glutaminergic, or by suppression of inhibitory control by neighboring neurons in the VP (Fig. 4.3A, neuron 4) [83].

To further investigate the VP function on behavior, we induced local reversible dysfunction by bicuculline injection into the VP while the monkeys were or were not performing tasks (Fig. 4.2C and D). First, we

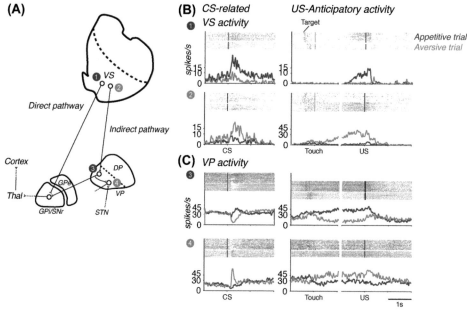

FIGURE 4.3   (A) Schematic view of direct and indirect pathways between the striatum and the pallidum. Two types of neurons in the VS project to the GPi (direct pathway) and the VP (indirect pathway). Then a signal is sent out to cortical areas via the thalamus. (B) Two examples of neurons in the VS showing response to CS (*left*: aligned on CS onset) and two examples of neurons with US anticipatory activity (*right*: aligned on screen touch and US onset, respectively). *Blue* and *red* raster and histogram represent activity in appetitive and aversive trials. (C) Two examples of the VP neurons showing CS-related and two others with US anticipatory activity [92]. The representations of raster and histograms are in the same format as in (B). *CS*, conditioned stimulus; *DP*, dorsal pallidum; *GPe*, external segment of the globus pallidus; *GPi*, internal segment of the globus pallidus; *SNr*, substantia nigra pars reticulata; *STN*, subthalamic nucleus; *Thal*, thalamus; *US*, unconditioned stimulus; *VP*, ventral pallidum; *VS*, ventral striatum.

evaluated the spontaneous behaviors induced by the bicuculline injection (Fig. 4.2C). The VP injections induced significantly compulsive grooming and licking or biting of the fingers (typical behavior induced by a stressful situation). During the tasks (Fig. 4.2D), the bicuculline injections induced mainly the release of the response bar before presentation of the targets (premature responses: red marks). These error responses were nonadaptive behavior because an identical trial was repeated if they made an error. Additionally, the monkeys showed no movement impairment or motivational problems because they could perform trials in the appetitive context. Thus, the monkeys showed reduction of active avoidance and increase of aberrant aversive reaction in an aversive context, perhaps because they were in an excessive aversive anticipation state. These injection results support the view that the VP contributes to avoidance behavior by encoding CS and enabling anticipation of outcome. These results strongly suggest that the VS—VP pathway, an indirect pathway in the basal ganglia, greatly contributes to aversive processing and avoidance behavior in an aversive context. Moreover, an aberrant function of this pathway could contribute to generating an anxious state potentially relevant to anxiety-related disorders.

ABNORMAL AVERSIVE PROCESSING COULD BIAS DECISION-MAKING TOWARD PATHOLOGICAL BEHAVIORS

Another important issue concerning anxiety is that an anxious state could be modulated by predictable and unpredictable future events. Predictability is crucial for making plans for a subsequent action, and an unpredictable situation creates a more uncomfortable and anxious state [84]. Thus, a situation of normal predictability (certainty) enables the subject to reduce the anxiety level, whereas unpredictability increases this level. The uncertainty of the potential aversive situation enhances the anxious state, and uncertainty is associated with increased bold responses in the ACC, amygdala, and insula [85]. In humans, the anxious state can induce hypervigilance for aversive stimuli as well as excessive aversive anticipation and avoidance behaviors [47]. Hypervigilance, which is excessive fear reactions in the face of aversive stimuli, can be seen in patients with anxiety disorders like OCD and PTSD [86—89]. Presumably, the hypervigilance for aversive stimuli can induce excessive reactions even in a context that is normally perceived as appetitive or neutral. Moreover, patients with anxiety show attention bias when they are exposed to a fear or threat stimulus [86]. This abnormal perceptual processing causes abnormal value-based decision-making, especially when negative value stimuli are present. Thus, the pathological aspects of anxiety and abnormal decision-making could have a close relationship in terms of fundamental mechanisms.

CONCLUSION

We have reviewed recent studies about the neuronal basis of positive and negative motivational behavior mainly in two structures of the basal ganglia, the VS and the VP, in nonhuman primates. The comprehension of this mechanism is vital, not only to open the door for diagnostic and therapeutic developments concerning psychiatric diseases that could be caused by dysfunctions inside the basal ganglia, but also to understand the effects of neurotransmitters such as dopamine and serotonin, which could influence the activity of these structures and how they induce their therapeutic effects [90]. Gaining such understanding is a challenge that we can now master, which will greatly help us to understand some psychiatric disorders that will be discussed in the fourth part of this book. Such disorders include abnormal decision-making, which can be seen in drug addiction, pathological gambling, and anxiety-related disorders.

Acknowledgments

This work was performed within the framework of the LABEX CORTEX (ANR-11-LABX-0042) of Université de Lyon, within the program "Investissements d'Avenir" (ANR-11-IDEX-0007) operated by the French National Research Agency. Y.S. is supported by the Japan Society for the Promotion of Science Postdoctoral Fellow for Research Abroad.

References

[1] Barberini CL, Morrison SE, Saez A, Lau B, Salzman CD. Complexity and competition in appetitive and aversive neural circuits. Front Neurosci 2012;6:170.

[2] Cybulska-Klosowicz A, Zakrzewska R, Kossut M. Brain activation patterns during classical conditioning with appetitive or aversive UCS. Behav Brain Res 2009;204:102—11.

[3] Gottfried JA, O'Doherty J, Dolan RJ. Appetitive and aversive olfactory learning in humans studied using event-related functional magnetic resonance imaging. J Neurosci 2002;22:10829—37.

[4] Jensen J, Smith AJ, Willeit M, Crawley AP, Mikulis DJ, Vitcu I, et al. Separate brain regions code for salience vs. valence during reward prediction in humans. Hum Brain Mapp 2007;28:294—302.

[5] Kobayashi S, Nomoto K, Watanabe M, Hikosaka O, Schultz W, Sakagami M. Influences of rewarding and aversive outcomes on activity in macaque lateral prefrontal cortex. Neuron 2006;51:861—70.

[6] Matsumoto M, Hikosaka O. Two types of dopamine neuron distinctly convey positive and negative motivational signals. Nature 2009;459:837—41.

[7] Plassmann H, O'Doherty JP, Rangel A. Appetitive and aversive goal values are encoded in the medial orbitofrontal cortex at the time of decision making. J Neurosci 2010;30:10799—808.

[8] Everitt BJ, Parkinson JA, Olmstead MC, Arroyo M, Robledo P, Robbins TW. Associative processes in addiction and reward. The role of amygdala-ventral striatal subsystems. Ann NY Acad Sci 1999;877:412−38.

[9] Grupe DW, Nitschke JB. Uncertainty and anticipation in anxiety: an integrated neurobiological and psychological perspective. Nat Rev Neurosci 2013;14:488−501.

[10] Cardinal RN, Daw N, Robbins TW, Everitt BJ. Local analysis of behaviour in the adjusting-delay task for assessing choice of delayed reinforcement. Neural Netw 2002;15:617−34.

[11] Kelley AE, Baldo BA, Pratt WE, Will MJ. Corticostriatal-hypothalamic circuitry and food motivation: integration of energy, action and reward. Physiol Behav 2005;86:773−95.

[12] Le Moal M, Koob GF. Drug addiction: pathways to the disease and pathophysiological perspectives. Eur Neuropsychopharmacol 2007;17:377−93.

[13] Mogenson GJ, Jones DL, Yim CY. From motivation to action: functional interface between the limbic system and the motor system. Prog Neurobiol 1980;14:69−97.

[14] O'Doherty JP, Buchanan TW, Seymour B, Dolan RJ. Predictive neural coding of reward preference involves dissociable responses in human ventral midbrain and ventral striatum. Neuron 2006;49:157−66.

[15] Ponseti J, Bosinski HA, Wolff S, Peller M, Jansen O, Mehdorn HM, et al. A functional endophenotype for sexual orientation in humans. NeuroImage 2006;33:825−33.

[16] Bray S, O'Doherty J. Neural coding of reward-prediction error signals during classical conditioning with attractive faces. J Neurophysiol 2007;97:3036−45.

[17] Knutson B, Bossaerts P. Neural antecedents of financial decisions. J Neurosci 2007;27:8174−7.

[18] Izuma K, Saito DN, Sadato N. Processing of social and monetary rewards in the human striatum. Neuron 2008;58:284−94.

[19] Alexander GE, DeLong MR, Strick PL. Parallel organization of functionally segregated circuits linking basal ganglia and cortex. Annu Rev Neurosci 1986;9:357−81.

[20] Parent A, Hazrati LN. Functional anatomy of the basal ganglia. I. The cortico-basal ganglia-thalamo-cortical loop. Brain Res 1995; 20:91−127.

[21] Haber SN. The primate basal ganglia: parallel and integrative networks. J Chem Neuroanat 2003;26:317−30.

[22] Yelnik J, Francois C, Percheron G. Spatial relationships between striatal axonal endings and pallidal neurons in macaque monkeys. Adv Neurol 1997;74:45−56.

[23] Cardinal RN, Parkinson JA, Hall J, Everitt BJ. Emotion and motivation: the role of the amygdala, ventral striatum, and prefrontal cortex. Neurosci Biobehav Rev 2002;26:321−52.

[24] Middleton FA, Strick PL. Basal ganglia output and cognition: evidence from anatomical, behavioral, and clinical studies. Brain Cogn 2000;42:183−200.

[25] Middleton FA, Strick PL. Basal-ganglia 'projections' to the prefrontal cortex of the primate. Cereb Cortex 2002;12:926−35.

[26] Middleton FA, Strick PL. New concepts about the organization of basal ganglia output. Adv Neurol 1997;74:57−68.

[27] Fuster JM. The prefrontal cortex—an update: time is of the essence. Neuron 2001;30:319−33.

[28] Hikosaka O, Wurtz RH. Modification of saccadic eye movements by GABA-related substances. I. Effect of muscimol and bicuculline in monkey superior colliculus. J Neurophysiol 1985;53:266−91.

[29] Chevalier G, Vacher S, Deniau JM, Desban M. Disinhibition as a basic process in the expression of striatal functions. I. The striato-nigral influence on tecto-spinal/tecto-diencephalic neurons. Brain Res 1985;334:215−26.

[30] Deniau JM, Chevalier G. Disinhibition as a basic process in the expression of striatal functions. II. The striato-nigral influence on thalamocortical cells of the ventromedial thalamic nucleus. Brain Res 1985;334:227−33.

[31] Mink JW. The basal ganglia: focused selection and inhibition of competing motor programs. Prog Neurobiol 1996;50:381−425.

[32] Redgrave P, Prescott TJ, Gurney K. The basal ganglia: a vertebrate solution to the selection problem? Neuroscience 1999;89: 1009−23.

[33] Hollerman JR, Tremblay L, Schultz W. Involvement of basal ganglia and orbitofrontal cortex in goal-directed behavior. Prog Brain Res 2000;126:193−215.

[34] Berridge KC. Motivation concepts in behavioral neuroscience. Physiol Behav 2004;81:179−209.

[35] Bromberg-Martin ES, Matsumoto M, Hikosaka O. Dopamine in motivational control: rewarding, aversive, and alerting. Neuron 2010;68:815−34.

[36] Kahnt T, Park SQ, Haynes JD, Tobler PN. Disentangling neural representations of value and salience in the human brain. Proc Natl Acad Sci USA 2014;111:5000−5.

[37] Cromwell HC, Schultz W. Effects of expectations for different reward magnitudes on neuronal activity in primate striatum. J Neurophysiol 2003;89:2823−38.

[38] Hassani OK, Cromwell HC, Schultz W. Influence of expectation of different rewards on behavior-related neuronal activity in the striatum. J Neurophysiol 2001;85:2477−89.

[39] Hollerman JR, Tremblay L, Schultz W. Influence of reward expectation on behavior-related neuronal activity in primate striatum. J Neurophysiol 1998;80:947−63.

[40] Schultz W, Apicella P, Scarnati E, Ljungberg T. Neuronal activity in monkey ventral striatum related to the expectation of reward. J Neurosci 1992;12:4595−610.

[41] Shidara M, Aigner TG, Richmond BJ. Neuronal signals in the monkey ventral striatum related to progress through a predictable series of trials. J Neurosci 1998;18:2613−25.

[42] Tremblay L, Hollerman JR, Schultz W. Modifications of reward expectation-related neuronal activity during learning in primate striatum. J Neurophysiol 1998;80:964−77.

[43] Tachibana Y, Hikosaka O. The primate ventral pallidum encodes expected reward value and regulates motor action. Neuron 2012;76:826−37.

[44] Hayes DJ, Northoff G. Identifying a network of brain regions involved in aversion-related processing: a cross-species translational investigation. Front Integr Neurosci 2011;5:49.

[45] Buchel C, Morris J, Dolan RJ, Friston KJ. Brain systems mediating aversive conditioning: an event-related fMRI study. Neuron 1998; 20:947−57.

[46] Carlson JM, Greenberg T, Rubin D, Mujica-Parodi LR. Feeling anxious: anticipatory amygdalo-insular response predicts the feeling of anxious anticipation. Soc Cogn Affect Neurosci 2011;6: 74−81.

[47] Nitschke JB, Sarinopoulos I, Mackiewicz KL, Schaefer HS, Davidson RJ. Functional neuroanatomy of aversion and its anticipation. NeuroImage 2006;29:106−16.

[48] Nitschke JB, Sarinopoulos I, Oathes DJ, Johnstone T, Whalen PJ, Davidson RJ, et al. Anticipatory activation in the amygdala and anterior cingulate in generalized anxiety disorder and prediction of treatment response. Am J Psychiatry 2009;166:302−10.

[49] Simmons A, Matthews SC, Stein MB, Paulus MP. Anticipation of emotionally aversive visual stimuli activates right insula. Neuroreport 2004;15:2261−5.

[50] Simmons A, Strigo I, Matthews SC, Paulus MP, Stein MB. Anticipation of aversive visual stimuli is associated with increased insula activation in anxiety-prone subjects. Biol Psychiatry 2006;60:402−9.

[51] Yang H, Spence JS, Devous Sr MD, Briggs RW, Goyal A, Xiao H, et al. Striatal-limbic activation is associated with intensity of anticipatory anxiety. Psychiatry Res 2012;204:123–31.

[52] Reynolds SM, Berridge KC. Emotional environments retune the valence of appetitive versus fearful functions in nucleus accumbens. Nat Neurosci 2008;11:423–5.

[53] Delgado MR, Jou RL, Ledoux JE, Phelps EA. Avoiding negative outcomes: tracking the mechanisms of avoidance learning in humans during fear conditioning. Front Behav Neurosci 2009; 3:33.

[54] Jensen J, McIntosh AR, Crawley AP, Mikulis DJ, Remington G, Kapur S. Direct activation of the ventral striatum in anticipation of aversive stimuli. Neuron 2003;40:1251–7.

[55] Worbe Y, Baup N, Grabli D, Chaigneau M, Mounayar S, McCairn K, et al. Behavioral and movement disorders induced by local inhibitory dysfunction in primate striatum. Cereb Cortex 2009;19:1844–56.

[56] Worbe Y, Epinat J, Feger J, Tremblay L. Discontinuous long-train stimulation in the anterior striatum in monkeys induces abnormal behavioral states. Cereb Cortex 2011;21(12):2733–41.

[57] Bryant CE, Rupniak NM, Iversen SD. Effects of different environmental enrichment devices on cage stereotypies and autoaggression in captive cynomolgus monkeys. J Med Primatol 1988;17: 257–69.

[58] McBride SD, Parker MO. The disrupted basal ganglia and behavioural control: an integrative cross-domain perspective of spontaneous stereotypy. Behav Brain Res 2015;276:45–58.

[59] Novak MA, Kinsey JH, Jorgensen MJ, Hazen TJ. Effects of puzzle feeders on pathological behavior in individually housed rhesus monkeys. Am J Primatol 1998;46:213–27.

[60] Pomerantz O, Paukner A, Terkel J. Some stereotypic behaviors in rhesus macaques (Macaca mulatta) are correlated with both perseveration and the ability to cope with acute stressors. Behav Brain Res 2012;230:274–80.

[61] Bhatia KP, Marsden CD. The behavioural and motor consequences of focal lesions of the basal ganglia in man. Brain 1994; 117(Pt 4):859–76.

[62] Levy R, Dubois B. Apathy and the functional anatomy of the prefrontal cortex-basal ganglia circuits. Cereb Cortex 2006;16:916–28.

[63] Grabli D, McCairn K, Hirsch EC, Agid Y, Feger J, Francois C, et al. Behavioural disorders induced by external globus pallidus dysfunction in primates: I. Behavioural study. Brain 2004;127: 2039–54.

[64] LeDoux JE. Evolution of human emotion: a view through fear. Prog Brain Res 2012;195:431–42.

[65] Gillan CM, Morein-Zamir S, Urcelay GP, Sule A, Voon V, Apergis-Schoute AM, et al. Enhanced avoidance habits in obsessive-compulsive disorder. Biol Psychiatry 2014;75:631–8.

[66] Stein MB, Paulus MP. Imbalance of approach and avoidance: the yin and yang of anxiety disorders. Biol Psychiatry 2009;66:1072–4.

[67] Hartley CA, Phelps EA. Anxiety and decision-making. Biol Psychiatry 2012;72:113–8.

[68] Paulus MP, Yu AJ. Emotion and decision-making: affect-driven belief systems in anxiety and depression. Trends Cogn Sci 2012; 16:476–83.

[69] Bolstad I, Andreassen OA, Reckless GE, Sigvartsen NP, Server A, Jensen J. Aversive event anticipation affects connectivity between the ventral striatum and the orbitofrontal cortex in an fMRI avoidance task. PLoS One 2013;8:e68494.

[70] Pohlack ST, Nees F, Ruttorf M, Schad LR, Flor H. Activation of the ventral striatum during aversive contextual conditioning in humans. Biol Psychol 2012;91:74–80.

[71] Schlund MW, Magee S, Hudgins CD. Human avoidance and approach learning: evidence for overlapping neural systems and experiential avoidance modulation of avoidance neurocircuitry. Behav Brain Res 2010;225:437–48.

[72] Richard JM, Berridge KC. Metabotropic glutamate receptor blockade in nucleus accumbens shell shifts affective valence towards fear and disgust. Eur J Neurosci 2011;33:736–47.

[73] Bravo-Rivera C, Roman-Ortiz C, Brignoni-Perez E, Sotres-Bayon F, Quirk GJ. Neural structures mediating expression and extinction of platform-mediated avoidance. J Neurosci 2014;34: 9736–42.

[74] Choi JS, Cain CK, LeDoux JE. The role of amygdala nuclei in the expression of auditory signaled two-way active avoidance in rats. Learn Mem 2010;17:139–47.

[75] Lazaro-Munoz G, LeDoux JE, Cain CK. Sidman instrumental avoidance initially depends on lateral and basal amygdala and is constrained by central amygdala-mediated Pavlovian processes. Biol Psychiatry 2011;67:1120–7.

[76] Ramirez F, Moscarello JM, LeDoux JE, Sears RM. Active avoidance requires a serial basal amygdala to nucleus accumbens shell circuit. J Neurosci 2015;35:3470–7.

[77] Amemori KI, Graybiel AM. Localized microstimulation of primate pregenual cingulate cortex induces negative decision-making. Nat Neurosci 2012;15:776–85.

[78] Pessiglione M, Schmidt L, Draganski B, Kalisch R, Lau H, Dolan RJ, et al. How the brain translates money into force: a neuroimaging study of subliminal motivation. Science 2007;316: 904–6.

[79] Laplane D, Baulac M, Widlocher D, Dubois B. Pure psychic akinesia with bilateral lesions of basal ganglia. J Neurol Neurosurg Psychiatry 1984;47:377–85.

[80] Laplane D, Levasseur M, Pillon B, Dubois B, Baulac M, Mazoyer B, et al. Obsessive-compulsive and other behavioural changes with bilateral basal ganglia lesions. A neuropsychological, magnetic resonance imaging and positron tomography study. Brain 1989;112(Pt 3):699–725.

[81] Calder AJ, Beaver JD, Davis MH, van Ditzhuijzen J, Keane J, Lawrence AD. Disgust sensitivity predicts the insula and pallidal response to pictures of disgusting foods. Eur J Neurosci 2007;25: 3422–8.

[82] Ho CY, Berridge KC. Excessive disgust caused by brain lesions or temporary inactivations: mapping hotspots of the nucleus accumbens and ventral pallidum. Eur J Neurosci 2014;40: 3556–72.

[83] Matsumura M, Tremblay L, Richard H, Filion M. Activity of pallidal neurons in the monkey during dyskinesia induced by injection of bicuculline in the external pallidum. Neuroscience 1995;65: 59–70.

[84] Grillon C, Baas JP, Lissek S, Smith K, Milstein J. Anxious responses to predictable and unpredictable aversive events. Behav Neurosci 2004;118:916–24.

[85] Sarinopoulos I, Grupe DW, Mackiewicz KL, Herrington JD, Lor M, Steege EE, et al. Uncertainty during anticipation modulates neural responses to aversion in human insula and amygdala. Cereb Cortex 2010;20:929–40.

[86] Bar-Haim Y, Lamy D, Pergamin L, Bakermans-Kranenburg MJ, van IJzendoorn MH. Threat-related attentional bias in anxious and nonanxious individuals: a meta-analytic study. Psychol Bull 2007;133:1–24.

[87] Mathews A, MacLeod C. Cognitive vulnerability to emotional disorders. Annu Rev Clin Psychol 2005;1:167–95.

[88] Mogg K, Bradley BP. A cognitive-motivational analysis of anxiety. Behav Res Ther 1998;36:809–48.

[89] Reinecke A, Cooper M, Favaron E, Massey-Chase R, Harmer C. Attentional bias in untreated panic disorder. Psychiatry Res 2001;185:387–93.

[90] Tremblay L, Worbe Y, Thobois S, Sgambato-Faure V, Feger J. Selective dysfunction of basal ganglia subterritories: from movement to behavioral disorders. Mov Disord 2015;30(9): 1155—70.

[91] Worbe Y, Sgambato-Faure V, Epinat J, Chaigneau M, Tande D, Francois C, et al. Towards a primate model of Gilles de la Tourette syndrome: anatomo-behavioural correlation of disorders induced by striatal dysfunction. Cortex 2013;49: 1126—40.

[92] Saga Y, Richard A, Sgambato-Faure V, Hoshi E, Tobler PN, Tremblay L. Ventral pallidum encodes contextual information and controls aversive behaviors. Cereb Cortex April 24, 2016. pii:bhw107.

5

Reward and Decision Encoding in Basal Ganglia: Insights From Optogenetics and Viral Tracing Studies in Rodents

J. Tian, N. Uchida, N. Eshel

Harvard University, Cambridge, MA, United States

Abstract

For the past several decades, electrophysiologists have taught us much about how the brain works by recording from neurons [1,2]. In most cases, however, we have not known what types of neuron we were recording from, or what those neurons were connected to in the rest of the brain. To truly understand a circuit, we need to record from neurons with knowledge of their cell type and patterns of connectivity. Furthermore, the ability to manipulate the activity of specific neural types provides an important means to test hypotheses of circuit function. With the advent of optogenetics, molecular techniques, and more sophisticated tracing protocols, we have finally begun to take these steps. Here we highlight advances in the physiology, function, and circuit mechanism of decision-making, focusing especially on the involvement of dopamine and its major recipient area, the striatum.

It is one of the peculiar features of most modern neurophysiology that the experimentalist seldom knows which type of neuron he or she is listening to. There is no indication of where these different sets of neurons are sending their information, let alone exactly what type of neuron they are. This is not science but rather natural history. Rutherford would probably have called it stamp collecting [3].

DOPAMINE

Decision-making is an iterative process. We list the possible choices, predict outcomes for each one, make our choice, and then experience the consequences. Finally, depending on whether the outcome was better or worse than expected, we update our predictions to make better decisions in the future. Dopamine, the neuromodulator famed for its role in Parkinson's disease [4]

and addiction [5], among many other diseases, is critical for this process.

Physiology

In the mid-1990s, Schultz et al. demonstrated that dopamine-producing neurons in the midbrain of monkeys exhibit a peculiar response to reward [6] (see Chapter 2 for details). When reward is delivered unexpectedly, dopamine neurons fire a burst of action potentials. If the monkeys are taught to expect reward, however, dopamine neurons no longer respond to that reward. Finally, if an expected reward is omitted, dopamine neurons pause their firing at the time reward is expected [7]. Together, these results suggest that dopamine neurons signal reward-prediction error, or the difference between expected and actual reward. When reward is greater than expected, dopamine neurons fire; if reward is the same as expected, there is no response; if reward is less than expected, there is a suppression of activity.

This finding has been replicated dozens of times, from rodents [8–11] to humans [12]. The error signal has been shown to vary with all of the features that determine reward value, including magnitude [13–15], probability [16], timing [7,17,18], and subjective preference [19]. Furthermore, the signal develops in lockstep with behavior over learning [20,21]. When dopamine bursting is removed, either through genetics [22] or pharmacology [23–25], animals and humans show impaired cue-dependent reward learning, implying that dopamine may play a causal role in this process (see Ref. [26] for review).

Although these findings were first reported with standard electrophysiology, new molecular techniques have allowed us to confirm and refine the prediction error model. Classically, for example, dopamine neurons were identified during recording using spike waveforms, firing rate, reward response, and other indirect physiologic features. However, these criteria have low sensitivity and specificity for dopamine identification [27], implying that previous studies may have recorded from a mixture of dopaminergic and nondopaminergic midbrain neurons. This is particularly important given the heterogeneity of the ventral tegmental area (VTA), which is primarily dopaminergic, but has substantial minorities of GABAergic and glutamatergic neurons [28]. To circumvent this issue, Cohen et al. [10] recorded from VTA in mice while using optogenetics to definitively tag neurons as dopaminergic or GABAergic (Fig. 5.1A–C). They found that dopamine neurons indeed showed phasic activations to reward-predictive cues and reward, similar to those found in nonhuman primates. Furthermore, the ability to record from identified dopamine neurons also allowed us to examine the computation that dopamine neurons perform without the risk of contamination by other cell types. In particular, Eshel et al. [29] demonstrated that reward expectation reduces phasic reward responses of optogenetically identified dopamine neurons in a purely subtractive fashion, an unusual computation in the brain but one that is consistent with classic reinforcement learning models.

In addition to phasic responses time-locked to salient events, baseline or tonic dopamine levels are thought to regulate various behaviors. For instance, tonic dopamine may regulate the vigor of responding or the pace of behavioral performance depending on the overall availability of reward [30]. Cohen et al. recorded the activity of optogenetically identified dopamine neurons in a task in which the overall value of the state was modulated over a relatively long time scale (>1 min) [31]. They found that dopamine neurons did not change their baseline firing rates depending on the state value; instead, they found that many serotonergic neurons in the dorsal raphe changed their baseline firing rates in the same task. A study in nonhuman primates made a similar finding [32]. These studies, using molecular identification of different cell types, are quickly refining our knowledge of how the neuromodulator systems interact during learning.

In contrast to the phasic activity of dopamine neurons in the VTA, GABA neurons showed a very different response: they ramped up activity from the onset of a reward-predicting cue to the time of expected reward [10]. This pattern of activity supports the idea that local GABA neurons provide dopamine neurons with an expectation signal. When reward is expected, the inhibitory GABAergic input might cancel excitatory reward input from elsewhere in the brain, eliminating dopamine neurons' response. Causal manipulations using optogenetics allowed us to test this hypothesis (see later) [29]. Note that it is only with the advent of advanced molecular techniques that we are able to dissociate intermingled cell types and assign possible functions to each one.

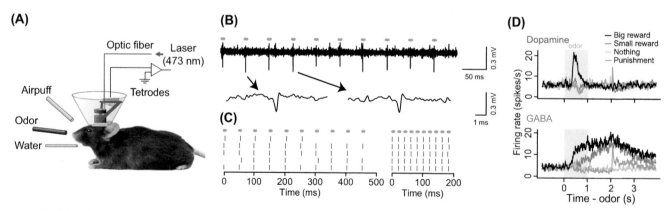

FIGURE 5.1   **Optogenetic tagging.** One of the major shortcomings of traditional extracellular recordings is the inability to identify the cell type being recorded. Here optogenetics was paired with electrophysiology to remove this limitation. (A) Transgenic mice were injected with a virus expressing the light-sensitive cation channel channelrhodopsin-2 (ChR2) and then implanted with tetrodes and a fiber optic, allowing for simultaneous light delivery and extracellular recording. Depending on the transgenic strain, ChR2 would be expressed in either dopamine or GABA neurons. (B) Voltage trace of an example identified dopamine neuron during 10 pulses of 20-Hz light stimulation (*cyan bars*). Two individual spikes are shown below. (C) Response from the same neuron to 20 (left) and 50 Hz (right) light stimulation. (D) Firing rate of one identified dopamine neuron (top) and one identified GABA neuron (bottom) during a classical conditioning task in which mice associated different odors with a big reward (black), a small reward (dark gray), nothing (light gray), or an air puff (orange). *Adapted with permission from Cohen JY, Haesler S, Vong L, Lowell BB, Uchida N. Neuron-type-specific signals for reward and punishment in the ventral tegmental area. Nature 2012;482:85–8. http://dx.doi. org/10.1038/nature10754.*

Recording from dopamine cell bodies is only one part of the puzzle. To understand the downstream effects of dopamine release, it is also important to record from dopamine axons in target regions. Here again, new techniques have paved the way. Gunaydin et al. [33] used fiber photometry along with genetically encoded calcium indicators to explore the role of dopamine afferents in the striatum for social behavior. They implanted transgenic mice expressing a calcium indicator (GCaMP) in neurons expressing tyrosine hydroxylase (TH) with an optical fiber in the nucleus accumbens and successfully detected activity in dopamine axons while mice interacted with one another. In particular, they found that increases in VTA—nucleus accumbens projection activity predicted social interaction. This finding—specific to both the cell type of origin and the downstream target—opens up new avenues for understanding the physiology of normal behavior.

Function

Understanding dopamine function requires both readout and control. New molecular techniques have allowed us to make headway on the latter as well. Several experimenters have used channelrhodopsin (ChR2) in transgenic mice to optogenetically activate VTA dopamine neurons and examine the effects on behavior. In the first study of its kind, Tsai et al. [34]

showed that mice preferred environments that had been paired with dopamine stimulation. Although previous studies had demonstrated conditioned place preference with electrical stimulation of dopamine-producing regions, this was the first experiment using selective dopamine neuron stimulation. Importantly, mice showed a place preference only when dopamine neurons were stimulated in a phasic fashion; tonic stimulation, which did not elicit dopamine transients in the striatum, had no behavioral effect. Later studies confirmed and extended these results, finding that optogenetic dopamine stimulation facilitates approach behavior in both mice and rats [35—37].

One elegant study [38] used optogenetic manipulation to test a crucial tenet of reinforcement learning theory (Fig. 5.2). In the 1960s, Kamin reported that if stimulus A already predicts a reward, then combining A with a new stimulus X fails to trigger an association between X and the reward [39]. In other words, if a reward was already fully predicted, pairing that reward with another stimulus would not cause a conditioned response. This phenomenon, called blocking, revealed the importance of prediction error. For associations to form, the outcome must be different from the expected.

We now know that dopamine neurons encode prediction error (see earlier). If this dopamine signal is indeed important for learning, then dopamine stimulation should be sufficient to reverse blocking. This is exactly

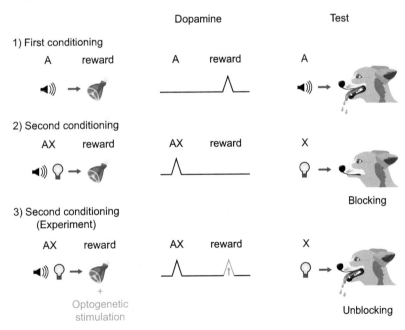

FIGURE 5.2 **The role of dopamine in blocking.** During initial conditioning, the animal learns to associate a tone with reward. After many pairings, a light is added to the tone, and this combination is paired with reward. But now the reward is fully predicted, so dopamine no longer responds to it. In the absence of dopamine, the animal fails to associate the light with the reward. However, if dopamine is artificially increased at the time of predicted reward, the animal learns the association. *Adapted with permission from Eshel N, Tian J, Uchida N. Opening the black box: dopamine, predictions, and learning. Trends Cogn Sci September 2013;17(9):430—431. http://dx.doi.org/10.1016/j.tics.2013.06.010 [Epub 2013 July, 3]. Based on the work of Steinberg EE, Keiflin R, Boivin JR, Witten IB, Deisseroth K, Janak PH. A causal link between prediction errors, dopamine neurons and learning. Nat Neurosci 2013. http://dx.doi.org/10.1038/nn.3413.*

what Steinberg et al. [38] found. They used ChR2 to stimulate dopamine neurons during the fully predicted reward, mimicking a prediction error. They found that such stimulation was sufficient to unblock learning, causing rats to respond to the previously blocked cue. Thus, dopamine prediction errors appear to be causally involved in associative learning. The temporal precision and cell-type specificity of optogenetics was crucial for this experiment to succeed.

Circuitry

Ultimately the goal of our field is not to understand dopamine neurons in isolation, but rather to understand the circuit in which they act. New techniques have allowed us a much better window into this circuit.

Anatomically, decades of experiments have attempted to trace inputs to dopamine neurons [40,41]. However, conventional tracing methods could not distinguish between dopaminergic and nondopaminergic neurons in the same region, and were susceptible to contamination by axons of passage. To resolve these issues, Watabe-Uchida et al. used a modified rabies virus to map and quantify all monosynaptic inputs to dopamine neurons in the whole mouse brain [42]. They found that dopamine neurons in both VTA and substantia nigra pars compacta (SNc) received direct inputs from a more diverse set of areas than previously believed (Fig. 5.3). Furthermore, although the two regions received many common inputs, there was also a topographical shift, with SNc dopamine neurons receiving more inputs from dorsal and lateral regions.

Studies have further refined the dopamine "inputome" by identifying monosynaptic inputs to subpopulations of dopamine neurons defined by their projection targets (e.g., striatum, cortex, amygdala, etc.) [43–45]. These studies show that even projection-specific dopamine populations receive inputs from many areas, and that most populations receive a similar set of inputs. There are some differences, however. Menegas et al. [45], for example, found that the ventral striatum is a major source of input to most dopamine neurons, but does not project to dopamine neurons that project to the posterior striatum. By compiling the entire list of direct inputs to specific populations of dopamine neurons, these findings provide a foundation with which to study how neural circuits regulate dopamine neuron activity.

One of the major inputs to VTA dopamine neurons are VTA GABA neurons. As mentioned earlier, these GABA neurons encode reward expectation, ramping up in response to cues predicting water reward [10]. Do dopamine neurons use this expectation signal to calculate prediction error? To find out, Eshel et al. used optogenetics to selectively stimulate or inhibit

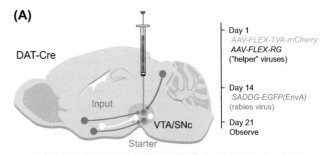

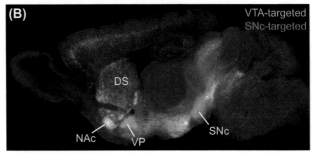

FIGURE 5.3 **Rabies virus tracing.** Compared with traditional anatomical tracing techniques, rabies virus has the advantage of labeling neurons that are synaptically connected to a population of neurons with a defined cell type. (A) Modified rabies virus was used to trace the monosynaptic inputs of dopamine neurons. The virus is pseudotyped with an avian virus envelope protein (EnvA) and therefore cannot infect mammalian cells unless they express a cognate receptor (e.g., TVA). In addition, the gene for the rabies virus envelope glycoprotein (RG), which is required for transsynaptic spread, is deleted. Therefore, transsynaptic transfer occurs only from neurons that exogenously express RG. By using dopamine transporter (DAT) −Cre mice [90], we can restrict the expression of TVA and RG proteins to dopamine neurons so that rabies virus labels only the monosynaptic inputs of dopamine neurons. (B) Sagittal slice of the mouse brain in which two types of rabies virus expressing different fluorescent proteins were injected into the VTA and SNc, respectively. Cyan neurons are monosynaptic inputs to VTA dopamine neurons. Red neurons are monosynaptic inputs to SNc dopamine neurons. The starter neurons were also labeled. *DS*, dorsal striatum; *NAc*, nucleus accumbens; *SNc*, substantia nigra pars compacta; *VP*, ventral pallidum; *VTA*, ventral tegmental area.

VTA GABA neurons while recording from dopamine neurons [29]. They found that stimulating VTA GABA neurons suppressed dopamine reward responses, as if reward was more expected. Furthermore, this suppression took the same arithmetic form—subtraction—as natural, cue-driven expectation. Conversely, inhibiting VTA GABA neurons increased dopamine reward responses, as if reward was less expected. Finally, bilaterally stimulating VTA GABA neurons while keeping rewards unchanged caused mice to learn new, reduced values for odor stimuli, consistent with prediction error theory. By combining recording, optogenetic stimulation, and behavior, these experiments dissected the dopamine circuit in ways that were never before possible, revealing a crucial role for VTA GABA neurons in putting the "prediction" in prediction error.

As mentioned before, dopamine neurons also receive long-range inputs from many areas. In addition to VTA GABA input, could these long-range inputs contribute to the computation of reward prediction error? For example, many neurons in the orbitofrontal cortex signal reward expectation. One 2011 study showed that lesions of the orbitofrontal cortex impair expectation-dependent suppressions of dopamine reward responses, suggesting that the orbitofrontal cortex is a source of the reward expectation signal [9]. Similarly, previous studies in primates showed that neurons in the lateral habenula (LHb) signal reward prediction errors (RPEs), but in an opposite way compared to dopamine neurons (i.e., they are inhibited when reward is greater than expected and excited when reward is less than expected) [46]. As a result, it has been proposed that RPE is already calculated in the LHb and that RPE signals are simply relayed to dopamine neurons via inhibitory neurons in the rostromedial tegmental area, rather than being calculated locally at dopamine neurons [47]. Tian and Uchida [48] tested this proposal by lesioning the habenula and examining dopamine responses (Fig. 5.4). They found that specific aspects of dopamine RPE responses were impaired by habenula lesions. Specifically, even after large lesions of the habenula, excitatory responses to reward-predictive cues

and reward were relatively unaffected. In contrast, inhibitory responses to reward omissions were impaired, even while other inhibitory responses (e.g., suppression by air-puff-predictive cues or air puff itself) were unchanged. These results suggested that the LHb is not the only source of inputs that support RPE signaling by VTA dopamine neurons. Future work will need to determine how all of the inputs dopamine neurons receive are combined to calculate RPE.

Although dopamine is classically known for its role in RPE, increasing evidence suggests that dopamine neurons are not a homogeneous population [49,50]. Rather, there is a complex system of dopamine subtypes, with differences in projections, inputs, cotransmission, and electrophysiologic and molecular features. Molecular techniques have become essential to dissect these differences. For example, Lammel et al. [51] combined slice recordings with rabies-mediated expression of ChR2 to distinguish dopamine neurons receiving inputs from the laterodorsal tegmentum versus the LHb. Those receiving inputs from the laterodorsal tegmentum tended to project to the nucleus accumbens lateral shell, and activating these inputs elicited conditioned place preference in mice. Conversely, dopamine neurons receiving inputs from the LHb tended to project to the medial prefrontal cortex and rostromedial tegmental

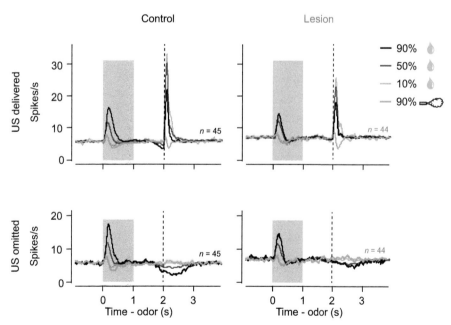

FIGURE 5.4 **Habenula lesions impair specific components of RPE signals in dopamine neurons.** Mice were trained to associate different odors with different probabilities of outcomes (a droplet of water reward or an aversive air puff). Odors were delivered for 1 s and the outcome was revealed following 1 s of delay. Dopamine neurons were identified using the optogenetic tagging method mentioned in the text. The lesion group was mice with bilateral electrolytic lesions in the habenula; the control group was mice with no operation or sham lesion. The average firing rates across all identified dopamine neurons were plotted. Compared to dopamine neurons in the control group, neurons in the lesion group tended to show weaker phasic responses. Notably, in habenula-lesioned animals, dopamine neurons were still able to encode positive prediction error, responding more strongly when reward was less predicted. However, for reward omission responses, dopamine neurons from habenula-lesioned animals were no longer able to distinguish omission of 90% reward from omission of 50% reward.

nucleus, and activating these inputs led to conditioned place aversion. These results suggest that VTA dopamine neurons are part of multiple circuits that mediate different behaviors, and that input- or projection-specific targeting will be one important method to understand these differences. Again, it is only with advanced molecular techniques that we are now able to gain traction on the physiology, function, and circuitry underlying dopamine's involvement in decision-making and reward.

STRIATUM

The striatum is one of the major targets of dopamine axons. Historically, the striatum was considered a predominantly motor structure because of the severe motor impairments that occur when the structure is dysfunctional, for example, in Parkinson's disease and Huntington disease. Mounting evidence now suggests that the striatum also plays important roles in learning and cognitive functions, including the ability to represent and flexibly update values, learn habits and other motor skills, select appropriate actions, and regulate the vigor of responding. The striatum consists of various types of neurons expressing specific molecular markers. Over the past several years, efforts have been made to tease apart how these different neurons contribute to these important functions. In the following section, we review some of these studies, focusing particularly on those using molecular tools. More comprehensive reviews on specific functions of the striatum are available elsewhere [52–56], also see Chapter 4.

Initiating Movement, Habits, and Motor Skills

Decades of work on Parkinson's disease and the anatomy of the basal ganglia have led to various hypotheses about how the striatal circuit regulates movement initiation and action selection. More than 95% of neurons in the striatum are medium spiny neurons (MSNs), which are inhibitory projection neurons. However, there are two largely nonoverlapping subtypes of MSNs, defined by how they link to the output structures of the basal ganglia (internal segment of the globus pallidus and the substantia nigra). Direct-pathway MSNs project directly to the output basal ganglia (BG) structures, whereas indirect-pathway MSNs project to the output BG structures via intermediate targets (external segment of the globus pallidus and subthalamic nucleus). Fortunately for molecular techniques, MSNs in the direct and indirect pathways express different dopamine receptors, D1 and D2, respectively. One class of models postulates that direct-pathway MSNs facilitate movements, whereas indirect-pathway MSNs inhibit them [57,58].

Dopamine activates direct-pathway MSNs via D1 dopamine receptors, whereas dopamine suppresses the activity of indirect-pathway MSNs via D2 dopamine receptors. Thus, in Parkinson's disease patients, in which striatal dopamine is depleted, the imbalance between indirect and direct pathway activation leads to difficulty in initiating movements. A different class of models, however, proposes that coactivation of both pathways is necessary for movement [59]. For instance, direct-pathway MSNs might select the desired action, whereas indirect-pathway MSNs inhibit competing, less desired actions.

Although these models have been influential in thinking about how the basal ganglia circuit regulates movement, it has been difficult to test them directly. How does the activity of each pathway relate to action initiation? Although spike waveforms and other physiologic properties have been successfully used to distinguish MSNs from other cell types (e.g., cholinergic neurons or fast-spiking GABAergic neurons), until recently there was no way to tell whether an MSN being recorded in a behaving animal was in the direct or indirect pathway. In particular, spike waveforms and other physiological properties are indistinguishable between the two types, and identification methods based on their projection targets (i.e., antidromic spikes) were known to be ineffective. Furthermore, specifically stimulating either pathway to test its causal role in behavior was impossible. The developments in molecular tools have made it possible, for the first time, to modulate and monitor the activity of direct and indirect pathways during behavior.

In some of the first studies using molecular techniques, ChR2 was expressed in direct- or indirect pathway-MSNs in the dorsal striatum, and each pathway was activated optogenetically in freely moving mice [60,61]. These studies revealed that activation of direct-pathway MSNs increased locomotor initiation, whereas activation of the indirect pathway decreased locomotion and increased freezing. Consistent with this, optogenetic stimulation of direct-pathway MSNs increases the spontaneous activity of motor cortex neurons, whereas stimulation of indirect-pathway MSNs either decreases or causes biphasic changes in the spontaneous activity in the motor cortex [62]. Interestingly, these effects were much diminished during movement. These experiments are generally consistent with the model of the direct pathway promoting movement initiation and indirect pathway inhibiting such initiation (Fig. 5.5, model 1).

This simple see–saw model is only part of the picture, however; other studies have used molecular techniques to reveal a complicated coordination between direct and indirect pathways. To monitor the activity

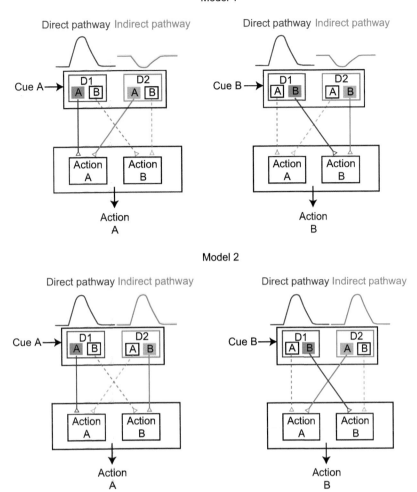

FIGURE 5.5 **Models of action selection in striatum.** Assume that to earn a reward, animals must take action A in response to cue A and action B in response to cue B. In model 1 (top), when cue A is presented, D1 neurons in the striatum are activated to promote action A and D2 neurons are inhibited to disinhibit action A. In model 2 (bottom), when cue A is presented, D1 neurons in the striatum are activated to promote action A, and D2 neurons are also activated, which suppresses competing action B. In both models, when cue B appears, a different set of D1 and D2 neurons work together to promote action B (top right, bottom right). Both models are consistent with the observations of optogenetic stimulation experiments in which activation of D1 neurons promotes movements and activation of D2 neurons suppresses movements. However, in recording experiments, D1 and D2 neurons seem to be coactivated during action selection, which suggests that model 2 might be more realistic. See main text for details.

of MSNs in each pathway, a genetically encoded calcium indicator (GCaMP3) was expressed in direct- versus indirect-pathway MSNs using transgenic mice expressing Cre-recombinase under the control of D1 and D2 dopamine receptor genes, respectively [63]. The dynamics of population activity were monitored through single optical fibers (fiber photometry). This study found concurrent activation of direct- and indirect-pathway MSNs immediately preceding the initiation of a contralateral movement. Similarly, at the single-neuron level, using the aforementioned optogenetic tagging method, Jin et al. recorded the activity of D1 and D2 MSNs in the dorsal striatum while mice executed a sequence of lever presses to get a reward

[64]. Consistent with activation of both pathways during movement initiation, a similar proportion of both types of neurons were activated at the start of a motor sequence. During the execution of a motor sequence, however, D1 MSNs tended to show sustained activation, whereas D2 MSNs tended to show sustained inhibition. Finally, another study showed that balanced activation of direct and indirect pathways is necessary for proper contraversive movements, as inhibition of either pathway or both pathways disrupted contraversive movements [65]. These studies together support the model that coordinated activation of both direct- and indirect-pathway MSNs is necessary for proper movement (Fig. 5.5, model 2).

The start/stop activities in MSNs are thought to play an important role in movement, in particular forming habits and acquiring motor skills [53,54]. But the above studies have showed that many MSNs remain active throughout the execution of habitual movements. In fact, a 2015 study showed that these activities are correlated with movement kinematics (e.g., speed or position) [66]. How different types of motor-related activities in direct and indirect pathways relate to the execution of motor behavior, in addition to initiation, remains to be further clarified.

Value Representations and Adaptive Decision-Making

In addition to the learning and execution of habitual and skilled motor behaviors, studies on the striatum have highlighted its role in flexible behavior. Accumulating evidence suggests that neurons in the ventral and dorsomedial striatum flexibly represent dynamically changing values associated with particular actions and the environment ("state"). In value-based decision-making paradigms, many neurons in the striatum modulate their activity according to the expectation of reward [67–73]. Some of these activities are related to the value of specific actions [67,70,72–74]. Furthermore, after a choice is made, some striatal neurons encode either the value of the chosen action or the actual reward received [70]. In addition to encoding the value of a specific action, some neurons in dorsomedial striatum encode the net expected reward of all available options, which is important for regulating response vigor [73].

The recording experiments mentioned earlier have indicated that MSNs represent value in various forms, but it remains unclear whether direct- and indirect-pathway MSNs make different contributions to this process. There are clues, however, from studies stimulating these two pathways during decision-making. Using a value-based decision-making task in mice, Tai et al. showed that unilateral optogenetic stimulation of D1 neurons in the dorsomedial striatum biases the animal's choices to the contralateral side, whereas stimulation of D2 neurons biases their choices to the ipsilateral side [75]. Thus, it is possible that these two populations carry inverse value functions, or might represent value in the same way, but have different mappings to behavioral output.

New molecular techniques have also begun to demystify the function of cholinergic neurons and GABAergic interneurons in the striatum. Cholinergic neurons constitute only about 1% of the total neurons in the striatum. They are known as tonically active neurons (TANs) and are characterized by 2- to 10-Hz tonic firing and wide spike waveforms. Classic primate physiology studies have shown that those neurons often show brief pauses and rebound excitation to cues associated with reward [76,77]. However, when cue values were modulated systematically by varying the probability of associated rewards, TANs did not encode the value of the cues [78], in contrast to midbrain dopamine neurons. What regulates these cholinergic interneurons during conditioning? A 2012 study using ChR2-assisted circuit mapping found that VTA GABAergic neurons selectively synapse onto cholinergic interneurons but not onto MSNs, even though the latter are the vast majority of neurons in the striatum [79]. Optogenetic stimulation of this projection caused brief pauses of striatal cholinergic interneurons, which resembles cholinergic interneurons' response to cues in classical conditioning [79].

Circuits

The aforementioned studies have begun to reveal how striatal neurons respond to various stimuli and contribute to behavior. However, how the activity of striatal neurons themselves is regulated remains less understood. The striatum receives excitatory inputs from sensory, motor, and associative cortical areas, as well as subcortical inputs from thalamic nuclei [80,81] and the amygdala. In addition, the striatum also receives neuromodulatory inputs from midbrain dopamine neurons and dorsal raphe serotonin neurons [82].

Among these inputs, synapses between cortical inputs and MSNs have been thought of as a site of learning-dependent plasticity. Learning to assign values to actions that lead to reward in a specific context is known as the credit assignment problem. It has been hypothesized that RPE signals from dopamine neurons modulate the plasticity of corticostriatal synapses, strengthening connections that promote rewarding actions and weakening connections that lead to aversive outcomes. The hypothesis is largely based on computational theories and slice physiology, and it is unknown whether it is actually implemented in striatal circuits. However, optogenetic manipulation of these circuits has provided some supporting evidence. In an auditory perceptual decision-making task, stimulation of auditory cortex neurons that are tuned to a specific frequency and project to striatum can bias animals' choices [83]. Moreover, after learning to associate a sensory stimulus with contralateral reward-seeking movements, the synapses between cortical neurons that encode this sensory information and ipsilateral striatal neurons are selectively potentiated [84]. These studies suggest that potentiation of specific corticostriatal synapses might mediate action value learning. It is still unknown what roles dopamine and serotonin might play in this process.

New molecular tools have also been used to examine the connectivity between striatum and its inputs. There are two main types of cortical neurons projecting to MSNs, namely intratelencephalic (IT) neurons and pyramidal tract (PT) neurons. Classical anatomical studies using electron microscopy found that IT neurons preferentially synapse onto D1 MSNs, whereas PT neurons are preferentially wired with D2 MSNs [85,86]. However, using ChR2-assited circuit-mapping techniques, Kress et al. found that both D1 and D2 MSNs receive inputs from both types of cortical projection neurons [87]. To examine the whole brain inputs to D1 and D2 MSNs, two studies used rabies virus to trace the monosynaptic inputs to D1 and D2 MSNs [88,89]. Interestingly, the inputs of D1 and D2 are largely the same across cortical and subcortical areas. Thus, it remains unclear how these two types of MSNs develop their different responses; to find out, future studies will need to use molecular tools to understand the cell types that send input to each population, as well as the role of local synaptic regulation.

CONCLUSIONS

All brain functions, including reward processing and decision-making, emerge from intricate interactions among neurons. The goal of neuroscience is to probe these interactions and understand how they control behavior. In this chapter, we have reviewed advances in optogenetic, molecular, and genetic techniques, and shown how these tools have been used to dissect the circuits underlying decision-making. Discoveries are happening at an unprecedented pace; as we rapidly uncover circuit structure and function, we can be sure about one thing: that our dopamine neurons and striatum are happily engaged.

References

[1] Hubel DH, Wiesel TN. Receptive fields, binocular interaction and functional architecture in the cat's visual cortex. J Physiol (Lond) 1962;160:106−54.

[2] Schultz W, Dayan P, Montague PR. A neural substrate of prediction and reward. Science 1997;275:1593−9.

[3] Crick F. The impact of molecular biology on neuroscience. Philos Trans R Soc Lond B Biol Sci 1999;354:2021−5. http://dx.doi.org/10.1098/rstb.1999.0541.

[4] Lotharius J, Brundin P. Pathogenesis of Parkinson's disease: dopamine, vesicles and alpha-synuclein. Nat Rev Neurosci 2002;3:932−42. http://dx.doi.org/10.1038/nrn983.

[5] Hyman SE, Malenka RC, Nestler EJ. Neural mechanisms of addiction: the role of reward-related learning and memory. Annu Rev Neurosci 2006;29:565−98. http://dx.doi.org/10.1146/annurev.neuro.29.051605.113009.

[6] Mirenowicz J, Schultz W. Importance of unpredictability for reward responses in primate dopamine neurons. J Neurophysiol 1994;72:1024−7.

[7] Hollerman JR, Tremblay L, Schultz W. Influence of reward expectation on behavior-related neuronal activity in primate striatum. J Neurophysiol 1998;80:947−63.

[8] Pan W-X, Schmidt R, Wickens JR, Hyland BI. Dopamine cells respond to predicted events during classical conditioning: evidence for eligibility traces in the reward-learning network. J Neurosci 2005;25:6235−42. http://dx.doi.org/10.1523/JNEUROSCI.1478-05.2005.

[9] Takahashi YK, Roesch MR, Wilson RC, Toreson K, O'Donnell P, Niv Y, et al. Expectancy-related changes in firing of dopamine neurons depend on orbitofrontal cortex. Nat Neurosci 2011;14:1590−7. http://dx.doi.org/10.1038/nn.2957.

[10] Cohen JY, Haesler S, Vong L, Lowell BB, Uchida N. Neuron-type-specific signals for reward and punishment in the ventral tegmental area. Nature 2012;482:85−8. http://dx.doi.org/10.1038/nature10754.

[11] Jo YS, Lee J, Mizumori SJY. Effects of prefrontal cortical inactivation on neural activity in the ventral tegmental area. J Neurosci 2013;33:8159−71. http://dx.doi.org/10.1523/JNEUROSCI.0118-13.2013.

[12] D'Ardenne K, McClure SM, Nystrom LE, Cohen JD. BOLD responses reflecting dopaminergic signals in the human ventral tegmental area. Science 2008;319:1264−7. http://dx.doi.org/10.1126/science.1150605.

[13] Tobler PN, Fiorillo CD, Schultz W. Adaptive coding of reward value by dopamine neurons. Science 2005;307:1642−5. http://dx.doi.org/10.1126/science.1105370.

[14] Bayer HM, Glimcher PW. Midbrain dopamine neurons encode a quantitative reward prediction error signal. Neuron 2005;47:129−41. http://dx.doi.org/10.1016/j.neuron.2005.05.020.

[15] Bayer HM, Lau B, Glimcher PW. Statistics of midbrain dopamine neuron spike trains in the awake primate. J Neurophysiol 2007;98:1428−39. http://dx.doi.org/10.1152/jn.01140.2006.

[16] Fiorillo CD, Tobler PN, Schultz W. Discrete coding of reward probability and uncertainty by dopamine neurons. Science 2003;299:1898−902. http://dx.doi.org/10.1126/science.1077349.

[17] Kobayashi S, Schultz W. Influence of reward delays on responses of dopamine neurons. J Neurosci 2008;28:7837−46. http://dx.doi.org/10.1523/JNEUROSCI.1600-08.2008.

[18] Fiorillo CD, Newsome WT, Schultz W. The temporal precision of reward prediction in dopamine neurons. Nat Neurosci 2008. http://dx.doi.org/10.1038/nn.2159.

[19] Lak A, Stauffer WR, Schultz W. Dopamine prediction error responses integrate subjective value from different reward dimensions. Proc Natl Acad Sci USA 2014;111:2343−8. http://dx.doi.org/10.1073/pnas.1321596111.

[20] Stuber GD, Klanker M, de Ridder B, Bowers MS, Joosten RN, Feenstra MG, et al. Reward-predictive cues enhance excitatory synaptic strength onto midbrain dopamine neurons. Science 2008;321:1690−2. http://dx.doi.org/10.1126/science.1160873.

[21] Flagel SB, Clark JJ, Robinson TE, Mayo L, Czuj A, Willuhn I, et al. A selective role for dopamine in stimulus-reward learning. Nature 2011;469:53−7. http://dx.doi.org/10.1038/nature09588.

[22] Zweifel LS, Parker JG, Lobb CJ, Rainwater A, Wall VZ, Fadok JP, et al. Disruption of NMDAR-dependent burst firing by dopamine neurons provides selective assessment of phasic dopamine-dependent behavior. Proc Natl Acad Sci USA 2009;106:7281−8. http://dx.doi.org/10.1073/pnas.0813415106.

[23] Frank MJ, Seeberger LC, O'reilly RC. By carrot or by stick: cognitive reinforcement learning in parkinsonism. Science 2004;306:1940−3. http://dx.doi.org/10.1126/science.1102941.

[24] Cools R, Altamirano L, D'Esposito M. Reversal learning in Parkinson's disease depends on medication status and outcome valence. Neuropsychologia 2006;44:1663−73. http://dx.doi.org/10.1016/j.neuropsychologia.2006.03.030.

[25] Santesso DL, Evins AE, Frank MJ, Schetter EC, Bogdan R, Pizzagalli DA. Single dose of a dopamine agonist impairs reinforcement learning in humans: evidence from event-related potentials and computational modeling of striatal-cortical function. Hum Brain Mapp 2009;30:1963–76. http://dx.doi.org/10.1002/hbm.20642.

[26] Schultz W. Updating dopamine reward signals. Curr Opin Neurobiol 2013;23:229–38. http://dx.doi.org/10.1016/j.conb.2012.11.012.

[27] Margolis EB, Lock H, Hjelmstad GO, Fields HL. The ventral tegmental area revisited: is there an electrophysiological marker for dopaminergic neurons? J Physiol (Lond) 2006;577:907–24. http://dx.doi.org/10.1113/jphysiol.2006.117069.

[28] Sesack SR, Grace AA. Cortico-Basal Ganglia reward network: microcircuitry. Neuropsychopharmacology 2010;35:27–47. http://dx.doi.org/10.1038/npp.2009.93.

[29] Eshel N, Bukwich M, Rao V, Hemmelder V, Tian J, Uchida N. Arithmetic and local circuitry underlying dopamine prediction errors. Nature 2015;525:243–6. http://dx.doi.org/10.1038/nature14855.

[30] Niv Y, Daw ND, Joel D, Dayan P. Tonic dopamine: opportunity costs and the control of response vigor. Psychopharmacology (Berl) 2007;191:507–20. http://dx.doi.org/10.1007/s00213-006-0502-4.

[31] Cohen JY, Amoroso MW, Uchida N. Serotonergic neurons signal reward and punishment on multiple timescales. Elife August 31, 2015;4:e10032. http://dx.doi.org/10.7554/eLife.10032.

[32] Hayashi K, Nakao K, Nakamura K. Appetitive and aversive information coding in the primate dorsal raphé nucleus. J Neurosci 2015;35:6195–208. http://dx.doi.org/10.1523/JNEUROSCI.2860-14.2015.

[33] Gunaydin LA, Grosenick L, Finkelstein JC, Kauvar IV, Fenno LE, Adhikari A, et al. Natural neural projection dynamics underlying social behavior. Cell 2014;157:1535–51. http://dx.doi.org/10.1016/j.cell.2014.05.017.

[34] Tsai H-C, Zhang F, Adamantidis A, Stuber GD, Bonci A, de Lecea L, et al. Phasic firing in dopaminergic neurons is sufficient for behavioral conditioning. Science 2009;324:1080–4. http://dx.doi.org/10.1126/science.1168878.

[35] Adamantidis AR, Tsai H-C, Boutrel B, Zhang F, Stuber GD, Budygin EA, et al. Optogenetic interrogation of dopaminergic modulation of the multiple phases of reward-seeking behavior. J Neurosci 2011;31:10829–35. http://dx.doi.org/10.1523/JNEUROSCI.2246-11.2011.

[36] Witten IB, Steinberg EE, Lee SY, Davidson TJ, Zalocusky KA, Brodsky M, et al. Recombinase-driver rat lines: tools, techniques, and optogenetic application to dopamine-mediated reinforcement. Neuron 2011;72:721–33. http://dx.doi.org/10.1016/j.neuron.2011.10.028.

[37] Kim KM, Baratta MV, Yang A, Lee D, Boyden ES, Fiorillo CD. Optogenetic mimicry of the transient activation of dopamine neurons by natural reward is sufficient for operant reinforcement. PLoS One 2012;7:e33612. http://dx.doi.org/10.1371/journal.pone.0033612.

[38] Steinberg EE, Keiflin R, Boivin JR, Witten IB, Deisseroth K, Janak PH. A causal link between prediction errors, dopamine neurons and learning. Nat Neurosci 2013. http://dx.doi.org/10.1038/nn.3413.

[39] Kamin L. Selective association and conditioning. In: Fundamental issues in associative learning; 1969. p. 42–64.

[40] Geisler S, Zahm DS. Afferents of the ventral tegmental area in the rat-anatomical substratum for integrative functions. J Comp Neurol 2005;490:270–94. http://dx.doi.org/10.1002/cne.20668.

[41] Phillipson OT. Afferent projections to the ventral tegmental area of Tsai and interfascicular nucleus: a horseradish peroxidase study in the rat. J Comp Neurol 1979;187:117–43. http://dx.doi.org/10.1002/cne.901870108.

[42] Watabe-Uchida M, Zhu L, Ogawa SK, Vamanrao A, Uchida N. Whole-brain mapping of direct inputs to midbrain dopamine neurons. Neuron 2012;74:858–73. http://dx.doi.org/10.1016/j.neuron.2012.03.017.

[43] Beier KT, Steinberg EE, DeLoach KE, Xie S, Miyamichi K, Schwarz L, et al. Circuit architecture of VTA dopamine neurons revealed by systematic input-output mapping. Cell 2015;162:622–34. http://dx.doi.org/10.1016/j.cell.2015.07.015.

[44] Lerner TN, Shilyansky C, Davidson TJ, Evans KE, Beier KT, Zalocusky KA, et al. Intact-brain analyses reveal distinct information carried by SNc dopamine subcircuits. Cell 2015;162:635–47. http://dx.doi.org/10.1016/j.cell.2015.07.014.

[45] Menegas W, Bergan JF, Ogawa SK, Isogai Y, Umadevi Venkataraju K, Osten P, et al. Dopamine neurons projecting to the posterior striatum form an anatomically distinct subclass. Elife 2015;4:e10032. http://dx.doi.org/10.7554/eLife.10032.

[46] Matsumoto M, Hikosaka O. Representation of negative motivational value in the primate lateral habenula. Nat Neurosci 2009;12:77–84. http://dx.doi.org/10.1038/nn.2233.

[47] Hong S, Jhou TC, Smith M, Saleem KS, Hikosaka O. Negative reward signals from the lateral habenula to dopamine neurons are mediated by rostromedial tegmental nucleus in primates. J Neurosci 2011;31:11457–71. http://dx.doi.org/10.1523/JNEUROSCI.1384-11.2011.

[48] Tian J, Uchida N. Habenula lesions reveal that multiple mechanisms underlie dopamine prediction errors. Neuron 2015;87:1304–16. http://dx.doi.org/10.1016/j.neuron.2015.08.028.

[49] Lammel S, Hetzel A, Häckel O, Jones I, Liss B, Roeper J. Unique properties of mesoprefrontal neurons within a dual mesocortico-limbic dopamine system. Neuron 2008;57:760–73. http://dx.doi.org/10.1016/j.neuron.2008.01.022.

[50] Lammel S, Lim BK, Malenka RC. Reward and aversion in a heterogeneous midbrain dopamine system. Neuropharmacology 2014;76 Pt B:351–9. http://dx.doi.org/10.1016/j.neuropharm.2013.03.019.

[51] Lammel S, Lim BK, Ran C, Huang KW, Betley MJ, Tye KM, et al. Input-specific control of reward and aversion in the ventral tegmental area. Nature 2012;491:212–7. http://dx.doi.org/10.1038/nature11527.

[52] Ding L, Gold JI. The basal ganglia's contributions to perceptual decision making. Neuron 2013;79:640–9. http://dx.doi.org/10.1016/j.neuron.2013.07.042.

[53] Graybiel AM, Grafton ST. The striatum: where skills and habits meet. Cold Spring Harb Perspect Biol 2015;7. http://dx.doi.org/10.1101/cshperspect.a021691.

[54] Jin X, Costa RM. Shaping action sequences in basal ganglia circuits. Curr Opin Neurobiol 2015;33:188–96. http://dx.doi.org/10.1016/j.conb.2015.06.011.

[55] Kable JW, Glimcher PW. The neurobiology of decision: consensus and controversy. Neuron 2009;63:733–45. http://dx.doi.org/10.1016/j.neuron.2009.09.003.

[56] Lee D, Seo H, Jung MW. Neural basis of reinforcement learning and decision making. Annu Rev Neurosci 2012;35:287–308. http://dx.doi.org/10.1146/annurev-neuro-062111-150512.

[57] Alexander GE, Crutcher MD. Functional architecture of basal ganglia circuits: neural substrates of parallel processing. Trends Neurosci 1990;13:266–71.

[58] DeLong MR. Primate models of movement disorders of basal ganglia origin. Trends Neurosci 1990;13:281–5.

[59] Mink JW. The basal ganglia: focused selection and inhibition of competing motor programs. Prog Neurobiol 1996;50:381–425.

[60] Bateup HS, Santini E, Shen W, Birnbaum S, Valjent E, Surmeier DJ, et al. Distinct subclasses of medium spiny neurons differentially regulate striatal motor behaviors. Proc Natl Acad Sci USA 2010;107:14845–50. http://dx.doi.org/10.1073/pnas.1009874107.

[61] Kravitz AV, Freeze BS, Parker PRL, Kay K, Thwin MT, Deisseroth K, et al. Regulation of parkinsonian motor behaviours by optogenetic control of basal ganglia circuitry. Nature 2010;466: 622—6. http://dx.doi.org/10.1038/nature09159.

[62] Oldenburg IA, Sabatini BL. Antagonistic but not symmetric regulation of primary motor cortex by basal ganglia direct and indirect pathways. Neuron 2015;86:1174—81. http://dx.doi.org/10.1016/j.neuron.2015.05.008.

[63] Cui G, Jun SB, Jin X, Pham MD, Vogel SS, Lovinger DM, et al. Concurrent activation of striatal direct and indirect pathways during action initiation. Nature 2013;494:238—42. http://dx.doi.org/10.1038/nature11846.

[64] Jin X, Tecuapetla F, Costa RM. Basal ganglia subcircuits distinctively encode the parsing and concatenation of action sequences. Nat Neurosci 2014;17:423—30. http://dx.doi.org/10.1038/nn.3632.

[65] Tecuapetla F, Matias S, Dugue GP, Mainen ZF, Costa RM. Balanced activity in basal ganglia projection pathways is critical for contraversive movements. Nat Commun 2014;5:4315. http://dx.doi.org/10.1038/ncomms5315.

[66] Rueda-Orozco PE, Robbe D. The striatum multiplexes contextual and kinematic information to constrain motor habits execution. Nat Neurosci 2015;18:453—60. http://dx.doi.org/10.1038/nn.3924.

[67] Cai X, Kim S, Lee D. Heterogeneous coding of temporally discounted values in the dorsal and ventral striatum during intertemporal choice. Neuron 2011;69:170—82. http://dx.doi.org/10.1016/j.neuron.2010.11.041.

[68] Kawagoe R, Takikawa Y, Hikosaka O. Expectation of reward modulates cognitive signals in the basal ganglia. Nat Neurosci 1998;1:411—6. http://dx.doi.org/10.1038/1625.

[69] Tremblay L, Hollerman JR, Schultz W. Modifications of reward expectation-related neuronal activity during learning in primate striatum. J Neurophysiol 1998;80:964—77.

[70] Lau B, Glimcher PW. Value representations in the primate striatum during matching behavior. Neuron 2008;58:451—63. http://dx.doi.org/10.1016/j.neuron.2008.02.021.

[71] Lauwereyns J, Watanabe K, Coe B, Hikosaka O. A neural correlate of response bias in monkey caudate nucleus. Nature 2002;418: 413—7. http://dx.doi.org/10.1038/nature00892.

[72] Samejima K, Ueda Y, Doya K, Kimura M. Representation of action-specific reward values in the striatum. Science 2005;310: 1337—40. http://dx.doi.org/10.1126/science.1115270.

[73] Wang AY, Miura K, Uchida N. The dorsomedial striatum encodes net expected return, critical for energizing performance vigor. Nat Neurosci 2013;16:639—47. http://dx.doi.org/10.1038/nn.3377.

[74] Ito M, Doya K. Distinct neural representation in the dorsolateral, dorsomedial, and ventral parts of the striatum during fixed- and free-choice tasks. J Neurosci 2015;35:3499—514. http://dx.doi.org/10.1523/JNEUROSCI.1962-14.2015.

[75] Tai L-H, Lee AM, Benavidez N, Bonci A, Wilbrecht L. Transient stimulation of distinct subpopulations of striatal neurons mimics changes in action value. Nat Neurosci 2012;15: 1281—9. http://dx.doi.org/10.1038/nn.3188.

[76] Aosaki T, Tsubokawa H, Ishida A, Watanabe K, Graybiel AM, Kimura M. Responses of tonically active neurons in the primate's striatum undergo systematic changes during behavioral sensorimotor conditioning. J Neurosci 1994;14:3969—84.

[77] Aosaki T, Kimura M, Graybiel AM. Temporal and spatial characteristics of tonically active neurons of the primate's striatum. J Neurophysiol 1995;73:1234—52.

[78] Morris G, Arkadir D, Nevet A, Vaadia E, Bergman H. Coincident but distinct messages of midbrain dopamine and striatal tonically active neurons. Neuron 2004;43:133—43. http://dx.doi.org/10.1016/j.neuron.2004.06.012.

[79] Brown MTC, Tan KR, O'Connor EC, Nikonenko I, Muller D, Lüscher C. Ventral tegmental area GABA projections pause accumbal cholinergic interneurons to enhance associative learning. Nature 2012;492:452—6. http://dx.doi.org/10.1038/nature11657.

[80] Giménez-Amaya JM, McFarland NR, De Las Heras S, Haber SN. Organization of thalamic projections to the ventral striatum in the primate. J Comp Neurol 1995;354:127—49. http://dx.doi.org/10.1002/cne.903540109.

[81] McFarland NR, Haber SN. Convergent inputs from thalamic motor nuclei and frontal cortical areas to the dorsal striatum in the primate. J Neurosci 2000;20:3798—813.

[82] Moore RY, Halaris AE, Jones BE. Serotonin neurons of the midbrain raphe: ascending projections. J Comp Neurol 1978;180: 417—38. http://dx.doi.org/10.1002/cne.901800302.

[83] Znamenskiy P, Zador AM. Corticostriatal neurons in auditory cortex drive decisions during auditory discrimination. Nature 2013; 497:482—5. http://dx.doi.org/10.1038/nature12077.

[84] Xiong Q, Znamenskiy P, Zador AM. Selective corticostriatal plasticity during acquisition of an auditory discrimination task. Nature 2015;521:348—51. http://dx.doi.org/10.1038/nature14225.

[85] Lei W, Jiao Y, Mar ND, Reiner A. Evidence for differential cortical input to direct pathway versus indirect pathway striatal projection neurons in rats. J Neurosci 2004;24:8289—99. http://dx.doi.org/10.1523/JNEUROSCI.1990-04.2004.

[86] Reiner A, Jiao Y, Del Mar N, Laverghetta AV, Lei WL. Differential morphology of pyramidal tract-type and intratelencephalically projecting-type corticostriatal neurons and their intrastriatal terminals in rats. J Comp Neurol 2003;457:420—40. http://dx.doi.org/10.1002/cne.10541.

[87] Kress GJ, Yamawaki N, Wokosin DL, Wickersham IR, Shepherd GMG, Surmeier DJ. Convergent cortical innervation of striatal projection neurons. Nat Neurosci 2013;16:665—7. http://dx.doi.org/10.1038/nn.3397.

[88] Guo Q, Wang D, He X, Feng Q, Lin R, Xu F, et al. Whole-brain mapping of inputs to projection neurons and cholinergic interneurons in the dorsal striatum. PLoS One 2015;10:e0123381. http://dx.doi.org/10.1371/journal.pone.0123381.

[89] Wall NR, De La Parra M, Callaway EM, Kreitzer AC. Differential innervation of direct- and indirect-pathway striatal projection neurons. Neuron 2013;79:347—60. http://dx.doi.org/10.1016/j.neuron.2013.05.014.

[90] Bäckman CM, Malik N, Zhang Y, Shan L, Grinberg A, Hoffer BJ, et al. Characterization of a mouse strain expressing Cre recombinase from the 3′ untranslated region of the dopamine transporter locus. Genesis 2006;44:383—90. http://dx.doi.org/10.1002/dvg.20228.

[91] Eshel N, Tian J, Uchida N. Opening the black box: dopamine, predictions, and learning. Trends Cogn Sci September 2013;17(9): 430—1. http://dx.doi.org/10.1016/j.tics.2013.06.010 [Epub 2013 July, 3].

I. ANIMAL STUDIES ON REWARDS, PUNISHMENTS, AND DECISION-MAKING

6

The Learning and Motivational Processes Controlling Goal-Directed Action and Their Neural Bases

L.A. Bradfield, B.W. Balleine
University of Sydney, Camperdown, NSW, Australia

Abstract

Research has made considerable progress in describing the learning and motivational processes controlling goal-directed action and their related neural circuitry. Broadly this circuitry spans various regions within the corticostriatal–thalamic loop, within which specific structures mediate differentiable aspects of goal-directed learning and performance. These aspects include the acquisition of action-outcome contingencies, the encoding and retrieval or incentive value, the matching of that value to specific outcome representations, and finally the integration of this information for action selection. Information from each of the structures that mediate these processes converges on the striatum, with the posterior dorsomedial striatum in particular hypothesized to represent the neural locus of goal-directed action. Here we discuss evidence from rodent studies regarding the neural circuits mediating each of these individual processes before considering how they are integrated within the striatum for successful goal-directed action selection.

The neural bases of the learning and motivational processes that control goal-directed actions are becoming increasingly better understood. Not only are new components of this neural network increasingly being identified, but the exact nature of the unique role played by each structure within this network, as well as the interactions between them, is becoming better recognized. As a consequence, a clear picture of how the brain encodes and expresses goal-directed learning is beginning to emerge. In this chapter we will separate these processes into several key psychological capacities and explore the distinct neural bases of each. We will then examine how these individual structures and circuits interact to form the larger neural system that underlies goal-directed action.

As has been made clear previously, an action is considered to be goal-directed if its performance respects: (1) the current value of its outcome and (2) the extant contingency between that action (A) and its outcome (O) (cf. Ref. [1]). In contrast, reflexive actions or habits are controlled by antecedent stimuli (S) rather than their relationship to, or the value of, their consequences and, as such, tend to be relatively inflexible or automatic. Although there is a rich literature concerning the behavioral and neural bases of habits, this chapter will focus solely on evidence from goal-directed actions particularly from studies in rodents.

The behavioral procedure most commonly used to determine whether an action is goal-directed, and the one that will be primarily relied on here, is outcome devaluation. Typically, animals are first trained to perform an action, such as a lever press response, to obtain an outcome, such as pellets. The value of that outcome is then reduced either by outcome-specific satiety (induced by free consumption of the outcome) or by taste aversion learning (in which the ingestion of the outcome is paired with illness induced by an injection of lithium chloride) (cf. Ref. [1]). If responding is primarily goal-directed, the animal should subsequently reduce its propensity to perform the action previously associated with the now-devalued outcome, in comparison either to alternate actions earning nondevalued outcomes or to animals that did not receive the devaluation treatment. Importantly, testing occurs in the absence of outcome delivery (i.e., in extinction) so that any feedback that could be gained from experiencing the outcome is not available. This ensures that the animal must rely on its knowledge of the A—O contingency, as well as its ability to retrieve the specific outcome of its actions—including its current value—to perform appropriately and reduce the performance of the now devalued action relative to other actions.

There are a number of psychological phenomena that underlie goal-directed action, specifically: the encoding of the A—O contingency, reward or incentive learning, outcome-specific retrieval, and action selection and initiation in the context of goal-directed performance. These four phenomena will be described later along with current evidence as to their neural bases. Generally, at present, A—O encoding is thought to depend primarily on the prelimbic cortex (PL) in rats (homologous to the medial frontal/anterior cingulate cortex in humans), and probably involves connections with the mediodorsal thalamus (MD) and dorsomedial striatum (DMS). This learning is influenced by an incentive learning process [2] via which the incentive (or reward) value of the outcome is altered largely by virtue of the motivational state of the animal when contact with that outcome occurs. Incentive learning processes are thought to rely on structures connected with the PL, the MD, and the DMS such as the basolateral nucleus of the amygdala (BLA) and the insular cortex (IC). Once encoded, the incentive value of the outcome is matched to a broader mental representation of the outcome, including all its sensory-specific characteristics, to drive goal-directed responding. Recent data suggest that, in the absence of the physical outcome, outcome-specific retrieval and attendant reward processing depend on the medial portion of the orbitofrontal cortex (mOFC). The mOFC, IC, and BLA all project strongly to the DMS but perhaps even more densely to the ventral striatum, particularly the nucleus accumbens (NAC) core, which is known to control the performance of goal-directed actions. The posterior region of the DMS (pDMS; homologous to the anterior caudate in humans) appears to be uniquely involved in both the learning and the expression of goal-directed actions, suggesting that the information provided by each of these other structures must converge on the pDMS or on the immediate circuits controlled by the pDMS and that constitute the neural locus of goal-directed action. This chapter will first examine the neural bases of each of these psychological phenomena separately and then examine how they interact as part of the broader network with a focus on the convergence of information on the pDMS to ultimately control goal-directed actions.

LEARNING "WHAT LEADS TO WHAT": THE NEURAL BASES OF ACTION—OUTCOME LEARNING

The acquisition of instrumental responding depends on encoding the relationship between the action and the outcome (A—O). There are now several lines of evidence that suggest a specific role for the PL in the acquisition, but not performance, of goal-directed actions. The role of the PL in acquisition was first demonstrated by Balleine and Dickinson [1], who examined the effects of pretraining excitotoxic PL lesions on a variation of the outcome devaluation procedure described earlier. Specifically, rats with either sham or excitotoxic lesions of the PL were trained to either press a lever or pull a chain for food pellets or a starch solution, respectively. Prior to testing, the rats were prefed either the food pellets or the starch solution to devalue that particular outcome by way of specific satiety. On a subsequent extinction test rats with sham lesions displayed sensitivity to outcome devaluation by selectively reducing the action that had been paired with the now devalued outcome relative to the action paired with a still valued outcome. By contrast, rats with PL lesions were insensitive to devaluation and performed both actions similarly on testing. This finding is robust and has been replicated several times [3,4]. However, these results contrast with the results of "reinforced" tests that are conducted in the same manner as the outcome devaluation test previously described except that each action delivers its related outcome. In these instances, PL lesions spare sensitivity to outcome devaluation [1,3]. This finding is important in demonstrating the specificity of PL involvement in A—O learning by ruling out alternate explanations such as an inability to encode changes in value or discriminate between levers or outcomes. Further experiments have suggested that this role for PL is limited to the acquisition of goal-directed actions but not their performance. For example, Ostlund and Balleine [5] found that, even when tested in extinction, rats that received posttraining PL lesions were as sensitive to outcome devaluation as sham controls.

A separate line of evidence showing that PL lesions abolish contingency degradation performance further supports the role of the PL in A—O learning [1,3]. Specifically, although outcome devaluation performance provides evidence of A—O contingency knowledge, intact degradation performance provides specific evidence that the experimental subjects are capable of encoding the causal relationship between the action and the outcome. For example, Balleine and Dickinson [1] again trained hungry rats to perform two actions (lever press and chain pull) for two separate outcomes (food pellets and starch solution). At the end of instrumental training each action earned each outcome with a probability of .05. In the next stage of the experiment, one of these outcomes was also delivered with a probability of .05 in each second that no response was recorded. Thus, in any one second of the session, the probability of receiving that specific outcome was equal, whether or not its associated action was performed. Importantly, this procedure maintains contiguity between the action and the outcome, because each action is rewarded at the same rate, but the causal efficacy of the action is degraded. This typically results in a reduction in the performance of the degraded action relative to an alternate, nondegraded action that continues uniquely to predict its outcome. Indeed, Balleine and Dickinson [1] demonstrated this precise result in sham animals, but found that animals with PL lesions reduced the performance of both actions. Corbit and Balleine [3] conducted a similar experiment and again found insensitivity to degradation. In that study the rats were found to increase performance on the degraded relative to the nondegraded action. This pattern of results was revealing; further experiments suggested that, although the PL rats clearly lacked the ability to learn the A–O associations necessary for normal degradation performance, their ability to recognize the outcome as a discriminative stimulus (rather than as a goal) was intact [3]. This means to say these rats retained the ability to retrieve a specific action based on the outcome, via an O–A association, resulting in increased performance on the degraded relative to the nondegraded action; the free outcome delivery acting to drive responding on its associated action. Consistent with this explanation, when animals were given a choice lever test in extinction, rats with PL lesions chose each lever to a similar degree, suggesting that, when the outcome was not available, neither O–A or A–O associations were accessible to PL-lesioned animals.

Lesions of the MD, a structure that has strong and reciprocal connections to the PL, have been found to produce a similar pattern of results. Specifically, Corbit et al. [6] found that excitotoxic lesions of the MD abolished outcome devaluation performance (although not on reinforced tests) and contingency degradation using designs similar to those described earlier. Later, Ostlund and Balleine [7] further demonstrated that, like the PL, posttraining MD lesions spared outcome devaluation, implying that this structure also plays a role in the acquisition but not the expression of A–O associations.

The similarity in the results produced by PL and MD lesions in these tasks, as well as the heavy interconnections between these regions, led Bradfield et al. [8] to examine whether a disconnection of these regions would also produce a deficit in outcome devaluation performance. This disconnection was complicated by the fact that not only does the PL project ipsilaterally to the MD, but also there are significant contralateral projections. To overcome this we employed a novel surgical procedure that was successful in disconnecting the contralateral PL–MD projections, illustrated in Fig. 6.1 ([8], see particularly Fig. 3 in that paper). This procedure consisted of performing an electrolytic lesion of the genu of the corpus callosum (CC), and injecting fluorogold into the MD in one hemisphere. We then examined the pattern of labeling in the PL contralateral to the injection site and found that, relative to controls, the amount of labeling was almost completely eliminated, suggesting that these projections had been severed [8]. We then conducted a behavioral experiment using electrolytic CC lesions in combination with sham, ipsilateral, or contralateral excitotoxic lesions of the PL and MD. This procedure ensured that the sham group had intact ipsilateral PL–MD pathways in each hemisphere, the ipsilateral group had an intact PL–MD pathway in one hemisphere, and the contralateral group had no intact PL–MD pathway in either hemisphere. Using procedures similar to those described, we found that only

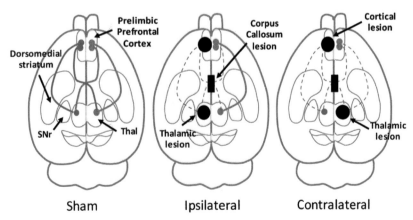

Sham   Ipsilateral   Contralateral

FIGURE 6.1 **Prelimbic cortex–mediodorsal thalamus disconnection.** Summary of the hypothesized influence of unilateral neurotoxic lesions of the prelimbic prefrontal cortex and the mediodorsal thalamus plus electrolytic lesion severing the crossed connections in the corpus callosum. Groups had no damage (sham); damage to the two structures ipsilaterally, leaving only one intrahemispheric connection intact; and damage contralaterally, severing all intra- and interhemispheric connections. Only the last manipulation is predicted to abolish the acquisition of goal-directed action. *SNr*, substantia nigra pars reticulata; *Thal*, thalamus.

the contralateral group was impaired in their outcome devaluation performance but that their discrimination between actions and outcomes was intact.

Although these studies suggest a role for the MD in goal-directed learning, they appear to contradict several other studies that have failed to find an effect of MD inactivation on outcome devaluation. For example, Pickens [9] described two relevant experiments, one in which MD lesions abolished outcome devaluation performance, and another for which performance was intact. The author suggested that MD lesions might impair outcome devaluation only when rats had to alter their strategy within a task: the rats that showed the impairment had previously experienced a Pavlovian devaluation task, whereas the rats that showed no impairment had not. However, there are a number of other methodological differences between this study and the studies that did find effects of MD lesions. In particular, Corbit et al. [10] and Bradfield et al. [8] employed two instrumental actions that earned two outcomes and examined devaluation within subjects by devaluing one outcome that was prefed to satiety. Pickens [9], on the other hand, employed only one action and one outcome and compared outcome devaluation performance between subjects that had received lithium chloride injections either paired or unpaired with the outcome. Although it is not clear which of these differences could account for the different results, it would be of future interest to assess whether MD involvement in the acquisition of A−O learning is limited to situations in which devaluation is assessed in choice.

There is another study, however, that used parameters similar to ours that also failed to find an effect of MD inactivation on outcome devaluation. Parnaudeau et al. [11] used the inhibitory hM4Di designer receptors exclusively activated by designer drugs (DREADDs) paired with intraperitoneal injections of clozapine-N-oxide (CNO) to inactivate the MD during instrumental training or outcome devaluation testing, and failed to find an effect of either. Like our prior studies [7,8,10], Parnaudeau et al. [11] also trained two instrumental actions to predict two outcomes and examined devaluation within subjects using sensory-specific satiety. We suspect, therefore, that the reason for the discrepancy might lie in the use of DREADDs versus permanent lesions. First, the extent of the DREADDs transfection reported by Parnaudeau et al. does not appear to be as extensive as the lesions reported by Balleine and colleagues [7,8,10], particularly in more caudal sections. Second, in a prior report, Parnaudeau et al. [11] found that the administration of CNO in awake mice induced a decrease in the firing rate of only 40% and that decrease was observed in only a third of MD neurons. Although this appeared to be sufficient to produce deficits in a number of other cognitive tasks, perhaps it is

not sufficient to produce a deficit in outcome devaluation in particular. Finally, the outcome devaluation effect in the group that received CNO during instrumental training—the group for which we might expect an impairment—does appear to be smaller than that observed in the control group, particularly in the first 2−4 min of the test. Although not significant, it is possible that a more complete inactivation might allow the effect to reach statistical significance.

HOW REWARDS ARE FORMED: NEURAL BASES OF INCENTIVE LEARNING

Incentive learning constitutes another process that contributes to the production of goal-directed action. Demonstrations of this effect have revealed that, contrary to the predictions of early researchers [12], alterations in motivational state (or "drive") are not sufficient to alter instrumental responding. Rather, the incentive value of the outcome must first be altered, and this is achieved through consummatory contact with the outcome within a particular motivational state, known as the incentive learning experience [17].

The earliest direct evidence for this effect as it relates to instrumental responding came from a series of experiments conducted by Dickinson and Dawson [13] in which they trained rats to press a lever and pull a chain, with one action earning a food pellet and the other a liquid sucrose solution. When tested on the two actions thirsty, for the first time, the rats failed to show any difference in performance of the two actions. If, however, prior to this test, the rats were allowed to drink a little of the sucrose solution or to eat a little of the food pellets when in the thirsty state, then, on test, they preferred the actions previously trained with sucrose relative to the other action. Balleine [2] extended this finding by examining changes in food deprivation. For the first of these experiments rats were food deprived, and half were exposed to the target food (either a food pellet or a maltodextrin solution) in the deprivation state, and half were not. All rats were then returned to free food and trained to lever press. For the final testing period, half of the rats that were preexposed to the target food and half of the nonpreexposed rats were again made hungry, then all rats were tested for their propensity to press the lever. Interestingly, in this test, only the group that had previous consummatory contact with the target outcome in a deprived state and were later tested in that state increased their instrumental responding. The group that was tested in a deprived state but not preexposed to the outcome in that state responded at levels that were indistinguishable from those of nondeprived controls. Together with that of Dickinson and Dawson [13], this finding demonstrates that in the absence of any incentive learning, shifts in motivational state are insufficient to influence instrumental responding. Rather the rats have to learn about the value of an outcome in a new motivational state before that value can influence performance.

Incentive learning is central to the production of goal-directed actions because it regulates the incentive, or reward, value of the outcome, as per the first of the "goal-directed" criteria outlined above. Indeed, all outcome devaluation procedures directly rely on incentive learning. During training animals usually learn to perform the instrumental actions in a deprived state, increasing the incentive value of the associated outcomes. However, if devaluation occurs by way of sensory-specific satiety, the value of the outcome loses value over the course of prefeeding as consummatory contact occurs in an increasingly sated state. Likewise, if taste aversion learning is employed, it has been demonstrated that the animal must have consummatory contact with the target outcome after its pairing with illness to have any effect on responding [14,15,16]. The current view of the structure of incentive learning is presented in Fig. 6.2A. Briefly, the sensory features of an outcome, such a pellet, must first become connected with a specific motivational structure; here, one sensitive to nutrient commodities (N) that is itself modulated by deprivation conditions. Activation of the nutrient center provokes activity in an appetitive affective structure (Ap), one outcome of which is an emotional response. In this view, it is the contiguous activity of the outcome representation and the emotional response that constitutes incentive learning. From this learning can be extracted more abstract values (such as good or bad—see Ref. [17] for details).

With regard to the neural bases of incentive learning (refer to Fig. 6.2B), previous findings suggest that the PL and MD are unlikely candidates for the mediation of incentive learning because they mediate goal-directed acquisition but not performance. Rather, considerable research has focused on the anterior IC and the BLA in incentive learning. The IC was first considered because of its role in gustatory (and therefore taste) processing, and the BLA was considered because of its role in regulating emotional feedback. Specifically, Balleine [17] argued that emotional responses to particular outcomes might form the bases of encoding outcome value in this manner.

In the same study in which they examined the effects of PL lesions, Balleine and Dickinson [1] investigated the effects of IC lesions using the same tasks. IC lesions, like PL lesions, were found to abolish outcome devaluation performance. However, unlike PL lesions, IC lesions left contingency degradation performance intact, suggesting that animals with IC lesions had successfully encoded the contingency between the action and the

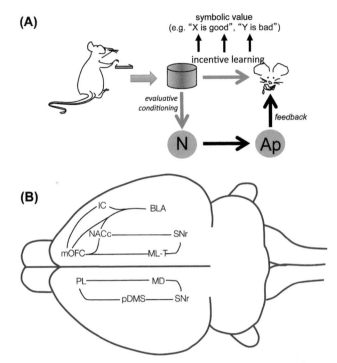

FIGURE 6.2 **Incentive learning and the neural basis of goal-directed action.** (A) Summary of the associate structure thought to underlie incentive learning. Rats learn to press a lever for an outcome, the value of which is determined by motivational and affective feedback in the form of an emotional response. The association of the sensory properties of the outcome with this emotional response constitutes incentive learning. In this example, a rat presses for a food pellet, the sensory properties of which, through evaluative conditioning, become able to activate nutrient detection processes and thus an appetitive affective center, one effect of which is the generation of a pleasant emotional response. Based on incentive learning more abstract values can be generated. (B) Summary of the neural circuits mediating incentive learning and instrumental performance (upper hemisphere) and goal-directed learning (lower hemisphere); see text for details. *N*, nutrient commodities; *Ap*, appetitive affective structure; *BLA*, basolateral amygdala; *IC*, insular cortex; *MD*, mediodorsal thalamus; *ML-T*, midline thalamus; *mOFC*, medial portion of the orbitofrontal cortex; *NACc*, nucleus accumbens core; *pDMS*, posterior dorsomedial striatum; *PL*, prelimbic cortex; *SNr*, substantia nigra pars reticulata.

outcome. In this, and a later study [18], the authors further found that IC lesions abolished sensitivity to an incentive learning treatment that, in sham controls, reduced responding on an action paired with an outcome that they had experienced in a sated state relative to an outcome experienced while hungry. Rats with IC lesions, on the other hand, performed each action indiscriminately. Importantly, Balleine and Dickinson [18] demonstrated that this effect was confined to situations in which testing occurred in extinction, suggesting that the IC plays a role in the memory for changes in incentive value rather than in its encoding. This was later confirmed by Parkes and Balleine [19]. They demonstrated that bilateral intra-IC infusions of the NMDA receptor antagonist ifenprodil, when made after devaluation (specific satiety) but prior to a choice test, were successful in abolishing the effect of outcome devaluation on performance. Together, these experiments suggest that the role of the IC is to retrieve the incentive value of the outcome prior to test to influence performance.

So where, then, is the incentive value of the instrumental outcome encoded? Balleine et al. [20] examined whether the BLA might perform this function. First they demonstrated that lesions of the BLA attenuated outcome devaluation performance. Unlike lesions of other structures previously discussed, BLA lesions attenuated performance whether tested in extinction or in reinforced tests. A further experiment showed that these deficits were not due to BLA lesions preventing discrimination between actions (BLA-lesioned rats were able to perform these actions in a distinct sequence to achieve reward). Parkes and Balleine [19] later demonstrated that although bilateral BLA infusions of ifenprodil abolished outcome devaluation performance when administered prior to devaluation treatment (specific satiety), similar infusions given after devaluation but prior to test left performance intact. This finding, together with previous results, suggests that the BLA encodes the alteration in incentive value that occurs during sensory-specific satiety, but is not involved in the retrieval of that value. A similar result was reported by Wassum et al. [21] in an experiment in which they trained rats in a low deprivation state and reexposed them to the outcome in a food-deprived state. In that case the increase in incentive value was blocked by the infusion of the opioid antagonist naloxone into the BLA but only if the infusion was given prior to incentive learning rather than test (see Refs. [21,22], for review).

Because the BLA and the IC share strong reciprocal connections [23] Parkes and Balleine [19] hypothesized that they form a circuit in which the incentive value of the instrumental outcome is encoded by the BLA, and then retrieved by the IC, to influence performance. They developed a novel, sequential disconnection procedure to test this hypothesis. All rats received unilateral infusions of ifenprodil into the IC in one hemisphere and BLA infusions in the alternate hemisphere, but the order of infusions was altered between groups. That is, one group received the BLA infusion prior to devaluation and the IC infusion prior to testing, and the remaining group received infusions in the opposite order. It was hypothesized that if the BLA were important for encoding incentive value, and the IC important for retrieving that value, then only the group that received the BLA infusion prior to devaluation and the IC infusion prior to test should show a deficit in devaluation performance (Fig. 6.2A). This is because, first, the BLA infusion prior to devaluation treatment should block encoding of the incentive value to be retrieved by the intact ipsilateral

IC on test and, second, the contralateral inactivation of the IC prior to the test should block any retrieval of incentive learning encoded by the contralateral BLA. By contrast, for the group that received the opposite arrangement, both BLAs were intact and able to encode alterations in incentive value during devaluation, and these values could be retrieved by each intact IC on testing. This was the observed result; the BLA infusion prior to devaluation and the contralateral IC infusion prior to test blocked the outcome devaluation effect, whereas the reversal of this order had no effect. This general circuit is summarized in the upper hemisphere of Fig. 6.2B.

DECIDING WHAT TO DO: THE NEURAL BASES OF OUTCOME RETRIEVAL AND CHOICE

How is information regarding the specific consequences of actions integrated with the value of those consequences so that we can choose actions appropriately? Although the IC retrieves the incentive value of a specific sensory event, when tested in extinction that value has to be linked to a specific mental representation of the instrumental outcome, including its sensory-specific characteristics. Work suggests that the integration of the retrieved value with the retrieved outcome of a specific instrumental action involves the mOFC [24]: a projection target of both the BLA and the IC [25] (see Fig. 6.2B). The mOFC also receives inputs from sensory areas, particularly olfactory regions that could carry sensory-specific information, as well as inputs from the posterior parietal cortex [25], which itself receives visual, auditory, and somatosensory information to influence planned movements. Thus, the mOFC could integrate these sensory inputs with information about available actions to retrieve the specific outcome representation to guide choice.

Our work has suggested that this role for mOFC is limited to situations in which the physical outcome is absent [24]. Specifically, we first showed that both pretraining excitotoxic lesions and posttraining mOFC inactivation by way of the hM4Di DREADDs abolished outcome devaluation performance in extinction but not reinforced tests. mOFC lesions further spared performance in paradigms in which the outcome was actually presented, such as contingency degradation. A final experiment demonstrated that animals with mOFC lesions, unlike shams, were unable to learn or be appropriately guided by specific inhibitory S—A—O contingencies, which rely on the retrieval of absent (but expected) outcomes. These findings confirmed predictions generated using simulations of reinforcement learning models in which we assumed that sham animals could infer "states" determined by the identity of absent outcomes, but animals with a dysfunctional mOFC could not.

The mOFC projects widely throughout the neural circuit controlling goal-directed actions, including the entire medial wall of the striatum [26]. These projections do, however, appear to be particularly strong in the ventromedial striatum, especially in the NAC core; see Chapter 1. It is of some interest, then, that this region has been specifically implicated in controlling the performance, but not learning, of goal-directed actions. Similar to IC and mOFC lesions, lesions of the NAC core have been found to abolish outcome devaluation performance but spare contingency degradation [6]. Again this suggests a role in the expression, rather than the learning, of goal-directed actions. This was confirmed by a later study [27] that found that μ opioid receptors within the NAC core underlie this effect. Specifically, bilateral intracore infusions of the μ opioid receptor antagonist D-Phe-Cys-Tyr-D-Trp-Arg-Thr-Pen-Thr-NH$_2$ (CTAP) were found to successfully abolish outcome devaluation performance. Because these infusions were administered after A−O learning and outcome devaluation had occurred, this finding suggests that the NAC core is not involved in encoding the A−O contingency or the incentive value of the instrumental outcome, but rather implies a specific role in the performance (see Refs. [21,6,28], for review); refer to Fig. 6.2B.

Like the mOFC, the NAC core also receives projections from BLA and IC. Furthermore, several studies have examined the impact of disconnecting the NAC core either from the BLA using contralateral lesions [29] or from the IC using contralateral lesions (experiment 1) or contralateral infusions (experiment 2) of the GABA$_A$ agonist muscimol into the IC and CTAP into the NAC core [30]. All of these manipulations were found to abolish the ability of animals to choose appropriately following outcome devaluation. The infusions in the latter experiment were again given after devaluation but prior to test, implicating the IC−NAC core circuit in goal-directed performance specifically.

Taken together with the findings regarding incentive learning, and as illustrated in the full circuit in the upper hemisphere in Fig. 6.2B, it appears that alterations in incentive value are encoded in the BLA and retrieved by the IC, and that this information is matched with the abstract mental representation of the outcome (including its sensory-specific characteristics) by the mOFC, which passes this information to the NAC core for the performance of goal-directed actions. Although no disconnection studies between the mOFC and any of the target structures (BLA, IC, NAC core) have yet been conducted, and thus any ideas about the functional circuitry involving the mOFC in this role are purely speculative, these studies are clearly of great potential interest particularly because they could point to a specific circuit that plays an important role in bringing A−O and reward information together to guide choice. How exactly this ventral performance circuit interacts with the learning processes supported by the dorsal striatum is another critical issue and we turn to this next.

HOW THESE CIRCUITS INTERACT: CONVERGENCE ON THE DORSOMEDIAL STRIATUM

To briefly recap, the processes controlling goal-directed actions that we have outlined here include first learning the A−O contingency, which is then integrated with the current incentive value of the outcome that has been matched with a representation of that outcome. Together these learning and reward processes are integrated to form action values, based on which distinct courses of goal-directed action can be compared. Importantly, although all of these learning and motivational processes are somewhat behaviorally and neurally separable, they are also interdependent, and we will now discuss views on how this interdependence is achieved.

As illustrated in Fig. 6.2B, a crucial circuit upon which these processes and neural inputs are likely to converge is that involving the pDMS. In rhesus monkeys, single-unit recordings from neurons in the anterior nucleus found that activity was closely correlated with the rate of instrumental learning [31] and, in humans, fMRI studies have revealed that caudate nucleus activity is modulated specifically by A−O contingency knowledge [32,33]. In rats the evidence is particularly revealing because, unlike all previously discussed structures, not only do both pre- and posttraining manipulations that inactivate the pDMS abolish outcome devaluation, they also impair contingency degradation without affecting discrimination between actions or outcomes [34,35]. Therefore, of all the structures that have been examined, the pDMS is the only one that is critical for both the learning and the expression of goal-directed actions. It appears likely, therefore, that the pDMS is the neural locus of goal-directed learning, and we will consider how each of the neural structures outlined here as important to various aspects of goal-directed action production makes contact with the pDMS.

The source of contact between the PL and the pDMS is straightforward, as there are strong ipsilateral and contralateral projections from layer 5 of the PL to the pDMS. Although there are no direct projections from the MD (the other structure specifically implicated in A−O learning) to the pDMS, the MD projects to the PL so it can influence the pDMS indirectly via that corticothalamic circuit. As discussed, A−O learning depends on the PL and MD but is not stored there: it becomes independent of the MD and PL once it is learned. It does not, however, become independent of the pDMS; therefore it is likely that A—O contingency information reaches the pDMS from the PL (and indirectly from the MD) to be stored in the pDMS. Studies in our laboratory are directly assessing these questions.

With regard to the more performance-related circuits, although a number of the studies outlined here have explored "minicircuits" (e.g., BLA—IC, BLA—NAC, and IC—NAC core), few have assessed how these circuits might make contact with the larger circuit involving the pDMS. With respect to this, it is of some interest that all of these structures, with the exception of the NAC core, project directly to the pDMS, although the relative strengths of these projections is quite varied.

In fact, of these connections, only one has been experimentally examined: that between the BLA and the pDMS [36]. For this study, each rat received a unilateral BLA lesion and a pDMS cannulation in the hemisphere that was either ipsilateral or contralateral to the lesion. They were then trained to press two levers for a common outcome (Polycose) to achieve a base rate of responding, followed by 2 days of training with specific A—O contingencies. For the 2 days that followed this initial A—O training, one group of rats received an infusion of muscimol into the pDMS just prior to being trained with the reverse of the A—O contingencies, whereas another group received an infusion of vehicle. All rats were then tested for outcome devaluation drug free. Rats that received ipsilateral muscimol infusions had encoded the reversed A—O contingencies, responding appropriately (i.e., nondevalued > devalued), whereas rats that received contralateral infusions responded according to the original A—O contingencies (devalued > nondevalued). This suggests that the infusion prevented A—O learning from occurring during reversal training without affecting initial A—O encoding. Although this finding might appear to be at odds with our previous assertion that the BLA encodes incentive value rather than A—O learning, it is clear that incentive value information provided by the BLA makes an important contribution to A—O learning during acquisition. Specifically, the prediction error signal driving selective learning (i.e., the difference between the reward that the animal predicts based on an action and that actually delivered) should be anticipated to be relatively weak in the face of a low incentive value, resulting in both a reduced asymptote and a relatively slower rate of A—O learning. As such, the rats in the experiment of Corbit et al. [36] would have strongly encoded the first set of contingencies when no infusions were given and incentive value encoding was free to influence A—O learning in the pDMS, but only weakly encoded the opposing contingencies after pDMS inactivation.

Thus, when pitted against each other on the final choice test, the more strongly encoded original contingencies were those that influenced choice.

A second experiment in this series showed that outcome devaluation was abolished in animals that had unilateral BLA lesions and received contralateral pDMS infusions of muscimol prior to test, whereas it was intact in animals that received ipsilateral infusions. This is consistent with our account of BLA function because the hemisphere containing the BLA lesion could not have encoded incentive value during devaluation, and the infused pDMS in the opposite hemisphere could not retrieve any incentive value on test that was encoded by the intact BLA.

We argued that the BLA—pDMS connection mediating these effects is likely to be direct. This is partly because the major indirect pathway by which BLA-encoded information could affect A—O learning is via the PL, and PL—BLA disconnections have been found to have no impact on goal-directed learning ([37], although no attempt to sever contralateral projections was made). It is possible, however, that the BLA conveys incentive value information directly to the pDMS during A—O acquisition to influence learning, but that a wider, indirect circuit is involved in the expression of goal-directed learning during decision-making. This interpretation would fit more readily with the results of disconnection studies involving BLA and IC [19] and BLA and NAC core [29] and, indirectly, the IC and NAC core [30] examining goal-directed performance. To speculate even further, it is possible that the direct IC—pDMS and mOFC—pDMS projections also contribute to A—O learning but not performance, whereas their influence on performance is indirect, via the NAC core.

Although posttraining manipulations of IC and mOFC have been found to abolish outcome devaluation, suggesting that each of these structures plays a specific role in goal-directed performance, a possible role for these structures in A—O learning has not been explicitly ruled out (i.e., no studies have inactivated IC or mOFC during acquisition but not test). If these structures do influence A—O learning in the pDMS via direct projections, this is likely to occur via the processes that have already been identified for these structures, rather than via a direct influence over A—O learning. For example, during A—O learning the IC might mediate the continual retrieval of the incentive value of the outcomes, which would contribute to any updates in the encoding of the A—O associations. By contrast, because mOFC inactivation has been found not to influence A—O learning or responding in the presence of the outcomes themselves, mOFC—pDMS connections are likely to have little role in learning excitatory A—O associations. However, the mOFC might perform another important function recently identified: learning an inhibitory association between the current action and any outcomes that are not currently being presented [38]. Indeed if this type of learning proves to also depend on the pDMS, then it is likely to require inputs from the mOFC to supply the representation of the absent outcome(s).

Although highly speculative, these ideas do produce a number of testable hypotheses. For example, a direct IC—pDMS disconnection can now be achieved using a combination of wheat germ agglutinin—Cre and Cre-dependent DREADDs and should affect goal-directed learning but not performance, whereas a direct disconnection of the IC—NAC core pathway should affect performance but not learning. If correct, a final question remains with regard to how the NAC core makes contact with the pDMS to enable goal-directed responding, as there are no direct connections between these structures. One possibility that has been suggested is that it does so via the series of spiraling networks linking the striato-ventral tegmental and striatonigral pathways, which direct a "feed-forward" flow of information from the NAC core (and shell) through to the DMS (and then to the dorsolateral striatum) [39]. A second or perhaps additional alternative hypothesis is that these circuits interact in the corticothalamic pathway where they come together in the cortical basal ganglia feedback network (see Refs. [28,40], for reviews of this suggestion). Such a network has already been proposed to support the influence of reward-related cues on goal-directed action, in which integration appears to involve the corticothalamic network, particularly that linking the MD and the OFC [28]. Nevertheless, it is clear that a network centered on the MD does not support the integration necessary for changes in performance following outcome devaluation, which probably involves some other, functionally undiscovered, region of the midline thalamus (see Fig. 6.2B).

CONCLUSION

The studies reviewed here indicate that substantial progress has been made in recent years toward identifying the neural bases of the learning and motivational processes that control goal-directed actions. It is clear, however, that there is still much to be explored. With the advent of new techniques, such as DREADDs and optogenetics (see Chapter 5), as well as the diligent use of existing techniques, it is likely that many of the remaining questions will be answered in the years to come. It is further possible, even likely, that new questions will arise with the use of these new techniques, and that exploring these questions will reveal the circuits outlined here to be considerably more complex

than currently characterized. Indeed, there are several layers of complexity that we are already aware of but that are beyond the scope of this chapter. For example, we have focused entirely on the circuits that underlie self-initiated goal-directed actions and ignored the rich literature examining the neural circuitry underlying goal-directed actions that are elicited by Pavlovian stimuli (i.e., Pavlovian—instrumental—transfer). Furthermore, we have not explored evidence demonstrating that parafascicular thalamic-controlled cholinergic interneurons in the pDMS control learning about any alterations in already-learned A—O contingencies but leave original contingencies unaffected [41]. Nevertheless, a great deal has been already achieved in the process of delineating these circuits, and it will be of significant interest to continue to add to this rapidly developing literature.

Acknowledgments

The preparation of this chapter was supported by Grant DP150104878 from the Australian Research Council and a Senior Principal Research Fellowship from the National Health and Medical Research Council of Australia to BWB.

References

[1] Balleine BW, Dickinson A. The role of incentive learning in instrumental revaluation by specific satiety. Anim Learn Behav 1998;26: 46—59.

[2] Balleine BW. Instrumental performance following a shift in primary motivation depends on incentive learning. J Exp Psychol Anim Behav Process 1992;18:236—50.

[3] Corbit LH, Balleine BW. The role of prelimbic cortex in instrumental conditioning. Behav Brain Res 2003;146:145—57.

[4] Killcross S, Coutureau E. Coordination of actions and habits in the medial prefrontal cortex of rats. Cereb Cortex 2003;13:400—8.

[5] Ostlund SB, Balleine BW. Lesions of medial prefrontal cortex disrupt the acquisition but not expression of goal-directed learning. J Neurosci 2005;25:7763—70.

[6] Corbit LH, Muir JL, Balleine BW. The role of the nucleus accumbens in instrumental conditioning: evidence of a functional dissociation between accumbens core and shell. J Neurosci 2001;21: 3251—60.

[7] Ostlund SB, Balleine BW. Differential involvement of the basolateral amygdala and mediodorsal thalamus in instrumental action selection. J Neurosci 2008;28:4398—405.

[8] Bradfield LA, Hart G, Balleine BW. The role of the anterior, mediodorsal, and parafascicular thalamus in instrumental conditioning. Front Syst Neurosci 2013;7:1—15.

[9] Pickens CL. A limited role for mediodorsal thalamus in devaluation tasks. Behav Neurosci 2008;122:659—76.

[10] Corbit LH, Muir JL, Balleine BW. Lesions of mediodorsal thalamus and anterior thalamic nuclei produce dissociable effects on instrumental conditioning in rats. Eur J Neurosci 2003;18: 1286—94.

[11] Parnaudeau S, Taylor K, Bolkan SS, Ward RD, Balsam PD, Kellendonk C. Mediodorsal thalamus hypofunction impairs flexible goal-directed behavior. Biol Psychiatry 2015;77:445—53.

[12] Hull CL. Principles of behavior. 1943. Appleton, New York.

[13] Dickinson A, Dawson GR. Pavlovian processes in the motivational control of instrumental performance. Q J Exp Psychol 1987;39B:113—34.

[14] Lopez M, Balleine BW, Dickinson A. Incentive learning following reinforcer devaluation is not conditional upon the motivational state during re-exposure. Q J Exp Psychol B 1992;45:265—84.

[15] Balleine BW, Dickinson A. Instrumental performance following reinforcer devaluation depends upon incentive learning. Q J Exp Psychol 1991;43B:279—96.

[16] Balleine BW, Dickinson A. Signalling and incentive processes in instrumental reinforcer devaluation. Q J Exp Psychol B 1992;45: 285—301.

[17] Balleine BW. Incentive processes in instrumental conditioning. In: Mowrer R, Klein S, editors. Handbook of contemporary learning theories. Hillsdale (NJ): LEA; 2001. p. 307—66.

[18] Balleine BW, Dickinson A. The effect of lesions of the insular cortex on instrumental conditioning: evidence for a role in incentive memory. J Neurosci 2000;20:8954—64.

[19] Parkes SL, Balleine BW. Incentive memory: evidence the basolateral amygdala encodes and the insular cortex retrieves outcome values to guide choice between goal-directed actions. J Neurosci 2013;33:8753—63.

[20] Balleine BW, Killcross AS, Dickinson A. The effect of lesions of the basolateral amygdala on instrumental conditioning. J Neurosci 2003;23:666—75.

[21] Wassum KM, Ostlund SB, Maidment NT, Balleine BW. Distinct opioid circuits determine the palatability and the desirability of rewarding events. Proc Natl Acad Sci USA 2009;106:12512—7.

[22] Laurent V, Morse AK, Balleine BW. The role of opioid processes in reward and decision-making. Br J Pharmacol 2015;172:449—59.

[23] Sripanidkulchai K, Sripanidkulchai B, Wyss JM. The cortical projection of the basolateral amygdaloid nucleus in the rat: a retrograde fluorescent dye study. J Comp Neurol 1984;229:419—31.

[24] Bradfield LA, Dezfouli A, Chieng B, van Holstein M, Balleine BW. Medial orbitofrontal cortex mediates outcome retrieval in partially observable task situations. Neuron December 16, 2015; 88(6):1268—80.

[25] Reep RL, Corwin JV, King V. Neuronal connections of orbital cortex in rats: topography of cortical and thalamic afferents. Exp Brain Res 1996;111:215—32.

[26] Hoover WB, Vertes RP. Collateral projections from nucleus reunions of thalamus to hippocampus and medial prefrontal cortex in the rat: a single and double retrograde fluorescent labeling study. Brain Struct Funct 2012;217:191—209.

[27] Laurent V, Leung B, Maidment N, Balleine BW. μ- and δ- opioid-related processes in the accumbens core and shell differentially mediate the influence of reward-guided and stimulus-guided decisions on choice. J Neurosci 2012;32:1875—83.

[28] Balleine BW, Morris R, Leung B. Thalamocortical integration of instrumental learning and performance and their disintegration in addiction. Brain Res December 13, 2014. http://dx.doi.org/10.1016/j.brainres.2014.12.023. pii:S0006—8993(14)01692-8.

[29] Shiflett MW, Balleine BW. At the limbic-motor interface: disconnection of basolateral amygdala from nucleus accumbens core and shell reveals dissociable components of incentive motivation. Eur J Neurosci 2010;32:1735—43.

[30] Parke SL, Bradfield LA, Balleine BW. Interaction of insular cortex and ventral striatum mediates the effect of incentive memory on choice between goal-directed actions. J Neurosci 2015;35:6464—71.

[31] Williams ZM, Eskander EN. Selective enhancement of associative learning by microstimulation of the anterior caudate. Nat Neurosci 2006;9:562—8.

[32] Liljeholm M, Tricomi E, O'Doherty JP, Balleine BW. Neural correlates of instrumental contingency learning: differential effects of action-reward conjunction and disjunction. J Neurosci 2011;31: 2474—80.

[33] Tanaka SC, Balleine BW, O'Doherty JP. Calculating consequences: brain systems that encode the causal effects of actions. J Neurosci 2008;28:6750—5.

[34] Yin HH, Ostlund SB, Knowlton BJ, Balleine BW. The role of the dorsomedial striatum in instrumental conditioning. Eur J Neurosci 2005;22:513—23.

[35] Yin HH, Knowlton BJ, Balleine BW. Blockade of NMDA receptors in the dorsomedial striatum prevents action-outcome learning in instrumental conditioning. Eur J Neurosci 2005;22:505—12.

[36] Corbit LH, Leung BK, Balleine BW. The role of the amygdala-striatal pathway in the acquisition and performance of goal-directed instrumental actions. J Neurosci 2013;33:17682—90.

[37] Coutureau E, Marchand AR, Di Scala G. Goal-directed responding is sensitive to lesions to the prelimbic cortex or basolateral nucleus of the amygdala but not to their disconnection. Behav Neurosci 2009;123:443—8.

[38] Laurent V, Balleine BW. Factual and counterfactual action-outcome mappings control choice between goal-directed actions in rats. Curr Biol 2015;25:1074—9.

[39] Haber SN, Fudge JL, McFarland NR. Striatonigrostriatal pathways in primates form an ascending spiral from the shell to the dorsolateral striatum. J Neurosci 2000;20:2369—82.

[40] Hart G, Leung BK, Balleine BW. Dorsal and ventral streams: the distinct role of striatal subregions in the acquisition and performance of goal-directed actions. Neurobiol Learn Mem 2014;108: 104—18.

[41] Bradfield LA, Bertran-Gonzalez J, Chieng B, Balleine BW. The thalamostriatal pathway and cholinergic control of goal-directed action: interlacing new with existing learning in the striatum. Neuron 2013;79:153—66.

7

Impulsivity, Risky Choice, and Impulse Control Disorders: Animal Models

T.W. Robbins, J.W. Dalley

University of Cambridge, Cambridge, United Kingdom

Abstract

The neural and neurochemical bases of various forms of impulsive behavior in rodents are reviewed. We distinguish three main forms of impulsivity: waiting impulsivity (based on premature responding in the five-choice serial reaction time task and delayed temporal discounting of reward), risky choice impulsivity (based on probabilistic discounting of reward), and stopping impulsivity (based on stop-signal inhibition). Some of these forms have trait-like characteristics. Dopamine- and serotonin-dependent functions of the nucleus accumbens are implicated in waiting impulsivity and risky choice impulsivity, as well as cortical structures projecting to the nucleus accumbens. For stopping impulsivity, dopamine-dependent functions of the dorsal striatum are implicated, as well as circuitry including the orbitofrontal cortex and dorsal prelimbic cortex. The noradrenaline reuptake blocker atomoxetine ameliorates both waiting impulsivity (reducing premature responding and enhancing tolerance to delayed gratification) and stopping impulsivity (speeding the stop-signal reaction time), effects that are mediated in the shell region and the prefrontal cortex, respectively. Differences between the contrasting forms of impulsive responding, as well as their commonalities, are highlighted, suggesting that impulsivity is not a unitary construct. Various applications to human neuropsychiatric disorders such as drug addiction and attention deficit/hyperactivity disorder are discussed.

Impulsivity, the tendency to respond prematurely or in an unduly risky fashion, is becoming an increasingly important construct in psychiatry. The emphasis in this working definition of impulsivity is on its potentially maladaptive nature; clearly there are occasions when it is advantageous to respond quickly and in a risky manner. Thus, impulsivity is a dimensional construct that can describe not only behavior of the healthy population, but also a range of psychiatric symptoms expressed in disorders as seemingly distinct as drug addiction, attention deficit/hyperactivity disorder

(ADHD), obsessive–compulsive disorder, mania, schizophrenia, and Parkinson's disease. From that perspective "impulsivity" may qualify for what has been described by the National Institute of Mental Health as a research domain criterion [1], cutting across traditional psychiatric categories. For this to be the case, impulsivity would also have to be tied to specific neurobehavioral processes, possibly also linked to genetic factors. There is every indication that impulsivity meets these requirements. Thus, the aforementioned definition also captures the fact that there is emphasis on anticipating goals or outcomes—at the beginning of response sequences prior to response initiation, when it is important, not only to select responses appropriately, but also to perform the response at the right time and place. Such processes have now been linked definitively to the functions of neural networks of the basal ganglia, including the nucleus accumbens and its contributions to reward-related, goal-directed responding. We have previously referred to this form of impulsive behavior as "waiting impulsivity" [2].

In clinical applications, impulsivity is often measured in terms of self-report, as in the Barratt Impulsiveness Scale (BIS) [3], which contains questions about everyday behavior such as whether individuals make comments "without thinking" and whether they switch jobs frequently or feel "restless in lectures." While such questionnaires are useful and sensitive clinical instruments, they are subjective and often do not reflect what is measured more objectively by other methods. In fact, measures of impulsive behavior frequently do not intercorrelate very highly, suggesting that impulsivity is not a unitary construct, a theme we will develop further later. However, the availability of several objective measures of impulsive responding that can be used across species has been an advantage for establishing the neural networks of various forms of impulsivity. In this chapter we will refer to several methods for measuring impulsive behavior, ranging from temporal and probabilistic discounting of reward to measures of "motor impulsivity" such as go/no-go performance, premature responding in the five-choice serial reaction time task (5CSRTT), and also differential reinforcement of low rates of responding (DRL schedules). We will also consider the stop-signal reaction time as another prominent example of requirement of response inhibition when the response has already been initiated. After an initial survey of these procedures, we will discuss how they have been used to define the involvement of discrete neural circuitries in control of impulsive behavior (Fig. 7.1).

DELAYED AND PROBABILISTIC DISCOUNTING: "IMPULSIVE CHOICE" IN DECISION-MAKING

Delayed gratification, or impulsive choice, procedures have been developed in animals including pigeons as well as rodents, but are also applicable to humans. Much early work [4−6] showed that individuals (pigeons, rats, or humans) have distinct discounting functions, hyperbolic in form, with regard to delay, magnitude, and probability of reinforcement, and that certain patient groups, notably drug abusers, discount at significantly steeper rates than healthy volunteers, in other words, exhibiting impulsive choice. Perry and Carroll [7] provide an excellent review of the relevance

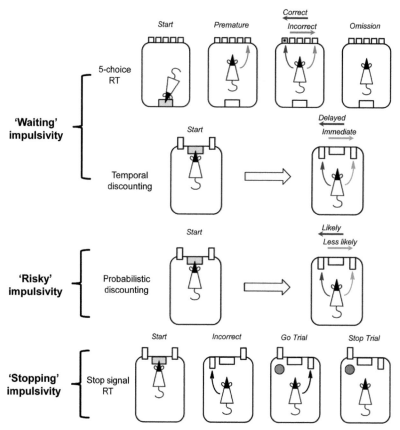

FIGURE 7.1　Translatable subtypes of impulsivity and their operational assessment in rodents. The schematic depicts the multifaceted nature of impulsivity and its hypothetical component dimensions of waiting, risk-taking, and stopping. "Waiting" impulsivity is assessed by the inadequate suppression of responses made prepotent by their association with reward, leading to a "premature" response on the five-choice serial reaction time task (or the analogous four-choice task in humans), and by a preference for instant, small-magnitude rewards, assessed by temporal discounting procedures. "Risky" impulsivity refers to the tendency to prefer less certain, large-magnitude rewards over smaller, but more certain rewards. "Stopping" impulsivity is widely assessed by stop-signal reaction time procedures in rodents and humans, a form of motor impulsivity requiring inhibition after, rather than before, response initiation. *RT*, reaction time.

of this form of impulsivity to drug abuse. Evenden [8] was perhaps the first to consider temporal discounting in comparison to other models of impulsivity. Cardinal [9] has reviewed, in considerable detail, the neural and neurochemical basis of behavior under delayed and probabilistic reinforcement; this review serves to update and extend that previous synthesis, for reasons of space.

In temporal discounting procedures, a choice is generally given between a small immediate reward and a delayed larger one. In this case, impulsive choice is reflected as a preference for the smaller, more immediate outcome. The resulting behavior can usually be modeled as a hyperbolic function in which the constant k can provide an index of temporal reward discounting and hence also of impulsive choice [4,5]. A parallel case is that of "probability discounting," in which the same general principles obtain, but in which the risky aspects of impulsive behavior are measured as preference for the larger, but less likely, option. This paradigm is best suited to human studies of decision-making under risk and is directly relevant to such paradigms as the Iowa Gambling Task [10] and the Cambridge Gambling Task [11]. Temporal and probabilistic discounting clearly overlap and engage similar processes in theoretical terms [12]; however, there is mounting evidence that they are in fact distinct and both contribute to complex decision-making processes [9].

There are further, major methodological problems for delayed discounting both for human and for experimental animal paradigms. Human temporal discounting of reward is generally measured by giving subjects hypothetical choices; for example, the possibility of receiving £X now and £Y after some elapsed time, often large numbers of days. Hence, actual behavior is not measured and one cannot be absolutely sure that people will actually do in practice what they say they will do in the abstract. The problem with the animal paradigms of temporal discounting is somewhat different; one has to be sure that the animal is able to bridge delays of many seconds between its response and an outcome; in other words to learn with delay of reinforcement, for which most species have severely limited capability. One way of ameliorating this difficulty is to provide an immediate conditioned reinforcer that is associated with the eventual large reward outcome.

Temporal and probabilistic discounting are clearly important potential components of decision-making under risk and uncertainty. However, also important is the capacity to accumulate relevant information to a point at which a decision can be made with confidence. The tendency to choose too soon under conditions when insufficient evidence has accumulated may be referred to as "reflection impulsivity." This is an important construct in developmental psychology, but has recently been employed in studies of adults. Animal tests of reflection impulsivity are certainly feasible but have not so far been widely employed [8].

PREMATURE RESPONDING IN THE 5CSRTT

The discovery that individual differences in the premature responding of rodents in the attentional setting of the 5CSRTT [13] is associated with an increased propensity for cocaine, nicotine, and sucrose reinforcement [14–16] has led to considerable interest in the nature of this potential behavioral endophenotype. In the 5CSRTT, rats or mice are trained to detect brief visual stimuli to earn food reinforcers by a nosepoke response in the illuminated location. Incorrect or premature nosepoke responses are punished by a time-out from reinforcement, signaled by dimming of the house lights. Individuals vary considerably in their tendency to respond prematurely, although the persistence of this tendency makes it a likely trait measure of impulsivity. The 5CSRTT essentially incorporates a DRL contingency in its design, and so it is likely that efficiency of performance in DRL schedules also reflects a tendency toward premature responding, although the typically longer intervals for which responding has to be inhibited make timing an even more important requirement than in the 5CSRTT. There are also elements of the requirement for go/no-go choice responding in the 5CSRTT, but this is clearly a distinct requirement, at the point of response selection.

The 5CSRTT is in fact based on a human analogue of the task, previously much used for examining effects of stressors, fatigue, and drugs on human performance, including sustained attention [13]. In 2013, the rodent 5CSRTT was adapted for a human version with four choices (4CSRTT), which lays emphasis on measuring premature responding as premature release of a response button prior to touching a screen at the location of a briefly presented visual target [17].

STOP-SIGNAL REACTION TIME

The stop-signal reaction time (SSRT) task is well established as a paradigm for measuring response inhibition and volitional control in humans [18]. Impairments in response inhibition are associated with impulsivity. However, the key point of the SSRT task, which distinguishes it from other measures of motor impulsivity, is that inhibition is applied after, rather than prior to, response initiation. Thus, we have previously distinguished "waiting impulsivity" from that exemplified in SSRT performance as "stopping impulsivity" [3]. Other than being more relevant following rather than prior to response initiation, "stopping inhibition" can be distinguished from "waiting inhibition" in terms of the underlying neural networks engaged by each task. Effective translation of the SSRT paradigm has been achieved in nonhuman primates, rats, and mice (eg, [19]), with often striking similarities in typical values of key variables such as SSRT that encourage the conclusion that similar processes are being engaged in animal and human studies.

NEURAL BASIS OF IMPULSIVITY

This review focuses on evidence gained from animal models of impulsivity, but the similarity of some of the procedures employed is useful for cross-species comparative studies of their neural substrates and homologies, using, for example, brain imaging methodology for humans. In general, it appears that corticostriatal pathways are especially implicated in various forms of impulsivity, although with considerable overlap of some key structures. An especially important consideration is the parallel nature of the corticostriatal parallel "loop" pathways, which are fed back via pallidal and thalamic nodes to the frontal cortex; waiting and stopping impulsivity appear, for example, to utilize distinct, though parallel, systems [3]. Moreover, choice and motor aspects of waiting impulsivity, while converging in the nucleus accumbens, also appear to utilize different systems. Finally, the different forms of impulsivity may be subject to distinct modes of chemical neuromodulation by the ascending monoaminergic systems, which may have implications for the treatment of such conditions as ADHD.

NEURAL SUBSTRATES OF WAITING IMPULSIVITY

5CSRTT Premature Responding

Premature responding on the 5CSRTT clearly depends on dopamine (DA)-dependent mechanisms in the nucleus accumbens. Initial evidence showed that the tendency of D-amphetamine to induce premature responding, whether the drug was infused systemically or centrally, depended on the integrity of the DA innervation of the nucleus accumbens [20]. Moreover, this premature responding could be blocked by infusions of intraaccumbens DA receptor antagonists such as the mixed D1/D2 antagonist α-flupenthixol and also the D1 receptor antagonist SCH22390 or the D2 receptor antagonist sulpiride. The nucleus accumbens is divided into two subregions, the shell and the core, but both are

implicated in this behavior. Effects of excitotoxic lesions of the nucleus accumbens on baseline premature responding have been quite inconsistent. Christakou et al. [21] found that lesions of the core region mainly increased premature responding following an error on the previous trial, suggesting that the arousal induced by frustrative nonreward may contribute to premature responding. This evidence of core involvement is further substantiated by effects of core lesions on responding under DRL schedules, which typically require inhibition of responding for periods of up to 18 s [22].

Murphy et al. [23] showed that core lesions exacerbated the effects of systemic D-amphetamine to elevate premature responding, whereas lesions of the shell actually reduced such responding following D-amphetamine. This evidence suggested a form of opponency in the control over responding by the core and shell subregions; further evidence in support of this opponency comes from the contrasting effects of deep brain stimulation on impulsivity [24] and divergent changes in electrically evoked DA release in these two regions in low- and high-impulsive rats [15].

Other evidence has shown that damage to the infralimbic (IL) cortex, anterior cingulate, and hippocampus can all enhance premature responding in the 5CSRTT, suggestive of top-down inhibitory influences [3]; these structures project to the shell (especially hippocampus and IL cortex) and the core accumbens subregions. It should also be noted that damage to the dorsomedial striatum can also enhance premature responding but often in the context of more general changes in performance [3].

Studies of individual differences in impulsive responding on the 5CSRTT have greatly bolstered these conclusions. Dalley et al. [25] originally noted that the large, though consistent, individual variation in premature responding to a single, fixed location was suggestive of impulsive trait. Studies (Dalley et al., unpublished findings) have shown that the 5CSRTT impulsivity indeed breeds true over four generations and is heritable, with a quantitative trait locus on chromosome 1 that includes genes for synaptic proteins and the GABA and mGluR5 receptors. The phenotype was quite specific; the high-impulsive rats were not overactive or disposed to stress reactions, nor did they exhibit impaired Pavlovian or instrumental learning. They did show a mild novelty preference and occasionally a mild attentional impairment [3].

Subsequent to screening, high-impulsive rats were also shown, however, to exhibit elevated self-administration of cocaine in a binge access paradigm [14]. Other studies reported similarly enhanced nicotine [15], but not opiate, self-administration [3]. The elevated cocaine self-administration occurred in high-impulsive rats even if punished with electric shock [26] and so

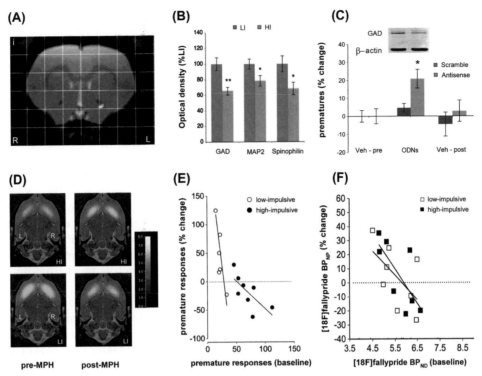

FIGURE 7.2   Neural mechanisms underlying trait-like impulsivity in rats. (A) In vivo magnetic resonance (MR) scan image showing reduced gray matter density in the left nucleus accumbens core of low- versus high-impulsive rats *(LI, HI)* on the five-choice serial reaction time task. (B) Reduced levels of glutamate decarboxylase *(GAD)*, the rate-limiting enzyme for GABA, and markers of dendritic spines [microtubule-associated protein-2 *(MAP2)*, spinophilin] in the left nucleus accumbens core of high-impulsive rats. (C) Increased impulsivity in low-impulsive rats following transient inactivation of GAD in the nucleus accumbens core by locally administered antisense oligonucleotides *(ODNs)*. (D) Coregistered MR structural scans (gray scale) and positron emission scans showing uptake of the selective, high-affinity D2/3 receptor antagonist [^{18}F]fallypride in the ventral striatum of low- versus high-impulsive rats prior to, and after, oral methylphenidate *(MPH)* administration. MPH increased uptake when baseline [^{18}F]fallypride binding was low, as in HI rats, but decreased binding when baseline binding was high, as in LI rats. (E) Rate-dependent effect of MPH on behavioral impulsivity. MPH had the effect of reducing impulsivity when baseline impulsivity was high (HI rats) but increased impulsivity when baseline impulsivity was low (LI rats). (F) Baseline-dependent effects of MPH on D2/3 receptors in the ventral striatum of LI and HI rats. MPH increased [^{18}F]fallypride uptake when baseline binding was low but had the opposite effect when baseline binding was high. *Veh,* vehicle; BP_{ND}, nondisplaceable binding potential. *Reproduced with permission from Caprioli D, Sawiak SJ, Merlo E, Theobald DE, Spoelder M, Jupp B, et al. Gamma aminobutyric acidergic and neuronal structural markers in the nucleus accumbens core underlie trait-like impulsive behavior. Biol Psychiatry 2014;75:115—23. and Caprioli D, Jupp B, Hong YT, Sawiak SJ, Ferrari V, Wharton L, et al. Dissociable rate-dependent effects of oral methylphenidate on impulsivity and D2/3 receptor availability in the striatum. J Neurosci. 2015;35:3747—55.*

could be described as compulsive—an important sign of drug addiction as distinct from recreational drug use. Of considerable significance was that D2/3 receptor binding was reduced in the ventral (nucleus accumbens) but not dorsal striatum of high-impulsive rats before they were exposed to cocaine, suggesting that low D2/3 binding was an endophenotype for stimulant abuse and relevant to the interpretation of analogous findings of Volkow et al. [27] of reduced D2/3 binding potentials in the dorsal striatum of drug abusers.

These findings were substantiated by several studies. Besson et al. [28] showed that mRNA for D2 receptors was reduced specifically in the shell region. Jupp et al. [29] used autoradiographic methods to show similarly that there was reduced D2/3 receptor protein expression in the shell, but not the core, accumbens subregions. This was accompanied by evidence of reduced DA

transporter binding in the shell and D1 receptor binding in the core region. Besson et al. [30] found that infusions of the D3/D2 antagonist nafodotride into the shell region increased premature responding, whereas infusions into the core region actually reduced such responding, again demonstrating apparent opponent actions of core and shell in the regulation of behavior. Together with evidence of increased DA release in the shell region [15], these findings point to elevated DA release in the shell region being associated with high impulsivity. Thus, low presynaptic levels of D2/3 receptors in the shell region will tend to enhance DA release and high extracellular levels would be sustained by the reduction in DA transporters, which would normally regulate such excessive overflow. So far, in vivo dialysis measures of accumbens DA have not revealed direct evidence of elevated levels of DA, but this failure may be a product of low temporal resolution and a failure to distinguish phasic DA release.

The involvement of the core subregion has been confirmed by a 2014 study in which magnetic resonance imaging was used to show a reduced signal in the core subregion of high-impulsive rats that correlated to a very high extent in the left core subregion with levels of impulsive responding [31]. Associated with this finding was reduced activity of GABA decarboxylase (GAD) in the left core (and not the shell) subregion of high-impulsive rats, as well as evidence of a reduction of synaptic proteins. The reduced GAD activity presumably reflects reductions in numbers of medium spiny output neurons or the density of dendritic spines in the core region. The causal role of the core subregion was shown by the effects of GAD antisense infused into that region in low-impulsive rats, which selectively increased levels of premature responding [31] (Fig. 7.2).

Therefore, it would appear that the shell and core subregions interact to some extent in the production of this form of impulsivity: elevated DA in the shell presumably enhances output of this structure, which is normally counteracted or gated by core output, but less effectively so in the high-impulsive rats.

An early observation was that cocaine intravenous (IV) self-administration appeared to reduce high levels of premature responding [14], raising the possibility that this impulsivity was regulated by drugs with similar actions on catecholamine transporters, such as methylphenidate, which are used in the treatment of ADHD. This modulation has been studied in much more detail. Caprioli et al. [32,33] have examined the effects of IV self-administered cocaine and also systemic methylphenidate on impulsive responding and striatal D2/3 receptor binding. These authors replicated the finding of diminished D2/3 receptors in the nucleus accumbens associated with high levels of impulsive responding, although this was shown only for the left hemisphere. Both IV cocaine and oral methylphenidate given subchronically reduced impulsivity in high-impulsive rats, whereas it increased in animals with low levels of impulsivity—a variation of the well-known rate dependency effect. Parallel effects were observed for D2/3 binding. Thus, for example, previously low levels of striatal D2/3 binding were increased by cocaine but previously high levels (associated with high impulsivity) were decreased. However, unlike the effects for behavioral impulsivity, these effects were found for both the ventral and the dorsal striatum. Moreover, these receptor changes were found not to covary predictably with behavioral changes, suggesting that other factors may be operating to reduce impulsivity. Nevertheless, it appears that blocking the reuptake of DA can have the apparently paradoxical effect of reducing impulsivity associated with increased DA release. An explanation of this, as well as other rate-dependent effects, is required, but may reflect an altered functional balance between tonic DA levels and phasic DA neuronal activity [34].

An obvious additional candidate for modulating impulsive responding is central noradrenaline (NA), as the NA transporter is also a target for methylphenidate and cocaine. In fact, the NA reuptake blocker atomoxetine is a very effective antiimpulsivity agent, reducing impulsivity at doses that do not affect attentional performance [35]. Remarkably, when infused into the shell [but not the core or the prefrontal cortex (PFC)] region it also dose-dependently reduces impulsive responding [36]. Presumably, the drug interacts with the sparse noradrenergic innervation of the basal ganglia, which is most pronounced for the shell region [37].

In addition to regulation by DA and NA, impulsivity is also modulated importantly by serotonin (5-HT) mechanisms. Infusion of 5-HT2A receptor antagonists (ketanserin or M100907) or a 5-HT1A agonist (8-OH-DPAT) into the medial PFC reduces impulsivity, perhaps via the IL–PFC pathway [38]. These effects are mimicked by infusion of M100907 directly into the nucleus accumbens. Moreover, impulsivity is enhanced by infusions of the 5-HT2C antagonist SB-242084 into the nucleus accumbens [39]. This is consistent with earlier evidence that profound forebrain depletion of 5-HT by infusions of the selective neurotoxin 5,7-dihydroxytryptamine also led to enhanced impulsive responding [40,41], which was further exacerbated by systemic treatment with SB-242084.

Mouse models have also augmented our understanding of impulsive responding in the 5CSRTT. Mice are capable of excellent performance in this task and a large number of genetic strains have now been studied in particular with respect to the premature response measure [42]. One particular finding of interest has been that the deletion of the gene controlling α-synuclein decreases impulsivity in mice [43].

In summary, premature responding in the 5CSRTT is governed by a network of brain regions including the PFC and nucleus accumbens (core and shell) and is prevalent in disorders including ADHD and drug addiction. While we have begun to construct a neural network for the mediation and modulation of this behavior, little is known of the relationship between neural activity in these regions [44]. Therefore, we investigated local field potential (LFP) oscillations in distinct subregions of the PFC and nucleus accumbens during performance of the 5CSRTT. The main findings showed that the power in γ frequency (50–60 Hz) LFP oscillations transiently increased in the PFC and nucleus accumbens both during the anticipation of a cue signaling the spatial location of a nosepoke response and, again, following correct responses, confirming our hypothesis that this form of impulsivity was related to reward anticipation [45]. γ oscillations were coupled to low-frequency δ oscillations in both regions; this coupling strengthened specifically when an error response was made. θ (7–9 Hz) LFP power in the PFC and nucleus accumbens increased during the waiting period and was also related to response outcome. Additionally, both γ and θ power were significantly affected by upcoming premature responses as rats waited for the visual cue to respond. In high-impulsive rats we found that impulsivity was associated with increased error signals following a nosepoke response, as well as reduced signals of previous trial outcome during the waiting period [45]. This evidence is consistent with the previous findings that high impulsivity often follows errors or activation produced by excessive DA release, as occurs following treatment with psychomotor-stimulant drugs.

Other sources of evidence tend to support our view of waiting impulsivity as exemplified by premature responding on the 5CSRTT. Human studies of impulsivity are consistent with low D2/3 binding in the midbrain predicting impulsivity as measured by the BIS; this would also lead to elevated DA release in terminal domains in the ventral striatum [46]. There has as of this writing been little parallel evidence in humans; however, the 4CSRTT described earlier has identified elevated premature responding in methamphetamine and recreational cannabis abusers [17]. Moreover, tryptophan depletion (which transiently reduces central 5-HT) significantly increased premature responding in human volunteers [47], analogous to the effects of forebrain 5-HT depletion in rats [40]. Further evidence for a role of the nucleus accumbens in impulsivity comes from converging evidence of other forms of waiting impulsivity, notably delayed discounting of reward.

Temporal Discounting of Reward

Nucleus Accumbens

Cardinal et al. [48] initially reported that selective bilateral excitotoxic lesions of the nucleus accumbens core region (but not the anterior cingulate or the medial PFC) resulted in more impulsive choice using a temporal discounting task. In general, this result has held up in subsequent studies using different variants of the delayed discounting procedure. Thus, Pothuizen et al. [22] also varied the uncertainty of reward and showed at certain parameters that core lesions similarly enhanced choice for the immediate outcome. Bezzina et al. [49] reported a selective reduction of the intercept of an indifference function in core-lesioned rats, which indicates a higher rate of delay discounting. One issue

of interpretation is whether these effects are confounded by effects on discrimination of magnitude of reinforcement [50] or on the memory for a response following delay of reinforcement [51]. However, Bezzina et al. [49] concluded that effects on reinforcement discrimination, while present, were insufficient to account for impulsive choice following core lesions. Moreover, da Costa Araujo et al. [52] concluded that the effects on delay of reinforcement were similarly insufficient to account for changes in discounting.

Delays in discounting tasks can be signaled by feedback stimuli following choice of the delayed outcome and apparently have an important influence on effects of drugs and lesions. Cardinal et al. [53] showed that D-amphetamine promoted choice for the larger, delayed reinforcer in the presence of the signaled delay, but impulsive choice in its absence, possibly consistent with other evidence that D-amphetamine enhances effects of conditioned reinforcers in their control over behavior [54]. As D-amphetamine is an indirect DA agonist, the potentiation of conditioned reinforcement through a DA-dependent mechanism in the nucleus accumbens is intuitively attractive, as its actions on temporal discounting are similarly mediated by this region. However, it appears that the situation may be more complex and depends on an interaction of DA and 5-HT mechanisms in the mesolimbic system [55].

Winstanley et al. [56] found, after extensive training in a delayed discounting procedure, that there was considerable variation in the extent of impulsive choice across a group of rats. Those rats that exhibited impulsive choice (ie, low k values) preferred the large delayed reinforcer to a greater extent after systemic D-amphetamine and this effect was reduced both by forebrain 5-HT depletion and by α-flupenthixol, the mixed D1/D2 receptor antagonist. In a later study [57], the 5-HT1A receptor agonist, 8-OH-DPAT, also antagonized the effects of D-amphetamine to reduce impulsive choice, whereas the 5-HT1A antagonist WAY 100635 potentiated it, presumably acting at 5-HT-inhibitory autoreceptors. Acute 8-OH-DPAT increased impulsive choice, an effect that depended on the integrity of the DA innervation of the nucleus accumbens [57]. This finding is consistent with the evidence that, at least under certain parameters, central 5-HT depletion can cause steep temporal discounting of reward [58]. Additionally, in vivo microdialysis has been used to measure DA and 5-HT changes selective to temporal discounting with appropriate yoked controls for the contribution of contingency. Findings indicated some selective changes in 3,4-dihydroxyphenylacetic acid, a primary DA metabolite, in the orbitofrontal cortex (OFC) and 5-HT in the medial PFC. In short, temporal discounting is powerfully modulated by both dopaminergic and serotoninergic influences, within the nucleus accumbens and various sectors of the PFC [59].

Amygdala and Corticolimbic Influences

Winstanley et al. [60] showed that rats with excitotoxic lesions of the basolateral nucleus of the amygdala exhibited a form of impulsive choice that was similar to that following lesions of the nucleus accumbens core, perhaps reflecting the innervation of this structure. Similar effects of hippocampal lesions on impulsive choice were also reported by Cardinal and Cheung [61].

By contrast, Winstanley et al. [60] also showed that similar, large lesions of the OFC produced a paradoxical preference for delayed food rewards. The fact that the delay did not devalue the large reward to such an extent in OFC-lesioned animals supported the suggestion by Schoenbaum et al. [62] that the OFC is involved in updating the incentive value of outcomes in response to devaluation. The effects of these OFC lesions thus contrasted with the lack of effect shown of lesions of the anterior cingulate or medial PFC previously reported by Cardinal et al. [48]. They also contrasted with the findings of a notable previous study by Mobini et al. [63] in which OFC lesions made prior to the acquisition of the delay-discounting choice procedure had had precisely the opposite effect: to increase impulsive choice. At the time of that publication, Winstanley et al. [60] explained the anomaly by citing evidence that the ability of OFC neurons to encode the expected value of reward occurred only relatively late in training. The tasks used also differed in that Mobini et al. [63] offered rats a choice between a one-pellet immediate and a two-pellet delayed reinforcer, whereas Winstanley et al. [60] gave them the option of a one-pellet immediate versus a four-pellet delayed reinforcer. Thus, an increase in reward magnitude discounting following OFC lesions could possibly mask any changes in delay discounting. For example, if small rewards are also delayed, this greatly increases preference for the large delayed reward. In fact, Kheramin et al. [64,65] had shown that OFC lesions and OFC DA depletion both appear to increase sensitivity to (or discounting of) both delay and magnitude of reward.

However, a later study, using a T-maze procedure to investigate impulsive choice, also found that OFC lesions produced impulsive choice in previously trained rats [66], providing another contrast with the findings of Winstanley et al. [60], and so it is clear that it is necessary to consider other possible variables to explain these apparent discrepancies. One such variable, which may have influenced the effects of OFC lesions in all of these studies, and, as described earlier [53], has been shown to modulate the effects of amphetamine, is the presence of a cue bridging the delay between response and large reinforcing outcome. Zeeb et al. [67] found that when the delay to the large reward was cued, transient inactivation of the OFC (using a baclofen/muscimol mixture)

increased impulsive choice in rats with low baseline levels of impulsivity. In contrast, the same treatment decreased impulsive choice in the uncued condition in highly impulsive animals. The effects of the cue were also found to be D2 receptor dependent, as intra-OFC infusion of a D2 receptor antagonist blocked its ameliorative effects on impulsive choice. Thus, when baseline levels of impulsive choice are high, as often occurs in rats with uncued delays, OFC lesions tend to reduce impulsive choice, as may have occurred in the Winstanley et al. [60] study. By contrast, when the large delayed reward is cued, as occurred in Mobini et al. [63] and may well have occurred in the T-maze study of Rudebeck et al. [66], inactivation of the OFC leads to impulsive choice.

Yet another factor influencing the outcome of these studies of OFC manipulations on impulsive choice is the precise site of the lesion or inactivation. All of the interventions within the OFC in the previous studies described above have centered on the ventrolateral OFC region to varying extents. However, in their study of delayed discounting, Mar et al. [68] explicitly employed discrete excitotoxic lesions of the medial OFC and ventrolateral OFC, as well as larger lesions including both structures. Ventrolateral OFC lesions induced stable patterns of impulsive choice, whereas medial OFC lesions after retraining had opposite effects. Entire OFC lesions initially replicated the effects of Winstanley et al. [60], but were transient, perhaps consistent with the interactive but opponent sequelae of the two component regions. These differential effects presumably reflect the possible role of the lateral OFC in cognitive control processes when reward-related responses require suppression, a hypothesis supported by the slowed reversal learning of lateral OFC-lesioned rats when the location of the reward contingencies was switched. By contrast, medial OFC-lesioned rats after retraining appear to be more sensitive to reward magnitude, which counteracts delay devaluation, similar to the effects of larger lesions in Kheramin et al. [63].

Overall, these findings are probably relevant to human findings for hypothetical temporal reward discounting, but there are apparent discrepancies. The increased impulsive choice following lateral OFC lesions in rats may be consistent with evidence in humans that the degree of activation during reward discounting of lateral OFC predicts the extent of deferral of gratification [69]. However, less consistent is the finding that humans showed impulsive choice in three separate paradigms following medial OFC damage [70].

The McClure et al. [69] functional imaging study in humans varied both delay and magnitude for reward and found evidence for two interactive systems of choice, an interaction of which may explain hyperbolic discounting. These systems consist of a "declarative, rational discounting" component including parietal and lateral OFC areas and a system more sensitive to immediate effects of highly salient rewards, focused on the medial OFC and nucleus accumbens. However, the effects of nucleus accumben regions to enhance impulsive choice reviewed earlier (eg, [48]) appear to be inconsistent with this formulation, possibly depending on the differences between hypothetical and actual reward discounting.

NEURAL SUBSTRATES OF PROBABILITY DISCOUNTING OF REWARD: RISKY CHOICE

Risky choice is that between more certain small rewards and less certain large rewards. Although probability discounting can be subsumed in the same theoretical framework as temporal discounting [12], it appears that the underlying neural substrates are distinct. Correlational studies in humans are consistent with evidence that the nucleus accumbens might also be implicated in probabilistic discounting in animals. Thus, for example, midbrain DA neurons innervating the nucleus accumbens fire in relation to reward probability in monkeys [71] and an increase in blood flow occurs in the human nucleus accumbens just prior to risk-taking decisions in human volunteers [72]. Moreover, lesions of the nucleus accumbens core [73] and lateral OFC [63] induce risk-averse choice in rats. Effects of inactivation of the nucleus accumbens with acute infusions of GABA agonists were broadly similar but distinct contributions of the shell and core subregions were obtained [74]. Manipulation of DA receptors in the nucleus accumbens has complex results; D1 receptor blockade reduces risky choice, much like lesions or inactivation; a D1 receptor agonist appears to optimize risky choice. D3 receptor antagonism has effects almost opposite to those of D1 receptor antagonism; surprisingly D2 receptors play no obvious role [75]. A parallel analysis of effects of DA receptor agents in the rat medial PFC showed that, again, D1 receptor blockade reduced risky choice, whereas D2 receptor blockade had the opposite effect [76]. These findings are consonant with the relationship of tonic fluctuations of DA release to several factors influencing risky choice in the nucleus accumbens, and contrast with the more obvious "reward metering" provided by DA release in the PFC, somewhat independent of precise contingencies [59,77]. These are striking results, as one would normally associate impulsivity with risk-taking as well as temporal discounting of reward; it suggests either that our notions of impulsivity have to be revised or that these structures are associated only with certain forms of impulsivity.

Following up the findings of dissociable effects of manipulations of medial OFC lesions on temporal discounting [68], Stopper et al. [78] investigated the contribution of the rat medial OFC to risk and delay-based decision-making, assessed with probabilistic and delay-discounting tasks. In well-trained rats, reversible inactivation of the medial OFC with baclofen/muscimol infusions increased risky choice on the probabilistic discounting task, irrespective of whether the odds of obtaining a larger/risky reward decreased (100–12.5%) or increased (12.5–100%) over the course of a session. This enhanced risky choice was associated with increased win–stay behavior, in which the rats exhibited an increased tendency to choose the risky option after being reinforced for the risky choice on the preceding trial. In contrast, acute medial OFC inactivation did not alter delay discounting. These findings suggest that the medial OFC plays a selective role in decisions involving reward uncertainty, possibly promoting the exploration of novel options when reward contingencies change. A comprehensive program of studies of the roles of other PFC and cingulate regions following their inactivation in probability discounting has highlighted the specific role of the medial OFC [79,80]. For example, inactivation of either the dorsal anterior cingulate cortex or the insula or the lateral OFC failed to affect probabilistic discounting (in the last case, unlike the results of Mobini et al. [63] following chronic lesions), although response latencies were slower following lateral OFC inactivation. By contrast, inactivation of the medial PFC (prelimbic) increased risky choice when reward probabilities were initially high, decreasing over the session, but had the opposite effect when the reward probabilities began low and increased subsequently [79]. This is perhaps consistent with a proposed role for adjusting decision biases in volatile situations. In their studies, St Onge et al. [80] showed that, whereas basolateral amygdala–accumbens circuits tended to drive choice toward the large, risky option, medial PFC–amygdala pathways promoted the opposite bias. Moreover, other structures, such as the lateral habenula, have been found to exert fundamental influences on probabilistic discounting, with inactivation essentially obliterating decisional factors influencing choice and leading to random responding [81].

A further complication applies when risky decision-making also involves possible punishment, eg, with electric shock, rather than reward omission as in the typical probability discounting paradigm. Setlow and colleagues have added punishing electric shock to the risk associated with high levels of reward. This paradigm appears more sensitive to manipulations of D2 receptors than probabilistic discounting, which simply depends on reward omission. Thus, Simon et al. [82] found that systemic D2 receptor agonists reduced risk-taking, whereas D1 agonists had no effect. Lower levels of D2 mRNA in the dorsal striatum were associated with greater risk-taking. D1 mRNA expression in the insula and shell subregion was positively associated with risk-taking. In an interesting echo of the findings of Dalley et al. [14] for premature responding, Mitchell et al. [83] reported that high-risk takers were predisposed to elevated cocaine self-administration (which also led to elevated risk expressed 6 weeks following abstinence). Presumably the neural substrates of these two phenotypes overlap a great deal in the nucleus accumbens. Similarly also, just as basolateral amygdala lesions lead to impulsive choice, such treatments also enhance risky choice, whereas lateral OFC lesions reduce such choice [84].

These studies of temporal discounting and probabilistic discounting are relevant to new developments in the measurement of gambling-like behavior in rats [85], reminiscent of decision-making in such tasks as the Iowa Gambling Task, which is much used in human neuropsychological studies of patients with PFC lesions. This test for "gambling rats" offers considerable promise for future studies relevant to compulsive gambling in humans. One highly relevant finding for this chapter is that motor impulsivity was positively correlated with poor decision-making under risk [86].

Overall, while it is clear that many of the mechanisms of decision-making cognition appear to overlap and recruit similar neural networks, there are important differences, and the neural networks themselves may overlap but also differ in precise function. There appears to be excellent translational potential to human studies of decision-making, eg, the Iowa Gambling Task [10] and the Cambridge Gambling Task [11]. However, we are still somewhat far from being able to define precisely what each node in a network contributes to the processing of the network as a whole.

MOTOR IMPULSIVITY: STOP-SIGNAL INHIBITION

Another paradigm that measures impulsivity, in perhaps the purest form of inhibition, is that required in the SSRT procedure [18]. Humans are trained to respond as quickly as possible in a choice reaction time task, for example, to cues to the left or right. On a proportion of trials (often 25%) a stop signal is presented that instructs the subject to cancel his or her response. This stop signal is offset from the imperative signal by a variable short delay, which means that the subject has often initiated a response before having to inhibit it. The task is often considered to reflect a race between a go process and a stop process for control over an already initiated response; for this race model to be valid the SSRT should be independent of the go reaction time. The SSRT task, which requires cancellation of an already-initiated

response, differs considerably from the well-known go/no-go paradigm, in which response inhibition or a rule has to be exerted prior to initiation of a response, at the stage of response selection [87]. Human neuropsychological studies, for example, using patients with PFC lesions [88], in parallel with neuroimaging studies [89], have been able to identify a neural network mediating stop-signal inhibition, which includes the right inferior frontal gyrus, the supplementary motor area, and the basal ganglia, especially the subthalamic nucleus. This network has been replicated to some extent using a rodent analogue of the SSRT task. Eagle et al. [90] found that selective cell body lesions of the rat lateral OFC and anterior cingulate, but not the medial PFC [91], produced major slowing of the SSRT. The lateral OFC site was interesting because it may represent the region homologous to that of the human inferior frontal cortex.

Particularly striking is the evidence that, whereas lesions of the rat caudate-putamen can be shown significantly to retard the SSRT [19], damage to the nucleus accumbens core, which affects premature responding on the 5CSRTT and disrupts the temporal discounting of reward, has no effect on SSRT [91]. This emphasizes that the construct of impulsivity can be fractionated into several forms, often utilizing distinct neural networks. A more recent study [92] has shown that D1 and D2 receptor antagonists infused into the caudate-putamen (and not the nucleus accumbens core) had opposite effects, the D2 receptor antagonist slowing SSRT and the D1 antagonist speeding it—this perhaps represents the opposed actions of the direct (D1-dominated) and indirect (D2-dominated) pathways and also helps to explain why a systemic general DA agonist such as L-dopa has no significant effect. The noradrenergic reuptake blocker atomoxetine, given either systemically or within the rat OFC [93], also exerts antiimpulsive effects, quite parallel to its effects on other forms of impulsivity including premature responding and temporal discounting [35]. Another surprising result is that manipulations of 5-HT (either via the selective 5-HT inhibitor citalopram or by gross and profound forebrain depletion) have no apparent effects on SSRT performance [94], arguing against the classic, but oversimplified, concept of this neurotransmitter as a mediator of response inhibition. By contrast, 5-HT is a major modulator of premature responding in the 5CSRTT [40], an index of waiting impulsivity [3].

CONCLUSION

Overall, it is evident that the construct of impulsivity has to be subdivided into several distinct forms; some of these reflect the requirement to wait before a response should be made at a particular time or to endure a delay to obtain a large reward (waiting impulsivity); some reflect risky choice behavior and some reflect the motor inhibition of a response that has already been initiated (stopping impulsivity). These forms are mediated by neural systems that often include overlapping elements, as well as some distinct circuits, notably within the frontostriatal systems. They are also mediated by distinct neuromodulatory influences, such as the 5-HT and D1/D2 receptor systems. Further advances may depend on using these paradigms to investigate genetic factors using mouse models; indeed some studies are indicating the potential of this approach [95,96].

These different forms of impulsivity have already been shown to have some neuropsychiatric significance, although we still do not have a complete picture of how they are all expressed in various disorders. Temporal discounting of reward has been shown to be generally steeper in many so-called impulsive–compulsive syndromes. SSRT has similarly been shown to be impaired in disorders as distinct as addiction, obsessive–compulsive disorder, trichotillomania, schizophrenia, and Parkinson's disease. Whether all of these behavioral deficits reflect common neural dysfunctions would seem unlikely, but the analysis has undoubted heuristic value in the continuing effort to refine our understanding of neuropsychiatric phenotypes.

Acknowledgments

The Behavioural and Clinical Neuroscience Institute is supported by a joint award from the MRC (UK) and Wellcome Trust.

References

[1] Cuthbert BN. The RDoC framework: facilitating transition from ICD/DSM to dimensional approaches that integrate neuroscience and psychopathology. World Psychiatry 2010;13:28–35.

[2] Patton JH, Stanford MS, Barratt ES. Factor structure of the Barratt impulsiveness scale. J Clin Psychol 1995;51:768–74.

[3] Dalley JW, Everitt BJ, Robbins TW. Impulsivity, compulsivity, and top-down cognitive control. Neuron 2011;69:680–94.

[4] Mazur JE. Theories of probabilistic reinforcement. J Expt Anal Behav 1989;51:87–99.

[5] Bradshaw CM, Szabadi E. Choice between delayed reinforcers in a discrete trials schedule. Q J Exp Psychol B 1992;44:1–16.

[6] Bickel WK, Marsch LA. Toward a behavioral economic understanding of drug dependence: delay discounting processes. Addiction 2001;96:73–86.

[7] Perry JL, Carroll ME. The role of impulsive behavior in drug abuse. Psychopharmacology (Berl) 2008;200:1–26.

[8] Evenden JL. Varieties of impulsivity. Psychopharmacology (Berl) 1999;146:348–61.

[9] Cardinal RN. Neural systems implicated in delayed and probabilistic reinforcement. Neural Netw 2006;19:1277–301.

[10] Bechara A, Damasio AR, Damasio H, Anderson SW. Insensitivity to future consequences following damage to human prefrontal cortex. Cognition 1994;50:7–15.

[11] Rogers RD, Everitt BJ, Baldacchino A, Blackshaw AJ, Swainson R, Wynne K, et al. Dissociable deficits in the decision-making cognition of chronic amphetamine abusers, opiate abusers, patients with focal damage to prefrontal cortex, and tryptophan-depleted normal volunteers: evidence for monoaminergic mechanisms. Neuropsychopharmacology 1999;20:322–39.

[12] Rachlin H, Logue AW, Gibbon J, Frankel M. Cognition and behavior in studies of choice. Psych Rev 1986;93:33–45.

[13] Robbins TW. The 5-choice serial reaction time task: behavioural pharmacology and functional neurochemistry. Psychopharmacology 2002;163:362–80.

[14] Dalley JW, Fryer TD, Brichard L, Robinson ESJ, Theobald DEH, Lääne K, et al. Nucleus accumbens D2/3 receptors predict trait impulsivity and cocaine reinforcement. Science 2007;315:1267–70.

[15] Diergaarde L, Pattij T, Poortvliet I, Hogenboom F, de Vries W, Schoffelmeer AN, et al. Impulsive choice and impulsive action predict vulnerability to distinct stages of nicotine seeking in rats. Biol Psychiatry 2008;63:301–8.

[16] Diergaarde L, Pattij T, Nawijn L, Schoffelmeer AN, De Vries TJ. Trait impulsivity predicts escalation of sucrose seeking and hypersensitivity to sucrose-associated stimuli. Behav Neurosci 2009; 123:794–803.

[17] Voon V, Irvine MA, Derbyshire K, Worbe Y, Lange I, Abbott S, et al. Measuring "waiting" impulsivity in substance addictions and binge eating disorder in a novel analogue of rodent serial reaction time task. Biol Psychiat 2013;75:148–55.

[18] Logan GD. On the ability to inhibit thought and action. A user's guide to the stop signal paradigm. In: Dagenbach D, Carr TH, editors. Inhibitory processes in attention, memory and language. San Diego, CA: Academic Press; 1994. p. 189–236.

[19] Eagle DM, Robbins TW. Inhibitory control in rats performing on the stop-signal reaction-time task: effects of lesions of the medial striatum and d-amphetamine. Behav Neurosci 2003;117: 1302–17.

[20] Cole BJ, Robbins TW. Amphetamine impairs the discriminative performance of rats with dorsal noradrenergic bundle lesions on a 5-choice serial reaction time task: new evidence for central dopaminergic-noradrenergic interactions. Psychopharmacology 1987;91:458–66.

[21] Christakou A, Robbins TW, Everitt BJ. Prefrontal cortical-ventral striatal interactions involved in affective modulation of attentional performance: implications for corticostriatal circuit function. J Neurosci 2004;24:773–80.

[22] Pothuizen HH, Jongen-Relo AL, Feldon J, Yee BK. Double dissociation of the effects of selective nucleus accumbens core and shell lesions on impulsive-choice behaviour and salience learning in rats. Eur J Neurosci 2005;22:2605–16.

[23] Murphy ER, Robinson ESJ, Theobald DEH, Dalley JW, Robbins TW. Contrasting effects of selective lesions of nucleus accumbens core or shell on inhibitory control and amphetamine-induced impulsive behaviour. Eur J Neurosci 2008;28:353–63.

[24] Sesia T, Temel Y, Lim LW, Blokland A, Steinbusch HW, Visser-Vandewalle V. Deep brain stimulation of the nucleus accumbens core and shell: opposite effects on impulsive action. Exp Neurol 2008;214:135–9.

[25] Dalley JW, Theobald DEH, Eagle DM, Passetti F, Robbins TW. Deficits in impulse control associated with tonically-elevated serotonergic function in rat prefrontal cortex. Neuropsychopharmacology 2002;26:716–28.

[26] Belin D, Mar AC, Dalley JW, Robbins TW, Everitt BJ. High impulsivity predicts the switch to compulsive cocaine-taking. Science 2008;320:1352–5.

[27] Volkow ND, Chang L, Wang GJ, Fowler JS, Ding YS, Sedler M, et al. Low level of brain dopamine D2 receptors in methamphetamine abusers: association with metabolism in the orbitofrontal cortex. Am J Psychiatry 2001;158:2015–21.

[28] Besson M, Pelloux Y, Dilleen R, Theobald DEH, Lyon A, Belin-Rauscent A, et al. Cocaine modulation of fronto-striatal expression of zif268, D2 and 5-HT2c receptors in high and low impulsive rats. Neuropsychopharmacology 2013;38(10):1963–73.

[29] Jupp B, Caprioli D, Saigal N, Reverte I, Shrestha S, Cumming P, et al. Dopaminergic and GABA-ergic markers of impulsivity in rats: evidence for anatomical localisation in ventral striatum and prefrontal cortex. Eur J Neurosci 2013;37:1519—28.

[30] Besson M, Belin D, McNamara R, Theobald DEH, Castel A, Beckett VL, et al. Dissociable control of impulsivity in rats by dopamine D2/3 receptors in the core and shell subregions of the nucleus accumbens. Neuropsychopharmacology 2010;35: 560—9.

[31] Caprioli D, Sawiak SJ, Merlo E, Theobald DE, Spoelder M, Jupp B, et al. Gamma aminobutyric acidergic and neuronal structural markers in the nucleus accumbens core underlie trait-like impulsive behaviour. Biol Psychiatry 2013;75:115—23.

[32] Caprioli D, Hong YT, Sawiak SJ, Ferrari V, Williamson DJ, Jupp B, et al. Baseline-dependent effects of cocaine pre-exposure on impulsivity and D(2/3) receptor availability in the rat striatum: possible relevance to the attention-deficit hyperactivity syndrome. Neuropsychopharmacology 2013;38(2013): 1460—71.

[33] Caprioli D, Jupp B, Hong YT, Sawiak SJ, Ferrari V, Wharton L, et al. Dissociable rate-dependent effects of oral methylphenidate on impulsivity and D2/3 receptor availability in the striatum. J Neurosci 2015;35:3747—55.

[34] Goto Y, Grace AA. The Yin and Yang of dopamine release: a new perspective. Neuropharmacology 2007;53:583—7.

[35] Robinson ESJ, Eagle DM, Mar AC, Bari A, Banerjee G, Jiang X, et al. Similar effects of the selective noradrenaline reuptake inhibitor atomoxetine on three distinct forms of impulsivity in the rat. Neuropsychopharmacology 2008;33:1028—37.

[36] Economidou D, Theobald DE, Robbins TW, Everitt BJ, Dalley JW. Norepinephrine and dopamine modulate impulsivity on the five-choice serial reaction time task through opponent actions in the shell and core sub-regions of the nucleus accumbens. Neuropsychopharmacology 2012;37:2057—66.

[37] Delfs JM, Zhu Y, Druhan JP, Aston-Jones GS. Origin of noradrenergic afferents to the shell subregion of the nucleus accumbens: anterograde and retrograde tract-tracing studies in the rat. Brain Res 1998;806:127—40.

[38] Winstanley CA, Chudasama Y, Dalley JW, Theobald DE, Glennon JC, Robbins TW. Intra-prefrontal 8-OH-DPAT and M100907 improve visuospatial attention and decrease impulsivity on the five-choice serial reaction time task in rats. Psychopharmacology 2003;167:304—14.

[39] Robinson ESJ, Dalley JW, Theobald DEH, Glennon JC, Pezze MA, Murphy ER, et al. Opposing roles for 5-HT2A and 5-HT2C receptors in the nucleus accumbens on inhibitory response control in the 5-choice serial reaction time task. Neuropsychopharmacology 2008;33:2398—406.

[40] Harrison AA, Everitt BJ, Robbins TW. Central 5-HT depletion enhances impulsive responding without affecting the accuracy of attentional performance: interactions with dopaminergic mechanisms. Psychopharmacology 1997;133:329—42.

[41] Winstanley CA, Theobald DEH, Dalley JW, Glennon JC, Robbins TW. 5-HT2A and 5-HT2C receptor antagonists have opposing effects on a measure of impulsivity: interactions with global 5-HT depletion. Psychopharmacology 2004;176:376—85.

[42] Sanchez-Roige S, Pena-Oliver Y, Stephens DN. Measuring impulsivity in mice: the five-choice serial reaction time task. Psychopharmacology 2012;219:253—70.

[43] Peña-Oliver Y, Buchman VL, Dalley JW, Robbins TW, Schumann G, Ripley TL, et al. Deletion of alpha-synuclein decreased impulsivity in mice. Genes Brain Behav 2012;11: 137—46.

[44] Donnelly NA, Holtzman T, Rich PD, Nevado-Holgado AJ, Fernando AB, Van Dijck G, et al. Oscillatory activity in the medial prefrontal cortex and nucleus accumbens correlates with impulsivity and reward outcome. PLoS One 2014;9:e111300.

[45] Donnelly NA, Paulsen O, Robbins TW, Dalley JW. Ramping single unit activity in the medial prefrontal cortex reflects the onset of waiting and not imminent impulsive actions. Eur J Neurosci 2015;41:1524—37.

[46] Buckholtz JW, Treadway MT, Cowan RL, Woodward ND, Li R, Ansari MS, et al. Dopaminergic network differences in human impulsivity. Science 2010;329:532.

[47] Worbe Y, Savulich G, Voon V, Fernandez-Egea E, Robbins TW. Serotonin depletion induces 'waiting impulsivity' on the human four-choice serial reaction time task: cross-species translational significance. Neuropsychopharmacology 2014;39:1519—26.

[48] Cardinal RN, Pennicott DR, Sugathapala CL, Robbins TW, Everitt BJ. Impulsive choice induced in rats by lesions of the nucleus accumbens core. Science 2001;292:2499—501.

[49] Bezzina G, Cheung TH, Asgari K, Hampson CL, Body S, Bradshaw CM, et al. Effects of quinolinic acid-induced lesions of the nucleus accumbens core on inter-temporal choice: a quantitative analysis. Psychopharmacology (Berl) 2007;195:71—84.

[50] Acheson A, Farra AM, Patak M, Hausknecht KA, Kieres AK, Choi S, et al. Nucleus accumbens lesions decrease sensitivity to rapid changes in the delay to reinforcement. Behav Brain Res 2006;173:217—28.

[51] Cardinal RN, Cheung THC. Nucleus accumbens core lesions retard instrumental learning and performance with delayed reinforcement in the rat. BMC Neurosci 2005;6:9.

[52] Da Costa Araujo S, Body S, Hampson CL, Langley RW, Deakin JFW, Anderson IM, et al. Effects of lesions of the nucleus accumbens core on inter-temporal choice: further observations with an adjusting-delay procedure. Behav Brain Res 2009;202: 272—7.

[53] Cardinal RN, Robbins TW, Everitt BJ. The effects of d-amphetamine, chlordiazepoxide, alpha-flupenthixol and behavioural manipulations on choice of signalled and unsignalled delayed reinforcement in rats. Psychopharmacology 2000;152:362—75.

[54] Taylor JR, Robbins TW. Enhanced behavioural control by conditioned reinforcers following microinjections of d-amphetamine into the nucleus accumbens. Psychopharmacology 1984;84: 405—12.

[55] Dalley JW, Roiser J. Dopamine, serotonin, impulsivity. Neuroscience 2012;215:42—58.

[56] Winstanley CA, Dalley JW, Theobald DEH, Robbins TW. Global 5-HT depletion attenuates the ability of amphetamine to decrease impulsive choice on a delay-discounting task in rats. Psychopharmacology 2003;170:320—31.

[57] Winstanley CA, Theobald DEH, Dalley JW, Robbins TW. Interactions between serotonin and dopamine in the control of impulsive choice in rats: therapeutic implications for impulse control disorders. Neuropsychopharmacology 2005;30:669—82.

[58] Mobini S, Chiang TJ, Al-Ruwaitea AS, Ho MY, Bradshaw CM, Szabadi E. Effect of central 5-hydroxytryptamine depletion on inter-temporal choice: a quantitative analysis. Psychopharmacology 2000;149:313—8.

[59] Winstanley CA, Theobald DEH, Dalley JW, Cardinal RN, Robbins TW. Double dissociation between serotonergic and dopaminergic modulation of medial prefrontal and orbitofrontal cortex during a test of impulsive choice. Cereb Cortex 2006;16: 106—14.

[60] Winstanley CA, Theobald DEH, Cardinal RN, Robbins TW. Contrasting roles of basolateral amygdala and orbitofrontal cortex in impulsive choice. J Neurosci 2004;24:4718—22.

[61] Cheung TH, Cardinal RN. Hippocampal lesions facilitate instrumental learning with delayed reinforcement but induce impulsive choice in rats. BMC Neurosci 2005;6:36.

[62] Schoenbaum G, Setlow B, Ramus SJ. A systems approach to orbitofrontal cortex function: recordings in rat orbitofrontal cortex reveal interactions with different learning systems. Behav Brain Res 2003;146:19—29.

[63] Mobini S, Body S, Ho MY, Bradshaw CM, Szabadi E, Deakin JF, et al. Effects of lesions of the orbitofrontal cortex on sensitivity to delayed and probabilistic reinforcement. Psychopharmacology (Berl) 2002;160:290−8.

[64] Kheramin S, Body S, Ho MY, Velazquez-Martinez DN, Bradshaw CM, Szabadi E, et al. Effects of quinolinic acid-induced lesions of the orbital prefrontal cortex on inter-temporal choice: a quantitative analysis. Psychopharmacology 2002;165:9−17.

[65] Kheramin S, Body S, Ho MY, Velazquez-Martinez DN, Bradshaw CM, Szabadi E, et al. Effects of orbital prefrontal cortex dopamine depletion on inter-temporal choice: a quantitative analysis. Psychopharmacology 2004;175:206−14.

[66] Rudebeck PH, Walton ME, Smyth AN, Bannerman DM, Rushworth MF. Separate neural pathways process different decision costs. Nat Neurosci 2006;9:1161−8.

[67] Zeeb FD, Floresco SB, Winstanley CA. Contributions of the orbitofrontal cortex to impulsive choice: interactions with basal levels of impulsivity, dopamine signalling, and reward-related cues. Psychopharmacology (Berl) 2010;211:87−98.

[68] Mar AC, Walker AL, Theobald DEH, Eagle DM, Robbins TW. Dissociable effects of lesions to orbitofrontal cortex subregions on impulsive choice in the rat. J Neurosci 2011;31:6398−404.

[69] McClure SM, Ericson KM, Laibson DI, Loewenstein G, Cohen JD. Time discounting for primary rewards. J Neurosci 2007;27: 5796−804.

[70] Sellitto M, Ciaramelli E, di Pellegrino G. Myopic discounting of future rewards after medial orbitofrontal damage in humans. J Neurosci 2010;30:16429−36.

[71] Fiorillo CD, Tobler PN, Schultz W. Discrete coding of reward probability and uncertainty by dopamine neurons. Science 2003; 299:1898−902.

[72] Kuhnen CM, Knutson B. The neural basis of financial risk taking. Neuron 2005;47:763−70.

[73] Cardinal RN, Howes NJ. Effects of lesions of the nucleus accumbens core on choice between small certain rewards and large uncertain rewards in rats. BMC Neurosci 2005;6:37.

[74] Stopper CM, Floresco SB. Contributions of the nucleus accumbens and its subregions to different aspects of risk-based decision making. Cogn Affect Behav Neurosci 2011;11:97−112.

[75] Stopper CM, Khayambashi S, Floresco SB. Receptor-specific modulation of risk-based decision-making by nucleus accumbens dopamine. Neuropsychopharmacology 2013;38:715−28.

[76] St Onge JR, Abhari H, Floresco SB. Dissociable contributions by prefrontal D1 and D2 receptors to risk-based decision making. J Neurosci 2011;31:8625−33.

[77] St Onge JR, Ahn S, Phillips AG, Floresco SB. Dynamic fluctuations in dopamine efflux in the prefrontal cortex and nucleus accumbens during risk-based decision making. J Neurosci 2012;32:16880−91.

[78] Stopper CM, Green EB, Floresco SB. Selective involvement by the medial orbitofrontal cortex in biasing risky, but not impulsive, choice. Cereb Cortex 2014;24:154−62.

[79] St Onge JR, Floresco SB. Prefrontal cortical contribution to risk-based decision making. Cereb Cortex 2010;20:1816−28.

[80] St Onge JR, Stopper CM, Zahm DS, Floresco SB. Separate prefrontal-subcortical circuits mediate different components of risk-based decision making. J Neurosci 2012;32:2886−99.

[81] Stopper CM, Floresco SB. What's better for me? Fundamental role for lateral habenula in promoting subjective decision biases. Nature Neurosci 2014;17:33−5.

[82] Simon NW, Montgomery KS, Beas BS, Mitchell MR, LaSarge CL, Mendez IA, et al. Dopaminergic modulation of risky decision-making. J Neurosci 2014;31:17460−70.

[83] Mitchell MR, Weiss VG, Beas BS, Morgan D, Bizon JL, Setlow B. Adolescent risk-taking, cocaine self-administration, and striatal dopamine signaling. Neuropsychopharmacology 2014;39: 955−62.

[84] Orsini CA, Trotta RT, Bizon JL, Setlow B, et al. Dissociable roles for the basolateral amygdala and orbitofrontal cortex in decision-making under risk of punishment. J Neurosci 2015;35:1368−79.

[85] Zeeb FD, Robbins TW, Winstanley CA. Serotonergic and dopaminergic modulation of gambling behavior as assessed using a novel rat gambling task. Neuropsychopharmacology 2009;34: 2329−43.

[86] Barrus MM, Hosking JG, Zeeb FD, Tremblay M, Winstanley CA. Disadvantageous decision-making on a rodent gambling task is associated with increased motor impulsivity in a population of male rats. J Psychiatry Neurosci 2015;40:108−17.

[87] Eagle DM, Bari A, Robbins TW. The neuropsychopharmacology of action inhibition: cross-species translation of the stop-signal and go/no-go tasks. Psychopharmacology 2008;199:439−56.

[88] Aron AR, Fletcher PC, Bullmore ET, Sahakian BJ, Robbins TW. Stop-signal inhibition disrupted by damage to right inferior frontal gyrus in humans. Nat Neurosci 2003;6:115−6.

[89] Aron AR, Poldrack R. Cortical and subcortical contributions to stop signal response inhibition: role of the subthalamic nucleus. J Neurosci 2006;26:2424−33.

[90] Eagle DM, Baunez C, Hutcheson DM, Lehmann O, Shah AP, Robbins TW. Stop-signal reaction-time task performance: role of prefrontal cortex and subthalamic nucleus. Cereb Cortex 2008; 18:178−88.

[91] Eagle DM, Robbins TW. Lesions of the medial prefrontal cortex or nucleus accumbens core do not impair inhibitory control in rats performing a stop-signal reaction time task. Behav Brain Res 2003;146:131−44.

[92] Eagle DM, Wong JC, Allan ME, Mar AC, Theobald DEH, Robbins TW. Contrasting roles for dopamine D1 and D2 receptor subtypes in the dorsomedial striatum but not the nucleus accumbens core during behavioural inhibition in the stop-signal task in rats. J Neurosci 2011;31:7349−56.

[93] Bari A, Mar AC, Theobald DEH, Elands SA, Oganya KC, Eagle DM, et al. Prefrontal and monoaminergic contributions to top-signal performance in rats. J Neurosci 2011;31(2011):9254−63.

[94] Eagle DM, Lehmann O, Theobald DEH, Peña Y, Zakaria R, Ghosh R, et al. Serotonin depletion impairs waiting but not stop-signal reaction time in rats: implications for theories of the role of the 5-HT in behavioural inhibition. Neuropsychopharmacology 2009;34:1311−21.

[95] Isles AR, Humby T, Walters E, Wilkinson LS. Common genetic effects on variation in impulsivity and activity in mice. J Neurosci 2004;24:6733−40.

[96] Dent CL, Isles AR, Humby T. Measuring risk-taking in mice; balancing the risk against reward and danger. Eur J Neurosci 2014;39:520−30.

[97] Caprioli D, Sawiak SJ, Merlo E, Theobald DE, Spoelder M, Jupp B, et al. Gamma aminobutyric acidergic and neuronal structural markers in the nucleus accumbens core underlie trait-like impulsive behavior. Biol Psychiatry 2014;75:115−23.

8

Prefrontal Cortex in Decision-Making: The Perception–Action Cycle

J.M. Fuster

University of California Los Angeles, Los Angeles, CA, United States

Abstract

The neural mechanisms of decision-making are understandable only in the structural and dynamic context of the perception–action (PA) cycle. The PA cycle is the biocybernetic processing of information that adapts the organism to its environment. That circular processing involves a variety of neural structures at several hierarchical levels, though with close functional interactions between them. At its lowest level, the PA cycle is largely reflex and automatic, and involves the vegetative and visceral structures of the hypothalamus and the autonomic nervous system. At intermediate levels, the cycle involves limbic structures supporting its emotional and value-assessing mechanisms. At the cortical level, under the commanding role of the prefrontal cortex, the PA cycle incorporates prefrontal cognitive components. The posterior cortex contributes to decision-making mainly information from perceptual memory and knowledge; the frontal cortex contributes mainly executive memory and knowledge. Its prefrontal sector contributes to decision-making predictive and preadaptive control through its top–down executive functions—especially attentional set, working memory, monitoring, and inhibitory control.

Of all the executive functions of the frontal lobe, decision-making is arguably the most valued, for it implicates the freedom of choice to adapt to the environment for self-benefit or the benefit of others. For this reason, in dynamic terms, decision-making can be considered a *preadaptive* function. Indeed, like other executive frontal-lobe functions—planning, set, working memory, inhibitory control, monitoring, and error prevention—decision-making has an essential future perspective; it is prospective by definition, because its objectives are in the future, however immediate or distant that future may be. Decision-making is an integral part of the predictive and preadaptive roles of the prefrontal cortex in behavior and in language, as reviewed by Fuster and Bressler [1].

In addition, like all other executive functions, decision-making is, in neurobiological terms dependent on the cortical and subcortical structures with which the prefrontal cortex is connected. All these structures constitute the neural substrate of the perception–action (PA) cycle, the circular biocybernetic processing of information—with feed-forward and feedback—that governs the relationships of the organism with its environment. The prefrontal cortex plays an integrative, coordinating, and commanding role in the PA cycle. To understand that role and the physiology of decision-making, it is necessary to view in the light of that cycle the evidence that has been acquired about the involvement of assorted neural structures in decision-making. That evidence makes it readily obvious that most of those structures participate critically in the perception or action aspects of the PA cycle in the broadest sense of the two words, "perception" and "action." In that sense, perception would include (1) the memory and knowledge by which sensation is converted into perception [2,3], (2) the biological drives of the organism ("biodrives"), and (3) the outcome, expected or real, rewarded or unrewarded, of present and previous decisions. At low levels, action in the PA cycle is automatic and reducible to reaction (e.g., a defensive reflex). At higher levels, action includes the enactment of any decision driven by, and giving rise to, probabilistic prediction from experience, in other words, Bayesian prediction of success or failure [4].

This chapter presents a mesoscopic view of the role of the prefrontal cortex in decision-making, accompanied by brief accounts of some data that not only substantiate that view but also provide insights into the neural mechanisms of decision-making. The chapter is divided into five parts to deal with the following subjects: (1) the PA cycle; (2) inputs to the PA cycle; (3) prediction and preparation toward decision; (4) execution of decision; and (5) feedback from decision (closure of the cycle).

THE PERCEPTION–ACTION CYCLE

The PA cycle is deeply rooted in biology. Essentially, it is the cybernetic processing of information that regulates the environmental adaptation of the organism. At its biological base, the cycle consists of the homeostatic mechanisms that maintain equilibrium in the internal milieu. Those mechanisms are involuntary and involve mainly the autonomic nervous system and the hypothalamus. In addition, at that low level, and regulated by the same structures, the cycle includes the sensory and motor interactions founded on biological drive that ensure self-preservation and the protection and propagation of the species. In higher organisms those structural components of the lower cycle include parts of the limbic brain and basal ganglia, notably the amygdala, the orbitofrontal cortex, the insula, and the nucleus accumbens. These lower parts of the cycle, interacting with the environment, will constitute the emotional component of the PA cycle, which in all mammalian species supports the biological evaluation of rewards and values in the external world, as well as the behavioral acts to attain them.

Biologist Jakob von Uexküll [5] was the first to remark on the cybernetic nature of what would become the PA cycle, though he used neither of these terms to describe it; instead he used naturalistic expressions that now appear somewhat archaic (Fig. 8.1). An environmental stimulus leads to an adaptive motor reaction, which induces a change in the environment, which in turn leads to further stimulus and reaction, and so on. At every step toward a goal, the appropriate action is guided by feedback from the environment. In a lower animal like the sea anemone, for example, the environmental changes take place in the chemical composition of the surrounding water, and the adaptive reactions consist of predatory movements of the tentacles to find the prey and to attract it to the mouth of the animal.

In the mammal, Uexküll notes, a new circuit appears inside the brain, made of connections (*red arrow* in lower Fig. 8.1) that flow from the action organ to the sensor ("mark") organ; those connections carry an internal feedback from effectors to sensors. At the highest level of the PA cycle, in the primate brain, that feedback will be carried by long fiber connections from frontal (executive) cortex to posterior (sensory) cortex. That will be a kind of "anticipated feedback," operating on the PA cycle *before* the action, and preparing the sensory and motor apparatus for that action and its expected consequences. It will contain "efferent copies" of the imminent motor act as well as the so-called "corollary

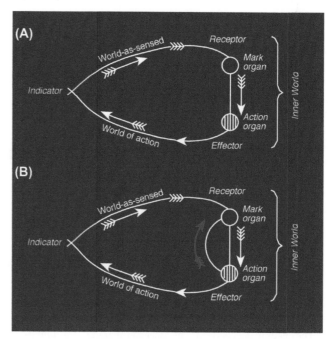

FIGURE 8.1 Uexküll's diagram of the organism's relation to its environment. The following are excerpts from his description of the diagram. (A) "The inner world is divided into two parts; one, which receives the impressions, is turned toward the world-as-sensed, and the other, which distributes the effects, is turned toward the world of action." (B) "In the highest animals, however, the creature's own action-rule penetrates further into the world-as-sensed, and there assumes direction and control… A new circle is introduced within the animal's own central organ, for the support of the external function-circle, and this connects the action organ with the mark-organ (*red arrow*)." *From Uexküll J. Theoretical biology. New York: Harcourt, Brace & Co; 1926. pp.155–157.*

discharge," which Teuber [6] construed as one of the fundamental preparatory functions of the prefrontal cortex. In this cortex, however, the function of its anticipatory output to the posterior cortex transcends the corollary discharge of Teuber, and supports the prospective cognitive functions that will construct the action. Those functions in the aggregate constitute the so-called cognitive control by the prefrontal cortex, which includes top–down attentional set and working memory. Based on value, risk, reward, and utility, both these functions, among others, will select from the neural substrates of perception, memory, and biodrive, the information that will go into any plan and any decision. The same functions will select from the neural substrates of action the information that will guide the execution of the plan or the decision.

Before approaching the dynamics of the PA cycle in decision-making, a brief description is needed of the cortical cognitive networks, which I call *cognits* and which harbor the memory and the knowledge that an individual has acquired through life and that are to enter decision-making [7]. Cognits consist of widely distributed neuronal networks of association cortex that extend beyond cytoarchitectonic areal boundaries; they are made of synaptic connections and fibers between neuronal groups representing discrete features of sensation and/or movement having co-occurred more or less simultaneously in life experience. It is believed that, by mechanisms still obscure, cortical cognits are made, and their synapses modulated, by the hippocampus. Because of the variety and multiplicity of potentially co-occurring events in one's life, one neuron or group of neurons practically anywhere in the cortex can be part of many cognits, and thus many memories or items of knowledge—the latter also called semantic memory. Cognits are hierarchically organized, from the most simple and concrete in sensory and motor areas up to the most abstract in higher association areas. Perceptual cognits are mainly distributed in the posterior cortex (parietal, temporal, and occipital), whereas executive cognits are mainly distributed in the frontal cortex; the dorsolateral prefrontal cortex is the highest level of the executive hierarchy of cognits.

Fig. 8.2A depicts schematically the cortical connective substrate of the PA cycle in the primate. In the diagram, the major connections of the cycle (all of them anatomically substantiated) are represented by thicker arrows, which run clockwise. Feedback connections are represented by thinner arrows, which run counterclockwise.

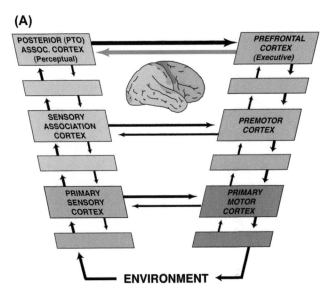

FIGURE 8.2A Cortical circuitry of the perception—action cycle, all bidirectional and anatomically substantiated in the primate. Unlabeled rectangles represent cortical areas intermediate between labeled areas or subareas of the same. The perceptual cognit hierarchies are on the left of the image, the executive hierarchy on the right. Main feedforward processing takes place through *thick arrows*, feedback processing through *thin arrows*; one exception, in the primate, is the *thick red arrow* from frontal to posterior cortex (the highest expression of Uexküll's internal feedback that carries cognitive-control signals, as in decision-making). *PTO*, parietal—temporal—occipital (cortex).

An exception is the *thick red, straight, arrow* that represents the internal flow of—anticipatory—cognitive control from the prefrontal to the posterior cortex (the mammalian equivalent, at high levels, of the counter-flowing *arrow* in Uexküll's schema). That "counterflow" serves all the prospective cognitive prefrontal functions—notably attentional set, working memory, and decision-making.

INPUTS TO THE PERCEPTION—ACTION CYCLE IN DECISION-MAKING

It is impossible to dissociate the cerebral dynamics of a decision from its antecedents and the cerebral inputs that convey them. Even present external sensory stimuli will be interpreted by past history and converted into perceptual cognits before they gain access to decision. Even binary decisions, such as those commonly investigated by behavioral physiologists and neuroeconomists will be determined by the concomitant confluence of inputs from myriad sources inside and outside the brain, many of them imponderable in the experimental setting. Consider briefly the diagram in Fig. 8.2B, schematically illustrating the principal categories of those sources, some with a cortical and others a subcortical substrate. The schema is intended to highlight very generally the multifactorial background to decision-making.

Not only are many the possible sources of information that go into a decision, but also they interact with one another in many ways to produce that decision (which in the schema is symbolized by a vector). The process toward decision converges from the past and, through the present, diverges toward many potential actions in the future, when the decision will be actually made and its consequences will occur. The PA cycle, symbolized over the schema with an image of the human brain, contains feedback at all levels. The diagram obviously does not take into account the computational complexities of the interactions between input factors, including the potential synergies and incompatibilities between them.

Subcortical inputs from the internal milieu and the realm of emotion arrive in the prefrontal cortex from limbic structures involved in the evaluation and motivational significance of external and internal stimuli. These inputs enter, for the most part, the orbital prefrontal cortex. This cortex, together with those subcortical structures, and the feedback into them from the internal and external environments, form the emotional PA cycle, the parallel counterpart to the cognitive PA cycle running through cortical structures that I have already described. The two cycles meet and interact with each other, whereby emotional information modulates cognitive information and vice versa. Strictly speaking, from

FIGURE 8.2B Schematic diagram of various categories of cognitive, emotional, and biological information converging on the present for the making of a decision with prospective consequences. *A*, action; *P*, perception.

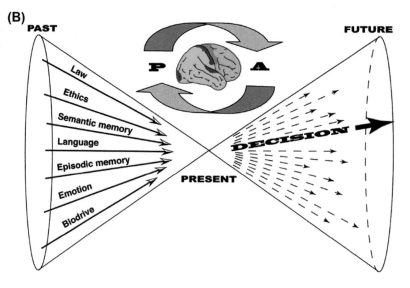

the point of view of systems neuroscience, the two cycles could be considered part of one and the same general PA cycle by which the organism adapts to the environment. Fig. 8.3 illustrates the major inputs to and outputs from the prefrontal cortex in both cognitive and emotional compartments of that cycle.

A large number of studies, some of them reviewed in other chapters of this volume, implicate structures of the emotional PA cycle component in decision-making. Briefly, there is now abundant evidence from lesion, electrophysiological, and imaging studies—that several of these structures, notably the amygdala, the insula,

the ventral striatum, the ventromedial prefrontal cortex, and the anterior cingulate cortex, are critically involved in the assessment of the value and motivational significance of present or expected stimuli or events [8–18]. A review by Saga and Tremblay [19] points to the critical role of the ventral striatum, clearly a component of the emotional cycle, in the control of aversive/appetitive behavior by the negative or rewarding feedback from decisions.

The emotional PA cycle has two major categories of output to decision. One is direct, reflexive, and automatic. It flows from those limbic and paralimbic

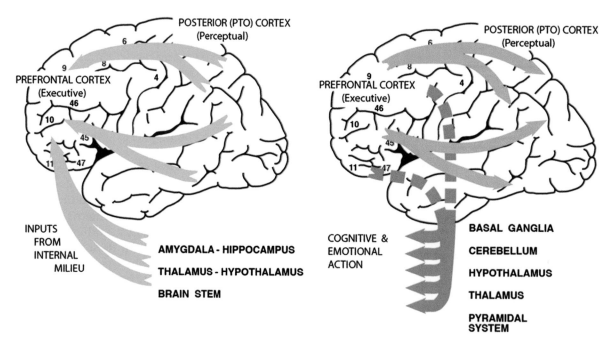

FIGURE 8.3 Major cerebral sources of input (left) and targets of output (right) of the cognitive and emotional perception–action cycles. Prefrontal areas are designated by Brodmann's numeration. *PTO*, parietal–temporal–occipital (cortex).

I. ANIMAL STUDIES ON REWARDS, PUNISHMENTS, AND DECISION-MAKING prefrontal structures into the hypothalamus and the autonomic nervous system. It will modulate homeostatic mechanisms that tend to maintain the internal milieu in equilibrium. Bertalanffy [20] invoked some of those mechanisms to substantiate his seminal theory of open adaptive systems.

The other major category of outputs from limbic and paralimbic prefrontal structures flows through orbital into lateral prefrontal cortex (review in Ref. [21]) and, from there, to the neocortex at large. The orbitofrontal cortex itself—which includes parts of the ventromedial and ventrolateral prefrontal cortex—has long been associated not only with the encoding of reward signals from many sources, but also with the outputs to decisions based on those signals [22—30]. In this manner, the affective and motivational information collected by the orbitofrontal cortex flows into cognitive and motor systems. It is by way of outputs to the lateral prefrontal and posterior cortex that the emotional PA cycle will reach the environment and at the same time modulate the operations of the cognitive PA cycle. Both cycles, cognitive and emotional, will continue to govern behavior in parallel and essentially by cybernetic loops of feed-forward, feedback, monitoring, and error correction.

PREDICTION AND PREPARATION TOWARD DECISION

Confirming neuropsychological evidence from lesion studies, imaging studies provide ample evidence that the anterior prefrontal cortex is involved in the formulation of plans of future behavior [31—36]. Some of these studies, for example, Spreng et al. [37], highlight the important fact that the same prefrontal area is activated by recall of a memory as by mentally including that memory in the imagination of a future plan. This is in line with the concept that a cortical network of long-term memory (a cognit) can be activated for its utilization in a future plan or decision—"memory of the future" [38]. Another important finding of that research [39] is that the imagining of a pleasant plan lights up limbic areas, notably the amygdala, that are involved in the signaling of reward or punishment. This finding points to the intersection, indeed interaction, between the cognitive and the emotional PA cycles that I have mentioned in the previous section.

To weigh onto decision-making, sensory inputs to the PA cycle arrive in executive frontal structures having passed two major "filters" or complementary selection processes (Fig. 8.2A): (1) one is the cortical processing of the sensory information of those inputs in posterior cortical—perceptual—hierarchies through cognits of perception and memory, and (2) the other is the cognitive-control functions of the prefrontal cortex

upon that sensory information in the posterior cortex. The two processes are simultaneous and intimately tied together in tandem. Here, we concern ourselves principally with the second, the top—down cognitive control.

The prefrontal cognitive control of decision-making has been effectively investigated by electrophysiological and functional imaging methods. It is not easy, however, to disambiguate the neural correlates of the decision control function from the correlates of other executive cognitive functions that precede and support choice. A useful procedure to that end is, by behavioral manipulation, to parse in time decision-making and its cognitive predecessors, notably attentional set and working memory. This can best be achieved experimentally by submitting the subject to performance of a working-memory task, in which sensory information must be attended to, retained for a period (delay) in working memory, and decided upon with an appropriate action. What follows pertains to the prefrontal role in these functions as revealed in working-memory tests.

Except for the automatic and immediate correction of so-called "prediction error" [40—42], attentional set is ordinarily the first and most rapid of the anticipatory cognitive functions of the prefrontal cortex in preparation for decision. Essentially, attentional set consists in the priming of the central and peripheral components of sensory and motor systems for an expected percept or motor action. It is therefore top—down attention and, like attention in all its forms, it has two complementary components, one inclusionary and the other exclusionary: the first can be characterized as focused selection and the second as inhibitory control. Both are exerted from the lateral prefrontal cortex. The first or inclusionary one modulates in the posterior cortex what has been called the "biased competition" [43] or "mixed selectivity" [44] between stimuli, whereas the exclusionary component of attention exerts inhibitory control of extraneous inputs and outputs through the inferior-lateral orbitofrontal cortex [45].

By neuroimaging, signals can be recorded in prefrontal and posterior cortex, indicating that perceptual attentional set occurs in these regions already in anticipatory expectation of a stimulus that will bear on a decision [46,47]. As we see next, in the delay-of-memory tasks, when the focus of attention shifts to working memory and preparatory motor set, electrical signals can be recorded from the prefrontal cortex that are indicative of that shift.

Working memory, such as attentional set, is by definition a prospective and predictive cognitive function: it is attention focused on the internal representation of a stimulus-activated item of memory or perception *for a prospective act* to solve a problem or to perform a task [48]. My colleagues and I have investigated the neuronal activity in the prefrontal cortex of the monkey during the working memory of cross-modal associations, whereby a stimulus of one sensory modality has to be retained for later associative matching to a stimulus of a different modality [49–51]. Before our recordings, the animal has established those associations in long-term memory by the learning of a cross-modal memory task. Thus, in one study, a cross-modal association was preestablished between auditory and visual stimuli [49]. Under those conditions, neurons were found in prefrontal cortex that not only reacted similarly to two associated stimuli of different modality, audition, and vision, but also retained that association between the two across the delay period of the working-memory task (Fig. 8.4).

In a working-memory task in which a monkey can predict the spatial direction of the motor response that the memorandum (a color) will call for, it is possible to discern in the lateral prefrontal and parietal cortices the electrical correlates of motor attentional set [52]. During the delay period, cells that are attuned to that motor response show increasing firing in advance of the response. That increase, most likely related to the priming of the motor apparatus, is proportional to the probability with which the monkey can correctly predict the direction of the response. This finding, in the context of the behavioral task (Fig. 8.5), can be appropriately characterized as a neural expression of Bayesian probability.

Reentry of neural excitation is part of the essence of the PA cycle at the mesoscopic level in preparation for decision, but now there is additional evidence that the reentry mechanism plays a fundamental role in working memory also at the microscopic level. For a long time, reentry has been a key feature of the most plausible

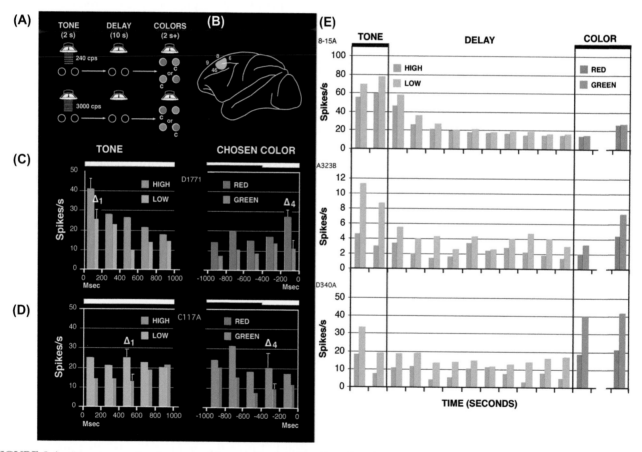

FIGURE 8.4 Cross-temporal and cross-modal association by cells of lateral prefrontal cortex. (A) Audiovisual working-memory task. Trial sequence: (1) 2-s tone, high pitch (3000 cps) or low pitch (240 cps); (2) 10-s delay; (3) two colors presented simultaneously, red and green; (4) monkey chooses red if tone before the delay was high pitch, green if it was low pitch. (B) Brain diagram, numbers indicating Brodmann's cytoarchitectonic areas; in *blue*, prefrontal region from which tone- and color-discriminating cells were recorded. (C and D) Firing frequency histograms of two cells at the times of the tone and of the matching color choice, one cell (C) attuned to the high-tone-red association and the other cell (D) to the low-tone-green association. The histograms are time-locked (0) with the tone onset and the manual color choice. Note the correlation of selective cell reactions to tones and colors in accord with the task rule. (E) Firing frequency histograms of three cells selective for low-tone-green. Note that throughout the memory period (10 s) after a low tone, the cells maintain their selective firing in anticipation of the choice of green. *From Fuster JM, Bodner M, Kroger JK. Cross-modal and cross-temporal association in neurons of frontal cortex, Nature 2000;405:347–51 with permission.*

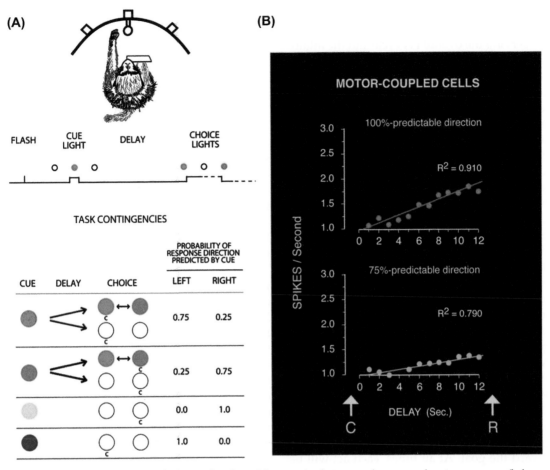

FIGURE 8.5 (A) Working-memory task with temporal and spatial separation between color cues and motor responses (below, contingencies between them). The monkey faces a panel with three stimulus—response buttons above a pedal, where the operant hand rests at all times except to respond to stimuli. A trial begins with a brief color display in the central button (the cue for the trial). A delay follows, at the end of which the two lateral buttons simultaneously turn either one red and the other green or both white. If they turn red and green, the animal must choose the color of the cue (*red* or *green*). If they turn white, the animal must choose left for red cue, right for green cue, right for yellow cue, or left for blue cue. Therefore, during the delay, if the cue has been yellow or blue, the animal can predict with certainty the response direction (right or left, respectively). If the cue has been red or green, the response side can be predicted with only 75% probability of success (left if red, right if green). Cue color and position of correct choice are changed at random from trial to trial. *c*, correct choice. (B) Accelerating activation of motor-coupled neurons during the delay. These cells were presumably involved in priming the motor system for the approaching motor response. Their accelerated firing was in accord with the predictability of that response: the steepness of the firing ramp during the delay was greater when that predictability was 100% than when it was 75%. *C*, cue; *R*, response; R^2, square of correlation coefficient. *From Quintana J, Fuster JM. From perception to action: temporal integrative functions of prefrontal and parietal neurons, Cereb Cortex 1999;9:213–21 with permission.*

computational models of working memory [53—59]. The fact that working memory is maintained by reentry between prefrontal and posterior cortices is supported by neurobehavioral studies in which one cortex is cooled (depressed) and unit activity is recorded from the other. Under these conditions, working-memory performance deteriorates, while neuronal activity in the noncooled cortex is depressed or otherwise altered [60—62].

The role of cross-cortical reentry in working memory is highlighted also by functional neuroimaging evidence. This evidence demonstrates that the maintenance of a sensory stimulus in memory, for ulterior action, activates simultaneously two cortical regions: one in the lateral prefrontal cortex and the other in a sensory area of posterior cortex corresponding to the modality of the memorandum (graphic metaanalysis in Fuster [63]); Ref. [64].

EXECUTION OF DECISION

Barring the execution of routine actions that are part of habit or overlearned sequences, the execution of a series of goal-directed actions, in the face of conflicting motives, risk, or uncertain outcome, is a continuous process of temporal integration under the cognitive control of the prefrontal cortex. That process can be broken down into a series of steps of mediated cross-temporal contingency, each of which can be construed as a subsidiary PA cycle serving the overall PA cycle and the attainment of the ultimate goal. Every step in that process requires the seriatim or simultaneous intervention of various executive prefrontal functions, notably attentional set and working memory, all of which in turn require the cooperation of prefrontal cortex with other cortical and subcortical structures. If the cross-temporal contingencies and the decision are relatively simple, neuroimaging studies point to a cascade of neural processing from anterior or polar prefrontal cortex toward the motor cortex and the basal ganglia [33,65–69].

If the decision is to be implemented by a more complex goal-directed sequence of acts, the cascading process recruits a succession of interleaved PA cycles of varying temporal span and complexity in terms of cognitive as well as contextual content. Some of the cycles will be unconscious and automatic; others will be conscious and novel. The mediation of multiple temporal contingencies toward a goal must consist of a series of short PA cycles nested within longer ones, all successively activating executive cognits of commensurate complexity.

Unless it is simple and automatic, each cycle will in turn recruit the executive prefrontal functions that will mediate the cross-temporal contingencies for continued progress toward the goal. At all times, each act will be modulated by the effects of previous acts on the environment and on the self, the latter to include the emotional and affective connotations of those effects entering the PA cycle through its limbic inputs.

At the lowest level of the executive apparatus, in the pyramidal system, a mechanism has been postulated that would be based on the prediction of the proprioceptive input resulting from movement. Based on fine structural and physiological features of motor neurons and their connectivity, Adams et al. [42] postulate that the motor system sends to muscles a "predictive code" whose function is to modulate action *in anticipation* of its predicted feedback. In effect, the pyramidal system would be dedicated to precorrection of error. Movements would then be produced essentially to meet expected feedback and thus to minimize predicted error. Friston and his coworkers [70,71] have generalized that local principle to the entire action systems of the brain. Clark [72], following similar thinking, speaks of the entire brain as a "predictive machine" that sends to the periphery "action-oriented-predictive coding." Its objective, again, would be to minimize error.

These theoretical positions are, of course, germane to the preadaptive concept of the prefrontal cortex, and therefore to its preparatory cognitive functions, as discussed in the previous section. But, given a more extended timeline, those ideas are also germane to the monitoring function of the prefrontal cortex, *after* the action, which is the topic of the next section.

FEEDBACK FROM DECISION: CLOSURE OF THE PERCEPTION–ACTION CYCLE

Once a decision has been executed in the form of a specific action or choice, inputs flow into the prefrontal cortex with information about the outcome of the action on the environment and on the self. Those inputs come from sensory receptors, from other regions of the cortex, and from subcortical structures, especially those in the limbic system that assess the value of an obtained reward or the adverse consequences of the action taken. The aggregate of that feedback into the prefrontal cortex from both cognitive and emotional outposts of the PA cycle can be characterized as so-called "monitoring." In essence, monitoring in that sense serves the verification of predicted outcome and the avoidance of prediction error, now and in the future. Hence, the prefrontal monitoring after decision can be considered a continuation of the preadaptive, predictive functions, before decision, which we have reviewed above. In essence, monitoring is the matching of results to expectations. With that matching the PA cycle is closed. Not surprisingly, as the bulk of the literature shows, the matching of results to expectations activates the same prefrontal areas that receive cognitive and emotional inputs to the decision process as well as those that receive outputs from it. Those structures close the PA cycle in cognition as well as in affect and emotion.

Accordingly, the orbitofrontal cortex is activated as a consequence of a rewarded decision as well as its expectation [24–26,73,74], or its absence when predicted [14,76–78]. The same is true for the insula [75,79–81], a paralimbic frontal structure in the emotional PA cycle and the recipient of precursor inputs to decision as well as outputs from it. The insula, like the ventral striatum [19], is activated not only by reward but also by aversive emotions, such as anxiety, disappointment, and regret.

Probably no prefrontal area is as clearly at the center of the monitoring circuits of decision-making as the anterior cingulate cortex. This part of the cortex integrates feed-forward loops that receive inputs to decision-making as well as feedback loops that collect neural influences from the outcome of decision. Thus, in the decision process the anterior cingulate cortex serves to monitor inputs to and outputs from the PA cycle. Some of the outputs are involved in the construction of action to avoid future error in both cognitive and emotional spheres.

The anterior cingulate cortex has been consistently shown to be a decision outcome predictor, whether that outcome is to be favorable or unfavorable, and relatedly also a predictor of possible conflict [82–93]. For these reasons, and on the basis of the results of additional experiments, some investigators attribute to the anterior cingulate cortex a role in the cognitive control of prospective action, at the transition from uncertainty or fear of failure to effective preventive action [79,94–99].

Shackman et al. [100], in a comprehensive review of cingulate cortices, reach the conclusion that the anterior of those cortices is critically situated at the interface between limbic structures and the lateral frontal cortex. Therefore its role would be critical, not only to monitor the emotional and cognitive outcomes of a decision, but also to integrate the information from both emotional and cognitive sectors, to integrate a correct subsequent decision. That integrated information closes the PA cycle by flowing from anterior cingulate into motor, premotor, and lateral prefrontal cortices through anatomically identified pathways [101].

References

[1] Fuster JM, Bressler SL. Past makes future: role of pFC in prediction. J Cogn Neurosci 2015;27:639–54.

[2] von Helmholtz H. Helmholtz's treatise on physiological optics [Translated from German by J.P.C. Southall]. Menasha (WI): The Optical Society of America; 1925 [G. Banta].

[3] Hayek FA. The sensory order. Chicago: University of Chicago Press; 1952.

[4] Jaynes ET. Bayesian methods: general background. In: Maximum-entropy and Bayesian methods in applied statistics. Cambridge: Cambridge University Press; 1986.

[5] Uexküll J. Theoretical biology. New York: Harcourt, Brace & Co.; 1926.

[6] Teuber HL. Unity and diversity of frontal lobe functions. Acta Neurobiol Exp 1972;32:625–56.

[7] Fuster JM. Cortex and memory: emergence of a new paradigm. J Cogn Neurosci 2009;21:2047–72.

[8] Volz KG, Schubotz RI, Von Cramon DY. Decision-making and the frontal lobes. Curr Opin Neurol 2006;19:401–6.

[9] Dolan RJ. The human amygdala and orbital prefrontal cortex in behavioural regulation. Philos Trans R Soc Lond B Biol Sci 2007;362:787–99.

[10] Kable JW, Glimcher PW. The neurobiology of decision: consensus and controversy. Neuron 2009;63:733–45.

[11] Glascher J, Adolphs R, Damasio H, Bechara A, Rudrauf D, Calamia M, et al. Lesion mapping of cognitive control and value-based decision making in the prefrontal cortex. Proc Natl Acad Sci USA 2012;109:14681–6.

[12] Jimura K, Chushak MS, Braver TS. Impulsivity and self-control during intertemporal decision making linked to the neural dynamics of reward value representation. J Neurosci 2013;33: 344–57.

[13] Dreher JC. Neural coding of computational factors affecting decision making. Prog Brain Res 2013;202:289–320.

[14] Sescousse G, Caldu X, Segura B, Dreher JC. Processing of primary and secondary rewards: a quantitative meta-analysis and review of human functional neuroimaging studies. Neurosci Biobehav Rev 2013;37:681–96.

[15] Vassena E, Krebs RM, Silvetti M, Fias W, Verguts T. Dissociating contributions of ACC and vmPFC in reward prediction, outcome, and choice. Neuropsychologia 2014;59:112–23.

[16] Jocham G, Furlong PM, Kroger IL, Kahn MC, Hunt LT, Behrens TE. Dissociable contributions of ventromedial prefrontal and posterior parietal cortex to value-guided choice. NeuroImage 2014;100:498–506.

[17] Amarante LM, Laubach M. For better or worse: reward comparison by the ventromedial prefrontal cortex. Neuron 2014;82: 1191–3.

[18] Sescousse G, Li Y, Dreher JC. A common currency for the computation of motivational values in the human striatum. Soc Cogn Affect Neurosci 2015;10:467–73.

[19] Saga Y, Tremblay L. Ventral striato-pallidal pathways involved in appetitive and aversive motivational processes.

[20] Bertalanffy LV. The theory of open systems in physics and biology. Science 1950;13:23–9.

[21] Fuster JM. The prefrontal cortex. 5th ed. London: Academic Press; 2015.

[22] Bechara A, Damasio H, Tranel D, Anderson SW. Dissociation of working memory from decision making within the human prefrontal cortex. J Neurosci 1998;18:428–37.

[23] Bechara A, Damasio H, Damasio AR, Lee GP. Different contributions of the human amygdala and ventromedial prefrontal cortex to decision-making. J Neurosci 1999;19:5473–81.

[24] Kringelbach ML. The human orbitofrontal cortex: linking reward to hedonic experience. Nat Rev Neurosci 2005;6:691–702.

[25] Roesch MR, Olson CR. Neuronal activity in primate orbitofrontal cortex reflects the value of time. J Neurophysiol 2005;94: 2457–71.

[26] Schultz W. Behavioral theories and the neurophysiology of reward. Annu Rev Psychol 2006;57:87–115.

[27] Wallis JD. Orbitofrontal cortex and its contribution to decision-making. Annu Rev Neurosci 2007;30:31–56.

[28] Zald DH. Orbitofrontal cortex contributions to food selection and decision making. Ann Behav Med 2009;38(Suppl. 1):S18–24.

[29] Levy DJ, Glimcher PW. The root of all value: a neural common currency for choice. Curr Opin Neurobiol 2012;22:1027–38.

[30] Metereau E, Dreher JC. The medial orbitofrontal cortex encodes a general unsigned value signal during anticipation of both appetitive and aversive events. Cortex 2015;63:42–54.

[31] Okuda J, Fujii T, Ohtake H, Tsukiura T, Tanji K, Suzuki R, et al. Thinking of the future and past: the roles of the frontal pole and the medial temporal lobes. NeuroImage 2003;19:1369–80.

[32] Addis DR, Wong AT, Schacter DL. Remembering the past and imagining the future: common and distinct neural substrates during event construction and elaboration. Neuropsychologia 2007;45:1363–77.

[33] Badre D. Cognitive control, hierarchy, and the rostro-caudal organization of the frontal lobes. Trends Cogn Sci 2008;12:193–200.

[34] Harrison LM, Bestmann S, Rosa MJ, Penny W, Green GG. Time scales of representation in the human brain: weighing past information to predict future events. Front Hum Neurosci 2011;26: 5–37.

[35] Schacter DL, Addis DR, Hassabis D, Martin VC, Spreng RN, Szpunar KK. The future of memory: remembering, imagining, and the brain. Neuron 2012;76:677–94.

[36] Wunderlich K, Dayan P, Dolan RJ. Mapping value based planning and extensively trained choice in the human brain. Nat Neurosci 2012;15:786–91.

[37] Spreng RN, Stevens WD, Chamberlain JP, Gilmore AW, Schacter DL. Default network activity, coupled with the fronto-parietal control network, supports goal-directed cognition. NeuroImage 2010;53:303–17.

[38] Ingvar DH. Memory of the future: an essay on the temporal organization of conscious awareness. Hum Neurobiol 1985;4: 127–36.

[39] Gerlach KD, Spreng RN, Madore KP, Schacter DL. Future planning: default network activity couples with frontoparietal control network and reward-processing regions during process and outcome simulations. Soc Cogn Affect Neurosci 2014;9:1942–51.

[40] Modirrousta M, Fellows LK. Dorsal medial prefrontal cortex plays a necessary role in rapid error prediction in humans. J Neurosci 2008;28:14000–5.

[41] Asaad WF, Eskandar EN. Encoding of both positive and negative reward prediction errors by neurons of the primate lateral prefrontal cortex and caudate nucleus. J Neurosci 2011;31:17772–87.

[42] Adams RA, Shipp S, Friston KJ. Predictions not commands: active inference in the motor system. Brain Struct Funct 2013; 218:611–43.

[43] Desimone R, Duncan J. Neural mechanisms of selective visual attention. Annu Rev Neurosci 1995;18:193–222.

[44] Rigotti M, Barak O, Warden MR, Wang XJ, Daw ND, Miller EK, et al. The importance of mixed selectivity in complex cognitive tasks. Nature 2013;497:585–90.

[45] Bari A, Robbins TW. Inhibition and impulsivity: behavioral and neural basis of response control. Prog Neurobiol 2013;108:44–79.

[46] Sylvester CM, Shulman GL, Jack AI, Corbetta M. Anticipatory and stimulus-evoked blood oxygenation level-dependent modulations related to spatial attention reflect a common additive signal. J Neurosci 2009;29:10671–82.

[47] Bollinger J, Rubens MT, Zanto TP, Gazzaley A. Expectation-driven changes in cortical functional connectivity influence working memory and long-term memory performance. J Neurosci 2010;30:14399–410.

[48] Baddeley A. Working memory. Philos Trans R Soc Lond B Biol Sci 1983;B302:311–24.

[49] Fuster JM, Bodner M, Kroger JK. Cross-modal and cross-temporal association in neurons of frontal cortex. Nature 2000; 405:347–51.

[50] Zhou Y-D, Fuster JM. Visuo-tactile cross-modal associations in cortical somatosensory cells. Proc Natl Acad Sci USA 2000;97: 9777–82.

[51] Wang X, Li SS, siao F, Lenz F, Bodner M, Zhou Y-D, et al. Differential roles of delay-periods neural activity in monkey dorsolateral prefrontal cortex in visual-haptic crossmodal workig memory. Proc Natl Acad Sci USA 2015;112:E214–9.

[52] Quintana J, Fuster JM. From perception to action: temporal integrative functions of prefrontal and parietal neurons. Cereb Cortex 1999;9:213–21.

[53] Zipser D, Kehoe B, Littlewort G, Fuster J. A spiking network model of short-term active memory. J Neurosci 1993;13:3406–20.

[54] Compte A, Brunel N, Goldman-Rakic PS, Wang XJ. Synaptic mechanisms and network dynamics underlying spatial working memory in a cortical network model. Cereb Cortex 2000;10: 910–23.

[55] Durstewitz D. Implications of synaptic biophysics for recurrent network dynamics and active memory. Neural Netw 2009;22: 1189–200.

[56] Verduzco-Flores S, Bodner M, Ermentrout B, Fuster JM, Zhou Y. Working memory cell's behavior may be explained by cross-regional networks with synaptic facilitation. PLoS One 2009;4: e6399.

[57] Tang W, Bressler SL, Sylvester CM, Shulman GL, Corbetta M. Measuring Granger causality between cortical regions from voxelwise fMRI BOLD signals with LASSO. PLoS Comput Biol 2012; 8:e1002513.

[58] Liang L, Wang R, Zhang Z. The modeling and simulation of visuospatial working memory. Cogn Neurodyn 2010;4:359–66.

[59] Wang M, Yang Y, Wang CJ, Gamo NJ, Jin LE, Mazer JA, et al. NMDA receptors subserve persistent neuronal firing during working memory in dorsolateral prefrontal cortex. Neuron 2013;77:736–49.

[60] Fuster JM, Bauer RH, Jervey JP. Functional interactions between inferotemporal and prefrontal cortex in a cognitive task. Brain Res 1985;330:299–307.

[61] Quintana J, Fuster JM, Yajeya J. Effects of cooling parietal cortex on prefrontal units in delay tasks. Brain Res 1989;503:100–10.

[62] Chafee MV, Goldman-Rakic PS. Inactivation of parietal and prefrontal cortex reveals interdependence of neural activity during memory-guided saccades. J Neurophysiol 2000;83:1550–66.

[63] Fuster JM. The prefrontal cortex. 4th ed. London: Academic Press; 2008.

[64] Sreenivasan KK, Curtis CE, D'Esposito M. Revisiting the role of persistent neural activity during working memory. Trends Cogn Sci 2014;18:82–9.

[65] O'Reilly RC, Noelle DC, Braver TS, Cohen JD. Prefrontal cortex and dynamic categorization tasks: representational organization and neuromodulatory control. Cereb Cortex 2002;12:246–57.

[66] Koechlin E, Ody C, Kouneiher F. The architecture of cognitive control in the human prefrontal cortex. Science 2003;302: 1181–5.

[67] Badre D, D'Esposito M. Functional magnetic resonance imaging evidence for a hierarchical organization of the prefrontal cortex. J Cogn Neurosci 2007;19:2082–99.

[68] Koechlin E, Summerfield C. An information theoretical approach to prefrontal executive function. Trends Cogn Sci 2007;11:229–35 [Ref ID: 3911].

[69] Azuar C, Reyes P, Slachevsky A, Volle E, Kinkingnehun SF, Kouneiher F, et al. Testing the model of caudo-rostral organization of cognitive control in the human with frontal lesions. NeuroImage 2014;84:1053–60.

[70] Friston K. Learning and inference in the brain. Neural Netw 2003;16:1325–52.

[71] Shipp S, Adams RA, Friston KJ. Reflections on agranular architecture: predictive coding in the motor cortex. Trends Neurosci 2013;36:706–16.

[72] Clark A. Whatever next? Predictive brains, situated agents, and future of cognitive science. Behav Brain Sci 2013;36:181–253.

[73] Rosenkilde CE, Bauer RH, Fuster JM. Single cell activity in ventral prefrontal cortex of behaving monkeys. Brain Res 1981; 209:375–94.

[74] Ramnani N, Elliott R, Athwal BS, Passingham RE. Prediction error for free monetary reward in the human prefrontal cortex. NeuroImage 2004;23:777–86.

[75] Hardin MG, Pine DS, Ernst M. The influence of context valence in the neural coding of monetary outcomes. NeuroImage 2009; 48:249–57.

[76] Mitchell JP, Schirmer J, Ames DL, Gilbert DT. Medial prefrontal cortex predicts intertemporal choice. J Cogn Neurosci 2011;23: 857–66.

[77] Smith DV, Hayden BY, Truong TK, Song AW, Platt ML, Huettel SA. Distinct value signals in anterior and posterior ventromedial prefrontal cortex. J Neurosci 2010;30:2490–5.

[78] Hampshire A, Chaudhry AM, Owen AM, Roberts AC. Dissociable roles of lateral orbitofrontal cortex and lateral prefrontal cortex during preference driven reversal learning. NeuroImage 2012;59:4102–12.

[79] Magno E, Foxe JJ, Molholm S, Robertson IH, Garavan H. The anterior cingulate and error avoidance. J Neurosci 2006;26: 4769–73.

[80] Mohr PN, Biele G, Krugel LK, Li SC, Heekeren HR. Neural foundations of risk-return trade-off in investment decisions. NeuroImage 2010;49:2556–63.

[81] Prévost C, Pessiglione M, Metereau E, Clery-Melin ML, Dreher JC. Separate valuation subsystems for delay and effort decision costs. J Neurosci 2010;30:14080—90.

[82] Knutson B, Cooper JC. Functional magnetic resonance imaging of reward prediction. Curr Opin Neurol 2005;18:411—7.

[83] Brown JW. Conflict effects without conflict in anterior cingulate cortex: multiple response effects and context specific representations. NeuroImage 2009;47:334—41.

[84] Brown JW, Braver TS. Risk prediction and aversion by anterior cingulate cortex. Cogn Affect Behav Neurosci 2007;7:266—77.

[85] Magno E, Simoes-Franklin C, Robertson IH, Garavan H. The role of the dorsal anterior cingulate in evaluating behavior for achieving gains and avoiding losses. J Cogn Neurosci 2009;21: 2328—42.

[86] Potts GF, Martin LE, Kamp SM, Donchin E. Neural response to action and reward prediction errors: comparing the error-related negativity to behavioral errors and the feedback-related negativity to reward prediction violations. Psychophysiology 2011:218—28.

[87] Alexander WH, Brown JW. Medial prefrontal cortex as an action-outcome predictor. Nat Neurosci 2011;14:1338—44.

[88] Bode S, He AH, Soon CS, Trampel R, Turner R, Haynes JD. Tracking the unconscious generation of free decisions using ultra-high field fMRI. PLoS One 2011;6:e21612.

[89] Forster SE, Brown JW. Medial prefrontal cortex predicts and evaluates the timing of action outcomes. NeuroImage 2011;55: 253—65.

[90] Kim B, Sung YS, McClure SM. The neural basis of cultural differences in delay discounting. Philos Trans R Soc Lond B Biol Sci 2012;367:650—6.

[91] Nee DE, Kastner S, Brown JW. Functional heterogeneity of conflict, error, task-switching, and unexpectedness effects within medial prefrontal cortex. NeuroImage 2011;54:528—40.

[92] Kuwabara M, Mansouri FA, Buckley MJ, Tanaka K. Cognitive control functions of anterior cingulate cortex in macaque monkeys performing a Wisconsin Card Sorting Test analog. J Neurosci 2014;34:7531—47.

[93] Michelet T, Bioulac B, Langbour N, Goillandeau M, Guehl D, Burbaud P. Electrophysiological correlates of a versatile executive control system in the monkey anterior cingulate cortex. Cereb Cortex 2015 [electronic].

[94] Brown JW, Braver TS. Learned predictions of error likelihood in the anterior cingulate cortex. Science 2005;307:1118—21.

[95] Lee D, Rushworth MF, Walton ME, Watanabe M, Sakagami M. Functional specialization of the primate frontal cortex during decision making. J Neurosci 2007;27:8170—3.

[96] Rushworth MF, Behrens TE. Choice, uncertainty and value in prefrontal and cingulate cortex. Nat Neurosci 2008;11: 389—97.

[97] Seo H, Lee D. Behavioral and neural changes after gains and losses of conditioned reinforcers. J Neurosci 2009;29:3627—41.

[98] Bissonette GB, Powell EM, Roesch MR. Neural structures underlying set-shifting: roles of medial prefrontal cortex and anterior cingulate cortex. Behav Brain Res 2013;250:91—101.

[99] Boorman ED, Rushworth MF, Behrens TE. Ventromedial prefrontal and anterior cingulate cortex adopt choice and default reference frames during sequential multi-alternative choice. J Neurosci 2013;33:2242—53.

[100] Shackman AJ, Salomons TV, Slagter HA, Fox AS, Winter JJ, Davidson RJ. The integration of negative affect, pain and cognitive control in the cingulate cortex. Nat Rev Neurosci 2011;12: 154—67.

[101] Morecraft RJ, Stilwell-Morecraft KS, Cipolloni PB, Ge J, McNeal DW, Pandya DN. Cytoarchitecture and cortical connections of the anterior cingulate and adjacent somatomotor fields in the rhesus monkey. Brain Res Bull 2012;87:457—97.

HUMAN STUDIES ON MOTIVATION, PERCEPTUAL, AND VALUE-BASED DECISION-MAKING

9

Reward, Value, and Salience

T. Kahnt[1], P.N. Tobler[2]

[1]Northwestern University Feinberg School of Medicine, Chicago, IL, United States;
[2]University of Zurich, Zurich, Switzerland

Abstract

Value and salience are key variables for associative learning, decision-making, and attention. In this chapter we review definitions of value and salience, and describe human neuroimaging studies that dissociate these variables. Value increases with the magnitude and probability of reward but decreases with the magnitude and probability of punishment, whereas salience increases with the magnitude and probability of both reward and punishment. Moreover, salience may be particularly enhanced in situations with probabilistic as opposed to safe outcomes. At the behavioral level, both value and salience independently accelerate behavior. At the neural level, value signals arise in striatum, orbitofrontal and ventromedial prefrontal cortex, and superior parietal areas, whereas magnitude-based salience signals arise in the anterior cingulate cortex and the inferior parietal cortex. By contrast, probability-based salience signals have been found in the ventromedial prefrontal cortex. In conclusion, the related nature of value and salience stresses the importance of disentangling both variables experimentally.

INTRODUCTION

One of the most adaptive functions for humans and animals is the ability to predict future outcomes. This ability allows us to prepare and execute appropriate actions to obtain rewards and avoid punishments. For example, the sight of fruit hanging from a tree allows us to approach the tree and harvest the fruit. Similarly, if we have learned about the likely consequences of this stimulus, the sight of a leopard in a tree allows for avoidance behavior. In other words, the appropriate use of information provided by predictive stimuli allows for survival-promoting, adaptive behavior. Such behavior is also beneficial in evolutionary terms, by providing an advantage over other members of the species who have learned the meaning of the stimuli less

well and who, as a consequence, get to eat less fruit and are more likely to fall prey to a leopard [1].

A hallmark of adaptive behavior is that it relates stimuli in a meaningful way to their predicted outcomes. In particular, the magnitude of the predicted outcome is an important factor that influences behavior. For example, the sight of a lot of fruit on a tree signals more reward and may increase the probability of approach compared to the sight of less fruit. Similarly, an adult leopard may be avoided more resolutely than a baby leopard. Despite the different responses they elicit, both the tree with plenty of fruit and the adult leopard will capture more attention and cognitive resources than the tree with just a few pieces of fruit and the baby leopard. Increased attention in turn will reduce the latency of approach and avoidance responses. These orderly relations between stimuli and predicted outcomes can be captured by two theoretical concepts: value and salience. Roughly speaking, value guides appropriate action selection in a magnitude-dependent fashion, whereas salience captures attention according to the overall importance of the stimulus.

Value arises from the intrinsic meaning of two different types of predicted outcomes: rewards and punishments. Rewards have positive value and are approached, whereas punishments have negative value and are avoided. Thus, value is a signed currency that varies in a monotonic fashion from negative to positive. In contrast, salience is an unsigned currency, and the more negative or the more positive the predicted outcome, the more salient it is and the more attention it will draw. Thus, in the scheme proposed here, salience relates to value such that the absolute value of the predicted outcome determines its salience, irrespective of whether the predicted outcome is a reward or a punishment [2,3]. For example, if the absolute value of a banana is the same as the absolute value of an intimidating display of a conspecific, then the two have equal salience. The fact that the definitions of value and salience are related to each other may explain why the literature has not always separated them in empirical studies [4,5].

The predictive relations between stimuli and outcomes have to be learned from experience. It should therefore come as no surprise that learning theory has much to say about predicted value and salience [6–8]. In particular, associative learning theory describes the mechanisms with which predictive stimuli become valuable [9] and salient [10–12]. Moreover, learning theory has provided several, not mutually exclusive, definitions of salience and suggested ways in which they could be combined [13].

This chapter reviews how value and salience are defined by learning theory, how they can be distinguished from each other experimentally, and how they are processed in the human brain to mediate adaptive behavior. Importantly, we focus on predictive (learned) value and salience rather than on experienced (outcome-related) value and salience (e.g., Ref. [14]). Moreover, this chapter discusses only human studies, and the reader is referred to different chapters of this book covering the neural processing of value and salience in animals (see Chapters 2–5 and also Ref. [15]).

VALUE

Definition and Behavioral Function

Value has been defined in many different ways and is an important concept for various scientific disciplines such as psychology, economics, and ecology. The importance of the concept arises from its various functions. Value guides choice behavior [16]. In particular, stimuli with positive value elicit approach, whereas stimuli with negative value elicit avoidance behavior. In addition to guiding decisions, value carrying rewards and punishments also elicit associative learning [9] and are associated with emotions, including pleasure and pain [17]. Here we consider learned value, that is, the value that a previously neutral stimulus has acquired through association with a reward or punishment. The computational and neural processes by which predictive stimuli acquire value through associative learning are discussed in different chapters of this book.

At the most general level, value can be defined as the capacity of a stimulus or goal to elicit effort exertion. We exert energy (i.e., work) to obtain stimuli with positive value and to avoid stimuli with negative value. In addition, we work harder with more highly valued stimuli than with less highly valued stimuli. Experimenters

can harness the monotonic relationship between behavior and predicted outcomes, and infer the value of the outcome by the strength, latency, or probability of the response elicited by the outcome predictive stimulus (i.e., the conditioned response). For example, stimuli predictive of larger or more likely rewards elicit stronger approach behavior in rats (e.g., as measured by their running speed down an alley) than stimuli predictive of smaller or less likely rewards [18]. Accordingly, experimenters can take the running speed elicited by the different predictive stimuli as a proxy for value.

Work or energy exertion in the above definition is meant in a broad sense and not restricted to specific types of learned actions. Still, it applies most naturally to instrumental conditioning, i.e., situations in which delivery of the outcome depends on whether a specific response is performed [8]. Nevertheless, also with Pavlovian conditioning, i.e., in situations in which the delivery of the outcome is independent of any response production, conditioned responses occur and scale with value. For example, dogs salivate more to stimuli that predict more meat powder [7], and thus, Pavlovian responses can also be taken as a proxy for value.

In the definition above, value is a subjective variable that has to be inferred from how individuals respond to stimuli. However, value directly relates to objective parameters that can be controlled experimentally. For instance, value increases with increasing reward *magnitude* and decreases with increasing punishment magnitude (Fig. 9.1A). Thus, a stimulus predicting two cherries is more valuable than a stimulus predicting one cherry. Similarly, value increases with increasing *probability* of reward and decreases with increasing probability of punishment (Fig. 9.1B). Accordingly, a stimulus predicting a cherry with 100% probability is more valuable than a stimulus predicting a cherry with 50%, which in turn is more valuable than a stimulus predicting a cherry with 0%. For simplicity, Fig. 9.1 depicts value as a linear function of magnitude and probability but empirically, this relation is often found to be nonlinear, at least for large magnitudes and probabilities close to 0 and 1 [16].

The value functions depicted in Fig. 9.1 span the whole range of rewards and punishments. This implies that a true value signal not only distinguishes rewards and punishments according to their positive versus negative valence but that a difference of two value units in the punishment domain is represented similarly to a difference of two units in the reward domain. Such value coding on a common scale is a requirement for making computations across the entire range of possible values and allows us to determine the integrated value of stimuli predicting mixed outcomes (i.e., losses and gains, rewards and punishments, or costs and benefits).

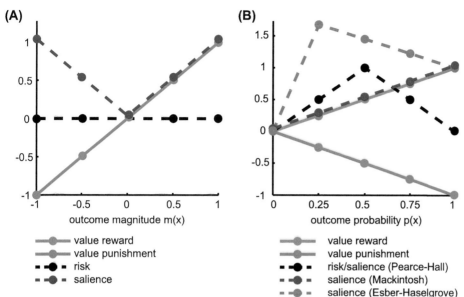

FIGURE 9.1 **Operational definitions of value and salience as a function of magnitude and probability.** (A) Positive value, negative value, risk, and salience of deterministic predicted outcomes of different magnitudes (−1, −0.5, 0, 0.5, 1 in arbitrary units). (B) Positive value, negative value, risk (salience according to the model by Pearce and Hall), salience according to the model by Mackintosh, and salience according to the model by Esber and Haselgrove of predicted outcomes (rewards and punishments) of different probabilities (0%, 25%, 50%, 75%, and 100%). Note, in (B), Esber−Haselgrove salience is generated using the reduced model (assuming complete learning) with parameters from [79], resulting in a peak at 0.25; in (A) and (B), risk is defined as variance, following the tradition of finance theory [19,98].

The definition of value as a function of magnitude and probability has been taken up in various economic theories suggesting different ways in which probability and magnitude should be, or are, combined during choice (expected value, expected utility, prospect; for review see Ref. [19]). For instance, by multiplying the magnitude of an outcome by the probability with which it occurs, the two variables are integrated to form the *expected value*. In neoclassical economic theory (e.g., Ref. [20]), value (or utility) has become a placeholder that can capture overt choice behavior (revealed preferences) if that behavior obeys a set of basic principles (axioms, such as transitivity: if one prefers A over B and B over C then one ought to prefer A over C). In other words, as long as choice behavior follows these principles, it can be described as if decision-makers maximize a hypothetical value function. Although research in behavioral economics, psychology, and neuroscience has repeatedly found empirical deviations from this normative neoclassical ideal (e.g., Ref. [16]), these deviations do not contradict the basic notion that value is a function of magnitude and probability.

Reinforcement learning theory also recognizes that the value of the predicted reward increases with magnitude and probability of occurrence. Accordingly, the strength of the predictive response increases with magnitude and probability. Value learning corresponds to the gradual adjustment of predictions until the predicted value reflects the experienced value (for review see Ref. [21]). At the end of each trial [9], or at each moment in time [22], the organism is thought to compute a prediction error that corresponds to the value difference between the actual outcome (or state) and the predicted outcome (or state). The prediction error is used to update the previous prediction to generate a more accurate prediction. The ultimate goal is to adjust the predictions such that they are proportional to the value of the outcome, both for magnitude and for probability. Thus, at the end of learning, value predictions correspond to the value functions depicted in Fig. 9.1.

So far we have considered one single type of value learning for a single use, namely stimulus-outcome learning for the prediction of rewards and punishments. However, it is likely that independent valuation systems can be employed in parallel that are differentially suited to handle variable versus stable environments. For example, a goal-directed learning system may be based on a mental model of the world, and represent the long-term value of outcomes resulting from a number of different future actions and states. Alternatively, a habitual learning system may represent experience-based value predictions of the immediate state without the need of a model. In other words, different value systems may use different mechanisms to learn different value predictions and to support habitual or goal-directed behavior [23].

Value Signals in the Human Brain

Functional neuroimaging has revealed predicted outcome magnitude and probability signals in the human striatum and midbrain, consistent with results from studies using single-cell recordings [24]. The striatal signals are reminiscent of the responses emitted by single dopamine neurons, which increase firing with both predicted reward magnitude and probability [25,26]. In one human neuroimaging study [27], subjects learned the meaning of visual stimuli that were graded in two dimensions: color (between *light yellow* and *dark red*) and shape (between one and five circles). One dimension was associated with gradations in reward magnitude, and the other with gradations in probability (across subjects, the dimension to probability or magnitude assignment and the increasing/decreasing direction of the dimensions were counterbalanced). fMRI data were acquired after learning was completed. In each trial, one stimulus was shown and subjects indicated with a button press where on the screen it was displayed. Stimulus-induced activations in the striatum increased with probability, magnitude, and their combination. Similar magnitude- and probability-related striatal activations have been reported for a variety of tasks [28–44]. For instance, with magnitude, striatal activity integrates punishment and reward value in response to stimuli predicting mixed outcomes [45].

While the aforementioned fMRI studies support the notion that a single decision variable capturing both probability and magnitude is encoded in these areas, some studies have shown spatial decomposition of probability and magnitude information within the striatum [27,32,41]. In one example, Yacubian and colleagues asked subjects to place a bet of one of two magnitudes (either €1.00 or €5.00) on gambles involving a winning probability of either 12.5% or 50% [41]. Activity was analyzed during an anticipatory period, before the outcome was revealed. Probability-related activations arose in more anterior and lateral regions of the ventral striatum, whereas magnitude-related activations arose in more posterior and medial regions (Fig. 9.2). Despite this segregation, it should be kept in mind that magnitude- and probability-related activations overlap substantially in large areas of the striatum, as would be expected from a value-coding region.

Moreover, fMRI activations related to the magnitude or probability of expected outcomes have also been reported in cortical regions, including anterior cingulate cortex (ACC) [14,31], posterior parietal cortex (PPC) [3,45], and superior frontal [31], lateral prefrontal (LPFC) [27,46–48], as well as medial (MPFC) and ventromedial prefrontal cortex (VMPFC) [37,38,42,46,48–50]. The responses in the orbitofrontal cortex (OFC) to rewards as well as to reward-predictive stimuli represent high versus low reward magnitude information in distributed patterns [3,51]. Thus, even though electrophysiological recordings in animals report heterogeneous signals in prefrontal regions, such that the firing rates of different value-coding neurons increase and decrease with magnitude [52], neuroimaging methods can nevertheless be used to study the representation of magnitude in such regions. Moreover, human OFC responses represent the value of rewards and punishments on a common monotonic scale [3], which enables computations across the entire value range. Common or combined coding of both components has been observed in ACC for error magnitude and probability [53], in medial OFC [31,54] and in LPFC for reward magnitude and probability [27]. By contrast, preferential activations to probability, rather than magnitude, arise in parts of the MPFC [37].

Different brain regions have been suggested to code value in the context of model-based (goal-directed) and model-free (habitual) behavior [55,56]. Specifically, a dissociation of these different value signals has been shown within the striatum, with activity in the caudate nucleus coding for model-based and in the putamen for model-free value signals [56]. However, both regions are functionally connected to the VMPFC, where they can be integrated to support value-based choice. This

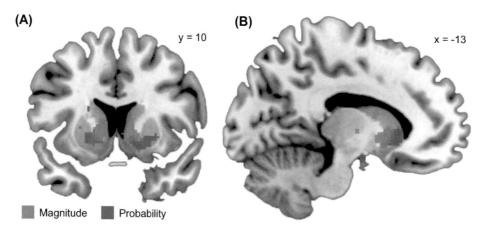

FIGURE 9.2 **Segregated encoding of probability and magnitude in the ventral striatum.** (A and B) Coronal and sagittal sections depicting magnitude- and probability-related fMRI activity (average of 98 subjects) in *green* and *blue*, respectively. *Adapted from Yacubian J, Sommer T, Schroeder K, Glascher J, Braus DF, Buchel C. Subregions of the ventral striatum show preferential coding of reward magnitude and probability. NeuroImage 2007;38:557–63.*

is in line with a large number of human neuroimaging studies reporting value-related activity in the VMPFC during value-based choice [57–66].

Moreover, evidence suggests that different regions in the ventral PFC are involved in coding the general versus specific value of predicted outcomes. General value signals are independent of the specific nature of the reward (e.g., cherries or elderberries) and can be used to compare and choose between alternative outcomes, but are unable to inform expectations about the specific identity of the outcome. In contrast, identity-specific representations conjointly represent information about both value (how good is it?) and outcome identity (what is it?). Human imaging work has linked activity in the VMPFC to general value signals [59,67–70]. In contrast, activity in the OFC has been shown to code for specific rewards [71,72], and neural value codes for different reward categories differ in this region [69].

In line with the fact that the medial and lateral OFC belong to largely separate networks [73,74], the central and lateral OFC represent the value of future rewards in the form of identity-specific value codes, whereas value representations in medial areas generalize across different outcome identities [68,75,76]. Such a dual representation of general and specific values in the PFC may allow for different forms of reward-related behaviors. Specifically, individuals may use general values to compare different options and to exert simple behaviors such as approach and avoidance responses. On the other hand, identity-specific values may support choices in line with the current needs of the organism, allow for outcome-specific responding, and enable the organism to flexibly update value predictions according to changes in the environment.

SALIENCE

Definition and Behavioral Function

At the most general level, salience can be defined as the capacity of a stimulus to direct attention. However, salience is an ill-defined concept in cognitive neuroscience. Part of the confusion lies in that different researchers have used the same term to refer to different ideas of what makes a stimulus salient. For instance, in visual neuroscience, the term is used to describe the perceptual salience (often saliency) of a stimulus, that is, its physical properties (i.e., the color, contrast, orientation, or luminance) that make it more likely to capture attention [77,78]. On the other hand, the field of associative learning and value-based decision-making uses the term to describe the acquired salience of a stimulus, that is, the importance that a stimulus has acquired through association with an incentive outcome. Perceptual salience is not a topic of this chapter and instead, we will focus only on acquired salience and define it as a function of the value of the stimulus.

More specifically, in line with associative learning theories, we define salience as the *absolute* (i.e., *unsigned*) *value* predicted by a stimulus (Fig. 9.1). In other words, both rewarding and punishing outcomes are salient, and the degree of salience is determined by the value of the predicted outcome [2,3]. Without loss aversion [16], a stimulus predicting the receipt of one cherry has the same salience as a stimulus predicting the loss of one cherry. As such, salience relates to the importance of a stimulus for motivated behavior independent of its valence. The key behavioral function is therefore to guide attention to stimuli that predict motivationally relevant outcomes and to facilitate neural processes related to planning and executing appropriate responses.

The definition of salience is straightforward for deterministic outcomes, where outcomes are predicted with no uncertainty. In the case of probabilistic stimuli, however, the definition of salience is more complicated. Specifically, probability affects salience not only by changing value (see earlier). In fact, the attention-capturing properties of a stimulus have historically been described as *associability*, a concept that is closely related to salience and is meant to describe how well associations can be formed with a stimulus. A key determinant of associability is the predictability of a stimulus, that is, the probability with which a stimulus predicts an outcome. Different mechanisms have been suggested to describe the relationship between predictability, uncertainty, and associability [10,12]. Specifically, the Mackintosh model [10] states that reliable predictors of outcomes (close to 100%) will acquire higher salience and consequently capture more attention than poor predictors (close to 0%), which have a low level of salience and will consequently be ignored (Fig. 9.1B). In contrast, the Pearce–Hall model [12] posits that a high level of salience (and thus attention) should be applied to unreliable predictors of outcomes such that learning of these is enhanced. In other words, unreliable predictors (50%) are assumed to have higher salience than reliable predictors (close to 0% and 100%, Fig. 9.1B).

The models of Mackintosh and Pearce and Hall have been combined by a third model of attention in associative learning [13]. The Esber–Haselgrove model [13] assumes that (1) stimuli acquire salience to the degree that they predict incentive outcomes (similar to Mackintosh), (2) both the occurrence and the nonoccurrence of an incentive outcome are salient events that are both partially predicted by a nondeterministic stimulus, and (3) salience is defined as the sum of the absolute values of all possible outcomes (occurrence and nonoccurrence of positive and negative outcomes) associated with a stimulus. Based on these assumptions, this model predicts that both high (100%) and low probability cues (0%) have a lower salience than intermediate predictors of outcomes (e.g., 50%), and that good predictors (100%) have a generally higher salience than poor predictors (0%) [79]. In other words, the Esber—Haselgrove model predicts a skewed, inverted U-shaped relationship between probability and salience (Fig. 9.1B).

Of note, defining salience as the sum of all (positive and negative) outcomes predicted by a stimulus is a fundamental idea, which can also be applied to deterministic predictors of outcomes. Specifically, in the case of compound stimuli, predicting multiple deterministic (or probabilistic) outcomes, salience can be defined as the absolute value of the average predicted outcome (global salience): a stimulus predicting both the receipt of one cherry and the loss of one cherry would have a global salience of zero. Alternatively, salience can be defined as the sum of the absolute values

of the individual predicted outcomes (elemental salience, Fig. 9.3C): a stimulus predicting both the receipt of one cherry and the loss of one cherry would have an elemental salience corresponding to two cherries. Supporting the fundamental assumption of the Esber—Haselgrove model, we found [45] that the positive and negative outcomes predicted by individual stimuli contribute independently to the salience of the compound stimulus (Fig. 9.3D). Specifically, elemental salience predicted the latency of value-based responses and did so significantly better than global salience.

Confounds in Operational Definitions of Salience

To test for behavioral and neural effects of salience, operational definitions of salience are required. From the preceding paragraphs, it should have become clear that the definition of salience is not as straightforward

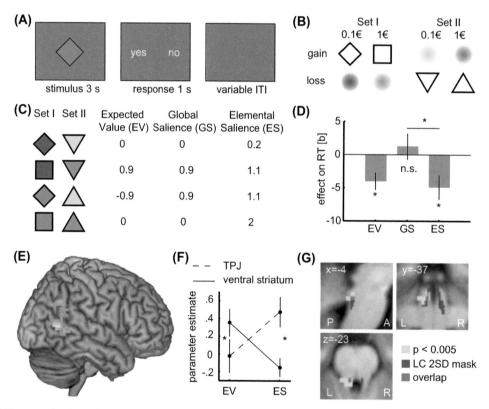

FIGURE 9.3 **Salience and value of compound stimuli.** (A—C) Trial structure and experimental stimuli. Subjects underwent classical conditioning with the single stimuli depicted in (B). These stimuli were then combined into compound stimuli [depicted in (C)] that independently varied as a function of expected value (*EV*, i.e., value 1 + value 2), global salience (*GS*, absolute value of EV, i.e., |value 1 + value 2|), and elemental salience (*ES*, sum of absolute values of single stimuli, i.e., |value 1| + |value 2|) (*ITI*, intertrial interval). (D) Response times (*RT*) in the choice task were independently predicted by EV and ES, but did not correlate with GS. (E) fMRI activity in the right temporoparietal junction (*TPJ*) was best correlated with elemental salience. (F) Double dissociation between value-related activity in the ventral striatum and salience-related activity in the TPJ. (G) Functional connectivity between the TPJ and the locus coeruleus (*LC*) was related to elemental salience. *Adapted from Kahnt T, Tobler PN. Salience signals in the right temporoparietal junction facilitate value-based decisions. J Neurosci 2013;33:863—69.*

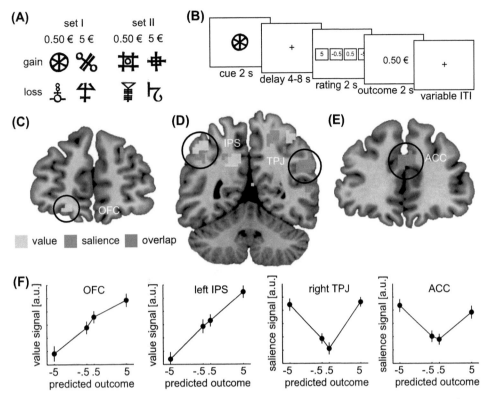

FIGURE 9.4 **Brain regions encoding value and salience.** (A and B) Design and conditions of the noninstrumental outcome prediction task. (C–E) Regions in the OFC, IPS, TPJ, and ACC, in which activity patterns correlate with the value *(yellow)* and salience *(pink)* of Pavlovian stimuli predicting rewarding and punishing outcomes of different magnitudes. (F) Value and salience signals from linear support vector regression models that were trained on multivoxel activity patterns to decode value and salience of the stimuli, respectively. *ACC,* anterior cingulate cortex; *IPS,* intraparietal sulcus; *OFC,* orbitofrontal cortex; *TPJ,* temporoparietal junction. *Adapted from Kahnt T, Park SQ, Haynes JD, Tobler PN. Disentangling neural representations of value and salience in the human brain. Proc Natl Acad Sci USA 2014;111:5000–05.*

as intuitively assumed. Moreover, in many cases, salience is confounded by value, magnitude, or probability [80]. Specifically, in most studies using only rewards or only punishments as outcomes, value and salience are perfectly correlated (Fig. 9.1). Accordingly, neural signals that increase with reward magnitude could be related to value or salience [3,81]. This is especially true if neural responses to value and salience are heterogeneous, i.e., if the activity of different neuronal populations increases and decreases with increasing value [52]. Thus, if value is defined by different levels of outcome magnitude, experimental designs need to include both appetitive and aversive outcomes to dissociate value and salience.

Similarly, when varying salience as a function of value operationalized by increases in probability, salience is likely to be confounded by probability and/or risk. As can be seen in Fig. 9.1B, to dissociate salience as defined in the Esber–Haselgrove model (see Fig. 9.1 and the section "Definition and Behavioral Function" for a description of the model) from coding of value and risk (i.e., salience as defined in the Pearce–Hall model), experimental designs need to vary probability in multiple steps from 0% to 100%. The critical

comparison is that between 100% and 0% cues, which have the same risk, but different levels of salience [79]. At the same time, however, to control for the different values of these cues, intermediate levels of probability, which have lower value but higher salience, need to be tested. Alternatively, to dissociate salience as defined in the Mackintosh model from value and risk, experimental designs need to cover multiple steps of probability from 0% to 100% and must include both positive and negative outcomes. Here, value increases and decreases with the probability of positive and negative outcomes, respectively, whereas salience increases with the probability of both appetitive and aversive outcomes.

Importantly, while it is theoretically possible to define salience in terms of absolute value based on magnitude, probability, or both (expected value), it is not clear whether these different operationalizations are computationally equivalent. Also, while both have been shown to correlate with response times, and thus appear to facilitate behavior and neural processing [3,45,79,82], it is unclear whether they are processed in the same brain regions and whether they subserve the same behavioral functions.

Salience Signals in the Human Brain

Given the numerous potential confounds of salience just described, experimental investigations of salience can be challenging. Accordingly, only relatively few imaging studies have comprehensively addressed how value versus salience is coded in the human brain. For instance, one study [82] used a simple value-based choice task involving images of snack items to dissociate value from salience. Specifically, subjects were asked to make a binding decision about whether they wanted to eat the currently displayed food item by choosing "Strong No," "No," "Yes," or "Strong Yes." Thus, by choosing "Yes" they could obtain rewards, whereas by choosing "No" they could avoid punishments. The choices were used to determine the value of the items by assigning -2, -1, $+1$, and $+2$ to the four answers, respectively. In line with the above magnitude-based definition, salience was defined as absolute value. The study identified several brain regions in which fMRI activity was correlated with salience, including the ACC extending to the supplementary motor area, the precentral gyrus, the insula, and the fusiform gyrus. In contrast, correlations with value were found in the VMPFC extending into the MPFC, in the precuneus, and in the ventral striatum.

A general problem of experiments involving choices is that aversive outcomes can be avoided. Specifically, the possibility of avoiding potential harm is likely to change the value associated with them [50]. Accordingly, it is unclear whether value and salience signals identified in studies involving choices reflect the same processes as in Pavlovian tasks, in which outcomes cannot be avoided. Moreover, decisions about stimuli with low absolute values are intrinsically more difficult because the values are closer to indifference than choices involving high absolute values, and salience signals in choice paradigms are therefore correlated with difficulty and nonspecifically enhanced processing. Finally, anticipatory outcome signals related to value and salience are related to neural processing of action selection and choice execution, preventing a straightforward interpretation of these signals.

To clearly dissociate neural representations of value and salience, we used a noninstrumental task, involving low- and high-magnitude monetary rewards and punishments [3]. In addition to dissociating value and salience, this task controlled for several additional confounds. First, because of the noninstrumental nature of the task, aversive outcomes could not be avoided, thus maintaining the salience and value of the aversive stimuli and at the same time preventing choice-related confounds. Second, we used abstract visual stimuli and associated them with rewarding ($+0.5$ or $+5$ euros) and punishing (-0.5 or -5 euros) outcomes using

classical conditioning (Fig. 9.4A). This controlled for preexisting associations of the stimuli, and thus ensured approximately equal value of the outcomes across subjects. Furthermore, instead of using univariate fMRI data analysis, we used multivoxel pattern analysis techniques [83,84], which are sensitive to distributed representations.

Our results revealed that fMRI activity patterns in the OFC, which had been assumed to be related to value [51,69], indeed represent value, and not salience (Fig. 9.4C). Moreover, activity patterns in the intraparietal sulcus (IPS) were also related to value (Fig. 9.4D), which is noteworthy because previously reported value signals in the IPS [85–87] had been reinterpreted as salience signals [81]. Thus, in the OFC and the PPC, distributed signals decreased with the magnitude of predicted punishments and increased with the magnitude of predicted rewards, reflecting a value signal. In contrast, we found salience signals in the ACC (Fig. 9.4E) and the inferior parietal cortex, extending into the temporoparietal junction (TPJ; Fig. 9.4D). Salience signals also emerged in more superior regions of the parietal cortex, overlapping with neural representations of value in the IPS (Fig. 9.4D). These results demonstrate that value and salience can be processed in the very same brain region, highlighting the importance of experimentally dissociating both variables.

Taken together, the results from this study using deterministic outcome magnitude to vary salience identified a network including the ACC and the TPJ that is involved in salience processing. Interestingly, the TPJ has been suggested to be part of a ventral attention network that is involved in mediating orientation of responses toward behaviorally relevant stimuli [88,89]. In parallel to results from studies on acquired salience discussed earlier, experiments found that the TPJ responds to behaviorally relevant stimuli [90], even if these stimuli are of low visual salience or unattended [91].

Our earlier study [45] converged with the subsequent one by showing salience signals in the TPJ. Specifically, similar to response times, fMRI signals in the TPJ were best explained by the summed salience of all positive and negative outcomes associated with the individual cues in the compound (Fig. 9.3E). Moreover, functional connectivity analyses suggested that salience signals in the TPJ were functionally linked to activity in the locus coeruleus (LC; Fig. 9.3G), a noradrenergic nucleus in the brain stem [45]. This finding is in line with the hypothesis that the TPJ's role in redirecting attention is supported by the activity of noradrenergic neurons in the LC [89], which directly innervate the superior temporal gyrus and inferior parietal cortex [92,93]. Noradrenaline has long been implicated in arousal and attention-related increases in behavioral performance. In particular, noradrenergic neurons in the LC respond to target cues with phasic activity increases [94], and the magnitude of this response is correlated with measures of task performance [95]. These findings contrast with striatal value signals that are presumably related to midbrain dopamine (Fig. 9.3F).

A few studies have investigated brain signals that correlate with salience, defined as a function of probability rather than magnitude. For instance, using an associative learning task with positive and negative outcomes (appetitive juice, money, salt water, and aversive pictures), one study showed that during learning, activity in the VMPFC was correlated with the salience of stimuli that predicted positive and negative outcomes with 50% contingency [96]. Because activity for stimuli predicting positive and negative outcomes had the same sign, and thus cannot be explained by value (Fig. 9.1B), this finding suggests that VMPFC activity correlates with the acquired salience of predicted outcomes (Fig. 9.5). However, even though this design controlled well for value, it included only stimuli that predicted the outcome with 50% contingency, and thus, it is unclear whether the observed VMPFC signals reflect salience or risk. Nevertheless, a similar finding of salience in the VMPFC has been reported in a study that

operationalized salience as 80% versus 20% probability of reward or punishment [97]. Thus, levels of salience were independent of risk. Moreover, the same study reported a correlation between value (defined as the product of probability and valence and thus independent of salience) and fMRI activity in the central OFC, close to where value signals (defined as magnitude) have been reported [3].

Taken together, these studies suggest a role for the VMPFC in salience processing, operationalized using probability. Supporting this idea, the firing patterns of neurons in the animal OFC, a region closely related to the human VMPFC, can be explained by salience as defined by the Esber–Haselgrove model [79]. By contrast, studies using the magnitude of positive and negative outcomes reliably identified the ACC [82] and the TPJ [3,45], whereas VMPFC activity has not been linked to salience defined in this way [82]. Given these contrasting findings, it appears that salience operationalized as a function of probability and magnitude is differently processed in the human brain (Fig. 9.5C). In theory, probability- and magnitude-based salience could play differential roles for learning and behavior, respectively.

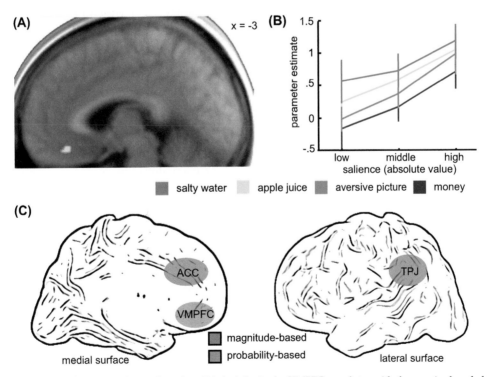

FIGURE 9.5 **Probability-based salience during learning.** (A) Activity in the VMPFC correlates with the acquired probability-based salience (absolute predicted value) of different aversive and appetitive probabilistic (50%) outcomes during associative learning. (B) Parameter estimates for low, middle, and high salience of different predicted outcomes (*green,* salty water; *yellow,* apple juice; *red,* aversive pictures; *blue,* money). (*Adapted from Metereau E, Dreher JC. The medial orbitofrontal cortex encodes a general unsigned value signal during anticipation of both appetitive and aversive events. Cortex 2015;63:42–54.*) (C) Brain regions representing probability- and magnitude-based salience are depicted in *cyan* and *magenta,* respectively. *ACC,* anterior cingulate cortex; *TPJ,* temporoparietal junction; *VMPFC,* ventromedial prefrontal cortex.

CONCLUSIONS

Value is a well-understood concept that increases with magnitude and probability of reward and decreases with magnitude and probability of punishment, whereas salience has been less well defined. We show that by defining salience as absolute value, salience can be given a precise and quantifiable interpretation, with respect to both magnitude and probability. Defined in this way, salience increases with magnitude and probability of both reward and punishment. The close relation of value and salience implies that they can be separated experimentally only if both rewards and punishments are used. Human neuroimaging studies that followed this approach revealed value signals in the striatum, OFC and VMPFC, and PPC. Typically, magnitude- and probability-based value signals colocalized. By contrast, magnitude-based salience signals arose in the TPJ and ACC, whereas probability-based salience signals were found in the VMPFC, suggesting different types of neural salience signals.

References

[1] Mayr E. Behavior programs and evolutionary strategies. Am Scientist 1974;62:650–9.

[2] Bromberg-Martin ES, Matsumoto M, Hikosaka O. Dopamine in motivational control: rewarding, aversive, and alerting. Neuron 2010;68:815–34.

[3] Kahnt T, Park SQ, Haynes JD, Tobler PN. Disentangling neural representations of value and salience in the human brain. Proc Natl Acad Sci USA 2014;111:5000–5.

[4] O'Doherty JP. The problem with value. Neurosci Biobehav Rev 2014;43:259–68.

[5] Schultz W. Updating dopamine reward signals. Curr Opin Neurobiol 2013;23:229–38.

[6] Dickinson A. Contemporary animal learning theory (problems in the behavioural sciences. New York: Cambridge University Press; 1981.

[7] Pavlov P. Conditioned reflexes: an investigation of the physiological activity of the cerebral cortex [Anrep GV, Trans.]. London: Oxford University Press; 1927.

[8] Thorndike E. Animal intelligence: experimental studies. New York: MacMillan; 1911.

[9] Rescorla RA, Wagner AR. A theory of Pavlovian conditioning: variations in the effectiveness of reinforcement and nonreinforcement. In: Black AH, Prokasy WF, editors. Classical conditioning II: current research and theory. New York: Appleton Century Crofts; 1972.

[10] Mackintosh NJ. Theory of attention - variations in associability of stimuli with reinforcement. Psychol Rev 1975;82:276–98.

[11] Mitchell CJ, Le Pelley ME. Attention and associative learning: from brain to behaviour. Oxford, UK: Oxford University Press; 2010.

[12] Pearce JM, Hall G. A model for Pavlovian learning: variations in the effectiveness of conditioned but not of unconditioned stimuli. Psychol Rev 1980;87:532–52.

[13] Esber GR, Haselgrove M. Reconciling the influence of predictiveness and uncertainty on stimulus salience: a model of attention in associative learning. Proc Biol Sci 2011;278:2553–61.

[14] Fujiwara J, Tobler PN, Taira M, Iijima T, Tsutsui K. Segregated and integrated coding of reward and punishment in the cingulate cortex. J Neurophysiol 2009;101:3284–93.

[15] Tobler PN, Dickinson A, Schultz W. Coding of predicted reward omission by dopamine neurons in a conditioned inhibition paradigm. J Neurosci 2003;23:10402–10.

[16] Kahneman D, Tversky A. Prospect theory - analysis of decision under risk. Econometrica 1979;47:263–91.

[17] Cabanac M. Pleasure: the common currency. J Theor Biol 1992;155:173–200.

[18] Bower G, Tarpold M. Reward magnitude and learning in a single-presentation discrimination. J Comp Physiol Psychol 1959;52:727–9.

[19] Tobler PN, Weber EU. Valuation of risky and uncertain choices. In: Glimcher PW, Fehr E, editors. Neuroeconomics. Oxford: Academic Press; 2013. p. 149–72.

[20] von Neumann J, Morgenstern O. Theory of games and economic behavior. 2nd ed. Princeton: Princeton University Press; 1947.

[21] Daw ND, Tobler PN. Value learning through reinforcement: the basics of dopamine and reinforcement learning. In: Glimcher PW, Fehr E, editors. Neuroeconomics. Oxford: Academic Press; 2013. p. 283–98.

[22] Sutton R, Barto A. Reinforcement learning: an introduction. Cambridge, MA: MIT Press; 1998.

[23] Dolan RJ, Dayan P. Goals and habits in the brain. Neuron 2013;80:312–25.

[24] Duzel E, Bunzeck N, Guitart-Masip M, Wittmann B, Schott BH, Tobler PN. Functional imaging of the human dopaminergic midbrain. Trends Neurosci 2009;32:321–8.

[25] Tobler PN, Fiorillo CD, Schultz W. Adaptive coding of reward value by dopamine neurons. Science 2005;307:1642–5.

[26] Fiorillo CD, Tobler PN, Schultz W. Discrete coding of reward probability and uncertainty by dopamine neurons. Science 2003;299:1898–902.

[27] Tobler PN, O'Doherty JP, Dolan RJ, Schultz W. Reward value coding distinct from risk attitude-related uncertainty coding in human reward systems. J Neurophysiol 2007;97:1621–32.

[28] Abler B, Walter H, Erk S, Kammerer H, Spitzer M. Prediction error as a linear function of reward probability is coded in human nucleus accumbens. NeuroImage 2006;31:790–5.

[29] Shenhav A, Greene JD. Moral judgments recruit domain-general valuation mechanisms to integrate representations of probability and magnitude. Neuron 2010;67:667–77.

[30] Tobler PN, Christopoulos GI, O'Doherty JP, Dolan RJ, Schultz W. Neuronal distortions of reward probability without choice. J Neurosci 2008;28:11703–11.

[31] Studer B, pergis-Schoute AM, Robbins TW, Clark L. What are the odds? The neural correlates of active choice during gambling. Front Neurosci 2012;6:46.

[32] Berns GS, Bell E. Striatal topography of probability and magnitude information for decisions under uncertainty. NeuroImage 2012;59:3166–72.

[33] Hsu M, Bhatt M, Adolphs R, Tranel D, Camerer CF. Neural systems responding to degrees of uncertainty in human decision-making. Science 2005;310:1680–3.

[34] Levy I, Snell J, Nelson AJ, Rustichini A, Glimcher PW. Neural representation of subjective value under risk and ambiguity. J Neurophysiol 2010;103:1036–47.

[35] Burke CJ, Tobler PN. Reward skewness coding in the insula independent of probability and loss. J Neurophysiol 2011;106:2415–22.

[36] Knutson B, Adams CM, Fong GW, Hommer D. Anticipation of increasing monetary reward selectively recruits nucleus accumbens. J Neurosci 2001;21:RC159.

[37] Knutson B, Taylor J, Kaufman M, Peterson R, Glover G. Distributed neural representation of expected value. J Neurosci 2005;25:4806–12.

[38] Tom SM, Fox CR, Trepel C, Poldrack RA. The neural basis of loss aversion in decision-making under risk. Science 2007;315:515–8.

[39] Preuschoff K, Bossaerts P, Quartz SR. Neural differentiation of expected reward and risk in human subcortical structures. Neuron 2006;51:381–90.

[40] Yacubian J, Glascher J, Schroeder K, Sommer T, Braus DF, Buchel C. Dissociable systems for gain- and loss-related value predictions and errors of prediction in the human brain. J Neurosci 2006;26:9530–7.

[41] Yacubian J, Sommer T, Schroeder K, Glascher J, Braus DF, Buchel C. Subregions of the ventral striatum show preferential coding of reward magnitude and probability. NeuroImage 2007; 38:557–63.

[42] Breiter HC, Aharon I, Kahneman D, Dale A, Shizgal P. Functional imaging of neural responses to expectancy and experience of monetary gains and losses. Neuron 2001;30:619–39.

[43] Christopoulos GI, Tobler PN, Bossaerts P, Dolan RJ, Schultz W. Neural correlates of value, risk, and risk aversion contributing to decision making under risk. J Neurosci 2009;29:12574–83.

[44] Delgado MR, Stenger VA, Fiez JA. Motivation-dependent responses in the human caudate nucleus. Cereb Cortex 2004;14: 1022–30.

[45] Kahnt T, Tobler PN. Salience signals in the right temporoparietal junction facilitate value-based decisions. J Neurosci 2013;33: 863–9.

[46] Plassmann H, O'Doherty J, Rangel A. Orbitofrontal cortex encodes willingness to pay in everyday economic transactions. J Neurosci 2007;27:9984–8.

[47] Tobler PN, Christopoulos GI, O'Doherty JP, Dolan RJ, Schultz W. Risk-dependent reward value signal in human prefrontal cortex. Proc Natl Acad Sci USA 2009;106:7185–90.

[48] Plassmann H, O'Doherty JP, Rangel A. Appetitive and aversive goal values are encoded in the medial orbitofrontal cortex at the time of decision making. J Neurosci 2010;30:10799–808.

[49] Kahnt T, Heinzle J, Park SQ, Haynes JD. The neural code of reward anticipation in human orbitofrontal cortex. Proc Natl Acad Sci USA 2010;107:6010–5.

[50] Kim H, Shimojo S, O'Doherty JP. Is avoiding an aversive outcome rewarding? Neural substrates of avoidance learning in the human brain. PLoS Biol 2006;4:e233.

[51] Kahnt T, Heinzle J, Park SQ, Haynes JD. Decoding the formation of reward predictions across learning. J Neurosci 2011;31: 14624–30.

[52] Kennerley SW, Dahmubed AF, Lara AH, Wallis JD. Neurons in the frontal lobe encode the value of multiple decision variables. J Cogn Neurosci 2009;21:1162–78.

[53] Brown JW, Braver TS. Risk prediction and aversion by anterior cingulate cortex. Cogn Affect Behav Neurosci 2007;7:266–77.

[54] Symmonds M, Bossaerts P, Dolan RJ. A behavioral and neural evaluation of prospective decision-making under risk. J Neurosci 2010;30:14380–9.

[55] McNamee D, Liljeholm M, Zika O, O'Doherty JP. Characterizing the associative content of brain structures involved in habitual and goal-directed actions in humans: a multivariate FMRI study. J Neurosci 2015;35:3764–71.

[56] Wunderlich K, Dayan P, Dolan RJ. Mapping value based planning and extensively trained choice in the human brain. Nat Neurosci 2012;15:786–91.

[57] Hare TA, Camerer CF, Rangel A. Self-control in decision-making involves modulation of the vmPFC valuation system. Science 2009;324:646–8.

[58] Kable JW, Glimcher PW. The neurobiology of decision: consensus and controversy. Neuron 2009;63:733–45.

[59] Levy DJ, Glimcher PW. The root of all value: a neural common currency for choice. Curr Opin Neurobiol 2012;22:1027–38.

[60] Wunderlich K, Rangel A, O'Doherty JP. Neural computations underlying action-based decision making in the human brain. Proc Natl Acad Sci USA 2009;106:17199–204.

[61] Barron HC, Dolan RJ, Behrens TE. Online evaluation of novel choices by simultaneous representation of multiple memories. Nat Neurosci 2013;16:1492–8.

[62] Behrens TE, Woolrich MW, Walton ME, Rushworth MF. Learning the value of information in an uncertain world. Nat Neurosci 2007;10:1214–21.

[63] Rushworth MF, Noonan MP, Boorman ED, Walton ME, Behrens TE. Frontal cortex and reward-guided learning and decision-making. Neuron 2011;70:1054–69.

[64] Clithero JA, Rangel A. Informatic parcellation of the network involved in the computation of subjective value. Soc Cogn Affect Neurosci 2013;9(9):1289–302.

[65] Bartra O, McGuire JT, Kable JW. The valuation system: a coordinate-based meta-analysis of BOLD fMRI experiments examining neural correlates of subjective value. NeuroImage 2013;76:412–27.

[66] Park SQ, Kahnt T, Rieskamp J, Heekeren HR. Neurobiology of value integration: when value impacts valuation. J Neurosci 2011;31:9307–14.

[67] Chikazoe J, Lee DH, Kriegeskorte N, Anderson AK. Population coding of affect across stimuli, modalities and individuals. Nat Neurosci 2014;17:1114–22.

[68] Howard JD, Gottfried JA, Tobler PN, Kahnt T. Identity-specific coding of future rewards in the human orbitofrontal cortex. Proc Natl Acad Sci USA 2015;112:5195–200.

[69] McNamee D, Rangel A, O'Doherty JP. Category-dependent and category-independent goal-value codes in human ventromedial prefrontal cortex. Nat Neurosci 2013;16:479–85.

[70] Chib VS, Rangel A, Shimojo S, O'Doherty JP. Evidence for a common representation of decision values for dissimilar goods in human ventromedial prefrontal cortex. J Neurosci 2009;29: 12315–20.

[71] Klein-Flugge MC, Barron HC, Brodersen KH, Dolan RJ, Behrens TE. Segregated encoding of reward-identity and stimulus-reward associations in human orbitofrontal cortex. J Neurosci 2013;33:3202–11.

[72] Sescousse G, Redoute J, Dreher JC. The architecture of reward value coding in the human orbitofrontal cortex. J Neurosci 2010; 30:13095–104.

[73] Kahnt T, Chang LJ, Park SQ, Heinzle J, Haynes JD. Connectivity-based parcellation of the human orbitofrontal cortex. J Neurosci 2012;32:6240–50.

[74] Ongur D, Price JL. The organization of networks within the orbital and medial prefrontal cortex of rats, monkeys and humans. Cereb Cortex 2000;10:206–19.

[75] Li Y, Sescousse G, Amiez C, Dreher JC. Local morphology predicts functional organization of experienced value signals in the human orbitofrontal cortex. J Neurosci 2015;35:1648–58.

[76] Sescousse G, Caldu X, Segura B, Dreher JC. Processing of primary and secondary rewards: a quantitative meta-analysis and review of human functional neuroimaging studies. Neurosci Biobehav Rev 2013;37:681–96.

[77] Bogler C, Bode S, Haynes JD. Decoding successive computational stages of saliency processing. Curr Biol 2011;21:1667–71.

[78] Itti L, Koch C. Computational modelling of visual attention. Nat Rev Neurosci 2001;2:194–203.

[79] Ogawa M, van der Meer MA, Esber GR, Cerri DH, Stalnaker TA, Schoenbaum G. Risk-responsive orbitofrontal neurons track acquired salience. Neuron 2013;77:251–8.

[80] Roesch MR, Calu DJ, Esber GR, Schoenbaum G. All that glitters… dissociating attention and outcome expectancy from prediction errors signals. J Neurophysiol 2010;104:587–95.

[81] Leathers ML, Olson CR. In monkeys making value-based decisions, LIP neurons encode cue salience and not action value. Science 2012;338:132–5.

[82] Litt A, Plassmann H, Shiv B, Rangel A. Dissociating valuation and saliency signals during decision-making. Cereb Cortex 2011;21: 95–102.

[83] Norman KA, Polyn SM, Detre GJ, Haxby JV. Beyond mind-reading: multi-voxel pattern analysis of fMRI data. Trends Cogn Sci 2006;10:424–30.

[84] Haynes JD, Rees G. Decoding mental states from brain activity in humans. Nat Rev Neurosci 2006;7:523–34.

[85] Platt ML, Glimcher PW. Neural correlates of decision variables in parietal cortex. Nature 1999;400:233–8.

[86] Peck CJ, Jangraw DC, Suzuki M, Efem R, Gottlieb J. Reward modulates attention independently of action value in posterior parietal cortex. J Neurosci 2009;29:11182–91.

[87] Louie K, Glimcher PW. Separating value from choice: delay discounting activity in the lateral intraparietal area. J Neurosci 2010;30:5498–507.

[88] Corbetta M, Shulman GL. Control of goal-directed and stimulus-driven attention in the brain. Nat Rev Neurosci 2002;3:201–15.

[89] Corbetta M, Patel G, Shulman GL. The reorienting system of the human brain: from environment to theory of mind. Neuron 2008;58:306–24.

[90] Geng JJ, Mangun GR. Right temporoparietal junction activation by a salient contextual cue facilitates target discrimination. NeuroImage 2011;54:594–601.

[91] Indovina I, Macaluso E. Dissociation of stimulus relevance and saliency factors during shifts of visuospatial attention. Cereb Cortex 2007;17:1701–11.

[92] Foote SL, Morrison JH. Extrathalamic modulation of cortical function. Annu Rev Neurosci 1987;10:67–95.

[93] Morrison JH, Foote SL. Noradrenergic and serotoninergic innervation of cortical, thalamic, and tectal visual structures in Old and New World monkeys. J Comp Neurol 1986;243:117–38.

[94] Aston-Jones G, Rajkowski J, Kubiak P, Alexinsky T. Locus coeruleus neurons in monkey are selectively activated by attended cues in a vigilance task. J Neurosci 1994;14:4467–80.

[95] Usher M, Cohen JD, Servan-Schreiber D, Rajkowski J, Aston-Jones G. The role of locus coeruleus in the regulation of cognitive performance. Science 1999;283:549–54.

[96] Metereau E, Dreher JC. The medial orbitofrontal cortex encodes a general unsigned value signal during anticipation of both appetitive and aversive events. Cortex 2015;63:42–54.

[97] Rothkirch M, Schmack K, Schlagenhauf F, Sterzer P. Implicit motivational value and salience are processed in distinct areas of orbitofrontal cortex. NeuroImage 2012;62:1717–25.

[98] O'Neill M, Schultz W. Coding of reward risk by orbitofrontal neurons is mostly distinct from coding of reward value. Neuron 2010; 68:789–800.

10

Computational Principles of Value Coding in the Brain

K. Louie, P.W. Glimcher

New York University, New York, NY, United States

Abstract

The notion of value is central to theoretical and empirical approaches to decision-making. In psychological and economic choice theory, value functions quantify the relationship between relevant decision information and choice behavior. Evidence for value coding in neural circuits suggests that value information is explicitly represented in brain activity and plays a critical role in the neurobiological choice process. Here, we review a research approach centered on the computations underlying neural value coding. As in sensation and perception, neural information processing in valuation and choice relies on core computational principles including contextual modulation and divisive gain control. The form of these computations reveals details about the nature of decision-related value information and the constraints inherent in computing with biological systems. Understanding value representation at the intermediate level of computation promises insight into decision-making at the level of both the underlying circuit architecture and the resulting choice behavior.

INTRODUCTION

Decision-making is a fundamental behavior for any organism that interacts with its environment. At its most complete, the decision process involves assessing the current state of the world, assessing the current state of the decision-maker, and selecting the optimal course of action. In higher order organisms, this multifaceted process can involve a host of cognitive functions such as attention, emotion, and memory in addition to the selection mechanism itself, complicating the neurobiological study of decision-making.

To approach this complexity, research has begun to focus on the aggregation of information guiding the choice process. Drawing from classic work in economics, an integrated decision variable—combining all relevant information into a common scale for comparison—forms the basis of standard neurobiological models of decision-making. In economic and psychological models, this quantity represents a subjective measure of satisfaction termed expected utility or decision utility [1,2]. In ecological foraging theory, this quantity represents an internal currency ultimately related to reproductive fitness [3]. In this chapter, we use the general term *value* to denote the integrated individual-specific subjective worth of a choice option as inferred from either neural or behavioral measurements.

Understanding how the brain processes value information is a key target of the neuroscience of decision-making. Neuroscience research spans multiple levels of analysis, from the detailed study of synaptic proteins to the examination of broad patterns of behavior. Between these extremes lies a key level of analysis, concerned with the computations that serve as a bridge between biological implementation and organism behavior [4]. In the fields of systems and cognitive neuroscience, an approach based on neural computation and information coding has led to significant advances in the understanding of sensory and motor circuits. Here, we outline a neural coding and computation framework for the study of value representation and decision-making (Fig. 10.1).

VALUE AND CHOICE BEHAVIOR

The idea of a summary decision variable is central to most formalized models of decision-making. In these models, value functions serve to integrate all relevant decision information into a single (although not necessarily consistent) metric for comparison and selection. In computer science, reinforcement learning (RL)

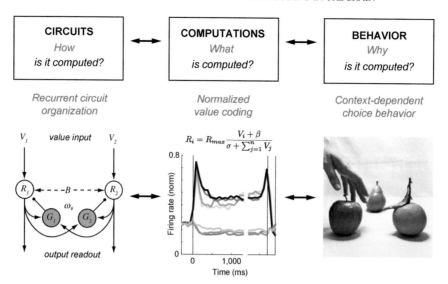

FIGURE 10.1 Computation as a bridge between mechanism and behavior. Information processing can be separated into distinct levels of description concerned with biological implementation, computational algorithm, and behavioral goal [4,116]. In the neurobiological study of decision-making, research is extending beyond documenting value-correlated neural activity to identifying the algorithms governing value coding in decision circuits. Understanding at the intermediate level of computation provides insight into both the organization of underlying decision circuitry and the nature of resulting value-guided choice behavior.

models assume that agents learn a value function of this kind based on the rewards experienced after a given action [5]. Action selection is then based on a comparison of the expected future rewards associated with different alternative actions. RL models precisely characterize reward-guided choice behavior in animals and humans and predict reward-related activity in diverse brain areas (see section "Value Coding in Decision Circuits"), empirically supporting the theoretical notion that value information plays a crucial role in the decision process [6,7] (see also Chapters 2, 9, and 10).

As noted above, value functions are also central to normative theories of decision-making, such as economic rational choice theory and ecological foraging theory [1,3], and to more behavioral or psychological theories like prospect theory [2]. Normative models aim to define optimal choices—assuming perfect knowledge and full accuracy—as benchmarks for examining empirical choice behavior. These optimal benchmarks require the existence of an underlying function defining the value of choice alternatives. Unlike the computational value functions used in RL, the value functions guiding empirical choosers cannot be analytically defined and must be inferred through observed choice behavior.

The development of a normative, value-based framework is the foundation for modern economic theory. Central to this development is the question of how people assign values in risky decisions, of which outcomes are probabilistic and drawn from known

distributions. Early theorists, like the mathematician Pascal, asserted that decisions were driven solely by the product of outcome probability and magnitude:

$$EV = px$$

where EV is the *expected value* of a prospect (x, p) offering $\$x$ with probability p [8]. However, this simple formulation fails to accurately describe empirically observed human behavior, in which the degree of risk significantly affects preference. For example, most choosers prefer a certain $\$49$ over a 50:50 chance of $\$100$ or nothing, suggesting that values depend on additional information beyond objective option properties alone.

A key breakthrough was the introduction of subjective, individual-specific parameters to value functions. In response to the inadequacy of expected value, Bernoulli proposed instead that choices are based on a transformed version of expected value termed *expected utility*:

$$EU = pu(x)$$

where the utility function $u(x)$ defines a transformation of reward magnitude [9]. Critically, the preference behavior of individual choosers can be defined by different parameters in the utility function. For example, given a utility function $u(x) = x^\alpha$, a risk-averse chooser behaves according to a value function with $\alpha < 1$ (Fig. 10.2). Subsequently, Samuelson and others developed an axiomatic framework around expected utility, defining a normative theory (often referred to as *rationality*) as choices consistent with

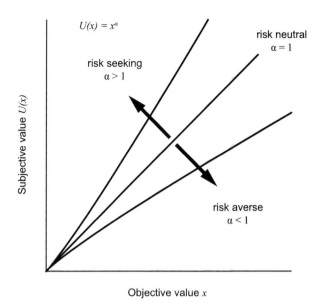

$U(x) = x^a$

risk neutral
$\alpha = 1$

risk seeking
$\alpha > 1$

risk averse
$\alpha < 1$

Subjective value $U(x)$

Objective value x

FIGURE 10.2 **Value functions derived from behavior.** Example value functions relating subjective value (utility) to objective value (outcome amount). In this case, the use of a parametric form allows the value function to characterize different attitudes to risky outcomes between individuals.

maximizing a subjective, internal representation of value [1,10,11].

A central biological question is whether the internal value signals posited by behavioral choice theories, ranging from that of Pascal to that of Kahneman and Tversky, are represented by specific underlying neural systems. The economic definition of rationality requires only that decision-makers choose "as if" they are maximizing a theoretical utility quantity. However, electrophysiological [12–14] and neuroimaging data [15–18] show that neural activity in a number of brain areas correlates with behaviorally inferred estimates of subjective value. Such findings suggest that behaviorally defined value functions may be explicitly coded in neural activity and play a critical role in the biological decision process [19].

VALUE CODING IN DECISION CIRCUITS

Reward-related responses are, of course, evident in many different brain areas. A key step in examining neural value coding is distinguishing decision-related value representation from other forms of reward-related activity. Various neural responses may appear closely correlated with value but instead encode other related forms of information [20]. For example, midbrain dopaminergic neurons reflect reward prediction error, a quantity encoding the difference between expected and received reward value [21,22] (see Chapter 2). Prediction errors are an integral component of computational RL models [5], and dopaminergic activity probably reflects a teaching signal used to update rather than represent value information.

A widely used approach to identifying decision-related value coding activity is to target for study neural circuits believed to play a causal role in the choice process itself. For example, electrophysiological studies have focused on sensorimotor circuits underlying action selection in frontal brain areas, posterior parietal cortex, and the basal ganglia [23–28]. In these brain regions, the simultaneous representation of multiple potential motor actions suggests that target selection and movement preparation represent an integrated decision mechanism [29]. Value modulation of activity in these circuits probably serves to guide the selection of actions, playing an integral role in the decision process [29–31].

While value coding has been examined in multiple brain regions, including the orbitofrontal cortex (Chapters 3, 10, and 16), the dorsolateral prefrontal cortex (Chapter 8), and the basal ganglia (Chapter 5), we specifically focus here on value representation in the monkey lateral intraparietal area (LIP). The LIP is a sensorimotor region of the posterior parietal cortex that has provided many of the seminal results in the neurophysiological study of decision-making (Fig. 10.3). Both anatomical connectivity and electrophysiological evidence support a role for LIP in saccadic action (eye movement) selection. Anatomically, LIP receives afferents from higher order sensory areas such as the extrastriate visual cortex and sends efferents to saccade execution areas including the frontal eye fields and superior colliculus [32,33]. Functionally, many individual LIP neurons respond to both visual stimulus presentation in a circumscribed region of visual space termed the response field (RF) and before and during eye movements to the same spatial location [34–36]. Critically, LIP responses are neither strictly sensory nor strictly motor in nature; in memory saccade tasks in which the sensory stimulus is only transiently presented and saccades are delayed, LIP neurons exhibit persistent spiking activity dissociated from both stimulus presence and active eye movements [37–39]. These findings suggest that LIP is involved in a higher order cognitive process related to the intention to execute an action [40]. In fact, intention-related activity appears to be a general feature of the posterior parietal cortex, with different effector-specific subregions representing the planning of different motor actions including saccades, grasps, and reaches [41].

Consistent with a role in decision-making, LIP activity is strongly correlated with decision-related information in a wide variety of saccade-related choice paradigms. In perceptual discrimination tasks using noisy random dot motion stimuli that provide gradual

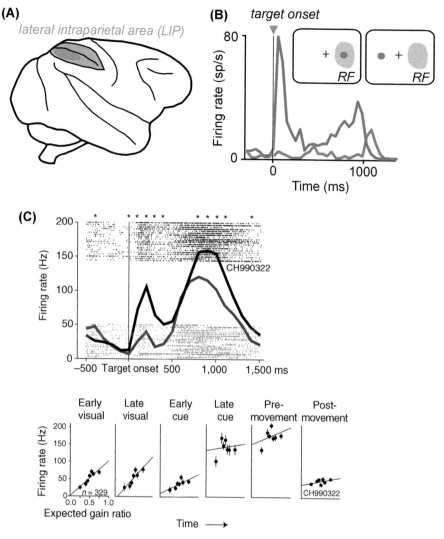

FIGURE 10.3   **Value representation in sensorimotor circuits.** (A) The lateral intraparietal area (LIP) in the monkey brain. Neurons in this sensorimotor region in the intraparietal sulcus respond to both visual stimulus presentation and saccadic eye movements. (B) Action-selective activity in LIP neurons. Individual LIP neurons exhibit visual and motor activity selective for a circumscribed region of visual space termed the response field (RF). (C) Value coding in LIP neurons. Example LIP neuron activity with activity correlated with reward magnitude (*gray, low; black, high*). *Data adapted with permission from Platt ML, Glimcher PW. Neural correlates of decision variables in parietal cortex. Nature 1999;400:233–38.* Regardless of the experimental manipulation used to control the behavioral significance of the intended action, LIP activity correlates with subjective value across a broad range of studies.

decision-related information, LIP responses appear to ramp (or gradually increase their activity) during stimulus presentation in a manner dependent on the strength of the motion signal [42–44]. Under these circumstances, individual LIP neuron firing rates increase if the presented motion information supports a decision to saccade toward the RF of the recorded neuron. This neural activity is consistent with an evolving decision signal representing the accumulated sensory evidence for the choice of a particular action [45]. When LIP activity is perturbed with either a brief motion pulse or electrical microstimulation, choice behavior shifts toward the stimulated RF target, suggesting a causal role in the decision process under at least some conditions

[46,47]. While the initial use of a motion-based task was motivated by the existence of direct afferents from motion-sensitive neurons to the posterior parietal cortex [44], decision-related activity in the LIP generalizes to different sources of choice information. For example, in monkeys trained on a probabilistic categorization task, LIP activity tracks the likelihood of reward for a particular saccade as indicated by dynamically changing cue configurations of various visual shapes [48,49].

A general framework for decision-related LIP activity emerged with the discovery of value representation during economic choice tasks [12,26,30]. In these paradigms, options differ in their associated rewards, and choices are driven by the subject's evaluation of potential outcomes. In contrast to perceptual tasks, in economic choices the relevant decision variable relies on internal subjective valuation rather than external sensory information. Across a number of studies, the activity of LIP neurons covaries with the value of saccade-associated outcomes (Fig. 10.3). For example, LIP activity increases with the magnitude of reward outcome and the probability of reinforcement associated with a saccade to the neural RF [26]. This reward coding is flexible and dynamic; in foraging environments, LIP activity tracks the trial-by-trial changing value of options as determined by recent reward history [28,50]. Moreover, LIP neurons simultaneously encode both reward magnitude and perceptual evidence in motion discrimination tasks with variable reinforcement [51], consistent with an overall subjective value measure integrating outcome and likelihood information [26,50,52].

As expected for a decision variable representing subjective value, LIP value coding incorporates the influence of all behaviorally relevant costs and benefits. In a temporal discounting task, LIP activity reflects the individual-specific decrement of reward value driven by imposed delays to reward [52]. In competitive games against opponents, LIP activity reflects the value of options determined by strategic interactions [50,53]. In social decision tasks, LIP activity reflects the behavioral value of conspecific visual targets as governed by features such as social status and reproductive salience [54]. Thus, firing rates in LIP represent a consistent value signal that reflects both the objective characteristics of rewards (e.g., delay to reinforcement) and the subjective weighting of relevant reward information (e.g., delay discount function). In terms of decision theory, this neural representation of subjective value is closely tied to the economic notion of expected utility derived solely from behavior.

While most extensively documented in the LIP, value coding appears to be a general feature of decision-related brain areas. Multiple brain structures necessary for normal saccade initiation are modulated by reward information, including the frontal eye fields, the dorsal striatum, the substantia nigra pars reticulata, and the superior colliculus [23,27,55–58]. In addition to the saccadic system, value representation extends to brain areas responsible for different motor effectors. For example, neurons in the medial intraparietal area of the posterior parietal cortex that represent specific arm movements are modulated by the value of associated reaches [59]. Similarly, the activity of reach-selective neurons in dorsal premotor cortex reflect an action value signal [25]. These findings suggest that decisions implemented via action selection occur in effector-specific pathways, where reward characteristics, internal subjective weighting, and motor cost information are combined into a subjective value signal to guide

choice. Additionally, decisions may also be made in an action-independent manner, with choices made over option identity without regard to motor implementation [60,61]. Neurons thought to underlie such goods-based decision-making, such as those in the orbitofrontal cortex, also show extensive modulation by value [62–65] (see also Chapter 10 with regard to humans).

CONTEXT DEPENDENCE IN BRAIN AND BEHAVIOR

A principal goal of systems neuroscience is elucidating the link between neural activity and behavior. While value coding occurs widely in the brain, the relationship between these signals and neural decision mechanisms is still unknown. Electrophysiological experiments done as of this writing have largely demonstrated neural activity that correlates with behavioral value measures. However, unlike theoretical value quantities, neural value representations are subject to biophysical constraints. These constraints include hard limitations like nonnegative spiking rates and maximum levels of neural activity and more general optimization concerns such as minimization of metabolic costs [66,67]. To represent information via spiking activity under biological constraints, neural systems must implement a realizable transformation between behaviorally relevant input variables and output firing rates. A critical question for decision neuroscience is the nature of this input–output function relating value information to firing rates.

The principles relating neural activity and information representation have been most thoroughly documented in the study of sensation and perception. This is true in large part because the environmental distribution of sensory quantities can be readily characterized, providing a metric for the capacity constraints and other limitations of information coding in sensory systems. It is now known that natural sensory signals are characterized by widespread statistical regularities in the distribution of environmental variables [68]. A fundamental principle of neural information coding that has emerged from theoretical and empirical work on sensory signal representation is the *efficient coding hypothesis*. Inspired by earlier theories of information transmission and channel capacity in the field of communication [69], Barlow proposed that sensory neurons implement coding strategies that minimize regularity-induced redundancies to maximize information under biological capacity constraints [70]. Under efficient coding, neural systems should adjust their representations to take advantage of known or learned regularities in the sensory environment. A large body of empirical evidence now supports the idea that neurons in sensory circuits adopt coding strategies adapted to characteristics of their input distributions [71–80].

While different coding strategies have been proposed to mediate efficient coding in sensory circuits, we focus here on a particular process relevant to both sensory and decision-related processing: *contextual modulation*. Neural and perceptual forms of contextual modulation are widely studied, but there are no standard definitions of either "context" or "modulation" in the literature. For explanatory purposes, we adopt here a broad conceptual definition of contextual modulation as the interaction of two classes of inputs to a given neuron or neural circuit [81]. The first class consists of the primary driving inputs to a neural system; in an organized hierarchy of cortical areas, these are typically feed-forward connections from earlier areas in the processing stream (Fig. 10.4). The second class consists of modulatory inputs that control the system response to the driving inputs; hierarchically, contextual inputs arise from recurrent interactions at the same level or feedback connections from later areas in the processing stream. Functionally, contextual control by modulatory connections implements a tuning of the neural input–output function defined for driving inputs.

The prototypical example of contextual modulation— *surround suppression*—occurs in visual cortical neurons, whereby responses to stimuli presented in a neuron's classical receptive field (CRF) are decreased by stimuli presented in a nearby region of visual space termed the nonclassical or extraclassical receptive field (nCRF) [82–85]. At the experimental level, sufficiently strong CRF input generates action potentials, whereas modulatory nCRF inputs cannot generate spiking activity on

their own, but act by refining responses to CRF stimulation. Inputs contemporaneous with the driving inputs, such as that elicited by nCRF stimuli in surround suppression, mediate a modulation by *spatial context*. However, neural responses are also sensitive to the recent history of past inputs, or the *temporal context*. Temporal context robustly affects neural processing in sensory systems, most commonly documented in the form of adaptation to recent sensory experience [86]. There are many psychophysical and electrophysiological parallels between spatial and temporal context effects, suggesting that they reflect a common information processing mechanism in neural circuits [87].

Contextual modulation of neural responses is a widespread phenomenon in sensory processing [81,88–90], governing neural responses to information ranging from low-level features such as stimulus orientation in early visual areas [91–93] to high-level features like invariant object identity in the medial temporal lobe [94]. While contextual modulation has been documented most extensively in visual pathways, analogous effects occur in circuits underlying other sensory modalities including audition [95–97] and somatosensation [80,98,99]. Moreover, multiple spatial and temporal contextual effects have been documented in perceptual psychophysics, linking contextual modulation in sensory circuits to observable behavioral phenomena [87].

Given its ubiquity, contextual modulation is thought to be a central mechanism of cortical function [87,88,100,101]. While many hypothesized roles have been proposed for contextual modulation [81,90], we focus here on its contribution to efficient coding. Given the finite dynamic range of biological neurons, both spatial [76,102–105] and temporal [74,75,106,107] contextual modulation can contribute to efficient coding by accounting for statistical regularities in the environment. Spatial regularities reflect the intrinsic structure in natural images, such as the strong correlation in luminance between neighboring points in space and the relative abundance of vertical and horizontal orientations. Temporal regularities reflect the short-term persistence of local statistics over time, which will change on a longer timescale in a dynamic world. These regularities introduce an intrinsic correlation in the outputs of linear encoding systems, reducing information densities in the sensory representation; contextual modulation implements nonlinear processing that can effectively decorrelate the outputs of a neural circuit and thus produce higher information densities. A key aspect of contextual modulation in this regard is that information is encoded in a relative manner, with contextual inputs controlling the transformation between driving input and spiking output;

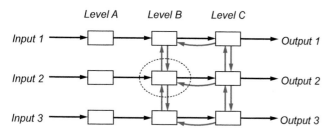

FIGURE 10.4 **Contextual modulation in information processing.** Contextual modulation can be broadly interpreted in terms of the various inputs in a hierarchical processing stream. In this schematic, information is processed in separate channels through successive layers of hierarchically organized areas via feed-forward inputs (*black*). These channels represent initially distinct information streams, for example, separate regions of visual space or individual saccadic eye-movement plans. Contextual modulation at a given area (*dotted*) can be mediated by recurrent inputs from the same hierarchical level or feedback inputs from higher level areas (*gray*).

this produces a coding of input information dependent on the surrounding spatiotemporal environment.

For the neurophysiology of decision-making, contextual modulation provides a framework for the transformation between behaviorally defined measures of value and their neural representation. Under contextual modulation, the neural representation of the value of a choice option should incorporate relevant information in the surrounding spatial or temporal value environment. While previous neurophysiological studies have largely examined value coding independent of context, more recent recordings from sensorimotor circuits like the LIP demonstrate a relative value code implemented via contextual modulation [51,108]. These studies find that, in addition to the standard activation by RF target value [26,28,52], LIP neurons exhibit suppression by the value of extra-RF targets (Fig. 10.5). For example, Rorie et al. recorded LIP neurons while monkeys performed a standard motion discrimination task [51]; additionally, the experimenters introduced variable reward outcomes (low or high) associated with the two alternative actions. In this task, LIP neurons show a multiplexed representation of all relevant variables including sensory, reward, and motor information. Crucially, LIP activity reflects the relative value of a saccade, with higher firing rates for a given RF target reward if presented with a low-valued extra-RF alternative. This relative value coding extends beyond binary decision scenarios to more diverse choice set architectures. Louie et al. examined LIP neurons while monkeys were presented single, double, and triple target arrays with varying reward contingencies [108]. In all configurations, LIP activity exhibits a consistent relative value coding with activity increased by RF target value and suppressed by the summed value of alternative, extra-RF targets. Moreover, the precise form of this representation links contextual value modulation to a specific and well-studied canonical computation, divisive normalization (see section "Neural Computations Underlying Value Representation").

Contextual modulation appears to be a fundamental characteristic of neural value coding [12,109]. The observed relative value coding in LIP is driven by the spatial context defined by the available choice alternatives. In this form of contextual modulation, the response of action-selective neurons is modulated by the value of alternative actions that alone are ineffective in driving neural activity. A similar contextual valuation occurs in monkey dorsal premotor cortex, where reach-selective neurons exhibit a relative value signal incorporating the rewards of both preferred and nonpreferred arm movements [25]. Furthermore, contextual value modulation is consistent with the effect of choice set size and target uncertainty in reducing firing rates in

downstream motor structures like the superior colliculus [110,111]. In addition to spatial context, value coding activity also depends on the recent reward environment or temporal context. Reward-related responses in monkey orbitofrontal cortex (OFC), which encode decision-related variables like the subjective value of option types and the value of chosen alternatives, adapt to the local statistics (e.g., range or variance) of recently received rewards [112,113]. Similar adaptive value coding occurs in the monkey anterior cingulate cortex [61] and midbrain dopaminergic nuclei [114] and extends to reward-related processing in the human brain [115]. Both the spatial and the temporal forms of contextual modulation function as a form of gain control, modifying the input—output function that relates value information to firing rates. Consideration of the precise form of this modulation offers insight into both the underlying circuit computations and how they may be implemented via circuit architecture.

NEURAL COMPUTATIONS UNDERLYING VALUE REPRESENTATION

While contextual modulation provides an empirical description of neural value coding, further understanding requires a quantification of how such modulation arises in decision-related circuits. To link circuit activity to behavioral phenomena like choice, proposals emphasize an intermediate-level approach focusing on information processing and computation [12,116]. This approach is inspired by the identification of certain canonical computations that appear in neural activity in diverse circuit architectures, brain areas, and animal species. Such canonical computations can arise from markedly different neural mechanisms, suggesting a conservation of information processing algorithm rather than specific biological implementation [117].

In the LIP, contextual value coding is consistent with a particular canonical computation, divisive normalization. Originally developed to describe response properties in primary visual cortex [118], divisive normalization is a nonlinear gain control algorithm widely observed in multiple sensory modalities and brain regions, including contrast gain control in the retina and thalamus [119,120], surround suppression in the middle temporal area [121,122], ventral stream responses to multiple objects [123], and gain control in the auditory cortex [124]. Furthermore, normalization characterizes neural activity related to higher order functions including visual attention [125,126] and multisensory integration [127], and aberrant normalization processes may underlie the unbalanced excitation and inhibition

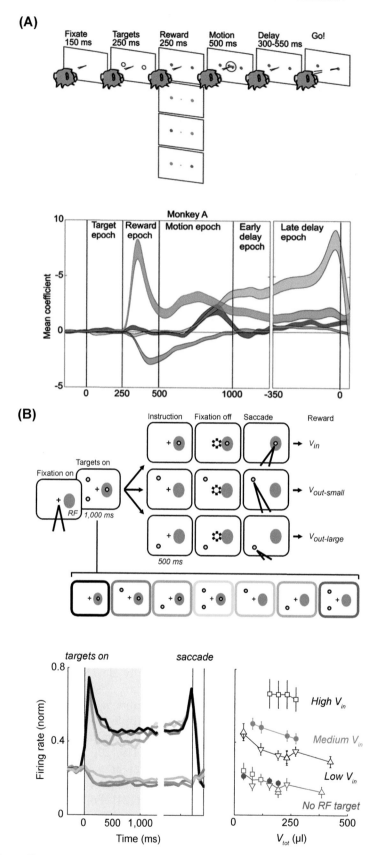

FIGURE 10.5   Relative value coding in monkey parietal cortex. (A) Contextual modulation of value information in a perceptual task. Top, standard motion discrimination task with different cued reward outcomes (low or high, for each target). Bottom, average regression weights in a

in pathophysiological conditions like autism and schizo-phrenia [128].

The key computation underlying normalization is a ratio between the direct driving inputs to a neuron and the inputs to a more broadly tuned pool of neurons. The general equation:

$$R_i = R_{\max} \frac{d_i^\alpha + \beta}{\sigma^\alpha + \sum_{j=1}^n w_j d_j^\alpha}$$

describes how the response of neuron i depends on both the direct input d_i to that neuron and the inputs to a larger normalization pool d_j. Additional parameters $R_{\max}$ and β determine the maximal response level and spontaneous baseline activity, respectively, whereas the parameter α governs exponentiation of the inputs. The semisaturation parameter σ governs the overall shape of the response function, describing how activity saturates with increasing driving input and determining the range of inputs with maximal response sensitivity. Under normalization, the direct drive to a specific neuron is divisively scaled by a term that sums over the inputs to a larger pool of neurons (typically including the neuron in question). In contextual modulation, this term mediates the effect of modulatory inputs via divisive gain control. The weight parameter w_j allows a precise tuning of modulatory inputs to reproduce empirical characteristics of contextual effects (e.g., visual extent of nCRF in surround suppression).

In a 2011 study (Fig. 10.6), quantification of LIP responses in varying reward scenarios reveals a relative value coding mediated by divisive normalization [108]. Neural responses depended on rewards associated with both RF and extra-RF target values, and were well approximated with a simple form of the normalization model:

$$R_i = R_{\max} \frac{V_i + \beta}{\sigma + \sum_{j=1}^n V_j}$$

In this value normalization algorithm, the response of a given neuron increases with the value of the target in its RF, consistent with previous studies of LIP value-related activity [26,28,50]. Relative value coding is mediated by a divisive term, which includes the summed value of all available alternatives, consistent with suppression driven by extra-RF option values [51]. Importantly, the examination of activity across a range of reward configurations enabled model comparison between divisive normalization and alternative relative value algorithms based on value difference or simple fractional value. In this comparison, divisive normalization far outperformed these alternative models in characterizing the nonlinear relationship between LIP activity and option values.

Normalized value coding carries a number of computational ramifications for how values are represented and choices are made. First, the hyperbolic ratio in the algorithm guarantees that neural activity eventually saturates as option value increases. This type of saturation is a key element of how normalization operates in the sensory domain, describing, for example, why V1 neural activity saturates with increasing contrast regardless of stimulus orientation [118,129,130]. Analogously, a compressive saturating value function ensures that the neural representation of the wide range of possible environmental rewards does not exceed the finite dynamic range of spiking neurons. Compressive utility functions model the risk-averse choice behavior seen in most human subjects [131,132], raising the possibility that decision behavior under risk may arise from the underlying biological mechanism. Second, the semisaturation term σ governs the sensitivity of the neural input—output function, determining the range of values that will elicit the largest changes in spiking activity. Intuitively, only values with magnitudes equivalent or larger than σ can elicit significant contextual suppression, producing a response continuum between absolute and relative value coding as a function of the value context defined by the choice set [108]. In sensory processing, changes in σ are proposed to mediate neural adaptation in response to changes in the temporal pattern of inputs [118], such as the lateral shifts in V1 response functions observed after changes in recent stimulus contrast [72,133]. In adaptive terms, σ specifies a weighted average of past inputs representing the expected input in the current environment. How value coding in LIP

single monkey showing the effect of response field (RF) value (*red*) and extra-RF value (*blue*) on lateral intraparietal area (LIP) activity; also shown are the effects of motion information strength (*black*) and choice (*green*). Activation by RF value and suppression by extra-RF value implements a relative reward code. *Adapted with permission from Rorie AE, Gao J, McClelland JL, Newsome WT. Integration of sensory and reward information during perceptual decision-making in lateral intraparietal cortex (LIP) of the macaque monkey. PLoS One 2010;5:e9308 #50.* (B) Contextual modulation of value information in a reward-guided task. Top, instructed reward task with variable reward configurations. Bottom left, average LIP population activity segregated by value context (total summed value of options). LIP activity is differentially modulated despite constant reward conditions (*top lines*, fixed reward target in RF; *bottom lines*, no reward target in RF). Bottom right, average LIP population activity during target presentation (0—1000 ms). *Red and blue points* correspond to data plotted on the left; *black points* represent additional data collected under varying RF value conditions. Consistent with a relative value code, LIP activity increases with increasing value in the RF (different lines) and decreases with increasing value outside the RF (V_{tot}). *Adapted with permission from Louie K, Grattan L, Glimcher PW. Reward value-based gain control: divisive normalization in parietal cortex. J Neurosci 2011;31:10627—39.*

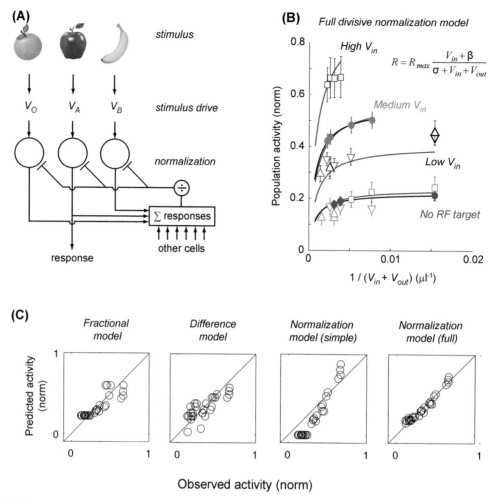

FIGURE 10.6 **Divisive normalization explains relative value coding.** (A) Circuit schematic of divisive normalization. The critical feature of the normalization computation is a pooling of responses over a large neuronal pool, which in turn serves as a divisive gain control on stimulus-driven responses (in this case, value inputs). (B) Lateral intraparietal area (LIP) activity characterized by the divisive normalization. The normalization model captures both activation by response field (RF) target value and suppression driven by the value of extra-RF targets. (C) Comparison of fractional, difference, and normalization models fits. Model predictions are plotted versus empirically recorded average LIP responses across different value conditions. *(B) and (C) are adapted with permission from Louie K, Grattan LE, Glimcher PW. Reward value-based gain control: divisive normalization in parietal cortex. J Neurosci 2011;31:10627—39.*

responds to changing reward statistics is unexplored, but the potential role of normalization in adaptive OFC responses to reward range and variance is an important target for future research. Finally, divisive normalization contributes to the efficient coding of sensory information in visual and auditory processing [76,134], effectively decorrelating neural responses despite significant redundancies in the original sensory information. Regularities in both spatial and temporal statistics can be accommodated by normalization via proper adjustment of the contextual weight and semisaturation terms, respectively. Normalized value coding may serve a similar role in decision-making, adjusting the limited capacity of neural value representations to the most likely rewards and reward scenarios [12].

TEMPORAL DYNAMICS AND CIRCUIT MECHANISMS

Identifying the role of divisive normalization in value coding sheds light on the intermediate algorithmic level of analysis in Marr's tripartite scheme of information processing. A key advantage of this tripartite framework is that knowledge of the underlying computation provides information about the implementation and behavioral levels of analysis [4,116]. One feature supporting the canonical nature of the normalization computation is that it can be implemented by multiple biophysical and network mechanisms [117]. However, the most prominent candidate mechanism for normalization in cortical circuits is recurrent inhibition. Extensive recurrent connectivity is a notable feature of cortical microcircuit organization [135] and feedback inhibition plays an important role in shaping cortical activity [136]. Sensory normalization processes are well characterized by models with lateral connectivity and divisive inhibition [137,138], suggesting that value normalization may reflect an underlying recurrent circuit organization.

Examining activity dynamics provides a key link between the normalization computation and its biophysical mechanism. Nonlinear dynamical systems theory is a fundamental approach in computational neuroscience, modeling data ranging from single-channel kinetics to the evolution of large-scale neural network activity [139,140]. The dynamics of decision-related LIP activity have been extensively studied in perceptual choice tasks, where they are thought to reflect an evidence-accumulation process [42—44] (see also Chapters 11 and 12). In contrast, few studies have specifically investigated the dynamics of value coding and value normalization; value coding has generally been examined in a static manner, assuming stationarity over long time intervals such as the delay period between target presentation and saccade initiation. However, the LIP displays a stereotyped temporal response during saccade choice tasks: a transient response to target presentation, a sustained intervening delay activity, and an action-selective increase tied to motor execution. Little is known in particular about decision coding during the initial phasic response, which is often presumed to be purely sensory driven.

Using a dynamical systems approach to examine value normalization, computational work finds significant value information during early LIP activity [141,142]. This model comprises a recurrent circuit with feed-forward excitation, lateral connectivity, and recurrent inhibition (Fig. 10.7), with paired excitatory (R) and inhibitory (G) neurons assigned to each choice option i. Time-dependent activity is implemented using a system of differential equations:

$$\tau \frac{dG_i}{dt} = -G_i + \sum_{j=1}^{N} \omega_{ij} R_j$$

$$\tau \frac{dR_i}{dt} = -R_i + \frac{V_i + \beta}{1 + G_i}$$

where R and G are neural firing rates, $i = 1, ..., N$ corresponds to individual choice options, τ is a timescale parameter, β describes baseline background input, and the parameter ω_{ij} weights the input R_j to the gain neuron G_i. In this model, feed-forward inputs to R neurons carry value information (V) about the associated option, mediating direct driving input. Contextual modulatory input to a given R neuron arrives via its paired inhibitory neuron, which receives lateral connections from a broader range of R neurons (the normalization pool). Together, this system of equations describes how a model neuron's response evolves as a function of time, value inputs, and the neural activity in the circuit.

The dynamic normalization model makes a number of predictions about the strength and time course of LIP activity. First, the equations have asymptotically stable and unique equilibrium points [141], indicating that neural activities eventually reach a steady state. This behavior matches the characteristic persistent activity of empirical LIP neurons, which maintain fixed firing rates over long delay periods. Second, from physiologically realistic initial conditions, model activity displays a characteristic phasic burst prior to reaching a lower level of equilibrium activity. These phasic-sustained dynamics are evident as a characteristic spiral pattern in the vector field of the dynamical system, which describes how the circuit approached equilibrium following the onset of value input. At the circuit level, these dynamics arise from the asymmetric influence of input: R neurons are driven directly and immediately by input, whereas G neurons receive pooled R inputs and respond only once R activity rises. The subsequent increase in the activity of the G neurons inhibits R neuron activity, driving the network toward its stable equilibrium. Crucially, value input to the model remains constant after onset, and the observed dynamics stem solely from the internal interactions of the normalization circuit. Together, these results suggest that the widely observed stereotyped early dynamics of LIP activity can arise from contextual gain control mechanisms at the circuit level.

In addition to these basic dynamics, the normalization model predicts time-varying aspects of value coding. At equilibrium, the model replicates the relative value coding observed during delay period activity in parietal and premotor cortices [25,51,108]. Specifically, the dynamic model predicts an activation by option value and a contextual suppression by the value of alternatives in steady-state activity. Beyond steady-state value coding, the model also exhibits a temporal evolution in how value information is encoded. In both single- and multiple-option scenarios, modulation of neural activity by RF target value is strongest early in the process, during the initial phasic transient. Furthermore, there are notable differences between direct (RF) and contextual (extra-RF) value modulations: activation by option value is stronger, rises earlier, and peaks earlier then suppression by value context. Critically, these predictions of the dynamic normalization model are consistent with both documented and newly examined aspects of LIP physiology [142]. RF target value most strongly affects neural activity early in the decision process in a number of previous studies

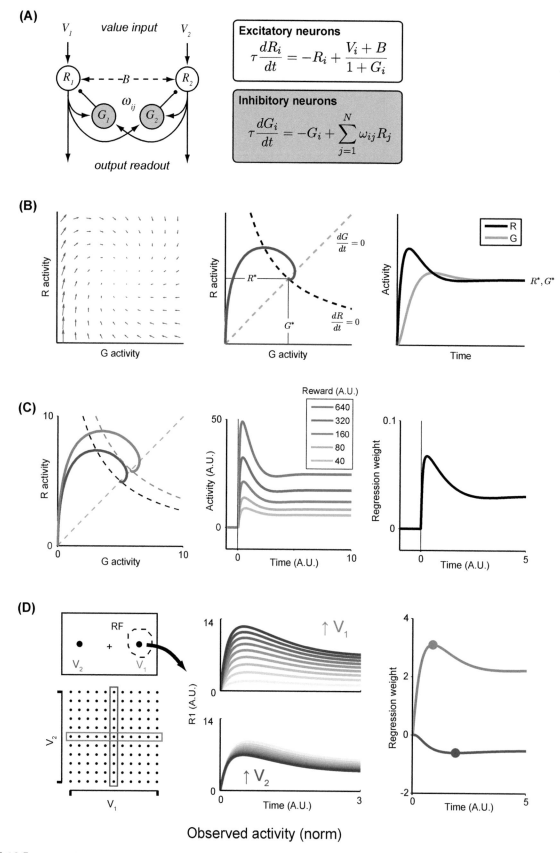

FIGURE 10.7 Dynamic normalization model. (A) Schematic of circuit organization corresponding to dynamic normalization model. The model consists of paired excitatory (*R*) and inhibitory (*G*) neurons for each choice option. Contextual value modulation is implemented via

[26,52], and the difference between direct and contextual value effects is evident in reward-related processing during perceptual decision tasks [51]. The ability of the dynamic normalization model to reproduce both equilibrium and time-varying properties of empirical electrophysiology data suggests that recurrent circuit architecture may play a key role in how value is processed in the brain.

computation. Furthermore, a computational understanding of value coding can also shed light on aspects of choice behavior. For example, several studies suggest that adaptive value-coding computations like normalization can reproduce context-dependent decision phenomena [143−146]. A key step for future research will be a full identification of circuit-level value-coding computations and their relationship to both brain organization and resulting choice behavior.

CONCLUSIONS

The study of decision-making spans multiple scientific disciplines, including diverse fields like computer science, ecology, economics, psychology, and neuroscience. These fields have historically pursued independent lines of inquiry, but research has begun to take an increasingly interdisciplinary approach. In particular, the neuroeconomic approach seeks to merge formal psychological and economic frameworks for choice behavior with neuroscientific characterization of brain activity. A central focus of current neuroeconomic research is value information, which sits at the core of formal choice models and extensively influences neural activity. Behavioral choice and the activity of underlying choice circuits correspond to the lowest and highest levels of Marr's tripartite framework for information processing, highlighting the importance of understanding value coding and decision-making at the crucial intermediate level of computation.

Value coding has now been demonstrated in diverse brain areas; current work is extending these findings beyond simply documenting value-correlated activity to identifying the computations employed to represent value information in neural circuits. Emerging research suggests that neural information processing for decision-making, as for perception, revolves around core computational principles like contextual modulation and divisive normalization. As reviewed above, identification of these computations can shed light on how the responsible neural circuits are organized; in particular, neural activity dynamics provides a key linkage between biological mechanism and

References

[1] von Neumann J, Morgenstern O. Theory of games and economic behavior. Princeton (NJ): Princeton University Press; 1944.

[2] Kahneman D, Tversky A. Prospect theory − analysis of decision under risk. Econometrica 1979;47:263−91.

[3] Stephens DW, Krebs JR. Foraging theory. Princeton (NJ): Princeton University Press; 1986.

[4] Marr D. Vision. San Francisco: W.H. Freeman; 1982.

[5] Sutton RS, Barto AG. Reinforcement learning: an introduction. Cambridge (MA): MIT Press; 1998.

[6] Lee D, Seo H, Jung MW. Neural basis of reinforcement learning and decision making. Annu Rev Neurosci 2012;35:287−308.

[7] Dayan P, Niv Y. Reinforcement learning: the good, the bad and the ugly. Curr Opin Neurobiol 2008;18:185−96.

[8] Pascal B. Great shorter works of Pascal. Philadelphia: The Westminster Press; 1948.

[9] Bernoulli D. Specimen theoriae novae de mensura sortis. Comment Acad Sci Imp Petropol 1738;5:175−92.

[10] Houthakker HS. Revealed preference and the utility function. Economica 1950;17:159−74.

[11] Samuelson PA. Foundations of economic analysis. Cambridge: Harvard University Press; 1947.

[12] Louie K, Glimcher PW. Efficient coding and the neural representation of value. Ann N Y Acad Sci 2012;1251:13−32.

[13] Schultz W. Neural coding of basic reward terms of animal learning theory, game theory, microeconomics and behavioural ecology. Curr Opin Neurobiol 2004;14:139−47.

[14] Glimcher PW. Making choices: the neurophysiology of visual-saccadic decision making. Trends Neurosci 2001;24:654−9.

[15] Bartra O, McGuire JT, Kable JW. The valuation system: a coordinate-based meta-analysis of BOLD fMRI experiments examining neural correlates of subjective value. NeuroImage 2013;76:412−27.

[16] Levy DJ, Glimcher PW. The root of all value: a neural common currency for choice. Curr Opin Neurobiol 2012;22:1027−38.

[17] Rangel A, Hare T. Neural computations associated with goal-directed choice. Curr Opin Neurobiol 2010;20:262−70.

recurrent connections driving a lateral inhibition. (B) Characteristic dynamics of the differential equation normalization model. Left, example single-option circuit vector field. *Arrows* indicate instantaneous rates of change in the activities of the (G, R) pair for a given constant value input. Middle, example system trajectory corresponding to the vector field on the left. *Dashed lines* indicate the nullclines of the two component equations; the intersection of the nullclines indicates the stable equilibrium point of the system. Right, activity of R and G neurons as a function of time. Note that despite a constant input, R activity exhibits significant transient early dynamics before achieving a stable equilibrium level. (C) Time-varying value coding. Left, single-option model trajectories under two different value conditions. Higher value input increases both equilibrium R activity and the extent of the initial transient peak. Middle, dynamic value modulation evident in R activity traces. Right, regression analysis of model R activity shows value coding is strongest during the initial transient. (D) Differential strength and timing of direct and contextual value coding. Left, schematic of varying value conditions tested in a two-option model. Middle, model R_1 activity as a function of varying either V_1 or V_2, showing earlier onset and stronger extent of direct value modulation. Right, multiple regression analysis showing effect of direct (*red*) and contextual (*blue*) information of model activity. RF, response field. *Adapted with permission from Louie K, LoFaro T, Webb R, Glimcher PW. Dynamic divisive normalization predicts time-varying value coding in decision-related circuits. J Neurosci 2014;34:16046−16057.*

[18] Kable JW, Glimcher PW. The neurobiology of decision: consensus and controversy. Neuron 2009;63:733–45.

[19] Glimcher PW. Foundations of neuroeconomic analysis. New York: Oxford University Press; 2011.

[20] O'Doherty JP. The problem with value. Neurosci Biobehav Rev 2014;43:259–68.

[21] Schultz W, Dayan P, Montague PR. A neural substrate of prediction and reward. Science 1997;275:1593–9.

[22] Hollerman JR, Schultz W. Dopamine neurons report an error in the temporal prediction of reward during learning. Nat Neurosci 1998;1:304–9.

[23] Ding L, Hikosaka O. Comparison of reward modulation in the frontal eye field and caudate of the macaque. J Neurosci 2006; 26:6695–703.

[24] Lau B, Glimcher PW. Action and outcome encoding in the primate caudate nucleus. J Neurosci 2007;27:14502–14.

[25] Pastor-Bernier A, Cisek P. Neural correlates of biased competition in premotor cortex. J Neurosci 2011;31:7083–8.

[26] Platt ML, Glimcher PW. Neural correlates of decision variables in parietal cortex. Nature 1999;400:233–8.

[27] Samejima K, Ueda Y, Doya K, Kimura M. Representation of action-specific reward values in the striatum. Science 2005;310: 1337–40.

[28] Sugrue LP, Corrado GS, Newsome WT. Matching behavior and the representation of value in the parietal cortex. Science 2004; 304:1782–7.

[29] Cisek P, Kalaska JF. Neural mechanisms for interacting with a world full of action choices. Annu Rev Neurosci 2010;33:269–98.

[30] Glimcher PW. The neurobiology of visual-saccadic decision making. Annu Rev Neurosci 2003;26:133–79.

[31] Hikosaka O, Kim HF, Yasuda M, Yamamoto S. Basal ganglia circuits for reward value-guided behavior. Annu Rev Neurosci 2014;37:289–306.

[32] Blatt GJ, Andersen RA, Stoner GR. Visual receptive field organization and cortico-cortical connections of the lateral intraparietal area (area LIP) in the macaque. J Comp Neurol 1990;299:421–45.

[33] Andersen RA, Brotchie PR, Mazzoni P. Evidence for the lateral intraparietal area as the parietal eye field. Curr Opin Neurobiol 1992;2:840–6.

[34] Andersen RA, Essick GK, Siegel RM. Neurons of area 7 activated by both visual stimuli and oculomotor behavior. Exp Brain Res 1987;67:316–22.

[35] Barash S, Bracewell RM, Fogassi L, Gnadt JW, Andersen RA. Saccade-related activity in the lateral intraparietal area. II. Spatial properties. J Neurophysiol 1991;66:1109–24.

[36] Barash S, Bracewell RM, Fogassi L, Gnadt JW, Andersen RA. Saccade-related activity in the lateral intraparietal area. I. Temporal properties; comparison with area 7a. J Neurophysiol 1991;66: 1095–108.

[37] Gnadt JW, Andersen RA. Memory related motor planning activity in posterior parietal cortex of macaque. Exp Brain Res 1988;70: 216–20.

[38] Bracewell RM, Mazzoni P, Barash S, Andersen RA. Motor intention activity in the macaque's lateral intraparietal area. II. Changes of motor plan. J Neurophysiol 1996;76:1457–64.

[39] Mazzoni P, Bracewell RM, Barash S, Andersen RA. Motor intention activity in the macaque's lateral intraparietal area. I. Dissociation of motor plan from sensory memory. J Neurophysiol 1996;76:1439–56.

[40] Andersen RA, Buneo CA. Intentional maps in posterior parietal cortex. Annu Rev Neurosci 2002;25:189–220.

[41] Andersen RA, Cui H. Intention, action planning, and decision making in parietal-frontal circuits. Neuron 2009;63:568–83.

[42] Churchland AK, Kiani R, Shadlen MN. Decision-making with multiple alternatives. Nat Neurosci 2008;11:693–702.

[43] Roitman JD, Shadlen MN. Response of neurons in the lateral intraparietal area during a combined visual discrimination reaction time task. J Neurosci 2002;22:9475–89.

[44] Shadlen MN, Newsome WT. Neural basis of a perceptual decision in the parietal cortex (area LIP) of the rhesus monkey. J Neurophysiol 2001;86:1916–36.

[45] Gold JI, Shadlen MN. The neural basis of decision making. Annu Rev Neurosci 2007;30:535–74.

[46] Hanks TD, Ditterich J, Shadlen MN. Microstimulation of macaque area LIP affects decision-making in a motion discrimination task. Nat Neurosci 2006;9:682–9.

[47] Huk AC, Shadlen MN. Neural activity in macaque parietal cortex reflects temporal integration of visual motion signals during perceptual decision making. J Neurosci 2005;25: 10420–36.

[48] Kira S, Yang T, Shadlen MN. A neural implementation of Wald's sequential probability ratio test. Neuron 2015;85:861–73.

[49] Yang T, Shadlen MN. Probabilistic reasoning by neurons. Nature 2007;447:1075–80.

[50] Dorris MC, Glimcher PW. Activity in posterior parietal cortex is correlated with the relative subjective desirability of action. Neuron 2004;44:365–78.

[51] Rorie AE, Gao J, McClelland JL, Newsome WT. Integration of sensory and reward information during perceptual decision-making in lateral intraparietal cortex (LIP) of the macaque monkey. PLoS One 2010;5:e9308.

[52] Louie K, Glimcher PW. Separating value from choice: delay discounting activity in the lateral intraparietal area. J Neurosci 2010; 30:5498–507.

[53] Seo H, Barraclough DJ, Lee D. Lateral intraparietal cortex and reinforcement learning during a mixed-strategy game. J Neurosci 2009;29:7278–89.

[54] Klein JT, Deaner RO, Platt ML. Neural correlates of social target value in macaque parietal cortex. Curr Biol 2008;18:419–24.

[55] Ikeda T, Hikosaka O. Reward-dependent gain and bias of visual responses in primate superior colliculus. Neuron 2003;39: 693–700.

[56] Roesch MR, Olson CR. Impact of expected reward on neuronal activity in prefrontal cortex, frontal and supplementary eye fields and premotor cortex. J Neurophysiol 2003;90:1766–89.

[57] Lau B, Glimcher PW. Value representations in the primate striatum during matching behavior. Neuron 2008;58:451–63.

[58] Handel A, Glimcher PW. Contextual modulation of substantia nigra pars reticulata neurons. J Neurophysiol 2000;83:3042–8.

[59] Kubanek J, Snyder LH. Reward-based decision signals in parietal cortex are partially embodied. J Neurosci 2015;35:4869–81.

[60] Padoa-Schioppa C. Neurobiology of economic choice: a good-based model. Annu Rev Neurosci 2011;34:333–59.

[61] Cai X, Padoa-Schioppa C. Neuronal encoding of subjective value in dorsal and ventral anterior cingulate cortex. J Neurosci 2012; 32:3791–808.

[62] Roesch MR, Olson CR. Neuronal activity related to reward value and motivation in primate frontal cortex. Science 2004;304: 307–10.

[63] Tremblay L, Schultz W. Relative reward preference in primate orbitofrontal cortex. Nature 1999;398:704–8.

[64] Wallis JD, Miller EK. Neuronal activity in primate dorsolateral and orbital prefrontal cortex during performance of a reward preference task. Eur J Neurosci 2003;18:2069–81.

[65] Padoa-Schioppa C, Assad JA. Neurons in the orbitofrontal cortex encode economic value. Nature 2006;441:223–6.

[66] Laughlin SB, de Ruyter van Steveninck RR, Anderson JC. The metabolic cost of neural information. Nat Neurosci 1998;1:36–41.

[67] Laughlin SB. Energy as a constraint on the coding and processing of sensory information. Curr Opin Neurobiol 2001;11:475–80.

[68] Simoncelli EP, Olshausen BA. Natural image statistics and neural representation. Annu Rev Neurosci 2001;24:1193—216.

[69] Shannon CE, Weaver W. The mathematical theory of communication. Urbana (IL): University of Illinois Press; 1949.

[70] Barlow HB. Possible principles underlying the transformation of sensory messages. In: Rosenblith WA, editor. Sensory communication. MIT Press; 1961.

[71] Laughlin S. A simple coding procedure enhances a neuron's information capacity. Z Naturforsch C 1981;36:910—2.

[72] Ohzawa I, Sclar G, Freeman RD. Contrast gain control in the cat visual cortex. Nature 1982;298:266—8.

[73] Smirnakis SM, Berry MJ, Warland DK, Bialek W, Meister M. Adaptation of retinal processing to image contrast and spatial scale. Nature 1997;386:69—73.

[74] Brenner N, Bialek W, de Ruyter van Steveninck R. Adaptive rescaling maximizes information transmission. Neuron 2000;26: 695—702.

[75] Fairhall AL, Lewen GD, Bialek W, de Ruyter van Steveninck RR. Efficiency and ambiguity in an adaptive neural code. Nature 2001;412:787—92.

[76] Schwartz O, Simoncelli EP. Natural signal statistics and sensory gain control. Nat Neurosci 2001;4:819—25.

[77] Lewicki MS. Efficient coding of natural sounds. Nat Neurosci 2002;5:356—63.

[78] Bonin V, Mante V, Carandini M. The statistical computation underlying contrast gain control. J Neurosci 2006;26:6346—53.

[79] Sharpee TO, Sugihara H, Kurgansky AV, Rebrik SP, Stryker MP, Miller KD. Adaptive filtering enhances information transmission in visual cortex. Nature 2006;439:936—42.

[80] Maravall M, Petersen RS, Fairhall AL, Arabzadeh E, Diamond ME. Shifts in coding properties and maintenance of information transmission during adaptation in barrel cortex. PLoS Biol 2007;5:e19.

[81] Phillips WA, Clark A, Silverstein SM. On the functions, mechanisms, and malfunctions of intracortical contextual modulation. Neurosci Biobehav Rev 2015;52:1—20.

[82] Hubel DH, Wiesel TN. Receptive fields and functional architecture of monkey striate cortex. J Physiol 1968;195:215—43.

[83] Blakemore C, Tobin EA. Lateral inhibition between orientation detectors in the cat's visual cortex. Exp Brain Res 1972;15:439—40.

[84] Levitt JB, Lund JS. Contrast dependence of contextual effects in primate visual cortex. Nature 1997;387:73—6.

[85] Cavanaugh JR, Bair W, Movshon JA. Nature and interaction of signals from the receptive field center and surround in macaque V1 neurons. J Neurophysiol 2002;88:2530—46.

[86] Kohn A. Visual adaptation: physiology, mechanisms, and functional benefits. J Neurophysiol 2007;97:3155—64.

[87] Schwartz O, Hsu A, Dayan P. Space and time in visual context. Nat Rev Neurosci 2007;8:522—35.

[88] Lamme VA, Super H, Spekreijse H. Feedforward, horizontal, and feedback processing in the visual cortex. Curr Opin Neurobiol 1998;8:529—35.

[89] Albright TD, Stoner GR. Contextual influences on visual processing. Annu Rev Neurosci 2002;25:339—79.

[90] Krause MR, Pack CC. Contextual modulation and stimulus selectivity in extrastriate cortex. Vis Res 2014;104:36—46.

[91] Gilbert CD, Wiesel TN. The influence of contextual stimuli on the orientation selectivity of cells in primary visual cortex of the cat. Vis Res 1990;30:1689—701.

[92] Cavanaugh JR, Bair W, Movshon JA. Selectivity and spatial distribution of signals from the receptive field surround in macaque V1 neurons. J Neurophysiol 2002;88:2547—56.

[93] Muller JR, Metha AB, Krauskopf J, Lennie P. Local signals from beyond the receptive fields of striate cortical neurons. J Neurophysiol 2003;90:822—31.

[94] Quiroga RQ, Reddy L, Kreiman G, Koch C, Fried I. Invariant visual representation by single neurons in the human brain. Nature 2005;435:1102—7.

[95] Calford MB, Semple MN. Monaural inhibition in cat auditory cortex. J Neurophysiol 1995;73:1876—91.

[96] Brosch M, Schreiner CE. Time course of forward masking tuning curves in cat primary auditory cortex. J Neurophysiol 1997;77: 923—43.

[97] Bartlett EL, Wang X. Long-lasting modulation by stimulus context in primate auditory cortex. J Neurophysiol 2005;94:83—104.

[98] Simons DJ. Temporal and spatial integration in the rat SI vibrissa cortex. J Neurophysiol 1985;54:615—35.

[99] Steinmetz PN, Roy A, Fitzgerald PJ, Hsiao SS, Johnson KO, Niebur E. Attention modulates synchronized neuronal firing in primate somatosensory cortex. Nature 2000;404:187—90.

[100] Salinas E, Thier P. Gain modulation: a major computational principle of the central nervous system. Neuron 2000;27:15—21.

[101] Gilbert CD, Sigman M. Brain states: top-down influences in sensory processing. Neuron 2007;54:677—96.

[102] Atick JJ. Could information theory provide an ecological theory of sensory processing? Network 1992;3:213—51.

[103] Rao RP, Ballard DH. Predictive coding in the visual cortex: a functional interpretation of some extra-classical receptive-field effects. Nat Neurosci 1999;2:79—87.

[104] Vinje WE, Gallant JL. Natural stimulation of the nonclassical receptive field increases information transmission efficiency in V1. J Neurosci 2002;22:2904—15.

[105] Felsen G, Touryan J, Dan Y. Contextual modulation of orientation tuning contributes to efficient processing of natural stimuli. Network 2005;16:139—49.

[106] Wainwright MJ. Visual adaptation as optimal information transmission. Vis Res 1999;39:3960—74.

[107] Wark B, Lundstrom BN, Fairhall A. Sensory adaptation. Curr Opin Neurobiol 2007;17:423—9.

[108] Louie K, Grattan LE, Glimcher PW. Reward value-based gain control: divisive normalization in parietal cortex. J Neurosci 2011;31:10627—39.

[109] Rangel A, Clithero JA. Value normalization in decision making: theory and evidence. Curr Opin Neurobiol 2012;22:970—81.

[110] Basso MA, Wurtz RH. Modulation of neuronal activity by target uncertainty. Nature 1997;389:66—9.

[111] Basso MA, Wurtz RH. Modulation of neuronal activity in superior colliculus by changes in target probability. J Neurosci 1998; 18:7519—34.

[112] Padoa-Schioppa C. Range-adapting representation of economic value in the orbitofrontal cortex. J Neurosci 2009;29:14004—14.

[113] Kobayashi S, Pinto de Carvalho O, Schultz W. Adaptation of reward sensitivity in orbitofrontal neurons. J Neurosci 2010;30: 534—44.

[114] Tobler PN, Fiorillo CD, Schultz W. Adaptive coding of reward value by dopamine neurons. Science 2005;307:1642—5.

[115] Cox KM, Kable JW. BOLD subjective value signals exhibit robust range adaptation. J Neurosci 2014;34:16533—43.

[116] Carandini M. From circuits to behavior: a bridge too far? Nat Neurosci 2012;15:507—9.

[117] Carandini M, Heeger DJ. Normalization as a canonical neural computation. Nat Rev Neurosci 2012;13:51—62.

[118] Heeger DJ. Normalization of cell responses in cat striate cortex. Vis Neurosci 1992;9:181—97.

[119] Shapley RM, Victor JD. The effect of contrast on the transfer properties of cat retinal ganglion cells. J Physiol 1978;285:275—98.

[120] Bonin V, Mante V, Carandini M. The suppressive field of neurons in lateral geniculate nucleus. J Neurosci 2005;25:10844—56.

[121] Britten KH, Heuer HW. Spatial summation in the receptive fields of MT neurons. J Neurosci 1999;19:5074—84.

[122] Rust NC, Mante V, Simoncelli EP, Movshon JA. How MT cells analyze the motion of visual patterns. Nat Neurosci 2006;9: 1421–31.

[123] Zoccolan D, Cox DD, DiCarlo JJ. Multiple object response normalization in monkey inferotemporal cortex. J Neurosci 2005;25:8150–64.

[124] Rabinowitz NC, Willmore BD, Schnupp JW, King AJ. Contrast gain control in auditory cortex. Neuron 2011;70:1178–91.

[125] Reynolds JH, Heeger DJ. The normalization model of attention. Neuron 2009;61:168–85.

[126] Itthipuripat S, Cha K, Rangsipat N, Serences JT. Value-based attentional capture influences context-dependent decision-making. J Neurophysiol 2015;114:560–9.

[127] Ohshiro T, Angelaki DE, DeAngelis GC. A normalization model of multisensory integration. Nat Neurosci 2011;14:775–82.

[128] Rosenberg A, Patterson JS, Angelaki DE. A computational perspective on autism. Proc Natl Acad Sci USA 2015;112(30): 9158–65.

[129] Albrecht DG, Geisler WS. Motion selectivity and the contrast-response function of simple cells in the visual cortex. Vis Neurosci 1991;7:531–46.

[130] Carandini M, Heeger DJ, Movshon JA. Linearity and normalization in simple cells of the macaque primary visual cortex. J Neurosci 1997;17:8621–44.

[131] Holt CA, Laury SK. Risk aversion and incentive effects: new data without order effects. Am Econ Rev 2005;95:902–4.

[132] Wu G, Gonzalez R. Curvature of the probability weighting function. Manage Sci 1996;42:1676–90.

[133] Ohzawa I, Sclar G, Freeman RD. Contrast gain control in the cat's visual system. J Neurophysiol 1985;54:651–67.

[134] Sinz F, Bethge M. Temporal adaptation enhances efficient contrast gain control on natural images. PLoS Comput Biol 2013;9:e1002889.

[135] Douglas RJ, Martin KA. Neuronal circuits of the neocortex. Annu Rev Neurosci 2004;27:419–51.

[136] Isaacson JS, Scanziani M. How inhibition shapes cortical activity. Neuron 2011;72:231–43.

[137] Wilson HR, Humanski R. Spatial frequency adaptation and contrast gain control. Vis Res 1993;33:1133–49.

[138] Carandini M, Heeger DJ. Summation and division by neurons in primate visual cortex. Science 1994;264:1333–6.

[139] Izhikevich EM. Dynamical systems in neuroscience: the geometry of excitability and bursting. Cambridge (MA): MIT Press; 2007.

[140] Wang XJ. Neural dynamics and circuit mechanisms of decision-making. Curr Opin Neurobiol 2012;22:1039–46.

[141] LoFaro T, Louie K, Webb R, Glimcher P. The temporal dynamics of cortical normalization models of decision-making. Lett Biomath 2014;1:209–20.

[142] Louie K, LoFaro T, Webb R, Glimcher PW. Dynamic divisive normalization predicts time-varying value coding in decision-related circuits. J Neurosci 2014;34:16046–57.

[143] Louie K, Khaw MW, Glimcher PW. Normalization is a general neural mechanism for context-dependent decision making. Proc Natl Acad Sci USA 2013;110:6139–44.

[144] Cheadle S, Wyart V, Tsetsos K, Myers N, de Gardelle V, Herce Castanon S, et al. Adaptive gain control during human perceptual choice. Neuron 2014;81:1429–41.

[145] Soltani A, De Martino B, Camerer C. A range-normalization model of context-dependent choice: a new model and evidence. PLoS Comput Biol 2012;8:e1002607.

[146] Hunt LT, Dolan RJ, Behrens TE. Hierarchical competitions subserving multi-attribute choice. Nat Neurosci 2014;17: 1613–22.

11

Spatiotemporal Characteristics and Modulators of Perceptual Decision-Making in the Human Brain

M.G. Philiastides, J.A. Diaz, S. Gherman
University of Glasgow, Glasgow, United Kingdom

Abstract

Perceptual decision-making is the process of choosing between two or more alternatives based on an evaluation and integration of sensory information. Converging evidence from electrophysiology, neuroimaging, and theoretical modeling work suggests that the decision process relies on a cascade of neural events. Sensory input is first encoded by the neural modules selective to the choice alternatives before it is passed on to a decision center, which compares the sensory outputs in a noisy process of gradual accumulation of evidence that ultimately leads to a decision. In this chapter we start out with an introduction to the general principles guiding perceptual decision-making. We then take a critical turn to look beyond sensory information as the decisive variable for the decision, and discuss additional factors that interact with, and contribute to, the decision process. Specifically, we review the influence of the following factors: prestimulus state, reward and punishment, speed—accuracy trade-off, learning and training, confidence, and neuromodulation. We show how these decision modulators can exert their influence at various stages of processing, in line with predictions derived from sequential-sampling models of decision-making.

INTRODUCTION

Perceptual decision-making is the act of selecting one option or course of action from a larger set of alternatives on the basis of available sensory information [1]. This process has frequently been modeled using sequential sampling models, such as the well-known drift-diffusion model [2], which assumes that the decision formation hinges on a noisy (stochastic) accumulation of incoming sensory information [3—6] (also see Chapter 12). More specifically, these computational accounts suggest that perceptual decisions involve an integrative mechanism whereby the difference in sensory evidence supporting the alternatives accumulates over time to a preset internal decision boundary, which ultimately determines the choice (Fig. 11.1).

Correspondingly, several nonhuman primate (NHP) electrophysiology studies have revealed patterns of single-unit activity that are in line with this integrative mechanism [7]. Using a visual motion direction-discrimination task [random-dot kinematogram (RDK) task], these studies revealed that, whereas sensory areas responsive to motion direction [such as the middle temporal area (MT)] encoded the amount of evidence for each alternative, higher-level regions known to orchestrate choice [such as the lateral intraparietal area (LIP), the frontal eye fields, and the superior colliculus] accumulated the evidence for the decision. Specifically, firing rates of individual neurons in these areas built up gradually over time at a rate proportional to the amount of evidence for the decision (i.e., difficulty of the task), eventually converging on a common firing level (decision boundary) as animals committed to a choice [7—9].

More recently, human studies using time-resolved electroencephalography (EEG) signals were able to measure the process of evidence accumulation on the scalp [10—12]. Philiastides et al. [11], using a visual binary categorization task (e.g., is the stimulus a face or a car/house), showed that population responses on the scalp (with a broad centroparietal topography) can capture activity that builds up gradually over time with a rate proportional to the amount of sensory evidence in the stimulus (Fig. 11.2A). The buildup rate of this accumulating activity was consistent with the properties of a drift-diffusion-like process as characterized by computational modeling (i.e., EEG buildup rate correlated with

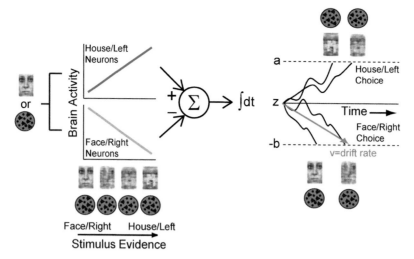

FIGURE 11.1   Representation (left) and integration (right) of sensory evidence in perceptual decision-making. Sensory areas/neurons encoding each of two decision alternatives (e.g., face/house or right/left motion) respond parametrically to the amount of evidence in the stimulus. For example, face-responsive regions respond stronger to clear than to noisy images of faces and even less to images of houses. Conversely, house-responsive regions respond stronger to clear than to noisy images of houses and even less to images of faces. The difference signal between the two competing areas or groups of neurons is subsequently integrated over time in the decision process from a starting point (z) toward one of two internally set decision boundaries (a or −b) representing the possible choices. The rate at which the evidence is accumulated [drift rate (v) in the diffusion model of simple decision-making] is proportional to the amount of stimulus evidence. *Adapted from Philiastides MG, Heekeren HR. Spatiotemporal characteristics of perceptual decision making in the human brain. In: Dreher J-C, Tremblay L, editors. Handbook of reward and decision making. Academic Press; 2009. p. 185−212.*

drift rate in the model; Fig. 11.2B) and it was additionally shown to predict participants' choice accuracy as in the NHP work described earlier (Fig. 11.2C).

Intriguingly, regions of the parietal and prefrontal cortices were linked to this accumulating activity in humans, via functional magnetic resonance imaging (fMRI). A study by Ploran et al. [13] recorded the time of recognition of noisy pictures that were revealed gradually over the course of several seconds. During this period the authors identified a gradual buildup in the

fMRI signal, peaking in correspondence with the time of recognition in a set of regions, the inferior temporal, frontal, and parietal regions. Similar patterns of activity were also reported using a face/house categorization task [14].

Heekeren et al. [15] directly tested whether a comparison operation is also at work (as shown in Fig. 11.1) using a similar face/house categorization task. A brain region in the posterior portion of the left dorsolateral prefrontal cortex (DLPFC) uniquely correlated with the

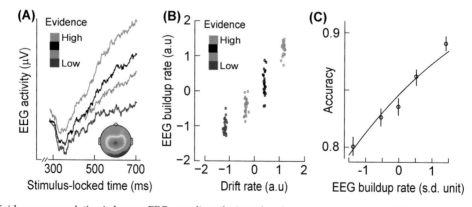

FIGURE 11.2   Evidence accumulation in human EEG recordings during a face/car categorization task. (A) Ramp-like activity consistent with a process of evidence accumulation recorded on the surface of the scalp. As sensory evidence increases so does the buildup rate (slope) of the evidence accumulation. Scalp topography depicts the spatial distribution of the accumulating activity. (B) The buildup rate of the EEG activity correlates with drift rate estimates from a diffusion model analysis of individual subject behavioral responses. (C) The buildup rate of the accumulating activity is predictive of participant's choice accuracy, even after accounting for task difficulty. *Adapted from Philiastides MG, Heekeren HR, Sajda P. Human scalp potentials reflect a mixture of decision-related signals during perceptual choices. J Neurosci 2014;34:16877−89.*

difference between the output signals of face- and house-responsive regions and it additionally predicted behavioral performance in the categorization task. This finding was replicated using a direction discrimination task and two different response modalities (i.e., eyes and hands) [16] and was shown to persist even when the perceptual decision was fully decoupled from the impending motor response [17].

Finally, Philiastides et al. [18] showed that hindering activity in the left DLPFC with transcranial magnetic stimulation resulted in behavioral impairments (relative to sham stimulation) that were attributed to changes in the efficiency of integration (drift rate) in a diffusion model fit to the behavioral data. Taken together these findings suggest that the left DLPFC plays a causal role in the integration of the outputs of lower-level sensory regions and uses a subtraction operation to compute perceptual decisions.

It is noteworthy that similar findings have been reported during decision-making in other domains (e.g., auditory [19,20] and somatosensory [21−23]), suggesting that the process of sensory evidence accumulation supporting perceptual decisions generalizes across sensory modalities. Nonetheless, in addition to the sensory input, decisions can also be affected by other factors (i.e., decision modulators), which have the capacity to influence different processing stages along the decision stream as highlighted in Fig. 11.1 (e.g., from early sensory encoding to the decision formation itself). In this chapter we will review the role of the following factors on perceptual decision-making: prestimulus state, reward and punishment, speed−accuracy trade-off, learning and training, confidence, and neuromodulation.

FACTORS AFFECTING PERCEPTUAL DECISION-MAKING

Prestimulus State

An important factor that can influence the course of the decision process is the state of the neural activity prior to any task-relevant sensory stimulation. For example, a strong link has been observed between the trial-to-trial fluctuations in prestimulus oscillatory activity and the behavioral outcome of the perceptual decision [24−27]. In one such study, Van Dijk et al. [27] asked participants to perform a simple visual discrimination task, which involved detecting whether a contrast difference was present in a target stimulus. Magnetoencephalography (MEG) recordings showed that spontaneous prestimulus oscillations in the alpha-frequency band, within occipitoparietal areas, were negatively correlated with subjects' performance, such that as alpha power increased, discrimination ability

decreased. Similar effects of prestimulus alpha activity have been shown in other studies [24−26,28] and are theorized to reflect variability in attentional processes exerting top-down modulatory influence on information processing [29].

The temporal locus of this influence on the decision-processing stream, however, remains an active topic of debate. Some studies suggest that prestimulus alpha power exerts its influence on early sensory processing. Mazaheri et al. [28] asked subjects to perform perceptual discriminations in which, on some trials, auditory and visual stimuli were presented simultaneously, with only one of the stimuli, cued in advance, requiring a perceptual judgment. On these trials, prestimulus alpha power was suppressed in the early sensory regions relevant to the target stimulus (i.e., early visual areas for visual stimuli and the supramarginal gyrus for auditory stimuli) and enhanced in the region relevant to the distractor sensory modality. The authors proposed that alpha activity might serve the role of gating or modulating information flow to sensory areas.

A different study by Lou et al. [30] directly sought to identify the temporal stage(s) of the decision process likely to be affected by prestimulus alpha activity during a face/car categorization task. Prestimulus alpha-band power was negatively correlated with the magnitude of an early EEG stimulus-discriminating component hypothesized to reflect early sensory evidence encoding [31−34], such that lower prestimulus alpha power was associated with higher absolute value of the discriminant component (Fig. 11.3A). Conversely, no correlation was found between prestimulus alpha power and a later decision-related discriminating component.

There has also been evidence to suggest a later influence directly affecting the evidence accumulation stage of the decision. Kelly and O'Connell [12] used an RDK task and found that prestimulus alpha power correlated positively with subjects' response time and negatively with the efficiency of decision formation, as indexed by the buildup rate of a centroparietal positivity potential in the stimulus-aligned EEG signal [35]. Their results are consistent with a top-down influence of prestimulus attentional fluctuations on the decision, whereby lower prestimulus alpha power (i.e., enhanced attention) leads to a more efficient evidence accumulation in the post-stimulus period.

Modulatory effects of prestimulus state on the perceptual decision have also been linked to spontaneous prestimulus oscillatory activity in the gamma-frequency band. Wyart and Tallon-Baudry [26] used a visual detection (present/absent) task in which the locations of the stimuli were precued with partial validity on each trial (i.e., the cues were correct on only 65% of the trials). They found that gamma-band activity in the lateral occipital cortex was predictive of subjects'

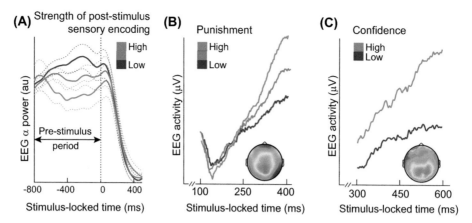

FIGURE 11.3   **Influence of different modulators on perceptual decision-making during a face/car categorization task.** (A) Prestimulus alpha power influences the encoding of early sensory evidence, whereby lower prestimulus alpha leads to a more reliable poststimulus sensory encoding. *(Adapted from Lou B, et al. Prestimulus alpha power predicts fidelity of sensory encoding in perceptual decision making. NeuroImage 2014;87: 242–51.)* (B) Modulating the level of punishment associated with incorrect perceptual decisions affects the efficiency of evidence accumulation in the decision, whereby higher punishment levels lead to a steeper rate of evidence integration. *(Adapted from Blank H, et al. Temporal characteristics of the influence of punishment on perceptual decision making in the human brain. J Neurosci 2013;33:3939–52.)* (C) Choice confidence during perceptual choices reflected in the process of decision formation, with more confident trials leading to a higher rate of evidence accumulation. *(Adapted from Gherman S, Philiastides MG. Neural representations of confidence emerge from the process of decision formation during perceptual choices. NeuroImage 2015; 106:134–43.)*

choices, increasing for detected versus undetected reports. The authors demonstrated that its perceptual impact resembled a decision bias at stimulus onset, thus making one response more likely than the other. Moreover, this did not appear to be due to changes in attentional focus, but rather to spontaneous predictions about the upcoming stimulus, potentially biasing the starting point of evidence accumulation.

Overall, it appears that variability in prestimulus activity can play an important role in altering the course of the decision, likely through top-down modulatory effects on information processing. It is possible that, at least in some circumstances, these internal fluctuations possess a volitional/adaptive component aimed at maximizing reward/performance; however, when and how these differ from spontaneous fluctuations remains to be explored.

Reward and Punishment

In a natural environment, the ultimate aim of any behavioral response is to try to maximize an organism's utility function (i.e., maximize rewards or minimize punishments) (also see Chapter 4). Value-based decision-making in humans, especially in the context of reinforcement learning and reward-related activity in dopaminergic systems [36] (also see Chapter 2), has already been studied extensively. Surprisingly, however, less has been done to explore the potential effects of reward and punishment on perceptual decision-making, whether on early encoding of sensory

information or on later decision-related processing and action selection.

Weil et al. [37] collected fMRI data while subjects performed an orientation discrimination task and were given monetary rewards for correct decisions. Greater rewards improved behavioral performance and increased fMRI activity in areas of the reward network (e.g., ventral striatum and orbitofrontal cortex). More importantly, however, the authors found that positive outcomes also led to increased activity in early visual areas (e.g., V2 and V3) at the time of reward delivery (presented auditorily) when no visual stimuli were being presented. Finally, rewarded trials led to improved performance on the subsequent trial and enhanced visual activity contralateral to the judged stimulus. In a related fMRI work, Schiffer et al. [38] used a face/house categorization task in which subjects received monetary rewards and punishments for correct and incorrect choices, respectively. At the time of outcome delivery, the authors found reward-predictive activity in visual areas associated with faces (i.e., fusiform face area) and houses (i.e., parahippocampal place area) when subjects made face and house choices, respectively. Not only these activations were dependent on the perceptual decision, they also covaried with activity in reward structures (e.g., ventral striatum), indicating an interaction between the human reward network and the early sensory cortex.

Similar results were obtained by Pleger et al. [39] in the somatosensory domain. The authors used fMRI and a tactile discrimination task in which subjects had to compare the frequency of two successive tactile stimuli applied to the same finger. The task was performed under different reward rates for each correct trial. Higher rewards enhanced behavioral performance and increased fMRI activity in both the primary somatosensory cortex (S1) and the ventral striatum. More importantly, however, these authors demonstrated that during reward delivery and in the absence of somatosensory stimulation, the S1 contralateral to the judged finger was reactivated and this reactivation was proportional to the amount of reward. Finally, they showed that reward magnitude on a particular trial influenced responses on the subsequent trial, with better behavioral performance and greater somatosensory responses for higher rewards. Taken together the results presented above clearly demonstrate that the systems involved in valuation interact with early sensory systems. More specifically, in situations in which the value of outcomes depends on decisions associated with different perceptual representations, reward and punishment signals are propagated back to sensory systems, in the form of a "teaching signal" whereby they can shape early sensory representations to optimize future choices and maximize reward.

In addition to the effects on the early sensory system, more recent studies focused on the influence of reward and punishment on the dynamics of the process of evidence accumulation during the decision itself. In an EEG study by Blank et al. [40] subjects performed a face/car categorization task, during which the level of punishment associated with incorrect choices changed in a block-wise fashion (three levels: low, medium, high). EEG activity discriminating between the three levels of punishment appeared mostly late in the trial. This activity exhibited a ramp-like response profile (Fig. 11.3B) and had a spatial topography consistent with the process of evidence accumulation defined earlier (compare Fig. 11.2A with Fig. 11.3B). Crucially, the buildup rate of this activity increased parametrically with the amount of punishment (Fig. 11.3B) and it was further predictive of the size of the behavioral improvements induced by punishment across participants. Finally, the trial-by-trial changes in prestimulus power in the alpha and gamma bands were good predictors of this accumulating activity, suggesting that different decision modulators (here prestimulus state and punishment) can interact to shape perceptual decisions. Similar results are obtained using manipulations of reward magnitude. These findings indicate that reward and punishment can exert a top-down influence (e.g., via attention and motivation) on the decision process itself, leading to more efficient integration of sensory evidence.

In another study, Green et al. [41] had participants perform an RDK task in blocks with identical duration but different reward/punishment schedules. Behavioral and diffusion modeling results indicated that subjects adjusted their decision boundaries (i.e., controlled the amount of accumulated evidence) to maximize the reward rate, consistent with earlier computational modeling work [42,43]. fMRI results from Green et al. [41] indicated that these changes in the decision boundaries were achieved by adjusting the connectivity within corticostriatal systems, responsible for accumulating sensory evidence, and cerebellar—striatal systems, responsible for temporal processing. These connectivity patterns were strongest for those individuals who obtained greater rewards by making greater adjustments in their decision boundaries. Similarly, Domenech and Dreher [44] showed decision boundary adjustments as a function of prior knowledge and the predictability of upcoming stimuli. These findings provide another source of evidence that in perceptual decision-making, reward and punishment can directly modulate the dynamics of the decision process itself.

Speed Versus Accuracy Trade-Off

The previous section offered evidence on how reward maximization in perceptual decision-making can be achieved by optimally adjusting the decision boundaries in the process of evidence accumulation to meet the payoff contingencies of the environment. Within the framework of sequential sampling models of decision-making, these boundary adjustments are thought to implement a speed—accuracy trade-off (SAT) [45] (also see Chapter 12 for additional discussion on SAT). More generally, SAT is considered a mechanism by which a reasonable balance between the competing demands of speed and accuracy can be achieved. While the likelihood of making a correct decision increases as information continues to accumulate over time, delaying decisions to ensure they are certainly correct may render them ineffective. Within this framework, studies have started to look at how SAT is implemented in the brain and how it influences the process of decision formation.

In an fMRI study by Forstmann et al. using the RDK task [46], subjects were instructed to increase either speed or accuracy on individual trials. As expected, subjects were fast but less accurate during the speed compared to the accuracy instruction and computational modeling analysis confirmed that this effect was driven by a reduction in the distance between the decision boundaries. fMRI analysis showed that the speed instruction was associated with increased activity (relative to the accuracy instruction) in the striatum and the pre-supplementary motor area, regions known to be involved in voluntary motor planning (see also van Maanen et al. [47]). Moreover, individual differences in the level of activation in these regions were correlated with individual variations in the decision boundaries estimated via a bounded-accumulation model. In a follow-up structural connectivity study, Forstmann et al. [48] provided additional evidence that decreases in decision boundaries are likely to be mediated via increased activation from cortex to striatum, releasing the motor system from inhibition (striatal hypothesis), rather than from decreased activation from cortex to the subthalamic nucleus (STN hypothesis). This evidence is consistent with a flexible mechanism of response caution adjustment in accordance to environmental (task) demands.

In a related study, using fMRI and the RDK task, Ivanoff et al. [49] showed that, in addition to the results described above, SAT also affected the activity of the lateral prefrontal cortex in a region known to be involved in evidence accumulation during the decision process. More specifically, the authors showed that emphasizing the speed rather than the accuracy of the perceptual decision lowered the amount of evidence-related activity in the lateral prefrontal cortex, consistent with a reduction in the decision boundary. Interestingly, this effect was not observed in sensory areas providing the evidence for the decision (e.g., MT), suggesting that SAT affects the dynamics of the decision process itself, rather than exerting an influence on early sensory processing.

More recently, Wenzlaff et al. [50] provided additional evidence on how the influence of SAT develops over time using MEG. The authors used a face/house categorization task and reported ramp-like activity consistent with the process of evidence accumulation that was modulated by SAT. Specifically, source analysis of this activity showed that supplementary motor areas and the medial precuneus increased their level of activation in the speed compared to the accuracy instruction. Increases in activity in these regions correlated with a reduction in the decision boundary (i.e., a negative correlation) as estimated by a diffusion model applied to the behavioral data, consistent with the results of the fMRI studies described earlier [46,49]. In addition, the authors reported that the level of activation in the DLPFC correlated positively with the model's boundary parameters as in the aforementioned Ivanoff et al. fMRI study [49].

Collectively, these neuroimaging studies offered strong support that SAT is implemented by changes in decision boundaries in (pre)motor and decision-related brain structures. However, evidence from NHPs challenged this view and provided evidence for an alternative account [51,52]. For example, Hanks et al. [52] found that during SAT, LIP neurons known to accumulate evidence for the decision did not exhibit a change in their decision boundaries. Instead, during the speed

instruction these neurons showed an increased initial firing rate (i.e., evidence-independent activation), ultimately enabling the animal to make a decision on the basis of less information. In contrast to the human neuroimaging results, these findings suggest that SAT could be mediated by changes in the amount of decision-related activity itself rather than through decision boundary adjustments.

Learning and Training

Another way to augment and optimize performance during perceptual choices is through learning. More specifically, training and experience are required to induce long-lasting improvements in our ability to make perceptual decisions based on ambiguous sensory information. This phenomenon is commonly referred to as perceptual learning [53] and though it has been studied extensively, especially in early vision [54,55], its effect on the decision-making process itself remains less well explored. The traditional view on the neural plasticity underlying learning has been that the influence of training is on the early parts of the perceptual system, such as in early visual cortex. An alternative view, however, is that learning in the context of perceptual decision-making can additionally affect higher-level areas responsible for driving the decision itself via enhanced cortical coupling with early sensory cortex.

The traditional view has been corroborated by human neuroimaging studies by Furmanski et al. [56], who used fMRI to measure neural signals in primary visual cortex before and after a month-long perceptual training period in which subjects learned to detect oriented patterns. They observed an increase in response in V1 that correlated with improvements in behavioral performance. Similarly, Jehee et al. [57] used an orientation discrimination task that was performed daily over the course of several weeks. They reported significant improvements in discrimination ability and corresponding enhancement in neural activity along the early visual areas (V1 to V4), albeit for only the trained orientations and locations. Correspondingly, single-cell recordings in NHPs by Yan et al. [58] showed that learning to discriminate visual contours led to strengthening and accelerating of neural responses in primary visual cortex, and that these changes correlated highly with behavioral performance.

Despite these results, evidence supporting the alternative view (i.e., late influences of learning on decision-making) has also been provided. For example, computational work [59] suggests that perceptual improvements are mediated by higher-level decision-related centers of the brain that learn to read out and reweight V1 inputs through training. In related work, Li et al. [60] asked subjects to decide whether a visually presented stimulus was radial or concentric by comparing the external sensory input to different internal decision criteria that they learned through training with feedback on separate sessions. After each training session, subjects performed the categorization task during fMRI. The authors used multivoxel pattern analysis of the fMRI signals to predict subjects' behavioral choices, and showed that category learning affects decision-related regions in frontal and higher occipitotemporal brain regions implicated in flexible adjustment of the decision criterion required for the task. These effects were not observed in regions responsible for early encoding of the stimulus or those controlling motor preparation and execution. Collectively, these studies provide evidence that perceptual learning effects extend beyond the perceptual system and can influence higher-level brain areas implicated in the decision process itself.

This interpretation is further supported by experiments in NHPs. Law and Gold [61] trained animals over several sessions to discriminate the direction of visual motion in an RDK task while activity in areas MT and LIP (thought to reflect the sites of sensory evidence encoding and evidence accumulation, respectively) was measured. The neurons in area LIP, but not MT, showed increased responsiveness to the decision evidence as a function of learning, reflected in steeper evidence-accumulation slopes. Correspondingly, there was a correlation between the neural responsiveness of these neurons and the performance on the task. This study provides evidence that perceptual learning does not change how sensory information is represented in the brain, but rather how sensory representations are interpreted, particularly by higher areas in the brain involved in decision-making.

In a follow up study, the same authors [62] showed that their results could be explained in terms of a reinforcement learning (RL) mechanism [63], whereby the connections between sensory neurons and the decision process are strengthened via a reward prediction error, gradually enhancing the readout of relevant information and improving perceptual sensitivity. This explanation was corroborated by work in humans by Kahnt et al. [64], who trained subjects on an orientation discrimination task over the course of 4 days. The authors explained behavioral improvements using an RL model that updated the decision evidence on every trial in accordance with a prediction error signal. Using fMRI, the authors showed that stimulus orientation was encoded in both early visual and higher cortical areas in lateral parietal and medial prefrontal cortices. However, only activity in the medial prefrontal cortex tracked the trial-by-trial changes in the decision variables estimated from the RL model. These findings suggest that a reinforcement-guided learning mechanism might be at work during both reward-related and perceptual learning.

Confidence

The previous section highlighted that the process of perceptual decision-making can undergo long-term optimization, which ultimately serves to fine-tune behavior and maximize reward. It was suggested that a potential mechanism by which perceptual learning is facilitated is the ability of a system to estimate the reliability of a perceptual decision [65]. In addition to providing an evaluation of the decision process itself, this sense of confidence in our judgments can also help predict choice outcome and ultimately motivate learning and inform future choices. Correspondingly, there has been growing interest in understanding the neural basis of decision confidence, in both the NHP and the human literature [66–68].

In particular, the relationship between confidence and the decision process has been the subject of some debate. Generally, decision confidence has been studied as a metacognitive (i.e., postdecisional) process. For instance, Fleming et al. [69] examined the spatial correlates of metacognition by asking subjects to perform a face/house categorization task and rate their confidence in their decisions after each choice. fMRI activity in the rostrolateral prefrontal cortex (RLPFC) was found to correlate with confidence at the time of rating, and was enhanced during confidence rating compared to a control task. Importantly, the strength of the relationship between RLPFC activity and confidence reports was predictive of subjects' ability to evaluate their own performance (i.e., metacognitive ability). Relatedly, Fleming et al. [70] demonstrated that metacognitive ability is correlated with the gray matter volume of the anterior portion of the prefrontal cortex, as well as the white matter projections to this region. While these studies have addressed important questions regarding the spatial characteristics of choice confidence in the postdecision period, it is not clear whether confidence might be arising earlier in time, during the decision process itself.

Evidence from animals suggests that information about confidence becomes available earlier in the decision stream [71,72]. Specifically, in the primate and rat brains, choice confidence appears to develop simultaneously with the decision process and is encoded by the same neural populations that form the decision. Only a handful of studies have tried to characterize choice confidence in the human brain as it develops in time during the decision process. In Zizlsperger et al. [73], subjects performed an RDK task and rated their confidence in the decision while EEG data were recorded. Event-related potentials locked to stimulus onset began to show a separation between high- and low-confidence trials soon after stimulus presentation, and almost concomitant with modulation by task difficulty, a finding consistent with a decisional account of confidence.

In line with this account, Gherman and Philiastides [10] used EEG to study the temporal characteristics of choice confidence during a delayed-response face/car categorization task that rewarded correct trials. On half of the trials, before indicating their choice, subjects were allowed to opt out of the task for a smaller but certain reward. This manipulation encouraged subjects to opt out of the decision when they were uncertain of their choice or ignore this option and commit to a choice when they were certain. A comparison between the neural signals for certain versus uncertain choices showed that discrimination between the two conditions increased gradually after the stimulus was presented, peaking before subjects initiated a response. These confidence-related signals exhibited a ramp-like response profile (Fig. 11.3C) and had a spatial topography consistent with the process of evidence accumulation defined earlier (compare Fig. 11.2A with Fig. 11.3C) [11]. Importantly, the accumulation rate of this activity was predictive of confidence on a trial-by-trial basis, even when difficulty effects were controlled for, offering strong evidence that confidence develops continuously as the decision process unfolds. Taken together, these studies suggest that confidence arises early after stimulus presentation and thus may help shape the course of the decision process itself.

One hypothesis as to how this can be achieved is that confidence may act as a learning signal, shaping subsequent decisions and optimizing performance. For instance, Hebart et al. [65] found that during an RDK task, fMRI activation increased with confidence in the ventral striatum. Using a connectivity analysis, they demonstrated a flow of information from areas of the brain correlating with the decision (i.e., parietal and prefrontal sites) to the ventral striatum, suggesting confidence computed within this region is derived from, and computed in parallel with, the ongoing decision signal. The authors proposed a potential role of the ventral striatum in confidence-driven learning, suggesting that introspective signals (e.g., the feeling of reward associated with a choice) serve to reinforce optimal behavior on subsequent choices. Thus, in this sense, confidence may be thought of as an implicit reward signal, which is being propagated back to the decision systems to optimize the dynamics of the decision process, possibly by means of an RL-like mechanism.

Neuromodulation

Although much effort is being invested in assessing the various influences on perceptual decision-making at the systems level, less attention has been devoted to understanding how neurotransmission affects the decision process. There has been some evidence pointing to a potential role of the neurotransmitter dopamine in the efficiency of information processing. MacDonald et al. [74] examined the influence of aging-related decline in dopaminergic activity on cognitive processing, using a speeded reaction-time task that relied on executive control. The authors demonstrated a tight link between the decrease in D1 receptor binding potential and intraindividual trial-to-trial variability in reaction time, a measure commonly used to indicate the precision of information processing [75]. Interestingly, another study by Ratcliff et al. [76] demonstrated that age-related decreases in processing efficiency during a brightness discrimination task were linked to the rate of decision-related evidence accumulation. Together, the two studies suggest that disruptions in dopaminergic activity can have an influence on the efficiency of decision formation; however, a direct link between the two is yet to be demonstrated empirically.

Additional evidence for the involvement of dopamine in the perceptual decision process comes from research on molecular genetics. The catechol-O-methyltransferase (COMT) protein, which metabolizes catecholamine neurotransmitters, has been shown to play a role in regulating prefrontal dopamine [77], and may affect processing in various decision-making centers. Saville et al. [78] demonstrated that Met/Met carriers of the COMT Val^{158}Met polymorphism showed greater intrasubject variability during an n-back reaction-time task, both in behavior (response time) and in the latency of the P3b event-related potential, compared against Val/Val carriers. The P3b component, which has often been associated with decision-related processes [79], has been argued to reflect the formation of the perceptual decision itself [11,35,80]. Together, these findings suggest a potential link between COMT-related individual differences in dopaminergic activity and decision formation.

Dopaminergic activity may also have an indirect influence on the perceptual decision, via its involvement in reward-related activity. Nagano-Saito et al. [81] tested the effect of reward-related dopaminergic activity on visual perceptual discrimination by temporarily hindering dopamine transmission in healthy subjects. Specifically, participants performed an RDK task in which cues presented prior to each trial informed them on the availability of a reward for a correct response. Behaviorally, availability of reward led to increased decision thresholds (as inferred using an accumulation-to-bound model), suggesting subjects prioritized accuracy over speed on these trials. During the anticipatory period leading to stimulus presentation, fMRI activation was greater for rewarded trials in the ventral striatum and prefrontal cortex, and interestingly, the magnitude of the activation was positively correlated with the threshold of the impending decision. Crucially, decreasing dopaminergic transmission eliminated both the blood oxygen level-dependent activation and its correlation with the decision threshold, pointing to a causal role in decision formation.

New insights on the neuromodulation of perceptual decision may also be gained from the study of pupil size changes across the time course of the decision process. Pupil size fluctuations in conditions of constant illumination are thought to reflect arousal state and have been shown to correlate with activity in the locus coeruleus, the center of the neuroadrenergic system [82] (for a review, see Aston-Jones and Cohen [83]). In one such study, Murphy et al. [84] demonstrated that spontaneous, stimulus-independent fluctuations in pupil diameter during an RDK task could be explained by the variability in the accumulation of decision evidence, a parameter derived by fitting a diffusion model to the behavioral data. Specifically, slow increases in pupil size, reflective of heightened arousal state, were associated with greater variability in the rate of evidence accumulation.

Similarly, De Gee et al. [85] measured pupil size fluctuations while subjects performed a visual detection task wherein they determined the presence (or absence) of low-contrast grating stimuli. Authors demonstrated that pupil dilation predicted subjects' behavior (i.e., whether they made a "present" vs "absent" response), and more importantly, it was best explained by a sustained component that persisted throughout the decision phase (i.e., starting at stimulus onset and ending when the subject made a response), suggesting pupil dilation contains information about the formation of the perceptual decision. As has been previously postulated by theoretical modeling work [86,87], it is plausible that the perceptual decision process may rely on noradrenergic modulation mediated by the locus coeruleus; however, this is still open to further investigation. The increase in availability of effective, transient, and noninvasive techniques for studying neuromodulatory systems in humans (e.g., amino acid challenge and depletion techniques, see Ref. [88]) may offer new opportunities for further understanding of their roles in perceptual decision-making.

CONCLUSION

In this chapter, we provided an overview of the general neurobiological principles guiding perceptual decision-making and reviewed the influence of various modulators on the process of decision formation. We also discussed how these influences could be understood in terms of changes in parameters of sequential sampling models of decision-making. Finally, we offered a general discussion on how the influence of these decision modulators can be thought of in the framework of reward maximization, whereby the perceptual decision-making system adjusts to adaptively optimize behavior. This perspective highlights possible future research directions into the role of reinforcement-guided learning, not only in reward- and value-based decision-making, but also in our understanding of perceptual decision-making.

References

[1] Heekeren HR, Marrett S, Ungerleider LG. The neural systems that mediate human perceptual decision making. Nat Rev Neurosci 2008;9:467−79.

[2] Ratcliff R. A theory of memory retrieval. Psychol Rev 1978;85:59.

[3] Usher M, McClelland JL. The time course of perceptual choice: the leaky, competing accumulator model. Psychol Rev 2001;108:550.

[4] Ratcliff R, Smith PL. A comparison of sequential sampling models for two-choice reaction time. Psychol Rev 2004;111:333.

[5] Ratcliff R, McKoon G. The diffusion decision model: theory and data for two-choice decision tasks. Neural Comput 2008;20: 873−922.

[6] Bogacz R. Optimal decision-making theories: linking neurobiology with behaviour. Trends Cognit Sci 2007;11:118−25.

[7] Gold JI, Shadlen MN. The neural basis of decision making. Annu Rev Neurosci 2007;30:535−74.

[8] Horwitz GD, Newsome WT. Separate signals for target selection and movement specification in the superior colliculus. Science 1999;284:1158−61.

[9] Kim JN, Shadlen MN. Neural correlates of a decision in the dorsolateral prefrontal cortex of the macaque. Nat Neurosci 1999;2: 176−85.

[10] Gherman S, Philiastides MG. Neural representations of confidence emerge from the process of decision formation during perceptual choices. NeuroImage 2015;106:134−43.

[11] Philiastides MG, Heekeren HR, Sajda P. Human scalp potentials reflect a mixture of decision-related signals during perceptual choices. J Neurosci 2014;34:16877−89.

[12] Kelly SP, O'Connell RG. Internal and external influences on the rate of sensory evidence accumulation in the human brain. J Neurosci 2013;33:19434−41.

[13] Ploran EJ, et al. Evidence accumulation and the moment of recognition: dissociating perceptual recognition processes using fMRI. J Neurosci 2007;27:11912−24.

[14] Tremel JJ, Wheeler ME. Content-specific evidence accumulation in inferior temporal cortex during perceptual decision-making. NeuroImage 2015;109:35−49.

[15] Heekeren HR, et al. A general mechanism for perceptual decision-making in the human brain. Nature 2004;431:859−62.

[16] Heekeren HR, et al. Involvement of human left dorsolateral prefrontal cortex in perceptual decision making is independent of response modality. Proc Natl Acad Sci USA 2006;103:10023−8.

[17] Filimon F, et al. How embodied is perceptual decision making? Evidence for separate processing of perceptual and motor decisions. J Neurosci 2013;33:2121−36.

[18] Philiastides MG, et al. Causal role of dorsolateral prefrontal cortex in human perceptual decision making. Curr Biol 2011;21:980−3.

[19] Noppeney U, Ostwald D, Werner S. Perceptual decisions formed by accumulation of audiovisual evidence in prefrontal cortex. J Neurosci 2010;30:7434—46.

[20] Kaiser J, Lennert T, Lutzenberger W. Dynamics of oscillatory activity during auditory decision making. Cereb Cortex 2007;17:2258—67.

[21] Romo R, Salinas E. Flutter discrimination: neural codes, perception, memory and decision making. Nat Rev Neurosci 2003;4:203—18.

[22] Preuschhof C, et al. Neural correlates of vibrotactile working memory in the human brain. J Neurosci 2006;26:13231—9.

[23] Spitzer B, Blankenburg F. Stimulus-dependent EEG activity reflects internal updating of tactile working memory in humans. Proc Natl Acad Sci USA 2011;108:8444—9.

[24] Haegens S, Handel BF, Jensen O. Top-down controlled alpha band activity in somatosensory areas determines behavioral performance in a discrimination task. J Neurosci 2011;31:5197—204.

[25] Thut G, et al. Alpha-band electroencephalographic activity over occipital cortex indexes visuospatial attention bias and predicts visual target detection. J Neurosci 2006;26:9494—502.

[26] Wyart V, Tallon-Baudry C. How ongoing fluctuations in human visual cortex predict perceptual awareness: baseline shift versus decision bias. J Neurosci 2009;29:8715—25.

[27] Van Dijk H, et al. Prestimulus oscillatory activity in the alpha band predicts visual discrimination ability. J Neurosci 2008;28:1816—23.

[28] Mazaheri A, et al. Region-specific modulations in oscillatory alpha activity serve to facilitate processing in the visual and auditory modalities. NeuroImage 2014;87:356—62.

[29] Foxe JJ, Snyder AC. The role of alpha-band brain oscillations as a sensory suppression mechanism during selective attention. Front Psychol 2011;2:154.

[30] Lou B, et al. Prestimulus alpha power predicts fidelity of sensory encoding in perceptual decision making. NeuroImage 2014;87:242—51.

[31] Philiastides MG, Sajda P. Temporal characterization of the neural correlates of perceptual decision making in the human brain. Cereb Cortex 2006;16:509—18.

[32] Philiastides MG, Ratcliff R, Sajda P. Neural representation of task difficulty and decision making during perceptual categorization: a timing diagram. J Neurosci 2006;26:8965—75.

[33] Philiastides MG, Sajda P. EEG-informed fMRI reveals spatiotemporal characteristics of perceptual decision making. J Neurosci 2007;27:13082—91.

[34] Ratcliff R, Philiastides MG, Sajda P. Quality of evidence for perceptual decision making is indexed by trial-to-trial variability of the EEG. Proc Natl Acad Sci USA 2009;106:6539—44.

[35] O'Connell RG, Dockree PM, Kelly SP. A supramodal accumulation-to-bound signal that determines perceptual decisions in humans. Nat Neurosci 2012;15:1729—35.

[36] Rangel A, Camerer C, Montague PR. A framework for studying the neurobiology of value-based decision making. Nat Rev Neurosci 2008;9:545—56.

[37] Weil RS, et al. Rewarding feedback after correct visual discriminations has both general and specific influences on visual cortex. J Neurophysiol 2010;104:1746—57.

[38] Schiffer A-M, et al. Reward activates stimulus-specific and task-dependent representations in visual association cortices. J Neurosci 2014;34:15610—20.

[39] Pleger B, et al. Reward facilitates tactile judgments and modulates hemodynamic responses in human primary somatosensory cortex. J Neurosci 2008;28:8161—8.

[40] Blank H, et al. Temporal characteristics of the influence of punishment on perceptual decision making in the human brain. J Neurosci 2013;33:3939—52.

[41] Green N, Biele GP, Heekeren HR. Changes in neural connectivity underlie decision threshold modulation for reward maximization. J Neurosci 2012;32:14942—50.

[42] Bogacz R, et al. Do humans produce the speed-accuracy trade-off that maximizes reward rate? Q J Exp Psychol (Hove) 2010;63:863—91.

[43] Drugowitsch J, et al. The cost of accumulating evidence in perceptual decision making. J Neurosci 2012;32:3612—28.

[44] Domenech P, Dreher JC. Decision threshold modulation in the human brain. J Neurosci 2010;30:14305—17.

[45] Bogacz R, et al. The neural basis of the speed-accuracy tradeoff. Trends Neurosci 2010;33:10—6.

[46] Forstmann BU, et al. Striatum and pre-SMA facilitate decision-making under time pressure. Proc Natl Acad Sci USA 2008;105:17538—42.

[47] van Maanen L, et al. Neural correlates of trial-to-trial fluctuations in response caution. J Neurosci 2011;31:17488—95.

[48] Forstmann BU, et al. Cortico-striatal connections predict control over speed and accuracy in perceptual decision making. Proc Natl Acad Sci USA 2010;107:15916—20.

[49] Ivanoff J, Branning P, Marois R. fMRI evidence for a dual process account of the speed-accuracy tradeoff in decision-making. PLoS One 2008;3:e2635.

[50] Wenzlaff H, et al. Neural characterization of the speed—accuracy tradeoff in a perceptual decision-making task. J Neurosci 2011;31:1254—66.

[51] Heitz RP, Schall JD. Neural mechanisms of speed-accuracy tradeoff. Neuron 2012;76:616—28.

[52] Hanks T, Kiani R, Shadlen MN. A neural mechanism of speed-accuracy tradeoff in macaque area LIP. Elife 2014;3.

[53] Goldstone RL. Perceptual learning. Annu Rev Psychol 1998;49:585—612.

[54] Recanzone GH, et al. Topographic reorganization of the hand representation in cortical area 3b owl monkeys trained in a frequency-discrimination task. J Neurophysiol 1992;67:1031—56.

[55] Recanzone GH, Schreiner CE, Merzenich MM. Plasticity in the frequency representation of primary auditory cortex following discrimination training in adult owl monkeys. J Neurosci 1993;13:87—103.

[56] Furmanski CS, Schluppeck D, Engel SA. Learning strengthens the response of primary visual cortex to simple patterns. Curr Biol 2004;14:573—8.

[57] Jehee JF, et al. Perceptual learning selectively refines orientation representations in early visual cortex. J Neurosci 2012;32:16747—53.

[58] Yan Y, et al. Perceptual training continuously refines neuronal population codes in primary visual cortex. Nat Neurosci 2014;17:1380—7.

[59] Zhang J-Y, et al. Rule-based learning explains visual perceptual learning and its specificity and transfer. J Neurosci 2010;30:12323—8.

[60] Li S, Mayhew SD, Kourtzi Z. Learning shapes the representation of behavioral choice in the human brain. Neuron 2009;62:441—52.

[61] Law C-T, Gold JI. Neural correlates of perceptual learning in a sensory-motor, but not a sensory, cortical area. Nat Neurosci 2008;11:505—13.

[62] Law C-T, Gold JI. Reinforcement learning can account for associative and perceptual learning on a visual-decision task. Nat Neurosci 2009;12:655—63.

[63] Sutton RS, Barto AG. Reinforcement learning: an introduction, vol. 1. Cambridge: MIT Press; 1998.

[64] Kahnt T, et al. Perceptual learning and decision-making in human medial frontal cortex. Neuron 2011;70:549—59.

[65] Hebart MN, et al. The relationship between perceptual decision variables and confidence in the human brain. Cereb Cortex 2016; 26:118–30.

[66] Fetsch CR, Kiani R, Shadlen MN. Predicting the accuracy of a decision: a neural mechanism of confidence. Cold Spring Harb Symp Quant Biol 2014;79:185–97.

[67] Yeung N, Summerfield C. Metacognition in human decision-making: confidence and error monitoring. Philos Trans R Soc Lond B Biol Sci 2012;367:1310–21.

[68] Fleming SM, Dolan RJ. The neural basis of metacognitive ability. Philos Trans R Soc Lond B Biol Sci 2012;367:1338–49.

[69] Fleming SM, Huijgen J, Dolan RJ. Prefrontal contributions to metacognition in perceptual decision making. J Neurosci 2012; 32:6117–25.

[70] Fleming SM, et al. Relating introspective accuracy to individual differences in brain structure. Science 2010;329:1541–3.

[71] Kiani R, Shadlen MN. Representation of confidence associated with a decision by neurons in the parietal cortex. Science 2009; 324:759–64.

[72] Kepecs A, et al. Neural correlates, computation and behavioural impact of decision confidence. Nature 2008;455:227–31.

[73] Zizlsperger L, et al. Cortical representations of confidence in a visual perceptual decision. Nat Commun 2014;5:3940.

[74] MacDonald SW, et al. Aging-related increases in behavioral variability: relations to losses of dopamine D1 receptors. J Neurosci 2012;32:8186–91.

[75] MacDonald SW, Hultsch DF, Bunce D. Intraindividual variability in vigilance performance: does degrading visual stimuli mimic age-related "neural noise"? J Clin Exp Neuropsychol 2006;28: 655–75.

[76] Ratcliff R, Thapar A, McKoon G. Aging and individual differences in rapid two-choice decisions. Psychon Bull Rev 2006;13:626–35.

[77] Tunbridge EM, et al. Catechol-o-methyltransferase inhibition improves set-shifting performance and elevates stimulated dopamine release in the rat prefrontal cortex. J Neurosci 2004;24: 5331–5.

[78] Saville CW, et al. COMT Val[158]Met genotype is associated with fluctuations in working memory performance: converging evidence from behavioural and single-trial P3b measures. NeuroImage 2014;100:489–97.

[79] Nieuwenhuis S, Aston-Jones G, Cohen JD. Decision making, the P3, and the locus coeruleus-norepinephrine system. Psychol Bull 2005;131:510–32.

[80] Twomey DM, et al. The classic P300 encodes a build-to-threshold decision variable. Eur J Neurosci 2015;42:1636–43.

[81] Nagano-Saito A, et al. From anticipation to action, the role of dopamine in perceptual decision making: an fMRI-tyrosine depletion study. J Neurophysiol 2012;108:501–12.

[82] Murphy PR, et al. Pupil diameter covaries with BOLD activity in human locus coeruleus. Hum Brain Mapp 2014;35:4140–54.

[83] Aston-Jones G, Cohen JD. An integrative theory of locus coeruleus-norepinephrine function: adaptive gain and optimal performance. Annu Rev Neurosci 2005;28:403–50.

[84] Murphy PR, Vandekerckhove J, Nieuwenhuis S. Pupil-linked arousal determines variability in perceptual decision making. PLoS Comput Biol 2014;10:e1003854.

[85] De Gee JW, Knapen T, Donner TH. Decision-related pupil dilation reflects upcoming choice and individual bias. Proc Natl Acad Sci USA 2014;111:E618–25.

[86] Shea-Brown E, Gilzenrat MS, Cohen JD. Optimization of decision making in multilayer networks: the role of locus coeruleus. Neural Comput 2008;20:2863–94.

[87] Gilzenrat MS, et al. Pupil diameter tracks changes in control state predicted by the adaptive gain theory of locus coeruleus function. Cogn Affect Behav Neurosci 2010;10:252–69.

[88] Biskup CS, et al. Amino acid challenge and depletion techniques in human functional neuroimaging studies: an overview. Amino Acids 2015;47:651–83.

[89] Philiastides MG, Heekeren HR. Spatiotemporal characteristics of perceptual decision making in the human brain. In: Dreher J-C, Tremblay L, editors. Handbook of reward and decision making. Academic Press; 2009. p. 185–212.

Perceptual Decision-Making: What Do We Know, and What Do We Not Know?

C. Summerfield, A. Blangero
University of Oxford, Oxford, United Kingdom

Abstract

Perceptual decisions occur when sensory inputs are converted to discrete categorical variables. The neural and computational mechanisms by which rodents, monkeys, and humans make perceptual decisions have been intensively studied since 1989. Here, we address five questions that continue to raise controversy, summarizing research that has asked (1) how information is integrated to form a categorical choice, (2) what computations are carried out in cortical neurons during perceptual decisions, (3) how perceptual decision-making can be studied using cognitive neuroimaging techniques, (4) how we decide when to draw our decisions to a conclusion, and (5) how perceptual decisions are biased by prior information. Despite considerable progress, these remain key open questions in psychology and the neurosciences.

INTRODUCTION

The term "perceptual decision-making" is a neologism—it first appeared in print in 1989, in the title of a now-classic article that describes neurophysiological recordings from motion-selective middle temporal (MT) area of the macaque monkey [1]. The paper showed, strikingly, that during discrimination of the motion direction of a random-dot kinetogram, the firing rates of individual neurons in the MT area can be more sensitive to the motion direction than the whole observer—a curiosity that neuroscientists have spent the subsequent years attempting to unravel [2]. In reporting this finding, however, the authors also laid the foundations for a remarkably successful new subfield within the neural sciences. Perceptual decision-making brings together historical work from psychophysics, the psychology of perceptual categorization, and mathematical modeling of response times, and marries them with state-of-the-art neural recording and cognitive neuroimaging techniques. The resulting observations have been argued to provide a unified neural and computational vision of how observers detect, discriminate, and categorize sensory information in the external world.

WHAT ARE PERCEPTUAL DECISIONS?

Perceptual decisions are sensorimotor acts: they allow us to classify sensory information in the immediate environment, and to select an appropriate course of action. As we cycle to work, the world passes by in a blur of objects, textures, and features. The field of perceptual decision-making attempts to understand the mechanisms by which the optic flow field is interpreted while zooming down a hill (to determine whether one should lean in a leftward or a rightward direction to stay on course) or the processes by which one determines that a passing face is that of a friend or a stranger (and thus perhaps stop for a chat or else hurry on by). In the laboratory, perceptual decision-making is usually studied using psychophysical methods, which allow the researcher to control tightly the parameters of sensory stimulation, and accurately measure an observer's responses and latencies. One widely studied paradigm, known as the a random-dot kinetogram or RDK task, requires an observer to classify the net direction of motion (e.g., leftward vs. rightward) in a cloud of random dots with variable fractional motion coherence (i.e., difficulty), signaling the decision with a saccadic eye movement or manual response [3]. More recently, new paradigms involving counting of punctate events (or "pulses") [4] or the integration of discrete stimulus information [5,6] have come to the fore. In conjunction with neural recordings, either at the single-cell level (via invasive neurophysiology) or at the whole-brain level (via cognitive neuroimaging) this psychophysical approach allows the researcher to measure how sensory input and neural activity collectively give rise to behavior. The widespread use of a simple, well-characterized paradigm, such as the RDK has brought the field converging explanatory power, in particular because the underlying neural substrates in the primate visual motion and saccadic systems have previously been well characterized at both the cellular and the systems levels. Nevertheless, the hope is that studying a simple model system will furnish general principles extending beyond elementary sensorimotor judgments, offering a general framework for understanding decision-making that provides a "window on cognition" [7].

AIMS AND SCOPE OF THIS CHAPTER

Since 1989, perceptual decision-making has blossomed into a vibrant and impactful field. An exhaustive survey of all the key theories and empirical findings is thus beyond the scope of this chapter. Instead, we have chosen to focus this commentary around five key questions that continue to provoke controversy in psychology and neuroscience. Our summary is far from exhaustive, and clear answers are not always forthcoming. Nevertheless, our hope is to use these questions to illustrate major themes, to consider relevant methodological issues, to chart where progress has been made, and to highlight where more research is needed.

DECISION OPTIMALITY

Decision-relevant sensory information is often variable, ambiguous, or partially observable. A recurrent theme in perceptual decision-making is whether observers make near-optimal decisions, that is, the best possible decisions given the level of uncertainty in the stimulus [8]. More formally, do choices resemble those of a "Bayesian" ideal observer, whose decisions respect the posterior probability of a given stimulus or category, given the available sensory input? Specifying how (or whether) decisions deviate from optimality is equivalent to the wider project of characterizing the computational mechanisms underlying perceptual choice; and as such, this issue subsumes the more specific questions raised below, and we give it a special introduction here.

The question of whether observers are optimal is theoretically fraught. Whether a decision is "optimal" or not depends on what exactly the agent is attempting to optimize. For example, if choices are differentially rewarded, then a policy that optimizes overall accuracy might fail to maximize reward rate. Moreover, perceptual decisions are limited not only by noise in the external stimulus (e.g., randomness of dot motion in the RDK), but also potentially by noise arising in the internal neural computations that underlie behavior. "Noise" in internal neural signals refers to residual or unexplained variability in activity, conditioned on some assumptions about how sensory information is encoded or processed. Whether a decision policy is deemed "optimal" thus depends on those assumptions [9]. For example, a given policy may be suboptimal if one assumes an unlimited processing capacity, but optimal under the assumption that cognitive resources are finite. Responses to this question are thus heavily assumption laden, and controversy regularly arises over how optimal decisions should be defined, complementing (and sometimes confusing) debate over how to interpret conflicting empirical results [10,11].

Q1: HOW IS INFORMATION INTEGRATED DURING PERCEPTUAL DECISION-MAKING?

The theoretical cornerstone of the field of perceptual decision-making is that decisions can be optimized through sequential sampling and integration. This follows from the law of large numbers, which states that the precision of an estimator will grow with the square root of the number of independent samples. In psychophysical tasks, such as the RDK, a more precise estimate of the direction of dot motion will be obtained after viewing and averaging information from several successive pairs of frames (or samples), rather than just a single pair.[1] Formally, when a single sample of information is available, our posterior belief in the (binary) category membership should be proportional to the relative log-likelihood that it is drawn from distribution A or B; where multiple samples are available, our posterior belief grows with the sum of sequential log-likelihoods accruing from successive samples [12].

Observers Sample Information Sequentially

This mathematical framework for understanding optimal choices, often referred to as the "sequential

---

[1] Note that in the RDK information about the direction of motion arises from the disparity in dot position over two or more successive frames.

probability ratio test" or SPRT, was first developed in the 1940s, and underpins almost all the models that have subsequently been proposed to account for perceptual choices (see Fig. 12.1A). Consider an observer discriminating motion direction in an RDK stimulus. When the viewing latency is determined by the experimenter (a fixed-response task), choice accuracy grows with the number of successive frame pairs (samples), as predicted by the sequential sampling framework [13]. Where the participant is free to make a response at any time (free-response task), decision latencies are well described by an approximation to the SPRT, known as the drift-diffusion model (or DDM), in which exact log

probabilities are replaced by a momentary input variable that is composed of signal and noise [14]. For example, in the RDK this input variable (the *drift rate*) will be close to zero on trials in which most dots are moving randomly, and larger when motion coherence is greater. Like the log-likelihood ratio in the SPRT, the drift rate reflects the relative evidence for one category over another, and it is integrated (by summation) until it reaches a criterion level or *bound*, at which point a decision is made. This class of sequential sampling model can capture many of the salient behavioral features of the task, including the rightward skew in reaction times, the relative latencies of error and correct trials, the

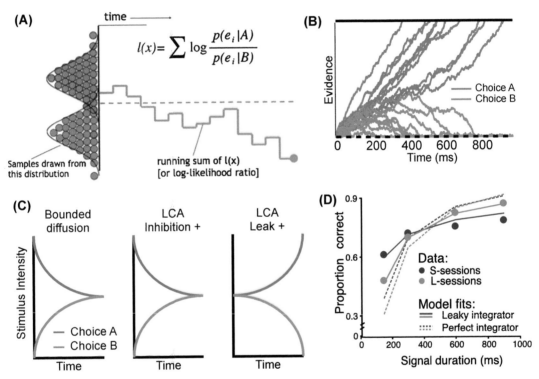

FIGURE 12.1 (A) Optimal decision models. Illustration of the sequential probability ratio test (SPRT). The SPRT describes the optimal mechanism by which an observer judges whether successive samples of information are drawn from one category (*green*) or another (*red*). Each sample is characterized by a momentary decision value (*vertical axis*) that is expressed as the log of the probability that it is drawn from the green or red category, and these log-likelihoods are summed over time (*horizontal axis*) to a fixed threshold. The cumulative decision variable $l(x)$ is depicted as the green trace (here, samples are drawn from the green category). The *dashed horizontal line* shows the equilibrium point (equivalent evidence for each category). See Ref. [12] for more details. (B) The leaky competing accumulator (LCA) model. Simulation of the amount of evidence over time during a two-alternative decision task. In the examples used here, option A (*green*) is selected every time (evidence reaching threshold), inhibiting choice B (*red*), which has less evidence. (C) Primacy and recency in perceptual decision-making. Schematic depiction of how the weight of evidence varies over time under bounded diffusion and LCA models, for a perceptual task in which an observer responds after a fixed interval. Each plot illustrates the result of a psychophysical reverse correlation analysis in which noisy evidence is averaged over time for choices favoring categories A (*red line*) and B (*green line*). The point at which the lines diverge indicates where evidence has the greatest influence on choices. In the diffusion model, early evidence has the greatest influence on choices ("primacy" effect), because late evidence arrives after an implicit bound has been crossed. In the LCA, the temporal profile of the weight of evidence depends on the parameter settings. Where inhibition is stronger than leak, a primacy effect is observed, similar to the diffusion model. When leak prevails, a "recency" effect is observed, with later evidence having more sway over choices. (*Adapted with permission from Tsetsos K, et al. Using time-varying evidence to test models of decision dynamics: bounded diffusion vs. the leaky competing accumulator model. Front Neurosci 2012;6:79.*) (D) Adaptive weighting of evidence over time. Human observers detected a small increment in luminance occurring in a stream of gray patches. Correct detections are shown for different stream durations on trials that occurred in a context comprising mostly short (*blue points*) or mostly long streams (*red points*). Observers adjusted their decision policy according to the context in a fashion that was predicted by the LCA (*solid lines*) but not a perfect integration (*dashed lines*). (*Reproduced with permission from Ossmy O, et al. The timescale of perceptual evidence integration can be adapted to the environment. Curr Biol 2013;23:981–86.*)

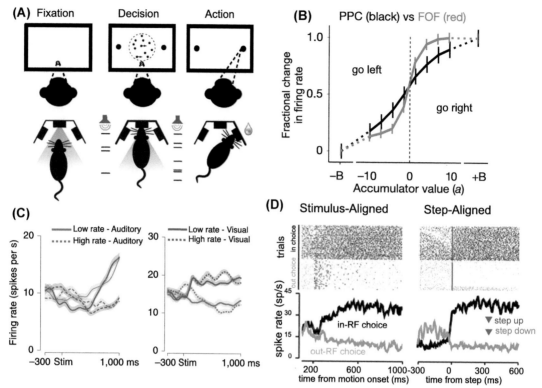

FIGURE 12.2   (A) Canonical paradigms. Illustration of the tasks used to investigate perceptual decision-making in the monkey (*upper row*) and rodent (*lower row*). Following a period of fixation, the animal has to decide whether a train of auditory pulses is more numerous on the left or the right (rodent) or whether a random dot kinetogram is moving to the right or the left (monkey). The response is made with a nosepoke (rodent) or saccade (monkey) to the corresponding side. (B) Distinct encoding of accumulator states in rodent parietal and prefrontal cortices. Firing rate change in the rodent posterior parietal cortex (PPC) and FOF, plotted against the state of the cumulative decision variable estimated from behavioral data. The relationship between PPC activity and accumulator state is graded, whereas that in FOF is steplike, indicative of a more categorical representation. *(Reproduced with permission from Hanks TD, et al. Distinct relationships of parietal and prefrontal cortices to evidence accumulation. Nature 2015;520:220–23.)* (C) Rodent PPC neurons show mixed selectivity. Average firing rates from two cells recorded in the PPC of a rodent performing a variant of the pulse task that involved both auditory (click) and visual (flash) stimuli. The *left* shows a cell that distinguishes low rate (*solid lines*) from high rate (*dashed lines*) stimuli, whereas the *right* shows a cell that distinguishes visual (*blue lines*) from auditory (*green lines*) stimuli. The study concluded that PPC neurons show mixed selectivity. *(Adapted with permission from Raposo D, et al. A category-free neural population supports evolving demands during decision-making. Nat Neurosci 2014;17:1784–92.)* (D) Steplike changes in lateral intraparietal region (LIP) activity during motion discrimination. Raster plots (*top*; each row is a raw spike train) and poststimulus time histograms (*bottom*; each line is an average firing rate) from LIP recordings in the macaque. *In-RF choice* refers to those trials in which the monkey saccaded into the response field of the LIP neuron. On the *left*, the activity is aligned on the onset of the motion dot stimulus, and after averaging, accumulation appears to be gradual. On the *right*, activity is aligned to a steplike change in firing rate in either a positive (*blue dots*) or negative (*red dots*) direction. The average accumulator state undergoes an abrupt change. *(Reprinted with permission from Fitzgerald JK, et al. Biased associative representations in parietal cortex. Neuron 2013;77:180–91.)*

relative benefit of stronger or weaker inputs over time, and the potential trade-off between speed and accuracy that is determined by the height of the bound (see Q4 below) [15].

However, Information Integration Is Not Lossless

The DDM proposes that humans sample and integrate information without loss, at least when psychophysical stimulation is brief (~1 or 2 s). In fixed-response tasks, time–accuracy curves grow to a fixed ceiling, as if observers fail to benefit fully from more prolonged streams of sensory information [16,17], but it is unclear whether

this occurs because early information is forgotten (recency effect) or because later information is ignored (primacy effect). When a psychophysical reverse correlation approach (in which noisy stimuli are averaged conditional on a subsequent response) was used in conjunction with a fixed-response version of the RDK paradigm, motion energy was found to have less impact later in the stimulation stream, as if integration halted when the cumulative decision variable crossed an implicit subjective bound [18]. However, viewing durations in this experiment were exponentially distributed, and so this primacy effect may have been triggered by the preponderance of short-latency trials. Indeed, in a later study by the same authors, the second of two successive fixed-duration motion pulses carried more weight in the decision, suggestive of a recency effect [19]. Other studies in which observers average information during rapid serial visual presentation (RSVP) have reported a recency bias [20]. In one study, recency was strongly dependent on the attentional resources available: dividing attention attenuated the influence of the earliest samples on choice, as if capacity limits gate the passage of momentary decision information into an accumulation process [21].

Both primacy and recency can be elegantly accounted for by a related model, known as the "leaky competing accumulator," in which cumulative decision variables for each choice compete via mutual inhibition (leading to "runaway" or attractor processes) but also leak back to equivalence over time (tempering these nonlinearities) (see Fig. 12.1B) [16]. A predominance of inhibition promotes primacy; recency occurs when the leak is more pronounced (see Fig. 12.1C) [22]. A particularly insightful experiment noted that in real world situations, in which decision information is available at unpredictable times, accuracy is not maximized by lossless integration, but by scaling the time constant of integration to the likely duration of the information pulse. Human observers who were instructed to detect long or short pulses of elevated luminance in an RSVP stream were able to adjust their integration window from trial to trial flexibly, scaling their "leak" to match the task demands (see Fig. 12.1D) [23].

Q1: Summary

Humans optimize perceptual decisions via sequential sampling and integration. However, information can be lost during integration. Where capacity is limited, information may be forgotten (or "leak" away). However, when the environment varies unpredictably, leaky integration can maximize decision sensitivity, and observers may flexibly adjust the time constant of integration to meet the demands of the task.

Q2: WHAT COMPUTATIONS DO CORTICAL NEURONS PERFORM DURING PERCEPTUAL DECISIONS?

In the macaque monkey, the RDK task (see Fig. 12.2A) engages cortical circuits that subserve motion detection, such as visual area MT, and saccadic control, such as the lateral intraparietal region (LIP) and the frontal eye fields (FEF). From there, information is routed to the

striatum and superior colliculus, where a saccade is initiated.[2]

In the Macaque, Lateral Intraparietal Region Firing Rates Grow During Information Integration

In area MT, firing rates depend on the quality of the evidence in the stimulus (i.e., percentage motion coherence), but activity remains broadly constant during stimulation. Many LIP exhibit persistent activity in advance of a saccade to a spatial target, and have been implicated in memory-based saccade planning [25]. When the monkey responds to the RDK stimulus with a saccade into its response field, firing rates in these neurons build up in a time-dependent fashion. The rate of buildup depends on the signal-to-noise ratio, and the saccadic response is initiated when firing rates reach a common maximum that is independent of motion coherence [26,27]. Similar phenomena are observed when recordings are made from FEF and other frontal areas [28,29]. In other words, LIP and FEF firing rates bear the two principal hallmarks of the cumulative decision variable proposed by the sequential sampling framework: a buildup that scales with the quality of incoming information, and a common termination point once the bound is breached. When a response is made to the alternative target (i.e., one that is outside the neuron's response field), firing rates correspondingly fall off from an initial burst at a rate that scales with percentage motion coherence, consistent with models, such as the DDM, which propose that the decision variable expresses relative evidence for the two competing choices. Microstimulation experiments have provided converging evidence that LIP acts as an integrator of motion signals encoded in MT [30,31].

In the Rodent, Posterior Parietal Cortex Neurons Show Similar Dynamics

Recent work has explored the use of rodent models of perceptual choice, opening the door to new interventions, such as optogenetic silencing of select classes of neurons [32]. Rats can report with a nosepoke whether the approximate number of pulses in an auditory click train was greater on the left or right hemifield (Poisson click task) (see Fig. 12.2A). In one study, fitting the data with an elaborate integration-to-bound model with free parameters for sensory and accumulation noise, leak, and adaptation revealed that decisions were "optimal," i.e., limited only by sensory noise [4].

---

[2] For brevity, here we focus on cortical regions. Integration-to-bound signals are also observed in the striatum, where motor and oculomotor signals may be gated before driving saccadic "burst" cells in the superior colliculus [24].

Neural recordings in the posterior parietal cortex (PPC) reveal a graded accumulation signal similar to those observed in the monkey, whereas neural activity in the prefrontal cortex represents the decision variable in a more categorical fashion [33] (see Fig. 12.2B). Indeed, in contrast to PPC lesions, unilateral inactivation of rat prefrontal regions induces a response bias best explained by heightened attraction to one decision boundary over another (i.e., a selective increase in leak for the contralesional choice) as if frontal regions categorized the output of a parietal accumulation signal [34].

However, Lateral Intraparietal Region and Posterior Parietal Cortex Neurons Exhibit Mixed Selectivity

One important limitation of these perceptual decision tasks is that the sensory, decision, and motor variables are confounded. Thus, in the RDK task, when the dots move toward the right, the monkey will typically decide that they are tending toward the right, and make a response toward the right. The frame of reference (sensory, decision, or motor) in which information is encoded is thus hard to discern. One might expect LIP neurons to encode decision information in the frame of reference of the response, because they are preselected on the basis of their saccadic response field target. Indeed, LIP neurons are sensitive to the monkey's decision even when the stimulus information was wholly ambiguous [27], and buildup activity emerges only when the response targets are revealed to the observer [35]. Nevertheless, the increasing firing rates are unlikely merely to reflect motor preparation, because the signal does not encode the fine parameters of saccadic control (e.g., acceleration, velocity, or accuracy). The claim has thus been made that buildup signals in posterior parietal neurons reflect a growing decision variable [26].

Nevertheless, reports aimed at disentangling these factors have emphasized the heterogeneity of the single-cell response in LIP. Although many neurons show build-to-threshold dynamics another large subpopulation is driven mainly by salient visual events [36,37]. Similarly. FEF is well known to contain a mixture of neurons encoding the motion stimulus and the saccadic response [38]. When the relevant response targets were introduced at variable times during motion stimulation, substantial subsets of LIP neurons showed selectivity for both motion direction and the target color (which indicated the relevant stimulus–response mapping) [35]. In a different study in which the selection criteria were relaxed to include cells that do not always show spatially specific persistent activity, individual LIP neurons were found to code for

target onset, motion onset, and saccadic choice with unique but characteristic motifs that mimic a diffusing decision variable only when averaged together [39,40]. During a multisensory version of the Poisson click task, posterior parietal neurons exhibit mixed selectivity for decision information and modality [41] (see Fig. 12.2C). Rather than coding for a decision variable, then, LIP and FEF code for a range of sensory-, task-, and response-related variables.

The continuous nature of the choice encoding in monkey LIP neurons has been challenged. Instead of averaging over trials the neuronal activity from the motion onset, the authors aligned the activity on a sudden change in the firing rate that happens at different times on each trial and showed that this original synchronization of activity explains better the choice selectivity of the majority of the recorded neurons. This result could mean that the usually observed ramping of activity during decision-making would be a by-product of trial averaging and that, by trial, the decision is made through a stochastic step (see Fig. 12.2D).

One related theme to emerge from recent work is that posterior parietal neurons encode category information with a low-dimensional code that rescales spontaneous firing patterns to variable levels, a highly redundant coding strategy that may facilitate robust readout at downstream processing stages [42,43]. However, as a rodent approaches the choice point in a T-maze, networks of PPC cells undergo striking state transitions that defy description with simple accumulation-to-bound models or low-dimensions categorical codes [44]. Most perplexingly, unilateral ablation of rodent PPC has a barely detectable effect on perceptual decisions in fixed-response versions of the Poisson click task [34]. Just how posterior parietal neurons contribute to perceptual decision-making, then, is an open question that continues to excite considerable debate.

Q2: Summary

The PPC has been studied as a model system for understanding perceptual decision-making. Parietal firing rates in rats and monkeys show information-dependent buildup to a fixed threshold during decisions about perceptual stimuli, mimicking the putative decision variable in the DDM. However, posterior neurons exhibit heterogeneous firing patterns and mixed selectivity for perceptual-, decision-, and task-related variables. An emerging body of work, then, has questioned the simple equivalence between the computations proposed by the integration-to-bound framework and findings of neurophysiological recordings in this region.

Q3: HOW CAN WE STUDY PERCEPTUAL DECISION-MAKING IN HUMANS?

One key goal in perceptual decision-making is to understand the dynamics of information integration, which unfolds on the millisecond timescale. However, a different question pertains to the location and extent of the functional brain network that underlies perceptual choices. Functional magnetic resonance imaging (fMRI), whose slowly varying measurements of blood oxygenation levels are poorly suited to approaching the former question, has been deployed to address the latter in humans [45,46]. However, research has been hampered by a lack of consensus over how decision information should be encoded in blood oxygen level-dependent (BOLD) signals [47]. By contrast, human magnetoencephalographic/electroencephalographic (MEG/EEG) signals have revealed signals with build-to-threshold dynamics that mirror those of single neurons [48].

It Is Unclear How BOLD Signals Encode Decision Variables

By analogy with monkey studies that have characterized LIP as integrating the noisy output of MT neurons coding for competing directions of motion, early human imaging studies searched for brain regions where the BOLD response correlated with the signal difference in category-selective extrastriate regions, such as those active during discrimination of faces and places. A wide swathe of cortex correlates with this difference, although only one dorsolateral region also shows stronger responses on "easy" than on "hard" decisions as individual LIP neurons do [49]. When categorizing faces and houses on the basis of their likelihood of predicting positive feedback, a comparable analysis yields activations in the ventromedial prefrontal cortex [50]. One framework for understanding these choices is provided by studies in value-guided decision-making, which have shown that medial and superior prefrontal signals scale with the value of a chosen relative to an unchosen option [51,52].

However, it seems unlikely that the BOLD signal, which indexes the combined responses of many millions of neurons with different selectivity, should encode decision variables in the same way as single posterior parietal neurons. Some authors have made the assumption that the BOLD signal reflects the total spiking activity occurring during (linear) evidence integration, such that stronger signals will be observed in free-response tasks when information is scarce and integration is prolonged [53,54]. Indeed, when human observers perform the RDK task, regions of the PPC show this "hard > easy" pattern, along with both medial prefrontal and anterior insular cortical zones well known to respond to decision entropy or conflict [55,56]. However, the above assumption holds only when integration times are variable; and yet activity in these regions encodes uncertainty in a variety of speeded and unspeeded choice situations. An alternative insight that helps explain this finding is offered by single-cell studies that have reported that the total firing rate across LIP neurons (considering both those that are selective for the saccadic target and those that are not) is greatest at low levels of motion coherence, because despite its shallower slope, the buildup signal on these trials is initiated at a higher level [39]. This finding is consistent with the view that selective single-neuron firing rates are normalized by the global levels of activity provided by the global LIP population, a computation that increases coding efficiency in neural circuits [57,58].

EEG Signals Capture Integration-to-Bound Dynamics During Perceptual Decisions

MEG/EEG has poor spatial resolution, but can capture the unfolding dynamics of decision processes with millisecond time resolution. EEG signals are typically averaged in response to punctate stimulus events, yielding transient evoked responses. However, when decision information is introduced gradually (for example, by embedding pulses of coherent motion in a fully random ongoing RDK stimulus), this evoked response is minimized, unmasking a build-to-threshold variable in the average EEG over centroparietal electrodes [59,60] (see Fig. 12.3A). Similar to the responses of single posterior parietal neurons, this signal scales with information quality and reaches a common threshold immediately prior to the response. Like the macroscopic BOLD signal, this signal is invariant to the response, and so must reflect the trajectory of the network state toward a threshold level of activation. The same authors have suggested that the well-known P300 response, a late-positive potential most often observed by averaging responses to surprising stimuli, may reflect a growing decision signal [61]. This finding has opened exciting new avenues to testing mechanistic models of perceptual decision-making in humans.

EEG Studies Suggest Discrete Computational Stages for Perceptual Decisions

Oscillatory activity in the beta band (15–30 Hz) exhibits a marked desynchronization over contralateral premotor cortex prior to a manual response. When humans indicate the motion direction in an RDK with a left- or right-handed button press, relative premotor

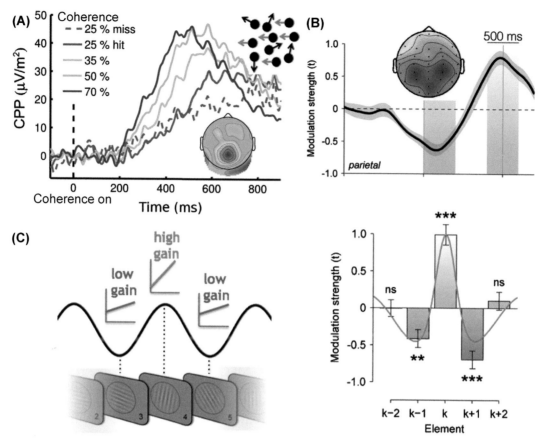

FIGURE 12.3   (A) Build-to-threshold signals in human EEG recordings. Average EEG activity recorded over centroparietal electrodes while humans detected various levels of coherent motion in a random-dot kinematogram display. When aligned to stimulus onset, a centroparietal positivity (CPP) builds up more rapidly when coherence is higher. *(Adapted with permission from Heekeren HR, et al. A general mechanism for perceptual decision-making in the human brain. Nature 2004;431:859—62.)* (B) Choice-predictive activity recorded over human parietal cortex. Humans categorized the average feature in a stream of discrete stimuli (oriented gratings) occurring in a 4-Hz stream. Choice-predictive signals were calculated for each sample by regressing residual variance in encoding its decision information in EEG signals against choices. Over the parietal cortex, choice-predictive activity was characterized by an early negativity (at 250 ms) and late positivity (500 ms). *(Adapted with permission from Wyart V, et al. Rhythmic fluctuations in evidence accumulation during decision making in the human brain. Neuron 2012;76:847—58.)* (C) Rhythmic gain control during perceptual choice. *Right:* At 500 ms following sample k, residual EEG activity positively predicted the weight given to sample k, but negatively predicted the weight given to samples $k-1$ and $k+1$. *Left:* Together, these findings are explained by a model in which gain varies rhythmically at approximately 2 Hz, acting as an oscillatory "bottleneck" that boosts processing of samples falling in the preferred phase of endogenous EEG activity over parietal cortex, and filtering those samples falling in the antipreferred phase of this rhythm. *(Adapted with permission from Wyart V, et al. Rhythmic fluctuations in evidence accumulation during decision making in the human brain. Neuron 2012;76:847—58, see for more details.)*

spectral power in the beta band shows the same integration-to-bound dynamics as broadband parietal MEG/EEG signals [62]. The isotropy between these two signals seems to mirror the homologous firing rate dynamics of macaque LIP and FEF signals during perceptual choice. However, can we distinguish the computations performed at parietal and premotor stages during perceptual choice?

One EEG study asked participants to average the information in a rapid stream of visual samples (tilted gratings), responding with a left- or right-button press. When the decision information provided was parametrically regressed against EEG signals over centroparietal sites, a buildup signal resembling that described earlier emerged following an initial negativity [6] (see

Fig. 12.3B). Similarly, lateralized premotor beta-band activity built up in a fashion that depended on the quality of the decision information, replicating previous findings. Critically, however, residual variance in broadband parietal EEG signals predicted the weight that would be given to each sample, i.e., how influential it was in the decision (multiplicative modulation), whereas the residuals of the beta-band signal predicted participants' bias to respond left or right (additive modulation). Further analyses suggest that category information passes through a rhythmic "bottleneck" over the parietal cortex, which gates entry into a cumulative decision variable (see Fig. 12.3C). In a follow-up study, the authors showed that under divided attention, the bottleneck constricts, leading to a leaky integration process [21].

The idea that the parietal and prefrontal cortices make distinct contributions to evidence integration is consistent with the rodent work cited earlier [34].

Q3: Summary

In fMRI studies, regions of the parietal, insular, and medial prefrontal cortices respond when perceptual decisions are "hard" rather than "easy." This follows from single-cell recordings during the RDK task, in which the collective firing rate of selective and nonselective neurons is greater when motion coherence is low. However, the link between decision variables and the BOLD signal remains unclear. In EEG recordings, build-to-threshold dynamics in average centroparietal signals are revealed when observers discriminate signals that emerge gradually from a noisy background. These broadband parietal signals may index the multiplicative gain of information processing, whereas lateralized beta-band activity over premotor cortex reflects an additive bias to respond left or right.

Q4: HOW DO OBSERVERS DECIDE WHEN TO DECIDE?

Computational models of perceptual decision-making propose that responses are initiated when the decision variable breaches a threshold level, or bound. When making binary category judgments of fixed difficulty via serial integration of log-likelihoods, a flat boundary ensures that the requisite accuracy is achieved with the fewest possible samples of information. The height of this bound controls the speed−accuracy trade-off; as the bound is raised, decisions become slower but more accurate [63].

It Is Unclear How Speed and Accuracy Are Traded Off at the Neural Level

One might thus expect that single-neuron responses in the LIP and FEF would build to a lower threshold when the monkey is rewarded for speed over accuracy. However, attempts to address this question have yielded perplexing results. In the LIP, firing rates build to a common value irrespective of whether the monkey works under speed or accuracy; under speed, however, activity is initiated at a higher level [64] (see Fig. 12.4A). Yet more confusingly, in the FEF, firing rates plateau at a higher level under speed than under accuracy, as if imposing a deadline on response raised, rather than lowering, the bound [65] (see Fig. 12.4B), a finding that is consistent with elevated BOLD under speed versus accuracy [66]. One interpretation of this finding is that LIP firing is itself integrated to threshold at a later subcortical stage, with steeper buildup (under high coherence

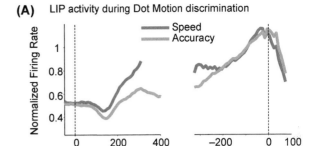

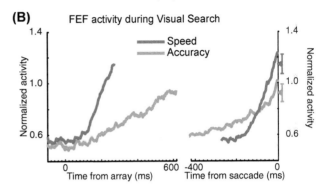

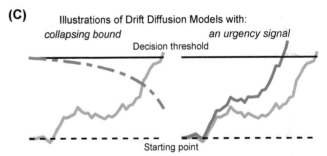

FIGURE 12.4 (A) Neural mechanisms of the speed−accuracy trade-off in the lateral intraparietal region (LIP). Normalized firing rates from cells in the LIP recorded while macaque monkeys discriminate a random-dot kinematogram stimulus under motivational regimes that emphasize speed (*purple trace*) and accuracy (*orange trace*). The *left* shows that LIP firing rates aligned to stimulus onset build up faster under speed pressure. The *right* shows firing rates aligned to the saccadic response: surprisingly, no difference is observed in the terminal firing rate achieved prior to a response. This is hard to reconcile with the view that LIP neurons terminate at a decision bound that can be raised and lowered to accommodate speed and accuracy conditions. (*Adapted with permission from Heitz RP, Schall JD. Neural mechanisms of speed-accuracy tradeoff. Neuron 2012;76:616−28.*) (B) Neural mechanisms of the speed−accuracy trade-off in the frontal eye fields (FEF). In the FEF, a similar pattern is observed for stimulus-aligned data (*left*) but termination is paradoxically higher under speed than under accuracy conditions (*right*). (*Adapted with permission from van Veen V, et al. The neural and computational basis of controlled speed-accuracy tradeoff during task performance. J Cogn Neurosci 2008;20: 1952−65.*) (C) Two mechanisms for increasing decision urgency. The *left* shows a standard diffusion process (*orange trace*) that terminates at either a fixed boundary (*dashed black line*) or a collapsing boundary (*dashed purple line*), in which the threshold decreases over time. The latter accelerates responding, particularly when decision information is weak or ambiguous. The *right* shows the same trace (*orange*) compared against a diffusion process that is boosted by a time-varying urgency signal (*purple*). This has the same effect on reaction time as the collapsing bound.

or speed stress) provoking earlier response. This theory takes a systems view of the perceptual decision circuit, casting LIP, FEF, and downstream subcortical circuits as links in a multistep accumulation process by which sensory information is gradually converted to a decision [67].

The Optimal Bound Collapses When Sensory Reliability Is Unknown

Computationally, the situation is more complex when the reliability of sensory information is unknown, for example, when variable coherence levels are intermingled across trials in the RDK task. Assuming that an optimal observer wishes to maximize the rate of return (i.e., of reward or positive feedback) per unit time, prolonged deliberation carries an opportunity cost, offsetting the benefit of a more precise decision variable. For example, when there is no motion coherence in the signal, the best policy is to respond as fast as possible (low bound) to receive random feedback; deliberation confers no benefit in this case. An ideal observer thus needs to estimate not only the direction of dot motion, but also the most likely coherence level, to know when to respond [68]. However, jointly estimating these quantities to determine the optimal height of the bound is computationally challenging.

Decisions May Be Driven by a Time-Varying Urgency Signal

One proposal states that observers solve this problem by monitoring the passage of time across the decision period. On average, longer integration times are associated with lower coherences, and so the time elapsed is a proxy for sensory reliability: if integration is prolonged but no conclusion has been reached, the decision threshold should be lowered. Under a mixture of difficulties, then, the optimal bound "collapses" as the epoch progresses, increasing the probability of a noisy "guess" as time goes on, and thereby scaling decision latencies to account for the opportunity cost associated with prolonged deliberation [69–71] (see Fig. 12.4C, left). How the decision bound is implemented at the neural level remains unknown, but many researchers find the notion of a time-varying bound biophysically implausible. One alternative proposal argues for a nonselective "urgency" signal that grows across the decision epoch, driving firing rates to threshold even where information is weak or ambiguous (see Fig. 12.4C, right). Indeed, time-dependent buildup is observed in LIP neurons even in the absence of sensory stimulation, when the monkey is estimating the passage of time [72]. This proposal remedies one oft-noted problem with the DDM, in

that it predicts a long tail to distribution of decision latencies (i.e., many trials with prolonged deliberation) that is rarely observed in empirical studies [17].

Evidence for urgency signals in humans and monkeys is currently mixed, however. One study noted that LIP firing rates build up even on 0% coherence trials, in a fashion that cannot be explained by noisy fluctuations in motion energy alone [73]. This buildup was faster for two-choice than for four-choice decisions, as if it reflected an urgency signal that was scaled by the relative posterior belief that the response will be correct. Other neural data recorded during the RDK task, and reaction times exhibited by the monkey, are captured by a model that computes the optimal cost of information accumulation and scales the urgency signal accordingly [70]. One interesting proposal goes further, suggesting that if decisions are brought to a close by an urgency signal, then the relative likelihood of either response can be calculated over a short, moving time window, tantamount to an extremely leaky accumulator [74]. However, evidence for this view has emerged in the context of a counting task in which the accumulated tallies remain visible to the observer, which may constitute a special case in which integration is less crucial [75]. Moreover, human studies involving discrimination of dot density or position have failed to find evidence for a collapsing bound; rather, the data are fit better by a model with flat bounds alone [76–78]. It thus remains unclear whether human and monkey decisions are driven to a close by an urgency signal.

Q4: Summary

When the reliability of sensory information is known, the SPRT is optimal for binary choices, but discriminating among choice alternatives with unknown sensory reliability is optimal under a collapsing bound. Optimal decision times can be inferred by a time-varying urgency signal that brings decisions about weak or ambiguous stimuli to a close. However, empirical evidence for urgency signals in humans and monkeys remains scarce.

Q5: HOW ARE PERCEPTUAL DECISIONS BIASED BY PRIOR BELIEFS?

All perceptual decisions are made in the context of past experience. When the context (e.g., an advance cue or a bias toward one alternative in the stimulation history) offers probabilistic information about which of two responses is more likely (e.g., leftward vs. rightward motion), observers are biased toward the favored response, and faster to respond correctly when sensory information is validly cued [79]. When a sensory reliability of the discriminanda is known, an ideal observer will combine the log-likelihood of the response conditioned on the stimulus and cue information, which is equivalent to adding an offset to the cumulative decision variable before integration begins. The optimal model thus predicts that probability cues will provoke fast errors on invalid trials, in which the decision variable originates close to the incorrect boundary [8].

Probability Cues Can Bias Both the Origin and the Slope of Information Integration

This finding is borne out by behavioral studies employing the RDK task and related psychophysical paradigms [80]. Consistent with this finding, differentially lateralized beta-band activity over premotor cortex, the aforementioned hallmark of information integration in human MEG/EEG signals, begins to build up in response to a probabilistic cue, ensuring a higher starting point when the RDK arrives [81]. Similarly, in a study combining modeling and imaging, individual differences in best-fitting model parameters for the origin of integration were predicted by BOLD signals in the prefrontal and parietal cortices [80].

Single-cell recordings in the PPC, however, have provided only weak evidence for an additive offset in firing rates when saccadic decisions indicate that the dots were moving in the more probable of two directions. Rather, when decisions about an RDK are biased by probabilistic cues, evidence builds up toward the expected choice with a steeper slope, as if prior knowledge scaled the urgency of the decision toward the corresponding boundary [82]. In fact, where the sensory reliability of the information is unknown, this is an optimal policy, because priors should bias signals most strongly when sensory information is weak, which is most likely when more time has passed without a decision. Early additive offsets in firing rate are seen, however, at downstream brain regions, such as the superior colliculus [83], further supporting the view that integration is a multistage process, with subcortical sites integrating growing signals from LIP and FEF.

Predictive Coding

Under the SPRT, optimal (Bayesian) integration of priors into the decision process reduces to the addition of the likelihood ratio (i.e., evidence conferred by the current sample) to the extant log posterior ratio (i.e., the relative evidence sampled thus far). One biologically plausible scheme for implementing this Bayesian integration across a variety of domains proposes that optimal perceptual inference occurs at nested stages of the sensory processing hierarchy, for example, in areas V1, V4, and IT in the ventral stream [84]. Under this framework, known as predictive coding, contextual information about an object or feature is encoded in higher processing stages, at which receptive fields have larger windows of integration in both space and time. This information flows backward to be optimally combined with information arriving at the immediately preceding cortical stage, with any discrepancies (prediction errors) being fed forward to adjust the posterior distribution over representations [85]. Thus, our belief about the presence of a feature or object (e.g., a leaf) will be driven not only by feed-forward evidence for that percept (e.g., the color green in V4 or an ovate shape in V2), but also by contextual signals that feed back from subsequent regions (e.g., an IT neuron signaling the presence of a tree), which in turn are informed by higher regions (e.g., a medial temporal lobe neuron indicating a forest scene). This scheme accounts for a host of salient neural phenomena at the cellular and systems levels, including repetition suppression, contextual facilitation and other extraclassical receptive field effects, the properties of end-stopped cells in early visual cortex, and the observed patterns of macroscopic BOLD and MEG/EEG signals to expected, unexpected, and unexpectedly omitted events across sensory domains [79]. Unlike the models based exclusively on integration of motion signals in the LIP, it also accounts for the large volume of "backward" or reentrant connections that link successive stages of the visual processing hierarchy [86]. Predictive coding offers a wider framework for understanding decision-making that is not limited to a simple model system, such as that for motion detection.

Adaptive Gain Control During Perceptual Decision-Making

Although integration of sensory likelihoods in psychophysical tasks is often argued to be optimal, in other domains humans are notoriously biased when incorporating prior information into a decision. Humans frequently neglect baseline probabilities or adjust their decision policy too sluggishly to maximize decision accuracy. In other tasks, humans are prone to confirmatory biases, prompting overconfident and often circular inferences [87]. These biases may emerge over the course of information integration. In one task involving averaging of visual features over successive discrete samples, human observers gave more weight to information that was consistent with the running tally of evidence accumulated thus far. This overweighting was accompanied by heightened gain of encoding of decision information in pupillometric signals and from EEG and fMRI activity in the parietal and medial prefrontal cortices [88]. This finding is consistent with the existence of "superstitious" biases in perceptual inference that are driven by local fluctuations in stimulation [89]. Under a Bayesian inference scheme, such as predictive coding, confirmatory biases occur when prior information is overweighted relative to sensory input, resulting in a runaway bias to favor existing preconceptions over new data [90]. Processing expected stimuli with heightened gain is suboptimal for binary classification, but may be efficient in a volatile world in which the information that is diagnostic for choices changes rapidly from moment to moment, and processing is a limited resource that should be focused on the most relevant sensory signals [10]. In other tasks in which observers compute the mean feature in a multielement array, a related bias to overweight the most commonly observed feature [91] and to adapt to environmental variability [92] may similarly reflect adaptive gain control mechanisms that maximize the efficiency of limited processing resources.

Q5: Summary

When the reliability of sensory information is known and resources are unlimited, prior information will optimally influence binary classification via an additive offset to information. In naturalistic settings, for example, discriminating an object or feature in a cluttered scene, optimal inference may rely on hierarchical predictive coding. However, human decisions often fail to properly account for prior information. A model in which the gain of information processing is adapted rapidly to give expected inputs more weight accounts for behavior in tasks involving decisions about discrete samples of information.

CONCLUSIONS AND FUTURE DIRECTIONS

Understanding how rodents, monkeys, and humans make decisions is a central concern in psychology and the neurosciences. Considerable progress has been made, but answers to the five questions outlined here remain elusive. In human work, future research would benefit from the establishment of more carefully grounded assumptions about how decision variables can be captured in the macroscopic signals measured with fMRI and MEG/EEG. Monkey researchers need to move beyond the simple notion that LIP and FEF neurons encode a growing decision variable, to an acknowledgment of the heterogeneity of the neural signal in these regions. New work studying perceptual decision-making in rodents offers considerable promise, and

opens the door to new imaging and interference techniques. How decision information is sampled, weighted, and integrated remain key open questions, and the use of a greater diversity of behavioral paradigms, including those involving integration of discrete samples of information, would help establish the domain generality of findings in all three species.

Acknowledgments

This work was funded by a European Research Council award to C.S. We thank Hannah Tickle and Konstantinos Tsetsos for comments on the manuscript.

References

[1] Newsome WT, et al. Neuronal correlates of a perceptual decision. Nature 1989;341:52–4.

[2] Nienborg H, et al. Decision-related activity in sensory neurons: correlations among neurons and with behavior. Annu Rev Neurosci 2012;35:463–83.

[3] Britten KH, et al. The analysis of visual motion: a comparison of neuronal and psychophysical performance. J Neurosci 1992;12:4745–65.

[4] Brunton BW, et al. Rats and humans can optimally accumulate evidence for decision-making. Science 2013;340:95–8.

[5] Yang T, Shadlen MN. Probabilistic reasoning by neurons. Nature 2007;447:1075–80.

[6] Wyart V, et al. Rhythmic fluctuations in evidence accumulation during decision making in the human brain. Neuron 2012;76:847–58.

[7] Shadlen MN, Kiani R. Decision making as a window on cognition. Neuron 2013;80:791–806.

[8] Bogacz R, et al. The physics of optimal decision making: a formal analysis of models of performance in two-alternative forced-choice tasks. Psychol Rev 2006;113:700–65.

[9] Beck JM, et al. Not noisy, just wrong: the role of suboptimal inference in behavioral variability. Neuron 2012;74:30–9.

[10] Summerfield C, Tsetsos K. Do humans make good decisions? Trends Cogn Sci 2015;19:27–34.

[11] Ma WJ. Organizing probabilistic models of perception. Trends Cogn Sci 2012;16:511–8.

[12] Wald A, Wolfowitz J. Bayes solutions of sequential decision problems. Proc Natl Acad Sci USA 1949;35:99–102.

[13] Wickelgren WA. Speed-accuracy tradeoff and information processing dynamics. Acta Psychol 1977;41:67–85.

[14] Ratcliff R. A theory of memory retrieval. Psychol Rev 1978;85:59–108.

[15] Ratcliff R, McKoon G. The diffusion decision model: theory and data for two-choice decision tasks. Neural Comput 2008;20:873–922.

[16] Usher M, McClelland JL. The time course of perceptual choice: the leaky, competing accumulator model. Psychol Rev 2001;108:550–92.

[17] Ditterich J. Stochastic models of decisions about motion direction: behavior and physiology. Neural Netw 2006;19:981–1012.

[18] Kiani R, et al. Bounded integration in parietal cortex underlies decisions even when viewing duration is dictated by the environment. J Neurosci 2008;28:3017–29.

[19] Kiani R, et al. Integration of direction cues is invariant to the temporal gap between them. J Neurosci 2013;33:16483–9.

[20] Tsetsos K, et al. Salience driven value integration explains decision biases and preference reversal. Proc Natl Acad Sci USA 2012;109:9659–64.

[21] Wyart V, et al. Neural mechanisms of human perceptual choice under focused and divided attention. J Neurosci 2015;35:3485–98.

[22] Tsetsos K, et al. Using time-varying evidence to test models of decision dynamics: bounded diffusion vs. the leaky competing accumulator model. Front Neurosci 2012;6:79.

[23] Ossmy O, et al. The timescale of perceptual evidence integration can be adapted to the environment. Curr Biol 2013;23:981–6.

[24] Ding L, Gold JI. Caudate encodes multiple computations for perceptual decisions. J Neurosci 2010;30:15747–59.

[25] Gnadt JW, Andersen RA. Memory related motor planning activity in posterior parietal cortex of macaque. Exp Brain Res 1988;70:216–20.

[26] Gold JI, Shadlen MN. The neural basis of decision making. Annu Rev Neurosci 2007;30:535–74.

[27] Roitman JD, Shadlen MN. Response of neurons in the lateral intraparietal area during a combined visual discrimination reaction time task. J Neurosci 2002;22:9475–89.

[28] Hanes DP, Schall JD. Neural control of voluntary movement initiation. Science 1996;274:427–30.

[29] Kim JN, Shadlen MN. Neural correlates of a decision in the dorsolateral prefrontal cortex of the macaque. Nat Neurosci 1999;2:176–85.

[30] Ditterich J, et al. Microstimulation of visual cortex affects the speed of perceptual decisions. Nat Neurosci 2003;6:891–8.

[31] Hanks TD, et al. Microstimulation of macaque area LIP affects decision-making in a motion discrimination task. Nat Neurosci 2006;9:682–9.

[32] Carandini M, Churchland AK. Probing perceptual decisions in rodents. Nat Neurosci 2013;16:824–31.

[33] Hanks TD, et al. Distinct relationships of parietal and prefrontal cortices to evidence accumulation. Nature 2015;520:220–3.

[34] Erlich JC, et al. Distinct effects of prefrontal and parietal cortex inactivations on an accumulation of evidence task in the rat. eLife 2015;4.

[35] Bennur S, Gold JI. Distinct representations of a perceptual decision and the associated oculomotor plan in the monkey lateral intraparietal area. J Neurosci 2011;31:913–21.

[36] Premereur E, et al. Functional heterogeneity of macaque lateral intraparietal neurons. J Neurosci 2011;31:12307–17.

[37] Leathers ML, Olson CR. In monkeys making value-based decisions, LIP neurons encode cue salience and not action value. Science 2012;338:132–5.

[38] Bruce CJ, Goldberg ME. Primate frontal eye fields. I. Single neurons discharging before saccades. J Neurophysiol 1985;53:603–35.

[39] Meister ML, et al. Signal multiplexing and single-neuron computations in lateral intraparietal area during decision-making. J Neurosci 2013;33:2254–67.

[40] Park IM, et al. Encoding and decoding in parietal cortex during sensorimotor decision-making. Nat Neurosci 2014;17:1395–403.

[41] Raposo D, et al. A category-free neural population supports evolving demands during decision-making. Nat Neurosci 2014;17:1784–92.

[42] Fitzgerald JK, et al. Biased associative representations in parietal cortex. Neuron 2013;77:180–91.

[43] Ganguli S, et al. One-dimensional dynamics of attention and decision making in LIP. Neuron 2008;58:15–25.

[44] Harvey CD, et al. Choice-specific sequences in parietal cortex during a virtual-navigation decision task. Nature 2012;484:62–8.

[45] Mulder MJ, et al. Perceptual decision neurosciences − a model-based review. Neuroscience 2014;277:872–84.

[46] Heekeren HR, et al. The neural systems that mediate human perceptual decision making. Nat Rev Neurosci 2008;9:467–79.

[47] O'Reilly JX, et al. How can a Bayesian approach inform neuroscience? Eur J Neurosci 2012;35:1169–79.

[48] Kelly SP, O'Connell RG. The neural processes underlying perceptual decision making in humans: recent progress and future directions. J Physiol Paris 2015;109:27–37.

[49] Heekeren HR, et al. A general mechanism for perceptual decision-making in the human brain. Nature 2004;431:859–62.

[50] Philiastides MG, et al. A mechanistic account of value computation in the human brain. Proc Natl Acad Sci USA 2010;107:9430–5.

[51] Boorman ED, et al. How green is the grass on the other side? Frontopolar cortex and the evidence in favor of alternative courses of action. Neuron 2009;62:733–43.

[52] Hunt LT, et al. Mechanisms underlying cortical activity during value-guided choice. Nat Neurosci 2012;15(3).

[53] Ho TC, et al. Domain general mechanisms of perceptual decision making in human cortex. J Neurosci 2009;29:8675–87.

[54] Basten U, et al. How the brain integrates costs and benefits during decision making. Proc Natl Acad Sci USA 2010;107:21767–72.

[55] Grinband J, et al. A neural representation of categorization uncertainty in the human brain. Neuron 2006;49:757–63.

[56] Botvinick MM, et al. Conflict monitoring and cognitive control. Psychol Rev 2001;108:624–52.

[57] Louie K, et al. Normalization is a general neural mechanism for context-dependent decision making. Proc Natl Acad Sci USA 2013;110(15):6139–44.

[58] Carandini M, Heeger DJ. Normalization as a canonical neural computation. Nat Rev Neurosci 2012;13:51–62.

[59] O'Connell RG, et al. A supramodal accumulation-to-bound signal that determines perceptual decisions in humans. Nat Neurosci 2012;15:1729–35.

[60] Kelly SP, O'Connell RG. Internal and external influences on the rate of sensory evidence accumulation in the human brain. J Neurosci 2013;33:19434–41.

[61] Twomey DM, et al. The classic P300 encodes a build-to-threshold decision variable. Eur J Neurosci 2015;42(1):1636–43.

[62] Donner TH, et al. Buildup of choice-predictive activity in human motor cortex during perceptual decision making. Curr Biol 2009;19:1581–5.

[63] Bogacz R, et al. The neural basis of the speed-accuracy tradeoff. Trends Neurosci 2010;33:10–6.

[64] Hanks T, et al. A neural mechanism of speed-accuracy tradeoff in macaque area LIP. eLife 2014;3.

[65] Heitz RP, Schall JD. Neural mechanisms of speed-accuracy tradeoff. Neuron 2012;76:616–28.

[66] van Veen V, et al. The neural and computational basis of controlled speed-accuracy tradeoff during task performance. J Cogn Neurosci 2008;20:1952–65.

[67] Heitz RP, Schall JD. Neural chronometry and coherency across speed-accuracy demands reveal lack of homomorphism between computational and neural mechanisms of evidence accumulation. Philos Trans R Soc Lond Ser B Biol Sci 2013;368:20130071.

[68] Deneve S. Making decisions with unknown sensory reliability. Front Neurosci 2012;6:75.

[69] Frazier P, Yu AJ. Sequential hypothesis testing under stochastic deadlines. Adv Neural Inf Process Syst 2008;20:465–72.

[70] Drugowitsch J, et al. The cost of accumulating evidence in perceptual decision making. J Neurosci 2012;32:3612–28.

[71] Simen P, et al. Reward rate optimization in two-alternative decision making: empirical tests of theoretical predictions. J Exp Psychol Hum Percept Perform 2009;35:1865–97.

[72] Janssen P, Shadlen MN. A representation of the hazard rate of elapsed time in macaque area LIP. Nat Neurosci 2005;8:234–41.

[73] Churchland AK, et al. Decision-making with multiple alternatives. Nat Neurosci 2008;11:693–702.

[74] Cisek P, et al. Decisions in changing conditions: the urgency-gating model. J Neurosci 2009;29:11560–71.

[75] Thura D, et al. Decision making by urgency gating: theory and experimental support. J Neurophysiol 2012;108:2912—30.

[76] Winkel J, et al. Early evidence affects later decisions: why evidence accumulation is required to explain response time data. Psychon Bull Rev 2014;21:777—84.

[77] Hawkins GE, et al. Revisiting the evidence for collapsing boundaries and urgency signals in perceptual decision-making. J Neurosci 2015;35:2476—84.

[78] Karsilar H, et al. Speed accuracy trade-off under response deadlines. Front Neurosci 2014;8:248.

[79] Summerfield C, de Lange FP. Expectation in perceptual decision making: neural and computational mechanisms. Nat Rev Neurosci 2014;15:745—56.

[80] Mulder MJ, et al. Bias in the brain: a diffusion model analysis of prior probability and potential payoff. J Neurosci 2012;32:2335—43.

[81] de Lange FP, et al. Prestimulus oscillatory activity over motor cortex reflects perceptual expectations. J Neurosci 2013;33:1400—10.

[82] Hanks TD, et al. Elapsed decision time affects the weighting of prior probability in a perceptual decision task. J Neurosci 2011; 31:6339—52.

[83] Dorris MC, Munoz DP. Saccadic probability influences motor preparation signals and time to saccadic initiation. J Neurosci 1998;18:7015—26.

[84] Kanai R, et al. Cerebral hierarchies: predictive processing, precision and the pulvinar. Philos Trans R Soc Lond Ser B Biol Sci 2015;370.

[85] Friston K. A theory of cortical responses. Philos Trans R Soc Lond Ser B Biol Sci 2005;360:815—36.

[86] Angelucci A, et al. Circuits for local and global signal integration in primary visual cortex. J Neurosci 2002;22:8633—46.

[87] Nickerson RS. Confirmation bias: a ubiquitous phenomenon in many guises. Rev Gen Psychol 1998;2:175—220.

[88] Cheadle S, et al. Adaptive gain control during human perceptual choice. Neuron 2014;81:1429—41.

[89] Yu AJ, Cohen JD. Sequential effects: superstition or rational behavior? Adv Neural Inf Process Syst 2009;21:1873—80.

[90] Jardri R, Deneve S. Circular inferences in schizophrenia. Brain 2013;136:3227—41.

[91] de Gardelle V, Summerfield C. Robust averaging during perceptual judgment. Proc Natl Acad Sci USA 2011;108:13341—6.

[92] Michael E, et al. Priming by the variability of visual information. Proc Natl Acad Sci USA 2014;111:7873—8.

13

Neural Circuit Mechanisms of Value-Based Decision-Making and Reinforcement Learning

A. Soltani[1], W. Chaisangmongkon[2,3], X.-J. Wang[3,4]

[1]Dartmouth College, Hanover, NH, United States; [2]King Mongkut's University of Technology Thonburi, Bangkok, Thailand; [3]New York University, New York, NY, United States; [4]NYU Shanghai, Shanghai, China

Abstract

Despite groundbreaking progress, currently we still know preciously little about the biophysical and circuit mechanisms of valuation and reward-dependent plasticity underlying adaptive choice behavior. For instance, whereas phasic firing of dopamine neurons has long been ascribed to represent reward-prediction error (RPE), only recently has research begun to uncover the mechanism of how such a signal is computed at the circuit level. In this chapter, we will briefly review neuroscience experiments and mathematical models on reward-dependent adaptive choice behavior and then focus on a biologically plausible, reward-modulated Hebbian synaptic plasticity rule. We will show that a decision-making neural circuit endowed with this learning rule is capable of accounting for behavioral and neurophysiological observations in a variety of value-based decision-making tasks, including foraging, competitive games, and probabilistic inference. Looking forward, an outstanding challenge is to elucidate the distributed nature of reward-dependent processes across a large-scale brain system.

INTRODUCTION

Animals survive in complex environments by learning to make decisions that avoid punishments and lead to rewards. In machine learning, a number of reinforcement learning (RL) algorithms have been developed to accomplish various tasks in terms of reward-optimization problems, ranging from sequential decision-making to strategic games to training multi-agent systems [1]. In neuroscience, models inspired by RL have been used both as normative descriptions of animals' behaviors and as mechanistic explanations of value-based learning processes (see also Chapters 2 and 4). Physiologically, such learning processes are thought to be implemented by changes in synaptic connections between pairs of neurons, which are regulated by reward-related signals. Over the course of learning, this synaptic mechanism results in reconfiguration of the neural network to increase the likelihood of making a rewarding choice based on sensory stimuli. The algorithmic computations of certain reinforcement models have often been translated to synaptic plasticity rules that rely on the reward-signaling neurotransmitter dopamine (DA).

There are two main theoretical approaches to derive plasticity rules that foster rewarding behaviors. The first utilizes gradient-descent methods to directly maximize expected reward—an idea known as policy gradient methods in machine learning [2,3]. Because neurons possess stochastic behaviors, many of these learning rules exploit the covariation between neural activity fluctuation and reward to approximate the gradient of reward with respect to synaptic weights. This class of learning rules has been implemented in both spiking models [4–7] and firing rate models [8–10] and has been shown to replicate operant-matching behavior in a foraging task [9], learn temporal spiking patterns [7], and generate an associative representation that facilitates perceptual decision [8,10]. The second approach is prediction-based, in which the agent estimates the values of encountered environmental states and learns decision policy that maximizes reward [1]. This idea encompasses several related algorithms, such as actor–critic models, which issue separate updates for value and policy, and temporal difference (TD) models that use a continuous prediction error signal over time [1,11]. Early work using prediction-based models aimed to explain the patterns of firing in DA neurons [12,13], whereas later work has expanded the framework to explain complex animal behaviors with more realistic spiking models [14,15].

Decision Neuroscience
http://dx.doi.org/10.1016/B978-0-12-805308-9.00013-0

In both gradient-based and prediction-based approaches, an essential element of successful learning is the reward-prediction error (RPE) signal—the difference between expected and actual reward that the agent receives. Such "teaching" signal, which derives from the Rescorla—Wagner model [16], is thought to be encoded by midbrain DA neurons. Different modeling studies interpret the DA modulatory signal differently, but the dominant idea is that DA signal dynamics mirror that of the TD error. In a series of classic experiments, Schultz et al. have shown that the firing patterns of DA neurons resembled TD error in a classical conditioning paradigm [17]. More recent experimental findings successfully used the TD framework to model DA neuron behaviors in more complex dynamic value-based choice tasks [18—21]. Experiments using optogenetics have begun to uncover the circuit logic in the ventral tegmental area that gives rise to the computation of an RPE signal [22,23] (see Chapter 5). On the other hand, some investigations into the response of DA neurons have revealed aspects of DA neural activity that do not match the theoretically defined RPE [24,25], leading others to provide alternative explanations for the observed patterns of DA response [26].

To examine the link between adaptive, value-based choice behavior and DA-dependent plasticity (Fig. 13.1), we first review various aspects of reward value representations in the brain and then discuss how such representations could be acquired through DA-dependent plasticity. The chapter will focus on a framework for reward-dependent learning in which the reinforcement signal is binary (DA release or no release) and learning does not involve any explicit optimization process. We show how this model can explain observed patterns of value representation and can account for some of the observations previously attributed to learning based on an error-driven signal and the RPE [27].

REPRESENTATIONS OF REWARD VALUE

To make good decisions, organisms must be able to assess environmental states and choose actions that are likely to yield reward. Accumulating evidence indicates that the primate brain achieves this goal by engaging multiple interconnected networks, especially in prefrontal cortex and basal ganglia, for computing and storing value estimates for sensory and motor events already experienced [28—30] (see Chapters 10, 14, and 16). In RL models, the notions of state value and action value are well distinguished. State value refers to the average reward that can be obtained in the future from the environmental state that the animal encounters at the present

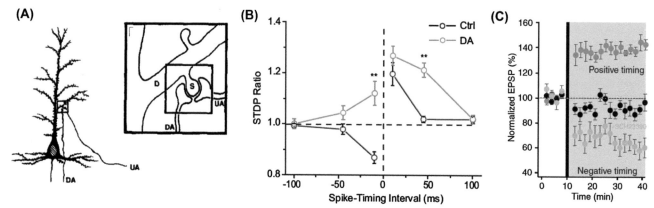

FIGURE 13.1 Modulation of synaptic plasticity by dopamine. (A) Triad configuration of dopamine (DA) and other synaptic inputs to pyramidal cells. Putative DA afferents terminate on the spine of a prefrontal pyramidal cell in the vicinity of an unidentified axon (UA). Inset shows an enlargement of axospinous synapses illustrated on the left. *(Adapted with permission from Goldman-Rakic PS. Cellular basis of working memory. Neuron 1995;14:477—85 with data published in Goldman-Rakic PS, Leranth C, Williams SM, Mons N, Geffard M. Dopamine synaptic complex with pyramidal neurons in primate cerebral cortex. Proc Natl Acad Sci USA 1989;86:9015—19.)* (B) Alteration of the spike-timing-dependent plasticity (STDP) window in hippocampal neurons with DA. *Black circles* show the STDP window under control (Ctrl) conditions and *red circles* show the same window when dopamine was present. In the presence of DA, positive spike timing includes a wider window for potentiation. With negative spike timing, dopamine reversed the direction of plasticity from depression to potentiation. *(Adapted with permission from Zhang JC, Lau PM, Bi GQ. Gain in sensitivity and loss in temporal contrast of STDP by dopaminergic modulation at hippocampal synapses. Proc Natl Acad Sci USA 2009;106(31):13028—33.)* (C) Dependence of the direction of plasticity on dopamine for cortical synapses on D1 receptor-expressing striatal, medium spiny neurons. During the STDP paradigm, when presynaptic spikes precede postsynaptic spikes (positive spike timing) long-term potentiation occurs (red). On the other hand, when presynaptic spikes follow postsynaptic spikes (negative spike timing) no changes in plasticity occur (black). When D1 receptors are blocked by SCH23390, negative timing results in long-term depression (blue). *(Adapted with permission from Surmeier DJ, Plotkin J, Shen W. Dopamine and synaptic plasticity in dorsal striatal circuits controlling action selection. Curr Opin Neurobiol 2009;19(6):621—8 with data published in Shen W, Flajolet M, Greengard P, Surmeier DJ. Dichotomous dopaminergic control of striatal synaptic plasticity. Science 2008;321(5890):848—51.)* D, dendrite; EPSP, excitatory postsynaptic potential; S, synapse.

time, whereas action value refers to the future reward expected from an action taken at a specific environmental state. Neurophysiological studies have shown that both state and action values are represented in a distributed fashion across the brain.

Signals related to reward expectancy have been abundantly found in striatum [31,32] and many subregions of prefrontal cortex [33]. More specifically, neurons in several areas of the brain encode the expected reward of the state environment, independent of actions, in accordance with the theoretical definition of state values. For example, the activity of some neurons in ventral striatum [34], dorsal anterior cingulate cortex [35], and amygdala [36] is correlated with the sum of values associated with different upcoming choice alternatives. Another type of action-independent reward expectation is termed "good-based value," meaning that the agent assigns different prices to different goods based on subjective preferences (e.g., one orange is worth more than two apples if the agent prefers orange). A series of neurophysiological studies in primates and rodents found that neurons in the orbitofrontal cortex play a prominent role in encoding the economic value of a chosen option in the unit of common currency [37,38]. The orbitofrontal cortex is known to be a part of the interconnected network that guides reward-related behaviors [39]. The network encompasses other areas that demonstrate similar chosen value encoding, such as striatum [40,41], medial prefrontal cortex [38], and ventromedial prefrontal cortex [42].

Signals related to values of specific actions are also prevalently observed across different brain areas. To name a few, the representation of action values was found in striatum [34,40,41,43,44] and posterior parietal cortex [45,46] (see also Chapter 14). These action-value neurons may exhibit correlations between firing activity and magnitude or probability of reward associated with different choices before the action is executed. The signals related to choice values were detected even in areas that directly regulate execution of movements, such as frontal eye field [47] and superior colliculus [48]. Furthermore, subsets of neurons in these areas are shown to encode the difference in values between two competing choice alternatives, which might reflect the underlying decision mechanism [34,38,46].

Overall, there is a diverse and heterogeneous representation of reward value throughout the brain. The implications of such diverse representation remain to be delineated. It is conceivable, for instance, that neurons in one brain area display value-related signal simply as a result of inputs from another area where such signal is computed. However, it is unclear what reward signal and DA-dependent learning mechanisms result in the formation of such diverse representation.

LEARNING REWARD VALUES

The prevalence of value representation across the brain begs the question of how the neurons learn to encode such properties in the first place. In RL models, a subnetwork responsible for value estimation is often referred to as the "critic" (Fig. 13.2A). An adaptive critic can be implemented simply as a single neuron (or a single pool of neurons) that receives synaptic connections from the sensory circuits, which represent environmental states [49]. The synapses are updated by a learning rule so that the activity of the critic neuron approximates the expected reward in each environmental state. The derivation of such learning rules can be accomplished by either an error-driven learning method [49] or a gradient-descent method [50]. The learning rule minimizes the difference between the actual reward and the system's reward prediction; hence, the error term is equivalent to the RPE. The idea of an adaptive critic forms the basis for TD learning, in which the expected reinforcement signal is defined by temporally discounted reward from all future time steps [49].

Early attempts in converting the TD learning algorithm into biophysical models aimed to understand the patterns of dopaminergic neuron firing [12,13,51,52]. Despite the partial success of these models in explanation of these patterns based on TD learning, there is experimental evidence indicating that other mechanisms are required to capture the exact pattern of experimental data. For example, a 2005 experiment suggested that the use of a long-lasting eligibility trace is necessary to replicate the dynamic behaviors of DA neurons as the agent is learning [53]. Interestingly, such memory traces are required to model human decisions in a sequential economic decision game [54]. These results indicate that in the neural circuits, RL may take place on a slower timescale (in the order of trial events), which is inconsistent with the rapid timescale proposed in previous theoretical studies [28,55].

Recent models have integrated more refined knowledge in machine learning to tackle difficult learning tasks, such as spatial navigation to solve a water maze [15,56] by expanding the actor—critic model for the case of continuous space and time [57]. In one spiking model, a population of critic neurons was trained by a reward-dependent Hebbian rule, which is a product of TD error and filtered pre- and postsynaptic spike trains. The synaptic kernel acts as an eligibility trace with a timescale of hundreds of milliseconds. As a result, the average firing rate of the critic population approximates the dynamic value function as the animal moves toward the target platform [15]. This is one example of many successful models that implement reward-dependent learning in spiking networks [4,6,14].

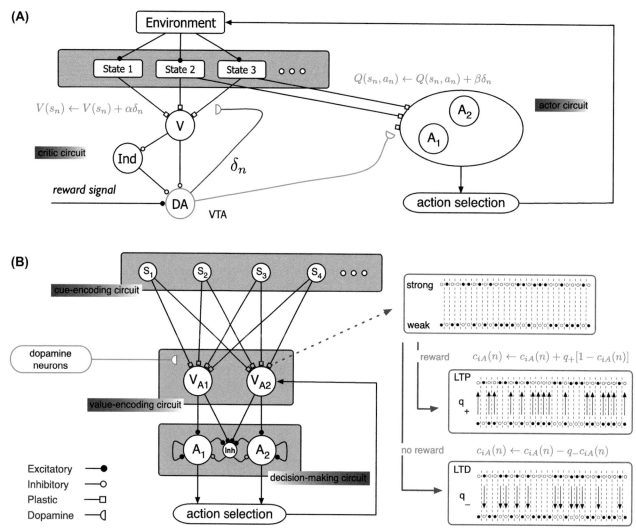

FIGURE 13.2 Comparison of two alternative models for value-based decision-making. (A) The actor–critic architecture of the temporal difference learning and use of the reward-prediction error (RPE). In this model, the critic circuit learns the reward value associated with different states of the environment, $V(s_n)$, and the actor circuit learns reward value associated with taking different actions (e.g., A_1 and A_2) in a given state, $Q(s_n, a_n)$. The critic circuit combines the reward signal and the output of state-value neurons (V pool) via a direct and indirect pathways (Ind pool) to compute the RPE. The RPE is then used to update value functions in the critic circuit and the action value functions in the actor circuit. (B) The architecture of our proposed network model with three distinct circuits: the cue-encoding, value-encoding, and decision-making circuits. The cue-encoding circuit represents sensory information pertinent for valuation and contains pools of sensory neurons that are selective for individual cues (shapes S_1 to S_4). They project to the value-encoding circuit, where neurons learn to encode reward values of the two alternative responses (action values V_{A1} and V_{A2}) through plastic synapses. At the end of the trial, only plastic synapses from sensory neurons selective for the presented cues onto action value-coding neurons selective for the chosen alternative are updated. The direction of update (long-term potentiation or depression, LTP and LTD) depends on the choice and a binary reward signal. If the choice of the model is rewarded, synapses in the depressed state transition to the potentiated state with a probability q_+; otherwise, the transition in the reverse direction occurs with a probability q_-. The decision-making circuit uses the output of the value-encoding circuit and, therefore, combines evidence from different sources to determine the model's choice on each trial. This circuit has two competing neural pools that are selective for two alternative choices (A_1 and A_2). Importantly, the finite number of synaptic states and the stochastic nature of plasticity allow synapses to transform reward signal into quantities important for value-based choice, but this transformation depends on the state of the synapses encoding reward and, therefore, on the existing representation of reward value in a given neuron. *DA*, dopamine; *VTA*, ventral tegmental area.

Nevertheless, there are a few serious caveats when applying an RPE framework in learning value-based decision-making. First, although much effort has been devoted to modeling the dynamics of DA neurons, little is understood about the temporal properties of DA release and the resulting intracellular signaling that determines synaptic plasticity in target neurons. One study suggests that the dynamics of DA release and re-uptake are slow and imprecise relative to the time resolution required for RPE signals in TD learning [58]. Second, RPE signals can be both positive and negative values, whereas the neural response of DA neurons can be only positive. Although depression in the baseline activity of DA neurons could act as a negative value, the baseline activity of DA is very low (about 2 Hz), which, in turn, limits the representation of negative errors. Therefore, the decreased activity might not result in a reversal in the direction of DA-dependent plasticity (Fig. 13.1) required by TD learning unless other assumptions are included [59]. Finally, the machinery required for the computation of RPE is expensive, as it requires representation of all possible actions at all time points (with relatively high time resolution) during a given trial [12].

One could ask whether the diverse representations of reward value described in the previous section could arise only from an RPE-based mechanism or whether other types of reward signal suffice to develop such representations. One possibility would be that reward signal is not sensitive to states and actions per se, and the recipient brain areas have a crucial role in translating a more general reward signal to diverse representation of reward value.

STOCHASTIC DOPAMINE-DEPENDENT PLASTICITY FOR LEARNING REWARD VALUES

To explain various aspects of reward-based choice behavior, we have proposed an alternative learning rule based on DA-dependent synaptic plasticity. In this model, we assume plastic synapses to be bounded, such that there are a limited number of synaptic states, each with a finite value of synaptic efficacy. There are three main assumptions underlying our learning rule [60]. First, learning is reward-dependent and Hebbian, meaning that it depends on the firing rates of presynaptic and postsynaptic neurons, but is modulated by the reward signal according to the experimental data on DA-dependent synaptic plasticity [61]. Second, the reward signal mediated by DA is binary, such that presence/absence of reward is signaled by release or no release of DA. This assumption can be extended to a graded reward signal reflecting size or uncertainty of reward. Third, DA-dependent plasticity is stochastic, such that synaptic modifications occur probabilistically when conditions for plasticity are met [62−64].

In a specific formulation of our model, plastic synapses are assumed to be binary [65,66] with two discrete states: potentiated ("strong") and depressed ("weak") (Fig. 13.2B). The Hebbian characteristic of learning implies that learning depends on the firing rates of pre- and postsynaptic neurons. The presynaptic neurons encode visual targets or, more generally, stimuli that precede or predict alternative responses. On the other hand, the postsynaptic neurons represent the value of alternative actions and could be the selective neurons in a putative decision-making network. Moreover, the presence of reward is signaled by a global DA release and results in long-term potentiation, whereas the absence of reward is signaled by the lack of DA release, which reverses the direction of synaptic plasticity from potentiation to depression. Because synaptic plasticity is stochastic, in potentiation instances, each synapse in the weak state has a probability q_+ to transition to the strong state. Similarly, in depression instances, each synapse in the strong state has a transition probability q_- to move to the weak state.

The information stored at a specific set of synapses between neurons encoding stimulus s and neurons encoding action A on trial n can be quantified as the fraction of these synapses in the strong state, $c_{sA}(n)$ (or equivalently, the remaining fraction, $1 - c_{sA}(n)$, in the weak state). Based on the aforementioned assumptions, at the end of each trial when reward feedback is received, the fraction of synapses in the strong state is updated as follows:

$$c_{sA}(n+1) = c_{sA}(n) + q_+(r, \nu_s, \nu_A)[1 - c_{sA}(n)]$$

in case of LTP

$$c_{sA}(n+1) = c_{sA}(n) - q_-(r, \nu_s, \nu_A)c_{sA}(n)$$ (13.1)

in case of LTD

where potentiation (q_+) and depression (q_-) rates, collectively called learning rates, are assumed to depend on the firing rates of pre- and postsynaptic neurons (ν_s and ν_A, respectively) and on the reward outcome ($r = 0$ or 1). More generally, the learning rates could also depend on the concentration of DA at the site of plasticity, allowing modifications of plasticity based on more graded response of DA neurons.

In a series of works, we have shown that a learning rule based on these assumptions enables the neural network to compute and estimate quantities that are crucial for performing various value-based decision-making tasks that require learning from reward feedback [55,60,67−69]. Importantly, in some cases, reward value computed by plastic synapses in our model using a binary reward signal resembles those computed by RL models using the RPE [55]. However, in our model, changes in reward value mainly depend on the state of plastic synapses receiving the reward signal, rather than the reward signal itself as in RL models.

FORAGING WITH PLASTIC SYNAPSES

In the first application of our learning rule described in Eq. (13.1), we show how reward-dependent and stochastic modifications in binary synapses allow these synapses to estimate reward value for possible choice alternatives in a value-based, decision-making task known as the "matching" task. In this task, which has been extensively used to study animals' response to reinforcement [70–73], the subject chooses between two options (color targets A and B), which are assigned different reward probabilities. Specifically, each target is baited with reward stochastically and independently of the other target, but if a reward is assigned to a target, it stays until harvested [46]. Moreover, the probability of baiting rewards on the two targets (i.e., reward schedule) changes between blocks of trials without any signal to the subject. The baiting probability ratios are randomly chosen from a set of ratios, whereas the overall baiting probability is fixed. Therefore, performing this task requires continuous estimation of reward value of the two choice options and dynamic shift of choice toward the better option.

We assumed that the neural circuit for solving this task should have a few components: sensory representation of choice options (color targets), plastic synapses that learn reward expected from choosing each target (action values), a set of neural populations for color to location remapping (because the color targets could appear on either side), and a decision-making (DM) network for choosing between the two options [60] (Fig. 13.2B). Because the two targets are identical except for the reward values, which have to be learned through reward feedback, it is reasonable to assume that target presentation leads to identical activation of presynaptic sensory neurons that project to the DM network. Therefore, the only difference in the inputs to the two competing neural populations (A and B) in the DM network is due to the efficacies of the plastic synapses. The average currents through plastic synapses selective for option A depend on the states of synapses, c_A (i.e., fraction of these synapses in the strong state), the number of plastic synapses (N_p), the presynaptic firing rate (r_{st}), and the peak conductance of the strong and weak states (g_+ and g_-) and their corresponding decay (τ_{syn}):

$$I_A \propto N_p r_{st}(c_A g_+ + (1-c_A)g_-)\tau_{syn} \quad (13.2)$$

Therefore, the difference in the average currents into neurons selective for the two options is equal to

$$I_A - I_B \propto (c_A - c_B)N_p r_{st}(g_+ - g_-)\tau_{syn} \quad (13.3)$$

Importantly, the choice behavior of the DM network is mainly a sigmoid function of the difference in the average input currents; therefore, the probability of choosing option A can be written as

$$P_A = \frac{1}{1 + e^{-\left(\frac{c_A - c_B}{\sigma}\right)}} \quad (13.4)$$

where σ determines the inverse sensitivity to the differential current. As a result, any of the factors in the right side of Eq. (13.3) can affect the difference in the overall synaptic currents and consequently change σ and, therefore, the choice behavior. Although σ in this model is equivalent to the temperature in RL models, it can be related to biophysical properties of the neural circuit involved in value-based DM.

Based on the learning rule defined in Eq. (13.1), the level of activity (high or low) in pre- and postsynaptic neurons determines the condition for synaptic plasticity, whereas the presence or absence of reward determines its direction. However, because both targets are present on each trial of the matching task, the presynaptic neurons are always active and therefore learning becomes independent of presynaptic activity. Moreover, we assume that the postsynaptic firing rate is low (respectively, high) for the neurons selective for the unchosen (respectively, chosen) target. Finally, assuming that learning happens only if the postsynaptic neurons are highly active, the learning rule in Eq. (13.1) can be simplified as

$$c_A(n+1) = c_A(n) + q_+[1 - c_A(n)],$$

A selected and rewarded

$$c_A(n+1) = c_A(n) - q_- c_A(n), \quad (13.5)$$

A selected but not rewarded

whereas synapses onto neurons selective for the unchosen target (B in this case) are not modified. We found that such a reward-dependent learning rule enables plastic synapses to estimate quantities crucial for matching behavior. More specifically, the steady state of synaptic strengths of the two sets of plastic synapses is a monotonic function of the returns from the two choices:

$$c_A^{ss} = \frac{q_+ R_A}{(q_+ - q_-)R_A + q_-} \quad (13.6)$$

where R_A is the return (i.e., average reward harvested per choice of A). In the special case in which $q_+ = q_-$, the steady state of c_A is equal to R_A. The steady state indicates what can be computed by plastic synapses over a long period of time. However, because these synapses have a limited number of states and bounded efficacy, the information stored in them is constantly rewritten [28]. Therefore, plastic synapses provide a local (in time) estimate of return (or a monotonic function of it). In other words, plastic synapses integrate reward feedback over time, but in a leaky fashion.

In order to see how the proposed learning rule enables the model to perform matching behavior we need to consider the interplay between learning and

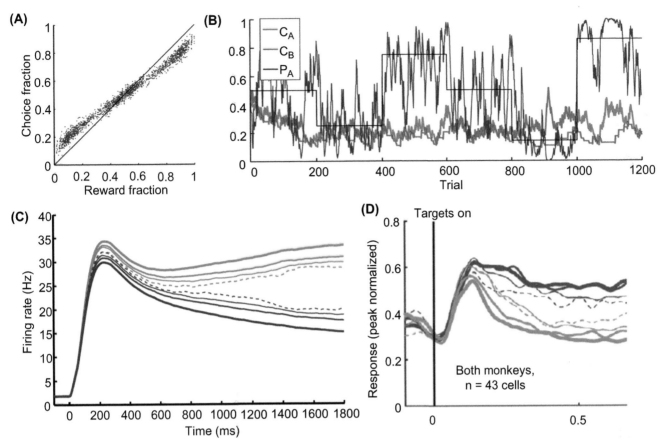

FIGURE 13.3 Behavioral and neural response of the model during the matching task. (A) The overall choice behavior of the model. Each point shows the fraction of trials within a given block on which a choice alternative is selected as a function of the fraction of reward on that choice. Matching law corresponds to the diagonal line, when the choice fraction matches the reward fraction. (B) Changes in synaptic strengths and the outcome choice behavior. Continuous update of plastic synapses selective to choices A and B, measured by synaptic strengths c_A and c_B, allows the model to track the reward fraction. P_A is the probability of choosing A and the *black line* shows the reward fraction in each block of the experiment. (C) The graded activity in the DM network during the matching task. Plotted is the average response of neurons in the two decision-making populations selective for the two targets. Activity is aligned on target onset and is shown separately for the choice that is the preferred (red) or nonpreferred (blue) target of the neurons. Moreover, trials are subdivided into four groups according to the difference between the strength of synapses onto the two pools of neurons selective to choices A and B: $[(c_A - c_B) = -0.05$ to -0.14 (*dashed*), 0 to -0.05 (*thin*), 0 to 0.05 (*normal*), 0.05 to 0.14 (*thick*)]. *((A)–(C) are adapted from Soltani A, Wang X-J. A biophysically-based neural model of matching law behavior: melioration by stochastic synapses. J Neurosci 2006;26:3731–44.)* (D) Average peak-normalized firing rates of lateral intraparietal area neurons over time. Activity is aligned on target onset and different colors correspond to trials when choice was into (blue) and out of (green) the neuron's receptive field. Trials are subdivided into four groups according to the local fractional income of the chosen target: *solid thick lines*, 0.75 to 1.0; *solid medium lines*, 0.5 to 0.75; *solid thin lines*, 0.25 to 0.5; *dotted thin lines*, 0 to 0.25. *(Adapted from Sugrue LP, Corrado GC, Newsome WT. Matching behavior and representation of value in parietal cortex. Science 2004;304:1782–7.)*

decision-making. On one hand, plastic synapses approximate return from each choice option (Eq. (13.6)). On the other hand, the difference between synaptic strengths, $c_A - c_B$ (Eq. (13.4)) determines the choice probability, which in turn modifies the returns (Fig. 13.3). This interplay between the synaptic strengths (or equivalently returns) and the choice probability gives rise to the dynamics underlying matching behavior. At each moment, the model selects the choice option with a larger return (say $R_A > R_B$) with a higher probability ($P_A > .5$). This then reduces the return on that option (without changing the return on the unselected option) because not every selection of A is accompanied by a reward. This

continues until the return on A falls below the return on the other options ($R_A < R_B$), which then causes the model to select choice B more often.

Because of the probabilistic nature of learning and DM in our model, there is always a limit for approaching perfect matching. The dynamic bias of choice toward the more rewarding option makes the model reach a choice probability that is generally smaller but close to the prediction of matching (i.e., undermatching). The extent of undermatching depends on the value of σ so that for a smaller value of σ, the steady state of the model is closer to the prediction of matching. Moreover, the model's estimate of reward values and ensuing choice behavior always fluctuate around a steady state due to ongoing learning. Therefore, perfect matching can be achieved by reducing the learning rates to zero however, such a solution hinders the model from adapting to changes in the reward schedule. Because the model selects the better option (in terms of return), our model acts according to the melioration principle [74,75], which states that the choice behavior should be biased toward the option with the higher return. However, decision is not deterministic in our model and, therefore, our model achieves matching through "probabilistic" melioration.

The model's choice and reward sequences can be used to quantify the dependence of choice on the history of reward by employing different methods, such as the "choice-triggered average of rewards" (CTA) [76]. Such analysis reveals that similar to experimental data and RL models based on RPE, the choice behavior of the model depends on previous reward outcomes in an exponential fashion or a variant of it (e.g., the sum of two exponentials). Interestingly, the time constant of the CTA is a function of the overall baiting probability (or the maximal reward rate) in the environment, such that it decreases as the overall baiting probability increases. In other words, the abundance of reward reduces the time constant of reward influence on choice. This happens because more rewarding instances increases the frequency of potentiation of plastic synapses, reducing the impact of more distant reward. The dependence of the time constant of reward integration on the overall reward rate in the environment enables the model to achieve matching over a wide range of learning rates.

Finally, the pattern of neural activity in the DM circuit of the model matches the graded neural response observed in the lateral intraparietal cortex during this task [46] (Fig. 13.3). This result supports the idea that DA-dependent learning based on a binary reward signal could account for the formation of value signal, as well as the resulting choice behavior.

RANDOM CHOICE AND COMPETITIVE GAMES

Many social interactions, often studied using games [77], require learning and predicting the behavior of other agents while behaving unpredictably to avoid getting exploited by other agents. Because game theory provides only a normative account of the average behavior (or equilibrium) in a given game, RL models are often used to describe the dynamic choice behavior. But how does the dynamics of choice depend on the underlying learning? To answer this question and explore the neural mechanisms of dynamic choice behavior during

competitive games, we incorporated our DA-dependent plasticity into a DM network and simulated the choice behavior during the game of matching pennies [55].

During the game of matching pennies, monkeys were required to freely choose between one of two visual targets (on the left and right sides of the fixation point) and were rewarded only if their choice matched the computer opponent's choice on a given trial (see also Chapter 21). The optimal strategy during this game was to choose the two targets randomly with equal probability [78,79]. However, the strategy or algorithm used by the computer opponent became more complex in three successive stages. In the first stage, the computer selected one of the two targets randomly, each with $p = .5$ (algorithm 0). In the second stage, the computer used the entire history of the animal's previous choices in a given session to predict the monkey's next choice (algorithm 1). The maximum reward could be obtained in algorithm 1 if the animal selected the two targets with equal probability and independently of its previous choices. In the final stage, the computer used the entire history of the animal's choice and reward in a given session to predict the monkey's choice on the next trial (algorithm 2). Therefore, the maximum performance could be obtained if the animal selected its targets, not only with equal probability and independently of its previous choices, but also independently of the combination of its previous choices and their outcomes.

Interestingly, animals showed specific patterns of choice during these three stages. During algorithm 0 when the computer opponent selected between two targets randomly, monkeys displayed a strong bias (choice bias) toward one of the two targets. These choice biases quickly diminished when the computer started to exploit the animal's preference for one of the targets in algorithm 1, but a new bias emerged. Specifically, the animal tended to repeat its decision if the choice was rewarded on the previous trial (win–stay strategy) and switch to the alternative target if the previous choice was not rewarded (lose–switch). Interestingly, the use of the win–stay–lose–switch (WSLS) strategy slowly increased over the period of many days. However, following the introduction of algorithm 2, the probability of WSLS strategy declined toward .5.

In many competitive games, it is important to select between choices stochastically to outsmart the opponent. Our model exhibits such probabilistic behavior because neural spike discharges are intrinsically stochastic. The network's choice varies from trial to trial but is more frequently biased toward the target with a stronger input determined by the recent reward history on the two targets. Similar to the matching game, the choice probability in this task is also a sigmoid function of the difference between the synaptic strengths as in

Eq. (13.4), where the value of σ determines the randomness of the choice behavior. Therefore, any of the biophysical factors appearing on the right side of Eq. (13.3) can change the value of σ and therefore, the desired stochastic behavior.

The specific form of the learning rule used for simulating foraging behavior (Eq. (13.5)) can also be applied to the game of matching pennies. However, for foraging we assumed that only synapses projecting to neurons selective for the chosen target are modified, making the learning rule "choice specific." Such choice-specific learning rule is equivalent to stateless Q-learning [1]. Alternatively, the general learning rule described by Eq. (13.1) can be simplified by assuming that synapses projecting to neurons selective to the unchosen target are also modified by the same amount as the synapses projecting to neurons selective for the chosen target, but in the opposite direction:

Right is selected and rewarded:

$$\begin{cases} c_R(t+1) = c_R(t) + (1 - c_R(t))q_r \\ c_L(t+1) = c_L(t) - c_L(t)q_r \end{cases}$$

(13.7)

Right is selected but not rewarded:

$$\begin{cases} c_R(t+1) = c_R(t) - c_R(t)q_n \\ c_L(t+1) = c_L(t) + (1 - c_L(t))q_n \end{cases}$$

where q_r and q_n are the learning rates in the rewarded and unrewarded trials, respectively. In game theory, learning rules that indiscriminately modify the value functions for chosen and unchosen actions are referred to as belief learning [80,81]. These results show the flexibility of our general learning rule in producing different update rules used in the RL, but with the advantage that our model connects learning rates to transition probabilities at the synaptic level.

The strong choice bias observed during algorithm 0 could be attributed to lack of incentive for the subject to adopt the equilibrium strategy (i.e., choosing the two targets with equal probability). An alternative explanation would be that reward-dependent learning causes the equilibrium strategy to be unstable and, therefore, unattainable. Analyzing the steady state of the model's choice behavior, we found that for specific values of learning rates, the equilibrium strategy becomes unstable, resulting in a strong bias toward one of the two choices.

This happens if $\frac{\sigma(q_r+q_n)}{2(q_r-q_n)} < 0.25$ when learning rates for rewarded trials are larger than unrewarded trials [i.e., $(q_r - q_n) > 0$]. On the other hand, when $(q_r - q_n) < 0$, an unstable equilibrium occurs when $\frac{\sigma(2-q_r-q_n)}{2(q_n-q_r)} < 0.25$.

Therefore, a biased choice behavior does not necessarily reflect insensitivity to reinforcement or feedback, but instead may result from a reward-dependent learning mechanism.

These results were obtained based on the assumption that the choice behavior is not intrinsically biased toward one of the choices. However, the DM network may have an intrinsic bias due to differences in pathways that drive the two competing pools in the DM circuit, resulting in preference for one target and shifting the steady state of choice behavior to a point rather than $p = .5$. Although we found that choice bias could emerge from learning when the computer opponent selects randomly, the question remains as to whether the same learning can mitigate an intrinsic bias in the network if the choice bias is penalized by the opponent (e.g., in algorithm 1). Interestingly, we found that such an intrinsic bias can be compensated through DA-dependent learning that modifies the strength of plastic synapses based on reward feedback during algorithm 1.

Finally, an important aspect of behavioral data described above was a slow change in the probability of WSLS over the course of many days [79]. Could such a change stem from an ongoing learning mechanism that tries to adjust the learning rate over time? To answer this question, we implemented a modified version of the metalearning algorithm [82]. The goal of a metalearning model is to maximize the long-term average of rewards, by comparing the medium-term and long-term running averages of the reward rate. We found that metalearning can account for the gradual change in the animal's strategy. Moreover, we found that, in addition to the initial condition and metalearning parameters, the probabilistic nature of this task contributes to the time course of the choice behavior. Therefore, metalearning and its stochastic nature can provide a mechanism for generating a diverse repertoire of choice behavior observed in various competitive games (see also Chapters 16 and 21).

PROBABILISTIC INFERENCE WITH STOCHASTIC SYNAPSES

In the third application of our DA-dependent plasticity rule (Eq. (13.1)), we show how such a learning mechanism enables plastic synapses to perform probabilistic inference. Probabilistic inference is the ability to combine information from multiple sources that are only partially predictive of alternative outcomes, as well as making inferences about the predictive power of individual sources. These tasks are challenging in naturalistic situations because often only a single action

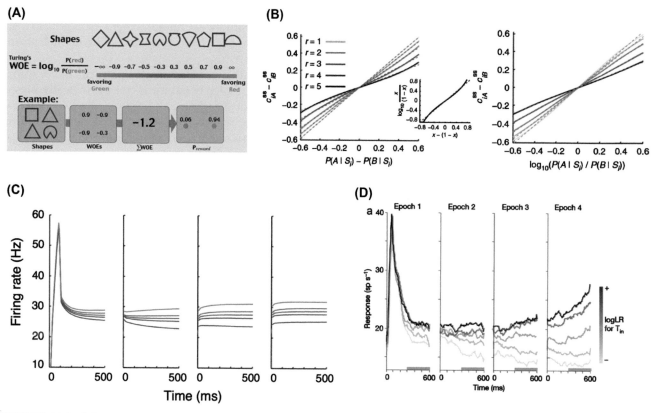

FIGURE 13.4   Behavioral and neural response of the model during the weather prediction task. (A) Reward assignment on each trial of the experiment. Reward is assigned based on the sum of the weight of evidence (WOE) associated with the presented shapes. (B) Difference in the steady state of the synaptic strengths as a function of the difference in the posteriors (left) and of the log posterior odds (right), for different learning rate ratios. *Dashed lines* show linear fits for the values of posterior between 0.2 and 0.8. The inset shows the relationship between $\log_{10}\left(\frac{x}{1-x}\right)$ and $x - (1 - x)$ over the same range, where $x = P(A|S_i)$ and $P(B|S_i) = 1 - P(A|S_i) = 1 - x$. (C) Effect of the log likelihood ratio (LR) on the firing rate of model neurons in the DM network. Plotted is the average population activity over many trials, computed for five quintiles of the log LR in each epoch (more red means larger log LR). The response is aligned to the onset of each shape in a given epoch. *((B) and (C) are adapted from Soltani A, Wang X-J. Synaptic computation underlying probabilistic inference. Nat Neurosci 2010;13:112–9.)* (D) Effect of the log LR on the population average firing rate in a monkey's lateral intraparietal area. Average responses are aligned to the onset of the shapes and extend 100 ms into the subsequent epoch. The averages were computed for five quintiles of log LR in each epoch (indicated by *shading*). *(Adapted from Yang T, Shadlen MN. Probabilistic reasoning by neurons. Nature 2007;447:1075–80.)*

outcome such as a binary reward feedback follows the presentation of multiple sources of information (or cues), and so it is unclear how presented cues should be associated with the final outcome.

An example of such a naturalistic situation is simulated in the so-called weather prediction task, in which a categorical choice (rain or sunshine) is predicted based on a number of given cues [83]. A 2007 study using a variant of this probabilistic categorization task suggested that even monkeys are capable of some forms of probabilistic inference and revealed neural correlates of this ability at the single-cell level [84]. In this task, four shapes precede a selection between two color targets on each trial [84] (Fig. 13.4A). These shapes were selected randomly from a set of 10 distinguishable shapes

$(S_i, i = 1, 2, ..., 10)$, each of which was allocated a unique weight of evidence (WOE) about the probability of reward assignment on one of the two choice alternatives $\left(\text{WOE} = \log_{10}\frac{P(A|S_i)}{P(B|S_i)}\right)$. The computer assigned a reward to one of the two alternatives with a probability that depended on the sum of the WOEs from all the shapes presented on a given trial.

We used the three-circuit network model (Fig. 13.2B) to simulate the choice behavior and neural activity during the weather prediction task to test whether our proposed learning rule (Eq. (13.1)) enables probabilistic inference. First, we found that the specific form of the learning rule allows plastic synapses to estimate variants of posteriors depending on the number of shapes presented simultaneously. When one shape is presented alone, the steady state of the synaptic strength for synapses selective for shape i and alternative A is

$$c_{iA}^{ss} = \frac{rP(A|S_i)}{1 + (r-1)P(A|S_i)} \qquad (13.8)$$

where $P(A|S_i)$ is the probability that A is assigned with reward, given shape i is presented, and r is the learning rate ratio ($r \equiv q_+/q_-$). Thus, when each cue is presented alone, the steady state is proportional to posteriors. When several shapes are presented together, the reward feedback is determined by all presented shapes, and synapses selective for all these shapes are updated. As a result, plastic synapses estimate the naïve posterior probability, $\widetilde{P}(A|S_i)$, or the posterior probability that a choice alternative is assigned with reward given that a cue is presented in any combination of cues.

Importantly, the decision circuit stochastically generates a categorical choice with a probability, which is a sigmoid function of the difference in the overall inputs (differential input, ΔI) to its selective pools [55,60,67,85]. We assume that cue-selective neurons fire at a similar rate, and, therefore, the differential input is determined by the sum of the difference in the synaptic strengths $\left(\Delta c_i^{ss} \equiv c_{iA}^{ss} - c_{iB}^{ss}\right)$ from the action value-coding neurons onto the decision neurons (Fig. 13.4B). Using Eq. (13.8) and replacing posteriors for naïve posteriors, we can compute the differential input:

$$\Delta I \propto \sum_i \Delta c_i^{ss} \propto \sum_i \frac{r\left(\widetilde{P}(A|S_i) - \widetilde{P}(B|S_i)\right)}{r + (r-1)^2 \widetilde{P}(A|S_i)\widetilde{P}(B|S_i)} \qquad (13.9)$$

This formula can be simplified by noting that $r \gg (r-1)^2 \widetilde{P}(A|S_i)\widetilde{P}(B|S_i) (\equiv k)$, which happens when $r \approx 1$ and the values of posterior probabilities are in an intermediate range $\left(2 \leq \left(\widetilde{P}(A|S_i)\right) \leq .8\right)$:

$$\Delta I \propto \sum_i \left(1 - \frac{k}{r}\right)\left(\widetilde{P}(A|S_i) - \widetilde{P}(B|S_i)\right) \qquad (13.10)$$

Using another simplification, $x - (1-x) \simeq \log_{10}\left(\frac{x}{1-x}\right)$ (true if $0.2 \leq x \leq 0.8$) we get:

$$\Delta I \propto \sum_i \left(1 - \frac{k}{r}\right)\log_{10}\frac{\widetilde{P}(A|S_i)}{\widetilde{P}(B|S_i)} \qquad (13.11)$$

Note that for smaller or larger values of posteriors, the choice behavior is deterministic and so our simplification does not affect the final calculation. Therefore, because of the convergence of sensory neurons onto action value-coding neurons, the latter naturally summate the currents through sets of plastic synapses related to presented cues. Subsequently, the outputs of action value neurons drive the decision circuit, and the resulting choice is a function of the sum of log naïve posterior odds. Thus, summation of currents through plastic synapses provides a natural mechanism for integrating evidence from different cues in terms of log posterior odds.

Overall, our model provides a solution to probabilistic inference based on a mixture of synaptic and neural mechanisms. On the one hand, modifications of synapses selective for the presented shapes enables plastic synapses to directly estimate naïve posteriors (but not posteriors). A consequence of such a learning rule is that the evidence a model assigns to a given cue is smaller than the WOE assigned to that cue. On the other hand, summation of currents evoked by presented cues causes the model to integrate evidence in terms of log posterior odds. This feature enables the model to perform cue combination near optimality (i.e., according to Bayes' rule), but only given equal priors. When priors are unequal, plastic synapses carry information about priors as well, and therefore, our model provides an answer different from what is expected by adding log prior odds to the summed log likelihood ratios. More specifically, the model accounts for a cognitive bias known as base-rate neglect, that is, a cue that is equally predictive of each outcome is perceived to be more predictive of the less probable outcome [86]. Moreover, it predicts a bias in making inference about a combination of cues which depends on the number of cues used for making inference. Interestingly, a recent study has provided evidence for both predictions of the model [86a].

Finally, neural activity in our model reproduces the main physiological observations from the lateral intraparietal area in the monkey experiment [84] (Fig. 13.4C and D). Overall, our results demonstrate that empirically observed neural correlates of probabilistic inference can rely on synaptic, rather than neuronal, computations. Despite the complexity of the weather prediction task, stochastic DA-dependent plasticity enables the model to perform the task using a binary "teaching" signal.

CONCLUDING REMARKS

Reward is one of the determining factors for choice behavior, and the neural correlates of reward value are observed in many brain areas. Consequently, linking the influence of reward on behavior and the measured neural response (i.e., representation of reward value) is one of the critical goals in the study of DM. It is generally believed that reward is signaled throughout the brain via DA [87–89], and DA-dependent plasticity provides the neural substrates for learning reward values; however, the neural mechanisms underlying these processes are not fully understood.

There are reasons for why identifying neural mechanisms underlying value-based learning and representation of reward is challenging. Dopaminergic neurons project to many brain areas, where DA influences both synaptic plasticity and neural excitability; the diverse effects remain poorly explored (see also Chapter 2). Furthermore, neural correlates of value-based behavior are often measured in one area at the time, not allowing for the exploration of the role of interactions between different areas in value-based DM. Similarly, computational models of value-based DM and learning often are not concerned with the plausibility of the learning rules, ignore important effects of DA on neuronal processes such as short-term synaptic plasticity and changes in neural excitability [69], overlook different types of DA receptors involved in dopaminergic modulations, and do not consider interactions between brain areas.

To move forward, the focus needs to shift to biological plausibility of proposed learning rules and future models of value-based DM. Learning should go beyond a local circuit and include interactions between various circuits in determining the choice behavior. Understanding the role of interactions between different brain areas in value-based DM may elucidate the origin and function of various types of value representation throughout the brain (see Chapters 10, 14, 16, and 24). The diverse representation of reward value throughout the brain also begs the question of investigating DA-dependent plasticity in those areas, and the role of local circuits in translating a "global" reward signal into meaningful information.

Acknowledgments

This work was supported by NIH Grant R01MH062349 to Xiao-Jing Wang.

References

[1] Sutton RS, Barto AG. Reinforcement learning: an introduction. Cambridge (MA): MIT Press; 1998.

[2] Williams RJ. Simple statistical gradient-following algorithms for connectionist reinforcement learning. Mach Learn 1992;8:229–56.

[3] Baxter J, Bartlett PL. Infinite-horizon policy-gradient estimation. Artif Intell 2001;15:319–50.

[4] Vasilaki E, Fremaux N, Urbanczik R, Senn W, Gerstner W. Spike-based reinforcement learning in continuous state and action space: when policy gradient methods fail. PLoS Comput Biol 2009;5:e1000586.

[5] Xie X, Seung HS. Learning in neural networks by reinforcement of irregular spiking. Phys Rev E 2004;69:1–10.

[6] Florian RV. Reinforcement learning through modulation of spike-timing-dependent synaptic plasticity. Neural Comput 2007;19:1468–502.

[7] Pfister JP, Toyoizumi T, Barber D, Gerstner W. Optimal spike-timing-dependent plasticity for precise action potential firing in supervised learning. Neural Comput 2006;18:1318–48.

[8] Engel TA, Chaisangmongkon W, Freedman DJ, Wang X-J. Choice-correlated activity fluctuations underlie learning of neuronal

[9] Loewenstein Y, Seung HS. Operant matching is a generic outcome of synaptic plasticity based on the covariance between reward and neural activity. Proc Natl Acad Sci USA 2006;103:15224–9.

[10] Roelfsema PR, van Ooyen A. Attention-gated reinforcement learning of internal representations for classification. Neural Comput 2005;17:2176–214.

[11] Sutton RS. Learning to predict by the methods of temporal differences. Mach Learn 1988;3:9–44.

[12] Montague PR, Dayan P, Sejnowski TJ. A framework for mesencephalic dopamine systems based on predictive Hebbian learning. J Neurosci 1996;16:1936–47.

[13] Suri RE, Schultz W. Temporal difference model reproduces anticipatory neural activity. Neural Comput 2001;13:841–62.

[14] Potjans W, Morrison A. A spiking neural network model of an actor-critic learning agent. Neural Comput 2009;339:301–39.

[15] Frémaux N, Sprekeler H, Gerstner W. Reinforcement learning using a continuous time actor-critic framework with spiking neurons. PLoS Comput Biol 2013;9:e1003024.

[16] Rescorla RA, Wagner AR. A theory of Pavlovian conditioning: variations in the effectiveness of reinforcement and nonreinforcement. In: Classical conditioning II: current research and theory; 1972. p. 64–9.

[17] Schultz W. Neuronal reward and decision signals: from theories to data. Physiol Rev 2015;95:853–951.

[18] Bayer HM, Glimcher PW. Midbrain dopamine neurons encode a quantitative reward prediction error signal. Neuron 2005;47:129–41.

[19] Nomoto K, Schultz W, Watanabe T, Sakagami M. Temporally extended dopamine responses to perceptually demanding reward-predictive stimuli. J Neurosci 2010;30:10692–702.

[20] Glimcher PW. Understanding dopamine and reinforcement learning: the dopamine reward prediction error hypothesis. Proc Natl Acad Sci USA 2011;108(Suppl. 3):15647–54.

[21] Enomoto K, et al. Dopamine neurons learn to encode the long-term value of multiple future rewards. Proc Natl Acad Sci USA 2011;108:15462–7.

[22] Cohen JY, Haesler S, Vong L, Lowell BB, Uchida N. Neuron-type-specific signals for reward and punishment in the ventral tegmental area. Nature 2012;482:85–8.

[23] Eshel N, Bukwich M, Rao V, Hemmelder V, Tian J, Uchida N. Arithmetic and local circuitry underlying dopamine prediction errors. Nature 2015;525:243–6.

[24] Fiorillo CD, Tobler PN, Schultz W. Evidence that the delay-period activity of dopamine neurons corresponds to reward uncertainty rather than backpropagating TD errors. Behav Brain Funct 2005;1:1–5.

[25] Fiorillo CD, Newsome WT, Schultz W. The temporal precision of reward prediction in dopamine neurons. Nat Neurosci 2008;11:966–73.

[26] Redgrave P, Gurney K. The short-latency dopamine signal: a role in discovering novel actions? Nat Rev Neurosci 2006;7:967–75.

[27] Daw ND, Tobler PN. Value learning through reinforcement: the basics of dopamine and reinforcement learning. In: Neuroeconomics. Academic Press; 2013. p. 283–98.

[28] Soltani A, Wang X-J. From biophysics to cognition: reward-dependent adaptive choice behavior. Curr Opin Neurobiol 2008;18:209–16.

[29] Lee D, Seo H, Jung MW. Neural basis of reinforcement learning and decision making. Annu Rev Neurosci 2012;35:287–308.

[30] Louie K, Glimcher PW. Efficient coding and the neural representation of value. Ann NY Acad Sci 2012;1251:13–32.

[31] Cromwell HC, Schultz W. Effects of expectations for different reward magnitudes on neuronal activity in primate striatum. J Neurophysiol 2003;89:2823–38.

category representations. Nat Commun 2015;6. http://dx.doi.org/10.1038/ncomms7454 [Article No: 6454].

[32] Hollerman JR, Tremblay L, Schultz W. Influence of reward expectation on behavior-related neuronal activity in primate striatum. J Neurophysiol 1998;80:947−63.

[33] Wallis JD, Kennerley SW. Heterogeneous reward signals in prefrontal cortex. Curr Opin Neurobiol 2010;20:191−8.

[34] Cai X, Kim S, Lee D. Heterogeneous coding of temporally discounted values in the dorsal and ventral striatum during intertemporal choice. Neuron 2011;69:170−82.

[35] Seo H, Lee D. Temporal filtering of reward signals in the dorsal anterior cingulate cortex during a mixed-strategy game. J Neurosci 2007;27:8366−77.

[36] Belova MA, Paton JJ, Salzman CD. Moment-to-moment tracking of state value in the amygdala. J Neurosci 2008;28:10023−30.

[37] Padoa-Schioppa C, Assad JA. Neurons in the orbitofrontal cortex encode economic value. Nature 2006;441:223−6.

[38] Sul JH, Kim H, Huh N, Lee D, Jung MW. Distinct roles of rodent orbitofrontal and medial prefrontal cortex in decision making. Neuron 2010;66:449−60.

[39] Öngür D, Price J. The organization of networks within the orbital and medial prefrontal cortex of rats, monkeys and humans. Cereb Cortex 2000;10:206−19.

[40] Lau B, Glimcher PW. Value representations in the primate striatum during matching behavior. Neuron 2008;58:451−63.

[41] Kim H, Sul JH, Huh N, Lee D, Jung MW. Role of striatum in updating values of chosen actions. J Neurosci 2009;29:14701−12.

[42] Kable JW, Glimcher PW. The neurobiology of decision: consensus and controversy. Neuron 2009;63:733−45.

[43] Samejima K, Ueda Y, Doya K, Kimura M. Representation of action-specific reward values in the striatum. Science 2005;310:1337−40.

[44] Lau B, Glimcher PW. Action and outcome encoding in the primate caudate nucleus. J Neurosci 2007;27:14502−14.

[45] Platt ML, Glimcher PW. Neural correlates of decision variables in parietal cortex. Nature 1999;400:233−8.

[46] Sugrue LP, Corrado GC, Newsome WT. Matching behavior and representation of value in parietal cortex. Science 2004;304:1782−7.

[47] Ding L, Hikosaka O. Comparison of reward modulation in the frontal eye field and caudate of the macaque. J Neurosci 2006;26:6695−703.

[48] Ikeda T, Hikosaka O. Reward-dependent gain and bias of visual responses in primate superior colliculus. Neuron 2003;39:693−700.

[49] Houk J, Davis J. Models of information processing in the basal ganglia. The MIT Press; 1994.

[50] Si J, Barto AG, Powell WB, Wunsch D. Handbook of learning and approximate dynamic programming (IEEE press series on computational intelligence). Wiley-IEEE Press; 2004.

[51] Montague P, Dayan P, Person C, Sejnowski T, et al. Bee foraging in uncertain environments using predictive Hebbian learning. Nature 1995;377:725−8.

[52] Suri RE, Schultz W. A neural network model with dopamine-like reinforcement signal that learns a spatial delayed response task. Neuroscience 1999;91:871−90.

[53] Pan W-X, Schmidt R, Wickens JR, Hyland BI. Dopamine cells respond to predicted events during classical conditioning: evidence for eligibility traces in the reward-learning network. J Neurosci 2005;25:6235−42.

[54] Bogacz R, McClure SM, Li J, Cohen JD, Montague PR. Short-term memory traces for action bias in human reinforcement learning. Brain Res 2007;1153:111−21.

[55] Soltani A, Lee D, Wang X-J. Neural mechanism for stochastic behavior during a competitive game. Neural Netw 2006;19:1075−90.

[56] Foster D, Morris R, Dayan P, et al. A model of hippocampally dependent navigation, using the temporal difference learning rule. Hippocampus 2000;10:1−16.

[57] Doya K. Reinforcement learning in continuous time and space. Neural Comput 2000;12:219−45.

[58] Montague PR, et al. Dynamic gain control of dopamine delivery in freely moving animals. J Neurosci 2004;24:1754−9.

[59] Potjans W, Diesmann M, Morrison A. An imperfect dopaminergic error signal can drive temporal-difference learning. PLoS Comput Biol 2011;7:e1001133.

[60] Soltani A, Wang X-J. A biophysically-based neural model of matching law behavior: melioration by stochastic synapses. J Neurosci 2006;26:3731−44.

[61] Reynolds JN, Wickens JR. Dopamine-dependent plasticity of corticostriatal synapses. Neural Netw 2002;15:507−21.

[62] Amit DJ, Fusi S. Dynamic learning in neural networks with material synapses. Neural Comput 1994;6:957−82.

[63] Fusi S. Hebbian spike-driven synaptic plasticity for learning patterns of mean firing rates. Biol Cybern 2002;87:459−70.

[64] Fusi S, Drew PJ, Abbott LF. Cascade models of synaptically stored memories. Neuron 2005;45:599−611.

[65] Petersen CC, Malenka RC, Nicoll RA, Hopfield JJ. All-or-none potentiation at CA3-CA1 synapses. Proc Natl Acad Sci USA 1998;95:4732−7.

[66] O'Connor DH, Wittenberg GM, Wang SS-H. Graded bidirectional synaptic plasticity is composed of switch-like unitary events. Proc Natl Acad Sci USA 2005;102:9679−84.

[67] Fusi S, Asaad WF, Miller EK, Wang X-J. A neural circuit model of flexible sensorimotor mapping: learning and forgetting on multiple timescales. Neuron 2007;54:319−33.

[68] Soltani A, Wang X-J. Synaptic computation underlying probabilistic inference. Nat Neurosci 2010;13:112−9.

[69] Soltani A, Noudoost B, Moore T. Dissociable dopaminergic control of saccadic target selection and its implications for reward modulation. Proc Natl Acad Sci USA 2013;110:3579−84.

[70] Herrnstein RJ. Relative and absolute strength of response as a function of frequency of reinforcement. J Exp Anal Behav 1961;4:267−72.

[71] Williams BA. Reinforcement, choice, and response strength. In: Atkison RC, Herrnstein RJ, Lindzey G, Luce RD, editors. Steven's handbook of experimental psychology. 2nd ed., vol. 2. New York: Wiley; 1988. p. 167−244.

[72] Gallistel CR. Foraging for brain stimulation: toward a neurobiology of computation. Cognition 1994;50:151−70.

[73] Herrnstein RJ, Rachlin H, Laibson DI. The matching law: papers in psychology and economics. Harvard UP; 1997.

[74] Herrnstein RJ, Vaughan WJ. Melioration and behavioral allocation. In: Staddon JER, editor. Limits to action: the allocation of individual behavior. New York: Academic; 1980. p. 143−76.

[75] Herrnstein RJ, Prelec D. Melioration: a theory of distributed choice. J Econ Perspect 1991;5:137−56.

[76] Corrado GS, Sugrue LP, Seung HS, Newsome WT. Linear-nonlinear-Poisson models of primate choice dynamics. J Exp Anal Behav 2005;84:581−617.

[77] Lee D. Game theory and neural basis of social decision making. Nat Neurosci 2008;11:404−9.

[78] Barraclough DJ, Conroy ML, Lee D. Prefrontal cortex and decision making in a mixed-strategy game. Nat Neurosci 2004;7:404−10.

[79] Lee D, Conroy ML, McGreevy BP, Barraclough DJ. Reinforcement learning and decision making in monkeys during a competitive game. Brain Res Cogn Brain Res 2004;22:45−58.

[80] Camerer CF. Behavioral game theory: experiments in strategic interaction. Princeton: Princeton Univ. Press; 2003.

[81] Lee D, McGreevy BP, Barraclough DJ. Learning and decision making in monkeys during a rock-paper-scissors game. Brain Res Cogn Brain Res 2005;25:416−30.

[82] Schweighofer N, Doya K. Meta-learning in reinforcement learning. Neural Netw 2003;16:5−9.

II. HUMAN STUDIES ON MOTIVATION, PERCEPTUAL, AND VALUE-BASED DECISION-MAKING

[83] Knowlton BJ, Squire LR, Gluck MA. Probabilistic classification learning in amnesia. Learn Mem 1994;1:106—20.

[84] Yang T, Shadlen MN. Probabilistic reasoning by neurons. Nature 2007;447:1075—80.

[85] Wang X-J. Probabilistic decision making by slow reverberation in cortical circuits. Neuron 2002;36:955—68.

[86] Gluck MA, Bower GH. From conditioning to category learning: an adaptive network model. J Exp Psychol Gen 1988;117:227—47.

[86a] Soltani A, Khorsand P, Guo CZ, Farashahi S, Liu J. Neural substrates of cognitive biases during probabilistic inference. Nat Commun 2016;7:11393.

[87] Schultz W. Multiple reward signals in the brain. Nat Rev Neurosci 2000;1:199—207.

[88] Vickery TJ, Chun MM, Lee D. Ubiquity and specificity of reinforcement signals throughout the human brain. Neuron 2011;72: 166—77.

[89] Clark AM. Reward processing: a global brain phenomenon? J Neurophysiol 2013;109:1—4.

[90] Goldman-Rakic PS. Cellular basis of working memory. Neuron 1995;14:477—85.

[91] Goldman-Rakic PS, Leranth C, Williams SM, Mons N, Geffard M. Dopamine synaptic complex with pyramidal neurons in primate cerebral cortex. Proc Natl Acad Sci USA 1989;86:9015—9.

[92] Zhang JC, Lau PM, Bi GQ. Gain in sensitivity and loss in temporal contrast of STDP by dopaminergic modulation at hippocampal synapses. Proc Natl Acad Sci USA 2009;106(31):13028—33.

[93] Surmeier DJ, Plotkin J, Shen W. Dopamine and synaptic plasticity in dorsal striatal circuits controlling action selection. Curr Opin Neurobiol 2009;19(6):621—8.

[94] Shen W, Flajolet M, Greengard P, Surmeier DJ. Dichotomous dopaminergic control of striatal synaptic plasticity. Science 2008; 321(5890):848—51.

SOCIAL DECISION NEUROSCIENCE

14

Social Decision-Making in Nonhuman Primates

M. Jazayari[1,2], S. Ballesta[1,2,3], J.-R. Duhamel[1,2]

[1]Centre National de la Recherche Scientifique, Bron, France; [2]Université Lyon 1, Villeurbanne, France; [3]The University of Arizona, Tucson, AZ, United States

Abstract

In primates, being able to predict the immediate and long-term consequences of their actions upon others requires keen cognitive abilities that are essential for individual fitness, as inappropriate social behavior can be synonymous with aggression or exclusion from the group. Behavioral and social neurosciences have turned their attention to nonhuman primates to gather new insights into the psychological and biological bases of normal and maladaptive social behavior. In this chapter, we describe experimental studies that have overcome the challenge of using real partners to investigate nonhuman primates' decision-making in a well-controlled social environment. We specifically focus on the neuronal mechanisms of social decision-making and summarize the actual state of our knowledge about the implications of various cortical and subcortical areas in the production of other-regarding behaviors that are naturally involved in the primate's prosocial, cooperative, or competitive interactions.

INTRODUCTION

Primates live in highly social environments characterized by strong competition and dominance hierarchies, but also by cooperation and empathy-driven behaviors that serve to promote social bonds and contribute to group members' fitness. The way we adapt to a dynamic, changing social environment depends on our ability to perceive social stimuli; to convey, but also sometimes conceal, our emotions and intentions; and to keep track of those of others. There has been a growing interest in the identification and characterization of the brain areas dedicated to the processing of social information and in their relation to other large-scale systems such as the cerebral networks involved in reward processing and emotions [1–3].

Compared to other vertebrates, human and nonhuman primates have a larger neocortex and this is to be directly correlated with the size and complexity of their social systems [4]. Furthermore group size has been shown to drive cortical gray matter expansion in key neocortical areas for social cognition [5]. Obviously, human social behavior has reached an unrivaled degree of sophistication in the course of primate evolution, thanks to some particularly developed cognitive abilities like theory of mind or recursive thinking. This should not lead us to minimize the relevance of comparative studies and of the parallels that can be drawn with nonhuman primate social cognitive abilities. There are pros and cons to the possible homologies a given nonhuman primate species affords with respect to brain architecture, behavioral repertoire, and ecological pressures. Like all models, the macaque monkey has some limitations but it has established itself over the years as a gold standard thanks to the wealth of accumulated knowledge on the neurobiology relevant to social decision-making, including the dopamine reward system, amygdala complex, temporal cortex, insula, and the anterior cingulate (ACC), ventromedial (VM), and orbitofrontal (OFC) subdivisions of the prefrontal cortex (Fig. 14.1). Human patients with VM or OFC lesions have been described as engaging in socially inappropriate behavior [6], lacking awareness of social norm violations [7], failing to adjust their communicative behavior according to the social partner [8], or showing reduced aversion to social inequity [9]. In monkeys, experimental lesions of the OFC similarly point to the role of these structures in a number of social domains, e.g., responding to and producing communicative facial expressions, affiliation and bonding, and aggression [10–12]. In addition, lesions of the ACC can reduce the interest of monkeys in social stimuli [13]. Altered social avoidance, inhibition, and aggression have been found following ablation of the amygdala complex [12,14]. In Chapter 15, Noonan et al. describe the striking

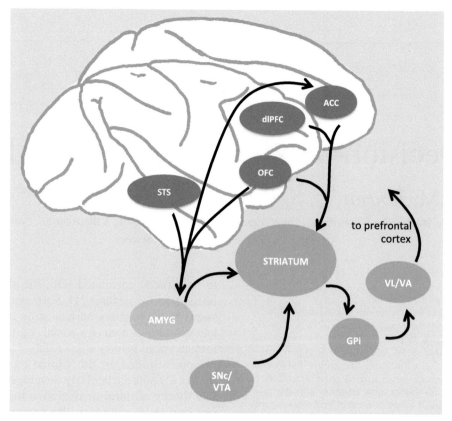

FIGURE 14.1 Simplified schematic representation of the primate reward and social decision-making network. Midbrain nuclei containing dopaminergic neurons: *AMYG*, amygdala; *GPi*, internal globus pallidus; *SNc/VTA*, substantia nigra pars compacta/ventral tegmental area; *STS*, superior temporal sulcus; *VL/VA*, ventral lateral and anterior thalamic nuclei. Subdivisions of the prefrontal cortex: *ACC*, anterior cingulate cortex; *dlPFC*, dorsolateral prefrontal cortex; *OFC*, orbitofrontal cortex.

similarities between the human and the macaque social brains using anatomical and functional brain imaging. In the following sections we draw on studies of macaque behavior and neuronal mechanisms that contribute to our understanding of the social brain. Our focus will be on studies of real dyadic interactions that have examined other-regarding behavior, many of which have relied on social decision-making paradigms as a model for natural prosocial, cooperative, or competitive interactions (see also Chapter 18 for a discussion of strategic reasoning algorithms revealed by simulated social interactions).

BEHAVIORAL STUDIES ON SOCIAL DECISION-MAKING

In their natural habitat, primates face demanding social challenges for which specialized cognitive competences are assumed to be advantageous, like the capacity to infer what others can hear or see [3,15,16] or to predict others' emotions or intentions [17]. Although higher forms of altruism may require uniquely human mentalizing and moral reasoning

abilities, much evidence supports an evolutionary continuity in the motivational and affective mechanisms that regulate affiliative behaviors [18–21]. Ethological observations show that nonhuman primates are sensitive to others' states, and their ubiquitous social play, grooming behavior, and hormonal correlates suggest an ability to conceive what is pleasant or unpleasant for others [22–24]. Pioneering experimental studies have shown that macaques can perceive and seek to alleviate their peers' distress [25,26], and more recent studies have shown emotional contagion in rodents and its promotion of helping behavior [27–29]. Studies in monkeys have evaluated their capacity to take into account the welfare of a conspecific using simple social dilemmas, such as choosing between one action leading to a reward for self only and another action leading to a reward for self as well as for a passive observer. Despite methodological differences, these experiments consistently show that monkeys often favor the benevolent option, to a degree that depends on factors such as kinship, dominance status, and preexisting social bonds among nonkin ([30–32], see also Chapter 19).

The question arises as to the proximal mechanisms of such prosociality. A strong unifying hypothesis is that prosocial behavior is shaped by the reinforcing value of certain social events. One proposal is that the mere sight of a conspecific receiving juice might recruit, through some form of mirror mechanism, the same brain reward circuit as an actual drop of juice [31]. Alternatively, it could be that juice allocation increases social attention received from the partner, which in turn initiates a positive loop of rewarding social interaction and promotes prosocial behavior through the monkeys' social attachment system [32,33]. Conversely, observing a conspecific being inflicted an aversive stimulus could be experienced as unpleasant, deactivate reward circuits, and negatively reinforce the antisocial option. The idea that prosocial actions generate their own rewards is related to the warm-glow hypothesis of human altruism: doing good makes us feel good [33,34]. Similarly, empathy

theories postulate that prosocial behavior directed to suffering individuals might primarily aim at alleviating vicariously experienced negative affect [35]. Empirical evidence for an empathy-based account is provided in a study by Ballesta and Duhamel [32], who challenged pairs of cynomolgus and rhesus macaques with social decisions involving appetitive and aversive outcomes while monitoring both animals' oculomotor behavior (Fig. 14.2). The results show that most monkeys take into account the fate of their peer when making choices that could cause outcomes like a drop of juice or an unpleasant air puff being delivered to their partner. Positive correlations between these two types of social decisions indicate that monkeys have coherent motivations in different social interactions. Importantly, benevolence was associated with enhanced social

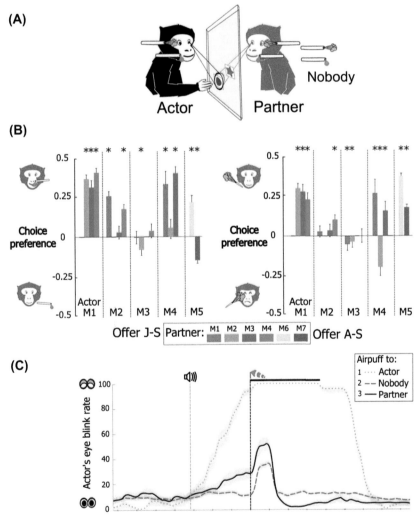

FIGURE 14.2 Evidence of rudimentary empathy during social decisions. (A) Pairs of monkeys sitting face to face made social decisions by touching images on a transparent touch slab. (B) In a majority of dyads, the monkey actor exhibits prosocial behavior, preferring to procure juice rewards and to refrain from administering unpleasant air puffs to the partner. (C) Affective-state matching observed through anticipatory eye blinking on trials in which the actor monkey chose to send an air puff to self or to its partner compared to nobody. *Adapted from Ballesta S, Duhamel J-R. Rudimentary empathy in macaques' social decision-making. Proc Natl Acad Sci USA 2015;112:15516–21.*

communication through deliberate mutual gaze and with empathic eye blinking in anticipation of the partner's being inflicted an aversive stimulus (Fig. 14.2C). Such enhanced eye blinking when watching a peer receiving an air puff is highly suggestive of induced negative affect in the observer, and in fact indifference or malevolence was associated with lower or suppressed such responses.

Theories of empathy, broadly defined as vicarious experiences of the affective states of others [19,21], postulate that shared representations serve as a proximate mechanism sustaining altruistic behaviors [18–20,36]. Our finding of similar behavioral responses to the direct experience of an aversive stimulus and to the observation of its impact upon a peer would thus appear consistent with simulation theories of empathy. Finally, it is noteworthy that the small population of tested macaques does not behave in a uniformly benevolent manner, but showed consistent partner preferences. Using manual and automatic scoring of spontaneous interactions in their living space [37], we assessed the social organization and affiliative behavior patterns of monkeys and found that benevolent ones spend more time and exchange more grooming with their partner than nonbenevolent monkeys. This is in accordance with ethological observations of social behaviors such as coalition building or reconciliations [38,39], which emphasize that nonhuman primates actively establish and preserve preferential social relationships. Together, these results show that macaques' responses to the welfare of others can be measured objectively, opening the way to the investigation of the neuronal processing during social interactions in a laboratory setting.

NEURONAL CORRELATES OF DECISION-MAKING IN A SOCIAL CONTEXT

The neural substrates underlying social behavior and motivation systems are likely to intersect for several obvious reasons. A first reason is that social information possesses intrinsic rewarding qualities. The tremendous success of smartphone photo exchange apps and online social networks is a patent testimony to the high value we assign to observing and learning about our conspecifics. This is true also of monkeys who spontaneously gaze at socially significant features of visual scenes [40–42] and are willing to partially sacrifice a primary reward such as fruit juice to be allowed to view images of conspecifics [43]. This has led to the view that, in addition to temporal lobe systems specialized for the processing of social information such as faces, and particularly the gaze direction, emotional expressions, and identity of faces [44–47], other regions participate in encoding the subjective value of such social stimuli [43].

Another obvious reason, and this is what the following section will focus on more specifically, is that vital resources come in limited amounts and thus generate different social adaptations to deal with competition and dominance status of peers. Access to goods (food) and services (grooming by a partner) in a social context implies that "others" can be used as comparison points and directly affect perception of one's own welfare.

Numerous electrophysiological experiments have investigated the neuronal correlates of motivated behaviors and reward processing in monkeys, notably in the dopamine reward system (see Ref. [48] for a review). The representation of various natural reinforcers, both appetitive [49,50] and aversive [51], has also been described in the OFC, and activity in this region is described as reflecting the subjective value of rewards during decision-making [52,53]. To determine whether neural encoding of motivational value reflects only the relation between a physiological state (e.g., blood glucose concentration) and a physical magnitude (number of calories supplied by a food item), or whether it encompasses a broader, more psychological dimension of value, the effect of social context on reward-related activity has been investigated in the OFC and in related structures.

In an attempt to extend and to investigate the role of the OFC from value-based decisions to socially motivated ones, Azzi et al. [54] trained monkeys to produce a manual response to visual cues predicting subsequent juice rewards for themselves and for passive observers (Fig. 14.3). The results showed that the spontaneous spike discharge activity of the OFC neuronal population is globally enhanced in a social context compared to a nonsocial one. Furthermore, value-sensitive neurons were found to respond to a fixed-size reward for self with different discharge rates depending on whether monkeys were working to obtain a reward for self only or for both self and a partner monkey. This is illustrated for a single unit and for a population of reward-value coding cells responding with a monotonic increase in discharge rate as a function of reward size during a series of nonsocial trials (left panels of Fig. 14.3B and C). In a social trial block, neuronal firing to a predicted reward for self was modified on trials in which an identical reward was provided to one of the monkey's partners, P1 or P2. Interestingly, these neuronal responses reflected the subjective value of rewards, as inferred from the behavior of the monkeys who, in this particular context, showed less motivation (i.e., performed fewer correct trials) on "shared" than on "unshared" reward trials. Finally, a session-by-session analysis further showed that OFC activity tracked the monkey's current social preferences, expressed in the relative willingness to procure rewards to one or the other partner. Together,

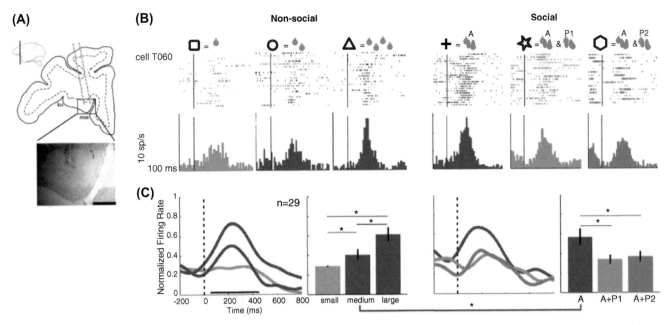

FIGURE 14.3 Social modulation of reward-value signals in the macaque OFC. (A) Single units were recorded in area 13 while monkeys worked to obtain rewards for self only or self plus a passive partner. (B) A positively coding value-sensitive single-unit responding as a function of reward size in a nonsocial context (left rasters and histograms) shows a reduced response in a social context when either partner P1 or partner P2 received a concomitant reward (right rasters and histograms). (C) Population activity of value-sensitive orbitofrontal cortex neurons. *Adapted from Azzi JCB, Sirigu A, Duhamel J-R. Modulation of value representation by social context in the primate orbitofrontal cortex. Proc Natl Acad Sci USA 2012;109: 2126—31.*

these results indicate that social influence of reward sharing and partner identity is taken into account in the valuation process operated by the OFC.

Such properties suggest that the OFC could be involved the neural computations taking place during social comparisons. However, the study by Azzi et al. did not address directly the issue of coding of subjective equity and inequity between one's own welfare and that of other individuals. People naturally compare their resources with those of others and, as Karl Marx noted, perceived reward differences are a major driving force in the pursuit of wealth and a source of perpetual frustration: "A house may be large or small; as long as the neighboring houses are likewise small, it satisfies all social requirement for a residence. But let there arise next to the little house a palace, and the little house shrinks to a hut. The little house now makes it clear that its inmate has no social position at all to maintain, or but a very insignificant one; and however high it may shoot up in the course of civilization, if the neighboring palace rises in equal or even in greater measure, the occupant of the relatively little house will always find himself more uncomfortable, more dissatisfied, more cramped within his four walls" [55]. Sensitivity to unequal distribution of rewards has been well documented in human adults [56] and children [57]. And in a similar and not surprising way it has been shown that also various nonhuman

primate species respond adversely to social inequity [58–62]. Báez-Mendoza et al. [63] investigated neuronal coding of inequity in the ventral striatum using an imperative reward-giving task in which monkeys experienced mainly four different reward conditions: reward to self, to the partner, to both animals, or to nobody. Behaviorally, monkeys exhibited unfavorable inequity aversion. At the single-cell level, as expected from prior studies [64–66], a sizeable proportion of striatal neurons encoded reward to self. However, a subset of striatal neurons were sensitive to reward inequity, coding either disadvantageous inequity (receiving less reward than the partner) or advantageous inequity (receiving more reward than the partner), and more rarely both forms of inequity.

So far we have described brain signals that track changes in subjective utility of own rewards as a function of the social context. Neurons in the striatum can also respond to another individual's actions, but only when resulting in own reward [67]; hence their "frame of reference," like in the OFC, remains self-centered and does not describe directly another individual's reward experience. Vicarious reward- and empathy-based accounts of prosocial behavior postulate the existence of mechanisms allowing an individual to perceive the welfare of others and potentially assign it a value based on self-experience. Such information could be encoded by other brain structures.

The amygdala is one such candidate, as it is implicated in various aspects of emotions, memory, and social information processing [1,68]. Primate amygdala neurons respond when animals observe emotional facial expressions of other monkeys [45] and engage in eye contact [45,69]. Furthermore the well-known effects of oxytocin (OT) on social approach, trust, and attachment in animals [70] and humans [71,72] could be mediated to a large extent via the connections of the amygdala with OT-rich structures such as the nucleus basalis of Meynert [73,74]. Chang et al. [75] reported that value-sensitive neurons in the macaque basolateral amygdala do not distinguish between obtained or observed rewards, showing similar response scaling as a function of reward size when a monkey chooses to grant juice to itself, its partner, or to itself and its partner, but not when the recipient was an inanimate agent (a drop of juice falling into a container). The authors thus argue for a motivational mirroring mechanism allowing the evaluation of rewards for self and others in a "common currency."

Social information about agency and reward destination also appears to be processed by the amygdala, as shown by preliminary findings from our laboratory. This region is known to be involved in the learning of associations between visual stimuli and reward or punishment [76]. Fig. 14.4 illustrates a response pattern that we recorded in the amygdala while two monkeys alternated in performing a reward allocation task. On some trials, monkeys responded to a visual cue associated with juice rewards to self, the partner, or nobody by a manual key press. This example neuron showed a brisk, short latency visual response to cues predicting rewards for the monkey's partner and for nobody (the same empty container condition as in the Chang et al. study), but it was inhibited by a cue predicting a reward for self. Interestingly, it responded much more strongly when the monkey was in the observer's than in the actor's role, hence providing information about the identity of the donor. Such signals could be interpreted as tracking the agent empowered with controlling access to the "resources," which in a certain sense can be considered as a proxy for social dominance [77]. Interestingly, the amygdala has been found to be part of a restricted set of brain areas showing a correlation between gray matter volume and position in the social hierarchy [78]. Other studies have shown that activity in the ACC distinguishes between self and other as the reward recipient [79], and medial prefrontal neurons were found to track rewards and errors made by monkey partners [80]. Taken together, these data indicate that the amygdala and the medial prefrontal cortex carry signals relevant for evaluating other's welfare, action, and social status.

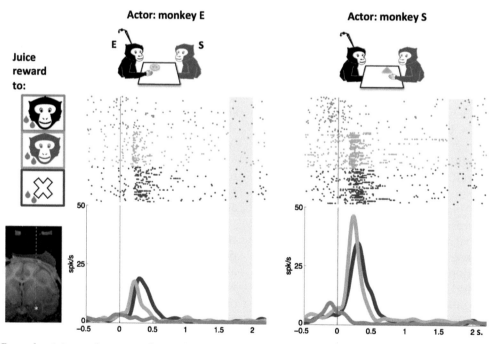

FIGURE 14.4   Reward recipient and agency coding in the macaque amygdala. Two monkeys alternated in performing an imperative task by touching a cue image predicting whether a juice reward would be granted to self, the partner, or nobody. Individual trial rasters (top) and mean spike density (bottom) for a single neuron recorded in the brain of monkey E, while either it or the partner monkey S performed the task, are shown. The neuron responds to cues predicting a reward to monkey S or to nobody, but not to monkey E, and responds more strongly when monkey E observed the task being performed rather than when performing it. Neuronal activity is aligned on cue onset, *blue-shaded rectangles* represent approximate time of outcome delivery.

These brain structures can thus provide key information about others that is required for empathy-based, prosocial behavior, but also, in other situations, for competition and cooperation. Haroush and Williams [81] used an iterated prisoner's dilemma game to investigate the capacity of monkeys to coordinate their actions. This game mimics a type of social interaction in which individuals must try to anticipate each other's move to pick the most beneficial, or least costly, course of action. When played multiple times, players can respond to each other's actions, retaliate against or forgive noncooperative behavior, and ideally, but not necessarily, converge on the mutually advantageous dominant strategy. The authors found that neuronal activity in the dorsal ACC accurately predicted other player's intention to cooperate or defect, and disruption of ACC activity by electrical stimulation reduced the preference for choosing the mutually beneficial option. In a context of virtual competition, in which monkeys played a video shooting game against each other, Hosokawa and Watanabe [82] identified neurons in the lateral prefrontal cortex (LPFC) that discriminated between task situations, responding more strongly to reward outcomes when playing against a monkey opponent than a computer opponent or no opponent at all. In a follow-up study, the same authors examined the neural mechanisms of competitive behavior under various reward contingencies. An interesting contrast was made between the usual outcome of a competition, in which the winner obtains the reward and the loser obtains nothing, and an egalitarian outcome in which both players are rewarded. Monkeys are obviously not natural-born collectivists, as their performance declined strikingly under the egalitarian reward condition [83]. At the neuronal level, the LPFC discriminated between the two competitive situations and showed little reward-related activity associated with winning in the egalitarian context. Together, these results suggest an important role for the ACC and LPFC in decision-making during competitive social interactions.

CONCLUSIONS AND PERSPECTIVES

The use of laboratory tasks mimicking real-world multiagent interactions of nonhuman primates is a relatively new direction in behavioral neuroscience. Although much work still has to be done, this approach highlights the role of several key players in social motivation and reward evaluation, and provides a fine-grained analysis of the neuronal computations performed by these structures. The picture remains patchy, in part because different experimental tasks have been used to study different areas. The specificity and the information flow among these areas are also far from

being elucidated. Nevertheless a rough sketch of the functional properties described in this chapter can be proposed. The OFC and striatum generate somewhat similar value signals that are weighted by social context, partner preference, and resource distribution and that may provide part of the motivational basis for social tolerance, approach, and avoidance. Value signals are also present in the amygdala, in which a class of cells responding to the value of both obtained and given rewards might be involved in affective state matching, social reward, and prosocial behavior. Finally, neurons in prefrontal areas like the ACC and LPFC, but also in the amygdala, distinguish between attributes of self and other, and represent information about agency and others' current intentions during social decision-making and could thus mediate complex decisions, such as whether and when to engage in cooperative or competitive behavior. The cognitive challenges that are typically being used in these experiments are simplified versions of the choices monkeys have to make when interacting with their conspecifics, and provide even more simplified models of complex human social cognitive processes. However, important features of primates' social decision-making are emerging, such as the fact that it does not necessarily follow a strict utility hypothesis and that everyday social decisions seem to rely, more than any other type of decisions, on affective processing. Indeed, it seems that nonhuman primates are able to create an elaborated representation of their social environment and thus express social preferences that reflect their desire to strengthen or weaken each unique social bond. Primates thus have expectations about the consequences of their behavior upon others' affect, but it is unclear to what extent nonhuman primates "know" about mental states of other live agents in the sense that humans do. Nevertheless, if one admits that a cognitive skill like theory of mind is not simply something one does or does not possess, but is assembled from more elementary abilities like social awareness and emotional empathy, monkeys clearly display some of these competences, at least in rudimentary form. Pursuing the study of the distal and ultimate bases of nonhuman primate's social decision-making will surely participate in fostering a unifying theory dealing with the roots of maladaptive and normal social behavior of humans.

References

[1] Cardinal RN, Parkinson JA, Hall J, Everitt BJ. Emotion and motivation: the role of the amygdala, ventral striatum, and prefrontal cortex. Neurosci Biobehav Rev 2002;26:321–52.
[2] Haber SN, Knutson B. The reward circuit: linking primate anatomy and human imaging. Neuropsychopharmacology 2010;35: 4–26.

[3] Rushworth MF, Mars RB, Sallet J. Are there specialized circuits for social cognition and are they unique to humans? Curr Opin Neurobiol 2013;23.

[4] Dunbar RIM, Shultz S. Evolution in the social brain. Science 2007; 317:1344—7.

[5] Sallet J, Mars RB, Noonan MP, Andersson JL, O'Reilly JX, Jbabdi S, et al. Social network size affects neural circuits in macaques. Science 2011;334:697—700.

[6] Bechara A, Damasio AR, Damasio H, Anderson SW. Insensitivity to future consequences following damage to human prefrontal cortex. Cognition 1994;50:7—15.

[7] Beer JS, John OP, Scabini D, Knight RT. Orbitofrontal cortex and social behavior: integrating self-monitoring and emotion-cognition interactions. J Cogn Neurosci 2006;18:871—9.

[8] Stolk A, D'Imperio D, di Pellegrino G, Toni I. Altered communicative decisions following ventromedial prefrontal lesions. Curr Biol 2015;25:1469—74.

[9] Moretti L, Dragone D, di Pellegrino G. Reward and social valuation deficits following ventromedial prefrontal damage. J Cogn Neurosci 2009;21:128—40.

[10] Babineau BA, Bliss-Moreau E, Machado CJ, Toscano JE, Mason WA, Amaral DG. Context—specific social behavior is altered by orbitofrontal cortex lesions in adult rhesus macaques. Neuroscience 2011;179:80—93.

[11] Machado CJ, Bachevalier J. Measuring reward assessment in a semi-naturalistic context: the effects of selective amygdala, orbital frontal or hippocampal lesions. Neuroscience 2007;148:599—611.

[12] Machado CJ, Bachevalier J. The impact of selective amygdala, orbital frontal cortex, or hippocampal formation lesions on established social relationships in rhesus monkeys (Macaca mulatta). Behav Neurosci 2006;120:761.

[13] Rudebeck PH, Buckley MJ, Walton ME, Rushworth MFS. A role for the macaque anterior cingulate gyrus in social valuation. Science 2006;313:1310—2.

[14] Amaral DG, Bauman MD, Capitanio JP, Lavenex P, Mason WA, Mauldin-Jourdain ML, et al. The amygdala: is it an essential component of the neural network for social cognition? Neuropsychologia 2003;41:517—22.

[15] Flombaum JI, Santos LR. Rhesus monkeys attribute perceptions to others. Curr Biol 2005;15:447—52.

[16] Santos LR, Nissen AG, Ferrugia JA. Rhesus monkeys, Macaca mulatta, know what others can and cannot hear. Anim Behav 2006;71:1175—81.

[17] Cheney D, Seyfarth R, Smuts B. Social relationships and social cognition in nonhuman primates. Science 1986;234:1361—6.

[18] Decety J. The neuroevolution of empathy: neuroevolution of empathy and concern. Ann NY Acad Sci 2011;1231:35—45.

[19] Preston SD, de Waal FB. Empathy: its ultimate and proximate bases. Behav Brain Sci 2002;25:1—20.

[20] de Waal FBM. The antiquity of empathy. Science 2012;336:874—6.

[21] Batson CD. These things called empathy: eight related but distinct phenomena. In: Decety J, Ickes W, editors. Soc. Neurosci. Empathy. Cambridge (MA): MIT Press; 2009. p. 3—15.

[22] Keverne EB, Martensz ND, Tuite B. Beta-endorphin concentrations in cerebrospinal fluid of monkeys are influenced by grooming relationships. Psychoneuroendocrinology 1989;14:155—61.

[23] Shimada M. Social object play among young Japanese macaques (Macaca fuscata) in Arashiyama, Japan. Primates 2006;47:342—9.

[24] Shutt K, MacLarnon A, Heistermann M, Semple S. Grooming in Barbary macaques: better to give than to receive? Biol Lett 2007; 3:231—3.

[25] Masserman JH, Wechkin S, Terris W. "Altruistic" behavior in rhesus monkeys. Am J Psychiatry 1964;121:584—5.

[26] Miller RE, Banks Jr JH, Kuwahara H. The communication of affects in monkeys: cooperative reward conditioning. J Genet Psychol 1966;108:121—34.

[27] Bartal IB-A, Decety J, Mason P. Empathy and pro-social behavior in rats. Science 2011;334:1427—30.

[28] Atsak P, Orre M, Bakker P, Cerliani L, Roozendaal B, Gazzola V, et al. Experience modulates vicarious freezing in rats: a model for empathy. PLoS One 2011;6:e21855.

[29] Ben-Ami Bartal I, Rodgers DA, Bernardez Sarria MS, Decety J, Mason P. Pro-social behavior in rats is modulated by social experience. Elife 2014;3:e01385.

[30] Massen JJM, van den Berg LM, Spruijt BM, Sterck EHM. Generous leaders and selfish underdogs: pro-sociality in despotic macaques. PLoS One 2010;5:e9734.

[31] Chang SWC, Winecoff AA, Platt ML. Vicarious reinforcement in rhesus macaques (Macaca mulatta). Front Neurosci 2011;5.

[32] Ballesta S, Duhamel J-R. Rudimentary empathy in macaques' social decision-making. Proc Natl Acad Sci USA 2015;112:15516—21.

[33] Preston SD. The origins of altruism in offspring care. Psychol Bull 2013;139:1305—41.

[34] Aknin LB, Hamlin JK, Dunn EW. Giving leads to happiness in young children. PLoS One 2012;7:e39211.

[35] Jackson PL, Meltzoff AN, Decety J. How do we perceive the pain of others? A window into the neural processes involved in empathy. NeuroImage 2005;24:771—9.

[36] Gallese V. Before and below "theory of mind": embodied simulation and the neural correlates of social cognition. Philos Trans R Soc B Biol Sci 2007;362:659—69.

[37] Ballesta S, Reymond G, Pozzobon M, Duhamel J-R. A real-time 3D video tracking system for monitoring primate groups. J Neurosci Methods 2014;234.

[38] Cords M. Post-conflict reunions and reconciliation in long-tailed macaques. Anim Behav 1992;44:57—61.

[39] Widdig A, Streich WJ, Tembrock G. Coalition formation among male Barbary macaques (Macaca sylvanus). Am J Primatol 2000; 50:37—51.

[40] Nahm FK, Perret A, Amaral DG, Albright TD. How do monkeys look at faces? J Cogn Neurosci 1997;9:611—23.

[41] Dahl CD, Wallraven C, Bülthoff HH, Logothetis NK. Humans and macaques employ similar face-processing strategies. Curr Biol 2009;19:509—13.

[42] Mosher CP, Zimmerman PE, Gothard KM. Videos of conspecifics elicit interactive looking patterns and facial expressions in monkeys. Behav Neurosci 2011;125:639—52.

[43] Watson KK, Platt ML. Social signals in primate orbitofrontal cortex. Curr Biol 2012;22:2268—73.

[44] Perrett DI, Hietanen JK, Oram MW, Benson PJ. Organization and functions of cells responsive to faces in the temporal cortex. Philos Trans R Soc Lond B Biol Sci 1992;335:23—30.

[45] Gothard KM, Battaglia FP, Erickson CA, Spitler KM, Amaral DG. Neural responses to facial expression and face identity in the monkey amygdala. J Neurophysiol 2007;97:1671—83.

[46] Sliwa J, Duhamel J-R, Pascalis O, Wirth S. Spontaneous voice-face identity matching by rhesus monkeys for familiar conspecifics and humans. Proc Natl Acad Sci USA 2011;108:1735—40.

[47] Sliwa J, Planté A, Duhamel J-R, Wirth S. Independent neuronal representation of facial and vocal identity in the monkey hippocampus and inferotemporal cortex. Cereb Cortex 2016;26:950—66.

[48] Schultz W. Neuronal reward and decision signals: from theories to data. Physiol Rev 2015;95:853—951.

[49] Rolls ET, Hornak J, Wade D, McGrath J. Emotion-related learning in patients with social and emotional changes associated with frontal lobe damage. J Neurol Neurosurg Psychiatry 1994;57: 1518—24.

[50] Rolls ET. The orbitofrontal cortex and reward. Cereb Cortex 2000; 10:284—94.

[51] Morrison SE, Salzman CD. The convergence of information about rewarding and aversive stimuli in single neurons. J Neurosci 2009; 29:11471—83.

[52] Tremblay L, Schultz W. Relative reward preference in primate orbitofrontal cortex. Nature 1999;398:704−8.

[53] Padoa-Schioppa C, Assad JA. Neurons in the orbitofrontal cortex encode economic value. Nature 2006;441:223−6.

[54] Azzi JCB, Sirigu A, Duhamel J-R. Modulation of value representation by social context in the primate orbitofrontal cortex. Proc Natl Acad Sci USA 2012;109:2126−31.

[55] Marx K. Wage labor and capital. Neue Rheinische Zeitung; 1849.

[56] Adams JS. Inequity in social exchange. Adv Exp Soc Psychol 1965;2.

[57] LoBue V, Nishida T, Chiong C, DeLoache JS, Haidt J. When getting something good is bad: even three-year-olds react to inequality. Soc Dev 2011;20:154−70.

[58] Brosnan SF. Justice-and fairness-related behaviors in nonhuman primates. Proc Natl Acad Sci USA 2013;110:10416−23.

[59] Brosnan SF, de Waal FBM. Monkeys reject unequal pay. Nature 2003;425:297−9.

[60] Burkart JM, Fehr E, Efferson C, Van Schaik CP. Other-regarding preferences in a non-human primate: common marmosets provision food altruistically. Proc Natl Acad Sci USA 2007;104: 19762.

[61] Massen JJM, Van Den Berg LM, Spruijt BM, Sterck EHM. Inequity aversion in relation to effort and relationship quality in long-tailed Macaques (*Macaca fascicularis*). Am J Primatol 2012;74:145−56.

[62] Proctor D, Williamson RA, de Waal FBM, Brosnan SF. Chimpanzees play the ultimatum game. Proc Natl Acad Sci USA 2013; 110:2070−5.

[63] Báez-Mendoza R, van Coeverden CR, Schultz W. A neuronal reward inequity signal in primate striatum. J Neurophysiol 2016;115:68−79.

[64] Hikosaka O, Sakamoto M, Usui S. Functional properties of monkey caudate neurons. III. Activities related to expectation of target and reward. J Neurophysiol 1989;61:814−32.

[65] Apicella P, Scarnati E, Ljungberg T, Schultz W. Neuronal activity in monkey striatum related to the expectation of predictable environmental events. J Neurophysiol 1992;68:945−60.

[66] Cromwell HC, Schultz W. Effects of expectations for different reward magnitudes on neuronal activity in primate striatum. J Neurophysiol 2003;89:2823−38.

[67] Báez-Mendoza R, Schultz W. The role of the striatum in social behavior. Front Neurosci 2013;7.

[68] Phelps EA. Emotion and cognition: insights from studies of the human amygdala. Annu Rev Psychol 2006;57:27−53.

[69] Mosher CP, Zimmerman PE, Gothard KM. Neurons in the monkey amygdala detect eye contact during naturalistic social interactions. Curr Biol 2014;24:2459−64.

[70] Insel TR, Young LJ. The neurobiology of attachment. Nat Rev Neurosci 2001;2:129−36.

[71] Guastella AJ, Mitchell PB, Mathews F. Oxytocin enhances the encoding of positive social memories in humans. Biol Psychiatry 2008;64:256−8.

[72] Kosfeld M, Heinrichs M, Zak PJ, Fischbacher U, Fehr E. Oxytocin increases trust in humans. Nature 2005;435:673−6.

[73] Freeman SM, Inoue K, Smith AL, Goodman MM, Young LJ. The neuroanatomical distribution of oxytocin receptor binding and mRNA in the male rhesus macaque (*Macaca mulatta*). Psychoneuroendocrinology 2014;45:128−41.

[74] Knobloch HS, Charlet A, Hoffmann LC, Eliava M, Khrulev S, Cetin AH, et al. Evoked axonal oxytocin release in the central amygdala attenuates fear response. Neuron 2012;73:553−66.

[75] Chang SWC, Fagan NA, Toda K, Utevsky AV, Pearson JM, Platt ML. Neural mechanisms of social decision-making in the primate amygdala. Proc Natl Acad Sci USA 2015;112:16012−7.

[76] Sanghera MK, Rolls ET, Roper-Hall A. Visual responses of neurons in the dorsolateral amygdala of the alert monkey. Exp Neurol 1979;63:610−26.

[77] Fruteau C, Voelkl B, Van Damme E, Noë R. Supply and demand determine the market value of food providers in wild vervet monkeys. Proc Natl Acad Sci USA 2009;106:12007−12.

[78] Noonan MP, Sallet J, Mars RB, Neubert FX, O'Reilly JX, Andersson JL, et al. A neural circuit covarying with social hierarchy in macaques. PLoS Biol 2014;12:e1001940.

[79] Chang SWC, Gariépy J-F, Platt ML. Neuronal reference frames for social decisions in primate frontal cortex. Nat Neurosci 2012;16: 243−50.

[80] Yoshida K, Saito N, Iriki A, Isoda M. Social error monitoring in macaque frontal cortex. Nat Neurosci 2012;15:1307−12.

[81] Haroush K, Williams ZM. Neuronal prediction of opponent's behavior during cooperative social interchange in primates. Cell 2015;160:1233−45.

[82] Hosokawa T, Watanabe M. Prefrontal neurons represent winning and losing during competitive video shooting games between monkeys. J Neurosci 2012;32:7662−71.

[83] Hosokawa T, Watanabe M. Egalitarian reward contingency in competitive games and primate prefrontal neuronal activity. Front Neurosci 2015;9:165.

15

Organization of the Social Brain in Macaques and Humans

M.P. Noonan[a], R.B. Mars[a], F.X. Neubert, B. Ahmed, J. Smith, K. Krug, J. Sallet[a]

University of Oxford, Oxford, United Kingdom

Abstract

Human social life has changed dramatically in the past 100 years, as first advances in transport and later the Internet allowed us to interact with a much greater and more diverse group of people. As a result, even the term "social networks" has a profound new meaning in the 21st century. The human species is now more connected than ever, and we live in a world in which, for better or worse, we can communicate our thoughts and intentions to vast numbers of our conspecifics, instantly. Yet while the apps behind this revolution are upgraded each year, the neural hardware that supports social behavior evolves over millennia. This chapter will explore the evidence that our social brain and the brains of our less-technology-savvy cousins may be surprisingly similar.

INTRODUCTION

Human social life has changed dramatically in the past 100 years, as first advances in transport and later the Internet allowed us to interact with a much greater and more diverse group of people. As a result, even the term "social networks" has a profound new meaning in the 21st century. The human species is now more connected than ever, and we live in a world in which, for better or worse, we can communicate our thoughts and intentions to vast numbers of our conspecifics, instantly. Yet while the apps behind this revolution are upgraded each year, the neural hardware that supports social behavior evolves over millennia. This chapter will explore the evidence that our social brain and the brains of our less-technology-savvy cousins may be surprisingly similar.

Over the course of primate evolution better social abilities may have helped primates cooperate among conspecifics and together deal with predators and prey. The advanced social abilities of humans and other primates have been related to the large increase in brain size. The ratio of brain to body size is correlated with the number of individuals per social group, a variable that indexes the social complexity of a species' life [20]. While social behavior is intricate and multifaceted in nature, the size of the individual's social network is a useful and well-validated index [30,40] correlating with emotional intelligence and mentalizing abilities [85]. Social network size (SNS) reflects not only species differences but also differences in brain structure between individuals of the same species. In humans, correlates are reported between measures of SNS and gray matter (GM) volume in the amygdala and subregions of the temporal and frontal cortex [6,36,90]. Furthermore, blood oxygen level-dependent (BOLD) activity in these regions, measured while subjects made social closeness judgments, also correlates with individuals' SNS [90]. However, while neuroimaging studies in humans can allude to a network of brain regions involved in social cognition, they cannot reveal the directionality of structure—function relationships.

It is well established that the social environment during the time of development has a causal influence on behavior and brain structure [14]. We showed that changes in the social environment could cause changes in *adult* brains of nonhuman primates [74] as well. We manipulated the size of groups in which animals were housed and related this to changes in the size of

---

[a] Authors contributed equally to this work.

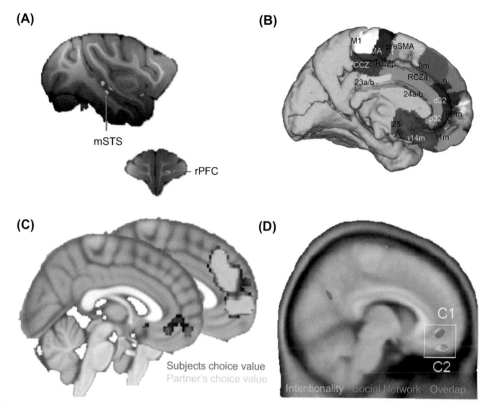

FIGURE 15.1   (A) Animals housed in large social groups had more gray matter volume in bilateral mid-superior temporal sulcus (mSTS) and rostral prefrontal cortex (rPFC). *(Adapted from Sallet J, Mars RB, Noonan MP, Andersson JL, O'Reilly JX, Jbabdi S, et al. Social network size affects neural circuits in macaques. Science 2011;334:697–700.)* (B) Subdivisions of the medial prefrontal cortex (MPFC). *(Adapted from Neubert FX, Mars RB, Sallet J, Rushworth MF. Connectivity reveals relationship of brain areas for reward-guided learning and decision-making in human and monkey frontal cortex. Proc Natl Acad Sci USA 2015;112:E2695–704.)* (C) MPFC blood oxygen level-dependent signal correlating modeled value of choices of a conspecific *(yellow)* and self-referential choices *(green). (Adapted from Nicolle A, Klein-Flügge MC, Hunt LT, Vlaev I, Dolan RJ, Behrens TE. An agent independent axis for executed and modeled choice in medial prefrontal cortex. Neuron 2012;75:1114–21.)* (D) Individual differences in gray matter volume correlate with differences in mentalizing abilities and social network size. *(Adapted from Lewis PA, Rezaie R, Brown R, Roberts N, Dunbar RI. Ventromedial prefrontal volume predicts understanding of others and social network size. Neuroimage 2011;57:1624–29.)*

specific brain areas (Fig. 15.1A). Using the same imaging techniques and analysis methods as in the aforementioned human studies, correlations with SNS were observed in a limited number of brain regions that resemble the human regions, particularly in the temporal lobe and the medial prefrontal cortex (MPFC). Collectively this body of work raises the possibility that the extent of similarity between macaque and human social brains is underestimated. However, testing this proposition is difficult, as functional imaging studies aimed at comparing activity profiles of areas in the human and macaque brain are limited by the complexity of the tasks that macaques can perform in the scanner. This has led some (e.g., Ref. [42]) to question whether these debates can be resolved with behavioral experiments alone. We propose that interpretations of different patterns of activation elicited by social tasks, or different deficits induced by specific lesions, can be better understood by establishing the foundations of the social brain in the two species by directly comparing regional changes in structure and connectivity.

This chapter focuses on brain structure and functional connectivity of two key areas of the social brain: the MPFC and the temporal cortex (particularly the temporal parietal junction, TPJ). Our approach is based on a combination of magnetic resonance imaging (MRI)-based methods. Structural MRI allows us to identify brain regions where individual differences in GM volume correlate with indices of sociocognitive factors. By contrast, diffusion-weighted imaging (DWI) and resting state functional MRI (rsfMRI) can be used to determine the connectivity-based organization of regions of interest [47,57,58,75].

MEDIAL PREFRONTAL CORTEX

The MPFC is associated with the control of social behavior [17]. Case histories describe patients with damage to this region as having "acquired sociopathy," manifesting social impairments such as increases in the expression of socially inappropriate behavior and aggression, as well as the tendency to misinterpret

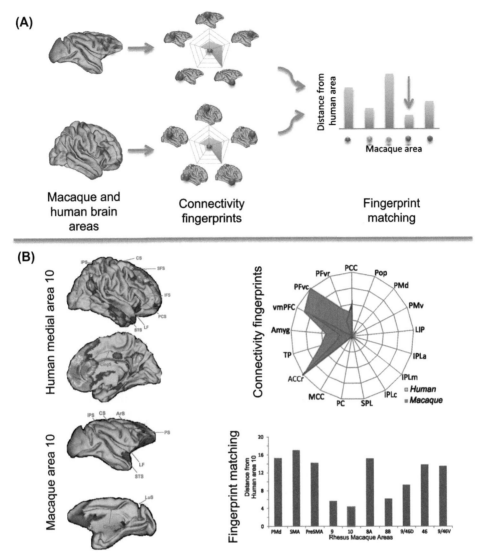

FIGURE 15.2 Connectivity-based comparison of human and macaque social areas in the medial prefrontal cortex. (A) Our strategy for identifying potential human/macaque homologs. We exploit the fact that each cortical area has a unique set of connections with the rest of the brain, termed its connectivity fingerprint (cf Ref. [63]). To identify the homolog of a human area (*red dot*) we chart its connectivity to a set of targets (*blue*) with known macaque homologies. We then define the connectivity fingerprint for a set of candidate areas in the macaque, with the same targets, and calculate a distance measure indicating how much each candidate's connectivity fingerprint differs from that of the human template. The macaque area with the smallest distance (indicated by the *arrow*) is the most likely candidate for between-species homology. (*Based on Mars RB, Verhagen L, Gladwin TE, Neubert FX, Sallet J, Rushworth MFS. Comparing brains by matching connectivity fingerprints (submitted for publication-b).)* (B) This approach was applied to area 10 in the medial frontal pole of the human brain. We determined its connectivity fingerprint (*top right*) using resting state fMRI (*top left*). The same was done for candidate areas in the macaque. The connectivity fingerprint of macaque area 10 best matched that of the human frontal pole, suggesting that even high-level areas share features between species. (*Adapted from Sallet J, Mars RB, Noonan MP, Neubert FX, Jbabdi S, O'Reilly JX, et al. The organization of dorsal frontal cortex in humans and macaques. J Neurosci 2013;33:12255–74.*)

other people's moods [31,32] and facial and vocal emotional expressions [28,32]. However, the MPFC is far from a unitary structure but refers to a collection of cytoarchitectonically distinct areas (Fig. 15.1B) with evidence that subregions support different computations [82]. A better understanding of the role of the MPFC in social cognition requires a careful examination of the subregions of this large cortical territory and an understanding of the distribution of sociocognitive function across these anatomical subregions (see Ref. [4,84]). This chapter will discuss which structural

subdivisions might contribute to social cognition for future investigations.

In the human neuroimaging literature, the MPFC has become synonymous with "theory of mind," which is the act of attributing thoughts and feelings to others [1]. This skill is thought to be particularly well developed in the human [77]. The cytoarchitectonic areas 9 medial and 10 medial (or Frontopolar cortex, medial subdivisions (FPm)) are the two regions of the MPFC that are most often reported in fMRI studies in which subjects attempt to infer the intentions or beliefs of others [1,59]. Other authors have emphasized a more ventral region, corresponding to cytoarchitectonic areas 11, 14, and p32 (Fig. 15.1C), as involved in mentalizing process [18,59]. Individual differences in GM volume in this ventromedial region also correlate with differences in mentalizing abilities, as do differences in SNS (Fig. 15.1D) [40,60].

Other parts of the MPFC have also been implicated in various aspects of social cognition. We find that GM volume in the dorsal anterior cingulate cortex (ACC) sulcus correlates with SNS both in humans and in monkeys [74,60]. Adjacent to it, the cortex of the cingulate gyrus is often shown to be recruited in tasks that require subjects to monitor the outcomes of other's decisions [2] or the reliability of information provided by a confederate [5]. An elegant study in monkeys by Rudebeck and colleagues [72] showed that macaques with anterior cingulate gyrus lesions (areas 24a,b and 32) lose interest in social stimuli. Intriguingly, in light of human neuroimaging results, lesions to the ventromedial cortex (area 11/14) do not induce the same social impairment [62].

A number of studies have investigated socioemotive functions in macaques [27,33] and the pattern is reminiscent of findings from human patients with MPFC lesions and human neuroimaging studies. However, when differences have been identified they have sometimes been thought to reflect uniquely human brain regions supporting uniquely human social behaviors [38,73]. Yet the uniqueness of human sociocognitive functions is still debated [29,42]. Direct between-species comparisons of function are challenging, but anatomical comparisons can provide the tools to establish homologies in the architecture of the social brain between the two species. Therefore, we have used different measures of connectivity to compare the organization of the MPFC between humans and macaques.

The connections of a brain region with the rest of the brain define a unique connectivity fingerprint [63], which can be used to compare regions across species ([51], see Fig. 15.2A [63], for description of technique). Using this technique, we found that the cortical areas composing human MPFC share similar connectivity patterns with areas of the macaque prefrontal cortex. Full regional comparisons are described in Refs. [75] and [58] but for the purpose of this chapter we focus on medial area 10(or FPm) to illustrate the principle of the comparative technique. We used diffusion MRI to identify the boundaries between cortical regions of the human MPFC and then used rsfMRI to compare the interactions or functional connectivity of these human areas with those of MPFC areas in the macaque. Analysis of the macaque data showed that area 10 is functionally coupled with other MPFC regions, with the temporal pole and cortex of the superior temporal sulcus (STS)

as well as with posterior cingulate regions (see Fig. 15.2B). Classic tracing studies in monkeys have revealed similar patterns of connections [10,45,65]. Human medial area 10 matched this profile best, showing functional connectivity with the ventral PFC and anterior temporal and posterior cingulate areas.

As well as identifying structural homologs between species, it is informative to study how different brain regions relate to one another. For instance, visual areas are often described as organized hierarchically and their position in this hierarchy provides clues to their function. Cluster analysis of the connectivity patterns of MPFC areas suggest that these regions can be separated into distinct networks of regions [3,57]. Based on their connectivity pattern, the cingulate (area 24) and dorsomedial cortex (medial area 9) could be grouped, and the ventromedial regions clustered together (area 11 and 14). Note that there is no consensus regarding areas 32 and 10. Collectively, it appears that the relationship between MPFC regions, i.e., their place in the cortical hierarchy, is also similar between the two species. These network divisions may reflect and define their different functional roles in learning from others' actions and comparing social choices. For example, Nicolle and colleagues [59] report that when subjects are asked to make self-referential choices in an fMRI scanner, signals from ventromedial prefrontal areas (32/25, 14, and 10) correlate with the choice of the agent, whereas signals from the rostral dorsomedial PFC (10/9) reflected the modeled choices of a conspecific. However, when the subjects chose on behalf of their partner the roles of these regions were swapped (Fig. 15.1C).

Finally, there are similarities across species in terms of the whole-brain networks that these clusters participate in. There is a distinct common pattern of connections for areas 9, 10, 32, 11, and 14 within the MPFC and the rest of the brain. These regions are monosynaptically connected with other regions of the MPFC, the temporal pole, and posterior cingulate cortex (PCC) [93]. These areas are referred to as the default mode network (DMN), a resting state network typically isolated in task-negative contrasts, with more BOLD signal in these regions when subjects are resting between task blocks in the scanner (see Refs. [9,69]). The DMN arguably reflects the default mode of social animals' brain function, that of coordinating behavior within a social context, which grows in demand with larger SNSs [79]. Moreover, the typical pattern of DMN brain activity partly reflects that seen during tasks of social cognition, mentalizing, and autobiographical memory [48].

Beyond the species similarities detailed above, there are also distinct differences between human and macaque prefrontal–temporal brain connectivity. Whereas functional connections between temporal cortex and lateral prefrontal cortex are stronger in humans compared to connections with MPFC in humans, the opposite pattern is observed in macaques [58]. Interestingly, reduced functional connectivity between frontal and temporal cortex has been observed in patients suffering from autism during the performance of a nonlinguistic theory of mind task, compared to a control group [35].

SUPERIOR TEMPORAL SULCUS AND TEMPORAL PARIETAL JUNCTION

As discussed previously, early human studies noted that GM in temporal cortex and the amygdala differed as a function of social complexity [7]. This modulation of amygdala GM by SNS was replicated in our macaque study [74] and subsequent human studies [36,40,60]. Interestingly, in macaques we found that amygdala GM was also associated with social status, an uncorrelated social variable [61]. However, the effect of SNS in amygdala GM was dorsal of the social status effect (see also Chapter 20 for the neural mechanisms underlying learning of social status in humans). This might reflect the diverse, dissociable nuclei of the amygdala, which have distinct connectivity profiles [24], and may support different roles in social cognition [6,56].

In the rest of the temporal cortex, there are many areas that respond to socially relevant stimuli—especially faces—and that correlate with SNS, in both humans and macaques. However, because the temporal cortex has undergone substantial reorganization since the last common ancestor [88], the identification of homologs between species is difficult. Some groups now use fMRI in awake macaque monkeys to give them face processing tasks similar to those given humans, to relate the different patches of activation in the two species, reporting potential homologs (e.g., Ref. [68]).

In the human brain, a prominent locus of socially related GM changes is in the posterior end of the superior temporal cortex, at the junction with the parietal and occipital cortex (cf Ref. [40]). It is known as the TPJ and has been shown to be active during higher order social reasoning tasks [78,82]. Indeed, TPJ GM changes are not correlated with SNS, but rather the ability to perform recursive social reasoning [40], which may enable complex chains of social inference ("I know that Mary thinks that John would like David to know that Gary wants..."). This ability may be unique, or at least more extensive, in humans [15]. Echoing this, the TPJ has been described as involved in "uniquely human social cognition" [77].

Despite the prominence of TPJ in the social literature, little is known about its precise anatomy. A number of authors have suggested that TPJ is not a separate region specifically involved in social cognition, but overlaps with another region also termed TPJ that is often reported when subjects reorient their attention in visual space. A number of studies either using metaanalyses of functional imaging studies [19] or testing both types of tasks in the same participants [54,81] could not resolve this controversy. However, despite similar loci of activity in the TPJ during processing of social information and reorienting of visual attention, the network of areas coactivated with TPJ in the two tasks is quite different, showing, among other regions, anterior MPFC during mentalizing task [11] but ventrolateral prefrontal cortex (VLPFC) during attentional reorienting [23]. This inspired Mars and colleagues to test the hypothesis that the large expanse of cortex termed "TPJ," as MPFC earlier, also consists of different subregions connected to different parts of cortex. Using a DWI tractography-based parcellation of TPJ, they indeed found that the posterior part of TPJ connects with nodes of the social brain, including anterior temporal cortex, PCC, and MPFC, whereas the anterior part of TPJ interacted with areas more often associated with attentional control, including the mid-cingulate cortex, anterior insula, and VLPFC [50]. Similar results were obtained by Bzdok et al. [12] using a connectivity-based parcellation of metaanalysis data.

Given the uncertainties of the anatomical properties of the TPJ and the fact that it is associated with activity during higher order social cognitive tasks that are thought to be uniquely human, the TPJ has not been linked to a monkey homolog. Because macaques cannot perform the same tasks that activate the human TPJ, a different approach was required. Taking an approach similar to that used when studying the human TPJ, Mars and colleagues [49] investigated whether TPJ can be matched to any macaque area based on anatomical grounds. They used a variant of the connectivity matching approach described earlier for frontal cortical areas and searched along the entire macaque brain for voxels that had a functional connectivity profile with the anterior, middle, and posterior cingulate cortex and the anterior insula/VLPFC similar to that of the human posterior TPJ. This approach identified the mid-STS (mSTS) as the macaque area with the most similar connectivity fingerprint (Fig. 15.3).

The area identified as anatomically similar to human TPJ was the same region that showed increased GM density in macaques living in larger social groups (see Ref. [74]). Interestingly, although identified from face processing in macaques, Perrett and colleagues [64] demonstrated that the cortex of the STS contains neurons that respond when a pair of eyes look in a particular location. If the eyes are not visible, then the neurons respond when the head is oriented in that direction. If the head is not visible, then the neurons respond to clues based on body posture. Thus, the mSTS seems to code for the focus of a conspecific's attention, rather than

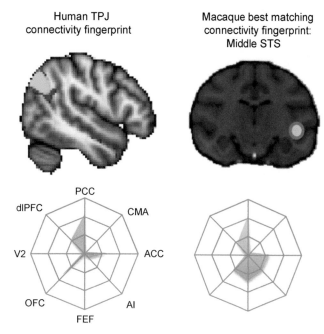

Human TPJ connectivity fingerprint

Macaque best matching connectivity fingerprint: Middle STS

FIGURE 15.3   Similarity between human temporoparietal junction (TPJ) and an area in the macaque superior temporal sulcus (STS). The human TPJ (*top left*) is active during mentalizing tasks suggested to be uniquely human. Using the techniques outlined in Fig. 15.2, we sought to establish whether this region nevertheless had an anatomical homolog in the macaque. Using resting state fMRI, we determined the connectivity fingerprint of the human TPJ (*bottom left*) and searched for the best matching connectivity fingerprint from a set of 32 candidate areas along the macaque STS and inferior parietal cortex. An area in the middle part of the STS (*top right*) had the most similar connectivity fingerprint (*bottom right*). A control analysis searched across all voxels in the macaque temporoparietal cortex using a more restricted connectivity fingerprint, and identified the same areas. *Adapted from Mars RB, Sallet J, Neubert FX, Rushworth MF. Connectivity profiles reveal the relationship between brain areas for social cognition in human and monkey temporoparietal cortex. Proc Natl Acad Sci USA 2013;110:10806—11.*

facial identity; this being more specifically supported by the inferotemporal cortex face patches system [25]. Such an ability might be considered a prerequisite of our human mentalizing ability. Without knowing the focus of another's attention, it is impossible to learn about his or her beliefs and desires.

A SOCIAL BRAIN NETWORK

We have now described the organization of the MPFC and TPJ in the social brain, but these areas do not operate in isolation. An outstanding question is how they interact within a larger "social brain network" to produce behavior. It has been argued that pathologies in social cognition might be the result of dysfunctional interactions between brain regions [8]. Our own work using rsfMRI in macaques established that functional connectivity between these two nodes was causally related to SNS [74]. Within this premise we examined

the contribution of the MPFC to larger distributed cortical networks, focusing on the DMN. In monkeys, and more recently in humans, we showed that changes in interareal couplings between the MPFC and the DMN relate to SNS [48,60].

While it is useful to characterize a network by its functional connectivity, the very nature of the method means we cannot address which anatomical white matter (WM) pathways support the transfer of neural information. Using DWI, a 2015 paper investigated the correlation between social network diversity and WM microstructural integrity in humans [55]. The results identify the corpus callosum, cingulum bundle, and hypothalamic pathways in a large group of subjects. Extending these findings, we found evidence that differences in the structural integrity of the cingulum bundle, extreme capsule (EmC), and arcuate fasciculus relate to SNS [60]. These WM pathways (Fig. 15.4) support frontal—temporal communication, and we have demonstrated with probabilistic tractography that the network connectedness of social GM nodes utilizes them.

An important next step will be to analyze the extent and connective profiles of WM bundles themselves. For example, whereas macaque tracing studies and human DWI studies show similar connections between superior temporal and lateral prefrontal cortex [26,66,80], the projection areas in the posterior part of the cortex may be unique to humans [22,86]. However, using DWI in both humans and macaques, Mars and colleagues [46] compared the course and cortical projections of WM fibers passing through the EmC. In both species, the EmC innervated a number of social brain regions in frontal cortical areas (areas 9 and 10) and superior temporal cortex, including the human TPJ. Notably, the authors observed some tracts that were not commonly reported in macaque studies. These tracts resemble those previously reported in the human, suggesting larger similarities between the species than originally thought.

The question remains from where the MPFC and mSTS receive their social information. The PCC and its associated neighbor, the precuneus (PCun), are prominent DMN nodes and often active in mentalizing and Theory of Mind (TOM) [44]. We observed GM changes as a function of SNS in the PCun/PCC [60]. The region is strongly connected to the MPFC and superior temporal cortex [9] and we report its interconnectedness to the social brain through some of the WM tracts found by Noonan et al. [60].

Gold standard tracing studies can further refine our understanding of the connectivity of these regions. For example, the macaque mSTS interconnects with adjacent cytoarchitectonic areas of the STS [83] (see also Fig. 15.5 [49]). It projects to the frontal lobes, including MPFC regions [41,45], but also to the inferior parietal lobe [71]. Note that projections to the frontal lobes seem to come

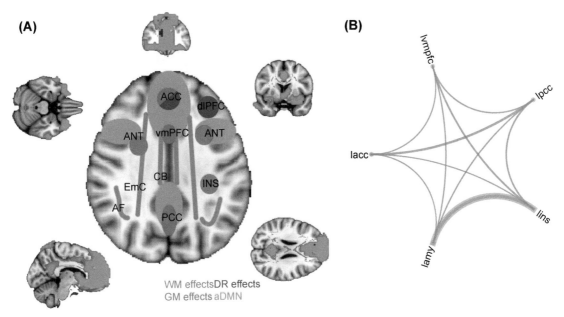

FIGURE 15.4 Summary figures of white matter (WM), gray matter (GM), and functional connectivity as a function of social network size (SNS) in humans. (A) Using diffusion-weighted imaging (DWI) we found that differences in the structural integrity of specific WM tracts—including cingulum bundle (CB), extreme capsule (EmC), arcuate fasciculus (AF), and corpus callosum—correlated with SNS measured over 30 days (*purple tracts*). Voxel-based morphology analysis demonstrated correlations between GM volume (*red-yellow*) and SNS in anterior cingulate cortex (ACC), ventromedial prefrontal cortex (vmPFC), and anterior temporal cortex. Finally, resting state fMRI demonstrated that the ACC and dorsolateral prefrontal cortex (dlPFC; *blue*) changed their functional contribution to the frontal component of the default mode network (aDMN; *green*) depending on SNS. (B) Probabilistic tractography seeded in the GM nodes, with the results of the DWI analysis used as waypoints, shows the proportion of tracts to reach each target seed in the left hemisphere. For each analysis the edges represent significant ($p < .05$) t-statistics against zero ($2.16 \leq t \leq 27.38$). (*Adapted from Noonan MP, Mars MB, Sallet J, Dunbar RIM, Fellows LK. The structural and functional brain networks that support human social networks (submitted for publication).*)

principally from the cortex on the dorsal bank of the STS. Fig. 15.5 illustrates the results of our work in which the retrograde tracer fluorogold was injected into the dorsal bank of the macaque mSTS. We show that this region receives monosynaptic input from both the dorsal visual area MT, linked to visual motion and depth perception [39], and the ventral visual pathway (area V4d), associated with object shape and color processing [70]. Altogether the connectivity of the macaque mSTS shows that this region is the nexus of several information streams.

SUMMARY AND PERSPECTIVES

We have described two brain regions associated with social cognition in both humans and macaques, the MPFC and the temporal cortex, in terms of their structure and their local/whole-brain connectivity. Despite the evolutionary expansion of the human brain and the increased complexity of our social networks, the results discussed in this chapter generally indicate that the brain networks supporting social cognition comprise similar building blocks in the human and in the macaque monkey. Although there are substantial differences in brain size and in the relative expansion of

certain human brain areas (e.g., Ref. [88]), our research suggests principal similarities between the two social brains. However, humans display some social behaviors that are quite different from macaque behaviors. Tomasello and colleagues, for instance, have argued that humans are unique in the collaborative nature of their social interactions [87]. The question, then, is what changed to allow us to display these behaviors, if most of the neural building blocks are similar?

Several authors have proposed computational processes that our brain might be uniquely capable of implementing, or at least of implementing much better. As discussed, the human ability to infer the mental states of others far outstrips any such ability in the macaque, and may be due to our ability to process information recursively ("I think, that he thinks, that I think, that he thinks...") [15]. Performance of such recursive tasks is indeed associated with activity in some of the regions highlighted earlier, in particular in the MPFC [16]. The hypothesis described earlier, that the mSTS region tracking social signals is anatomically similar to the human TPJ region associated with attributing belief states to others might be seen in this light, as suggesting that we process similar sensory information about others in a way similar to that of macaques but to a deeper extent (cf Ref. [21]).

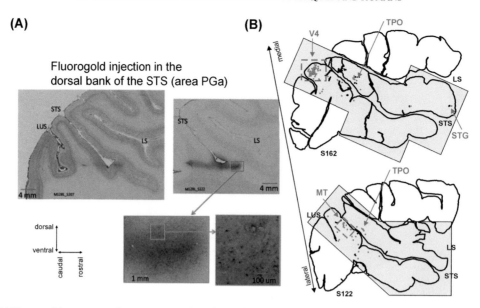

FIGURE 15.5 (A) Fluorogold, a retrograde tracer, was injected into the dorsal bank of a macaque's superior temporal sulcus (STS). Using Brainsight (Rogue Research, Montreal, QC, Canada), Mars et al. [49] showed that the injection site corresponded to monkey temporoparietal junction or mid-superior temporal sulcus. The panels illustrate the injection site and the stained cells at different levels of magnification on a parasagittal slice. (B) Parasagittal stained sections were examined under a light microscope using Neurolucida (MBF Bioscience, Williston, VT, USA). Fluorogold-stained cells (*red dots*) were found in dorsal V4 and middle temporal area (MT), indicating monosynaptic projections to the mSTS from both the dorsal (area MT) and the ventral visual pathways (area V4d). The *gray area* represents the area investigated for fluorogold-stained cells. *LS*, lateral sulcus; *LUS*, lunate sulcus.

While we have emphasized the similarities between species using MRI-based methods, we acknowledge that some disparity might be explained by methodological difference. For instance rsfMRI is usually recorded from humans at rest, yet macaques are usually scanned under anesthesia. General anesthesia is characterized by a decrease in spiking activity that results in a reduction in coupling between brain regions [89]. Therefore the difference between connectivity parameters, such as high-degree and betweenness centrality of the MPFC, in anesthetized macaques and awake humans should be considered carefully [53]. Despite this, our results continue to be upheld when examined with other techniques. For instance, the dissociation of human and macaque long-range connections to lateral prefrontal is mimicked in tracer and DWI data [10,67].

Finally, ending the discussion at the smallest scale, the cellular basis of experience-dependent plasticity following changes in the social environment inspires several hypotheses [34,52,76]. For instance blocking the synthesis of new neurons in subgranular and subventricular zones (neurons that will then integrate into the dentate gyrus of the hippocampus and the main olfactory bulb, respectively) had detrimental consequences for the social behavior of juvenile mice but it did not affect the social behavior of adult animals [92]. Increasing or decreasing the synaptic efficacy in the mouse MPFC markedly caused upward or downward change of the animal position in the dominance hierarchy [91] (see also Chapter 20). A study by Keifer et al. [37] showed that structural MRI changes were correlated with spine density. Findings that relate structural variations to differences in social behaviors may suggest that patients suffering from neurological and psychiatric disorders, as characterized by alteration of their social behaviors, will show corresponding changes in brain structure. Indeed changes in temporal and prefrontal cortex have also been reported in a monkey model of autism [43], whereas patients with Asperger syndrome have smaller prefrontal and temporal minicolumns than control subjects [13].

References

[1] Amodio DM, Frith CD. Meeting of minds: the medial frontal cortex and social cognition. Nat Rev Neurosci 2006;7:268–77.
[2] Apps MA, Ramnani N. The anterior cingulate gyrus signals the net value of others' rewards. J Neurosci 2014;34:6190–200.
[3] Averbeck BB, Seo M. The statistical neuroanatomy of frontal networks in the macaque. PLoS Comput Biol 2008;4:e1000050.
[4] Behrens TE, Hunt LT, Rushworth MF. The computation of social behavior. Science 2009;324:1160–4.
[5] Behrens TE, Hunt LT, Woolrich MW, Rushworth MF. Associative learning of social value. Nature 2008;456:245–9.
[6] Bickart KC, Hollenbeck MC, Barrett LF, Dickerson BC. Intrinsic amygdala-cortical functional connectivity predicts social network size in humans. J Neurosci 2012;32:14729–41.
[7] Bickart KC, Wright CI, Dautoff RJ, Dickerson BC, Barrett LF. Amygdala volume and social network size in humans. Nat Neurosci 2011;14:163–4.

[8] Buckholtz JW, Meyer-Lindenberg A. Psychopathology and the human connectome: toward a transdiagnostic model of risk for mental illness. Neuron 2012;74:990—1004.

[9] Buckner RL, Andrews-Hanna JR, Schacter DL. The brain's default network: anatomy, function, and relevance to disease. Ann NY Acad Sci 2008;1124:1—38.

[10] Burman KJ, Reser DH, Yu HH, Rosa MG. Cortical input to the frontal pole of the marmoset monkey. Cereb Cortex 2011;21:1712—37.

[11] Burnett S, Blakemore SJ. Functional connectivity during a social emotion task in adolescents and in adults. Eur J Neurosci 2009; 29:1294—301.

[12] Bzdok D, Langner R, Schilbach L, Jakobs O, Roski C, Caspers S, et al. Characterization of the temporo-parietal junction by combining data-driven parcellation, complementary connectivity analyses, and functional decoding. Neuroimage 2013;81:381—92.

[13] Casanova MF, Buxhoeveden DP, Switala AE, Roy E. Asperger's syndrome and cortical neuropathology. J Child Neurol 2002;17: 142—5.

[14] Champagne FA, Curley JP. How social experiences influence the brain. Curr Opin Neurobiol 2005;15:704—9.

[15] Corballis MC. The recursive mind: the origins of human language, thought, and civilization. Princeton: Princeton University Press; 2011.

[16] Coricelli G, Nagel R. Neural correlates of depth of strategic reasoning in medial prefrontal cortex. Proc Natl Acad Sci USA 2009;106:9163—8.

[17] Damasio AR. Descartes' error: emotion, reason, and the human brain. Putman Publishing; 1994.

[18] De Martino B, O'Doherty JP, Ray D, Bossaerts P, Camerer C. In the mind of the market: theory of mind biases value computation during financial bubbles. Neuron 2013;79:1222—31.

[19] Decety J, Lamm C. The role of the right temporoparietal junction in social interaction: how low-level computational processes contribute to meta-cognition. Neuroscientist 2007;13:580—93.

[20] Dunbar RI, Shultz S. Evolution in the social brain. Science 2007; 317:1344—7.

[21] Emery NJ. The eyes have it: the neuroethology, function and evolution of social gaze. Neurosci Biobehav Rev 2000;24:581—604.

[22] Forkel SJ, Thiebaut de Schotten M, Kawadler JM, Dell'Acqua F, Danek A, Catani M. The anatomy of fronto-occipital connections from early blunt dissections to contemporary tractography. Cortex 2014;56:73—84.

[23] Fox MD, Corbetta M, Snyder AZ, Vincent JL, Raichle ME. Spontaneous neuronal activity distinguishes human dorsal and ventral attention systems. Proc Natl Acad Sci USA 2006;103:10046—51.

[24] Freese JL, Amaral DG. Neuroanatomy of the primate amygdala. In: Whalen PJ, Phelps EA, editors. The human amygdala. New York: The Guildford Press; 2009.

[25] Freiwald WA, Tsao DY. Functional compartmentalization and viewpoint generalization within the macaque face-processing system. Science 2010;330:845—51.

[26] Frey S, Campbell JSW, Pike GB, Petrides M. Dissociating the human language pathways with high angular resolution diffusion fiber tractography. J Neurosci 2008;28:11435—44.

[27] Gariepy JF, Watson KK, Du E, Xie DL, Erb J, Amasino D, et al. Social learning in humans and other animals. Front Neurosci 2014;8:58.

[28] Heberlein AS, Padon AA, Gillihan SJ, Farah MJ, Fellows LK. Ventromedial frontal lobe plays a critical role in facial emotion recognition. J Cogn Neurosci 2008;20:721—33.

[29] Heyes C. Animal mindreading: what's the problem? Psychon Bull Rev 2015;22:313—27.

[30] Hill RA, Bentley RA, Dunbar RI. Network scaling reveals consistent fractal pattern in hierarchical mammalian societies. Biol Lett 2008;4:748—51.

[31] Hornak J, Bramham J, Rolls ET, Morris RG, O'Doherty J, Bullock PR, et al. Changes in emotion after circumscribed surgical lesions of the orbitofrontal and cingulate cortices. Brain 2003;126: 1691—712.

[32] Hornak J, Rolls ET, Wade D. Face and voice expression identification in patients with emotional and behavioural changes following ventral frontal lobe damage. Neuropsychologia 1996; 34:247—61.

[33] Izquierdo A, Suda RK, Murray EA. Comparison of the effects of bilateral orbital prefrontal cortex lesions and amygdala lesions on emotional responses in rhesus monkeys. J Neurosci 2005;25: 8534—42.

[34] Johansen-Berg H, Della-Maggiore V, Behrens TE, Smith SM, Paus T. Integrity of white matter in the corpus callosum correlates with bimanual co-ordination skills. Neuroimage 2007;36(Suppl. 2): T16—21.

[35] Kana RK, Keller TA, Cherkassky VL, Minshew NJ, Just MA. Atypical frontal-posterior synchronization of Theory of Mind regions in autism during mental state attribution. Soc Neurosci 2009;4: 135—52.

[36] Kanai R, Bahrami B, Roylance R, Rees G. Online social network size is reflected in human brain structure. Proc Biol Sci/R Soc 2012;279:1327—34.

[37] Keifer Jr OP, Hurt RC, Gutman DA, Keilholz SD, Gourley SL, Ressler KJ. Voxel-based morphometry predicts shifts in dendritic spine density and morphology with auditory fear conditioning. Nat Commun 2015;6:7582.

[38] Koechlin E. Frontal pole function: what is specifically human? Trends Cogn Sci 2011;15:241. author reply 243.

[39] Krug K, Cicmil N, Parker AJ, Cumming BG. A causal role for V5/MT neurons coding motion-disparity conjunctions in resolving perceptual ambiguity. Curr Biol 2013;23:1454—9.

[40] Lewis PA, Rezaie R, Brown R, Roberts N, Dunbar RI. Ventromedial prefrontal volume predicts understanding of others and social network size. Neuroimage 2011;57:1624—9.

[41] Luppino G, Calzavara R, Rozzi S, Matelli M. Projections from the superior temporal sulcus to the agranular frontal cortex in the macaque. Eur J Neurosci 2001;14:1035—40.

[42] Lurz RW. Mindreading animals. Cambridge (Massachusetts): MIT Press; 2011.

[43] Machado CJ, Whitaker AM, Smith SE, Patterson PH, Bauman MD. Maternal immune activation in nonhuman primates alters social attention in juvenile offspring. Biol Psychiatry 2015;77:823—32.

[44] Mar RA. The neural bases of social cognition and story comprehension. Annu Rev Psychol 2011;62:103—34.

[45] Markov NT, Ercsey-Ravasz MM, Ribeiro Gomes AR, Lamy C, Magrou L, Vezoli J, et al. A weighted and directed interareal connectivity matrix for macaque cerebral cortex. Cereb Cortex 2014; 24:17—36.

[46] Mars RB, Foxley S, Jbabdi S, Sallet J, Noonan MP, Neubert FX, et al. The extreme capsule fiber complex in humans and macaques: a comparative diffusion MRI tractography study (submitted for publication-a).

[47] Mars RB, Jbabdi S, Sallet J, O'Reilly JX, Croxson PL, Olivier E, et al. Diffusion-weighted imaging tractography-based parcellation of the human parietal cortex and comparison with human and macaque resting-state functional connectivity. J Neurosci 2011;31: 4087—100.

[48] Mars RB, Neubert FX, Noonan MP, Sallet J, Toni I, Rushworth MF. On the relationship between the "default mode network" and the "social brain". Front Hum Neurosci 2012;6:189.

[49] Mars RB, Sallet J, Neubert FX, Rushworth MF. Connectivity profiles reveal the relationship between brain areas for social cognition in human and monkey temporoparietal cortex. Proc Natl Acad Sci USA 2013;110:10806—11.

III. SOCIAL DECISION NEUROSCIENCE

[50] Mars RB, Sallet J, Schuffelgen U, Jbabdi S, Toni I, Rushworth MF. Connectivity-based subdivisions of the human right "temporo-parietal junction area": evidence for different areas participating in different cortical networks. Cereb Cortex 2012;22:1894–903.

[51] Mars RB, Verhagen L, Gladwin TE, Neubert FX, Sallet J, Rushworth MFS. Comparing brains by matching connectivity fingerprints (submitted for publication-b).

[52] May A. Experience-dependent structural plasticity in the adult human brain. Trends Cogn Sci 2011;15:475–82.

[53] Miranda-Dominguez O, Mills BD, Grayson D, Woodall A, Grant KA, Kroenke CD, et al. Bridging the gap between the human and macaque connectome: a quantitative comparison of global interspecies structure–function relationships and network topology. J Neurosci 2014;34:5552–63.

[54] Mitchell JP. Activity in right temporo-parietal junction is not selective for theory-of-mind. Cereb Cortex 2008;18:262–71.

[55] Molesworth T, Sheu LK, Cohen S, Gianaros PJ, Verstynen TD. Social network diversity and white matter microstructural integrity in humans. Social Cogn Affect Neurosci 2015;10(9).

[56] Mosher CP, Zimmerman PE, Gothard KM. Response characteristics of basolateral and centromedial neurons in the primate amygdala. J Neurosci 2010;30:16197–207.

[57] Neubert FX, Mars RB, Sallet J, Rushworth MF. Connectivity reveals relationship of brain areas for reward-guided learning and decision-making in human and monkey frontal cortex. Proc Natl Acad Sci USA 2015;112:E2695–704.

[58] Neubert FX, Mars RB, Thomas AG, Sallet J, Rushworth MF. Comparison of human ventral frontal cortex areas for cognitive control and language with areas in monkey frontal cortex. Neuron 2014; 81:700–13.

[59] Nicolle A, Klein-Flügge MC, Hunt LT, Vlaev I, Dolan RJ, Behrens TE. An agent independent axis for executed and modeled choice in medial prefrontal cortex. Neuron 2012;75:1114–21.

[60] Noonan MP, Mars MB, Sallet J, Dunbar RIM, Fellows LK. The structural and functional brain networks that support human social networks (submitted for publication).

[61] Noonan MP, Sallet J, Mars RB, Neubert FX, O'Reilly JX, Andersson JL, et al. A neural circuit covarying with social hierarchy in macaques. PLoS Biol 2014;12:e1001940.

[62] Noonan MP, Sallet J, Rudebeck PH, Buckley MJ, Rushworth MF. Does the medial orbitofrontal cortex have a role in social valuation? Eur J Neurosci 2010;31:2341–51.

[63] Passingham RE, Stephan KE, Kotter R. The anatomical basis of functional localization in the cortex. Nat Rev Neurosci 2002;3:606–16.

[64] Perrett DI, Hietanen JK, Oram MW, Benson PJ. Organization and functions of cells responsive to faces in the temporal cortex. Philos Trans R Soc Lond B Biol Sci 1992;335:23–30.

[65] Petrides M. The orbitofrontal cortex: novelty, deviation from expectation, and memory. Ann NY Acad Sci 2007;1121:33–53.

[66] Petrides M, Pandya DN. Association fiber pathways to the frontal cortex from the superior temporal region in the rhesus monkey. J Comp Neurol 1988;273:52–66.

[67] Petrides M, Pandya DN. Distinct parietal and temporal pathways to the homologues of Broca's area in the monkey. PLoS Biol 2009;7: e1000170.

[68] Pinsk MA, Arcaro M, Weiner KS, Kalkus JF, Inati SJ, Gross CG, et al. Neural representations of faces and body parts in macaque and human cortex: a comparative FMRI study. J Neurophysiol 2009;101:2581–600.

[69] Raichle ME, Snyder AZ. A default mode of brain function: a brief history of an evolving idea. Neuroimage 2007;37:1083–90. discussion 1097-1089.

[70] Roe AW, Chelazzi L, Connor CE, Conway BR, Fujita I, Gallant JL, et al. Toward a unified theory of visual area V4. Neuron 2012;74: 12–29.

[71] Rozzi S, Calzavara R, Belmalih A, Borra E, Gregoriou GG, Matelli M, et al. Cortical connections of the inferior parietal cortical convexity of the macaque monkey. Cereb Cortex 2006;16:1389–417.

[72] Rudebeck PH, Buckley MJ, Walton ME, Rushworth MF. A role for the macaque anterior cingulate gyrus in social valuation. Science 2006;313:1310–2.

[73] Rushworth MF, Noonan MP, Boorman ED, Walton ME, Behrens TE. Frontal cortex and reward-guided learning and decision-making. Neuron 2011;70:1054–69.

[74] Sallet J, Mars RB, Noonan MP, Andersson JL, O'Reilly JX, Jbabdi S, et al. Social network size affects neural circuits in macaques. Science 2011;334:697–700.

[75] Sallet J, Mars RB, Noonan MP, Neubert FX, Jbabdi S, O'Reilly JX, et al. The organization of dorsal frontal cortex in humans and macaques. J Neurosci 2013;33:12255–74.

[76] Sampaio-Baptista C, Khrapitchev AA, Foxley S, Schlagheck T, Scholz J, Jbabdi S, et al. Motor skill learning induces changes in white matter microstructure and myelination. J Neurosci 2013; 33:19499–503.

[77] Saxe R. Uniquely human social cognition. Curr Opin Neurobiol 2006;16:235–9.

[78] Saxe R, Wexler A. Making sense of another mind: the role of the right temporo-parietal junction. Neuropsychologia 2005;43: 1391–9.

[79] Schilbach L, Eickhoff SB, Rotarska-Jagiela A, Fink GR, Vogeley K. Minds at rest? Social cognition as the default mode of cognizing and its putative relationship to the "default system" of the brain. Conscious Cogn 2008;17:457–67.

[80] Schmahmann JD, Pandya DN. Fiber pathways of the brain. Oxford: Oxford University Press; 2006.

[81] Scholz J, Triantafyllou C, Whitfield-Gabrieli S, Brown EN, Saxe R. Distinct regions of right temporo-parietal junction are selective for theory of mind and exogenous attention. PLoS One 2009;4:e4869.

[82] Schurz M, Radua J, Aichhorn M, Richlan F, Perner J. Fractionating theory of mind: a meta-analysis of functional brain imaging studies. Neurosci Biobehav Rev 2014;42:9–34.

[83] Seltzer B, Pandya DN. Intrinsic connections and architectonics of the superior temporal sulcus in the rhesus monkey. J Comp Neurol 1989;290:451–71.

[84] Stanley DA, Adolphs R. Toward a neural basis for social behavior. Neuron 2013;80:816–26.

[85] Stiller J, Dunbar RIM. Perspective-taking and memory capacity predict social network size. Soc Netw 2007;29:93–104.

[86] Thiebaut de Schotten M, Dell'Acqua F, Valabregue R, Catani M. Monkey to human comparative anatomy of the frontal lobe association tracts. Cortex 2012;48:82–96.

[87] Tomasello M. The ultra-social animal. Eur J Soc Psychol 2014;44: 187–94.

[88] Van Essen DC, Dierker DL. Surface-based and probabilistic atlases of primate cerebral cortex. Neuron 2007;56:209–25.

[89] Vincent JL, Patel GH, Fox MD, Snyder AZ, Baker JT, Van Essen DC, et al. Intrinsic functional architecture in the anaesthetized monkey brain. Nature 2007;447:83–6.

[90] Von Der Heide R, Vyas G, Olson IR. The social network-network size is predicted by brain structure and function in the amygdala and paralimbic regions. Soc Cogn Affect Neurosci 2014;9(12).

[91] Wang F, Zhu J, Zhu H, Zhang Q, Lin Z, Hu H. Bidirectional control of social hierarchy by synaptic efficacy in medial prefrontal cortex. Science 2011;334:693–7.

[92] Wei L, Meaney MJ, Duman RS, Kaffman A. Affiliative behavior requires juvenile, but not adult neurogenesis. J Neurosci 2011;31: 14335–45.

[93] Yeterian EH, Pandya DN, Tomaiuolo F, Petrides M. The cortical connectivity of the prefrontal cortex in the monkey brain. Cortex 2012;48:58–81.

16

The Neural Bases of Social Influence on Valuation and Behavior

K. Izuma

University of York, York, United Kingdom

Abstract

An individual's beliefs, attitudes, and behaviors are influenced by other people, and social influence has been a primary research area throughout the history of social psychology. In this chapter, among a wide variety of forms of social influences, I will focus on social neuroscience studies investigating the following two forms of social influence: (1) observer effect (increased prosocial tendency in front of other people) and (2) social conformity (adjusting one's attitude or behavior to those of a group). I will first review studies investigating how one's concern for reputation formed by other people affects prosocial behavior and discuss how reputation processing is impaired in individuals with autism. Second, I will outline research on social conformity and especially highlight available evidence suggesting the link between social conformity and reward-based learning (reinforcement learning).

INTRODUCTION

Social influence is ubiquitous in human societies. It takes a wide variety of forms, including obedience, conformity, persuasion, social loafing, social facilitation, deindividuation, observer effect, bystander effect, and peer pressure. Research on social influence has a long history in social psychology, and an experiment on social facilitation effect that was conducted in 1898 by Triplett [1] is often considered the first social psychological experiment (see also Ref. [2]). Since then, social influence has fascinated scholars in various fields.

Since 2005, researchers in the emerging and growing fields of social neuroscience, neuroeconomics, and neuromarketing have begun to explore the neural bases of such complex social phenomena using methods in cognitive neuroscience including functional magnetic resonance imaging (fMRI), electroencephalography (EEG), and transcranial magnetic stimulation (TMS).

In this chapter, I will cover social neuroscience and neuroeconomics studies investigating the following two forms of social influence: (1) the effect of others' presence on prosocial behavior (observer effect) and (2) the effect of others' opinion on an individual's preference (social conformity). Both the observer effect and social conformity can be considered primary examples of social influence, as they represent two of the most relevant and ubiquitous forms of social influence in our everyday social lives.

THE EFFECT OF MERE PRESENCE OF OTHERS ON PROSOCIAL BEHAVIOR

The Neural Bases of Social Reward Processing

It has been widely known in social science research that the mere presence of other people increases an individual's tendency to behave prosocially, a phenomenon called the "observer effect" (also known as "audience effect"). For example, people donate more money when they know their decision is observed by other people [3–5]. In experimental economics, participants were usually more willing to share some benefits (i.e., money) in economic games (e.g., dictator game, public-goods game) if they believed that other people were able to know the participants' decisions, compared to the situation in which they believed that their decision remained completely confidential [6–8].

It has been considered that people act prosocially in front of other people because they want to get a good reputation (or avoid a bad reputation) from other people [9,10]. Thus, the observer effect involves processing of an important social reward for humans, good reputation or social approval from other people, which is a powerful incentive for a variety of human social behaviors [9].

Decision Neuroscience
http://dx.doi.org/10.1016/B978-0-12-805308-9.00016-6

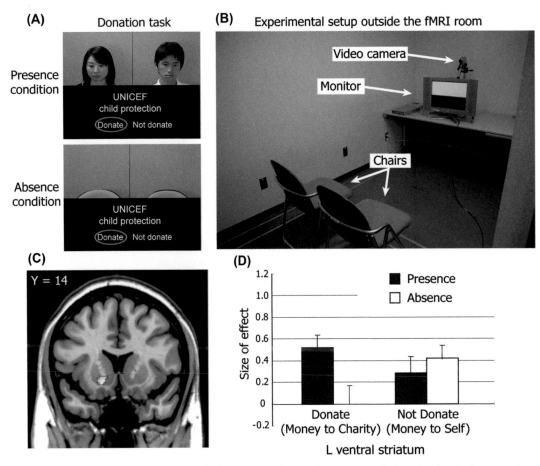

FIGURE 16.1   (A) The donation task in Izuma et al. [24]. In each trial, participants were asked to decide whether to make a donation to a charity. In one condition, their decisions were constantly observed by two other people (top). In the other condition, they engaged in the same donation task while no one was explicitly observing (bottom). (B) Before each subject entered the fMRI scanner room, all of them saw the experimental setup in the scanner control room. Participants were told that two other people would sit in chairs and watch the subject's decisions, which would be displayed on the monitor. Furthermore, they were told that inside the fMRI scanner, they would see the faces of two observers through the video camera. However, in reality, a prerecorded video of two observers was used throughout the experiment so that their facial expression was kept neutral at all times. (C) The area in the ventral striatum in which activity was significantly modulated by the participant's decision (donate or not) and the presence of observers. (D) Activation pattern in the left (L) ventral striatum. Activation was especially higher when making a donation in front of observers (expecting high social reward) and when not donating in the absence of observers (expecting monetary reward without social punishment). *(A), (C), and (D) are adapted from Izuma K, Saito DN, Sadato N. Processing of the incentive for social approval in the ventral striatum during charitable donation. J Cogn Neurosci 2010;22:621–31.*

Because understanding one's own reputation involves thinking about what other people think of us, it requires the ability to form a metarepresentation (e.g., thinking about thinking), which is closely related to the ability called "theory of mind" or "mentalizing" [11]. To date, there is no clear evidence that nonhuman animals have theory of mind (e.g., Ref. [12]), and accordingly the effect of observers on prosocial behavior may be unique to humans. Consistent with this idea, whereas even 5-year-old children showed the observer effect in humans [13–16], adult chimpanzees were insensitive to the presence of a conspecific [15,16], indicating that chimpanzees are not interested in increasing their reputations.

In the past decade, social neuroscientists have investigated the neural bases of how people make decisions based on the social reward [10,17]. Izuma et al. [18] first explored whether the seemingly more complex reward of one's good reputation shares the same neural bases as other tangible rewards such as money. Participants in this study were first asked to introduce themselves in front of a video camera, and they were led to believe that other people were going to form impressions of them based on their video-recorded self-introduction. While their brains were scanned by fMRI, they were presented with the results of impression evaluations (i.e., their reputations formed by other people). In a separate fMRI session, the same participants also performed a simple gambling task in which they could obtain monetary rewards. It was found that the ventral striatum, a part of the brain's reward system, was commonly activated by both the concrete reward of money and the more abstract social reward of good reputation. Other fMRI studies similarly demonstrated that social rewards of positive reputation [19–21] or being liked by others [22,23] activated the reward-related brain regions including the striatum and the ventromedial prefrontal cortex (VMPFC). These studies illustrated that an important social reward for humans, a good reputation, activates the same brain regions as do tangible rewards.

In a follow-up fMRI study, Izuma et al. [24] further tested if the striatum encodes the value of social reward during a real social decision-making situation in front of other people [24]. Participants in this experiment were asked to make a donation to real charitable organizations inside an fMRI scanner. Moreover, in one condition, their decisions were constantly observed through a video camera by two other people (confederates; Fig. 16.1A, top) who engaged in an impression formation task in an fMRI control room (Fig. 16.1B). In the other condition, the subject performed the same task while no one was watching (Fig. 16.1A, bottom). If the striatum encoded the value of both social and monetary rewards during donation decisions, its activity should be modulated by both their decision (keeping $5 for themselves or donating it to a charity) and the presence/absence of another person. The study first showed that participants actually made more donations if their decisions were observed by other people (i.e., observer effect) [24], replicating the findings of the aforementioned studies [3–5]. The study further revealed that, as predicted, the striatal activity during decision-making was significantly modulated by participants' decisions as well as by observer manipulation (Fig. 16.1C and D). Specifically, the striatal activity was greater when participants made donations in front of observers (high social reward is expected) as well as when participants kept the money for themselves while nobody was watching (monetary reward without social cost of bad reputation) (see Fig. 16.1D). The findings indicate that the decision utility of both social and monetary rewards is represented in the same regions within the striatum during prosocial decision-making.

Although common striatal activations between social and monetary rewards seem to support the notion that a single neural circuit is responsible for representing reward values of both social and nonsocial rewards (i.e., the "extended common currency schema" [25]), results from single-cell-recording studies of nonhuman animals suggest that this view may be too simplistic. For example, although there exist some neurons in a rat's ventral striatum, which encode both natural nutritive reward (e.g., food or juice) and drug reward (e.g.,

cocaine), these neuronal populations are largely distinct [26–28]. Thus, even among nonsocial rewards, a neural circuit may not be shared. Furthermore, using juice reward and social reward, such as images of female perinea, a monkey single-cell-recording study by Watson and Platt [29] showed that only 2% of neurons in the orbitofrontal cortex responded to both types of rewards. Similarly, in the ventral striatum, populations of neurons encoding juice and social information are largely distinct [30]. Thus, although the same reward-related brain regions were activated by both social and nonsocial rewards in human fMRI studies (e.g., Refs. [18,24,31]), neuronal populations are likely to be largely distinct, and the similarities and differences between the neural mechanisms of social and nonsocial reward-based decision-making are an important avenue for future research.

It should be noted that, as mentioned earlier, there is an important difference between the social reward of reputation and other tangible rewards; understanding one's own reputation (and thus its reward value) requires theory of mind. Consistent with this idea, regardless of whether reputations formed by others were good or neutral, perceiving one's reputation activates the medial prefrontal cortex (MPFC), a region implicated in theory of mind [11], whereas perceiving another person's reputation or obtaining a monetary reward does not [18]. Furthermore, using the same observer paradigm (Fig. 16.1A and B), Izuma et al. [32] further showed that the MPFC activity was heightened when participants were asked to answer whether they follow various social norms (e.g., "If I could get into a movie without paying and be sure I was not seen I would probably do it.") while their answers were constantly observed by other individuals, suggesting that the MPFC is activated when there is a strong situational demand for processing an individual's own reputation in the eyes of others.

The Observer Effect in Autism

While the human cognitive capacity to derivatively manipulate impressions of oneself formed by other people seems to far outstrip the abilities of other animals, not all humans seem capable of it [33]. Autism spectrum disorders (ASDs) are a class of neurodevelopmental disorders characterized by a profound difficulty in social communication and interaction [34]. People with ASD are considered to be impaired in theory of mind or mentalizing abilities [35,36], and it was suggested that people with ASD might have a specific difficulty in taking into account what other people think of them and modifying their behaviors accordingly [37].

To test this idea, Izuma et al. [38] compared the susceptibility to the observer effect between high-functioning adults with ASD and age- and IQ-matched normal control participants. This study involved a simple observer paradigm, and each participant was asked to make real donations in the presence or absence of a male observer (confederate), who was sitting about 3 feet behind the participant. To control for nonspecific effects of the presence of another person, such as heightened arousal, all participants also performed a simple attention task called a continuous performance test in the presence or absence of the observer. The results revealed that normal control participants made significantly more donations in the presence of the observer compared to when they were alone in the room, whereas this effect of the observer was completely absent in individuals with autism, and they did not show any increase in the number of donations in front of the observer [38] (Fig. 16.2A, left). However, both groups showed enhanced performance on the continuous performance test in the presence of the observer relative to his absence [38], demonstrating a typical social facilitation effect (Fig. 16.2A, right). More recently, Cage et al. [39] also

found reduced propensity to act prosocially in front of others in individuals with ASD.

A similar finding was demonstrated by Chevallier et al. [40] using a different paradigm. In this study, adolescents (about 13 years of age on average) with ASD and typically developing participants were first asked to rate a set of drawings of a face. In the control condition, participants were simply asked to rate two of the drawings again. In the experimental condition, the experimenter told them that she herself had drawn the two pictures, and participants were asked to rate them in front of the experimenter. The results showed that, whereas typically developing controls inflated their ratings in front of the experimenter, there was no difference in ratings in the ASD group [40] (Fig. 16.2B).

It is important to note that there are at least two possible mechanisms that explain the lack of behavioral modification based on self-reputation in individuals with ASD [38]. One idea is that their deficit is mainly cognitive, and they just lack the cognitive capacity to represent what others would think of them if they behaved in a certain way. The other idea is that their deficit is mainly motivational. They can process

FIGURE 16.2 (A) The effect of the observer's presence on donation (left) and continuous performance test (CPT; right) tasks. (*Adapted from Izuma K, Matsumoto K, Camerer CF, Adolphs R. Insensitivity to social reputation in autism. Proc Natl Acad Sci USA 2011;108:17302–07.*) (B) The effect of the drawer's presence on picture evaluations. *ASD*, autism spectrum disorder; *TD*, typically developing. (*Adapted from Chevallier C, Molesworth C, Happe F. Diminished social motivation negatively impacts reputation management: autism spectrum disorders as a case in point. PLoS One 2012;7:e31107.*)

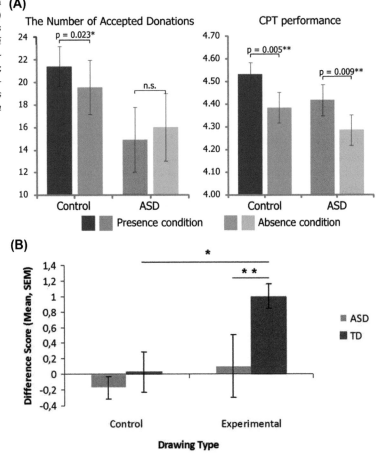

self-reputation in the eyes of observers but simply do not care about it. Whereas these two possibilities should be carefully dissociated in future research, as of this writing, available evidence seems to suggest that the social deficit in ASD is mainly explained by their specific deficit in social motivation [41].

Chevallier et al. [42] directly measured the social motivation of individuals with ASD by a questionnaire assessing self-reported pleasure in various situations including social interactions. Whereas children with ASD showed no difference in self-reported pleasure from physical (e.g., pleasure of eating, sounds, etc.) and other sources (e.g., intellectual pleasure) compared to normal control participants, they showed a significantly lower level of social pleasure (e.g., pleasure of being with people, talking with people, doing things with people, etc.) [42]. It suggests that for children with ASD, social situations are not as enjoyable and motivating as for typically developing children.

Similarly, several studies suggest that social stimuli that are considered to be rewarding for normal individuals, such as smiling faces, may not be rewarding for individuals with ASD. Individuals with ASD typically show reduced interest in social stimuli such as faces and eye contact [43–45]. While typically developing children showed increased pupillary diameter in response to happy faces with direct gaze compared to averted gaze, children with ASD showed no such difference [46], which suggests that individuals with ASD may have reduced sensitivity to the reward value of a smiling face with a direct gaze. An fMRI study further showed that the striatum is less sensitive to socially rewarding stimuli (smiling faces) in ASD compared to normal control participants [47] (see also Refs. [48,49]). Importantly, these past fMRI studies of ASD used simple social stimuli (e.g., smiling faces), and no study as of this writing has investigated the neural responses to the more complex social reward, one's reputation, in ASD. Such a future study is important to further inform the lack of the observer effect and social impairment in general in individuals with ASD.

THE EFFECT OF OTHERS' OPINIONS ON VALUATION

The Neural Bases of Social Conformity

People tend to align their attitudes, beliefs, and behavior to other people, a phenomenon known as "social conformity." While social psychological research on social conformity has a long history (e.g., Refs. [50,51]), only recently have researchers started investigating the neural mechanisms of how people's attitudes are influenced by others' opinions using cognitive neuroscience

methods such as fMRI, EEG, and TMS [52–54]. Nonetheless, social neuroscience research on conformity has generated largely consistent findings highlighting roles played especially by a posterior part of the medial frontal cortex (pMFC; Fig. 16.3A) and brain regions related to reward processing such as the striatum and VMPFC.

The majority of past social neuroscience studies on social conformity used tasks in which participants reported their personal preference for various stimuli such as faces, music, and T-shirts. Klucharev et al. [55] first asked female participants to rate the attractiveness of each female face while their brains were scanned by an fMRI. After they provided their ratings, an average group rating for the face was presented. Unknown to participants, group opinions were experimentally manipulated so that they matched with the participant's rating in some trials and did not match in other trials. After the scanning, participants were asked to rate the same faces again to see whether they changed their ratings after viewing group opinions. Behaviorally, participants showed a typical social conformity effect; their second ratings increased when group ratings for the same faces were higher than their first ratings, whereas their second ratings decreased when group ratings were lower. Their fMRI results revealed that the pMFC and insula were significantly activated when their ratings conflicted with group ratings, whereas the ventral striatum was activated when their ratings matched with group ratings (Fig. 16.3B). Striatal activity in response to agreement with group ratings or experts' ratings was also reported in two other fMRI studies [56,57] (Fig. 16.3C).

Both the pMFC and the ventral striatum are known to be involved in behavioral adjustment following positive or negative outcomes (i.e., reward-based learning) (e.g., Refs. [58–63]). Specifically, the pMFC is activated by a variety of aversive outcomes (e.g., negative performance feedback and errors) [62,64], while the ventral striatum is activated by a variety of positive outcomes (e.g., money and social reward) [17,60]. Based on this apparent similarity of brain regions involved between reinforcement learning and social conformity, Klucharev and his colleagues proposed that they might share the same neural mechanisms (Fig. 16.4) [55].

There are now three separate lines of evidence supporting the link between reinforcement learning and social conformity, in terms of brain structures engaged (pMFC and striatum), and two other lines of evidence (from pharmacological and genetics studies) pointing out the link between dopamine/serotonin and social conformity. First, studies indicate the causal role of the pMFC in both preference change based on group opinions and behavioral adjustment based on reward and punishment. Past monkey neurophysiological studies showed that neurons in the pMFC are especially

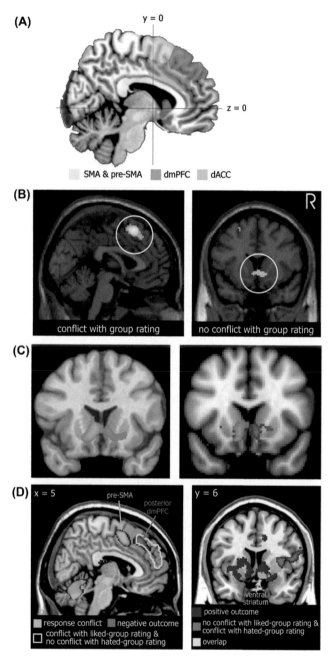

responsive to negative outcomes, which led to subsequent behavioral change compared to negative outcomes, which were not followed by behavioral change [58,63]. Similarly, Klucharev et al. [55] found that the magnitude of the pMFC activity in response to the mismatch between one's own and the group opinions was related to subsequent changes in attractiveness ratings. Furthermore, Shima and Tanji [58] showed that pharmacological inactivation of pMFC neurons impaired monkeys' behavioral adjustment following negative outcomes. Similarly, Klucharev et al. [65] demonstrated that social conformity is significantly reduced by downregulating the pMFC using TMS.

Furthermore, Izuma and Adolphs [66] provided evidence that the pMFC and striatum regions activated during a social conformity task overlap with the regions activated by negative and positive monetary outcomes, respectively. In this study [66], in addition to the social conformity task (a task similar to the previous studies by Klucharev et al. [55,65]), the same participants performed the Monetary Incentive Delay Task (MIDT) [67] and the Multi-Source Interference Task (MSIT) [68]. The MIDT is a simple speeded-response task in which participants have to press a key as soon as a white square appears on the center of the screen to obtain monetary reward. This task was intended to localize brain areas responsive to positive and negative outcomes. If they press the key before the white square disappears, they will get a monetary reward (positive outcome), whereas if they are too slow, they get no reward (negative outcome). The MSIT is conceptually similar to a variety of cognitive interference tasks such as the Stroop task, Eriksen flanker task, and Simon task and was intended to localize brain areas sensitive to response conflict [68]. Izuma and Adolphs [66] replicated the previous finding [55] and found that the pMFC is activated when there is a conflict between participants' (university students) rating of a T-shirt and the rating of a group (fellow students), whereas the ventral striatum is activated when the participant's opinion matches with other students' opinion. In addition, the study further demonstrated that the pMFC region activated by social conflict with group opinion overlaps with the areas activated by negative outcomes during the MIDT, whereas the ventral striatum regions activated by the match

FIGURE 16.3   (A) Posterior medial frontal cortex (pMFC), which consists of the dorsal anterior cingulate cortex (dACC), dorsomedial prefrontal cortex (dmPFC), and supplementary motor area (SMA) and pre-SMA. (B) The left shows the pMFC regions activated by conflict with group rating, whereas the right shows the ventral striatum regions activated when there is no conflict with group rating. (*Adapted and modified from Klucharev V, Hytonen K, Rijpkema M, Smidts A, Fernandez G. Reinforcement learning signal predicts social conformity. Neuron 2009;61:140–51.*) (C) Striatal regions activated by agreement with others' opinions relative to disagreement. (*Adapted from Nook EC, Zaki J. Social norms shift behavioral and neural responses to foods. J Cogn Neurosci 2005;27:1412–26 (left) and Campbell-Meiklejohn DK, Bach DR, Roepstorff A, Dolan RJ, Frith CD. How the opinion of others affects our valuation of objects. Curr Biol 2010;20:1165–70 (right).*) Green regions in the right show activation of the same contrast at a reduced statistical threshold. (D) The left shows the pMFC regions activated both by conflict with the opinion of a liked group (fellow students) and by no conflict with the

opinion of a hated group (sex offenders). *Cyan* and *red* regions represent areas sensitive to response conflict (as identified by the MSIT) and negative outcome (as identified by the MIDT), respectively. *Blue* areas show regions in the right activated by positive outcome (monetary gain, as identified by the MIDT). *Red* areas show regions activated both by no conflict with a liked group and by conflict with a hated group. *Yellow* areas indicate overlaps. *MIDT*, Monetary Incentive Delay Task; *MSIT*, Multi-Source Interference Task. (*Adapted and modified from Izuma K, Adolphs R. Social manipulation of preference in the human brain. Neuron 2013;78:563–73.*)

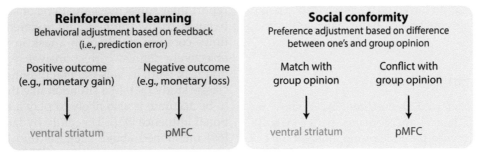

FIGURE 16.4   Schematic illustration of the mechanisms and brain regions involved in reinforcement learning (left) and social conformity (right). *pMFC*, posterior medial frontal cortex.

with group opinion overlap with the areas activated by positive outcomes (Fig. 16.3D). Striatum overlap between agreement with others' opinion and materialistic reward was also reported in another study [57]. Importantly, response conflict-related regions identified with the MSIT (pre-supplementary motor area) were more posterior to the pMFC regions activated by social conflict with group opinion and the MIDT (posterior dorsomedial prefrontal cortex; Fig. 16.3D, left). Thus, although the mismatch with group opinion can be considered as one type of cognitive conflict, its neural bases are likely to be different from those of conflict at the response level.

The study [66] further showed that activities of the pMFC and ventral striatum do not simply depend on the difference between one's own and the group opinions, but also on one's attitude toward the group, an effect predicted by balance theory [69]. In addition to the opinions of fellow university students, participants in this study were also presented with an opinion of sex offenders in different trials. Both behavioral and fMRI results revealed that the effect of sex offenders' opinion was the complete opposite of that of fellow students. Behaviorally, participants changed their opinions when their ratings matched with sex offenders, whereas ratings tended to stay the same when their opinions differed from sex offenders. The fMRI results showed that the same pMFC region was activated when the participant's opinion matched with sex offenders, and the ventral striatum was activated when they did not match (Fig. 16.3D).

Several EEG studies also suggest a functionally similar role of the pMFC between social conformity and reward-based learning. It is well established in EEG research that negative performance feedback induces a negative event-related potential signal, called feedback-related negativity (FRN), which occurs approximately 250 ms after receiving the feedback [70]. It is typically observed at central to frontal–central scalp regions and considered to be generated from the pMFC (i.e., anterior cingulate cortex) [71]. Three independent research groups [72–74] commonly showed that the mismatch between

one's own and the group opinion generates an FRN-like signal occurring between 200 and 400 ms after the mismatch was made apparent, and more recently, this finding was further replicated [75].

Second, two pharmacological studies [76,77] demonstrated the roles played by dopamine and serotonin in social conformity, both of which have been implicated in reward-based learning and decision-making [78–80]. In an experiment, Campbell-Meiklejohn et al. [76] gave one group of participants a drug called methylphenidate, which increases extracellular dopamine in the brain [81], whereas those in a control group received a placebo. Participants in both groups completed the same social conformity task as Klucharev et al. [55] (except that they rated the trustworthiness of faces instead of attractiveness). The results demonstrated that those who received the drug were influenced by group opinions more strongly than those who received a placebo [76], suggesting that increased dopamine in the brain enhances the social conformity effect. The social conformity effect is also enhanced by a drug citalopram, called the selective serotonin reuptake inhibitor (SSRI), especially when group ratings were lower than participants' ratings [77], suggesting that social conformity is influenced by serotonin.

Finally, the last piece of evidence comes from a study that demonstrated the effect of genetic polymorphism on an individual's tendency to conform to other people [82]. The gene *COMT*, encoding catechol-*O*-methyltransferase, is known to affect dopamine levels in the prefrontal cortex (PFC) [83], and the COMT Val158Met polymorphism modulates COMT enzymatic activity such that carriers of the Met allele presumably have greater PFC dopamine availability. It has been previously reported that during a reward-based learning task, Met allele carriers' behaviors are influenced by negative feedback (i.e., higher learning rate) more than those of noncarriers of the Met allele [84]. Similarly, Deuker et al. [82] demonstrated that Met allele carriers are influenced by others' opinions more than carriers of at least one Val allele, suggesting that the same genetic polymorphism is related to adjustment in behavior based on negative feedback and adjustment in judgment based on mismatch with group opinion.

Summary and Future Directions

Although, as reviewed above, there are studies suggesting common neural mechanisms between social conformity and reward-based learning, none of the evidence is conclusive. As I previously discussed potentially different neural circuits (neuronal populations) involved in social and nonsocial rewards in the striatum, activation overlaps between social conformity and positive/negative outcomes [66] (Fig. 16.3C) do not necessarily mean that the same neuronal circuit is responsible for both. Similarly, an FRN-like signal in response to the mismatch between one's own and group opinions [72–74] does not necessarily indicate that the same population of neurons is responsible for the social conflict signal and FRN. Furthermore, it is yet to be determined whether dopamine and serotonin play similar roles in both reward-based learning and social conformity, and the effect of gene polymorphisms in reinforcement learning versus social conformity should also be elucidated further (see Ref. [85]). It should be noted that three studies [86–88] demonstrated that social conformity is also modulated by oxytocin, which has been implicated in mediating a variety of social and affiliative behaviors [89,90]. The involvement of oxytocin in social conformity points to the possibility that the neural bases of social conformity are at least partially different from and potentially more complex than those of reinforcement learning. Importantly, so far, only one study [66] directly compared neural responses to social conflict with group opinion and negative (and positive) outcome within the same sample of individuals (Fig. 16.3C), and similarity and difference in the neural mechanisms between social conformity and reinforcement learning should be directly tested in future research (e.g., comparing EEG responses or the effects of methylphenidate, i.e., dopamine; SSRI, i.e., serotonin; oxytocin; and gene polymorphisms to negative performance feedback and social conflict with group opinion).

It is also important to acknowledge that social conformity is not a uniform phenomenon. Others' opinions affect not only an individual's valuation of stimuli, but also memory (e.g., Refs. [82,91]) and decision-making (e.g., Refs. [92–94]). Furthermore, people conform to others for different reasons, and if motivations are different, underlying neural processes are likely to differ. People may conform because they want to obtain social approval or avoid social disapproval (i.e., normative influence [95]). Although a majority of social conformity studies reviewed above asked participants to report their subjective valuations of stimuli (e.g., attractiveness, trustworthiness, and preference), people sometimes conform to others in a task in which there is an objectively defined correct answer (e.g., a mental rotation task [96]), as was the line judgment task in Asch's classic study [50]. In a situation like this, the motivation to be accurate is also likely to play a role (i.e., informational influence [95]). The study by Izuma and Adolphs [66] mentioned above suggests that the motivation to maintain cognitive consistency [69,97,98] is also important. Thus, the motivations for conformity seem to depend on many different factors that influence one or more of these different types of motivation, including who other people are, whether other people are actually present in an experiment, whether a task has a correct answer (and if so, how difficult the task is), and so on (see also Ref. [99] for a similar discussion). It is important to systematically manipulate these factors in future research to reveal the neural mechanisms underlying this complex social phenomenon.

CONCLUDING REMARKS

Despite its long history of research in social psychology, the investigation of the neural bases of social influence is still in its infancy. Currently available evidence suggests that reward-related brain regions (i.e., striatum and VMPFC) and theory of mind-related brain regions (i.e., MPFC) play key roles in prosocial behavior based on social reward (i.e., the observer effect), whereas the pMFC and striatum are the important brain regions in social conformity. Furthermore, past research on each of the two forms of social influence has generated a focused hypothesis on its neural mechanisms: similarity and difference between social and nonsocial reward-based decision-making (observer effect) and similarity and difference between social conformity and reinforcement learning. These ideas should be further elucidated in future studies. Finally, in addition to the observer effect and social conformity, researchers have also started investigating the neural bases of other forms of social influence, such as persuasion (e.g., Refs. [100,101]), obedience (e.g., Ref. [102]), and social influence on value-based decision-making (e.g., Refs. [92–94]). Uncovering the neural mechanisms of various forms of social influences is an essential step toward the complete understanding of human social behaviors.

References

[1] Triplett N. The dynamogenic factors in pacemaking and competition. Am J Psychol 1898;9:507–33.
[2] Stroebe W. The truth about triplett (1898), but nobody seems to care. Perspect Psychol Sci 2012;7:54–7.

[3] Bereczkei T, Birkas B, Kerekes Z. Public charity offer as a proximate factor of evolved reputation-building strategy: an experimental analysis of a real-life situation. Evol Hum Behav 2007; 28:277—84.

[4] Bereczkei T, Birkas B, Kerekes Z. Altruism towards strangers in need: costly signaling in an industrial society. Evol Hum Behav 2010;31:95—103.

[5] Satow KL. Social approval and helping. J Exp Soc Psychol 1975; 11:501—9.

[6] Andreoni J, Petrie R. Public goods experiments without confidentiality: a glimpse into fund-raising. J Public Econ 2004;88:1605—23.

[7] Hofman E, McCabe K, Shachat K, Smith VL. Preferences, property rights, and anonymity in bargaining games. Game Econ Behav 1994;7:346—80.

[8] Rege M, Telle K. The impact of social approval and framing on cooperation in public good situations. J Pub Econ 2004;88: 1625—44.

[9] Baumeister RF. A self-presentational view of social phenomena. Psychol Bull 1982;91:3—26.

[10] Izuma K. The social neuroscience of reputation. Neurosci Res 2012;72:283—8.

[11] Amodio DM, Frith CD. Meeting of minds: the medial frontal cortex and social cognition. Nat Rev Neurosci 2006;7:268—77.

[12] Call J, Tomasello M. Does the chimpanzee have a theory of mind? 30 years later. Trends Cogn Sci 2008;12:187—92.

[13] Fujii T, Takagishi H, Koizumi M, Okada H. The effect of direct and indirect monitoring on generosity among preschoolers. Sci Rep 2015;5:9025.

[14] Leimgruber KL, Shaw A, Santos LR, Olson KR. Young children are more generous when others are aware of their actions. PLoS One 2012;7:e48292.

[15] Engelmann JM, Herrmann E, Tomasello M. Five-year olds, but not chimpanzees, attempt to manage their reputations. PLoS One 2012;7:e48433.

[16] Engelmann JM, Herrmann E, Tomasello M. The effects of being watched on resource acquisition in chimpanzees and human children. Anim Cogn 2016;19:147—51.

[17] Izuma K. Social reward. In: Toga AW, editor. Brain mapping: an encyclopedic reference, vol. 3. Oxford: Elsevier; 2015. p. 21—3.

[18] Izuma K, Saito DN, Sadato N. Processing of social and monetary rewards in the human striatum. Neuron 2008;58:284—94.

[19] Korn CW, Prehn K, Park SQ, Walter H, Heekeren HR. Positively biased processing of self-relevant social feedback. J Neurosci 2012;32:16832—44.

[20] Ito A, Fujii T, Ueno A, Koseki Y, Tashiro M, Mori E. Neural basis of pleasant and unpleasant emotions induced by social reputation. Neuroreport 2011;22:679—83.

[21] Kawasaki I, Ito A, Fujii T, Ueno A, Yoshida K, Sakai S, Mugikura S, Takahashi S, Mori E. Differential activation of the ventromedial prefrontal cortex between male and female givers of social reputation. Neurosci Res 2016;103:27—33.

[22] Cooper JC, Dunne S, Furey T, O'Doherty JP. The role of the posterior temporal and medial prefrontal cortices in mediating learning from romantic interest and rejection. Cereb Cortex 2014;24:2502—11.

[23] Davey CG, Allen NB, Harrison BJ, Dwyer DB, Yucel M. Being liked activates primary reward and midline self-related brain regions. Hum Brain Mapp 2010;31:660—8.

[24] Izuma K, Saito DN, Sadato N. Processing of the incentive for social approval in the ventral striatum during charitable donation. J Cogn Neurosci 2010;22:621—31.

[25] Ruff CC, Fehr E. The neurobiology of rewards and values in social decision making. Nat Rev Neurosci 2014;15:549—62.

[26] Carelli RM, Ijames SG, Crumling AJ. Evidence that separate neural circuits in the nucleus accumbens encode cocaine versus "natural" (water and food) reward. J Neurosci 2000;20:4255—66.

[27] Carelli RM, Wondolowski J. Selective encoding of cocaine versus natural rewards by nucleus accumbens neurons is not related to chronic drug exposure. J Neurosci 2003;23:11214—23.

[28] Bowman EM, Aigner TG, Richmond BJ. Neural signals in the monkey ventral striatum related to motivation for juice and cocaine rewards. J Neurophysiol 1996;75:1061—73.

[29] Watson KK, Platt ML. Social signals in primate orbitofrontal cortex. Curr Biol 2012;22:2268—73.

[30] Klein JT, Platt ML. Social information signaling by neurons in primate striatum. Curr Biol 2013;23:691—6.

[31] Sescousse G, Li Y, Dreher JC. A common currency for the computation of motivational values in the human striatum. Soc Cogn Affect Neurosci 2015;10:467—73.

[32] Izuma K, Saito DN, Sadato N. The roles of the medial prefrontal cortex and striatum in reputation processing. Soc Neurosci 2010; 5:133—47.

[33] Frith U, Frith C. The social brain: allowing humans to boldly go where no other species has been. Philos Trans R Soc Lond B Biol Sci 2010;365:165—76.

[34] Frith U. Autism: explaining the enigma. 2nd ed. Oxford: Blackwell; 2003.

[35] Frith U. Mind blindness and the brain in autism. Neuron 2001;32: 969—79.

[36] Baron-Cohen S. Mindblindness: an essay on autism and theory of mind. Cambridge (MA): MIT Press; 1995.

[37] Frith CD, Frith U. The self and its reputation in autism. Neuron 2008;57:331—2.

[38] Izuma K, Matsumoto K, Camerer CF, Adolphs R. Insensitivity to social reputation in autism. Proc Natl Acad Sci USA 2011;108: 17302—7.

[39] Cage E, Pellicano E, Shah P, Bird G. Reputation management: evidence for ability but reduced propensity in autism. Autism Res 2013;6:433—42.

[40] Chevallier C, Molesworth C, Happe F. Diminished social motivation negatively impacts reputation management: autism spectrum disorders as a case in point. PLoS One 2012;7:e31107.

[41] Chevallier C, Kohls G, Troiani V, Brodkin ES, Schultz RT. The social motivation theory of autism. Trends Cogn Sci 2012;16:231—9.

[42] Chevallier C, Grezes J, Molesworth C, Berthoz S, Happe F. Brief report: selective social anhedonia in high functioning autism. J Autism Dev Disord 2012;42:1504—9.

[43] Dawson G, Meltzoff AN, Osterling J, Rinaldi J, Brown E. Children with autism fail to orient to naturally occurring social stimuli. J Autism Dev Disord 1998;28:479—85.

[44] Osterling J, Dawson G. Early recognition of children with autism — a study of 1st birthday home videotapes. J Autism Dev Disord 1994;24:247—57.

[45] Sasson NJ, Dichter GS, Bodfish JW. Affective responses by adults with autism are reduced to social images but elevated to images related to circumscribed interests. PLoS One 2012;7:e42457.

[46] Sepeta L, Tsuchiya N, Davies MS, Sigman M, Bookheimer SY, Dapretto M. Abnormal social reward processing in autism as indexed by pupillary responses to happy faces. J Neurodev Disord 2012;4:17.

[47] Scott-Van Zeeland AA, Dapretto M, Ghahremani DG, Poldrack RA, Bookheimer SY. Reward processing in autism. Autism Res 2010;3:53—67.

[48] Dichter GS, Felder JN, Green SR, Rittenberg AM, Sasson NJ, Bodfish JW. Reward circuitry function in autism spectrum disorders. Soc Cogn Affect Neurosci 2012;7:160—72.

[49] Kohls G, Schulte-Ruther M, Nehrkorn B, Muller K, Fink GR, Kamp-Becker I, et al. Reward system dysfunction in autism spectrum disorders. Soc Cogn Affect Neurosci 2013;8:565—72.

[50] Asch SE. Effects of group pressure upon the modification and distortion of judgment. In: Guetzkow H, editor. Groups, leadership and men. Pittsburgh: Carnegie Press; 1951. p. 177—90.

[51] Sherif M. The psychology of social norms. New York: Harper; 1936.

[52] Izuma K. The neural basis of social influence and attitude change. Curr Opin Neurobiol 2013;23:456–62.

[53] Schnuerch R, Gibbons H. A review of neurocognitive mechanisms of social conformity. Soc Psychol 2014;45:466–78.

[54] Stallen M, Sanfey AG. The neuroscience of social conformity: implications for fundamental and applied research. Front Neurosci 2015;9:337.

[55] Klucharev V, Hytonen K, Rijpkema M, Smidts A, Fernandez G. Reinforcement learning signal predicts social conformity. Neuron 2009;61:140–51.

[56] Nook EC, Zaki J. Social norms shift behavioral and neural responses to foods. J Cogn Neurosci 2015;27:1412–26.

[57] Campbell-Meiklejohn DK, Bach DR, Roepstorff A, Dolan RJ, Frith CD. How the opinion of others affects our valuation of objects. Curr Biol 2010;20:1165–70.

[58] Shima K, Tanji J. Role for cingulate motor area cells in voluntary movement selection based on reward. Science 1998;282:1335–8.

[59] Matsumoto M, Matsumoto K, Abe H, Tanaka K. Medial prefrontal cell activity signaling prediction errors of action values. Nat Neurosci 2007;10:647–56.

[60] Delgado MR. Reward-related responses in the human striatum. Ann NY Acad Sci 2007;1104:70–88.

[61] Schultz W, Tremblay L, Hollerman JR. Changes in behavior-related neuronal activity in the striatum during learning. Trends Neurosci 2003;26:321–8.

[62] Ridderinkhof KR, Ullsperger M, Crone EA, Nieuwenhuis S. The role of the medial frontal cortex in cognitive control. Science 2004;306:443–7.

[63] Kawai T, Yamada H, Sato N, Takada M, Matsumoto M. Roles of the lateral habenula and anterior cingulate cortex in negative outcome monitoring and behavioral adjustment in nonhuman primates. Neuron 2015;88:792–804.

[64] Hikosaka O, Isoda M. Switching from automatic to controlled behavior: cortico-basal ganglia mechanisms. Trends Cogn Sci 2010;14:154–61.

[65] Klucharev V, Munneke MAM, Smidts A, Fernandez G. Downregulation of the posterior medial frontal cortex prevents social conformity. J Neurosci 2011;31:11934–40.

[66] Izuma K, Adolphs R. Social manipulation of preference in the human brain. Neuron 2013;78:563–73.

[67] Knutson B, Westdorp A, Kaiser E, Hommer D. FMRI visualization of brain activity during a monetary incentive delay task. NeuroImage 2000;12:20–7.

[68] Bush G, Shin LM. The multi-source interference task: an fMRI task that reliably activates the cingulo-frontal-parietal cognitive/attention network. Nat Protoc 2006;1:308–13.

[69] Heider F. The psychology of interpersonal relations. New York: Wiley; 1958.

[70] Miltner WHR, Braun CH, Coles MGH. Event-related brain potentials following incorrect feedback in a time-estimation task: evidence for a "generic" neural system for error detection. J Cogn Neurosci 1997;9:788–98.

[71] Gehring WJ, Willoughby AR. The medial frontal cortex and the rapid processing of monetary gains and losses. Science 2002; 295:2279–82.

[72] Chen J, Wu Y, Tong GY, Guan XM, Zhou XL. ERP correlates of social conformity in a line judgment task. BMC Neurosci 2012:13.

[73] Kim BR, Liss A, Rao M, Singer Z, Compton RJ. Social deviance activates the brain's error-monitoring system. Cogn Affect Behav Neurosci 2012;12:65–73.

[74] Shestakova A, Rieskamp J, Tugin S, Ossadtchi A, Krutitskaya J, Klucharev V. Electrophysiological precursors of social conformity. Soc Cogn Affect Neurosci 2013;8:756–63.

[75] Schnuerch R, Trautmann-Lengsfeld SA, Bertram M, Gibbons H. Neural sensitivity to social deviance predicts attentive processing of peer-group judgment. Soc Neurosci 2014;9:650–60.

[76] Campbell-Meiklejohn DK, Simonsen A, Jensen M, Wohlert V, Gjerloff T, Scheel-Kruger J, et al. Modulation of social influence by methylphenidate. Neuropsychopharmacology 2012;37: 1517–25.

[77] Simonsen A, Scheel-Kruger J, Jensen M, Roepstorff A, Moller A, Frith CD, et al. Serotoninergic effects on judgments and social learning of trustworthiness. Psychopharmacology 2014;231: 2759–69.

[78] Schultz W. Updating dopamine reward signals. Curr Opin Neurobiol 2013;23:229–38.

[79] Cools R, Nakamura K, Daw ND. Serotonin and dopamine: unifying affective, activational, and decision functions. Neuropsychopharmacology 2011;36:98–113.

[80] Rogers RD. The roles of dopamine and serotonin in decision making: evidence from pharmacological experiments in humans. Neuropsychopharmacology 2011;36:114–32.

[81] Volkow ND, Wang G, Fowler JS, Logan J, Gerasimov M, Maynard L, et al. Therapeutic doses of oral methylphenidate significantly increase extracellular dopamine in the human brain. J Neurosci 2001;21:RC121.

[82] Deuker L, Muller AR, Montag C, Markett S, Reuter M, Fell J, et al. Playing nice: a multi-methodological study on the effects of social conformity on memory. Front Hum Neurosci 2013;7:79.

[83] Yavich L, Forsberg MM, Karayiorgou M, Gogos JA, Mannisto PT. Site-specific role of catechol-O-methyltransferase in dopamine overflow within prefrontal cortex and dorsal striatum. J Neurosci 2007;27:10196–209.

[84] Frank MJ, Moustafa AA, Haughey HM, Curran T, Hutchison KE. Genetic triple dissociation reveals multiple roles for dopamine in reinforcement learning. Proc Natl Acad Sci USA 2007;104: 16311–6.

[85] Falk EB, Way BM, Jasinska AJ. An imaging genetics approach to understanding social influence. Front Hum Neurosci 2012;6:168.

[86] Huang Y, Kendrick KM, Zheng H, Yu R. Oxytocin enhances implicit social conformity to both in-group and out-group opinions. Psychoneuroendocrinology 2015;60:114–9.

[87] Stallen M, De Dreu CK, Shalvi S, Smidts A, Sanfey AG. The herding hormone: oxytocin stimulates in-group conformity. Psychol Sci 2012;23:1288–92.

[88] Edelson MD, Shemesh M, Weizman A, Yariv S, Sharot T, Dudai Y. Opposing effects of oxytocin on overt compliance and lasting changes to memory. Neuropsychopharmacology 2015;40:966–73.

[89] Insel TR. The challenge of translation in social neuroscience: a review of oxytocin, vasopressin, and affiliative behavior. Neuron 2010;65:768–79.

[90] Meyer-Lindenberg A, Domes G, Kirsch P, Heinrichs M. Oxytocin and vasopressin in the human brain: social neuropeptides for translational medicine. Nat Rev Neurosci 2011;12:524–38.

[91] Edelson M, Sharot T, Dolan RJ, Dudai Y. Following the crowd: brain substrates of long-term memory conformity. Science 2011; 333:108–11.

[92] Engelmann JB, Moore S, Monica Capra C, Berns GS. Differential neurobiological effects of expert advice on risky choice in adolescents and adults. Soc Cogn Affect Neurosci 2012;7:557–67.

[93] Meshi D, Biele G, Korn CW, Heekeren HR. How expert advice influences decision making. PLoS One 2012;7:e49748.

[94] Chung D, Christopoulos GI, King-Casas B, Ball SB, Chiu PH. Social signals of safety and risk confer utility and have asymmetric effects on observers' choices. Nat Neurosci 2015;18:912–6.

[95] Deutsch M, Gerard HB. A study of normative and informational social influences upon individual judgement. J Abnorm Psychol 1955;51:629–36.

[96] Berns GS, Chappelow J, Zink CF, Pagnoni G, Martin-Skurski ME, Richards J. Neurobiological correlates of social conformity and independence during mental rotation. Biol Psychiatry 2005;58: 245–53.

[97] Gawronski B, Strack F, editors. Cognitive consistency: a fundamental principle in social cognition. 1st ed. New York: Guilford Press; 2012.

[98] Izuma K. Attitude change and cognitive consistency. In: Toga AW, editor. Brain mapping: an encyclopedic reference, vol. 3. Oxford: Elsevier; 2015. p. 247–50.

[99] Toelch U, Dolan RJ. Informational and normative influences in conformity from a neurocomputational perspective. Trends Cogn Sci 2015;19:579–89.

[100] Chua HF, Ho SS, Jasinska AJ, Polk TA, Welsh RC, Liberzon I, et al. Self-related neural response to tailored smoking-cessation messages predicts quitting. Nat Neurosci 2011;14:426–7.

[101] Falk EB, Berkman ET, Mann T, Harrison B, Lieberman MD. Predicting persuasion-induced behavior change from the brain. J Neurosci 2010;30:8421–4.

[102] Cheetham M, Pedroni AF, Antley A, Slater M, Jancke L. Virtual milgram: empathic concern or personal distress? Evidence from functional MRI and dispositional measures. Front Hum Neurosci 2009;3:29.

17

Social Dominance Representations in the Human Brain

R. Ligneul, J.-C. Dreher

Institute of Cognitive Science (CNRS), Lyon, France

Abstract

Social dominance refers to relationships wherein the goals of one individual prevail over the goals of another individual in a systematic manner. Dominance hierarchies have emerged as a major evolutionary force to drive dyadic asymmetries in a social group. Understanding how the brain detects, represents, implements, and monitors social dominance hierarchies constitutes a fundamental topic for social neuroscience as well as a major challenge for the future of clinical psychiatry. In this chapter, we argue that the emergence of dominance relationships is learned incrementally, by accumulating positive and negative competitive feedback associated with specific individuals and other members of the social group. We consider such emergence of social dominance as a reinforcement learning problem inspired by neurocomputational approaches traditionally applied to nonsocial cognition. We also report how dominance hierarchies induce changes in specific brain systems, and we review the literature on interindividual differences in the appraisal of social hierarchies, as well as the underlying modulations of the cortisol, testosterone, and serotonin/dopamine systems that mediate these phenomena.

INTRODUCTION

Social Hierarchies in Health and Well-Being

Social dominance refers to situations in which an "individual or a group controls or dictates others' behavior, primarily in competitive situations" [1,2]. Social dominance hierarchies influence access to resources and mating partners and therefore constitute a potent biological force binding together social behavior, well-being, and evolutionary success. The concept of social dominance is most often applied to "learned relationships," shaped by a history of social victories and defeats within dyads

of individuals [3]. Together with other forms of power, social dominance asymmetries constitute a pivotal concept for understanding social organizations and predicting individual behaviors.

Many animal studies indicate that iterated social defeats can trigger maladaptive social avoidance, behavioral inhibition, elevated glucocorticoid levels, and higher vulnerability to addiction, anxiety, or depression [4–7]. Epidemiological approaches in humans have subsequently confirmed that suffering from a chronically low socioeconomic status or enduring transient status-lowering threats facilitates both somatic and psychiatric disorders [8,9]. Unfortunately, it has long been difficult to disentangle the specific contributions of stress, socioeconomic status, and social dominance on the human brain. In particular, very little is known about the cerebral mechanisms governing the progressive establishment of social dominance hierarchies and associated neurobehavioral changes through real-life interactions (for a review of such mechanisms in nonhuman primates and of genetic mechanisms in zebra finches, see Chapters 15 and 28).

In this chapter, we will first consider the learning of social dominance as an incremental process, allowing us to develop a neurocomputational approach to a key decision problem (i.e., to initiate or not a competitive interaction). Second, we will review important interindividual differences that naturally derive and influence social dominance hierarchies. Third, we will highlight the tight relationship of social dominance with stress and neuroplasticity [6] by reporting its effects on the hypothalamic–pituitary–adrenal (HPA)/hypothalamic–pituitary–gonadal (HPG) axes as well as on the serotonin and dopamine systems [10].

Social Hierarchies as a Major Evolutionary Pressure and Pivotal Feature of Societies

In folk psychology, dominance is often considered a fundamental motive of social organisms. However, while many primatologists agree that social ranks correlate positively with offspring production in many primate and nonprimate species, effect sizes are usually small and many counterexamples exist, indicating that subordinate individuals often achieve decent reproductive success [11,12]. Moreover, dominance hierarchies spontaneously reemerge, even if only subordinate or only dominant individuals are put together to form a new social group at each generation [13,14], implying that the social environment dynamically tunes individuals' brains to promote the best behavioral strategy given the social context, through synaptic and epigenetic plasticity mechanisms. Dominance and subordination can thus be better described as life-history strategies [15,16], because both constitute adaptations to the social environment and both can increase evolutionary fitness.

The importance of social dominance for domain-general cognition is primarily rooted in the so called "social brain hypothesis," which stipulates that the need to optimize behavior within complex social environments largely constrained the evolution of the primate brain. This theory was first outlined in the pioneering study of Alison Jolly in lemurs [17], and it was popularized by Humphrey [18], Byrne and Whiten [19], and Dunbar [20], who provided correlational evidence for the coevolution of social complexity and various markers of brain development. In evolved animals, other group members constitute stochastically behaving entities guided by hidden mental states and obeying complex sets of rules. Consequently, predicting their actions and improving our transactions with them requires very elaborated computations which have long been overlooked in cognitive neurosciences. A similar argument also applies to simpler behaviors: the neural mechanism that enables human subjects to avoid selecting a blue square associated with electric shock delivery in the lab might have partly been sculpted, throughout evolution, to enable efficient avoidance of aggressive dominant individuals within one's social group. The possibility that "social life provided the evolutionary context of primate intelligence" [17] is thus key to foreseeing the importance of social dominance for domain-general, high level human cognition.

Previously, a large neural circuit activated when people make decisions in social settings has been identified using functional magnetic resonance imaging (fMRI) in humans. Key components include the orbitofrontal cortex, the ventromedial prefrontal cortex (VMPFC), the dorsomedial and dorsolateral prefrontal cortex (DLPFC), parts of the superior temporal sulcus (STS)

including a region near the temporoparietal junction (TPJ), and the anterior cingulate gyrus. In sum, there is now extensive evidence that social decision-making relies on many "nonsocial" subcortical and brain stem circuits. Social decision-making also overtakes the canonical cortical network of social cognition encompassing the medial prefrontal cortex (MPFC), the STS, and the TPJ (Fig. 17.1A).

Since 2005, a number of fMRI studies have investigated the perception of social ranks based on noncompetitive cues, such as wealth [23,24], postures [25], uniforms [23], facial traits [26], and celebrity, height, or intelligence [27,28]. Completing the pioneering work of Zink et al. [29], these studies have also demonstrated the engagement of a large brain network involved in social hierarchy processing, including the amygdala, hippocampus, striatum, ventrolateral prefrontal cortex (VLPFC), rostromedial prefrontal cortex (RMPFC), inferior parietal lobule (IPL), and the fusiform gyrus (Fig. 17.1B–C). Although these perceptual processes linked with social dominance raise many important questions (for a review see [30]), we will focus in this chapter on the neurocomputational processes that underlie the *learning* of social dominance statuses (SDSs).

LEARNING SOCIAL DOMINANCE HIERARCHIES

In what follows, we will propose that *one's own* dominance status is learned incrementally by accumulating the numerous competitive feedbacks (victories and defeats) obtained against other group members. Put simply, individuals who experience on average negative feedback following competitive encounters will develop an adaptive subordinate profile, which may take them away from social conflicts by promoting submission, and vice versa for dominant individuals. Importantly, the same principle can also be applied to the rapid update of *others'* SDS during competitive interactions. These assumptions allowed us to develop a neurocomputational approach to characterize the emergence of social dominance relationships through time and provide computational neuroscientists with an adequate framework to test quantitative and mechanistic hypotheses about this process [21,31].

Reinforcement Learning Approaches to Social Cognition

Mathematical models are increasingly used in social neuroscience as they can probe, simultaneously, several cognitive processes which would otherwise not be separable [21,31] (see also Chapters 18 and 19). Three main

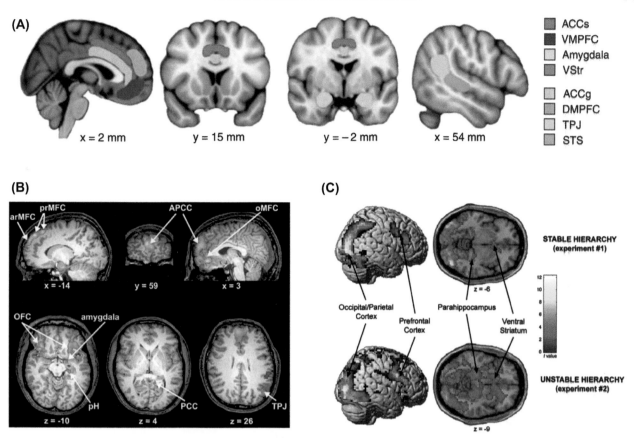

FIGURE 17.1   (A) **The functional neuroanatomy of social behavior.** Primary colors denote brain regions activated by reward and valuation, frequently identified in studies of social interaction within the frame of reference of the subject's own actions: anterior cingulate cortex sulcus (*ACCs*), ventromedial prefrontal cortex (*VMPFC*), amygdala, and ventral striatum (*VStr*). Pastels denote brain regions activated by considering the intentions of another individual: anterior cingulate cortex gyrus (*ACCg*), dorsomedial prefrontal cortex (*DMPFC*), temporoparietal junction (*TPJ*), and superior temporal sulcus (*STS*). *(From Behrens TEJ, Hunt LT, Rushworth MFS. The computation of social behavior. Science 2009;324:1160–64, with permission.)* (B) **Cerebral substrates of social comparison processes.** Comparative judgments about the height or the intelligence of others activate specifically the anterior prefrontal cortex, the amygdala, and the TPJ. *(From Lindner M, Hundhammer T, Ciaramidaro A, Linden DEJ, Mussweiler T. The neural substrates of person comparison—an fMRI study. Neuroimage 2008;40:963–71, with permission.)* (*APCC*, anterior paracingulate cortex; *prMFC*, posterior medial prefrontal cortex; *arMFC*, anterior portion of the rostral medial frontal cortex; *OFC*, orbitofrontal cortex; *oMFC*, orbital medial prefrontal cortex) (C) **Statistical maps of the brain regions more engaged in the comparison "superior player > inferior player."** Compared to inferior individuals, the perception of superior individuals elicited stronger activations in many brain regions including the dorsolateral prefrontal cortex, the medial prefrontal cortex, the striatum, and the occipitoparietal cortices. *(From Zink CF et al. Know your place: neural processing of social hierarchy in humans. Neuron 2008;58:273–83, with permission.)*

subsystems have been unraveled by this approach: the VMPFC, together with the ventral striatum, may be responsible for observational learning and reward prediction error signaling [32,33]. Second, the anterior cingulate cortex, the DLPFC, and the IPL may compute prediction errors elicited when others' behaviors deviate from our predictions about them. Third, the RMPFC and TPJ/posterior STS may compute the updating of one's own and others' mental states, [33,34] as well as the degree of interpersonal influence during social interactions, that is, the degree to which one's own actions and utilities are determined by others' skills and strategies [35].

Within iterated competitive games, rmPFC activity was often shown to encode second-order variables (i.e., variables inferred from other learned variables) [31,34,36]. For example, in the inspector game or in the

matching pennies task respectively used in Hampton et al. [34] and Seo et al. [36], a player able to take into account the influence of his or her past choices over the evolution of another's strategy will be able to increase his or her chances of winning to the detriment of the other, because this second-order information enables the player to better predict the other's choice (see Chapter 18 from Lee et al.). Given that the ability to control others' behaviors and outcomes is at the core of the social dominance concept, it would be tempting to characterize such player as "mentally dominating" his or her opponent. Interestingly, if the "subordinate" opponent starts to play on a purely random basis, the opportunity for interpersonal control disappears, possibly contributing to the frequency of inconsistent or irrational social behaviors in humans. Moreover, the consequences of

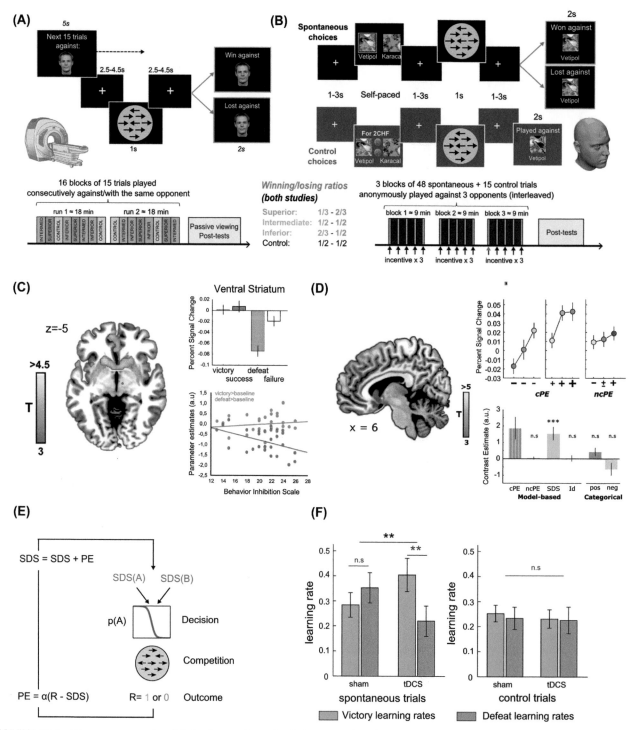

FIGURE 17.2   **The emergence of social dominance through reinforcement learning**. (A) **Typical trial and experiment time courses for the fMRI experiment**. During 15 trials of a "miniblock," subjects played against (or with) the same player in the competitive (or control) situation. The competitive task required subjects to evaluate a series of *stationary arrows*, indicating which direction the majority of these *arrows* pointed (*left or right*). The task was performed against one of three virtual opponents implicitly associated with three frequencies of winning and losing. To succeed in the competition, subjects were instructed to answer accurately and faster than their opponent. (B) **Typical trial and experiment time courses of the brain stimulation experiment**. Subjects performed a similar perceptual task but now opponents were marked by visual symbols and artificial names rather than a face photograph. Subjects could now choose which opponent to defy among two alternatives (three opponents per block), in two types of trials designed to distinguish dominance-based (spontaneous) and reward-based (control) choices. In half of the subjects, the rostromedial prefrontal cortex (RMPFC) was monitored with the excitatory anodal electrode of the transcranial direct current stimulation (tDCS) apparatus (magenta; the reference electrode on the vertex is in blue). (C) **Striatal encoding of competitive defeats**. Competition-specific outcome signals revealed by the interaction competitive victory and control failure > competitive defeat and control success

this interaction between social dominance and mentalization of interpersonal influence within strategic games has also been explored based on evolutionary reinforcement learning models [37], which showed that mentalizing abilities may indeed promote social status when the rate of cooperation among group members is low.

Finally, neuroimaging studies indicate that the RMPFC (as well as the TPJ and the medial STS) is generally more engaged when subjects are invited to compete against other humans compared to computers [29,38,39,40–42], suggesting that the RMPFC might be involved in representing the mental states of others and/or the dominance relationship emerging between two individuals. Competitive interactions also tend to lead to higher activity of the RMPFC compared to cooperative interactions, which indicates a possible competition specificity of the cognitive processes implemented in this structure [40,43].

The Emergence of Dominance and Subordination by Reinforcement Learning

According to the pioneering theory of Bernstein [3], dominance relationships are learned progressively, by integrating the positive and negative competitive feedback associated with specific individuals and other members of the social group taken as a whole. A competitive outcome thus provides information about others' behaviors (which underlies the representation of objective social hierarchies) and information about oneself (which may lead to adjustments in subjective social status and related variables such as self-esteem). In other words, the former information may foster target- or dyad-specific dominance behaviors while the latter may foster target-independent behavioral profiles of dominance and subordination with individuals. Although both can be described by a similar reinforcement learning scheme, the time constants (and learning rate) associated with those two processes might be different, as self-representations intuitively appear less volatile than representations about others, in most people.

By learning social dominance representations, individuals can anticipate their probability of winning versus losing and therefore decide whether they should carry on fighting or disengage from a confrontation in order to limit the physical and social costs associated with a defeat. Outside of agonistic interactions, monitoring such probability can also prevent the escalation of social conflicts for which the risks of losing outweigh the expected benefits of winning. Interestingly, in the cost–benefit trade-off that underlies the decision to compete against another conspecific, the cost is typically associated with a property of the opponent (i.e., his or her strength or skill, which translates into a given probability of losing and a given effort to be exerted when trying to win), whereas the benefit usually refers to the external resource at stake (a notable exception being social competitive play, in which no such external incentive is present). The anticipated energy costs associated with competitive interactions and external resources motivating the social conflict are thus pivotal to making optimal decisions. Yet we will here restrict the problem to winning-losing probability estimation, which already provide strong empirical evidence to the aforementioned reinforcement-learning model.

In a recent study, we have induced an implicit dominance hierarchy in men through a competitive game involving three opponents of different strengths, complemented by a noncompetitive control condition (Fig. 17.2A). Using fMRI, we first observed specific responses to social defeats in the ventral striatum and other subcortical regions, which were correlated with trait inhibition across subjects (Fig. 17.2C; see also "Interindividual Differences Resulting From Social Status and Personality"). Second, and more importantly, model-based analyses highlighted the functions of the rostromedial cortices in tracking the dominance status of opponents (i.e., anticipated winning-losing probabilities). More specifically, the RMPFC encoded opponent-specific prediction errors and appeared to monitor the probability of winning against each player in a dynamic fashion, throughout the competitive task (Fig. 17.2D).

These findings were obtained by applying a classical Rescorla–Wagner rule [Eq. (17.1); Fig. 17.2E] tracking

were observed in the bilateral ventral striatum. The amplitude of defeat-related deactivations was correlated with the behavioral inhibition personality trait across subjects. (D) **Encoding of competitive prediction errors (cPE) in the RMPFC**. Analyses showed that the activity changes observed in the RMPFC encoded a signed competitive prediction error, which did not reflect winning or losing per se, the identity of the opponent, or interactions of these two factors. ncPE, noncompetitive prediction error; SDS, social dominance status (Id, opponent type). (E) **Overview of the computational model**. Our reinforcement learning algorithm assumed that decisions are taken probabilistically (softmax policy) according to the value of each available opponent. Once the competition occurred, the value of the selected opponent was updated for the next trial proportional to the prediction error elicited by the outcome [i.e., $(R - SDS)$ with victory $R = 1$ and defeat $R = 0$] multiplied by the learning rate α. (F) **Effects of RMPFC tDCS on the parameters governing social dominance learning**. Whereas average learning rates related to defeats and victories were balanced in the sham group, stimulating the RMPFC using anodal tDCS induced a significant imbalance in the learning rates, with more weight placed on victories and less weight place on defeats. *From Ligneul R, Obeso I, Ruff CC, Dreher JC. Dynamical representation of dominance relationships in the human medial prefrontal cortex. Current Biology, in press.*

the probability of winning in such agonistic interactions, by simply initializing P_{win} (or SDS) at 0.5 and monitoring the outcome R (0 for defeat, 1 for victory);

$$SDS(t+1) = SDS(t) + \alpha * (R - SDS(t)) \qquad (17.1)$$

The prediction error term $R-SDS(t)$ multiplied by the learning rate α allows updating the anticipated chances of winning (or SDS) in the next encounter at $t+1$. Also called momentary social dominance status, or SDS, this "probability estimate could then be used to decide whether one should defy another conspecific, according to a decisional policy such as the probabilistic softmax rule. In real-life competitive settings, when the chances of winning are deemed too low to initiate or continue the fight, the decision-maker may start to submit, hence meeting the criterion of a dominance relationship [3,44]. Moreover, generalizing or averaging of target-specific SDSs over all members of the group would naturally underlie the emergence of chronically dominant or subordinate personality profiles.

Next, using an adapted version of our competitive perceptual decision-making task (Fig. 17.2B), we demonstrated that transcranial direct current stimulation (tDCS) applied over the RMPFC exerted a causal influence over social dominance learning, reflected in higher learning rates associated with victories and lower learning rates associated with defeats (Fig. 17.2F). This result paralleled a study in mice in which viral injections, producing an increase or decrease in overall MPFC activity, led to increases or decreases in the dominance ranks of animals [45].

In the past, it has been difficult to ascertain whether neural network covarying with learning of social dominance was a cause or a consequence of the emergence of social dominance in humans although original experiments in animals suggested a profound impact of dominance on brains and bodies. For example, it was shown that selective lesions of the amygdala or the administration of a serotonergic antidepressant can induce changes in the behavioral expression of dominance in monkeys [46,47] and human patients with lesions of the MPFC are also impaired in their ability to make social dominance attributions based on the narrative description of diverse social interaction [48]. Beyond those classical findings, brain stimulation techniques such as tDCS shall thus constitute an important tool for the study of causality in fine-grained social dominance behaviors among healthy human subjects.

INTERINDIVIDUAL DIFFERENCES AND SOCIAL DOMINANCE

Cognitive neuroscientists tend to rely on the convenient assumption that human brains are largely similar across genders, ethnicities, sexes, ages, and social groups. This assumption facilitates the generalization of observations typically made in a few dozens of students to the whole human population or, at least, to the whole student population. Impeding predictive power and reproducibility of empirical findings, this assumption is unfortunately often invalid. Therefore, the systematic study of interindividual differences has begun, so that many neuroscientists now routinely report differences in personality traits, gray matter volumes, functional activities, or connectivity estimates to better explain the behavioral variability of their subjects [49]. In this perspective, social dominance holds a strong potential to explain variability observed within social groups, which would otherwise be envisioned as homogeneous (such as psychology students participating in social learning experiments). Indeed, the study of social dominance promises more than simply *accounting* for interindividual differences in behavior and physiology: it may also offer mechanistic explanations for the *emergence* of neural, behavioral, cognitive, and social variability. For example, the existence of clear-cut dominance hierarchies in pure strains of rodents (i.e., genetically identical) stresses that the behavioral and physiological features of dominant and subordinate animals derive largely from experience and adaptation. Moreover, although social dominance is a universal principle structuring social groups, it is also highly dependent on the culture, the gender, and the personality of the participants.

Intercultural Differences in the Appraisal of Social Dominance

Human cultures vastly differ in how they value personality traits related to social dominance. For example, the construction of self in collectivistic east Asian cultures tends to be more interdependent upon other group members than in individualistic societies in which the construction of self appears more as a quest for independence and autonomy with respect to other group members [50]. As Sedikides et al. [51] wrote, "in individualistic cultures, the relevant dimension [of self-construal] is agency, defined as a concern with personal effectiveness and social dominance. In collectivistic cultures, however, the relevant dimension is communion, defined as a concern with personal integration and social connection." Acknowledging cultural differences in the promotion of *person-centric* versus *normative-contextual* models of self-construal is thus crucial to avoid a partial, Western-centric conception of behaviors related to social dominance. For example, an influential study demonstrated that European American children are more motivated to solve anagrams when they could choose the category of problem to be solved compared to when their mother or the experimenter chose it for them, whereas Asian American children displayed the opposite pattern [52]. This finding indicates that, early in development, the motivational attitudes toward dominant others (here, the experimenter or the parent) vary greatly across cultures. It is also consistent with the fact that high external locus of control (i.e., the feeling that one's own life is controlled by others) is much more strongly correlated with trait anxiety and negative emotions in individualistic compared to collectivistic societies [53].

In the only available neuroimaging study (as of this writing) probing intercultural differences in the appraisal of social dominance (Fig. 17.3A), Freeman and coauthors demonstrated that the neural correlates of social dominance expressed by body postures were reversed in the ventral striatum and the MPFC when comparing American (more activity in response to dominant postures) and Japanese subjects (more activity to subordinate postures). Interestingly, these responses to dominant and subordinate postures were also correlated with the behavioral tendency of the subjects: a stronger response to dominant postures predicted (self-reported) dominant behaviors typical of Western cultures, whereas a stronger response to subordinate cues predicted the opposite behaviors. Although we firmly believe that most of those

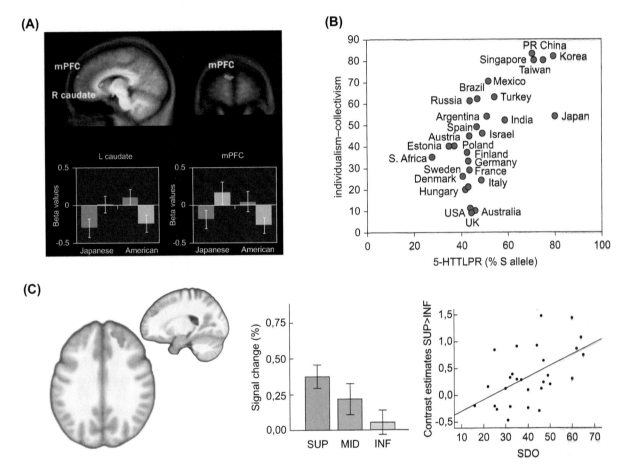

FIGURE 17.3 **Brain-related interindividual differences in the appraisal of social hierarchies.** (A) **Whole-brain analysis testing a status display × culture interaction effect.** In Americans, the medial prefrontal cortex (*mPFC*) and the caudate nucleus exhibited reliably stronger blood oxygen level-dependent (BOLD) responses to dominant stimuli relative to subordinate stimuli; in the Japanese, these same regions exhibited the opposite pattern, showing reliably stronger BOLD responses to subordinate stimuli relative to dominant stimuli. (*From Freeman JB, Rule NO, Adams RB, Ambady N. Culture shapes a mesolimbic response to signals of dominance and subordination that associates with behavior. Neuroimage 2009;47:353–59, with permission.*) (B) **Correlation analysis between Hofstede's individualism–collectivism index and frequency of S allele carriers of 5-HTTLPR across 29 nations.** Collectivistic nations showed higher prevalence of S allele carriers. (*From Chiao JY, Blizinsky KD. Culture-gene coevolution of individualism–collectivism and the serotonin transporter gene. Proc Biol Sci 2010;277:529–37, with permission.*) (5-HTTLPR, serotonin-transporter-linked polymorphic region). (C) **The relationship between political orientation and the neural sensitivity to competitive ranks.** In one of our experiments, the right anterior dorsolateral prefrontal cortex encoded social rank as induced by a prior competitive task against three opponents. In addition, the sensitivity of this brain region to social rank was strongly correlated with the social dominance orientation across subjects, thereby indicating that subjects more prone to legitimizing and reinforcing social inequalities are also more sensitive to competitive hierarchies [67]. *INF*, inferior; *LPP*, late positive potential; *MID*, middle; *SDO*, social dominance orientation; *SUP*, superior; *tDCS*, transcranial direct current stimulation.

inter-individual differences derive from the reinforcement-learning process described above, a cross-cultural genetic analysis revealed that endogenous serotonin reuptake capacity covaried with the individualistic—collectivistic opponency and the steepness of social hierarchies from one country to another (Fig. 17.3B) ([54]; see also "Dopamine, Serotonin, and Social Hierarchies in Rodents and Nonhuman Primates"). Located in the promoter region of the serotonin transporter, this polymorphism is also predictive of stress-coping strategies and resilience [55]. Cross-cultural differences in the perception of social dominance may thus be partly based on polymorphisms of genes engaged in serotoninergic (but also dopaminergic) transmission. Yet, as we will see, these polymorphisms may actually act by altering the learning process which is a the core of social dominance relationships.

Interindividual Differences Resulting From Social Status and Personality

By definition, the existence of a social dominance hierarchy means that different members of a given group experience different social environments and attribute different motivational values to specific social behaviors. Social hierarchies define the type of social dilemma most often faced by individuals and the range of options available to solve them. Thus, it is reasonable to expect that different social ranks would turn into different patterns of brain activity and—possibly—different brain anatomies.

A paper by Noonan, Sallet, Mars, and coauthors studied 36 macaques living in groups of two to seven members in which social dominance hierarchies could be reliably assessed [56] (see also Chapter 15). Their analyses showed that gray matter volumes markedly and reproducibly differed in several brain regions as a function of social rank. Higher ranked animals had more gray matter in the hippocampus, the amygdala, and the serotonergic brainstem, as well as in the medial temporal sulcus and the rostral prefrontal cortex. Because the last two were also correlated with the size of the group in which those animals were housed, the authors suggested that they might be "linked to the social cognitive processes that are taxed by life in more complex social networks and that must also be used if an animal is to achieve a high social status." In addition, lower ranked animals had more gray matter in three regions of the basal ganglia involved in habit learning and aversive processing: the dorsal striatum, the caudate nucleus, and the posterior putamen. As of this writing, the exact mechanisms that drive these correlations are unclear, but the ability of endogenous and drug-induced serotonin release to alter gray matter volumes and/or to stimulate neurogenesis in several brain regions offers promising perspectives [57,58]. In addition,

the pervasive effects of chronic stress on brain circuitry should be taken into account, as it is well known that subordinate individuals tend to be more stressed than dominant individuals and more prone to develop stereotypical behaviors, especially in captivity settings (see "Stress Asymmetries Paralleling Social Hierarchy Rank Have Adverse Consequences on Adrenocortical, Reproductive, and Neural Systems").

In humans, the existence of neuroanatomical correlates of social ranks per se remains underexplored. Some developmental neuroimaging studies indicate that, even after correction for several confounding factors, higher parental socioeconomic status still predicts higher prefrontal cortical thickness in children [59] and increased gray matter volumes in several brain regions, including prefrontal but also occipital, parietal, and limbic areas [60]. Behavioral sensitivity to social dominance expressed by facial traits or induced by competitive games may also be predicted by neuroanatomical variations in the insula and other regions [61,62].

In addition to the (sustained) neuroanatomical signature of social hierarchies, humans also *process* social events differently, depending on their own social standing. For example, Ly and coauthors demonstrated that low-status subjects had stronger striatal responses when presented with low-status faces, whereas the opposite was true for high-status subjects [63]. In one of our experiments [64], we found that the sensitivity of the ventral striatum to social defeats in a competitive perceptual decision-making game was correlated with the behavioral inhibition personality trait [65], often linked with social subordination and anxiety [66]. Because more inhibited individuals had more salient deactivations in response to defeats in this structure, one could infer that the repeated experience of social defeats not only lowers social status and social dominance, but also heightens the overall sensitivity of the motivational system to threats and negative events (Fig. 17.2C). Moreover, in another experiment [67], we observed that the sensitivity of the right anterior prefrontal cortex to social rank of neutral faces was strongly correlated with the Social Dominance Orientation questionnaire (Fig. 17.3E), which reflects the degree to which one envisions social hierarchy and economic inequalities as legitimate and necessary phenomena [68]. Deciphering the neurocognitive mechanisms involved in the appraisal of social hierarchy may thus help us understand real-world political divides [69].

NEUROCHEMICAL APPROACHES TO SOCIAL DOMINANCE AND SUBORDINATION

The neurochemical processes involved in the emergence, maintenance, and consequences of social dominance hierarchies by reinforcement-learning are key to translating fundamental social neurosciences into new therapeutic options and to improve our understanding of psychosocial disorders. Those disorders are largely mediated by pharmacological modulation of plastic, stress-sensitive systems such as hormonal and monoaminergic signaling. All "culture-like" features of animal societies can influence the dynamical form taken by a social hierarchy and the manifold individual profiles composing it. Nonetheless, the hormonal and neural systems that underlie the variable expressions of dominance across species, groups, and individuals have been highly conserved through evolution. Consequently, the acute and long-term consequences of social defeats are now widely studied because of their ability to trigger robust anxiety- and depression-like symptoms in animals, thereby providing a useful translational model of affective disorders [70].

Stress Asymmetries Paralleling Social Hierarchy Rank Have Adverse Consequences on Adrenocortical, Reproductive, and Neural Systems

In the attempt to explain dominance hierarchies in rodents, nonhuman primates, and humans, cortisol and testosterone have long played the first roles. Relatively easy to quantify, they reproducibly covary with social rank across species and experimental conditions. They are often jointly investigated because they interact at the physiological and behavioral levels. Cortisol and testosterone are the end products of two hormonal axes reciprocally inhibiting each other: the HPA axis and the HPG axis, respectively [71,72] (Fig. 17.4A). Moreover, exposure to stressors activates a chain of endocrine reactions, including secretion of glucocorticoids by the adrenal glands, which reallocate energy resources necessary to adapt rapidly to the stressor. In the long term, high levels of glucocorticoids can however disrupt an essential negative feedback loop, hence leading to immune function suppression as well as impairments in hippocampal and prefrontal functioning [73,74]. On the other hand, testosterone largely contributes to muscle mass, male secondary sexual characteristics, and reactive aggressive behaviors, which are relevant to predict the onset and the outcome of agonist interactions [6,72].

Modern ethology has reported a great variability between and even within species, regarding whether high- or low-ranking animals are the ones who are the most stressed in a dominance hierarchy. Many factors influence such rank-associated stress in stratified mammalian societies. Such factors include, but are not limited to, species-level variations in style of breeding system (cooperative/competitive), social and mating systems, housing, despotic versus egalitarian hierarchy style [75], and hierarchy stability within species [76]. In despotic hierarchies, resource access is skewed markedly and dominant positions are attained through aggression and intimidation, whereas in "egalitarian" hierarchies resource distribution is more equal and dominance is attained with the support of subordinate individuals. A general concept to help resolve these differences in the relationship between rank and stress across species is that it is the rank that experiences the most physical and psychological stressors that tends to display the most severe stress-related response.

In primates, glucocorticoid levels are often higher in subordinate males whenever a dominance hierarchy is stabilized and testosterone levels are generally independent of social rank [76,77]. However, higher-ranking males tend to experience higher testosterone and glucocorticoid (stress hormone) levels than lower-ranking males whenever their dominance rank is threatened (i.e., in a period of social instability) [76,77]. Together with the impact of living conditions (i.e., captivity, semi-captivity, free ranging, access to resources, size of the groups, etc.), this phenomenon probably explains the variability in empirical findings between and within species regarding whether high- or low-ranking animals endure more stress in a dominance hierarchy [6]. Ultimately, the reason why a psychosocial stressor is experienced as such by a given individual may depend on the amount of control exerted over its termination and the predictability of its occurence.

In humans, absolute dominance ranks have little meaning because of the multidimensional nature of social success in our species. It has been found that low socioeconomic status (SES) is reliably associated with a disruption of endogenous circadian fluctuations in cortisol levels, suggesting that cortisol might be linked with social hierarchy in our species as well [78,79]. The influence of parental SES and parent education in this phenomenon [80] may suggest the existence of a transgenerational epigenetic mechanism as observed in stress-related disorders [81], as the relationship holds after controlling for many confounding factors including the offspring's actual SES. This finding is also consistent with the well-established observation that uncontrollable psychosocial stressors involving a real or possible social subordination component invariably induce stress and cortisol release in humans [82,83].

The exact cognitive role of testosterone in humans is still debated. Although early studies proposed that testosterone plays a role in reactive aggression rather than aggression per se, studies proposed that it has an important function to establish social status in both men and women [71,84]. Yet, testosterone does not

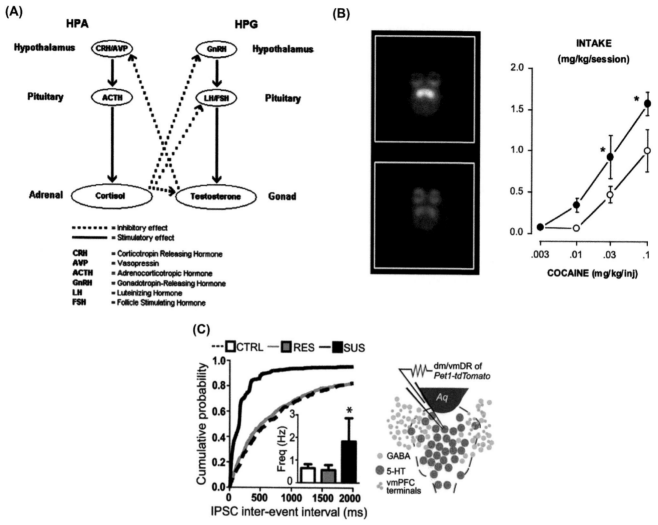

FIGURE 17.4   **Hormonal and neuromodulatory bases of social dominance.** (A) **Complexities of the cortisol—testosterone relationship involved in the maintenance of social dominance.** The hypothalamic—pituitary—adrenal (*HPA*) and hypothalamic—pituitary—gonadal (*HPG*) axes are represented with the brain structures involved, hormonal cascades, and functional interrelations. *(From Terburg D et al. The testosterone-cortisol ratio: a hormonal marker for proneness to social aggression. Int J L Psychiatry 2009;32:216—23, with permission.)* (B) **Role of dopamine in the emergence of social dominance and facilitation of cocaine addiction in subordinate individuals.** [^{18}F]FCP binding potential increases in dominant monkeys (*left*). (*Right*) Mean intake of cocaine per session for dominant (*white symbols*) and subordinate (*black symbols*) monkeys, as a function of the cocaine concentration in the self-administered solution. *(From Morgan D et al. Social dominance in monkeys: dopamine D2 receptors and cocaine self-administration. Nat Neurosci 2002:169—74.)* (*FCP*, fluoroclebopride) (C) **Involvement of serotonin neurons in the behavioral consequences of social defeats.** Optogenetic targeting showed that the serotonin neurons of susceptible (*SUS*) mice showing anxiety-like symptoms following social defeats were more inhibited than control (*CTRL*) or resilient (*RES*) mice showing no such symptoms. *5-HT*, serotonin; *vmPFC*, ventromedial prefrontal cortex. (*dm/vmDR*, dorsomedial/ventromedial dorsal raphe; *IPSC*, inhibitory post synaptic potential) *(From Challis C, et al. Raphe GABAergic neurons mediate the acquisition of avoidance after social defeat. J Neurosci 2013;33:13978—88. 13988a, with permission.)*

correlate linearly with socioeconomic status in humans, for two reasons. First, the aggressive tendencies of high-testosterone individuals are generally counterselected in many social organizations. Second, the net behavioral effects of testosterone on dominance-related behaviors depend upon cortisol levels: indeed, when resting cortisol levels are high, the positive association seen between dominance behaviors and testosterone is lost or even reversed [10,84]. A plausible role for testosterone would thus be to regulate the salience of and the

reactivity to social threats as a function of dominance ranks [85], whereas glucocorticoids would modulate the ability to shift flexibly between a "salience" network, which supports rapid but rigid decisions, and an "executive control" network, which supports flexible, elaborate social decisions (see also Chapter 30). According to this hypothesis, one may expect that the high and disruptive cortisol levels observed in chronic stress diminish the social and biological value of testosterone-mediated dominance behaviors because of the loss in behavioral flexibility. Moreover, transient fluctuations in cortisol levels seem causally involved in the adaptive memorization of social dominance relationships induced by competitive encounters [86], hence implying that disruption of cortisol signaling leads to imprecise representations of social dominance relationships.

While the HPG and HPA axes certainly play an important role in the implementation of proximal dominance behaviors, such as the arbitration of the "flee—fight—think" dilemma elicited by any social conflict, their functional physiology seems incompatible with the implementation of higher-order cognitive processes modulated by learned social hierarchies. The phenomena reported above are more likely to be mediated by central dopaminergic and serotonergic systems. Indeed, their modes of release in the forebrain enable also a refined coding of social information. To date, only a limited literature has investigated their roles in the social hierarchies of humans and nonhumans because of the methodological constraints associated with the measurement of central neuromodulators.

Dopamine, Serotonin, and Social Hierarchies in Rodents and Nonhuman Primates

In rodents, the emergence of an avoidant, subordinate behavior following social defeat is causally mediated by plasticity in the ventral tegmental area (VTA; containing dopamine neurons) occurring during and after a competitive interaction with negative outcome. More precisely, the sensitization of dopamine (DA) neurons occurs only in "susceptible" mice, which display a subordinate behavioral pattern following social defeats [87], and transient light stimulation of the VTA 1 day after social competition can reinstate avoidance and anhedonia symptoms induced by social defeats in most mice (Chaudhury et al. [88]). Suppression of dopaminergic firing may thus doubly contribute to the emergence of submissive behavior by inhibiting reward-related processes and by exacerbating the avoidance of subsequent social contacts with others, especially when they are dominant.

An outstanding question is whether postsynaptic sites to dopaminergic neurons are modulated by DA during social interaction. Does DA encode the presence/absence of dominant individuals? Does it encode social prediction errors (see "Social Hierarchies as a Major Evolutionary Pressure and Pivotal Feature of Societies") as in nonsocial settings? In dominant rats, microdialysis experiments showed that the imminence of a social conflict induces a strong release of DA in the nucleus accumbens [89,90]. In monkeys, no study has investigated dopaminergic firing per se as of this

writing, but the perception of dominant individuals was shown to interfere with a reward valuation process typically controlled by DA neurons [91,92], and one study showed that neurons in the ventral striatum—intensely innervated by DA neurons—may encode the experience of social subordination and dominance during social conflicts over reward [93]. These findings are highly relevant for clinicians, because the emergence of dominance hierarchies within groups of monkeys induced reversible changes in D2/D3 receptor availability, which mediates detrimental behavioral changes in subordinate individuals, including enhanced susceptibility to cocaine addiction (Fig. 17.4B) [94,95]. Interestingly, a later study extended this finding to humans, using a subjective social status questionnaire [96]. Beyond the evidence they provide for an involvement of DA in learning social dominance relationships, these fluctuations of D2/D3 receptors may also explain why recreational dopaminergic drugs are able to artificially upregulate self-esteem, self-confidence, and social dominance in the short term; why they result in degenerate social behavior patterns when used for a longer term; and why the experience of juvenile social stress, dominance motives, and low SES predispose to psychostimulant usage [5,97].

Compared to DA, the exact roles played by serotonin in reinforcement learning are less clear and constitute an active area of research. Nonetheless, this neurotransmitter is undoubtedly involved in the adaptive regulation of many aspects of social and nonsocial behaviors [98]. An influential theory has long maintained that serotonin (5-HT) might implement the coupling between the anticipation of aversive events and behavioral inhibition [99,100], which strongly resonates with situations of social subordination in which one has to inhibit the decision to compete for resources in front of threatening and powerful conspecifics. An outsider theory—which recently gained strong empirical support from electrophysiological recordings and optogenetic manipulation of 5-HT neurons [101,102]—proposed that serotonin firing might instead promote patience and cognitive control within both appetitive and aversive contexts [103]. Interesting, this second theory also resonates with dominance relationships, as dominant individuals tend to be impulsive decision-makers, whereas subordinates typically have to "wait their turn," because of the core importance of pecking orders in any dominance hierarchy (for both nutritional and social resources).

To date, the most striking demonstration of the causal role played by serotonin in the establishment of social hierarchies comes perhaps from the study of Raleigh and collaborators [46]. In this series of experiments performed in 12 groups of three vervet monkeys, the authors showed that the "enhancement of serotonin signaling" by a selective reuptake inhibitor and the suppression of serotonin signaling by a nonselective serotonin antagonist could induce dominance or subordination, respectively, in treated monkeys. More recently, it was confirmed that social defeats trigger sensitization of GABAergic neurons in the main serotonergic dorsal raphe nucleus (DRN) irrigating the forebrain [104]. Mirroring the aforementioned effect in the VTA [87,88], this sensitization phenomenon was visible only in the "susceptible" mice, which displayed an avoidant, subordinate-like behavioral pattern following social defeats (Fig. 17.4C). In humans, evidence supporting a role of 5-HT in social dominance is still sparse, but it was shown that enhancing 5-HT level through antidepressant medications or tryptophan supplementation (i.e., the precursor of 5-HT biosynthesis) might increase the frequency of dominance-related behaviors in everyday life [105,106]. Finally, in line with the definition of social dominance as an asymmetry of control over social stressors and social rewards, strong empirical evidence has emphasized that the reciprocal connections between the serotonergic DRN and the MPFC are crucial to adapt behavior in front of controllable stressors [87,107].

The joint involvement of DA and 5-HT in social dominance thus suggests (1) that the learning and decision-making processes controlled by these neuromodulators are central to the emergence of social hierarchies in mammals, (2) that social dominance might affect domain-general learning and decision-making through its influence on those neuromodulatory systems, and (3) that these neuromodulatory systems might have been sculpted throughout evolution to facilitate high flexibility in social behaviors, as required in species forming a dominance hierarchy. However, more research is needed to elucidate their exact computational roles, because no study has investigated how DA and 5-HT neurons react to conspecifics of different social ranks nor how they would implement the decision to compete or not against others.

CONCLUSION

The study of social dominance promises much more than simply accounting for interindividual differences in behavior and neurophysiology. Indeed, social dominance processes may offer mechanistic explanations for the emergence of such differences. Animal research indicates that social dominance affects serotonergic and dopaminergic neuromodulatory pathways responsible for behavioral and neural plasticity. It also affects the anatomical and functional properties of several brain structures traditionally linked with social and nonsocial perception, learning, and decision-making. It is the very nature of dominance hierarchies to shape

social behaviors and to promote the coexistence of various profiles within a single social group. In this perspective, the development of refined computational models and social learning tasks probing social dominance in humans (coupled with neuroimaging) may help us understand and treat specific psychosocial disorders that seem particularly prevalent in humans relative to other apes, such as pathological aggression, social anxiety, schizophrenia, psychopathy, and some forms of depression.

Acknowledgments

This work was performed within the framework of the LABEX CORTEX (ANR-11-LABX-0042) of Université de Lyon, within the program "Investissements d'Avenir" (ANR-11-IDEX-0007) operated by the French National Research Agency (ANR). JCD was also supported by Grant ANR-14-CE13-006 and by the European Institute for Advanced Study Fellowship Programme at the Hanse-Wissenschaftskolleg.

References

[1] Wilson R, Keil F. The MIT Encyclopedia of the cognitive sciences. Comput Linguist, vol. 26. Cambridge (Massachusetts): MIT Press; 1999.

[2] Ligneul R. Bases cérébrales des processus de compétition et de hiérarchisation sociales. 2014 [Doctoral dissertation], http://www.theses.fr/s101849.

[3] Bernstein IS. Dominance: the baby and the bathwater. Behav Brain Sci 1981;4:419–57.

[4] Price J. The dominance hierarchy and the evolution of mental illness. The Lancet 1967;290:243–6.

[5] Sandi C, Haller J. Stress and the social brain: behavioural effects and neurobiological mechanisms. Nat Rev Neurosci 2015;16: 290–304.

[6] Sapolsky RM. The influence of social hierarchy on primate health. Science 2005;308:648–52.

[7] Sloman L, Gilbert P. Subordination and defeat: an evolutionary approach to mood disorders and their therapy. Lawrence Erlbaum Associates; 2000.

[8] Singh-Manoux A, Marmot MG, Adler N. Does subjective social status predict health and change in health status better than objective status? Psychosom Med 2005;67:855–61.

[9] Marmot M, Wilkinson R. Social Determinants of health. Oxford University Press; 2005. At: http://books.google.com/books?id=AmwiS8HZeRIC&pgis=1.

[10] Mehta PH, Josephs RA. Testosterone and cortisol jointly regulate dominance: evidence for a dual-hormone hypothesis. Horm Behav 2010;58:898–906.

[11] Cowlishaw G, Dunbar RIM, Street G. Dominance rank and mating success in male primates. Anim Behav 1991:1045–56. http://dx.doi.org/10.1016/S0003-3472(05)80642-6.

[12] Ellis L. Dominance and reproductive success among nonhuman animals: a cross-species comparison. Ethol Sociobiol 1995;16: 257–333.

[13] McGuire MT, Brammer GL, Raleigh MJ. Resting cortisol levels and the emergence of dominant status among male vervet monkeys. Horm Behav 1986;20:106–17.

[14] van der Kooij MA, Sandi C, Kooij M, Van Der A, Sandi C. The genetics of social hierarchies. Curr Opin Behav Sci 2015;2:52–7.

[15] McNamara JM, Houston AI. State-dependent life histories. Nature 1996;380:215–21.

[16] Wolf M, van Doorn GS, Leimar O, Weissing FJ. Life-history trade-offs favour the evolution of animal personalities. Nature 2007;447:581–4.

[17] Jolly A. Lemur social behavior and primate intelligence. Science 1966;153:501–6.

[18] Humphrey NK. The social function of intellect. 1976. p. 303–17. http://dx.doi.org/10.2307/375925.

[19] Whiten A, Byrne R. Machiavellian intelligence hypothesis. Machiavellian Intell 1988. At: http://www.rm-f.net/~pennywis/MITECS/Articles/whiten.html.

[20] Dunbar RIM. Neocortex size as a constraint on group size in primates. J Hum Evol 1992;22:469–93.

[21] Behrens TEJ, Hunt LT, Rushworth MFS. The computation of social behavior. Science 2009;324:1160–4.

[22] Frith CD. The social brain?. 2007. http://dx.doi.org/10.1098/rstb.2006.2003.

[23] Chiao JY, et al. Neural representations of social status hierarchy in human inferior parietal cortex. Neuropsychologia 2009;47:354–63.

[24] Kumaran D, Melo HL, Duzel E. The emergence and representation of knowledge about social and nonsocial hierarchies. Neuron 2012;76:653–66.

[25] Marsh AA, Blair KS, Jones MM, Soliman N, Blair RJR. Dominance and submission: the ventrolateral prefrontal cortex and responses to status cues. J Cogn Neurosci 2009;21:713–24.

[26] Todorov A, Engell AD. The role of the amygdala in implicit evaluation of emotionally neutral faces. Soc Cogn Affect Neurosci 2008;3:303–12.

[27] Lindner M, Hundhammer T, Ciaramidaro A, Linden DEJ, Mussweiler T. The neural substrates of person comparison—an fMRI study. Neuroimage 2008;40:963–71.

[28] Farrow TFD, et al. Higher or lower? The functional anatomy of perceived allocentric social hierarchies. Neuroimage 2011;57:1552–60.

[29] Zink CF, et al. Know your place: neural processing of social hierarchy in humans. Neuron 2008;58:273–83.

[30] Watanabe N, Yamamoto M. Neural mechanisms of social dominance. Front Neurosci 2015;9.

[31] Dunne S, Doherty JPO. Insights from the application of computational neuroimaging to social neuroscience. Curr Opin Neurobiol 2013;23:1–6.

[32] Burke CJ, Tobler PN, Baddeley M, Schultz W. Neural mechanisms of observational learning. Proc Natl Acad Sci USA 2010;107:14431–6.

[33] Behrens TEJ, Hunt LT, Woolrich MW, Rushworth MFS. Associative learning of social value. Nature 2008;456:245–9.

[34] Hampton AN, Bossaerts P, O'Doherty JP, Doherty JPO. Neural correlates of mentalizing-related computations during strategic interactions in humans. Proc Natl Acad Sci USA 2008;105:6741–6.

[35] Renes RA, van Haren NEM, Aarts H, Vink M. An exploratory fMRI study into inferences of self-agency. Soc Cogn Affect Neurosci 2014. Nsu106.

[36] Seo H, Cai X, Donahue CH, Lee D. Neural correlates of strategic reasoning during competitive games. Science 2014;364:340–5.

[37] Devaine M, Hollard G, Daunizeau J. Theory of mind: did evolution fool us? PLoS One 2014;9.

[38] Fareri DS, Delgado MR. Differential reward responses during competition against in- and out-of-network others. Soc Cogn Affect Neurosci 2013;9:412–20.

[39] Katsyri J, Hari R, Ravaja N, Nummenmaa L. The opponent matters: elevated fMRI reward responses to winning against a human versus a computer opponent during interactive video game playing. Cereb Cortex 2012;23:2829–39.

[40] Le Bouc R, Pessiglione M. Imaging social motivation: distinct brain mechanisms drive effort production during collaboration versus competition. J Neurosci 2013;33:15894–902.

[41] Polosan M, et al. An fMRI study of the social competition in healthy subjects. Brain Cogn 2011;77:401–11.

[42] Chaminade T, et al. How do we think machines think? An fMRI study of alleged competition with an artificial intelligence. Front Hum Neurosci 2012;6:103.

[43] Decety J, Jackson PL, Sommerville JA, Chaminade T, Meltzoff AN. The neural bases of cooperation and competition: an fMRI investigation. Neuroimage 2004;23:744–51.

[44] Rowell TE. The concept of social dominance. Behav Biol 1974;11:131–54.

[45] Wang F, et al. Bidirectional control of social hierarchy by synaptic efficacy in medial prefrontal cortex. Science 2011;334:693–7.

[46] Raleigh MJ, McGuire MT, Brammer GL, Pollack DB, Yuwiler A. Serotonergic mechanisms promote dominance acquisition in adult male vervet monkeys. Brain Res 1991;559:181–90.

[47] Bauman MD, Toscano JE, Mason WA, Lavenex P, Amaral DG. The expression of social dominance following neonatal lesions of the amygdala or hippocampus in rhesus monkeys (Macaca mulatta). Behav Neurosci 2006;120:749–60.

[48] Karafin MS, Tranel D, Adolphs R. Dominance attributions following damage to the ventromedial prefrontal cortex. J Cogn Neurosci 2004;16:1796–804.

[49] Kanai R, Rees G. The structural basis of inter-individual differences in human behaviour and cognition. Nat Rev Neurosci 2011;12:231–42.

[50] Cross SE, Madson L. Models of the self: self-construals and gender. Psychol Bull 1997;122:5–37.

[51] Sedikides C, Gaertner L, Toguchi Y. Pancultural self-enhancement. J Pers Soc Psychol 2003;84:60–79.

[52] Iyengar SS, Lepper MR. Rethinking the value of choice: a cultural perspective on intrinsic motivation. J Pers Soc Psychol 1999;76:349–66.

[53] Cheng C, Cheung S-F, Chio JH, Chan MS. Cultural meaning of perceived control: a meta-analysis of locus of control and psychological symptoms across 18 cultural regions. Psychol Bull 2013;139.

[54] Chiao JY, Blizinsky KD. Culture-gene coevolution of individualism-collectivism and the serotonin transporter gene. Proc Biol Sci 2010;277:529–37.

[55] Homberg JR, Lesch K-PP. Looking on the bright side of serotonin transporter gene variation. Biol Psychiatry 2011;69:513–9.

[56] Noonan MP, et al. A neural circuit covarying with social hierarchy in macaques. PLoS Biol 2014;12:e1001940.

[57] Frodl T, et al. Reduced gray matter brain volumes are associated with variants of the serotonin transporter gene in major depression. Mol Psychiatry 2008;13:1093–101.

[58] Mahar I, Bambico FR, Mechawar N, Nobrega JN. Stress, serotonin, and hippocampal neurogenesis in relation to depression and antidepressant effects. Neurosci Biobehav Rev 2014;38:173–92.

[59] Lawson GM, Duda JT, Avants BB, Wu J, Farah MJ. Associations between children's socioeconomic status and prefrontal cortical thickness. Dev Sci 2013;16:641–52.

[60] Yeo RA, et al. The impact of parent socio-economic status on executive functioning and cortical morphology in individuals with schizophrenia and healthy controls. Psychol Med 2014;44:1257–65.

[61] Getov S, Kanai R, Bahrami B, Rees G. Human brain structure predicts individual differences in preconscious evaluation of facial dominance and trustworthiness. Soc Cogn Affect Neurosci 2014. http://dx.doi.org/10.1093/scan/nsu103.

[62] Santamaria-Garcia H, Burgaleta M, Sebastian-Galles N. Neuroanatomical markers of social hierarchy recognition in humans: a combined ERP/MRI study. J Neurosci 2015;35:10843–50.

III. SOCIAL DECISION NEUROSCIENCE

[63] Ly M, Haynes MR, Barter JW, Weinberger DR, Zink CF. Subjective socioeconomic status predicts human ventral striatal responses to social status information. Curr Biol 2011;21:794−7.

[64] Ligneul R, Obeso I, Ruff CC, Dreher JC. Dynamical representation of dominance relationships in the human medial prefrontal cortex. Current Biology, in press

[65] Carver CS, White TL. Behavioral inhibition, behavioral activation, and affective responses to impending reward and punishment: the BIS/BAS Scales. J Pers Soc Psychol 1994;67:319−33.

[66] Keltner D, Gruenfeld DH, Anderson C. Power, approach, and inhibition. Psychol Rev 2003;110:265−84.

[67] Ligneul R, Girard R, Dreher J-C. Social brains and divides: the interplay between political orientation and the neural sensitivity to social ranks. 2016 [Submitted for publication].

[68] Pratto F, Sidanius J, Stallworth LM, Malle BF. Social dominance orientation: a personality variable predicting social and political attitudes. J Pers Soc Psychol 1994;67:741−63.

[69] Stanton SJ, Beehner JC, Saini EK, Kuhn CM, Labar KS. Dominance, politics, and physiology: voters' testosterone changes on the night of the 2008 United States presidential election. PLoS One 2009;4:e7543.

[70] Chaouloff F. Social stress models in depression research: what do they tell us? Cell Tissue Res 2013. http://dx.doi.org/10.1007/s00441-013-1606-x.

[71] Dreher J-C. Neuroimaging evidences of gonadal steroid hormone influences on reward processing and social decision-making in humans. Brain Mapp 2015;3. Elsevier Inc.

[72] Terburg D, et al. The testosterone-cortisol ratio: a hormonal marker for proneness to social aggression. Int J Law Psychiatry 2009;32:216−23.

[73] Sapolsky RMM, Krey LC, McEwen BS. Glucocorticoid-sensitive hippocampal neurons are involved in terminating the adrenocortical stress response. Proc Natl Acad Sci USA 1984;81:6174−7.

[74] Arnsten AFT. Stress signalling pathways that impair prefrontal cortex structure and function. Nat Rev Neurosci 2009;10:410−22.

[75] Thierry B, Singh M, Kaumanns W. Macaque societies: a model for the study of social organization. Cambridge University Press,; 2004. 41.

[76] Gesquiere LR, et al. Life at the top: rank and stress in wild male baboons. Science 2011;333:357−60.

[77] Sapolsky RM, Ray JC. Styles of dominance and their endocrine correlates among wild olive baboons (Papio anubis). Am J Primatol 1989;18:1−13.

[78] Cohen S, et al. Socioeconomic status is associated with stress hormones. Psychosom Med 2006;68:414−20.

[79] Agbedia OO, et al. Blunted diurnal decline of cortisol among older adults with low socioeconomic status. Ann NY Acad Sci 2011;1231:56−64.

[80] Desantis AS, Kuzawa CW, Adam EK. Developmental origins of flatter cortisol rhythms: socioeconomic status and adult cortisol activity. Am J Hum Biol 2015;27:458−67.

[81] Yehuda R, et al. Parental posttraumatic stress disorder as a vulnerability factor for low cortisol trait in offspring of holocaust survivors. Arch Gen Psychiatry 2007;64:1040−8.

[82] Dickerson SS, Kemeny ME. Acute stressors and cortisol responses: a theoretical integration and synthesis of laboratory research. Psychol Bull 2004;130:355−91.

[83] Wirth MM, Welsh KM, Schultheiss OC. Salivary cortisol changes in humans after winning or losing a dominance contest depend on implicit power motivation. Horm Behav 2006;49:346−52.

[84] Montoya ER, Terburg D, Bos PA, van Honk J. Testosterone, cortisol, and serotonin as key regulators of social aggression: a review and theoretical perspective. Motiv Emot 2012;36:65−73.

[85] Hermans EJ, Ramsey NF, van Honk J. Exogenous testosterone enhances responsiveness to social threat in the neural circuitry of social aggression in humans. Biol Psychiatry 2008;63:263−70.

[86] Cordero MI, Sandi C, Cordero I, Sandi C. Stress amplifies memory for social hierarchy. Front Neurosci 2007;1:175−84.

[87] Cao JL, et al. Mesolimbic dopamine neurons in the brain reward circuit mediate susceptibility to social defeat and antidepressant action. Journal Neurosci 2010;30(49):16453−8.

[88] Chaudhury D, et al. Rapid regulation of depression-related behaviours by control of midbrain dopamine neurons. Nature 2013;493:532−6.

[89] Tidey JW, Miczek KA. Social defeat stress selectively alters mesocorticolimbic dopamine release: an in vivo microdialysis study. Brain Res 1996;721:140−9.

[90] Ferrari PF, et al. Accumbal dopamine and serotonin in anticipation of the next aggressive episode in rats. Eur J Neurosci 2003;17:371−8.

[91] Deaner RO, et al. Monkeys pay per view: adaptive valuation of social images by rhesus macaques. Curr Biol 2005;15:543−8.

[92] Klein JT, Platt ML. Social information signaling by neurons in primate striatum. Curr Biol 2013;23:691−6.

[93] Santos GS, Nagasaka Y, Fujii N, Nakahara H. Encoding of social state information by neuronal activities in the macaque caudate nucleus. Soc Neurosci 2012;7:42−58.

[94] Morgan D, et al. Social dominance in monkeys: dopamine D2 receptors and cocaine self-administration. Nat Neurosci 2002:169−74. http://dx.doi.org/10.1038/nn798.

[95] Nader MA, et al. Social dominance in female monkeys: dopamine receptor function and cocaine reinforcement. Biol Psychiatry 2012;72:414−21.

[96] Martinez D, et al. Dopamine type 2/3 receptor availability in the striatum and social status in human volunteers. Biol Psychiatry 2010;67:275−8.

[97] Sinha R. Chronic stress, drug use, and vulnerability to addiction. Ann NY Acad Sci 2008;1141:105−30.

[98] Kiser D, Steemers B, Branchi I, Homberg JR. The reciprocal interaction between serotonin and social behaviour. Neurosci Biobehav Rev 2012;36(2):786−98.

[99] Dayan P, Huys QJM. Serotonin in affective control. Annu Rev Neurosci 2009;32:95−126.

[100] Dayan P, Huys Q. Serotonin's many meanings. 2015. p. 3−5. http://dx.doi.org/10.7554/eLife.06346.

[101] Miyazaki KW, et al. Optogenetic activation of dorsal raphe serotonin neurons enhances patience for future rewards. Curr Biol 2014;24:2033−40.

[102] Miyazaki KW, Miyazaki K, Doya K. Activation of dorsal raphe serotonin neurons is necessary for waiting for delayed rewards. J Neurosci 2012;32:10451−7.

[103] Doya K. Modulators of decision making. Nat Neurosci 2008;11:410−6.

[104] Challis C, et al. Raphe GABAergic neurons mediate the acquisition of avoidance after social defeat. J Neurosci 2013;33:13978−88. 13988a.

[105] Moskowitz DS, Pinard G, Zuroff DC, Annable L, Young SN. The effect of tryptophan on social interaction in everyday life: a placebo-controlled study. Neuropsychopharmacology 2001;25:277−89.

[106] Tse WS, Bond AJ. Serotonergic intervention affects both social dominance and affiliative behaviour. Psychopharmacol (Berl) 2002;161:324−30.

[107] Amat J, et al. Medial prefrontal cortex determines how stressor controllability affects behavior and dorsal raphe nucleus. Nat Neurosci 2005;8:365−71.

[108] Freeman JB, Rule NO, Adams RB, Ambady N. Culture shapes a mesolimbic response to signals of dominance and subordination that associates with behavior. Neuroimage 2009;47:353−9.

CHAPTER

18

Reinforcement Learning and Strategic Reasoning During Social Decision-Making

H. Seo, D. Lee

Yale University, New Haven, CT, United States

Abstract

Finding an efficient learning algorithm in a complex social environment is challenging. When there is a sufficient amount of information about the intentions of other decision-makers and decision-makers are cognitively capable of predicting the outcomes of their actions more accurately using that information, a complex learning algorithm might be advantageous. However, a simpler heuristic learning algorithm might work better with limited information. Previous studies have found that humans and other primates apply a mixture of multiple learning algorithms during various social decision-making tasks. In addition, neural activity related to multiple learning algorithms has been identified in different cortical areas and basal ganglia, suggesting that various learning algorithms for social decision-making might be implemented throughout the brain. By contrast, the processes for evaluating the performance of various learning algorithms and switching between them might be more closely tied to specific regions in the prefrontal cortex.

INTRODUCTION

A problem of decision-making consists of selecting an action from a number of alternatives in a particular environment to achieve an outcome most desirable to the decision-maker. Studies of decision-making have taken various forms, depending on the specific types of the environment in question. For example, mathematical models of decision-making in economics, such as the expected utility theory, often focus on how a rational decision-maker would behave when the environment is stochastic and the outcomes of his or her choices are uncertain [1,2]. In addition, game theory provides the framework to analyze the behaviors of multiple rational self-interested decision-makers interacting in a social setting [2]. During social interactions, the outcome of a decision-maker's action is influenced by the actions of other decision-makers. For example, during a rock–paper–scissors game, choosing paper leads to win or loss depending on whether the opponent chooses rock or scissors. The strategies expected for a group of rational and self-interested decision-makers or players are referred to as a Nash equilibrium [3]. Formally, a set of strategies for all the players in a game is a Nash equilibrium when it is not possible for any player to increase his or her payoff by changing strategies individually. In game theory, selecting one particular action exclusively is referred to as a pure strategy, whereas a mixed strategy refers to choosing multiple actions stochastically. A Nash equilibrium can include mixed strategies, as in many competitive games. For example, there is a unique Nash equilibrium for the rock–paper–scissors game, which is to choose all three options with equal probabilities. Any deviation from this can be exploited by the opponent.

Although such normative theories of decision-making make predictions for rational decision-makers, they are often violated when the real human behaviors are examined. This suggests that some of the assumptions in these normative theories are wrong. For decision-making with probabilistic outcomes, the prospect theory proposes that decision-makers might evaluate the outcomes of their choices as gains and losses relative to a particular reference point, and that values of different outcomes might be weighted according to a nonlinear function of the outcome probability, rather than the probability itself [4]. People might also behave differently in social settings compared to the predictions of game theory, not only because it is cognitively

Decision Neuroscience
http://dx.doi.org/10.1016/B978-0-12-805308-9.00018-X

demanding to identify a Nash equilibrium strategy, but also because they might be altruistic and sensitive to the payoffs of others [5,6].

Standard economic theories, such as the expected utility theory and game theory, often ignore another important aspect of decision-making in real life, namely, the fact that humans and animals constantly change their strategies through experience according to the outcomes of their previous choices and other information they receive from the environment. For example, in the reinforcement learning theory [7], the expected values of future rewards expected from a particular state of the environment and from a particular action in a given state of the environment are referred to as the state and action value functions, respectively. Value functions are updated iteratively as the decision-maker interacts with his or her environment, and they determine the likelihood of selecting a given action at each time step. The reinforcement learning theory, rooted in dynamic programming and models of sequential Markov decision processes [8], has been applied very successfully to understand the neural mechanisms of value-based decision-making. Several review papers have focused on this topic [9—11].

How accurately the reinforcement learning theory can account for an animal's behavior can be tested using many different behavioral tasks. For example, nonstationary N-armed bandit tasks in which the reward probability changes over time for each of N different alternatives has been used in many previous studies [12—14]. More recently, a multistage sequential decision-making task has also been used to characterize the exact nature of learning algorithms [14,15]. Another fruitful area of research is to investigate how the strategies are dynamically modified during repeated games [16]. In fact, a number of studies have begun to accumulate converging evidence that humans and nonhuman primates might employ a mixture of learning algorithms to improve their decision-making tasks in both social and nonsocial contexts. In this chapter, we will discuss the neural substrates of reinforcement learning during social interactions. In particular, the neural substrates for simple, model-free reinforcement learning for social decision-making might be broadly distributed in multiple cortical and subcortical structures [11]. By contrast, the mechanisms responsible for arbitrating and switching between algorithms might be more localized in the medial prefrontal cortex [17].

MODEL-FREE VERSUS MODEL-BASED REINFORCEMENT LEARNING

The reinforcement learning theory is based on Markov decision processes, in which a combination of an action and a particular state of the environment entirely determines the probability of getting a particular amount of reward as well as how the state will change [7,8]. The goal of learning is to discover an optimal policy or action for each state that will maximize the expected value of future rewards. A large number of learning algorithms developed in this framework can be roughly divided into two different categories. First, in a simple or model-free reinforcement learning algorithm, value functions, namely, estimates for future rewards, are updated entirely according to the difference between the reward expected from the current value functions and the reward actually received by the agent. This difference is referred to as reward-prediction error. Although neural activity related to reward-prediction errors was originally identified in the brain-stem dopamine neurons [18], this has been also observed in many brain areas including the striatum and anterior cingulate cortex [11] (see also Chapters 2, 4, and 5). Because the value functions are continually updated through experience in model-free reinforcement learning algorithms, the process of selecting an action in a given state is relatively simple. On the other hand, if the property of the agent's environment (i.e., its transition probability) or the value of reward from different states changes abruptly, then the entire value functions have to be relearned gradually as the agent visits each state of the new environment repeatedly and incorporates the value of new reward from each state. Therefore, model-free reinforcement learning algorithms are relatively inflexible, and often considered to be analogous to habit learning [11,19] (see also Chapter 16).

Second, model-based reinforcement learning algorithms maintain a dynamic model of the agent's environment, which is updated through experience. This implies that the process of selecting an action would be more complex than in model-free reinforcement learning, because the value functions must be calculated iteratively using the current knowledge of the environment and reward from it. Such learning algorithms, however, have the advantage that when the environment or reward values change suddenly, value functions and hence actions appropriate in the new environment can be updated much more quickly than in model-free reinforcement learning algorithms, without having to experience the outcomes of actions in the new environment. This is analogous to the notion of goal-directed behaviors [11,20,21]. How these different types of reinforcement learning algorithms are implemented in the brain remains poorly understood, but this is an active area of research [14,15,22]. As described later, these two different types of reinforcement learning algorithms can be also used during dynamic social interactions [16,23].

NEURAL CORRELATES OF MODEL-FREE REINFORCEMENT LEARNING DURING SOCIAL DECISION-MAKING

Model-free reinforcement learning algorithms can be implemented using only a small number of free parameters for a wide variety of decision-making contexts, including iterative social interactions. For example, the value function in a model-free reinforcement learning model can be updated according to the following:

$$Q_{t+1}(a) = Q_t(a) + \alpha [r_t - Q_t(a)],$$

where $Q_t(a)$ denotes the value function for action a at time t, and α the learning rate. The probability of choosing an action is often related to its value function via a logistic transformation as follows:

$$p_t(a_i) = \exp\{\beta Q_t(a_i)\} \Big/ \sum_j \exp\{\beta Q_t(a_j)\}$$

where $p_t(a_i)$ denotes the probability of choosing the ith action a_i at time t, and β the inverse temperature that controls the randomness of actions. Namely, all actions will be chosen randomly and equally often when $\beta = 0$, and the action with the maximum value function will be chosen deterministically when $\beta = \infty$. Such model-free reinforcement learning models can parsimoniously account for the behaviors of human players in simple repeated games studied in a laboratory setting, better than the predictions based on the Nash equilibrium [24,25].

Similarly, previous studies have also found that choices of monkeys trained to play a matching pennies game against a computer opponent in an oculomotor decision-making task show the patterns expected for a model-free reinforcement learning algorithm [26,27]. Namely, although the matching pennies task required the animals to choose randomly and independently regardless of previous choices and their outcomes, monkeys were more likely to choose the same target rewarded in previous trials. The tendency to use a model-free reinforcement learning algorithm itself might not be learned, because this was not beneficial to the animal. In addition, signals necessary for model-free reinforcement learning algorithms are distributed broadly across many different cortical areas, consistent with the results obtained from the studies on individual, nonsocial decision-making tasks [11]. In particular, in model-free reinforcement learning algorithms, the value function for a given action is a weighted average of the rewards previously obtained from the same action. Therefore, neural activity related to the animal's previous choices and outcomes might play an important role in model-free reinforcement learning algorithm.

Such signals were identified in many regions of the brain, including the anterior cingulate cortex, dorsolateral prefrontal cortex, posterior parietal cortex, and striatum, in both primates (Fig. 18.1) [26–29] and rodents [12,30]. In addition, the time constant of such choice and reward signals, namely, the rate at which neural activity related to the animal's choice and its outcome decays, varies across neurons and follows a power-law distribution [31], suggesting that they might provide the substrate for adjusting the learning rate according to the dynamics of the decision-maker's environment [32]. Similarly, neuroimaging studies in humans have shown that the signals related to reinforcement and punishment can be decoded from almost the entire brain [33]. Therefore, the neural circuitry responsible for model-free reinforcement learning might be distributed broadly in the corticostriatal network and might contribute to behavioral improvement during iterative social interactions.

NEURAL CORRELATES OF HYBRID REINFORCEMENT LEARNING DURING SOCIAL DECISION-MAKING

Model-based reinforcement learning would be particularly advantageous during social interaction, because decision-makers can adjust their strategies based on information about the intentions of others. In economics, this is known as belief learning, but is formally equivalent to model-based reinforcement learning in which the value functions are updated based on information about the expected behaviors of other decision-makers [16]. Previous studies have shown that human choice behaviors during repeated social interactions are more consistent with a mixture of model-free and model-based reinforcement learning. Namely, they do not rely exclusively on a model-free or a model-based reinforcement learning algorithm, but instead their behaviors display features predicted by both algorithms [34,35].

In general, prediction errors would vary depending on the exact algorithms used to estimate the value functions. Previous studies, however, have found that error signals derived from model-free and model-based reinforcement learning algorithms are colocalized in the ventral striatum during repeated games [35], as in a nonsocial, multistage decision-making task [14]. This suggests that the ventral striatum might be involved in updating the value functions regardless of the exact nature of the learning algorithm used to update the value functions. Similarly, the neural substrates responsible for model-free and model-based reinforcement learning during social interactions might at least partially overlap

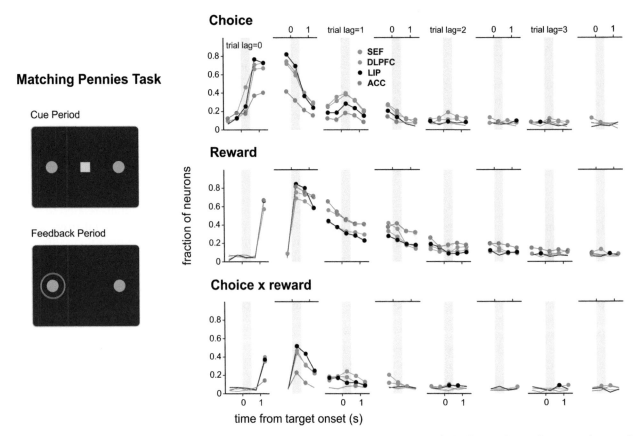

FIGURE 18.1 Neural activity related to choices and outcomes during a matching pennies task. (Left) Visual stimuli presented to monkeys during the cue and feedback periods of a matching pennies task [26]. (Right) Time course of signals related to the animal's choice (top) and to the reward (middle) and of their interaction, which is equivalent to the choice of the computer opponent (bottom). Each data point indicates the fraction of neurons in four different cortical areas (*ACC*, dorsal anterior cingulate cortex; *DLPFC*, dorsolateral prefrontal cortex; *LIP*, lateral intraparietal area; *SEF*, supplementary eye field) that significantly modulated their activity according to the animal's choice and outcome in the current and previous three trials [29]. The results are shown for a series of nonoverlapping 0.5-s bins, and the *colored disks* indicate that the fraction was significantly higher than the chance level (0.05).

at the cortical level. For example, single-neuron recordings in monkeys performing a computer-simulated rock–paper–scissors task identified signals related to hypothetical outcomes from unchosen actions in the orbitofrontal and dorsolateral prefrontal cortex (Fig. 18.2) [36]. Moreover, these signals related to hypothetical outcomes were often found in the same neurons encoding actual outcomes from chosen actions [36]. In addition, although some neuroimaging studies found that the activity in the rostral anterior cingulate cortex was related to the error signals from model-based reinforcement learning [35], neurons in the anterior cingulate cortex often encode reward prediction errors in model-free reinforcement learning [27,37]. On the other hand, the brain areas responsible for monitoring the performance of various types of learning algorithms and regulating them might be more specialized. For example, during social interaction, model-free and model-based learning algorithms would rely on the information about reward history and social interactions. The reliability of these two different types of information

was found in two different regions of the anterior cingulate cortex [38].

During social interactions, the ability to predict the actions of other players based on their beliefs and intentions plays an important role, and this is referred to as theory of mind [39]. Inferences based on the theory of mind often take a recursive form, because it is often necessary to take into account the fact that players can change their behaviors based on their predictions of what the other players might do. Such recursive reasoning during social decision-making might rely on the medial frontal cortex [40–43].

ARBITRATION AND SWITCHING BETWEEN LEARNING ALGORITHMS

If the statistical properties of the decision-maker's environment are fully known, and if the decision-maker has sufficient computational resources to estimate the likelihood and desirability of different

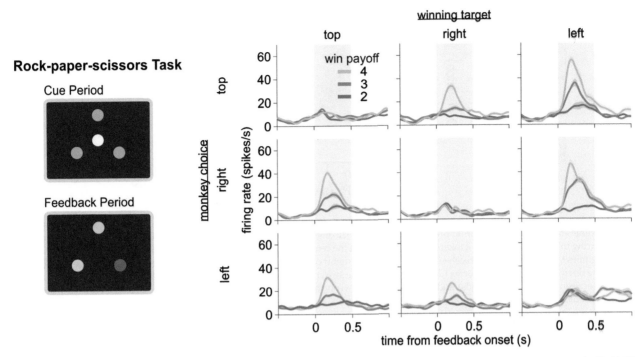

FIGURE 18.2 Neural activity related to hypothetical payoffs in the orbitofrontal cortex during a rock–paper–scissors task. (Left) Visual stimuli presented to monkeys during the cue and feedback periods of a rock–paper–scissors task [36]. (Right) Activity of an example neuron in the orbitofrontal cortex during the feedback period of a rock–paper–scissors task. *Lines* correspond to spike density functions of a single neuron sorted by the position of the target chosen by the animal (*rows*), the position of the winning target (*columns*), and the magnitude of reward available from the winning target (*colors*). Thus, this neuron modulated its activity according to the hypothetical winning payoff only when the animal did not choose the winning target, as shown by the panels off the diagonal (i.e., monkey choice ≠ winning target).

outcomes from each action using that knowledge, model-based reinforcement learning algorithms must perform better than model-free algorithms. However, in some cases, model-free reinforcement learning algorithms might be more robust and efficient, because the knowledge of the environment might not be sufficiently accurate and the decision-makers might not have the time or cognitive ability to use such knowledge for predicting the possible outcomes from alternative actions. As a result, the reliability of these different learning algorithms might vary with a particular task at hand and with the amount of the decision-maker's experience and knowledge. A 2014 study found that the activity of the lateral prefrontal cortex is modulated by the reliabilities of both model-free and model-based reinforcement learning algorithms [15]. Furthermore, the same study also found that the accuracy of the algorithm that was more reliable at any given time was encoded in the lateral prefrontal cortex and right frontopolar cortex, suggesting that these areas might be involved in arbitrating between these different learning algorithms [15]. By contrast, it has been shown that the reliability of recursive inference about the possible strategies of other decision-makers during

social decision-making was reflected in the medial frontal cortex [41]. Therefore, whether the process of arbitrating between algorithms differs for social and nonsocial decision-making remains to be further investigated.

For both social and nonsocial decision-making, model-free reinforcement learning might be often chosen as a default, given that this algorithm would be computationally simpler and also more reliable when the decision-maker does not have sufficient knowledge of the environment [15]. Therefore, the use of more complex learning algorithms might require a switching from or inhibiting of the processes involved in simple algorithms, such as model-free reinforcement learning algorithms. It was found that during a matching pennies game against a computerized opponent, monkeys often strategically switched from a model-free reinforcement learning algorithm to exploit the vulnerability in the opponent's strategies (Fig. 18.3). Moreover, neural signals related to switching from the simple reinforcement learning algorithm were found more frequently in the dorsomedial prefrontal cortex than in the lateral prefrontal cortex or basal ganglia [17].

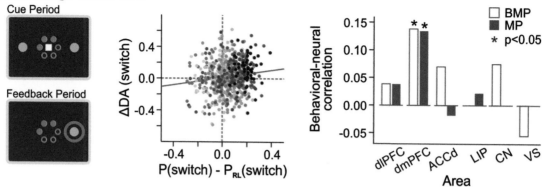

FIGURE 18.3   Correlation between neural activity and behavioral switching from a simple reinforcement learning algorithm during a biased matching pennies task. (Left) Visual stimuli presented to monkeys during the cue and feedback periods of a token-based biased matching pennies task [17]. During this task, the number of tokens (*red disks* shown at the center of the screen) was adjusted according to the payoff matrix of a biased matching pennies task, and the animal received juice reward only after it collected six tokens. (Middle) Relationship between the strength of neural signals in the dorsomedial prefrontal cortex (*dmPFC*) related to the switch in the animal's choice (*abscissa*; ΔDA, change in the decoding accuracy for choice between stay vs. switch trials) and the difference between the actual probability of behavioral switching, P(switch), and the same probability predicted by the simple reinforcement learning algorithm, P_{RL}(switch), (*ordinate*) are shown for alternative sequences of choices and outcomes in the last two trials (color coded). (Right) The correlation coefficient for the two quantities shown in the middle for different brain regions (*ACCd*, dorsal anterior cingulate cortex; *BMP*, biased matching pennies; *CN*, caudate nucleus; *dlPFC*, dorsolateral prefrontal cortex; *LIP*, lateral intraparietal area; *MP*, matching pennies; *VS*, ventral striatum). *Black histograms* show the results from the matching pennies task.

CONCLUSIONS

Dynamic changes in choices and decision-making strategies observed in human and animal subjects in a variety of contexts can be parsimoniously accounted for by the reinforcement learning theory. This framework has also been successfully adopted in a large number of neurobiological studies to characterize the functions of multiple cortical areas and basal ganglia. In addition, there is a wide variety of reinforcement learning algorithms. For example, for model-free reinforcement learning algorithms, value functions are updated exclusively based on the outcomes of previously chosen actions, whereas they can be adjusted according to the decision-maker's knowledge of the environment for model-based reinforcement learning algorithms. For complex decision-making tasks, including social interactions, it is likely that multiple learning algorithms operate in parallel, and their reliabilities might be continuously assessed. Studies have begun to elucidate the brain mechanisms involved in evaluating the performance of various learning algorithms and arbitrating and switching between learning algorithms at the time of action selection. Although this line of research has greatly advanced our understanding of the brain functions related to social learning and decision-making, the analytical framework and behavioral tasks used in the laboratories so far are still relatively simple. In particular, in real-life social interactions, there is much more uncertainty about the intentions and knowledge of other players, and even the payoff matrix that defines a specific decision-making problem might not be fully known. Accordingly, little is known about how the brain learns to identify the statistical structure of a new decision-making problem in social settings, and this might be an important topic for future studies.

References

[1] Bernoulli D. Exposition of a new theory on the measurement of risk. Econometrica 1954;22:23–36.
[2] von Neumann J, Morgenstern O. Theory of games and economic behavior. Princeton: Princeton University Press; 1944.
[3] Nash JF. Equilibrium points in n-person games. Proc Natl Acad Sci USA 1950;36:48–9.
[4] Kahneman D, Tversky A. Prospect theory: an analysis of decision under risk. Econometrica 1979;47:263–92.
[5] Fehr E, Schmidt KM. A theory of fairness, competition, and cooperation. Q J Econ 1999;114:817–68.
[6] Seo H, Lee D. Neural basis of learning and preference during social decision-making. Curr Opin Neurobiol 2012;22:990–5.
[7] Sutton RS, Barto AG. Reinforcement learning: an introduction. Cambridge: MIT Press; 1998.
[8] Puterman ML. Markov decision processes: discrete stochastic dynamic programming. New York: Wiley; 1994.
[9] Bornstein AM, Daw ND. Multiplicity of control in the basal ganglia: computational roles of striatal subregions. Curr Opin Neurobiol 2011;21:374–80.
[10] Ito M, Doya K. Multiple representations and algorithms for reinforcement learning in the cortico-basal ganglia circuit. Curr Opin Neurobiol 2011;21:368–73.
[11] Lee D, Seo H, Jung MW. Neural basis of reinforcement learning and decision making. Annu Rev Neurosci 2012;35:287–308.
[12] Kim H, Sul JH, Huh N, Lee D, Jung MW. Role of striatum in updating values of chosen actions. J Neurosci 2009;29:14701–12.
[13] Walton ME, Behrens TE, Buckley MJ, Rudebeck PH, Rushworth MF. Separable learning systems in the macaque brain and the role of orbitofrontal cortex in contingent learning. Neuron 2010;65:927–39.

[14] Daw ND, Gershman SJ, Seymour B, Dayan P, Dolan RJ. Model-based influences on humans' choices and striatal prediction errors. Neuron 2011;69:1204–15.

[15] Lee SW, Shimojo S, O'Doherty JP. Neural computations underlying arbitration between model-based and model-free learning. Neuron 2014;81:687–99.

[16] Camerer CF. Behavioral game theory: experiments in strategic interaction. Princeton: Princeton University Press; 2003.

[17] Seo H, Cai X, Donahue CH, Lee D. Neural correlates of strategic reasoning during competitive games. Science 2014;346:340–3.

[18] Schultz W. Predictive reward signal of dopamine neurons. J Neurophysiol 1998;80:1–27.

[19] Daw ND, Niv Y, Dayan P. Uncertainty-based competition between prefrontal and dorsolateral striatal systems for behavioral control. Nat Neurosci 2005;8:1704–11.

[20] Tolman EC. Cognitive maps in rats and men. Psychol Rev 1948;55:189–208.

[21] Balleine BW, Dickinson A. Goal-directed instrumental action: contingency and incentive learning and their cortical substrates. Neuropharmacology 1998;37:407–19.

[22] Doll BB, Duncan KD, Simon DA, Shohamy D, Daw ND. Model-based choices involve prospective neural activity. Nat Neurosci 2015;18:767–72.

[23] Lee D. Game theory and neural basis of social decision making. Nat Neurosci 2008;11:404–9.

[24] Mookherjee D, Sopher B. Learning behavior in an experimental matching pennies game. Games Econ Behav 1994;7:62–91.

[25] Erev I, Roth AE. Predicting how people play games: reinforcement learning in experimental games with unique, mixed strategy equilibria. Am Econ Rev 1998;88:848–81.

[26] Barraclough DJ, Conroy ML, Lee D. Prefrontal cortex and decision making in a mixed-strategy game. Nat Neurosci 2004;7:404–10.

[27] Seo H, Lee D. Temporal filtering of reward signals in the dorsal anterior cingulate cortex during a mixed-strategy game. J Neurosci 2007;27:8366–77.

[28] Seo H, Barraclough DJ, Lee D. Lateral intraparietal cortex and reinforcement learning during a mixed-strategy game. J Neurosci 2009;29:7278–89.

[29] Donahue CH, Seo H, Lee D. Cortical signals for rewarded actions and strategic exploration. Neuron 2013;80:223–34.

[30] Sul JH, Kim H, Huh N, Lee D, Jung MW. Distinct roles of rodent orbitofrontal and medial prefrontal cortex in decision making. Neuron 2010;66:449–60.

[31] Bernacchia A, Seo H, Lee D, Wang XJ. A reservoir of time constants for memory traces in cortical neurons. Nat Neurosci 2011;14:366–72.

[32] Behrens TE, Woolrich MW, Walton ME, Rushworth MF. Learning the value of information in an uncertain world. Nat Neurosci 2007;10:1214–21.

[33] Vickery TJ, Chun MM, Lee D. Ubiquity and specificity of reinforcement signals throughout the human brain. Neuron 2011;72:166–77.

[34] Camerer C, Ho TH. Experience-weighted attraction learning in normal form games. Econometrica 1999;67:827–74.

[35] Zhu L, Mathewson KE, Hsu M. Dissociable neural representations of reinforcement and belief prediction errors underlie strategic learning. Proc Natl Acad Sci USA 2012;109:1419–24.

[36] Abe H, Lee D. Distributed coding of actual and hypothetical outcomes in the orbital and dorsolateral prefrontal cortex. Neuron 2011;70:731–41.

[37] Matsumoto M, Matsumoto K, Abe H, Tanaka K. Medial prefrontal cell activity signalling prediction errors of action values. Nat Neurosci 2007;10:647–56.

[38] Behrens TE, Hunt LT, Woolrich MW, Rushworth MF. Associative learning of social value. Nature 2008;456:245–9.

[39] Premack D, Woodruff G. Does the chimpanzee have a theory of mind? Behav Brain Sci 1978;4:515–26.

[40] Coricelli G, Nagel R. Neural correlates of depth of strategic reasoning in medial prefrontal cortex. Proc Natl Acad Sci USA 2009;106:9163–8.

[41] Yoshida W, Seymour B, Friston KJ, Dolan RJ. Neural mechanisms of belief inference during cooperative games. J Neurosci 2010;30:10744–51.

[42] Suzuki S, Harasawa N, Ueno K, Gardner JS, Ichinohe N, Haruno M, et al. Learning to simulate others' decisions. Neuron 2012;74:1125–37.

[43] Griessinger T, Coricelli G. The neuroeconomics of strategic interactions. Curr Opin Behav Sci 2015;3:73–9.

19

Neural Control of Social Decisions: Causal Evidence From Brain Stimulation Studies

G. Ugazio, C.C. Ruff

University of Zurich, Zurich, Switzerland

Abstract

Social decisions are among the most important choices in our life. They are often proposed to rely on functionally specialized neural circuitry, based on correlated neural activity observed with neuroimaging. However, neuroimaging studies usually do not allow conclusions about whether the identified neural activity causally controls behavior, rather than being a consequence of it. This gap is now being bridged by brain stimulation studies that test the causal relationship between neural activity and three different types of processes underlying social decisions: social emotions, social cognition, and social behavioral control. Here we critically review this evidence and propose future steps that may help to advance our understanding of how the brain implements social decisions.

INTRODUCTION

Like most animal species, humans have evolved to live in groups [1,2], because doing so ensures better protection against predators and higher chances of survival [3—5]. However, these benefits come coupled with challenges, as group life requires constant interactions with other group members and also competition for fundamental resources such as food or access to mating partners [6]. To deal with these challenges of group life, most animal species have evolved strong dominance hierarchies [7—9] that constrain the need for constant social decision-making, thereby limiting the potential for within-group conflicts. In the case of humans, such hierarchies are complemented by emotional and cognitive skills that allow control of spontaneous behavior as well as maintenance of elaborated social codes of conduct [10—12], such as legal, moral, and social norms [13—15]. The need for these skills may have been a major force driving the evolution of the human brain, an assumption often referred to as the Social Brain Hypothesis [16,17]. This hypothesis proposes that the cortical enlargement in the human brain mainly reflects the evolution of neurocognitive processes beneficial for cohabiting and reproducing in large social groups. This theory motivates one of the central questions in social neuroscience: What are these neurocognitive processes governing our social behavior [18—20]?

Social neuroscience studies trying to answer this question usually focus on three aspects of social decisions that have each been found to recruit activity in somewhat different sets of brain areas (for reviews see Refs. [21—23]). The first group of studies examines *social emotions* [24], i.e., emotions that are experienced vicariously (such as empathy or compassion) or emotions that result from imposing externalities on others (such as guilt or shame, cf. also Chapters 17 and 20). The presence or strength of these emotions is often found to correlate with activity in the insula and the ventromedial prefrontal cortex [25]. A second group of studies investigates *social cognition* [26—28], mainly the ability to take the perspective of others or to attribute intentionality to the consequences of others' actions (often jointly referred to as theory of mind (ToM) [29]). These studies have routinely documented correlated activity in the temporoparietal junction (TPJ), the medial—frontal cortex, the posterior cingulate cortex, and the prefrontal cortex [30]. Finally, a third set of studies examines mechanisms of *social behavior control* that are at least partially independent of basic emotions and cognitions, for instance, the control of selfish impulses or compliance with social norms [31]. These studies often report correlated activity in various parts of the prefrontal cortex [32,33].

Based on these findings, several theoretical accounts have now proposed a core "social brain" comprising all brain areas that together govern our social behavior. However, the purely correlative nature of neuroimaging makes it impossible to assess whether activity in these brain areas is indeed required to control social behavior, rather than just correlating with it (e.g., it is not clear whether the activity is a consequence rather than the cause of the behavior). Moreover, many neuroimaging studies leave it unclear as to what degree the observed activity is indeed specific to social behavior, rather than just relating to some aspect of the choice situation that would also be present for comparable nonsocial decisions (for instance, the activity may relate to the risk [34] or arousal [35] present in both social and nonsocial situations).

For this reason, noninvasive brain stimulation methods are now increasingly used to test whether exogenous manipulation of neural activity indeed specifically affects social decisions, in the way suggested by the related neuroimaging results. Some of these approaches may even arbitrate between opposing interpretations of a set of neuroimaging findings [36]. In general, these approaches may generate a more mechanistic understanding of the neural functions underlying social decisions and may allow disentangling which of the observed neural functions are indeed specifically dedicated to social decisions, as assumed by the social brain hypothesis [16]. However, the use of these methods is not without its complications and limits. In the following, we will first briefly outline the possibilities and pitfalls associated with the two most commonly used methods of noninvasive brain stimulation in humans, before reviewing studies that have employed these methods to study social emotions, cognitions, and control of behavior.

NONINVASIVE BRAIN STIMULATION METHODS USED FOR STUDYING SOCIAL DECISIONS

Transcranial magnetic stimulation (TMS) [37,38] and transcranial electrical stimulation (tES) [39,40] are the two brain stimulation methods most commonly used for studying social decision-making. Both methods are truly noninvasive and can be applied to any healthy person when used according to specific guidelines [41,42]. Their effects are completely reversible and last only for minutes (for detailed methodological reviews see Refs. [43–45]). Because of these properties, both methods are now routinely used to measure how temporary changes in brain function affect social decisions in controlled laboratory settings. While both TMS and tES are often used with similar purpose, their effects on neural function, and thus possibly behavior, are rather different (see Fig. 19.1).

TMS uses magnetic pulses to induce action potentials in the targeted neural populations with good spatial precision (around 1 cm [46]). This artificial activity can modulate neural function and behavior both during the actual pulse application and in the minutes following the stimulation when the area recovers its normal function [46,47]. TMS can in principle have either excitatory or inhibitory effects, depending on the specifics of the stimulation protocol [44,48]. However, most studies to date have employed physiologically disruptive protocols to inhibit neural activity [47,49,50], creating what is often called a temporary "virtual lesion" in the stimulated brain area [51,52]. TMS is therefore mainly used to disrupt spatially distinct patterns of neural activity, to provide a proof of concept that the affected neural processes are causally involved in bringing about behavior. Major shortcomings of TMS for the study of social choices are that many routine TMS protocols may take minutes to apply ([52], but see Ref. [48]) and may generate potentially distracting noises and tactile sensations at the scalp. Both of these properties may modulate arousal and therefore also the behavior under study. These effects need to be properly accounted for by means of comparisons with control groups and may set some limits on the experimental settings that this technique can be used in.

tES, in contrast, does not directly generate action potentials in the stimulated neurons but instead induces neuroplastic effects by means of weak electric currents that modulate the sensitivity of neurons to driving input (for review see Ref. [53] or [42]). The most common form of electric stimulation—transcranial direct current stimulation (tDCS)—changes the excitability of neurons by means of direct currents, but other waveforms are also possible (e.g., oscillatory or noise stimulation, see Refs. [54,55]). The effect on the stimulated tissue depends on the polarity of the applied current: Anodal stimulation generally increases the excitability of the stimulated neurons, whereas cathodal stimulation decreases neuronal excitability [45,56]. The currents are applied by means of electrodes (typically several square centimeters in size) for a period of several minutes, but the effects outlast this period of stimulation [42]. tDCS is therefore not as spatially focal and temporally precise as TMS pulse protocols. Importantly, however, tDCS is relatively easy to deliver to several people at the same time (see Fig. 19.1D), in a manner that is hardly noticeable to the subjects. This allows testing the effects of brain stimulation during actual social interactions and to compare the effects of real stimulation with placebo-like sham stimulation without major worries about possible confounds associated with side effects of stimulation.

One common shortcoming of both TMS and tES is that they can typically be applied only to superficial areas of the neocortex (but see Refs. [57,58] for proposals

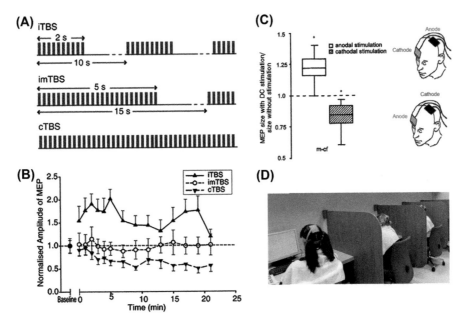

FIGURE 19.1 (A) A schematic illustration of some of the most common transcranial magnetic stimulation *(TMS)* protocols: single pulse *(sp)* TMS, low- (1 Hz) and high-frequency (10 Hz) repetitive TMS, and different sequences of theta burst stimulation *(TBS; c*, continuous; *i*, intermittent; *im*, intermediate). *(Taken from Dayan NC, Buch ER, Sandrini M, Cohen LG. Noninvasive brain stimulation: from physiology to network dynamics and back. Nat Neurosci 2013;16:839. http://dx.doi.org/10.1038/nn.3422.)* (B) The opposite effects of different TBS protocols: cTBS *(downward-pointing triangles)* decreases motor cortex excitability from a prestimulation baseline, whereas iTBS *(upward-pointing triangles)* increases motor cortex sensitivity, as revealed by the amplitude of motor-evoked potentials *(MEPs)* resulting from an spTMS over the motor cortex. *(Taken from Huang, YZ, Edwards MJ, Rounis E, Bhatia KP, Rothwell JC. Theta burst stimulation of the human motor cortex, Neuron 2005;45:p. 202.)* (C) The opposite effects of different tDCS montages over the motor cortex, with anodal *(top)* increasing and cathodal *(bottom)* decreasing motor cortex excitability, again measured with MEP changes following an spTMS as described above. *(Adapted from Nitsche MA, Paulus W. Excitability changes induced in the human motor cortex by weak transcranial direct current stimulation. J Physiol 2000;527:633–39.)* (D) A tDCS lab allowing for simultaneous testing of several participants, an ideal setup to study social interactions. *(All figures used with permissions.)*

to extend these methods). This precludes routine stimulation of potentially relevant subcortical or deep cortical structures, such as the ventromedial prefrontal cortex (but see Ref. [59]), ventral striatum [60], or insula [61]. This limitation has led brain stimulation researchers to mainly focus on two cortical areas that have been routinely implicated in social decision-making: the lateral prefrontal cortex (LPFC; [21]) and the TPJ [29]. In the following, we will discuss these studies, grouped by whether they investigated social emotions, social cognition, or social behavioral control.

BRAIN STIMULATION STUDIES OF SOCIAL EMOTIONS

Social emotions such as empathy result from observing, imagining, or anticipating emotional states of others. For instance, simply watching a person suffering may generate a vicarious feeling of pain in the observer [62]. A number of imaging studies (for a review see Ref. [63]) have shown that the presence of such vicarious emotions activates cortical limbic structures, such as the anterior insula, anterior cingulate cortex,

posterior parietal cortex, or TPJ, whereas the saliency of these emotions may relate to neural activity in the left dorsolateral prefrontal cortex (DLPFC).

To establish whether the neural activity in the left DLPFC is indeed causally necessary for modulating the representation of vicarious pain, Boggio et al. [64] had participants rate how uncomfortable and in pain they felt when watching images of other people experiencing aversive situations. Participants performed this rating task twice on different sets of images, once before tDCS and once after anodal tDCS over one of three different areas: the left primary motor cortex, the DLPFC, and the occipital cortex. Following anodal tDCS of the left DLPFC, the unpleasantness and discomfort/pain ratings given by participants were significantly decreased compared to both the preceding baseline and the sham placebo stimulation. Importantly, no significant changes were observed following tDCS over the two other brain areas. This demonstrates that the left DLPFC has a causal role in modulating the vicarious experience of observed pain in others.

In a similar vein, Kuehne et al. [65] used anodal tDCS to investigate whether the left DLPFC also has a causal role in modulating emotional saliency in a different social decision context: moral judgment. In this study, the participants completed a moral judgment task [66] that required them to decide whether it is morally acceptable to sacrifice the life of a person to save a larger group of people. The two choice options in these dilemmas are thought to differentially rely on emotions, with the decision to sacrifice the person relating to abstract cognitive mechanisms and the judgment that sacrificing the person is morally inacceptable guided by affective responses [66]. Kuehne et al. [65] found that increasing neural excitability in the left DLPFC by means of anodal tDCS led participants to judge sacrificing the person as less appropriate, compared to the moral judgments expressed by participants receiving either sham or cathodal stimulation. These results not only demonstrate that alterations in left DLPFC activity can change moral judgments but also suggest a causal role of the left DLPFC in integrating emotionally salient information associated with a choice option into moral decisions.

This conclusion is in line with the results of a previous TMS study [67] investigating the causal role of the DLPFC (now in the right hemisphere) for related moral decisions. In this study, the authors reasoned that if the DLPFC is causally necessary for integrating emotional information in moral judgment, then disrupting the DLPFC activity by means of repetitive TMS should reduce the saliency of emotional information in moral judgments. This expected pattern was confirmed by the results, as disrupting the right DLPFC with repetitive TMS resulted in increased utilitarian choices to sacrifice one person for the benefit of several others, compared to the responses given by the control group undergoing sham TMS. Jointly with the study by Kuehne et al. [65], these findings provide converging evidence that both the right and the left DLPFC play a causal role in integrating emotional information into complex social moral judgments.

A different aspect of social emotions was investigated by Silani et al. [36]. In this study, the scientists investigated the neural mechanisms of understanding the emotional states of others. To this end, they used TMS over the right supramarginal gyrus (rSMG) to establish if neural activity in this brain area is causally necessary to overcome the so-called emotional egocentricity bias (EEB): the tendency of using the self as a reference point to perceive the world and gain information about others' emotional states. The presence of an EEB was found to correlate with increased activation in the rSMG. When generating a temporary disruption of the rSMG with repetitive TMS, the authors found a substantial increase in EEB, meaning that participants were less able to ignore their own feelings at the moment of interpreting the emotional states of others. This suggests that the neural activity in the rSMG does not just indicate the presence of a conflict between oneself

and the empathically experienced state of another person, but rather is causally necessary for resolving this conflict. In other words, these brain stimulation findings show that the TPJ is specifically required to avoid vicarious emotions that are biased by one's own inner state (Fig. 19.2).

The studies reviewed so far point to important causal connections between neural activity in the DLPFC and TPJ with the modulation and representation of social emotions. The same brain areas have also been implicated in brain stimulation studies of social cognition, as reviewed in the next section.

BRAIN STIMULATION STUDIES OF SOCIAL COGNITION

To efficiently interact with our peers, we rely on several social cognitive mechanisms that enable us to take the perspective of others and to understand their intentions—an ability often referred to as ToM. This cognitive function has been shown to activate a network of brain areas, among them the TPJ [68]. To establish whether this activation indeed reflects a causal role of TPJ activity in attributing intentionality to others' actions, Young et al. [69] employed a moral task in which participants had to judge whether a fictitious agent was morally culpable for his actions. These actions were described as resulting in consequences that were either negative (the agent harming the peer) or neutral (the peer was unharmed by the agent's actions), and the agent could have intentions that were either negative (e.g., intending to do harm) or neutral. When the TPJ was disrupted with repetitive TMS, participants grounded their moral judgments more on the consequences of the agent's actions rather than on his intentions. This was particularly evident for scenarios in which the agent acted with the intention of causing harm but the consequences were neutral. This suggests that the right TPJ has a causal role in interpreting the intentions behind other people's actions and incorporating this understanding in the moral evaluation of the observed actions.

The potential roles of the DLPFC and TPJ in integrating cognitions and emotions into moral choices were directly compared in a study by Jeurissen et al. [70]. This study again used moral dilemmas requiring judgments on the moral permissibility of sacrificing the life of a person to save a larger number of lives (as already discussed in the previous section on social emotions). However, in this study the dilemmas were divided into two classes: personal moral dilemmas in which sacrificing the life of a person is a direct consequence of the action judged and thus may elicit strong affective responses [66], or impersonal moral dilemmas

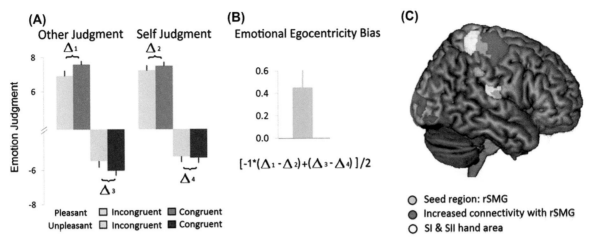

FIGURE 19.2 (A) The emotional egocentricity bias (EEB), the tendency to be affected by one's own emotional state when judging the emotional states of others. This is evident in the fact that participants rate others' projected emotional states (other judgment) as more pleasant or more unpleasant when these are congruent with their own current emotional state. No such effects are seen when participants rate their own projected emotional states (self judgment). (B) Disruption of the right supramarginal gyrus (rSMG) with repetitive transcranial magnetic stimulation (rTMS) results in a decreased ability to dissociate from one's own emotional state, as revealed by an increase in EEB relative to that observed during sham TMS (the zero line). ((A) and (B) adapted from Silani G, Lamm C, Ruff CC, Singer T. Right supramarginal gyrus is crucial to overcome emotional egocentricity bias in social judgments. J Neurosci 2013;33(39):p. 15473.) (C) The exact stimulation site targeted with repetitive TMS (labeled "seed region") (Adapted from Silani G, Lamm C, Ruff CC, Singer T. Right supramarginal gyrus is crucial to overcome emotional egocentricity bias in social judgments. J Neurosci 2013;33(39):p. 15470.)

in which the loss of a life would be as a side effect of the action, thus presumably causing less affect [66]. TMS was given over either the right DLPFC or the right TPJ, in the form of three TMS pulses (separated by 150 ms) starting at either of four different time points (1.5, 2, 2.5, or 3 s) after the response screen. For personal moral dilemmas, right DLPFC TMS given 2.5 s after stimulus onset led participants to consider it more acceptable to sacrifice the one person in personal moral dilemmas, compared to when TMS was delivered to the right TPJ. The opposite pattern of results was observed for impersonal moral dilemmas: Participants considered it more acceptable to sacrifice the one person after receiving TMS over the right TPJ (again starting at 2.5 s after stimulus onset) compared to the DLPFC. While this evidence suggests somewhat different roles for the TPJ and DLPFC in different types of moral decisions, perhaps related to integrating affective responses and more cognitive judgments into choice, it is still unclear which specific mechanisms could explain both the differential effects of TMS over these two areas and the selective effect of the stimulation for only one of the four time points at which TMS was delivered.

Another aspect of ToM is the capacity for taking the perspective of others, which requires dissociating one's point of view from that of other people, a phenomenon that may be related to attention reorientation [71]. Perspective taking can be seen as the cognitive analogue of the ability to overcome the EEB described earlier [36]. Interestingly, studies investigating the

neural underpinnings of either perspective taking [72] or attention reorientation between just visual stimuli [73,74] repeatedly identified correlated neural activity in the right TPJ, suggesting some functional overlap. However, it is currently debated whether different subareas in the TPJ may be differentially involved in just visual attention reorientation or perspective taking [75,76].

To test the specificity of TPJ function for just the social domain, Krall et al. [71] used continuous theta burst stimulation (cTBS) to disrupt neural activity in the right TPJ while participants sequentially completed two tasks: a visual cueing paradigm measuring basic attention reorientation and a false-belief cartoon task indexing ToM. The authors could therefore test if the right TPJ is causally necessary not only for ToM but also for the basic capacity of reorienting visual attention. Both attentional reorienting and ToM abilities were significantly impaired after right TPJ cTBS, compared with a control group receiving stimulation over the vertex, a brain area rarely found to be involved in social decision-making. Moreover, the size of the effect resulting from cTBS on the two tasks was positively correlated, suggesting that the right TPJ plays a crucial overarching role in implementing these two cognitive functions, at least at the functional resolution afforded by TMS. This study therefore extends the results of an antecedent study by Costa et al. [77], which had already shown that a similar TMS protocol over the

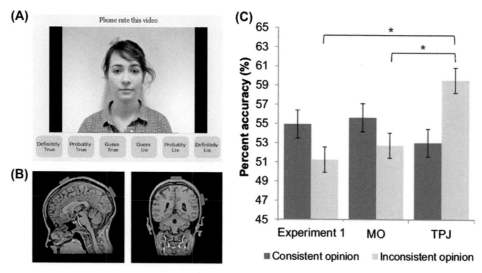

FIGURE 19.3   (A) A screenshot of an example trial of the lie-detection task used to measure attention reorientation bias in Sowden et al. [80]. (B) The tDCS electrode montage used in this study superimposed on a structural magnetic resonance image. The *red* electrode corresponds to the anode and the *blue* electrode corresponds to the cathode. (C) The tDCS-induced changes in accuracy of deception detection for trials with inconsistent opinions, following stimulation of the right TPJ but not of a control site (i.e., a mid-occipital region (MO)). *TPJ*, temporoparietal junction. *Original figure from Sowden S, Wright GRT, Banissy MJ, Catmur C, Bird, G. Transcranial current stimulation of the temporoparietal junction improves lie detection. Curr Biol 2015:2448.*

right TPJ affects performance of the false-belief cartoon task [78].

To further refine which aspect of ToM is implemented by the TPJ, Santiesteban et al. [79] applied anodal, cathodal, or sham tDCS over this brain region while participants completed three tasks: (1) an imitation control task measuring how well participants can identify and control motor representations evoked by oneself and others, (2) a task estimating the capacity to inhibit their own perspective and shift attention to the perspective of others, and (3) a self-referential task in which participants had to attribute mental states either to themselves or to others. This last task differed from the first two in that it required only the basic ability to attribute mental states to other people, but not the ability to distinguish between the perspectives held by oneself and others. Increased neural excitability in the right TPJ, by means of anodal tDCS, resulted in better performance for both the imitation control task (faster responses) and the perspective-taking task (higher accuracy), but no stimulation effect was found for the self-referential task. Thus, these results suggest that neural activity in the right TPJ is causally necessary for distinguishing between the perspectives of oneself and others, but not for ToM in general, as the basic capacity for attributing mental states to others was not affected.

The role of the TPJ in perspective taking was further documented in the context of lie detection [80]. In this task participants had to detect deception in statements made by others, who had an opinion

that was either similar to or different from that of the participant. When judging video recordings of these statements, the participants were significantly less accurate in detecting deception for statements in which the opinion of the watched person was inconsistent with their own. More importantly, this inaccuracy in detecting deception for opinions conflicting with their own was diminished when anodal tDCS was applied to the right TPJ. Consistent with the previous studies, this evidence confirms that neural activity in the right TPJ is causally necessary for ToM, specifically for the function of inhibiting the self-oriented point of view to more accurately represent the point of view of others (Fig. 19.3).

So far we have discussed studies that suggest a causal involvement of the DLPFC in processing social emotions, and a functional role for the right TPJ in assessing emotions, intentions, and perspectives held by others. In the following section we review studies that investigate which brain regions may causally implement social behavior even if social cognition and emotions remain unaltered.

BRAIN STIMULATION STUDIES OF SOCIAL BEHAVIORAL CONTROL

Many aspects of our social behavior can be seen as social types of impulse control, because they require resistance to temptations that appear beneficial in the short run but are actually suboptimal in the long term. One of the brain areas often associated with this ability is the right LPFC [81]. One study targeted this brain structure with TMS to investigate the neural basis underlying the ability to build a good reputation [82]. Creating a good reputation can be essential for a person's success in society, but it requires considerable effort and time investments as well as the readiness to incur costs. In the study, Knoch et al. [82] asked participants to take the role of responder in an iterated version of the trust game [83]. In this game, each response provides signals about one's trustworthiness and therefore contributes to the buildup of a reputation. Disruption of the right, but not the left, LPFC with low-frequency TMS impaired participants' ability to build a positive reputation, as it led to a decrease in returns of the investments received by participants. In other words, participants were less able to resist the temptation of keeping the money they received from the opponent, even if by doing so they harmed their reputation of trustworthiness and thus their possible future earnings. Interestingly, the stimulation did not affect participants' beliefs about the behavioral norms for behavior in this task, and it did not change their behavior in conditions where reputation concerns were absent. This suggests that impaired right LPFC activity results in a lower ability to resist the temptation of behaving selfishly in the trust game, at the expense of deteriorating the participant's social reputation.

Another study [84] tested if the right DLPFC has a related causal role in regulating self-interest when a person is faced with the decision of whether to incur a cost in order to enforce reciprocal fairness. The participants took the role of the responder in the classic ultimatum game [85], requiring them to decide whether to accept a monetary split proposed by the other player or to reject it, in which case both players did not obtain anything. For unfair offers, the participants were thus faced with a conflict between the selfish interest of keeping the little money they were offered and the social motive of punishing the other player by rejecting the offer in order to enforce the fairness norm. Using repetitive TMS, the study showed that disruption of the right, but not the left, DLPFC reduced subjects' willingness to reject unfair offers. This behavioral effect was successfully replicated in a second independent experiment [86] in which the neural involvement of the DLPFC was decreased using a different stimulation method (cathodal tDCS).

Consistent with the aforementioned study, these findings therefore support a causal role for the DLPFC in resisting the selfish economic temptation to accept unfair offers. Also in convergence with Knoch et al. [82], these behavioral effects were clearly dissociated from effects on the fairness beliefs held by the participants, which were unaffected by the stimulation. This general point was again confirmed by Buckholtz et al. [87], who employed hypothetical choice scenarios to examine the effects of inhibitory DLPFC TMS on choices of whether to punish norm transgressions, as well as ratings of the hypothetical actor's blameworthiness for these norm violations. Disrupting DLPFC activity with repetitive TMS led only to a reduction of punishment for wrongful acts without affecting the corresponding blameworthiness ratings. Taken together, all these studies therefore suggest that at least some neural mechanisms in the DLPFC relate to the regulation of behavior based on unaltered beliefs, rather than to representation of these beliefs themselves.

The neural mechanisms underlying this ability were further investigated by Baumgartner et al. [88], who repeated the ultimatum game setup inside an fMRI scanner following application of the TMS over the DLPFC. This allowed comparison of neural activity elicited by the fair and unfair offers between participants receiving stimulation and those receiving sham stimulation. Participants with disrupted DLPFC activity due to TMS again complied less with the fairness norm (i.e., they punished less) than those with unaltered DLPFC activity due to sham TMS. More importantly, TMS participants displayed significantly lower activity in, and connectivity between, the right DLPFC and the posterior ventromedial prefrontal cortex. These effects were present only for unfair offers, not for fair offers in which there was no conflict between economic self-interest and the fairness norm. Based on these findings the authors propose that participants' willingness for costly fairness norm enforcement causally depends on the neural activity in the right DLPFC, as well as on the functional connectivity between this brain area and the ventromedial PFC.

All the studies discussed so far used brain stimulation to inhibit DLPFC excitability and therefore disrupt behavior. This provides interesting causal evidence, but may be of more limited practical use. Ruff et al. [89] used anodal and cathodal tDCS to either increase or decrease neural excitability in the right LPFC. This area was chosen as it had responded strongly to the presence of sanction threats during exchanges in a modified dictator/ultimatum game. Its activity was increased when the participants had to choose how much money to transfer to opponents who could punish for norm transgressions, compared to opponents who could not react to the transfer [90]. Ruff et al. [89] now tested whether this neural activity indeed controls behavioral reactions to such punishment threats. To this end, they measured both voluntary norm compliance (transfers in a dictator game in which the opponent cannot punish) and sanction-induced norm compliance (changes in transfers from the voluntary compliance when sanction threats are present, i.e., when the opponent can punish). The results revealed that both types of norm compliance could be modified with tDCS, in opposite ways: sanction-induced norm compliance was increased by anodal tDCS and decreased by cathodal tDCS, whereas voluntary compliance was reduced by anodal tDCS and enhanced by cathodal tDCS (see Fig. 19.4). This pattern of results suggests that anodal (cathodal) tDCS rendered participants more (less) sensitive to the presence of sanction threats, causing larger (smaller) adjustments of behavior to the external incentives. Interestingly, and in clear congruence with the results discussed earlier, tDCS affected only choices and did not alter several beliefs about the fairness norm and the possible opponent reactions. This again shows that the LPFC mechanism affected by the tDCS controls behavior based on external incentives and internal beliefs, rather than representing these incentives and beliefs themselves. More generally, the study by Ruff et al. [89] shows that brain stimulation methods can be used not only to disrupt social behavior and associated neural function, but also to enhance it.

The finding that reducing neural excitability in the LPFC leads to reduced sanction-induced norm compliance was replicated in a study by Strang et al. [91], in which the neural activity in the right LPFC was disrupted using repetitive TMS. Moreover, that study also reported that disrupting the

contralateral brain region, i.e., left LPFC, did not produce any behavioral change, confirming that sanction-induced norm compliance appears to rely specifically on activity in the right LPFC.

BRAIN STIMULATION EVIDENCE FOR SOCIAL-SPECIFIC NEURAL ACTIVITY

As discussed in the preceding sections, numerous studies have now shown that neural activity in the DLPFC and TPJ is causally involved in guiding behavior based on social emotions and cognitions. However, only very few of these studies have established social specificity, by demonstrating that the brain stimulation affects neural mechanisms that are selectively expressed during social behavior. Such demonstrations would be important to test the original social brain hypothesis [17].

An ideal framework for such tests may be parallel choice settings in which the only element changing across conditions is the presence or absence of a human opponent. Spitzer et al. [90], for instance, identified socially specific neural correlates of social norm compliance by comparing neural activity when participants interacted with another person versus when they

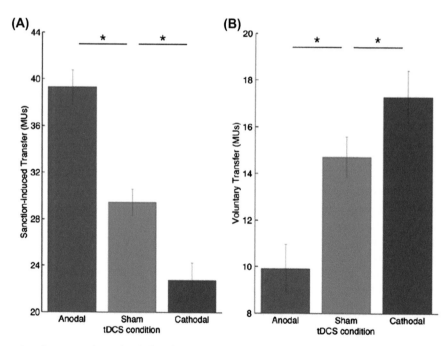

FIGURE 19.4   Changes in voluntary and sanction-induced norm compliance (measured by transfer of monetary units, *MUs*) following anodal or cathodal transcranial direct current stimulation *(tDCS)* over the right lateral prefrontal cortex (LPFC). Increased neural excitability (following anodal stimulation) led to (B) a decrease in voluntary fairness norm compliance and to (A) an increase in compliance with the fairness norm in the presence of a sanction threat. The opposite pattern was observed when neural excitability was inhibited by means of cathodal tDCS. *Original figure from Ruff CC, Ugazio G, Fehr E. Changing social norm compliance with noninvasive brain stimulation. Science 2013;342(6157):p. 483. http://dx.doi.org/10.1126/science.1241399.*

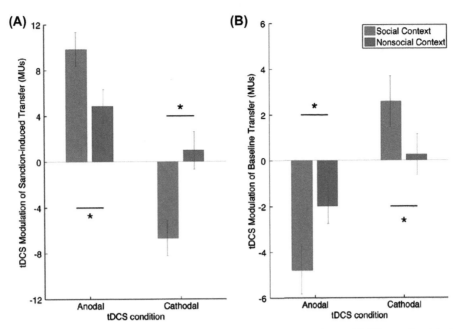

FIGURE 19.5 Changes in norm compliance following transcranial direct current stimulation *(tDCS)* over the right lateral prefrontal cortex were stronger in a social context *(green bars)* than when participants interacted with a computer algorithm *(purple bars)*, both (A) for sanction-induced norm compliance and (B) for voluntary transfers in the absence of sanction threats. These data suggest a specific causal involvement of this brain region during social interactions. *MUs,* monetary units. *Original figure from Ruff CC, Ugazio G, Fehr E. Changing social norm compliance with noninvasive brain stimulation. Science 2013;342(6157):p. 484. http://dx.doi.org/10.1126/science.1241399.*

performed the exact same task, but now against a computer algorithm. In this study, participants performed the same monetary allocation game also used in the Ruff et al. study discussed above [89]. Importantly, the right LPFC that was later targeted with tDCS was more strongly activated by the presence of sanction threats when participants interacted with human opponents than with the computer algorithms. Using a similar approach, Ruff et al. [89] also investigated the effects of LPFC tDCS when participants interacted with a computer algorithm that fully matched the punishment probabilities and magnitudes of human participants from previous sessions. This demonstrated that social context largely determined the strength of the tDCS effects, as the effect of tDCS was significantly more pronounced in the social condition (see Fig. 19.5). These findings thus indicate that the neural activity in the right LPFC is indeed causally necessary for behavioral control only during social interactions with other humans, rather than reflecting other aspects of the choice situation that may be similar for nonsocial choices (such as risk assessment, response selection, etc.). A similarly social-specific causal role of the right LPFC was also identified by Knoch et al. [84], who showed that TMS increased acceptance of unfair offers only from a human opponent, not from a computer. Thus, previous studies have mainly established social specificity for DLPFC control of compliance with fairness norms, whereas

such demonstrations are still crucially lacking for other domains of social behavior.

CONCLUSIONS

In this chapter, we have outlined how brain stimulation studies are starting to establish causal links between neural activity in the DLPFC and the TPJ and various aspects of social behavior (emotions, cognitions, control of behavior). For some of these aspects—specifically perspective taking and compliance with the fairness norm—these links are supported by converging evidence across behavioral paradigms and methods; moreover, there are some demonstrations that the affected neural mechanism may indeed be specific to the social aspects of behavior. While these results provide a promising starting point for a more mechanistic causal understanding of the neural mechanisms steering our social behavior, they have not yet addressed a whole range of fundamental questions that will guide research in the coming years.

For instance, many models of the processes contributing to social behavior have been specified at a purely conceptual level, without a quantitative formalization. This makes it difficult to investigate how brain stimulation affects these processes, beyond statements that they are "modulated" or "disrupted."

Promising avenues in this direction therefore include combining brain stimulation methods with computational models of social learning and choice processes [92–95], which more explicitly formulate distinct neural computations that may be mediated by the stimulated brain area. Conversely, while the effects of brain stimulation methods are well validated with respect to physiological responses of the motor system [37,40], much less is known about how these methods exactly affect neural processing outside the motor system. It may therefore be very promising to combine brain stimulation and neuroimaging methods with behavioral measures [96,97] to obtain a full picture of the causal chain of events that links brain-stimulation effects on neural processing with the corresponding behavioral effects.

At a more general level, it remains fundamentally unclear why the same two brain areas (DLPFC and TPJ) appear involved in a wide range of aspects of social behavior. One explanation for this puzzling generality may be that different aspects of behavior in fact rely on different subareas of these large regions, or different neural populations within these areas. However, this interpretation is hard to assess based on the existing literature, partially because not all studies report their precise stimulation coordinates, and partially because the brain stimulation methods may not have enough spatial resolution to differentially stimulate such subareas [98]. It may thus be interesting to attempt to dissociate different functional contributions of such subareas within the same study, using optimized methods.

A second possible explanation for the general involvement of DLPFC and TPJ may relate to the fact that both areas are heavily interconnected with several brain networks and are thought to play the role of information integration "hubs" [87,99–101]. Perhaps these regions contribute to many different aspects of behavior simply because they integrate and modulate information across different sets of brain areas for the different behavioral contexts. This perspective may also offer an explanation for the functional lateralization in the PFC and TPJ for control of social behavior [74,102,103], as patterns of functional connectivity may differ between analogous regions in the different hemispheres ([88], see also Ruff et al. [104] for similar demonstrations in the visual system). One pragmatic way to start investigating this issue again is the combination of brain stimulation and neuroimaging methods; this may reveal which brain networks are affected by stimulation of the same brain area in different task contexts and how these context-dependent brain network changes relate to distinct aspects of social behavior

[96,97]. Such an approach may help to shift the focus of social neuroscience away from the function of single brain areas toward a true network perspective of the social brain.

References

[1] Wilson EO. The social conquest of earth. New York: Norton; 2012.
[2] Aronson E. The social animal. New York: Palgrave Macmillan; 1980.
[3] Alcock J. Animal behavior. 3rd ed. Sunderland (MA): Sinauer Associates; 2001.
[4] Wilson EO. Sociobiology. abridged ed. Cambridge (MA): Belknap Press of Harvard University Press; 1980.
[5] Tomasello M. The ultra-social animal. Eur J Social Psychol 2014; 44:187–94.
[6] Alcock J. Animal behavior: an evolutionary approach. Sunderland (MA): Sinauer Associates; 2005.
[7] Mueller AE, Thalmann U. Origin and evolution of primate social organization: a reconstruction. Biol Rev Camb Philos Soc 2000; 75(3):405–35.
[8] Pusey AE, Packer C. The ecology of relationships. In: Krebs JR, Davies NB, editors. Behavioural ecology: an evolutionary approach. Oxford: Blackwell Science; 1997. p. 254–83.
[9] Huntingford F, Turner AK. Animal conflict. London: Chapman and Hall; 1987.
[10] Tomasello M. Human culture in evolutionary perspective. In: Gelfand M, editor. Advances in culture and psychology. Oxford: Oxford University Press; 2011.
[11] Dunbar RI, Schultz S. Evolution in the social brain. Science 2007; 317:1344–7.
[12] Barton R, Dunbar IM. Machiavellian intelligence II. In: Whiten A, Byrne R, editors. Cambridge: Cambridge University Press; 1997. p. 240–63.
[13] Bicchieri C. The grammar of society: the nature and dynamics of social norms. , New York: Cambridge University Press; 2006.
[14] Elster J. The cement of society. Cambridge University Press; 1989.
[15] Axelrod R. The evolution of cooperation. New York: Basic Books; 1984.
[16] David-Barrett T, Dunbar RIM. Processing power limits social group size: computational evidence for the cognitive costs of sociality. Proc R Soc B 2013;280(1765):20131151. http://dx.doi.org/10.1098/rspb.2013.1151.
[17] Dunbar RIM. The social brain hypothesis. Evol Anthrop 1998; 6(5):562–72. http://dx.doi.org/10.1002/(SICI)1520-6505(1998)6: 5<178::AID- EVAN5>3.0.CO;2–8.
[18] Decety J, Cacioppo JT. The oxford handbook of social neuroscience. Oxford: Oxford University Press; 2011. Driver J, Blankenburg F, Bestmann S, Vanduffel W, Ruff CC. Concurrent brain-stimulation and neuroimaging for studies of cognition. Trends Cogn Sci 2009; 13:319–27.
[19] Adolphs R. The social brain: neural basis of social knowledge. Annu Rev Psychol 2009;60:693–716. http://dx.doi.org/10.1146/annurev.psych.60.110707.163514.
[20] Maibom HL. Social systems. Philos Psychol 2007;20(5):557–78.
[21] Lieberman MD. Social cognitive neuroscience: a review of core processes. Annu Rev Psychol 2007;58:259–89.
[22] Dolan RJ. Emotion, cognition, and behavior. Science 2002; 298(5596):1191–4.
[23] Adolphs R. The neurobiology of social cognition. Curr Opin Neurobiol 2001;11(2):231–9. http://dx.doi.org/10.1016/S0959-4388(00)00202-6.

[24] Reeck C, Ames DR, Ochsner KN. The social regulation of emotion: an integrative, cross-disciplinary model. Trends Cogn Sci 2015:47—63. http://dx.doi.org/10.1016/j.tics.2015.09.003.

[25] van den Bos W, Guroglu B. The role of the ventral medial prefrontal cortex in social decision making. J Neurosci 2009;29(24):7631—2. http://dx.doi.org/10.1523/JNEUROSCI.1821-09.2009.

[26] Apperly I. Mindreaders: the cognitive basis of "theory of mind". Hove, East Sussex: Psychology Press; 2010.

[27] Hamilton DL, editor. Social cognition. New York (NY): Psychology Press; 2005.

[28] Gordon RM. 'Radical' simulationism. In: Carruthers P, Smith PK, editors. Theories of theories of mind. Cambridge: Cambridge University Press; 1996.

[29] Saxe R. Theory of mind (neural basis). Encyclopedia of consciousness. 2009.

[30] Dodell-Feder D, Koster-Hale J, Bedny M, Saxe R. fMRI item analysis in a theory of mind task. NeuroImage 2010;55(2):705—12. http://dx.doi.org/10.1016/j.neuroimage.2010.12.040.

[31] Montague PR, Lohrenz T. To detect and correct: norm violations and their enforcement. Neuron 2007;56:14—8. http://dx.doi.org/10.1016/j.neuron.2007.09.020.

[32] Buckholtz JW, Marois RE. The roots of modern justice: cognitive and neural foundations of social norms and their enforcement. Nat Neurosci 2012;15:655—61.

[33] Sanfey AG, Rilling JK, Aronson JA, Nystrom LE, Cohen JD. The neural basis of economic decision-making in the ultimatum game. Science 2003;300(5626):1755—8. http://dx.doi.org/10.1126/science.1082976.

[34] Mohr PNC, Biele G, Heekeren HR. Neural processing of risk. J Neurosci 2010;30(19):6613—9. http://dx.doi.org/10.1523/JNEUROSCI.0003-10.2010.

[35] Lindquist KA, Wager TD, Kober H, Bliss-Moreau E, Feldman Barrett L. The brain basis of emotion: a meta-analytic review. Behav Brain Sci 2012;35(3):121—43. http://dx.doi.org/10.1017/S0140525X11000446.

[36] Silani G, Lamm C, Ruff C, Singer T. Right supramarginal gyrus is crucial to overcome emotional egocentricity bias in social judgments. J Neurosci 2013;33(39):15466—6476.

[37] Walsh V, Pascual-Leone A. Transcranial magnetic stimulation: a neurochronometrics of mind. Cambridge (Massachusetts): MIT Press; 2003.

[38] Barker AT, Jalinous R, Freeston IL. Non-invasive magnetic stimulation of human motor cortex. Lancet 1985;325:1106—7.

[39] Ruffini G, Wendling F, Merlet I, Molaee-Ardekani B, Mekonnen A, Salvador R, et al. Transcranial current brain stimulation (tCS): models and technologies. IEEE Trans Neural Syst Rehabil Eng 2013;21(3):333—45.

[40] Nitsche MA, Paulus W. Excitability changes induced in the human motor cortex by weak transcranial direct current stimulation. J Physiol 2000;527:633—9.

[41] Rossi S, Hallett M, Rossini PM, et al. Safety, ethical considerations, and application guidelines for the use of transcranial magnetic stimulation in clinical practice and research. Clin Neurophysiol 2009;120:2008—39.

[42] Nitsche MA, Liebetanz D, Antal A, Lang N, Tergau F, Paulus W. Modulation of cortical excitability by weak direct current stimulation—technical, safety and functional aspects. Suppl Clin Neurophysiol 2003;56:255—76.

[43] Parkin BL, Ekhtiari H, Walsh VF. Non-invasive human brain stimulation in cognitive neuroscience: a primer. Neuron 2015;87(5):932—45. http://dx.doi.org/10.1016/j.neuron.2015.07.032.

[44] Dayan NC, Buch ER, Sandrini M, Cohen LG. Noninvasive brain stimulation: from physiology to network dynamics and back. Nat Neurosci 2013;16:838—44. http://dx.doi.org/10.1038/nn.3422.

[45] Nitsche MA, Fricke K, Henschke U, Schlitterlau A, Liebetanz D, Lang N, et al. Pharmacological modulation of cortical excitability shifts induced by transcranial direct current stimulation in humans. J Physiol 2003;553:293—301. http://dx.doi.org/10.1113/jphysiol.2003.049916.

[46] Riehl M. TMS stimulator design. In: Wassermann EM, Epstein CM, Ziemann U, Walsh V, Paus T, Lisanby SH, editors. Oxford handbook of transcranial stimulation. Oxford: Oxford University Press; 2008. p. 13—23, 25—32.

[47] Ko JH, Monchi O, Ptito A, Bloomfield P, Houle S, Strafella AP. Theta burst stimulation-induced inhibition of dorsolateral prefrontal cortex reveals hemispheric asymmetry in striatal dopamine release during a set-shifting task: a TMS-PET study. Eur J Neurosci 2008;28:2147—55.

[48] Huang YZ, Edwards MJ, Rounis E, Bhatia KP, Rothwell JC. Theta burst stimulation of the human motor cortex. Neuron 2005;45:201—6.

[49] Chen R, Classen J, Gerloff C, Celnik P, Wassermann EM, Hallett M, et al. Depression of motor cortex excitability by low-frequency transcranial magnetic stimulation. Neurology 1997;48:1398—403.

[50] Waldvogel D, van Gelderen P, Muellbacher W, Ziemann U, Immisch I, Hallett M. The relative metabolic demand of inhibition and excitation. Nature 2000;406:995—8.

[51] Quartarone A, Bagnato S, Rizzo V, Morgante F, Sant'angelo A, Battaglia F, et al. Distinct changes in cortical and spinal excitability following high-frequency repetitive TMS to the human motor cortex. Exp Brain Res 2005;161:114—24.

[52] Maeda F, Keenan JP, Tormos JM, Topka H, Pascual-Leone A. Modulation of corticospinal excitability by repetitive transcranial magnetic stimulation. Clin Neurophysiol 2000;111:800—5.

[53] Jacobson L, Koslowsky M, Lavidor M. tDCS polarity effects in motor and cognitive domains: a meta-analytical review. Exp Brain Res 2012;216(1):1—10. http://dx.doi.org/10.1007/s00221-011-2891-9.

[54] Antal A, Boros K, Poreisz C, Chaieb L, Terney D, Paulus W. Comparatively weak after-effects of transcranial alternating current stimulation (tACS) on cortical excitability in humans. Brain Stimul 2008;1:97—105.

[55] Terney D, Chaieb L, Moliadze V, Antal A, Paulus W. Increasing human brain excitability by transcranial high-frequency random noise stimulation. J Neurosci 2008;28:14147—55.

[56] Nitsche MA, Schauenburg A, Lang N, Liebetanz D, Exner C, Paulus W, et al. Facilitation of implicit motor learning by weak transcranial direct current stimulation of the primary motor cortex in the human. J Cogn Neurosci 2003;15(4):619—26.

[57] Ceccanti M, Inghilleri M, Attilia ML, Raccah R, Fiore M, Zangen A, et al. Deep TMS on alcoholics: effects on cortisolemia and dopamine pathway modulation. A pilot study. Can J Physiol Pharmacol 2015;93(4):283—90. http://dx.doi.org/10.1139/cjpp-2014-018.

[58] Isserles M, Shalev AY, Roth Y, Peri T, Kutz I, Zlotnick E, et al. Effectiveness of deep transcranial magnetic stimulation combined with a brief exposure procedure in post-traumatic stress disorder — a pilot study. Brain Stimul 2013;6:377—83.

[59] Chib VS, Yun K, Takahashi H, Shimojo S. Noninvasive remote activation of the ventral midbrain by transcranial direct current stimulation of prefrontal cortex. Transl Psychiatry 2013;3:e268. http://dx.doi.org/10.1038/tp.2013.44.

[60] Báez-Mendoza R, Schultz W. The role of the striatum in social behavior. Front Neurosci 2013;2013(7):233. http://dx.doi.org/10.3389/fnins.2013.00233.

[61] Singer T, Seymour B, O'Doherty J, Kaube H, Dolan RJ, Frith CD. Empathy for pain involves the affective but not sensory components of pain. Science 2004;303(5661):1157–62.

[62] Lamm C, Nusbaum HC, Meltzoff AN, Decety J. What are you feeling? Using functional magnetic resonance imaging to assess the modulation of sensory and affective responses during empathy for pain. PLoS One 2007;2:e1292. http://dx.doi.org/10.1371/journal.pone.000129.

[63] Lamm C, Decety J, Singer T. Meta-analytic evidence for common and distinct neural networks associated with directly experienced pain and empathy for pain. NeuroImage 2011;54(3):2492–502. http://dx.doi.org/10.1016/j.neuroimage.2010.10.014.

[64] Boggio PS, Zaghi S, Fregni F. Modulation of emotions associated with images of human pain using anodal transcranial direct current stimulation (tDCS). Neuropsychologia 2009;47(1):212–7.

[65] Kuehne TM, Heimrath K, Heinze H, Zaehle T. Transcranial direct current stimulation of the left dorsolateral prefrontal cortex shifts preference of moral judgments. PLoS One 2015;10(5):e0127061. http://dx.doi.org/10.1371/journal.pone.0127061.

[66] Greene JD, Sommerville RB, Nystrom LE, Darley JM, Cohen JD. An fMRI investigation of emotional engagement in moral judgment. Science 2001;293:2105–8. http://dx.doi.org/10.1126/science.1062872.

[67] Tassy S, Oullier O, Duclos Y, Coulon O, Mancini J, Deruelle C, et al. Disrupting the right prefrontal cortex alters moral judgement. Soc Cogn Affect Neurosci 2012;7(3):282–8. http://dx.doi.org/10.1093/scan/nsr008.

[68] Saxe R, Carey S, Kanwisher N. Understanding other minds: linking developmental psychology and functional neuroimaging. Annu Rev Psychol 2004;55:87–124.

[69] Young L, Camprodon J, Hauser M, Pascual-Leone A, Saxe R. Disruption of the right temporoparietal junction with transcranial magnetic stimulation reduces the role of beliefs in moral judgments. Proc Natl Acad Sci USA 2010;107:6753–8. http://dx.doi.org/10.1073/pnas.0914826107.

[70] Jeurissen D, Sack AT, Roebroeck A, Russ BE, Pascual-Leone A. TMS affects moral judgment, showing the role of DLPFC and TPJ in cognitive and emotional processing. Front Neurosci 2014;8:18. http://dx.doi.org/10.3389/fnins.2014.00018 [Epub 2014/03/05].

[71] Krall SC, Volz LJ, Oberwelland E, Grefkes C, Fink GR, Konrad K. The right temporoparietal junction in attention and social interaction: a transcranial magnetic stimulation study. Hum Brain Mapp 2015. http://dx.doi.org/10.1002/hbm.23068.

[72] Jackson PL, Meltzoff AN, Decety J. Neural circuits involved in imitation and perspective-taking. NeuroImage 2006;31:429–39.

[73] Corbetta M, Patel G, Shulman GL. The reorienting system of the human brain: from environment to theory of mind. Neuron 2008;58:306–24.

[74] van Overwalle F. Social cognition and the brain: a meta-analysis. Hum Brain Mapp 2009;30:829–58.

[75] Carter RM, Bowling DL, Reeck CC, Huettel SA. A distinct role of the temporal-parietal junction in predicting socially guided decisions. Science 2012;337:109–11.

[76] Eickhoff SB, Laird AR, Grefkes C, Wang LE, Zilles K, Fox PT. Coordinate-based activation likelihood estimation meta-analysis of neuroimaging data: a random-effects approach based on empirical estimates of spatial uncertainty. Hum Brain Mapp 2009;30:2907–26. http://dx.doi.org/10.1002/hbm.20718.

[77] Costa A, Torriero S, Oliveri M, Caltagirone C. Prefrontal and temporo-parietal involvement in taking others' perspective: TMS evidence. Behav Neurol 2008;19:71–4.

[78] Wimmer H, Perner J. Beliefs about beliefs: representation and the containing function of wrong beliefs in young children's understanding of deception. Cognition 1983;13:103–28.

[79] Santiesteban I, Banissy MJ, Catmur C, Bird G. Enhancing social ability by stimulating right temporoparietal junction. Curr Biol 2012;22:2274–7.

[80] Sowden S, Wright GRT, Banissy MJ, Catmur C, Bird G. Transcranial current stimulation of the temporoparietal junction improves lie detection. Curr Biol 2015:2447–51.

[81] Rilling JK, Sanfey AG. The neuroscience of social decision-making. Annu Rev Psychol 2011;2011(62):23–48. http://dx.doi.org/10.1146/annurev.psych.121208.131647.

[82] Knoch D, Schneider F, Schunk D, Hohmann M, Fehr E. Disrupting the prefrontal cortex diminishes the human ability to build a good reputation. Proc Natl Acad Sci USA 2009;106(49):20895–9.

[83] Camerer CF. Behavioral game theory: experiments in strategic interaction. Princeton (NJ): Princeton Univ Press; 2003.

[84] Knoch D, Pascual-Leone A, Meyer K, Treyer V, Fehr E. Diminishing reciprocal fairness by disrupting the right prefrontal cortex. Science 2006;314(5800):829–32.

[85] Gueth W, Schmittberger R, Schwarze B. An experimental analyses of ultimatum bargaining. J Econ Behav Organ 1982;3:367–88.

[86] Knoch D, Nitsche MA, Fischbacher U, Eisenegger C, Pascual-Leone A, Fehr E. Studying the neurobiology of social interaction with transcranial direct current stimulation—the example of punishing unfairness. Cereb Cortex 2008;18(9):1987–90.

[87] Buckholtz JW, Martin JW, Treadway MT, Jan K, Zald DH, Jones O, et al. From blame to punishment: disrupting prefrontal cortex activity reveals norm enforcement mechanisms. Neuron 2015:1369–80. http://dx.doi.org/10.1016/j.neuron.2015.08.023.

[88] Baumgartner T, Knoch D, Hotz P, Eisenegger C, Fehr E. Dorsolateral and ventromedial prefrontal cortex orchestrate normative choice. Nat Neurosci 2011;14(11):1468–74.

[89] Ruff CC, Ugazio G, Fehr E. Changing social norm compliance with noninvasive brain stimulation. Science 2013;342(6157):482–4. http://dx.doi.org/10.1126/science.1241399.

[90] Spitzer M, Fischbacher U, Herrnberger B, Gron G, Fehr E. The neural signature of social norm compliance. Neuron 2007;56:185–96.

[91] Strang S, Gross J, Schuhmann T, Riedl A, Weber B, Sack A. Be nice if you have to—the neurobiological roots of strategic fairness. Social Cogn Affective Neurosci 2014;10(6):790–6. http://dx.doi.org/10.1093/scan/nsu114.

[92] Hampton AN, Bossaerts P, O'Doherty JP. Neural correlates of mentalizing-related computations during strategic interactions in humans. Proc Natl Acad Sci USA 2008;105:6741–6. http://dx.doi.org/10.1073/pnas.0711099105.

[93] Behrens TE, Hunt LT, Woolrich MW, Rushworth MF. Associative learning of social value. Nature 2008;456:245–9. http://dx.doi.org/10.1038/nature07538.

[94] Charness G, Rabin M. Social preferences: some simple tests and a new model. Mimeo: University of California at Berkeley; 2000.

[95] Fehr E, Schmidt K. A theory of fairness, competition, and cooperation. Q J Econ 1999;114:817–68.

[96] Driver J, Blankenburg F, Bestmann S, Vanduffel W, Ruff CC. Concurrent brain-stimulation and neuroimaging for studies of cognition. Trends Cogn Sci 2009;13:319–27.

[97] Ruff CC, Driver J, Bestmann S. Combining TMS and fMRI: from "virtual lesions" to functional-network accounts of cognition. Cortex 2009;45:1043–9 [special issue on TMS and Cognition].

[98] Datta A, Bansal V, Diaz J, Patel J, Reato D, Bikson M. Gyri-precise head model of transcranial direct current stimulation: improved

spatial focality using a ring electrode versus conventional rectangular pad. Brain Stimul 2009;2:201–7.

[99] McKell Carter RM, Huettel SA. A nexus model of the temporal-parietal junction. Trends Cogn Sci 2013;17(7):328–36. http://dx.doi.org/10.1016/j.tics.2013.05.007.

[100] Duncan J. An adaptive coding model of neural function in prefrontal cortex. Nat Rev Neurosci 2001;2(11):820–9.

[101] Miller EK, Cohen JD. An integrative theory of prefrontal cortex function. Annu Rev Neurosci 2001;24:167–202.

[102] Kucyi A, Hodaie M, Davis KD. Lateralization in intrinsic functional connectivity of the temporoparietal junction with salience-and attention-related brain networks. J Neurophysiol 2012;108(12):3382–92. http://dx.doi.org/10.1152/jn.00674.2012.

[103] Levy B, Wagner AD. Cognitive control and right ventrolateral prefrontal cortex: reflexive reorienting, motor inhibition, and action updating. Ann NY Acad Sci 2011:40–62.

[104] Ruff CC, Blankenburg F, Bjoertomt O, Bestmann S, Weiskopf N, Deichmann R, et al. Hemispheric differences in frontal and parietal influences on human occipital cortex: direct confirmation with concurrent TMS-fMRI. J Cogn Neurosci 2009;21:1146–61.

20

The Neuroscience of Compassion and Empathy and Their Link to Prosocial Motivation and Behavior

G. Chierchia, T. Singer

Max Planck Institute for Human Cognitive and Brain Sciences, Leipzig, Germany

Abstract

Empathy enables us to connect with one another at an emotional level. However, this might not be enough to promote prosociality. For instance, it has often been argued that empathically suffering with others does not necessarily motivate us to help them, neither conceptually nor empirically. To fill this gap, a tradition in psychology has highlighted the role of empathic concern or compassion, and developments in social neuroscience have made this proposal increasingly clear. Indeed, empathy and compassion have been shown to tap on dissociable neurobiological mechanisms, as well as on different affective and motivational states. More specifically, while empathy for pain engages a network of brain areas centered around the anterior insula and anterior midcingulate cortex, areas associated with negative affect, compassionate states have been associated with activity in the medial orbitofrontal cortex and ventral striatum, and come with feelings of warmth, concern, and positive affect. Most intriguingly, much like any motor ability, it has also been shown that empathy and compassion can be trained; whereby compassion training has been associated with a number of intrapersonal and interpersonal benefits, ranging from increases in psychological well-being and health to increased cooperation, trust, and tolerance.

INTRODUCTION

There are several approaches to investigating prosocial behavior, and depending on formation or interest, different researchers have highlighted different factors. For instance, behavioral economists have shown how the presence of economic synergies, repeated interactions, or the threat of punishment can, in different ways, promote cooperation in a host of environments [1,2], whereas biologists have shown that prosocial behaviors, defined as costly gestures that benefit others [3], can be evolutionarily adaptive under given conditions [4]. Notably, however, both of these approaches have traditionally explained prosocial behavior by considering the direct/short-term or indirect/long-term benefits that organisms, or their genes, can gain by acting prosocially, that is, they focus on the *outcomes* of prosocial behavior. However [5], thanks to developments in social neuroscience and neuroeconomics, we also have a better understanding of the *motivational and emotional precursors* of cooperative and prosocial behavior.

This chapter will particularly focus on empathy and compassion. We will begin by defining what we intend with the word "empathy" and we will differentiate this from a number of other fundamental processes of social cognition. In section "The Neural Substrates of Empathy," we will continue to what we know about the neurobiological underpinnings of empathy. In the third section, "The Psychological and Neural Bases of Compassion," we will introduce the notion of compassion and discuss the similarities and differences between this and empathy. Specifically, we will suggest that, in contrast to common intuition, empathy may be neither sufficient nor necessary for prosocial motivation to emerge. We will thus illustrate why compassion seems necessary to fill the missing gap between empathy and prosocial motivation. We will review evidence that, indeed, empathy and compassion are related to different subjective feelings and distinct neural correlates. Finally, we will review evidence relating social emotions, such as empathy and compassion, to prosocial behavior.

Decision Neuroscience
http://dx.doi.org/10.1016/B978-0-12-805308-9.00020-8

THE "TOOLKIT" OF SOCIAL COGNITION

As a social species, humans are continuously required to navigate complex social signals that help them orient their behavior in relation to others. Better understanding of the cognitive and neural bases of these social behaviors can be important for various reasons. For instance, in addition to its intrinsic theoretical and scientific interest, it could be fundamental for informing policy makers as to which types of legal, corrective, or organizational institutions humans are more likely to respond to better with regard to improving social communication and cooperation. Furthermore, it may be important for diagnosing and treating clinical populations with social or emotional impairments, such as disorders in the autistic spectrum [6], alexithymia [7], psychopathy [8], the behavioral variant of frontotemporal dementia [9], or focal neurological damage [10].

As mentioned, this chapter particularly focuses on empathy, henceforth intended as the ability to share and understand the emotions of others, while not confusing these emotions with one's own [11–14]. At least dating back to Adam Smith's *Theory of Moral Sentiments,* fellow-feeling, or the ability to "change places in fancy" with another and share his or her "passions," has been considered a fundamental regulator of social behavior [15] and since 2005 there has been a 300% increase in scientific publications containing the word "empathy" [16]. Notoriously, however, there is still relatively little agreement on its precise definition [17]. In what follows, we will elaborate on the definition of empathy provided earlier, while disambiguating it from apparently related but distinct notions. In fact, perhaps one of the most important contributions of social neuroscience has consisted precisely in providing more fine-grained definitions and empirical evidence for several important socioaffective and sociocognitive processes, which had previously been typically lumped together under the broad term of empathy, though they are actually distinct.

Mentalizing

One important distinction made in recent years by social neuroscientists is the one between mentalizing (or "theory of mind") and empathizing. Whereas we have defined empathy as sharing and understanding the "feelings" of others [11], mentalizing is defined as the ability to attribute beliefs and intentions (or propositional knowledge) to others [14,18,19]. Naturally, in everyday life, empathizing with others and understanding their beliefs, thoughts, and intentions may be correlated phenomena, as both help to understand the minds of others. However, these two routes of social cognition have been shown to involve distinct mechanisms [12,20,21].

For instance, from an ontogenetic perspective, it has been suggested that empathy and mentalizing develop at different ages in (normally developing) infants [20]. In fact, the ability to attribute "false beliefs" to others, which is a hallmark of mentalizing, has been shown to develop around the age of 3 or 4, when tested with explicit tasks requiring verbal answers [22], and in younger ages when prompted with implicit tasks [23]. On the other hand, important precursors of empathy, such as emotional contagion, are often held to be innate [16], and signatures of empathic concern have been reported in children of 8–10 months [24]. Moreover, subjects with psychopathic traits have been associated with impairment in empathy but a spared ability to attribute intentions to others (i.e., thus allowing one to manipulate others) [8,16,25]. Conversely, subjects on the autistic spectrum often respond appropriately to affective states of others while frequently failing in mentalizing tasks [7,20,26,27]. Finally, from the neuroscientific perspective, while mentalizing has especially been associated with activity in the medial prefrontal cortex and the posterior superior temporal sulcus, together with the adjacent temporoparietal junction [18] (see also Chapters 18 and 20), empathic brain responses have been related to the anterior insula (AI) and anterior cingulate cortex (ACC), as we will further illustrate in our section on empathy [5,13].

Taken together, these distinct development curves, clinical double dissociations, and differential neural patterns suggest that empathy and mentalizing represent two distinct routes of social cognition.

Mimicry and Emotional Contagion

Despite these differences, one aspect that empathy and mentalizing have in common is that both require distinguishing one's own beliefs (i.e., when mentalizing) or affective states (i.e., when empathizing) from those of others. In other words, both processes require a self–other distinction [11,12]. Correspondingly, the definition of empathy provided above distinguishes it from nonconscious/automatic forms of "motor or affective resonance" with others, such as mimicry and emotional contagion, in which subjects are unaware that their affective states were triggered by the corresponding states in others. In what follows we briefly illustrate each in turn.

Mimicry refers to the observed tendency subjects have to involuntarily "synchronize" their facial expressions, postures, or vocalizations to those of others [28]. Some of these responses can occur roughly

300—400 ms after stimulus onset (i.e., the observed face) and are thus unlikely to be a result of deliberative processes, but rather of preattentive ones. Relatedly, a corpus of studies has revealed the existence of neurons that fire both when performing a motor program and when observing others execute the same program [29]. Such neurons, with potential "mirror-like" properties, have been observed in a widespread network of areas comprising the inferior parietal lobe, the inferior frontal gyrus, and the supplementary motor/ventral premotor motor areas [30,31], and they have often been held to contribute to understanding the actions of others. These brain areas, however, are clearly distinct from those generally observed in tasks specifically designed to induce empathy, such as the AI and ACC [13]. Moreover, such processes remain dissociable from empathy in that they need not regard affective states (as in the case of neuronal "mirror responses" to action observation), and/or they do not require subjects to understand that their states were triggered by others (as in mimicry).

Relative to motor mimicry, emotional contagion relates more specifically to affective state. However, as for mimicry, emotional contagion requires neither self—other distinction nor awareness. One compelling example is contagious crying: the observation that when babies hear the sound of other babies crying (relative to their own cry, white noise, or the crying sound of an infant chimpanzee), they are more likely to cry (or display facial expressions of distress) in turn [32,33], a phenomenon observed in babies as young as 18 h of age [32]. Notably, nonnutritive sucking and heart rate were decreased even when babies heard cries of others in their sleep [33,34], thus strengthening the notion that this form of contagion is nonconscious. In line with this, even in adults it has been shown that sad faces with varying pupil diameter selectively elicit proportional pupil contractions in observers (relative to neutral, happy, or angry faces) [35] and that smaller pupils in sad faces made the faces "look sadder." Finally, a 2014 study by Engert and colleagues [36] found that physiological signatures of stress (such as cortisol levels) were increased even when participants merely observed others in stressful situations. This suggests that emotional contagion can also take place at the hormonal level.

Last, it is worth noting that although mimicry and contagion are related, they are conceptually distinct [14]. For instance, exhibiting affect-laden facial expressions is not always accompanied by corresponding emotional states [37] and, vice versa, there can be emotion contagion without motor mimicry (i.e., as when imagining others in pain without corresponding facial expressions).

THE NEURAL SUBSTRATES OF EMPATHY

Since 2005, social neuroscientists have demonstrated that observing (or imagining) given emotions or sensations in others activates similar neural networks as when subjects experience the same affective states first hand. Such "shared networks of empathy" have now been proven in many domains [12], from disgust [38], to touch [39], to rewards [40] (see Refs. [12,14] for reviews). Here, however, we will focus on empathy for pain paradigms as those are the earliest [5] and by far most studied empathy paradigms in the field of social neuroscience (for reviews see Refs. [12,14]).

In such paradigms neural activity is compared when subjects directly receive painful stimulations (i.e., small electric shocks) and when they observe the delivery of painful stimulations to others. One metaanalysis [13] comprising 32 studies on empathy for pain revealed that a core network comprising the bilateral AI and a posterior portion of the ACC (pACC) was consistently activated across studies, and that some differences between studies could be explained by relatively minor differences in experimental design. These regions are part of the "pain matrix" [41], a set of areas typically recruited when subjects receive painful stimulations themselves. Furthermore, the AI has been involved in integrating interoceptive information into more abstract emotional experiences and decision-making [42,43], whereas the pACC—which is strongly connected to the AI—has been involved in conflict monitoring and cognitive control, integrating negatively valenced stimuli (such as pain) and mediating behavioral adjustments [44]. These areas could thus serve the dual function of representing one's own affective states and those of others [14,16,45] and they are also optimally positioned to relay empathy-related signals to decision-making processes, thus potentially mediating prosocial motivation.

Importantly, however, when we see others suffer, we do not always empathize with them. If we did, it would be difficult to explain how so frequently humans voluntarily inflict pain on one another, in the form of "allegedly corrective punishments" [2,46], as a preemptive attack [47], or out of mere spite [48]. Indeed, a corpus of studies has shown that empathic responses to pain are not universal, and they can be modulated by a number of factors, such as the perceived intensity of the observed pain [49,50], who is suffering [51,52], why he or she is suffering, and contextual reappraisal [53,54], on the available attentional resources [55], as well as on (i.e., personality) characteristics of the empathizer (see [11,12,14] for reviews). Moreover, and importantly, as we will illustrate later, even when empathic responses are present, they may not per se be sufficient to induce prosocial behavior.

THE PSYCHOLOGICAL AND NEURAL BASES OF COMPASSION

A Missing Link Between Empathy and Prosocial Motivation

One could be tempted to suggest that empathy, prosocial motivation, and behavior are correlated; if not even that they represent the same phenomena. However, there are many forms of "prosocial or altruistic" behaviors that need not co-occur with genuine prosocial motivation. For instance, subjects may "act" prosocially because they have learned (and maybe internalized) certain (pro)social norms and suffer when they—or others—deviate from them [56], because they aim to avoid the guilt of not helping [57], or, relatedly, because they want to preserve a positive image, both with regard to themselves [58], and with regard to real [59] or imagined [60] others. They may also act altruistically to elicit positive reciprocity from others [61] or avoid various forms of punishment that they might incur by acting selfishly [62]. These examples clearly show that genuine prosocial motivation is only one of the many routes to prosocial behavior.

Moreover, even when we do empathize with others, this alone may not be sufficient to induce prosocial motivation and behavior. In fact, it has been argued [17,63—65] that empathy can play out in at least two different ways: on one hand, the ability to empathize with others in pain could motivate subjects to help them, in order to relieve them of their pain. On the other, empathizers could be also motivated to relieve others of their sufferings just because, by doing so, they relieve *themselves* of their vicariously felt pain [66], thus reducing their own "negative arousal" [17,67]. Interestingly, this understanding interprets empathy-based prosociality as stemming from an egoistic motivation rather than just an altruistic one.

Notably, however, a number of experiments by Batson and colleagues (see Ref. [17] for a review) have been able to dissociate between these two motives. For instance, in one study [68] participants observed others receiving small (but uncomfortable) electric shocks and were subsequently offered the possibility of helping the victims by receiving shocks in their place. The authors then manipulated how "easy it was to escape" the "stressful" situation without helping. Specifically, in an "easy escape" condition, subjects were informed that (regardless of whether they decided to help or not) they would cease to observe the targets receiving shocks, whereas in the "difficult escape" condition, they knew they would have continued to observe the others suffering. The authors then manipulated the level of empathy by informing subjects that their confederates were similar or dissimilar to them. Notably,

the authors found that, across both empathy conditions, subjects offered to help the confederates when escape was difficult. However, subjects in the low-empathy condition ceased to help their confederates when escape opportunities were easy, whereas subjects in the high-empathy condition helped similarly in both cases. On the basis of this and similar studies [17,63—65], the literature has emphasized an important difference between two possible consequences of empathy. Namely, if empathy leads only to (empathic) distress, then the final goal of one's motivation will become only to help themselves—and helping others may be (at best) instrumental to this. Conversely, if empathy leads to compassion (or empathic concern), then subjects will have the ultimate goal of helping others, and any relief to themselves that they may incur in the process of doing so is accidental. In synthesis, compassion could "fill the gap" between empathy and prosocial behavior.

The Psychological Bases of Compassion

Compassion has been defined as concern for the sufferings of others combined with the motivation to increase their welfare [69]. Such a definition distinguished compassion from empathy in at least two ways. First, we reviewed evidence earlier showing that the vicarious affective states of the empathizer are *isomorphic* (both at the subjective and at the neural level) to those experienced by those they are empathizing with [11,14]. This is not true, however, for the affective states related to compassion, which have been instead characterized as positive feelings of concern, associated with a sense of "warmth" or "soft-heartedness" [17,67]. Second, and in contrast to empathy, the definition of compassion includes a motivational component and specifically a genuine prosocial motivation to alleviate the suffering of others.

For instance, a mother who witnesses her child crying because it has just injured its knee might first have an empathic response, and thus vicariously feel the pain of her child. Indeed, this might contribute to alarming the mother of the need of her child. If, however, the mother does not transform this empathic response into empathic concern or compassion, but rather into empathic distress, she may start to engage in a rather maladaptive response and, for instance, start to cry together with her child instead of helping. In contrast, motherly care typically implies a nearly opposite response (i.e., of concern) to the distress of her child, associated with comforting and helping behaviors, rather than distress. In other words, empathy entails feeling *with* others and can lead to empathic distress, whereas compassion leads to feeling *for* others and it is associated with prosocial motivation and behavior [14,63,70,71].

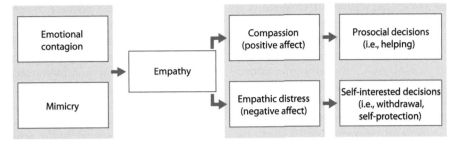

FIGURE 20.1   **The relationship between frequent precursors and possible consequences of empathy.** *Adapted with permission from Klimecki O, Singer T. Empathy from the perspective of social neuroscience. Cambridge University Press; 2011.*

In further support of this differentiation, several studies have suggested that empathic distress and compassion are differentially related to perspective taking [72–74]. For instance, when subjects were asked to "take the perspective of others" when empathizing with them, this was associated with increased concern and altruistic motivation, whereas taking their own perspective (i.e., putting themselves in the shoes of others) when empathizing was associated with withdrawal and distress. In line with this, one fMRI study [74] that used this manipulation found that, relative to taking "the other's perspective," adopting a "self-perspective" while watching pain-related videos (of subjects undergoing a painful auditory stimulation) increased feelings of distress and was related to enhanced activation in more sensory regions of the insula and the amygdala. It follows that empathizing from a first-person perspective might often induce distress rather than compassion (for an overview of possible precursors and consequences of empathy see Fig. 20.1).

In what follows, we will review a number of empirical studies that have dissociated compassion and empathic distress in terms of their psychological and neural characteristics. We will then conclude this chapter by reviewing evidence that, indeed, compassion is linked with prosocial behavior.

The Neural Bases of Compassion

As mentioned in our section on empathy, empathic responses may vary as a function of several important factors [49]. Among these the top-down processes of contextual reappraisal and perspective taking seemed particularly intriguing, as they possibly implied that empathic responses could be regulated for the better, perhaps, to tilt the balance away from empathic distress and toward compassion. Bolstering this notion, Kim and colleagues [76] showed that instructing subjects to take a "compassionate" attitude when observing images of sad faces resulted in activation of reward-related neural regions, such as the ventral tegmental area and substantia nigra, whereas taking a stance of "unconditional

love" toward individuals with mental disabilities was associated with enhanced activation in the caudate and putamen, the middle insula, and a dorsal portion of the ACC [77], areas also involved in romantic and maternal love [78].

Results such as these suggested not only that compassion could be regulated, but that it could maybe also be trained [63,71,79–82]. In fact, contemplative traditions have long engaged in meditation practices aimed to extend, through training, feelings of love and compassionate concern "unconditionally," that is, beyond the friends and family members it is frequently restricted to, as well as toward otherwise "undeserving" others. Indeed, a number of brain imaging studies in naïve or expert meditators have begun to investigate the impact of compassion training on affective responses to the suffering of others.

For instance, Lutz and colleagues [79] provided one of the first cross-sectional studies demonstrating enhanced neural responses to distressed sounds in the middle insula of expert meditators (relative to novice meditators). This cross-sectional approach was extended to naïve nonmeditators that were being taught practices of loving kindness and compassion in a longitudinal intervention design, whereas a control group received memory training. Subjective experiences and brain responses were recorded while subjects viewed video sets depicting other people suffering [63,71]. Notably, before training, the videos induced self-reported negative affect in both groups, and this was accompanied by activation in the empathy for pain-related network described earlier (namely, AI and pACC) [13]. However, after training, only the compassion group exhibited an increase in positive affect (and no change in negative affect) as well as enhanced activity in the putamen, ventral tegmental area, pallidum, and medial orbitofrontal cortex, areas previously implicated in reward, love, or affiliation [78,83,84].

Notably, a different study by the same group [71] extended this work by first training participants in empathy and then in loving kindness and compassion. Whereas compassion training resulted in activation in the same compassion-related brain network observed before and was again associated with positive feelings of concern and warmth, the short-term form of empathy training increased negative affect in response to the painful videos and was associated with increased response in AI and dorsal ACC, the aforementioned empathy for pain network (Fig. 20.2). In line with the view that compassion operates mainly through the activation of a motivational system associated with positive feelings of concern, care, and warmth, a 2015 study by Engen and Singer [70] compared the impact of different emotion-regulation strategies in a group of Buddhist expert meditators. More specifically, the study showed that empathic responses to the same pain-related videos used before [71] could be modulated in two potentially beneficial but distinct ways; whereas compassion

maximally increased positive affect (Fig. 20.3, top right) (even relative to a pain-unrelated video—"watch-neutral" condition), cognitive reappraisal most effectively decreased negative affect (Fig. 20.3, bottom right). Moreover, the two regulation strategies were associated with differential patterns of neural activity: whereas engaging in compassion elicited activity in the same compassion-related network observed by Klimecki et al. [63,71] (red-scaled brain activation), engaging in cognitive reappraisal recruited a frontoparietal network associated with cognitive top-down control (blue-scaled activation).

Taking these findings together, it follows that (1) empathy-related and compassion-related networks are dissociable on the subjective and neuronal levels;

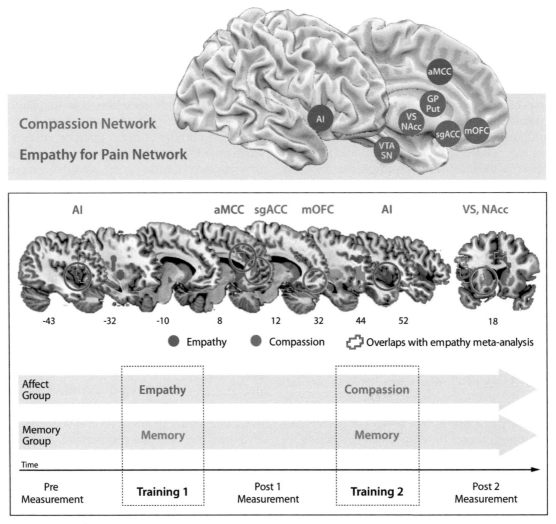

FIGURE 20.2 **Dissociable neural networks underlying compassion and empathy.** Neural plasticity after either empathy (in blue) or compassion training (in red) (relative to memory training) is presented. While watching empathy-inducing videos, empathy training was associated with heightened neural activity in the anterior insula (AI) and anterior middle cingulate cortex (aMCC). However, subsequent compassion training elicited increased activity in the subgenual anterior cingulate cortex (sgACC), medial orbitofrontal cortex (mOFC), ventral striatum (VS), and nucleus accumbens (NAcc). GP, globus pallidus; Put, putamen; SN, substantia nigra; VTA, ventral tegmental area. *Adapted with permission from Singer T, Klimecki OM. Empathy and compassion. Curr Biol 2014;24:R875–8.*

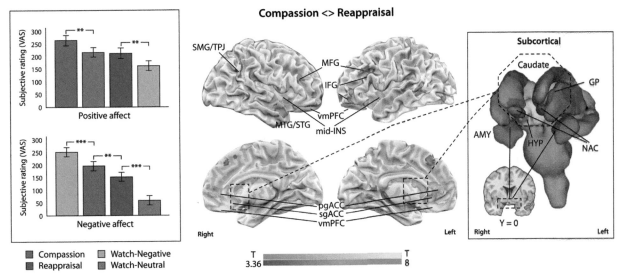

FIGURE 20.3 **Compassion-based emotion regulation upregulates experienced positive affect and associated neural networks.** Regions that exhibited enhanced activity for compassion meditation as opposed to cognitive reappraisal are shown in red, whereas results of the opposite contrast (cognitive reappraisal > compassion) are in blue. *AMY*, amygdala; *GP*, globus pallidus; *HYP*, hypothalamus; *IFG and MFG*, inferior frontal and middle frontal gyrus; *mid-INS*, midinsula; *MTG and STG*, middle and superior temporal gyrus; *NAC*, nucleus accumbens; *pgACC and sgACC*, perigenual and subgenual anterior cingulate cortex; *SMG*, supramarginal gyrus; *TPJ*, temporoparietal junction; *vmPFC*, ventromedial prefrontal cortex; *VAS*, visual analogue scale. *Adapted, with permission from Engen HG, Singer T. Compassion-based emotion regulation up-regulates experienced positive affect and associated neural networks. Soc Cogn Affect Neurosci 2015.*

(2) empathic responses are highly malleable and even short-term practices can tilt empathic responses either toward negative affect (i.e., empathic distress) or toward compassion (i.e., empathic concern); and (3) compassion may represent an adaptive emotion regulation strategy that works mainly through the endogenous generation of positive affect and through a neural system associated with affiliation and care. This could be important for everyone, but especially for subjects who are often more frequently confronted with the sufferings of others, such as physicians or nurses, for which empathic distress can easily lead to "burnout" [85]. It is also worth noting that enhanced compassion could also affect abilities that go beyond empathy for pain. For instance, Mascaro and colleagues [86] found that compassion training (as opposed to a group of subjects that took part in a health discussion forum) increased subjects' ability to infer emotional states from facial expressions.

Last, these benefits of empathic concern and empathic accuracy would seem futile if they did not translate into prosocial behavior, though—as we will briefly review in the next and last paragraphs—evidence suggests they do.

Compassion and Prosocial Behavior

A number of studies have demonstrated the beneficial effect of empathic concern on prosocial behavior. Among the early studies, Batson and colleagues [17] informed subjects about given misfortunes of confederates (i.e., a hard break-up) and then gave them the opportunity to help them while providing ratings of their feelings along two scales: a personal distress scale, composed of negatively valenced items (i.e., "how worried/disturbed/upset/distressed…do you feel?") and an empathic concern scale, which was generally composed of positively valenced items (i.e., "how soft-hearted/warm/compassionate…do you feel?"). The authors found that increases in self-reports of empathic concern were correlated with helping behavior, but that increases in personal distress ratings often predicted withdrawal. This line of evidence supports the notion that empathic distress and empathic concern have distinct motivational qualities that are observable through different (prosocial or selfish) behaviors.

In line with this, Leiberg and colleagues [81] showed that short-term compassion training increased altruistic moves in a multiplayer computer game intended to measure prosocial behavior (see Fig. 20.4, left). Similarly, in a double-blind study (in which participants were unaware that they were being tested), Condon and colleagues [82] showed that an 8-week compassion-training program increased helping behavior in an ecological setting (i.e., getting up to free a chair for a woman in crutches). Furthermore, a number of studies have adopted economic games to probe prosocial behavior, in which participants take part in social interactions with (typically) real monetary outcomes [87–89]. Indeed, it has been shown that when subjects experience empathic concern for their counterparts, they more frequently cooperate in one-shot cooperation dilemmas [90] and games

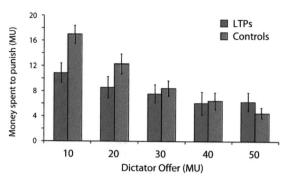

FIGURE 20.4   Compassion and prosocial behavior. (*Left*) The effect of short-term compassion training, as opposed to memory training, on a helping behavior in an ecological prosocial game. Short-term compassion training, but not memory training, increased helping. (*Right*) Money spent by victims of fairness violations to punish the violators (i.e., reduce their monetary payoff). Recipients were either long-term compassion practitioners (*LTPs*, in blue) or controls (in red). LTPs were less likely to punish when they were the direct victims of the violations. *MU*, monetary units. *Adapted with permission from left: Leiberg S, Klimecki O, Singer T. Short-term compassion training increases prosocial behavior in a newly developed prosocial game; 2011 and right: McCall C, Steinbeis N, Ricard M, Singer T. Compassion meditators show less anger, less punishment, and more compensation of victims in response to fairness violations. Front Behav Neurosci 2014;8:424.*

involving trust [91] (in which, in both cases, it would be "economically rational" to defect from cooperation). Even more impressively, empathic concern led to increased cooperation even when subjects knew their counterparts had already defected [92]. This form of "unconditional cooperation" appears rather characteristic of compassion in that it differs from the more frequent forms of reciprocal or conditional cooperation that are typically observed in economic games and in the field [46,93]. In line with this, a 2014 study by McCall and colleagues [94] showed that long-term practitioners of compassion meditation exhibited decreased negative reciprocity as well as less anger toward counterparts who had treated them unfairly. However, when third parties were treated unfairly, then long-term practitioners punished in the same degree as controls. This suggests that while compassion can decrease anger-based punishment, it need not decrease norm-based punishment intended to regain justice and equality (see Fig. 20.4, right).

In the previous section we reviewed evidence on the neural basis of compassion [64,70,71], whereas in the preceding paragraph we reviewed behavioral evidence linking compassion to prosocial behavior. As regards the direct link between compassion/concern-related neural activity and prosocial behavior, the field is still in its infancy, though preliminary findings are promising. For instance, in an fMRI study by Hein and colleagues [52], male soccer fans exhibited increased AI activity when observing fans of the same team receive (small) electric shocks. The magnitude of this activity correlated with ratings of empathic concern and, importantly, it predicted the degree to which participants later helped their in-group members (i.e., by relieving

them of half of their shock and receiving it themselves). Interestingly, when participants observed fans of a rival team receiving shocks, not only were AI responses decreased (relative to the corresponding measure for in-group members), but activity in reward-related regions (i.e., the nucleus accumbens) was observed (proportional to their reported negative attitudes toward the rival team). Indeed, subjects helped the out-group soccer fans less than the in-group ones. Tying these results together, AI activity positively predicted helping, whereas accumbens activity while watching out-group members suffer predicted withdrawal from help.

In this context reward-related activity in the ventral striatum is associated with feelings of "Schadenfreude" [95], thus also of a lack of empathy and a lack of prosocial behavior. Note that in the aforementioned compassion-related studies [63,70,71], the ventral striatum was part of a larger compassion-related network and thus its activation is likely to be predictive of engagement in prosocial behavior rather than of withdrawal (for a review of how reward-related neural activations may be related to prosociality see Ref. [96]). This point makes it clear that we cannot easily take activation of one area in the brain as a predictive marker for pro- or antisocial behavior (see discussion about reverse inference problems [97]). In fact, the functional role of a given neural region is likely to vary, depending on other concomitant neural activations and on the context. In line with this, a 2013 study by Weng et al. [80] compared compassion training to a form of reappraisal training (in which subjects learned how to reinterpret personally stressful events to decrease negative affect). The researchers found that, relative to the reappraisal, compassion training led to increased redistribution of money toward victims that had been treated unfairly. Intriguingly, helping in this case was predicted by increased connectivity between the dorsolateral prefrontal cortex and the nucleus accumbens, suggesting that prosociality requires cognitive upregulation of positive affect in favor of the victims. From these findings it follows that, upon witnessing others suffer, prosocial behaviors could emerge through the context-dependent interplay of neural regions associated with empathic responses (i.e., AI), cognitive regulation (i.e., dorsolateral prefrontal cortex), and compassion and positive affect (i.e., striatum, ventromedial prefrontal cortex/orbitofrontal cortex).

CONCLUSION

In this chapter we have reviewed theoretical, psychological, and neuroscientific work focusing on social emotions such as empathy and compassion and their relationship to prosocial behavior. We have seen that there are many forms of prosocial behaviors but that genuine prosocial motivation may underlie only a subset of these. We also reviewed ample evidence that compassion more than empathy is associated with prosocial motivation and behavior. Specifically, empathizing with the sufferings of others can lead to two possible outcomes, personal distress or compassion, each with different affective, motivational, and neural fingerprints. In fact, whereas the first type of response is perceived as negative, and may actually motivate subjects to withdraw from the person in need, compassion is perceived as pleasant and it has been linked to prosocial behavior. We also reviewed evidence suggesting that compassion can be trained and that this can lead subjects to better cope with empathic distress, to increased trust, and to a higher tolerance to the various "defections" that one may perceive on a daily basis. The intriguing next step would be to probe the extent to which longer periods of mental training on beneficial qualities such as loving kindness and compassion could affect behavior and brain structure in the long run.

In addition to this, the reviewed findings could have important implications for economics. In fact, while "social preference" models [98,99] have made their way into mainstream behavioral economics, they are still typically considered as stable attitudes that vary mostly between individuals [100,101]. The results based on either long-term compassion practitioners or short-term compassion training suggest a more malleable picture, in which social attitudes can change, thus varying within individuals. Finally, the beneficial effects of compassion training could have important implications for education programs, as well as for

the rehabilitation of clinical cases related to social cognition and affective dysregulation.

References

[1] Camerer CF. Behavioural studies of strategic thinking in games. Trends Cogn Sci 2003;7:225–31.
[2] Fehr E, Gächter S. Altruistic punishment in humans. Nature 2002;415:137–40.
[3] West SA, Griffin AS, Gardner A. Social semantics: altruism, cooperation, mutualism, strong reciprocity and group selection. J Evol Biol 2007;20:415–32.
[4] Nowak MA. Five rules for the evolution of cooperation. Science 2006;314:1560–3.
[5] Singer T, Seymour B, O'Doherty J, Kaube H, Dolan RJ, Frith CD. Empathy for pain involves the affective but not sensory components of pain. Science 2004;303:1157–62.
[6] Silani G, Bird G, Brindley R, Singer T, Frith C, Frith U. Levels of emotional awareness and autism: an fMRI study. Soc Neurosci 2008;3:97–112.
[7] Bird G, Silani G, Brindley R, White S, Frith U, Singer T. Empathic brain responses in insula are modulated by levels of alexithymia but not autism. Brain 2010;133:1515–25.
[8] Decety J, Chen C, Harenski C, Kiehl KA. An fMRI study of affective perspective taking in individuals with psychopathy: imagining another in pain does not evoke empathy. Front Hum Neurosci 2013;7.
[9] Cerami C, Dodich A, Canessa N, Crespi C, Marcone A, Cortese F, et al. Neural correlates of empathic impairment in the behavioral variant of frontotemporal dementia. Alzheimer's Dement 2014; 10:827–34.
[10] Shamay-Tsoory SG. The neural bases for empathy. Neuroscientist 2011;17:18–24.
[11] De Vignemont F, Singer T. The empathic brain: how, when and why? Trends Cogn Sci 2006;10:435–41.
[12] Bernhardt BC, Singer T. The neural basis of empathy. Annu Rev Neurosci 2012;35:1–23.
[13] Lamm C, Decety J, Singer T. Meta-analytic evidence for common and distinct neural networks associated with directly experienced pain and empathy for pain. NeuroImage 2011;54:2492–502.
[14] Singer T, Lamm C. The social neuroscience of empathy. Ann NY Acad Sci 2009;1156:81–96.
[15] Smith A. The theory of moral sentiments. Penguin; 2010.
[16] Decety J. The neural pathways, development and functions of empathy. Curr Opin Behav Sci 2015;3:1–6.
[17] Batson CD, Ahmad N, Powell AA, Stocks EL, Shah J, Gardner WL. Prosocial motivation. Handb Motiv Sci 2008: 135–49.
[18] Frith CD, Frith U. The neural basis of mentalizing. Neuron 2006; 50:531–4.
[19] Goldman AI. Simulating minds: the philosophy, psychology, and neuroscience of mindreading. New York: Oxford University Press; 2006.
[20] Singer T. The neuronal basis and ontogeny of empathy and mind reading: review of literature and implications for future research. Neurosci Biobehav Rev 2006;30:855–63.
[21] Kanske P, Böckler A, Trautwein F, Singer T. Dissecting the social brain: introducing the EmpaToM to reveal distinct neural networks and brain-behavior relations for empathy and theory of mind. NeuroImage 2015;122.
[22] Wellman HM, Cross D, Watson J. Meta-analysis of theory-of-mind development: the truth about false belief. Child Dev 2001:655–84.

[23] Baillargeon R, Scott RM, He Z. False-belief understanding in infants. Trends Cogn Sci 2010;14:110–8.

[24] Davidov M, Zahn-Waxler C, Roth-Hanania R, Knafo A. Concern for others in the first year of life: theory, evidence, and avenues for research. Child Dev Perspect 2013;7:126–31.

[25] Marsh AA, Finger EC, Fowler KA, Adalio CJ, Jurkowitz ITN, Schechter JC, et al. Empathic responsiveness in amygdala and anterior cingulate cortex in youths with psychopathic traits. J Child Psychol Psychiatry 2013;54:900–10.

[26] Singer T, Snozzi R, Bird G, Petrovic P, Silani G, Heinrichs M, et al. Effects of oxytocin and prosocial behavior on brain responses to direct and vicariously experienced pain. Emotion 2008;8:781.

[27] Jones AP, Happé FGE, Gilbert F, Burnett S, Viding E. Feeling, caring, knowing: different types of empathy deficit in boys with psychopathic tendencies and autism spectrum disorder. J Child Psychol Psychiatry 2010;51:1188–97.

[28] Hatfield E, Rapson RL, Le Y-CL. 2 emotional contagion and empathy. Soc Neurosci Empathy 2011;19.

[29] Rizzolatti G, Sinigaglia C. The functional role of the parieto-frontal mirror circuit: interpretations and misinterpretations. Nat Rev Neurosci 2010;11:264–74.

[30] Van Overwalle F. Social cognition and the brain: a meta-analysis. Hum Brain Mapp 2009;30:829–58. http://dx.doi.org/10.1002/hbm.20547.

[31] Molenberghs P, Cunnington R, Mattingley JB. Brain regions with mirror properties: a meta-analysis of 125 human fMRI studies. Neurosci Biobehav Rev 2012;36:341–9.

[32] Martin GB, Clark RD. Distress crying in neonates: species and peer specificity. Dev Psychol 1982;18:3.

[33] Dondi M, Simion F, Caltran G. Can newborns discriminate between their own cry and the cry of another newborn infant? Dev Psychol 1999;35:418.

[34] Field T, Diego M, Hernandez-Reif M, Fernandez M. Depressed mothers' newborns show less discrimination of other newborns' cry sounds. Infant Behav Dev 2007;30:431–5.

[35] Harrison NA, Singer T, Rotshtein P, Dolan RJ, Critchley HD. Pupillary contagion: central mechanisms engaged in sadness processing. Soc Cogn Affect Neurosci 2006;1:5–17.

[36] Engert V, Plessow F, Miller R, Kirschbaum C, Singer T. Cortisol increase in empathic stress is modulated by emotional closeness and observation modality. Psychoneuroendocrinology 2014;45:192–201.

[37] Laird JD, Alibozak T, Davainis D, Deignan K, Fontanella K, Hong J, et al. Individual differences in the effects of spontaneous mimicry on emotional contagion. Motiv Emot 1994;18:231–47.

[38] Jabbi M, Bastiaansen J, Keysers C. A common anterior insula representation of disgust observation, experience and imagination shows divergent functional connectivity pathways. 2008.

[39] Lamm C, Silani G, Singer T. Distinct neural networks underlying empathy for pleasant and unpleasant touch. Cortex 2015;70.

[40] Varnum MEW, Shi Z, Chen A, Qiu J, Han S. When "Your" reward is the same as "My" reward: self-construal priming shifts neural responses to own vs. friends' rewards. NeuroImage 2014;87:164–9.

[41] Derbyshire SWG. Exploring the pain "neuromatrix". Curr Rev Pain 2000;4:467–77.

[42] Craig AD. How do you feel—now? The anterior insula and human awareness. Nat Rev Neurosci 2009;10.

[43] Lindquist KA, Wager TD, Kober H, Bliss-Moreau E, Barrett LF. The brain basis of emotion: a meta-analytic review. Behav Brain Sci 2012;35:121–43.

[44] Shackman AJ, V Salomons T, Slagter HA, Fox AS, Winter JJ, Davidson RJ. The integration of negative affect, pain and cognitive control in the cingulate cortex. Nat Rev Neurosci 2011;12:154–67.

[45] Engen HG, Singer T. Empathy circuits. Curr Opin Neurobiol 2013;23:275–82.

[46] Gächter S. Conditional cooperation: behavioral regularities from the lab and the field and their policy implications, na. 2007.

[47] Simunovic D, Mifune N, Yamagishi T. Preemptive strike: an experimental study of fear-based aggression. J Exp Soc Psychol 2013;49:1120–3.

[48] Zizzo DJ, Oswald AJ. Are people willing to pay to reduce others' incomes? Ann Econ Stat 2001:39–65.

[49] Hein G, Singer T. I feel how you feel but not always: the empathic brain and its modulation. Curr Opin Neurobiol 2008;18:153–8.

[50] V Saarela M, Hlushchuk Y, Williams AC, Schurmann M, Kalso E, Hari R. The compassionate brain: humans detect intensity of pain from another's face. Cereb Cortex 2007;17:230–7 (New York, NY 1991).

[51] Singer T, Seymour B, O'Doherty JP, Stephan KE, Dolan RJ, Frith CD. Empathic neural responses are modulated by the perceived fairness of others. Nature 2006;439:466–9.

[52] Hein G, Silani G, Preuschoff K, Batson CD, Singer T. Neural responses to ingroup and outgroup members' suffering predict individual differences in costly helping. Neuron 2010;68:149–60.

[53] Decety J, Echols S, Correll J. The blame game: the effect of responsibility and social stigma on empathy for pain. J Cogn Neurosci 2010;22:985–97.

[54] Lamm C, Meltzoff AN, Decety J. How do we empathize with someone who is not like us? A functional magnetic resonance imaging study. J Cogn Neurosci 2010;22:362–76. http://dx.doi.org/10.1162/jocn.2009.21186.

[55] Gu X, Han S. Attention and reality constraints on the neural processes of empathy for pain. NeuroImage 2007;36:256–67.

[56] Akerlof GA, Kranton RE. Economics and identity. Q J Econ 2000:715–53.

[57] Chang LJ, Smith A, Dufwenberg M, Sanfey AG. Triangulating the neural, psychological, and economic bases of guilt aversion. Neuron 2011;70:560–72.

[58] Dana J, Weber RA, Kuang JX. Exploiting moral wiggle room: experiments demonstrating an illusory preference for fairness. Econ Theor 2007;33:67–80.

[59] Hoffman E, McCabe K, Smith VL. Social distance and other-regarding behavior in dictator games. Am Econ Rev 1996;86:653–60.

[60] Haley KJ, Fessler DMT. Nobody's watching?: Subtle cues affect generosity in an anonymous economic game. Evol Hum Behav 2005;26:245–56.

[61] Hardy CL, Van Vugt M. Nice guys finish first: the competitive altruism hypothesis. Pers Soc Psychol Bull 2006;32:1402–13.

[62] Fehr E, Gächter S. Fairness and retaliation: the economics of reciprocity. J Econ Perspect 2000:159–81.

[63] Klimecki OM, Leiberg S, Ricard M, Singer T. Differential pattern of functional brain plasticity after compassion and empathy training. Soc Cogn Affect Neurosci 2014;9:873–9.

[64] Singer T, Klimecki OM. Empathy and compassion. Curr Biol 2014;24:R875–8.

[65] Eisenberg N. Empathy-related emotional responses, altruism, and their socialization. 2002.

[66] Maner JK, Luce CL, Neuberg SL, Cialdini RB, Brown S, Sagarin BJ. The effects of perspective taking on motivations for helping: still no evidence for altruism. Personal Soc Psychol Bull 2002;28:1601–10.

[67] Batson CD, Fultz J, Schoenrade PA. Distress and empathy: two qualitatively distinct vicarious emotions with different motivational consequences. J Pers 1987;55:19–39.

[68] Batson CD, Duncan BD, Ackerman P, Buckley T, Birch K. Is empathic emotion a source of altruistic motivation? J Pers Soc Psychol 1981;40:290.

[69] Goetz JL, Keltner D, Simon-Thomas E. Compassion: an evolutionary analysis and empirical review. Psychol Bull 2010;136:351.

[70] Engen HG, Singer T. Compassion-based emotion regulation up-regulates experienced positive affect and associated neural networks. Soc Cogn Affect Neurosci 2015;10(9).

[71] Klimecki OM, Leiberg S, Lamm C, Singer T. Functional neural plasticity and associated changes in positive affect after compassion training. Cereb Cortex 2013;23:1552−61 (New York, NY 1991).

[72] Ruby P, Decety J. How would you feel versus how do you think she would feel? A neuroimaging study of perspective-taking with social emotions. J Cogn Neurosci 2004;16:988−99.

[73] Batson CD, Early S, Salvarani G. Perspective taking: imagining how another feels versus imaging how you would feel. Personal Soc Psychol Bull 1997;23:751−8.

[74] Lamm C, Batson CD, Decety J. The neural substrate of human empathy: effects of perspective-taking and cognitive appraisal. J Cogn Neurosci 2007;19:42−58.

[75] Klimecki O, Singer T. Empathy from the perspective of social neuroscience. Cambridge University Press; 2011.

[76] Kim J-W, Kim S-E, Kim J-J, Jeong B, Park C-H, Son AR, et al. Compassionate attitude towards others' suffering activates the mesolimbic neural system. Neuropsychologia 2009;47: 2073−81.

[77] Beauregard M, Courtemanche J, Paquette V, St-Pierre ÉL. The neural basis of unconditional love. Psychiatry Res Neuroimaging 2009;172:93−8.

[78] Bartels A, Zeki S. The neural correlates of maternal and romantic love. NeuroImage 2004;21:1155−66.

[79] Lutz A, Brefczynski-Lewis J, Johnstone T, Davidson RJ. Regulation of the neural circuitry of emotion by compassion meditation: effects of meditative expertise. PLoS One 2008;3:e1897.

[80] Weng HY, Fox AS, Shackman AJ, Stodola DE, Caldwell JZ, Olson MC, et al. Compassion training alters altruism and neural responses to suffering. Psychol Sci 2013;24:1171−80.

[81] Leiberg S, Klimecki O, Singer T. Short-term compassion training increases prosocial behavior in a newly developed prosocial game. 2011.

[82] Condon P, Desbordes G, Miller WB, DeSteno D. Meditation increases compassionate responses to suffering. Psychol Sci 2013; 24:2125−7.

[83] Vrticka P, Andersson F, Grandjean D, Sander D, Vuilleumier P. Individual attachment style modulates human amygdala and striatum activation during social appraisal. PLoS One 2008;3: e2868.

[84] Knutson B, Adams CM, Fong GW, Hommer D. Anticipation of increasing monetary reward selectively recruits nucleus accumbens. J Neurosci 2001;21:RC159.

[85] Adriaenssens J, De Gucht V, Maes S. Determinants and prevalence of burnout in emergency nurses: a systematic review of 25 years of research. Int J Nurs Stud 2015;52:649−61.

[86] Mascaro JS, Rilling JK, Negi LT, Raison CL. Compassion meditation enhances empathic accuracy and related neural activity. Soc Cogn Affect Neurosci 2013;8:48−55.

[87] Camerer C. Behavioral game theory: experiments in strategic interaction. 2003. New York.

[88] Singer T, Steinbeis N. Differential roles of fairness- and compassion-based motivations for cooperation, defection, and punishment. Ann NY Acad Sci 2009;1167:41−50.

[89] Fehr E, Fischbacher U. The nature of human altruism. Nature 2003;425:785−91.

[90] Batson CD, Moran T. Empathy-induced altruism in a prisoner's dilemma. Eur J Soc Psychol 1999;29:909−24.

[91] Eimontaite I, Nicolle A, Schindler I, Goel V. The effect of partner-directed emotion in social exchange decision-making. Front Psychol 2013;4.

[92] Batson CD, Ahmad N. Empathy-induced altruism in a prisoner's dilemma II: what if the target of empathy has defected? Eur J Soc Psychol 2001;31:25−36.

[93] Kocher MG, Cherry T, Kroll S, Netzer RJ, Sutter M. Conditional cooperation on three continents. Econ Lett 2008;101:175−8.

[94] McCall C, Steinbeis N, Ricard M, Singer T. Compassion meditators show less anger, less punishment, and more compensation of victims in response to fairness violations. Front Behav Neurosci 2014;8:424.

[95] Takahashi H, Kato M, Matsuura M, Mobbs D, Suhara T, Okubo Y. When your gain is my pain and your pain is my gain: neural correlates of envy and schadenfreude. Science 2009;323:937−9.

[96] Bhanji J, Delgado M. The social brain and reward: social information processing in the human striatum. Wiley Interdiscip Rev Cogn Sci 2014;5.1:61−73.

[97] Poldrack R. Inferring mental states from neuroimaging data: from reverse inference to large-scale decoding. Neuron 2011;72: 692−7.

[98] Kollock P. Social dilemmas: the anatomy of cooperation. Annu Rev Sociol 1998:183−214.

[99] Camerer CF, Fehr E. When does "economic man" dominate social behavior? Science (80-.) 2006;311:47−52.

[100] Van Lange PAM, De Bruin E, Otten W, Joireman JA. Development of prosocial, individualistic, and competitive orientations: theory and preliminary evidence. J Pers Soc Psychol 1997;73:733.

[101] Yamagishi T, Mifune N, Li Y, Shinada M, Hashimoto H, Horita Y, et al. Is behavioral pro-sociality game-specific? Pro-social preference and expectations of pro-sociality. Organ Behav Hum Decis Process 2013;120:260−71.

HUMAN CLINICAL STUDIES INVOLVING DYSFUNCTIONS OF REWARD AND DECISION-MAKING PROCESSES

21

Can Models of Reinforcement Learning Help Us to Understand Symptoms of Schizophrenia?

G.K. Murray, C. Tudor-Sfetea, P.C. Fletcher

University of Cambridge, Cambridge, United Kingdom

Abstract

We aim to show that the principles and empirical observations relating to reward and reinforcement learning may be useful in considering schizophrenia. Though this is a complex illness in which the primary symptoms are expressed in terms of bizarre beliefs and perceptions, there is reason to suppose that the underlying deficits reside in brain systems known to be linked to learning, reward, and motivational processes. We consider two very different symptoms that are characteristic of schizophrenia. The first, delusions, refers to apparently irrational beliefs that arise and are held with extraordinary tenacity in the face of contradictory evidence. There is emerging evidence that such beliefs emerge from abnormal associative processes perhaps arising as a consequence of disturbed dopaminergic function. The second symptom, anhedonia, refers to the apparent lack of enjoyment in life. In fact, there is now good evidence that anhedonia may relate more to a motivation deficit and, as such, can be considered in the context of the same system as that implicated in delusions. We highlight the evidence in favor of these perspectives, showing that reinforcement learning may be key to understanding schizophrenia.

INTRODUCTION

Schizophrenia can be a profoundly disabling condition. It is more common than many people think, having an estimated lifetime prevalence of 0.7−1% [1]. It might seem rather odd to include a consideration of schizophrenia in a book such as this one because the condition has not traditionally been linked directly to reward processing. In the main [2], it is defined and diagnosed according to the presence of certain experiences such as hallucinations (perceptions in the absence of stimuli; e.g., the sufferer may hear voices talking critically about him or her) and delusions (bizarre beliefs; e.g., the patient may come to believe that he or she is the victim of intrigue and conspiracy, even to the extent that his or her life is in danger). It is also accompanied by so-called negative symptoms, which are essentially defined by the absence of desirable functions or behaviors. For example, thinking may be disturbed and disorganized, making communication very difficult. Motivation may be markedly reduced and the sufferer may become withdrawn, often reporting difficulty enjoying previously pleasurable activities (anhedonia). Thus, whereas the diagnostic process does not invoke notions of disordered reward processing directly nor is a consideration of reward processing explicitly within the compass of standard treatments, there are nevertheless a number of aspects of psychopathology in schizophrenia that can fruitfully be considered within the framework of reward and reinforcement learning developed elsewhere in this book. Moreover, given extensive evidence from molecular imaging studies implicating dysregulated striatal dopamine in schizophrenia and other psychoses, combined with the evidence linking striatal dopamine to reinforcement learning and motivation, there are reasons, beyond descriptive considerations of the signs and symptoms, for suggesting that reinforcement learning is worth scrutinizing in psychosis.

We aim to show here that the principles and insights that have emerged from considerations of reward processing in animals and humans may be very important in coming to understand schizophrenia. We will illustrate this with respect to one characteristic positive symptom (delusional beliefs) and one negative symptom (anhedonia). We will begin by reviewing the history of empirical and theoretical observations of reward processing in schizophrenia before relating this to current neuroscientific insights into the brain and the possible neurochemical bases for such processing. We will consider how these observations relate to neurochemically based explanations of schizophrenia, specifically those invoking disruptions of dopaminergic and glutamatergic circuitry together with the possibility that the illness may arise from an impaired interaction between these two systems. We will then develop ideas of how disruptions in reward processing may provide an explanation for delusional beliefs and for anhedonia.

REWARD PROCESSING IN SCHIZOPHRENIA: A HISTORICAL PERSPECTIVE

While, as we have noted above, the traditional conceptualization of schizophrenia has not focused primarily on reward or reinforcement learning, in fact, both motivational deficits and dysfunction of associative learning (closely related to reinforcement learning) were proposed in schizophrenia long before the discovery of dopamine. For example, Bleuler described avolition and loosening of associations (due to a failure of any unifying concept of purpose or goal) as core features of schizophrenia [3].

Robbins proposed reinforcement learning abnormalities in schizophrenia when he drew parallels between stereotyped behaviors in stimulant-treated rats and schizophrenic stereotyped patterns of thought and behavior [4]. He argued that stereotypy-inducing properties of stimulants were related to their reinforcement-enhancing properties, and argued the case for such stimulant-induced behavior as a model for psychotic illness. In the same year, Miller proposed that dopamine disruption in the striatum could lead to abnormal associative learning processes, which in turn could lead to psychotic symptoms. Miller considered early evidence suggesting that recognition of the association of related features in the environment by humans and animals occurred in the basal ganglia via a dopamine-dependent process. Furthermore, he argued that basal ganglia dopamine overactivity in schizophrenia could lead to a lowering of the required threshold for acceptance of conclusions [5]. This is reminiscent of more recent arguments that schizophrenia patients "jump to conclusions" in probabilistic reasoning tasks [6]. Over the course of the 1970s, 1980s, and 1990s, in a series of theoretical articles, Miller developed his theory of the dopaminergic basis of positive psychotic symptoms acting via disruption of mesostriatal reward learning processes, to address specific psychotic and manic symptoms, and developments in the understanding of different dopamine receptor classes [7–10].

Meanwhile, Gray, Hemsley, and colleagues were working on a related theory, one that focused more directly on associative aspects of learning rather than on the rewarding and motivational aspects [11–14]. Gray drew on emerging evidence from studies of experimental animals that interactions between the hippocampus and the mesoaccumbens pathway are critical in allowing organisms to ignore irrelevant stimuli. While the importance of mismatches ("prediction error"; see Ref. [15]) between expectation and outcome has now been recognized in work on associative and reinforcement learning, this was not a component explicitly recognized in the models of Gray and his colleagues. However, in emphasizing the likelihood of a central dopamine-dependent cognitive disturbance in psychosis and in considering how this could contribute to a failure of "the integration of past regularities of experience with current stimulus recognition, learning and action," they clearly posited a model, which might now be discussed in terms of a disruption in prediction-error signaling.

Gray and Hemsley focused particularly on the classical behavioral associative learning paradigms of latent inhibition [16] and Kamin blocking [17]. They proposed that "Disruption of LI or KB would have the effect that, in accounting for the occurrence of a significant event [a US (unconditioned stimulus)], a schizophrenia patient would be likely to attribute causal efficacy to stimuli that normal individuals would ignore, either because they were familiar and had previously occurred without being followed by a significant event, or because the event was already predicted by other associations." [12]. One of the very attractive features of Gray's model, particularly to those who feel frustrated by the fact that a number of models of psychosis are very narrow in their focus, is that it encompassed four levels of explanation—neuroanatomy, chemistry, cognitive process, and symptoms. This is an important attribute in generating hypotheses that are testable and in providing an explanation that bridges the gaps between the physiological and the mental. Indeed, it foreshadows the development of cognitive neuropsychiatry, which adheres to this principle [18,19] and which has flourished since 1995.

The important work of these researchers sets the background for a number of fruitful avenues of research. Common to these developments is the drive to appeal to advances in cognitive neuroscience of learning and reward processing to inspire and constrain theories of schizophrenia. Later in this chapter, we will consider the importance of these developments with regard to delusions and anhedonia but it is worth noting at this stage that one very good example of this approach was exemplified by the work of Shitij Kapur [20], who drew on contemporary models of reward processing [21] to propose an influential model of how the bizarre beliefs that characterize schizophrenia might be considered in terms of disrupted incentive salience. We return to this later but first make some relevant background observations.

DOPAMINE, SCHIZOPHRENIA, AND REINFORCEMENT LEARNING

Subcortical monoamine function is important in both motivation and learning of goal-directed associations [22,23]. Perhaps dysfunction of this system could explain certain features of schizophrenia. It was previously thought that noradrenaline was the key neurotransmitter mediating these reward-related processes, and Stein and Wise were, as far as we are aware, the first to propose that damage to the noradrenergic reward system could explain both the core symptoms of schizophrenia and its progressive and worsening course [24]. However, as evidence has subsequently demonstrated the importance of another monoamine transmitter, dopamine, as opposed to noradrenaline, in reward processing, and as there is extensive evidence implicating dopamine dysfunction in schizophrenia, then the ascending dopamine system is a much more promising candidate. We now briefly and selectively review the extensive, and sometimes contradictory, literature on dopamine dysfunction in schizophrenia.

There is good evidence that a change in dopamine regulation explains, at least in part, the clinical picture associated with schizophrenia. Drugs that upregulate dopamine function (such as amphetamines, which enhance the release of dopamine and block its reuptake and breakdown) can lead to delusions and hallucinations in apparently healthy people. It seems likely that it is the dopaminergic effects of amphetamines that produce these symptoms because, in animals, the key effects of amphetamine administration (hyperactivity followed by stereotypy) are also produced by pure dopamine agonists. Furthermore, pure dopamine agonists in humans (e.g., L-dopa treatment in Parkinson's disease) can produce psychosis. On the other hand, drugs that block dopamine (as is the case with all established antipsychotic drugs) can dampen and remove the symptoms [25]. Furthermore, Creese and colleagues [26] and Seeman and colleagues [27,28] showed a very strong correlation (0.9) between the typical daily dose used to treat psychosis and the dose required to block a dopamine agonist (that is, drugs that need a large dose to block the dopamine agonists also require a large dose to produce an antipsychotic effect). While there is a clear correlation between a drug's antipsychotic dose and its dopamine-blocking properties, no such relationship exists with its serotonin-, noradrenaline-, or histamine-blocking properties. Also relevant is that the antipsychotic drug flupenthixol has two forms: the α-isomer (which blocks dopamine) produces a significantly better reduction in the symptoms of schizophrenia than does the β-isomer (which does not block

dopamine) [29]. Interested readers should note that this field is very well reviewed by McKenna [30].

These observations have led to the dopamine hypothesis of schizophrenia, which, simply stated, is that schizophrenia is caused by upregulation of dopamine. The dopamine hypothesis has been one of the major route maps in guiding research into schizophrenia and has engendered many experiments and observations. Unfortunately, the direct evidence in favor of the hypothesis is difficult to collect and does not lead to straightforward conclusions. A review of this evidence goes beyond the scope of the current chapter but is very well reviewed by McKenna [30], by Abi-Dargham [31−35], and by Howes [36−38].

It does seem that there is little in the way of direct evidence for increased levels of dopamine or its metabolites. An alternative view, that the upregulation is caused not by increased levels of the neurotransmitter, but by increased sensitivity of the receptors, also lacks clear support. In vivo nuclear medicine studies have shown inconclusive findings concerning D2 receptor availability in schizophrenia [39], although there is genetic evidence implicating variation in the D2 receptor gene in conferring a (slightly) increased risk for schizophrenia [40]. Increased striatal dopamine release in schizophrenia has been reliably documented by studies examining the dopamine response to amphetamine [41]. Abnormalities of presynaptic dopamine function have now been confirmed by several research teams, using various radioligands to assess differing stages of presynaptic dopamine pathways; nevertheless, in each study, while there have been group differences between patients and controls, the overlap with the measurements in the healthy population suggest that, whatever is abnormal about dopamine in schizophrenia, it is not straightforward.

THE POSSIBLE IMPORTANCE OF GLUTAMATE

It is generally accepted that the symptoms of schizophrenia are unlikely to be explained by a single neurotransmitter deficit. Attention has also focused on glutamate, the major excitatory transmitter of the brain [42,43]. When we block glutamatergic action through one of its key receptors—the NMDA receptor—we can produce, in healthy subjects, a set of symptoms that resemble schizophrenia [44] and, in patients, a worsening of the particular symptoms that tend to trouble them most when they are ill. The effects of NMDA blockade have been observed in recreational users of phencyclidine ("PCP" or "angel dust") and in people who have received the dissociative anesthetic ketamine.

Such observations have been important in the emergence of the NMDA receptor hypofunction model of schizophrenia (see, for example, Krystal and colleagues [45]). More recent work exploring the impact of ketamine on reinforcement learning processes has explicitly related its effects to the emergence [46] and persistence [47] of abnormal beliefs.

In addition to the impact of glutamate receptor blocking agents, further evidence to support glutamatergic pathology in schizophrenia has emerged from magnetic resonance spectroscopic studies, which show that there are alterations in glutamate levels discernible even in the early stages of schizophrenia [48]. There are clear interactions between glutamate (via NMDA) and dopamine systems. Laruelle et al. [41] proposes that glutamate exerts dual NMDA-dependent control of mesocortical and mesolimbic dopamine firing (see also the chapter by Haber in this volume). This consists of an "activating system" involving direct stimulation of dopamine neurons projecting from the ventral tegmental area back to the cortex together with bisynaptic stimulation of dopamine neurons projecting from the ventral tegmental area to the ventral striatum. It is complemented by a "brake system" in which glutamatergic stimulation of GABAergic inhibitory interneurons or GABA striatotegmental neurons produces an overall inhibitory effect on mesolimbic dopamine. In short, there appears to be a state of glutamate–dopamine balance. One can imagine that disruptions in this balance may lead to dysregulated mesocortical and mesolimbic dopamine firing and may appreciably reduce information flow between cortex and subcortical structures. Again, this is likely to be a simplistic model but one worth exploring. For example, administration of ketamine does indeed upregulate dopamine [49]. A fuller model recognizes that dopamine neurons modulate the NMDA effects on GABAergic neurons (as in the brake system just alluded to) and that this modulation differs for D1 and D2 receptors. Thus, D1 facilitates NMDA transmission on GABAergic neurons. D2, on the other hand, has an inhibitory effect. Laruelle points out that this has connotations for both the pathophysiology and the treatment of schizophrenia. Excess D2 receptor stimulation in schizophrenia would inhibit glutamate participation in the ventral striatal brake system. If glutamate function is already reduced, excess D2 activity would worsen this situation. Together, these two factors would appreciably reduce information flow between cortex and subcortical structures.

In short, a working hypothesis at present is that schizophrenia symptoms emerge from perturbed interaction between glutamatergic and dopaminergic systems. However, as has been compellingly argued [18], the extent to which this observation constitutes an explanation for the condition itself is questionable. After all, a person with schizophrenia expresses the signs and symptoms of his or her illness at a complex level, a level that feels far removed from neurochemistry. How does increased dopamine produce a strange belief? Why might life feel flat and unexciting as a consequence of the sorts of neurochemical perturbations described? We believe that both of these apparently disparate phenomena may be understood in terms of reward/reinforcement learning. The growing understanding of the role of these two key neurotransmitter systems—dopaminergic and glutamatergic—in reinforcement learning may provide us with a firmer footing on which to base our cognitive understanding of the symptoms of schizophrenia. In the following sections we try to show how this may be so by considering these two apparently disparate symptoms of the disease—anhedonia and delusions—and by examining the emerging evidence that they could arise from deficits in the brain systems underpinning reinforcement learning. We begin by considering the evidence for a reinforcement processing deficit in schizophrenia before speculating on how the presence of such a deficit might help to explain the emergence and nature of these symptoms.

STUDIES OF REWARD PROCESSING/REINFORCEMENT LEARNING IN PSYCHOSIS: BEHAVIORAL STUDIES

We now review evidence for whether there are reinforcement learning deficits in psychotic illness. One of the main challenges in answering this question is the lack of available behavioral measures to assess reward processing and incentive motivational processes in humans; the consequence has been that, to date, such processes have been addressed only indirectly in behavioral studies. Patients with schizophrenia display a range of abnormalities of classic associative learning phenomena including Kamin blocking and latent inhibition [12,50], and these abnormalities are responsive to short-term neuroleptic treatment, which would be consistent with a dopaminergic mechanism. In addition, reward-based decision-making on the Iowa Gambling Test (IGT) has been shown to be impaired in psychosis [51] and this effect also is sensitive to medication status [52]. Furthermore, there have been some failures to replicate the case–control difference [53], and it should be remembered that the IGT requires several cognitive processes in addition to reward sensitivity.

Generally, in the 1980s and 1990s, the cognitive focus in studies of schizophrenia tended toward processes such as attention, executive function, and memory, whereas the study of reward processing and motivation was relatively neglected. However, before this, in the 1950s and 1960s, a considerable number of studies examined the effects of either reward or punishment on patients with schizophrenia; however, the focus of these studies was rather different from that of studies in the modern era. As reviewed by Buss and Lang [54], an important question at the time was whether what today we would call cognitive deficits (or neuropsychological or neurocognitive deficits) in schizophrenia were purely the results of motivational deficits such that patients failed to engage with the experimenter and therefore any impairments in performance in comparison to a control group could be attributed to the motivational state of patients as opposed to reflecting genuine impairments in functioning. To examine this question, a number of studies were conducted by various authors, in which, for example, reaction times were measured without reinforcement. Then, the effect of adding reinforcement, either positive or negative verbal reinforcement or, typically, electric shock, white noise, or money, was examined. In fact, Skinner himself, together with Ogden Lindsley, conducted conditioning experiments in patients with psychosis over a number of years [55,56]. It should be remembered that these studies were conducted in the era prior to the use of modern diagnostic criteria. For example, Topping and O'Connor [57] found that, whereas healthy controls improved on a "serial anticipation task," patients with paranoid schizophrenia worsened, and other patients with schizophrenia were unchanged in performance. In contrast, reaction time performance did improve after electric shock [58].

In more recent years, Gold and colleagues have been at the forefront of behavioral studies of reward processing in schizophrenia. They have focused on studies of patients with chronic schizophrenia, almost all of whom were taking antipsychotic medication, and have examined various aspects of reward processing, usually in relation to negative symptoms (see also the anhedonia section later). Patients show deficits in rapid learning on the basis of trial-to-trial feedback, as shown by tasks such as reversal learning [59,60]. This is consistent with previous studies of such learning in schizophrenia [61,62] and with a study of our own, using a comparable manipulation, in first-episode psychosis. We observed that in early psychosis deficits were subtle, and modestly correlated with negative symptoms [63]. Interestingly, in some studies there has been intact sensitivity to reward, but evidence of impairment in decision-making in relation to attainment of reward [60,64,65].

A deficit related to prediction errors may also be at the root of another phenomenon observed in schizophrenia: patients can display impairments in learning from positive outcomes, yet learn normally from negative outcomes [66,67]. In a study that could be considered an early forerunner of this work, Garmenzy [68] examined sensitivity to positive and negative reinforcement, showing that schizophrenia patients were differentially sensitive to verbal punishment (WRONG) and to reward (RIGHT).

A differential learning deficit, if present, may be due to a failure to generate or use positive-prediction errors for positive outcomes (note that some studies have not confirmed this dissociation [69]), or it may be due to failure in representing the value of the alternative responses in the decision-making process [70]. In an attempt to differentiate between the two, Gold and colleagues [71] used a probabilistic reinforcement learning task involving gain and loss avoidance, designed in such a way that both were associated with the generation of positive-prediction errors. Computational modeling pointed to impairment in the representation of value during decision-making.

One limitation of these studies in patients with chronic schizophrenia is that nearly all the patients are taking antipsychotic medication, and such dopamine antagonist medication has been shown to affect reward processing in healthy volunteers [72,73]. Another study of ours showed a failure to modulate behavior in response to a motivational manipulation in patients with first-episode psychosis [74]. Importantly, when we carried out an analysis restricted to a subset of patients—those free of dopamine antagonist medication—the deficit was still observed.

STUDIES OF REWARD PROCESSING/REINFORCEMENT LEARNING IN PSYCHOSIS: NEUROIMAGING STUDIES

As noted above, one of the difficulties in studying reward processing in humans has been the difficulty of finding appropriate behavioral tests with which to investigate the processes, which are subtle and often indirectly expressed in terms of attendant behavior. To a large extent, the advent of functional neuroimaging reinvigorated this area of research. This is partly because the dependent measure, brain activation, may under certain circumstances offer a more sensitive and specific index of reward-related processing, and of learning, than behavioral measures.

In line with behavioral findings, dopamine-mediated striatal systems responsible for reinforcement learning and predicting cues that lead to rewards seem to be impaired, as well as frontal cortex-driven mechanisms associated with generating, updating, and maintaining value representations. The first fMRI study of reward anticipation in a psychotic disorder came from Juckel et al., who examined 14 unmedicated schizophrenia patients as they performed the Monetary Incentive Delay Task, which in controls produces robust ventral striatal activation in the anticipation of reward (compared to the anticipation of neural feedback). Patients recruited ventral striatum to a significantly less extent than did controls [75]; this finding has been shown to hold in several subsequent studies (e.g., Refs. [76] and [77]), including in drug-naïve, first-episode patients [78]. These differences may be partially normalized by second-generation antipsychotic treatment [79–81].

In the first fMRI study of reward learning in psychosis, we demonstrated abnormalities in patient brain responses correlating with reward-prediction error in the dopaminergic midbrain (Fig. 21.1), in striatal and limbic regions, and in cortical regions such as the dorsolateral prefrontal cortex [82]. The results could not solely be explained by medication effects and are consistent with subsequent studies (e.g., [83]) using different psychological paradigms. Using a partially overlapping sample of patients, we examined causal learning, as opposed to reward learning, in early psychosis and found abnormal activation in a network of midbrain, striatal, and frontal regions [50]. Interestingly, the severity of the dysfunction of activation in the lateral prefrontal cortex correlated with the severity of delusions (Fig. 21.2).

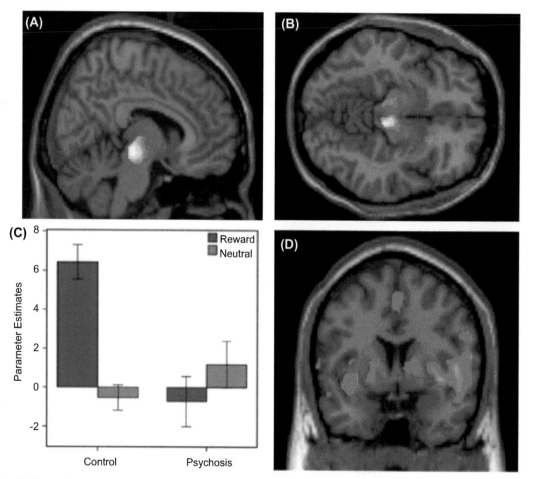

FIGURE 21.1   Differences between psychosis patients and controls in the relationship between prediction error and brain response (A, B) in the midbrain and ventral striatum and (D) in a wider network including the insula, cingulate, and dorsal striatum. (C) Patients showed attenuated responses associated with prediction error in salient trials and augmented prediction error responses in irrelevant trials (chart shows prediction error parameter estimates in the right substantia nigra). *Taken from Murray GK, Corlett PR, Clark L, Pessiglione M, Blackwell AD, Honey G, et al. How dopamine dysregulation leads to psychotic symptoms? Abnormal mesolimbic and mesostriatal prediction error signalling in psychosis. Mol Psychiatry 2008; 13:239.*

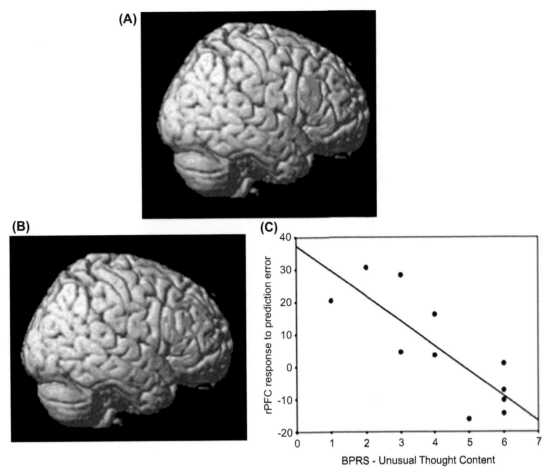

FIGURE 21.2 Causal learning brain prediction error signal disruption in psychosis. (A) The region of the right lateral prefrontal cortex (rPFC) in which there was a significant relationship between prediction-error changes (as estimated by changing behavioral response) and activation in controls (shown in red). Superimposed on this is the region (shown in yellow) in which there were group differences in this relationship. Specifically, there was an attenuation of this rPFC—prediction error relationship in the patient group. (B) A surface rendering showing the region of right lateral PFC in which there was a significant inverse correlation between PFC prediction-error response and symptom scores (unusual thought content on Brief Psychiatric Rating Scale, BPRS). (C) Symptom scores plotted against the extent to which lateral PFC activity correlated with our estimate of trial-by-trial prediction error; individuals in whom this relationship was most strongly positive showed lower scores on this symptom. *Taken from Corlett PR, Murray GK, Honey GD, Aitken MR, Shanks DR, Robbins TW, et al. Disrupted prediction-error signal in psychosis: evidence for an associative account of delusions. Brain 2007;130(Pt 9):2387—400.*

Consistent with behavioral findings and the "aberrant salience" hypothesis, Jensen and colleagues [84] employed an aversive-conditioning paradigm, using an unpleasantly loud noise, and they found that medicated patients showed elevated ventral striatal activity to a conditioned stimulus (CS) preceding neutral events compared to a CS preceding aversive events. CS-related responses in the midbrain to neutral events were correlated with delusion severity in another aversive-conditioning study [85]. Diaconescu and colleagues [86] used an appetitive classical conditioning task and found that dampened activation to reward CS in striatal areas was associated with weaker connectivity between the ventral striatum and the orbitofrontal cortex, the opposite of the pattern found in healthy controls. Instrumental-conditioning tasks yield similar results,

as shown, for instance, by Schlagenhauf and colleagues [87], who found reduced ventral striatal activation in a reversal learning task. Although studies have not shown evidence of striatal learning activity in schizophrenia relative to controls, a nuanced picture is suggested by the fact that some "negative" studies have shown relationships at the interindividual level between striatal activation and the severity of anhedonia [88,89].

One paradox has been to reconcile the finding that although there is an excessive dopaminergic striatal activity in schizophrenia, neuroimaging studies have generally shown impaired striatal signaling during learning studies. A study by Bernacer and colleagues [90] provided a possible explanation by showing that administration of methamphetamine to healthy volunteers (which floods the striatum with dopamine) leads

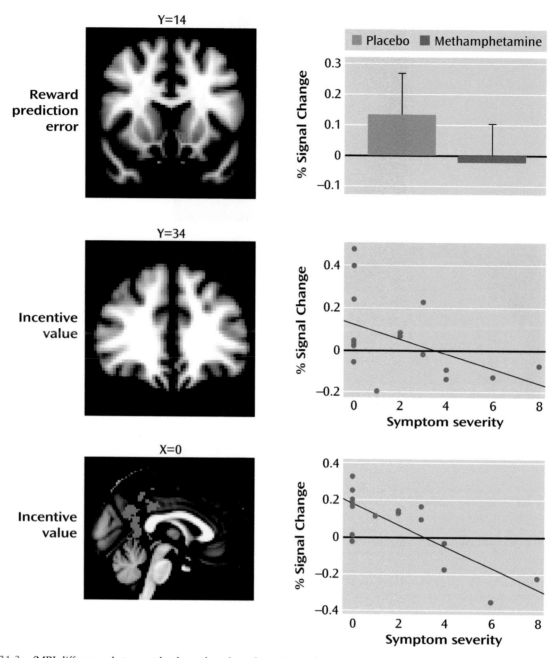

FIGURE 21.3 fMRI differences between placebo and methamphetamine and correlation with psychotic symptoms. The *top* image shows a cluster in the left ventral caudate nucleus indicating a significant difference for reward-prediction error signal; the bar graph indicates the drug effect on the signal change extracted for that particular cluster. Error bars indicate standard deviation. The *middle* image shows a methamphetamine-disrupted incentive value signal in the ventromedial prefrontal cortex; as shown in the graph, the signal change extracted from this cluster in the methamphetamine visit correlated with symptom severity. In the *bottom* image and graph, a large significant cluster was found in the posterior cingulate when the incentive value signal in the methamphetamine condition was correlated with symptom severity at a whole brain level. *Taken from Bernacer J, Corlett PR, Ramachandra P, McFarlane B, Turner DC, Clark L, et al. Methamphetamine-induced disruption of frontostriatal reward learning signals: relation to psychotic symptoms. Am J Psychiatry 2013;170(11):1326–34.*

to a disruption of striatal prediction error-associated activity. Interestingly, there was an association with the degree to which the drug induced mild psychosis and to which it disrupted the ventromedial prefrontal and posterior cingulate cortices' expected value signal, providing evidence linking dopaminergic disruption,

abnormal brain learning signaling, and psychosis (Fig. 21.3).

Taken together, these imaging and behavioral studies provide preliminary evidence for physiological abnormalities in learning about and anticipating rewards and punishments in psychotic illness. This appears to be accompanied by an impaired ability to modulate behavior in response to incentives.

CAN AN UNDERSTANDING OF REWARD AND DOPAMINE HELP US TO UNDERSTAND SYMPTOMS OF SCHIZOPHRENIA?

Delusions

We have summarized evidence that reward and reinforcement processing is abnormal in psychosis, perhaps secondary to dopamine, and/or glutamate, dysfunction, which appears to be linked to a network of subcortical (midbrain, striatum, and hippocampus/amygdala) and cortical dysfunction. Here, we consider more specifically whether such deficits may relate to, and perhaps explain, one of the key symptoms of schizophrenia: delusions. It is clinically very important to think about whether such abnormalities account for any aspects of patients' mental experience or behavior outside of the research laboratory. Patients with psychotic illnesses do not present to their doctor for the first time and complain that their dopamine or glutamate neurotransmitter systems, or indeed their reinforcement learning systems, are dysfunctional. More common is to describe unusual experiences or beliefs, which tend to be accompanied by varying levels of insight. In addition there is often a decline in occupational, educational, or social function. Thus, the encounter between a patient and a clinician very directly prompts the question of how disturbances at a physiological level relate to disturbances in the mental experience of the patient. This question is important philosophically and scientifically, but also is of very direct relevance to health care, if a mental experience is to be treated with a pharmacological (or, indeed, any other) intervention. As discussed earlier, the first scientist to make the case clearly and repeatedly for relating the emergence of psychotic symptoms to dopamine-driven associative and reinforcement learning was Robert Miller. In his 1976 article he wrote:

> The process of acquiring the associations necessary for learning a conditioned response in an experimental animal depends on the presence of dopamine. In human schizophrenic patients, an excessive supply of cerebral dopamine may facilitate the acquisition of associations between "units of information", to the point where unrelated features are associated and treated as if they are meaningful combinations: this process can be terminated by administering dopamine antagonists [5].

Miller's proposition, which has gathered widespread interest in recent years, remains speculative as opposed to definitive, but at the very least it is compatible with subjective reports from patients. One of a number of patients interviewed by McGhie [91] stated "I had very little ability to sort the relevant from the irrelevant—completely unrelated events became intricately connected in my mind." Similarly, another patient quoted by Mattusek [92] recounted the experiences as follows: "One is much clearer about the relatedness of things, because one can overlook the factuality of things." Another is quoted by Mattusek as reporting: "Out of these perceptions came the absolute awareness that my abilities to see connections had been multiplied many times over."

In regard to these quotations, it is well recognized that in many cases, a period of psychosis is preceded by a prodromal period, also known as a predelusional state, delusional mood, or delusional atmosphere, during which an individual experiences changes to his or her thought processes and emotions, but is not frankly ill [93]. A person developing a psychotic illness may feel suspicious, may have a sense that the world is changing in subtle but important ways, or may think he or she is on the verge of making a critically important breakthrough or insight. Such a patient may be feeling a new awareness of such subtleties (a patient previously cared for by one of the authors of this chapter used to refer to his very specific and personal experience of the world, and of detecting subtle changes that others could not, as "The Sensitivity"). In these early stages in the development of a psychotic illness, patients frequently experience abnormal thoughts, for example, that a neighbor or the police might be monitoring their behavior, but this thought may be initially no more than a suspicion and is not firmly or completely held. Additionally or alternatively they may have a sense of seeing the world in a new way, and observing that what appear to be innocuous events actually may have some other hidden but important meaning. As Kurt Schneider stated [94], "Often in these vague delusional moods, perceptions gain this sense of significance not yet defined." For example, one of the patients quoted by Mattusek (above) also stated:

> At ordinary times I might have taken pleasure in watching the dog, but [previously] would never have been so captivated by it.

Such experiences, in the early phase of a disorder, of seeing connections or reading possible meanings into innocuous events, do not represent psychosis, but have been termed psychosis-like experiences, or attenuated psychotic symptoms. A parsimonious explanation of the physiology of the early psychotic experience is that it will be closely related to, though perhaps of less severity than, the physiology of frank psychosis. Indeed, recent evidence suggests that patients with such attenuated symptoms, termed an at-risk mental state, do indeed have similar dopaminergic abnormalities in the striatum compared to patients with established illness, as detected by in vivo molecular imaging [95].

In the absence of a firm understanding of how delusions relate to any putative underlying physiological disturbance, it may be easier to conceptualize, at least initially, how physiological abnormalities could lead to the very early, mild symptoms. Indeed, one possibility is that is a clearer relationship between neurochemical and neurophysiological dysfunction and these initial symptoms than between chronic delusions and the underlying physiological state. After all, a chronic delusion is a belief that has been firmly held for a long time. The delusion may have been *formed* in the context of a profound physiological disturbance, but it is surely possible that it could, like other beliefs, persist, long after the physiological (or psychological) precipitants have subsided or faded. An account of the emergence of psychosis appealing to the dysfunction of reinforcement and associative learning can be outlined as follows. Excessive or dysregulated dopamine transmission, in which bursts of dopamine release, for example (irrespective of the etiology of the release of dopamine), coincide with unrelated thoughts or perceptions, may result in the coinciding thoughts or perceptions being invested with a sense of value, importance, and association. Extending this idea, and relating it to contemporary view of dopamine's role in normal psychological function, Shitij Kapur [20] appealed to Berridge and Robinson's concept of incentive, or motivational, salience [21], to explain why an objectively irrelevant experience could, in the mind of a patient with a disturbed dopamine system, become seen as being of the utmost significance. Recall that incentive salience, which is mediated by dopamine, is, according to Berridge and Robinson, that property of an event, object, or thought to capture attention and drive goal-directed behavior because of associations with reward (or punishment). Such feelings of significance in an unusual context would be very likely to preoccupy the mind of the person experiencing them, and would naturally be subjected to a good deal of reflection and possibly rumination. Thus, the speculation that the early symptoms of referential ideas and delusional atmosphere could emerge secondary to abnormalities in reward and motivational processing does seem plausible.

The next stage in the argument requires an understanding of how, in time, an unusual thought or experience may become or give rise to an unshakable false belief—a delusion. How could this happen? Two possibilities have been presented [96]. Under one account, if an experience is unusual enough, then by the exercise of normal reasoning processes, the protagonist may arrive at a delusional belief to resolve the uncertainties raised by the unusual and perplexing experience. This is sometimes known as a "one-stage theory," as it requires only one abnormality—abnormal experience. For example, the experience of hearing a voice in the third person commenting upon one's actions could be explained by the idea that a radio transmitter has been implanted in one's brain or inner ear, and that the voices are radio broadcasts being picked up by such a transmitter. Here, after seeing connections in many events for prolonged periods, eventually a patient may propose a unifying theory to explain a host of otherwise puzzling phenomena. Once an explanation has been reached, it is rewarding to the patient, as it provides relief from anxiety and perplexity [97].

Under an alternative "two-stage" account, an additional "lesion" is required: abnormal reasoning. Proponents of this model argue that even if one is subjected to very unusual experiences, such as auditory hallucinations, such experiences could be understood in an insightful manner, for example, as being evidence of illness, or one could simply refrain from making any inferences from unusual experiences (and thus, perhaps, remain in a prodromal state without moving to a psychotic state) and that to jump to the conclusion that a transmitter had been implanted in one's brain must indicate a massive failure in the normal process of evaluating evidence to arrive at such a far-fetched belief. Indeed there is some evidence, though as yet not conclusive, that patients with delusions do jump to conclusions on the basis of little evidence [6,98–100].

However, it has also been argued [101] that it may be possible to account for the bizarre perceptions, experiences, and beliefs that characterize schizophrenia within one explanatory framework and without the need to posit separate deficits in perception/experience and belief formation. This idea, which draws on hierarchical Bayesian accounts of brain function, considers models and inferences about the world in terms of a hierarchy of inferences such that failures of inference at lower levels engender prediction errors, which are fed forward and demand inferences at higher levels to cope with new or unexpected information. According to this view, a persistent deficit in prediction-error-related signaling could lead to the requirement for ever higher and more radical changes in inference leading to both bizarre experiences, when expressed lower in the hierarchy, and apparently irrational beliefs at the higher levels. This perspective has been developed and extended in recent years [102,103]. Though critical perspectives have been articulated [104] (see Ref. [105] for response), there is agreement that this hypothesis seeking to explain the emergence of delusions offers a credible explanation, although as yet, despite some evidence [47,106], it must be extended if it is to explain their characteristic persistence and content.

Anhedonia

The term anhedonia has been used in a number of ways and its use is by no means confined to schizophrenia. Indeed, it is probably more thought of in the context of depressive illness. We should remember, however, that it may affect between one-third and one-half of people with schizophrenia [107]. The term was coined by Ribot [108] (see William James. "The Sick Soul." for a discussion) who saw anhedonia as a decreased capacity for pleasure in a way that is analogous to the decreased capacity for feeling that is found in analgesia. James was graphic in his description of the condition, thinking of it as *passive joylessness and dreariness, discouragement, dejection, lack of taste and zest and spring* (and thus incorporating both loss of pleasure and loss of interest). A number of demographic observations with respect to the occurrence of anhedonia in schizophrenia have been made and are worth briefly recounting here. First, it tends to predominate in older patients in which the condition is more long-standing [109,110]. Stoichet and Oakley [111] suggested that it occurs at the early stages, perhaps being progressive, and may also associate with a poor overall prognosis. In an important study, Katsanis et al. [112] used the Chapman anhedonia scale [113] to evaluate patients with schizophrenia, their first-degree relatives, and control subjects. The patients scored more strongly on both social (e.g., "I attach very little importance to having close friends"; "There are few things more tiring than to have a long, personal discussion with someone") and physical (e.g., "I have seldom enjoyed any kind of sexual experience"; "One food tastes as good as another to me") anhedonia than did relatives or controls. Some would tend to group anhedonia with the negative symptoms of schizophrenia (social withdrawal, apathy, self-neglect) but there is some evidence (e.g., Ref. [107]) that the condition dissociates from other negative symptoms. Romney and Candido [114] used a factor analysis in people with schizophrenia and observed that anhedonia, which emerged as trait- rather than state-like, loaded more heavily on depression than other symptoms (positive or negative), although there was a correlation between anhedonia and the negative symptoms of social and emotional withdrawal. In keeping with this, Berenbaum and Oltmanns [115] showed that, in particular, patients who have the so-called blunted affect (i.e., not apparently expressing emotion) showed anhedonia compared to controls. They suggested that we should be wary of confusing anhedonia with emotional blunting, concluding that "Anhedonia is best thought of as a characteristic of depression, whereas affective blunting which encompasses anhedonia is a typical feature of the negative deficit syndrome in schizophrenia."

Clearly, the picture is a confusing one and, added to this, it is worth noting that nowadays the vast majority of people diagnosed with schizophrenia are, or have been, on antipsychotic medicines that have antidopaminergic action. The extent to which anhedonia arises from the treatment as opposed to the condition is not clear.

Attempts to understand the underlying causes or mechanisms of anhedonia have had to contend with the fact that it is a somewhat nebulous concept with a usage that can refer to a number of signs and behaviors in patients. Silver and Shlomo [116] have questioned the generally held notion that anhedonia is fundamentally a lack of, or an inability to obtain, pleasure or enjoyment. Using the Snaith−Hamilton Pleasure Scale in patients with schizophrenia, they observed first that about 41% of patients scored normally and, importantly, while there was a moderate positive correlation with negative symptoms (though the mean levels of negative symptoms did not significantly differ between those people with anhedonia and those without), the actual subjective measure scores for lack of enjoyment were less than would be predicted from observations of patients' apparent states of enjoyment and pleasure. In short, it may be that there is an appearance of anhedonia in schizophrenia but that this does not match the subjective state. This is in keeping with the work of Iwase et al. [117] suggesting that patients experience more pleasure in a rewarding stimulus than is shown by their facial expression. Both studies relate to the conclusions of Berenbaum and Oltmanns [115] stated earlier.

Specifically, it seems that, when it comes to ostensibly rewarding stimuli, people with anhedonia may not appear so motivated to obtain them and may not show marked signs of their enjoyment, but such evidence as there is does not point strongly toward a true reduction in actual enjoyment. At this stage, therefore, it may well be helpful to appeal to the more refined models of reward learning and processing that have emerged from the field of experimental and theoretical neuroscience just as we have discussed in relation to delusions earlier. Specifically, there is an important distinction between the anticipation and the consummation of reward (for a discussion of this field see Ref. [118]). Given that dopaminergic function (in particular, that of mesolimbic dopaminergic circuitry) may be important to modulate the anticipation of reward (see also Saga and Tremblay, this volume) and perhaps in engendering the motivation for reward and in selecting the actions that may be necessary to attain it, and given the studies cited earlier, it is worth scrutinizing the possibility that anhedonia in schizophrenia represents a deficit in the anticipation of, and quest for, reward rather than in the pleasure that rewards bring. As we see below, there is some empirical evidence in favor of this.

An important observation came from Gard et al. [119], who explicitly distinguished between anticipatory and consummatory pleasure in an elegant study evaluating how people with schizophrenia both predict the future experience of pleasure and enjoy this anticipation. In an initial study, they used an experience-sampling technique in which patients carried a pager and a notebook. When paged, they wrote down what they were doing and rated how pleasurable it was. They also noted whether they were anticipating any forthcoming pleasure. The results of this study indicated no difference between patients and controls in consummatory pleasure but a clear reduction in anticipatory pleasure. There was also evidence that they tended to be less likely to be engaged in a goal-directed activity at any time. In a complementary experiment, they found comparable results, shown in Fig. 21.4A, using a scale that attempts to measure experience of pleasure more specifically, the

Temporal Experience of Pleasure Scale (TEPS). Their conclusion that anhedonia is actually a failure to anticipate rather than experience pleasure was supported by the observation that Chapman ratings of anhedonia significantly correlated (negatively) with anticipatory but not consummatory pleasure. However, it should be noted that results using the TEPS have not been consistent. For example, a previously unpublished additional analysis of data from a study of ours involving 22 patients with schizophrenia and 20 controls [120] demonstrates deficiencies in TEPS consummatory and anticipatory pleasure (Fig. 21.4B); furthermore, Strauss and colleagues noted a deficit in TEPS consummatory pleasure in schizophrenia but intact anticipatory pleasure [121], and they discuss the limitations of the ingenious attempt by the TEPS developers to distinguish anticipatory and consummatory pleasure in a cross-sectional questionnaire setting.

Whereas it remains uncertain whether there is any primary consummatory pleasure deficit in schizophrenia and whether this contributes to motivational problems, there is an argument that if consummatory pleasure is intact, then motivational deficits in schizophrenia may be, at least in part, secondary to deficits in generating, maintaining, and updating mental representations of value [70]. As discussed earlier, if, as some studies have suggested, learning to act on the basis of positive outcomes is relatively impaired in schizophrenia, but learning from negative outcomes is intact, this would be a "reinforcement learning recipe" for anhedonia and avolition [67].

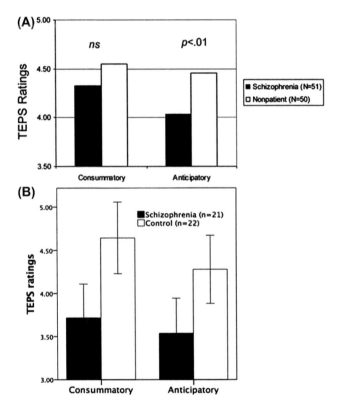

FIGURE 21.4 (A) "Consummatory" and "anticipatory" pleasure in a US sample of patients with schizophrenia and controls as measured by the Temporal Experience of Pleasure Scale (TEPS) rating scale. Patients report comparable degrees of consummatory pleasure but less anticipatory pleasure *(Taken from Gard DE, Kring AM, Gard MG, Horan WP, Green MF. Anhedonia in schizophrenia: distinctions between anticipatory and consummatory pleasure. Schizophr Res 2007;93(1−3):253−60)* (B) "Consummatory" and "anticipatory" pleasure in a UK sample of patients with schizophrenia and controls as measured by the TEPS rating scale. These data are previously unpublished. In contrast to the data in (A), we found that both consummatory and anticipatory pleasure were reduced in schizophrenia. The study is described in Arrondo et al. [77], Chuang et al. [120], and Segarra et al. [123].

SUMMARY

While schizophrenia is not conventionally considered in terms of reward and reinforcement learning, we believe that there are very good grounds for maintaining and adding to the body of work that has brought models of associative learning and rewards processing to psychiatric models of the illness. This is especially relevant now, given the advances in our neuroscientific understanding of reward learning and motivation and the consequent development of elegant models of how such processes might be invoked to explain symptoms of schizophrenia [20].

We have tried to show that, given the clear importance of mesolimbic circuitry, in particular, the dopaminergic system (and its interactions with the glutamate system), in schizophrenia, there are prima facie good grounds for supposing that the processes in which this system is most firmly implicated—the learning about and attainment of rewards—are abnormal in schizophrenia. As we have seen, a growing body of empirical evidence supports this possibility.

Furthermore, a consideration of such processes provides a very useful framework in which to relate specific symptoms of schizophrenia to the underlying physiological disturbances and, in doing so, helping to bridge the explanatory gap that greatly limits our ability to understand and treat schizophrenia.

Acknowledgments

This work was supported by the University of Cambridge, the Bernard Wolfe Health Neuroscience Fund, the Wellcome Trust, the UK National Institute of Health Research, the Brain and Behavior Research Foundation, the Medical Research Council, the Isaac Newton Trust, and the Cambridgeshire and Peterborough NHS Foundation Trust.

References

[1] McGrath J, Saha S, Chant D, Welham J. Schizophrenia: a concise overview of incidence, prevalence, and mortality. Epidemiol Rev 2008;30.

[2] APA, diagnostic and statistical manual of mental disorder-IV. Washington, DC: American Psychiatric Association; 1994.

[3] Bleuler E. Dementia Praecox or the group of schizophrenias. New York: International University Press; 1911/1950.

[4] Robbins TW. Relationship between reward-enhancing and stereotypical effects of psychomotor stimulant drugs. Nature 1976;264(5581):57–9.

[5] Miller R. Schizophrenic psychology, associative learning and the role of forebrain dopamine. Med Hypotheses 1976;2(5):203–11.

[6] Garety PA, Hemsley DR, Wessely S. Reasoning in deluded schizophrenic and paranoid patients. Biases in performance on a probabilistic inference task. J Nerv Ment Dis 1991;179(4):194–201.

[7] Miller R. Striatal dopamine in reward and attention: a system for understanding the symptomatology of acute schizophrenia and mania. Int Rev Neurobiol 1993;35:161–278.

[8] Miller R. Major psychosis and dopamine: controversial features and some suggestions. Psychol Med 1984;14(4):779–89.

[9] Chouinard G, Miller R. A rating scale for psychotic symptoms (RSPS): part II: subscale 2: distraction symptoms (catatonia and passivity experiences subscale 3: delusions and semi-structured interview (SSCI-RSPS). Schizophr Res 1999;38(2–3):123–50.

[10] Miller R. The time course of neuroleptic therapy for psychosis: role of learning processes and implications for concepts of psychotic illness. Psychopharmacology (Berl) 1987;92(4):405–15.

[11] Gray JA. Dopamine release in the nucleus accumbens: the perspective from aberrations of consciousness in schizophrenia. Neuropsychologia 1995;33(9):1143–53.

[12] Gray JA. Integrating schizophrenia. Schizophr Bull 1998;24(2):249–66.

[13] Gray JA, Feldon J, Rawlins JNP, Smith AD. The neuropsychology of schizophrenia. Behav Brain Sci 1991;14(1):1–19.

[14] Gray JA, Joseph MH, Hemsley DR, Young AM, Warburton EC, Boulenguez P, et al. The role of mesolimbic dopaminergic and retrohippocampal afferents to the nucleus accumbens in latent inhibition: implications for schizophrenia. Behav Brain Res 1995;71(1–2):19–31.

[15] Schultz W, Dickinson A. Neuronal coding of prediction errors. Annu Rev Neurosci 2000;23:473–500.

[16] Lubow RE. Latent inhibition: effects of frequency of nonreinforced preexposure of the CS. J Comp Physiol Psychol 1965;60(3):454–7.

[17] Kamin LJ. "Attention-like" processes in classical conditioning. In: Jones MR, editor. Miami symposium on the prediction of behavior, 1967: aversive stimulation. Coral Gables, Florida: University of Miami Press; 1968. p. 9–31.

[18] Frith CD. The cognitive neuropsychology of schizophrenia. Hove: Taylor and Francis; 1992.

[19] David AS. A method for studies of madness. Cortex 2006;42(6):921–5.

[20] Kapur S. Psychosis as a state of aberrant salience: a framework linking biology, phenomenology, and pharmacology in schizophrenia. Am J Psychiatry 2003;160(1):13–23.

[21] Berridge KC, Robinson TE. What is the role of dopamine in reward: hedonic impact, reward learning, or incentive salience? Brain Res Brain Res Rev 1998;28(3):309–69.

[22] Crow TJ. Catecholamine-containing neurones and electrical self-stimulation. 2. A theoretical interpretation and some psychiatric implications. Psychol Med 1973;3(1):66–73.

[23] Stein L. Self-stimulation of the brain and the central stimulant action of amphetamine. Fed Proc 1964;23:836–50.

[24] Stein L, Wise CD. Possible etiology of schizophrenia: progressive damage to the noradrenergic reward system by 6-hydroxydopamine. Science 1971;171(975):1032–6.

[25] Kapur S, Remington G. Dopamine D(2) receptors and their role in atypical antipsychotic action: still necessary and may even be sufficient. Biol Psychiatry 2001;50(11):873–83.

[26] Creese I, Burt DR, Snyder SH. Dopamine receptors and average clinical doses. Science 1976;194(4264):546.

[27] Seeman P, Lee T. Antipsychotic drugs: direct correlation between clinical potency and presynaptic action on dopamine neurons. Science 1975;188(4194):1217–9.

[28] Seeman P, Lee T, Chau-Wong M, Wong K. Antipsychotic drug doses and neuroleptic/dopamine receptors. Nature 1976;261(5562):717–9.

[29] Johnstone EC, Crow TJ, Frith CD, Carney MW, Price JS. Mechanism of the antipsychotic effect in the treatment of acute schizophrenia. Lancet 1978;1(8069):848–51.

[30] McKenna PJ. Schizophrenia and related syndromes. 2nd ed. Routledge; 2007.

[31] Abi-Dargham A. Do we still believe in the dopamine hypothesis? New data bring new evidence. Int J Neuropsychopharmacol 2004;7(Suppl. 1):S1–5.

[32] Toda M, Abi-Dargham A. Dopamine hypothesis of schizophrenia: making sense of it all. Curr Psychiatry Rep 2007;9(4):329–36.

[33] Guillin O, Abi-Dargham A, Laruelle M. Neurobiology of dopamine in schizophrenia. Int Rev Neurobiol 2007;78:1–39.

[34] Abi-Dargham A. The dopamine hypothesis of schizophrenia. 2012 [cited July 30, 2015]; Available from: www.schizophreniaforum.org/for/curr/AbiDargham/default.asp.

[35] Abi-Dargham A. Schizophrenia: overview and dopamine dysfunction. J Clin Psychiatry 2014;75(11):e31.

[36] Howes OD, Kapur S. The dopamine hypothesis of schizophrenia: version III—the final common pathway. Schizophr Bull 2009;35(3):549–62.

[37] Bonoldi I, Howes OD. The enduring centrality of dopamine in the pathophysiology of schizophrenia: in vivo evidence from the prodrome to the first psychotic episode. Adv Pharmacol 2013;68:199–220.

[38] Howes O, McCutcheon R, Stone J. Glutamate and dopamine in schizophrenia: an update for the 21st century. J Psychopharmacol 2015;29(2):97–115.

[39] Howes OD, Kambeitz J, Kim E, Stahl D, Slifstein M, Abi-Dargham A, et al. The nature of dopamine dysfunction in schizophrenia and what this means for treatment. Arch Gen Psychiatry 2012;69(8):776–86.

[40] Schizophrenia Working Group of the Psychiatric Genomics, C. Biological insights from 108 schizophrenia-associated genetic loci. Nature 2014;511(7510):421–7.

[41] Laruelle M, Kegeles LS, Abi-Dargham A. Glutamate, dopamine, and schizophrenia: from pathophysiology to treatment. Ann NY Acad Sci 2003;1003:138–58.

[42] Kim JS, Kornhuber HH, Schmid-Burgk W, Holzmuller B. Low cerebrospinal fluid glutamate in schizophrenic patients and a new hypothesis on schizophrenia. Neurosci Lett 1980;20(3): 379—82.

[43] Stone JM, Morrison PD, Pilowsky LS. Glutamate and dopamine dysregulation in schizophrenia—a synthesis and selective review. J Psychopharmacol 2007;21(4):440—52.

[44] Pomarol-Clotet E, Honey GD, Murray GK, Corlett PR, Absalom AR, Lee M, et al. Psychological effects of ketamine in healthy volunteers. Phenomenological Study. Br J Psychiatry 2006;189:173—9.

[45] Krystal JH, Abi-Saab W, Perry E, D'Souza DC, Liu N, Gueorguieva R, et al. Preliminary evidence of attenuation of the disruptive effects of the NMDA glutamate receptor antagonist, ketamine, on working memory by pretreatment with the group II metabotropic glutamate receptor agonist, LY354740, in healthy human subjects. Psychopharmacology (Berl) 2005; 179(1):303—9.

[46] Corlett PR, Honey GD, Aitken MR, Dickinson A, Shanks DR, Absalom AR, et al. Frontal responses during learning predict vulnerability to the psychotogenic effects of ketamine: linking cognition, brain activity, and psychosis. Arch Gen Psychiatry 2006;63(6):611—21.

[47] Corlett PR, Cambridge V, Gardner JM, Piggot JS, Turner DC, Everitt JC, et al. Ketamine effects on memory reconsolidation favor a learning model of delusions. PLoS One 2013;8(6):e65088.

[48] Chiappelli J, Hong LE, Wijtenburg SA, Du X, Gaston F, Kochunov P, et al. Alterations in frontal white matter neurochemistry and microstructure in schizophrenia: implications for neuroinflammation. Transl Psychiatry 2015;5:e548.

[49] Kegeles LS, Abi-Dargham A, Zea-Ponce Y, Rodenhiser-Hill J, Mann JJ, Van Heertum RL, et al. Modulation of amphetamine-induced striatal dopamine release by ketamine in humans: implications for schizophrenia. Biol Psychiatry 2000;48(7):627—40.

[50] Corlett PR, Murray GK, Honey GD, Aitken MR, Shanks DR, Robbins TW, et al. Disrupted prediction-error signal in psychosis: evidence for an associative account of delusions. Brain 2007;130(Pt 9):2387—400.

[51] Ritter LM, Meador-Woodruff JH, Dalack GW. Neurocognitive measures of prefrontal cortical dysfunction in schizophrenia. Schizophr Res 2004;68(1):65—73.

[52] Beninger RJ, Wasserman J, Zanibbi K, Charbonneau D, Mangels J, Beninger BV. Typical and atypical antipsychotic medications differentially affect two nondeclarative memory tasks in schizophrenic patients: a double dissociation. Schizophr Res 2003;61(2—3):281—92.

[53] Cavallaro R, Cavedini P, Mistretta P, Bassi T, Angelone SM, Ubbiali A, et al. Basal-corticofrontal circuits in schizophrenia and obsessive-compulsive disorder: a controlled, double dissociation study. Biol Psychiatry 2003;54(4):437—43.

[54] Buss AH, Lang PJ. Psychological deficit in schizophrenia: I. Affect, reinforcement, and concept attainment. J Abnorm Psychol 1965;70:2—24.

[55] Lindsley OR. Operant conditioning methods applied to research in chronic schizophrenia. Psychiatr Res Rep Am Psychiatr Assoc 1956;5:118—39. discussion, 140-53.

[56] Lindsley OR, Skinner BF. A method for the experimental analysis of the behavior of psychotic patients. Am Psychol 1954;9: 419—20.

[57] Topping GG, O'Connor N. The response of chronic schizophrenics to incentives. Br J Med Psychol 1960;33:211—4.

[58] Rosenbaum G, Mackavey WR, Grisell JL. Effects of biological and social motivation on schizophrenic reaction time. J Abnorm Psychol 1957;54(3):364—8.

[59] Waltz JA, Gold JM. Probabilistic reversal learning impairments in schizophrenia: further evidence of orbitofrontal dysfunction. Schizophr Res 2007;93(1—3):296—303.

[60] Gold JM, Waltz JA, Prentice KJ, Morris SE, Heerey EA. Reward processing in schizophrenia: a deficit in the representation of value. Schizophr Bull 2008;34(5).

[61] Elliott R, McKenna PJ, Robbins TW, Sahakian BJ. Neuropsychological evidence for frontostriatal dysfunction in schizophrenia. Psychol Med 1995;25(3):619—30.

[62] Pantelis C, Barnes TR, Nelson HE, Tanner S, Weatherley L, Owen AM, et al. Frontal-striatal cognitive deficits in patients with chronic schizophrenia. Brain 1997;120(Pt 10):1823—43.

[63] Murray GK, Cheng F, Clark L, Barnett JH, Blackwell AD, Fletcher PC, et al. Reinforcement and reversal learning in first-episode psychosis. Schizophr Bull 2008;34(5).

[64] Heerey EA, Bell-Warren KR, Gold JM. Decision-making impairments in the context of intact reward sensitivity in schizophrenia. Biol Psychiatry 2008;64(1):62—9.

[65] Heerey EA, Gold JM. Patients with schizophrenia demonstrate dissociation between affective experience and motivated behavior. J Abnorm Psychol 2007;116(2):268—78.

[66] Waltz JA, Frank MJ, Wiecki TV, Gold JM. Altered probabilistic learning and response biases in schizophrenia: behavioral evidence and neurocomputational modeling. Neuropsychology 2011;25(1):86—97.

[67] Strauss GP, Frank MJ, Waltz JA, Kasanova Z, Herbener ES, Gold JM. Deficits in positive reinforcement learning and uncertainty-driven exploration are associated with distinct aspects of negative symptoms in schizophrenia. Biol Psychiatry 2011;69(5):424—31.

[68] Garmezy N. Stimulus differentiation by schizophrenic and normal subjects under conditions of reward and punishment. J Pers 1952;20(3):253—76.

[69] Somlai Z, Moustafa AA, Keri S, Myers CE, Gluck MA. General functioning predicts reward and punishment learning in schizophrenia. Schizophr Res 2011;127(1—3):131—6.

[70] Strauss GP, Waltz JA, Gold JM. A review of reward processing and motivational impairment in schizophrenia. Schizophr Bull 2014;40(Suppl. 2):S107—16.

[71] Gold JM, Waltz JA, Matveeva TM, Kasanova Z, Strauss GP, Herbener ES, et al. Negative symptoms and the failure to represent the expected reward value of actions: behavioral and computational modeling evidence. Arch Gen Psychiatry 2012; 69(2):129—38.

[72] Abler B, Erk S, Walter H. Human reward system activation is modulated by a single dose of olanzapine in healthy subjects in an event-related, double-blind, placebo-controlled fMRI study. Psychopharmacology (Berl) 2007;191(3):823—33.

[73] Pessiglione M, Seymour B, Flandin G, Dolan RJ, Frith CD. Dopamine-dependent prediction errors underpin reward-seeking behaviour in humans. Nature 2006;442(7106):1042—5.

[74] Murray GK, Clark L, Corlett PR, Blackwell AD, Cools R, Jones PB, et al. Incentive motivation in first-episode psychosis: a behavioural study. BMC Psychiatry 2008;8:34.

[75] Juckel G, Schlagenhauf F, Koslowski M, Wustenberg T, Villringer A, Knutson B, et al. Dysfunction of ventral striatal reward prediction in schizophrenia. Neuroimage 2006;29(2): 409—16.

[76] Hagele C, Schlagenhauf F, Rapp M, Sterzer P, Beck A, Bermpohl F, et al. Dimensional psychiatry: reward dysfunction and depressive mood across psychiatric disorders. Psychopharmacology (Berl) 2015;232(2):331—41.

[77] Arrondo G, Segarra N, Metastasio A, Ziauddeen H, Spencer J, Reinders NR, et al. Reduction in ventral striatal activity when anticipating a reward in depression and schizophrenia: a replicated cross-diagnostic finding. Front Psychol 2015;6:1280.

[78] Nielsen MO, Rostrup E, Wulff S, Bak N, Lublin H, Kapur S, et al. Alterations of the brain reward system in antipsychotic naive schizophrenia patients. Biol Psychiatry 2012;71(10): 898—905.

[79] Schlagenhauf F, Juckel G, Koslowski M, Kahnt T, Knutson B, Dembler T, et al. Reward system activation in schizophrenic patients switched from typical neuroleptics to olanzapine. Psychopharmacology (Berl) 2008;196(4):673–84.

[80] Juckel G, Schlagenhauf F, Koslowski M, Filonov D, Wustenberg T, Villringer A, et al. Dysfunction of ventral striatal reward prediction in schizophrenic patients treated with typical, not atypical, neuroleptics. Psychopharmacology (Berl) 2006; 187(2):222–8.

[81] Nielsen MO, Rostrup E, Wulff S, Bak N, Broberg BV, Lublin H, et al. Improvement of brain reward abnormalities by antipsychotic monotherapy in schizophrenia. Arch Gen Psychiatry 2012;69(12):1195–204.

[82] Murray GK, Corlett PR, Clark L, Pessiglione M, Blackwell AD, Honey G, et al. Substantia nigra/ventral tegmental reward prediction error disruption in psychosis. Mol Psychiatry 2008; 13(3). 239, 267–76.

[83] Morris RW, Vercammen A, Lenroot R, Moore L, Langton JM, Short B, et al. Disambiguating ventral striatum fMRI-related bold signal during reward prediction in schizophrenia. Mol Psychiatry 2011;17(3).

[84] Jensen J, Willeit M, Zipursky RB, Savina I, Smith AJ, Menon M, et al. The formation of abnormal associations in schizophrenia: neural and behavioral evidence. Neuropsychopharmacology 2008;33(3):473–9.

[85] Romaniuk L, Honey GD, King JR, Whalley HC, McIntosh AM, Levita L, et al. Midbrain activation during Pavlovian conditioning and delusional symptoms in schizophrenia. Arch Gen Psychiatry 2010;67(12):1246–54.

[86] Diaconescu AO, Jensen J, Wang H, Willeit M, Menon M, Kapur S, et al. Aberrant effective connectivity in schizophrenia patients during appetitive conditioning. Front Hum Neurosci 2011;4:239.

[87] Schlagenhauf F, Huys QJ, Deserno L, Rapp MA, Beck A, Heinze HJ, et al. Striatal dysfunction during reversal learning in unmedicated schizophrenia patients. Neuroimage 2014;89:171–80.

[88] Waltz JA, Kasanova Z, Ross TJ, Salmeron BJ, McMahon RP, Gold JM, et al. The roles of reward, default, and executive control networks in set-shifting impairments in schizophrenia. PLoS One 2013;8(2):e57257.

[89] Dowd EC, Barch DM. Pavlovian reward prediction and receipt in schizophrenia: relationship to anhedonia. PLoS One 2012;7(5): e35622.

[90] Bernacer J, Corlett PR, Ramachandra P, McFarlane B, Turner DC, Clark L, et al. Methamphetamine-induced disruption of frontostriatal reward learning signals: relation to psychotic symptoms. Am J Psychiatry 2013;170(11):1326–34.

[91] McGhie A. Attention and perception in schizophrenia. Prog Exp Pers Res 1970;5:1–35.

[92] Mattusek P. Studies in delusional perception. Arch Psychiatr Z Neurol 1952;189:279–318. In: Cutting J, Shepherd M, editors. The clinical roots of the schizophrenia concept. Cambridge: Cambridge University Press; 1987. p. 96.

[93] Berrios GE. The history of mental symptoms: descriptive psychopathology since the nineteenth century. Cambridge: Cambridge University Press; 1996.

[94] Schneider K. Clinical psychopathology. New York: Grune and Stratton; 1959.

[95] Howes OD, Montgomery AJ, Asselin M, Murray RM, Grasby PM, McGuire PK. Pre-synaptic striatal dopamine synthesis capacity in subjects at risk of psychosis. Schizophrenia Bull 2007;33(2):371.

[96] Langdon R, Coltheart M. The cognitive neuropsychology of delusions. Mind Lang 2000;15(1):184–218.

[97] Maher BA. Delusional thinking and perceptual disorder. J Individual Psychol 1974;30:98–113.

[98] Fine C, Gardner M, Craigie J, Gold I. Hopping, skipping or jumping to conclusions? Clarifying the role of the JTC bias in delusions. Cognit Neuropsychiatry 2007;12(1):46–77.

[99] Peters E, Garety P. Cognitive functioning in delusions: a longitudinal analysis. Behav Res Ther 2006;44(4):481–514.

[100] Menon M, Mizrahi R, Kapur S. 'Jumping to conclusions' and delusions in psychosis: relationship and response to treatment. Schizophr Res 2008;98(1–3):225–31.

[101] Fletcher PC, Frith CD. Perceiving is believing: a Bayesian approach to explaining the positive symptoms of schizophrenia. Nat Rev Neurosci 2009;10(1):48–58.

[102] Corlett PR, Taylor JR, Wang XJ, Fletcher PC, Krystal JH. Toward a neurobiology of delusions. Prog Neurobiol 2010;92(3):345–69.

[103] Adams RA, Stephan KE, Brown HR, Frith CD, Friston KJ. The computational anatomy of psychosis. Front Psychiatry 2013;4:47.

[104] Griffiths O, Langdon R, Le Pelley ME, Coltheart M. Delusions and prediction error: re-examining the behavioural evidence for disrupted error signalling in delusion formation. Cogn Neuropsychiatry 2014;19(5):439–67.

[105] Corlett PR, Fletcher PC. Delusions and prediction error: clarifying the roles of behavioural and brain responses. Cogn Neuropsychiatry 2015;20(2):95–105.

[106] Corlett PR, Krystal JH, Taylor JR, Fletcher PC. Why do delusions persist? Front Hum Neurosci 2009;3:12.

[107] Loas G, Boyer P, Legrand A. Anhedonia and negative symptomatology in chronic schizophrenia. Compr Psychiatry 1996;37(1): 5–11.

[108] Ribot. The psychology of emotions. London: Walter Scott; 1897.

[109] Harrow M, Grinker RR, Holzman PS, Kayton L. Anhedonia and schizophrenia. Am J Psychiatry 1977;134(7):794–7.

[110] Watson CG, Jacobs L, Kucala T. A note on the pathology of anhedonia. J Clin Psychol 1979;35(4):740–3.

[111] Shoichet RP, Oakley A. Notes on the treatment of anhedonia. Can Psychiatr Assoc J 1978;23(7):487–92.

[112] Katsanis J, Iacono WG, Beiser M. Anhedonia and perceptual aberration in first-episode psychotic patients and their relatives. J Abnorm Psychol 1990;99(2):202–6.

[113] Chapman LJ, Chapman JP, Raulin ML. Scales for physical and social anhedonia. J Abnorm Psychol 1976;85(4):374–82.

[114] Romney DM, Candido CL. Anhedonia in depression and schizophrenia: a reexamination. J Nerv Ment Dis 2001;189(11):735–40.

[115] Berenbaum H, Oltmanns TF. Emotional experience and expression in schizophrenia and depression. J Abnorm Psychol 1992; 101(1):37–44.

[116] Silver H, Shlomo N. Anhedonia and schizophrenia: how much is in the eye of the beholder? Compr Psychiatry 2002;43(1):65–8.

[117] Iwase M, Yamashita K, Takahashi K, Kajimoto O, Shimizu A, Nishikawa T, et al. Diminished facial expression despite the existence of pleasant emotional experience in schizophrenia. Methods Find Exp Clin Pharmacol 1999;21(3):189–94.

[118] Berridge KC, Robinson TE. Parsing reward. Trends Neurosci 2003;26(9):507–13.

[119] Gard DE, Kring AM, Gard MG, Horan WP, Green MF. Anhedonia in schizophrenia: distinctions between anticipatory and consummatory pleasure. Schizophr Res 2007;93(1–3):253–60.

[120] Chuang JY, Murray GK, Metastasio A, Segarra N, Tait R, Spencer J, et al. Brain structural signatures of negative symptoms in depression and schizophrenia. Front Psychiatry 2014;5:116.

[121] Strauss GP, Wilbur RC, Warren KR, August SM, Gold JM. Anticipatory vs. consummatory pleasure: what is the nature of hedonic deficits in schizophrenia? Psychiatry Res 2011;187(1–2):36–41.

[122] Murray GK, Corlett PR, Clark L, Pessiglione M, Blackwell AD, Honey G, et al. How dopamine dysregulation leads to psychotic symptoms? Abnormal mesolimbic and mesostriatal prediction error signalling in psychosis. Mol Psychiatry 2008;13:239.

[123] Segarra N, Metastasio A, Ziauddeen H, pencer J, Reinders NR, Dudas RB, et al. Abnormal frontostriatal activity during unexpected reward receipt in depression and schizophrenia: relationship to anhedonia. Neuropsychopharmacology January 20, 2016. http://dx.doi.org/10.1038/npp.2015.370. Advance online publication.

22

The Neuropsychology of Decision-Making: A View From the Frontal Lobes

A.R. Vaidya, L.K. Fellows

McGill University, Montreal, QC, Canada

Abstract

A growing literature argues that specific frontal subregions contribute to various aspects of reward learning and value-based decision-making. This chapter reviews work on the effects of focal frontal lobe damage on these behaviors. These studies help to define the necessary contributions of these subregions to decision-making, and provide evidence as to the dissociability of component processes. There is clear evidence of distinct contributions of ventromedial, lateral, and dorsomedial frontal lobes to decision processes. We argue that the ventromedial frontal lobe is required for optimal learning from reward under dynamic conditions, and contributes to specific aspects of value-based decision-making. We discuss how these functions may reflect a common role for this region in guiding attention to reward-predictive features of complex stimuli. We also review work demonstrating a necessary contribution by the dorsomedial frontal lobe in decision-making, seemingly related to a role in representing action-value expectations.

INTRODUCTION

Decision-making, the process of choosing between options on the basis of their subjective value, is at the core of flexible, goal-oriented behavior. Shoppers at the mall, reviewers reading grants, and voters in the polling booth are faced with the same fundamental challenge of making optimal choices in complex environments that provide varying benefits and costs, risks and opportunities. Psychologists and economists have studied value-based decision-making for decades, but it has only recently become a focus for neuroscientists [1–6]. For clinicians, this neglect is surprising: neurologists and psychiatrists have long identified poor judgment as a feature of conditions ranging from dementia to drug addiction [7–11]. Preliminary efforts to translate decision neuroscience measures to clinical populations have found, for example, that decision-making difficulties are common in healthy aging and predict increased mortality [12], consistent with the idea that the behaviors captured under this rubric are important in real life. Component behaviors related to decision-making, such as impulsivity and apathy, feature prominently in efforts to shift mental health research toward a neurobehavioral framework [13], reflecting how frequently decision-making is affected in mental illness. Even "normal" decision-making can be maladaptive. For example, the obesity epidemic, a leading public health challenge, may be the expression of a mismatch between our decision neurobiology and an environment replete with calorie-rich foods [14–17]. A better understanding of the brain basis of decision-making thus has wide potential applications to neurological and psychiatric disorders, as well as maladaptive variations in behavior in the normal range, and has implications for public health, ethics, and the law.

Regions within the frontal lobe are believed to make important contributions to decision-making. However, central questions about the behavioral constructs and brain substrates of the set of processes that together produce this complex behavior remain unresolved. Neuropsychological studies have been influential from the beginnings of decision neuroscience [10]. This chapter reviews this literature, summarizing the initial work that helped to define current neural models of decision-making, and then focusing on more recent studies and increasingly specific hypotheses about the roles of frontal subregions in the component processes of decision-making.

LESION EVIDENCE IN HUMANS

Experimental neuropsychology is at the interface between fundamental and clinically applied research in human subjects. On one hand, studies of clinical populations can shed unique light on the functions of the healthy brain, by demonstrating the effects of loss of function. Studies of patients with focal injury, in particular, are a powerful means of testing whether particular component processes are dissociable and defining the conditions under which a given anatomical substrate is necessary for a given process [18–20]. Such work can support claims of necessity: showing that a given region must be intact to support a given process, i.e., that it is *necessary* for that process [19]. These loss-of-function experiments are an important complement to many of the other methods of cognitive neuroscience, such as fMRI or electroencephalography, that provide evidence that activity in a region is associated with a process, but cannot directly address whether this activity is causal [21,22].

Finally, lesion studies are particularly well positioned to support the translation of fundamental findings. Tasks shown to be sensitive measures of a given process may be useful as diagnostic tests in the clinic, or could serve to assess the effects of rehabilitation interventions. Studies relating the real world symptoms of patients with focal brain injury to experimental task performance can shed light on the ecological validity of decision component process measures, in health and disease. Both applications may be useful in other conditions in which the neural substrates are unknown or diffuse: indeed, many current approaches to analyzing behavioral disorders in psychiatry and neurodegenerative disorders draw on frameworks arising in part from experimental neuropsychology.

Lesion Study Designs

Lesion studies in humans are in part "experiments of nature": obviously, the lesions themselves are not produced experimentally. Rather, patients with lesions are invited to participate in an experiment, thus allowing designs driven by behavior or by lesion location. Behavior-driven designs select patients who show a particular behavior, whether a clinical symptom, such as impulsivity, or impaired performance on a particular experimental task. The goal of such a study could be to establish the necessary brain substrates for the behavior, in which case lesion location would be the dependent measure. In contrast, lesion-driven designs select participants for inclusion in the experimental group on the

basis of damage to a particular brain region or network, and then assess the effect of that damage on task performance. This tests the hypothesis that a given region is necessary for performance of a given task. The more specific the behavioral measure, and the more specific the lesion location, the stronger the conclusions. Behavioral specificity can be enhanced by thoughtful inclusion of control tasks, to build a case for a specific deficit. In this way, the dissociability of two putative processes can be established by asking whether there are instances in which performance of one task is impaired, whereas performance of another is intact. Lesion specificity can be harder to "design in," because the experimenter will be limited by the extent of damage in the pool of available participants. Systematic efforts to identify patients suitable for this kind of research can address some of these practical limitations [23]. Including lesion control groups provides stronger, more anatomically specific evidence: for example, patients with frontal damage affecting a region of interest (ROI), say the orbitofrontal cortex (OFC), could be compared to those with frontal damage sparing that predefined region, providing a control for nonspecific effects of brain damage in general, and frontal damage more specifically.

An alternate strategy to identify the damage that is most strongly related to the behavioral effect is termed "voxel-based lesion–symptom mapping" [24,25]. This method tests the effect of damage to a particular voxel on task performance, comparing task performance of those subjects with damage to that voxel (or cluster of voxels) against that of those without damage to that voxel (or cluster) in a large sample of patients with damage that variably overlaps. The result is a statistical map of whatever brain regions were adequately represented in the sample, showing the extent to which damage to a given voxel is reliably related to poor performance. This massively univariate approach requires a large sample size with adequate lesion coverage and variance in lesion overlap to provide sufficient power to detect effects [26].

With this as background, we turn now to the application of lesion methods to study frontal lobe contributions to human decision-making. The focus will be primarily on lesion studies that address major models of decision-making. We will critically review the extent to which these models are supported by this form of loss-of-function evidence, with the aim of showing where there is strong converging support, where such evidence is simply lacking, and where there are findings from lesion studies that fail to support these models.

DECISION-MAKING AND THE FRONTAL LOBES

Decision-making refers to a complex set of processes. We focus on value-based decision-making, that is, motivated behavior that concerns the subjective reward associated with options. Conceptually, this can be distinguished from perceptual decision-making (deciding what a stimulus is) or instructed action selection (e.g., deciding to press a left or right button based on a cue). A range of cortical, subcortical, and neuromodulatory systems have been implicated in value-based decision-making. Here, we focus on the frontal lobes, leaving the review of subcortical and neuromodulatory contributions to other chapters in this volume.

The frontal lobes clearly serve many functions. There are cytoarchitectonic and connectivity differences across this large region of the human brain, and it is not sensible to treat this area as having a single common function. Over the history of neuropsychology, different approaches have been taken to subdividing the frontal lobes, reflecting different research questions. Early work focused on distinctions between left and right frontal lobes, principally testing hemispheric differences rather than regional distinctions within the frontal lobes [27]. More recent work has tended to ignore hemispheric distinctions, while imposing subregional boundaries at varying degrees of resolution [28,29]. These boundaries can be justified on cytoarchitectonic grounds [30], or connectivity [31], or on observations of functional distinctions [32,33]. For practical reasons discussed above, lesion studies generally have a fairly coarse anatomical resolution. Fig. 22.1 shows an overlap image from 54 patients with frontal lobe damage tested in work from

our group [34]. Lesion masks define lateral frontal, ventromedial frontal (VMF), and dorsomedial frontal (DMF) areas, demonstrating how these patients were classified according the regional extent of their brain injury. Fig. 22.1 shows that, with sufficient sample size, it is possible to have good coverage of these sectors and to distinguish between them. It also demonstrates the natural limitations of this approach, showing how often damage can extend across regional boundaries and highlighting that some frontal regions are more heavily represented, because common causes of focal damage (stroke, tumor resection, aneurysm rupture) are more likely to affect some areas than others.

COMPONENT PROCESSES OF DECISION-MAKING

Overview

Just as the frontal lobes are too vast and varied to be treated as a single entity, decision-making is too broad a construct to be meaningfully related to the brain. As in cognitive neuroscience more generally, the major approach has been to define simpler component processes that can more plausibly be linked to brain mechanisms. The challenge here is to define the "right" processes; that is, to "carve behavior" at its neurobiologically relevant "joints." Various strategies have been employed for this purpose: behavioral frameworks have been borrowed from decision psychology [35], behavioral economics [36,37], animal reward learning research [3], and ecology [38]. Increasingly, computational modeling has been brought to bear in an effort to formalize and distinguish the ideas emerging from

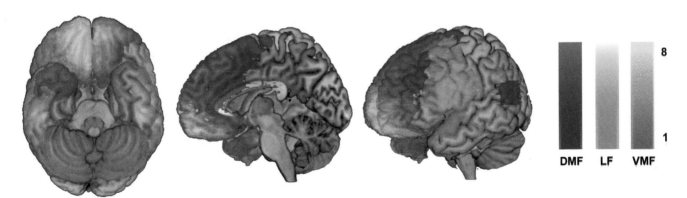

FIGURE 22.1 Lesion-overlap images from 54 patients with frontal lobe damage due to stroke, tumor resection, or aneurysm rupture. Lesion masks were manually registered to the Montreal Neurological Institute brain and classified as primarily affecting one of three regions of interest, as indicated by the three color scales shown. Each color scale also indicates the number of patients with damage in each voxel, with brighter colors indicating areas damaged in more patients. This sample included 28 patients with ventromedial frontal damage (VMF), 17 with dorsomedial frontal damage (DMF), and 13 with lateral frontal damage (LF). The extent to which damage extends across predefined regional boundaries, and in some cases outside the frontal lobes, can also be appreciated in this real sample.

these varied disciplines [39—41]. These various approaches have yielded interesting insights. No particular framework has emerged as clearly superior in explanatory power for the brain basis of decision-making, so for the moment it seems reasonable to consider evidence from all of them. In the absence of a definitive consensus model, we start with a schematic overview of the scope of decision component processes. Fig. 22.2 summarizes many of the processes that have been studied to date. Space prevents a full discussion of this literature in its entirety, but we will discuss many of the issues most central to the field today, or that seem particularly promising avenues for future progress.

Option Identification

Broadly, one can conceptualize decision-making as beginning at the level of option identification, that is, defining the possibilities upon which decision-making mechanisms will act. In the laboratory, this is often a trivial step, with the choices explicitly defined by the experiment (i.e., choose card deck A or card deck B). In the real world, however, identifying potential options, whether observable in the immediate environment, as at the grocery store, or existing in a more abstract state, as in a personal goal to eat better, may be an absolutely critical stage of decision-making. As one example, decision outcomes may be very different when a person is perusing the dessert menu in a restaurant if he or she is restricting decision-making to the menu alone or if he or she is comparing the value of any of those desserts against the option of

better long-term health resulting from adhering to a diet, i.e., choosing no dessert. This is an example of the importance of "framing" in decision-making: options are generally compared with one another, so the option set that is initially considered is very important in determining the final outcome.

Option identification has been little studied in decision neuroscience, but it would seem to have points of contact with other perspectives on frontal lobe function, such as planning, future thinking, problem-solving, and the ability to hold abstract goals in mind in the face of distraction. Creativity and fluency might also be relevant constructs when options must be generated in novel situations. There is a lesion literature implicating the frontal lobes in these various abilities [42—46]. The nature of much of this work makes specific structure—function claims difficult, although there have been efforts to distinguish right and left frontal contributions (e.g., Ref. [47]). The point is to study how people approach ill-defined problems, thus the results can be difficult to interpret in terms of very specific process deficits. A detailed review of this literature is beyond the scope of this chapter, but as our understanding of the later phases of decision-making improves, efforts to make contact with this literature are likely to be fruitful, particularly if the goal is to understand the brain basis of real-world decision-making [48].

Value Construction

Subjective value is an explanatory concept that captures the influence of "reward" on behavior [3,49]. It can be measured by asking people to make a value rating or indicate how much they are willing to pay for a particular outcome. Relative value can also be inferred from the choices people make, on the assumption that they generally act to maximize value [50]. Subjective value is a central concept in economics, and the finding that the fMRI blood oxygen level-dependent signal in specific regions scales with behavioral measures of this construct, and can predict choice, has been very influential in decision neuroscience [51,52]. One interpretation of the fMRI work on this topic is that the ventromedial prefrontal cortex (VMPFC) or the OFC regions within VMF serve as a general "value-meter" or "value look-up table" [53]. However, there are difficulties with that perspective. First, there is an inherent circularity in even defining subjective value as a construct: choice behavior reveals underlying subjective value, so value is simply whatever drives choice. Further, "value" is often confounded with a range of other, potentially dissociable processes, such as autonomic responses and salience, that may pose interpretation problems for correlational studies [53,54]. Even if subjective value

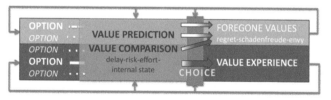

FIGURE 22.2 Schematic of decision-making processes. Options are generated or identified in the environment, the value of each option is derived and compared, a choice is enacted to select the desired option, and the outcome of the choice is experienced. In some cases, outcomes of the unchosen options can also be observed (i.e., if your dining partner chooses the other dessert on the menu and then describes how delicious it is). Outcome information, whether experienced or inferred, is compared with predictions, and if there are discrepancies, this serves as feedback to drive learning, influencing future option generation or identification, value predictions, and likelihood of a choice. Context is important in all these putative phases of decision-making. The *blue shading* distinguishes between concrete, real options observable in the world and eventually related to real experienced outcomes (*darker blue*) and abstract or imagined prospects with indirect outcome information (*paler blue*).

can be experimentally isolated, the nature of the information yielding these value-related signals is not clear, and may well vary substantially across experiments or conditions. Finally, there is only limited evidence for the claim that the value-related activity in the VMPFC, at least, is critical in guiding decisions, and the mechanisms by which this region might influence choice are not clear.

It is unlikely that any prefrontal region is needed for simply seeking reward: most views on the role of the VMPFC propose that it becomes necessary when value assessment is difficult: i.e., for making value predictions or comparisons within the current "context" of a choice. However, defining what is meant by this statement is a work in progress (reviewed in Ref. [55]). It has been applied to a range of observations, from classic decision biases (e.g., the framing effect [56]) to particular aspects of reinforcement learning [57–59]. Further, given the evidence of value-related signals in other regions, the focus on the VMPFC as "the" key region involved in value influences over choice may not be justified. At the moment, it is not clear what information is used to generate the VMPFC value signal, whether and how activity in that region comes to influence choice, nor whether the same information contributes to the value signals seen in the lateral OFC and DMF and what critical roles these areas may play in choice.

Lesion studies have addressed these questions. A useful starting point is to ask whether damage to brain region(s) that carry value-related signals impair (1) the ability to make subjective value ratings and (2) the ability to choose based on relative value. If impairments are found, are they general to all decision problems, or specific in some way?

We tested the hypothesis that the VMF is necessary for relative value comparison in several studies. We found that damage involving this region leads to less consistent preference judgments compared to healthy controls [50,60,61]. Two of these studies used very simple pairwise preference tasks, and found effects across a range of stimulus categories, from foods to social and even esthetic judgments, in entirely hypothetical choices. Another study used bundles of real food items (chocolate bars, juice), and again found that VMF damage was associated with less internally consistent value-based choice [50]. This supports the claim that the VMF (and, importantly, not other frontal regions) plays a critical role in value-based choice, but does not resolve the precise contribution of this region. Qualitatively, these findings suggest that whatever the role of the VMF in these simple preference judgments, it may be domain-general. However, the effects are relatively mild, and none of these studies had sufficient power to strongly test whether there might be specific categories of decisions spared or especially sensitive to the effects of VMF damage.

Provocatively, a study by our group found that patients with VMF damage were able to make value ratings of art stimuli presented one at a time, were as consistent as controls in rating the same stimuli a second time within the same test session, and made choices that were consistent with these a priori values to the same degree as healthy controls [34]. This poses a challenge to the simplest view of the VMF as uniquely critical in representing an integrated value signal, at least for visual art. However, a closer look at those data suggests that VMF patients may be relying on different information than controls when making value assessments. In particular, these patients showed more variance in their assessment of artwork that healthy controls and other prefrontal-damaged groups consistently rated as having low value. Another study provides insight into one potential mechanism for this difference [62]. That study used social stimuli: the faces of (unknown) politicians. Participants rated the faces on two attributes known to predict voting decisions: attractiveness and competence. Patients with VMF damage were able to rate these characteristics. However, there were differences when it came time to vote, i.e., to choose their preferred candidate in pairwise "elections." The choices of controls and patients with frontal damage sparing this region could be predicted by both attractiveness and competence ratings, as though they combined these two pieces of information to make their choice. In contrast, those with VMF damage could be predicted *only* by attractiveness ratings. Furthermore, this study had sufficient regional power to provide preliminary evidence that damage to the lateral OFC was most strongly associated with this effect.

One explanatory model that can reconcile these findings proposes that VMF plays a role in integrating value across relevant features, perhaps by biasing attention to the features that matter most in a given context. Variability in the attended features from choice to choice could lead to the observed variability in preference judgments in those with VMF damage. However, the consistent within-session value ratings of those with VMF damage observed in [34] do not fit readily with that account. It may be that the art stimuli used in that experiment were different from other stimuli, in that the value of art is in some sense "consumed" in the looking, whereas other stimuli (photos of foods, for example) are instead prompting expectations of future value outcomes. However, esthetic judgments of visual stimuli, including art, lead to robust increases in fMRI signal in the VMF across studies [63]. While there is a case to be made for a role for regions within the VMF in value prediction, rather than experienced value [64], there is also considerable evidence that OFC activity reflects received as well as predicted reward value from studies of both humans and nonhuman primates [54]. Further work is needed to fully reconcile these, and related, perspectives [53].

Learning Value

The ability to predict value is not only important for "one-off" decisions, but is also crucial in learning, with the difference between predicted and experienced reward believed to be a key signal driving the flexible optimization of behavior [65]. Lesion studies in human patients and animal models have repeatedly pointed to a critical role for the frontal lobes in associative learning. Notably, frontal damage does not disrupt learning value associations when contingencies are deterministic and unchanging. Instead, the frontal lobes appear to be critical in settings in which value associations are uncertain and require dynamically updating option values based on recent experience.

Reversal learning is one relatively simple task that appears to tap into this ability. Typically this task consists of two stimulus options (often cards, or objects in animal studies) that are paired with reward or no reward, with these reward contingencies reversing multiple times throughout the experiment. Several studies in both human patients [66,67] and nonhuman primates [68–70] have shown that damage to the VMF, but not other prefrontal subregions, impairs performance on this task, although the precise region within the VMF driving this effect is up for debate [71,72]. While VMF damage does not affect initial learning in these fully deterministic tasks, subjects tend to make more errors following reversals. This deficit was initially attributed to a failure of response inhibition, or perseveration [70], but more recent work indicates that reversal learning deficits result from impairments in learning the reward value of stimulus options. VMF damage in humans impairs learning in similar experiments in which stimuli are probabilistically associated with reward [73], even before reversals occur [57].

The Iowa Gambling Task is another much-studied task that is sensitive to VMF damage [10,74]. Although more complex than typical reversal learning paradigms in that it features four choices, and simultaneous wins and losses, this task shares with reversal learning tasks the requirement to flexibly update behavior in response to shifting reward contingencies. When the requirement to overcome an initial bias to the high-reward, but ultimately high-risk, options is removed from the task, VMF patients perform as well as healthy controls [75].

Interestingly, the role of VMF in learning about reward value is not general. Lesion work in macaques has identified a double dissociation in "stimulus-value" and "action-value" learning related to lesions of the OFC and the anterior cingulate sulcus, respectively

[76]. This same dissociation has been shown in humans with focal VMF or DMF damage [77]. These findings argue for dissociable roles for different prefrontal subregions in updating the reward value of actions and stimulus options. Of course, regions within the DMF have also been implicated in various aspects of action selection, error processing, and conflict monitoring as well; it is not yet clear how those ideas relate to this action-value account, although some interesting proposals have been made [78–81].

Foraging theories provide a different perspective on decision-making in dynamic environments, addressing how organisms optimally harvest reward while minimizing risk and effort. Here, the focus is on the trade-off between exploiting current choices or exploring other prospects [82]. Imaging work has suggested a role for the frontal polar cortex in facilitating exploratory choices [83,84]. Lesion work in humans provides support for this view. Damage to this region was found, remarkably, to increase the fidelity between choice and cumulative reward history in a dynamic stimulus–reward learning task; these patients were less sensitive to short-term trends in the pattern of rewards than were healthy controls [85]. Other work has shown that frontal pole damage disrupts the ability to multitask, i.e., to flexibly shift between realizing two or more goals [86]. A subsequent study from the same group demonstrated a specific deficit in prospective memory related to damage to the frontal pole: these patients had difficulty remembering to interrupt one ongoing task to perform a second task at a regular interval [87]. Together, these findings suggest a more general role for the frontal pole in shifting away from an ongoing behavior (i.e., exploring rather than exploiting) based on internal signals, whether short-term dips in reward or simply the passage of time.

Finally, the link between decisions and learning has also been explored in social decisions in games in which the behavior of one player must be taken into account by the other. One study tested how damage to the insula or VMF affected learning in a computerized version of the "ultimatum game" [88]. Participants chose between accepting or rejecting offers that split a predetermined amount of money. Critically, if the offer was rejected, no one received any money. This task required subjects to consider the fairness of the deal and adapt their choices based on the range of offers. Patients with insula damage were slow to adapt their decisions based on previous offers, suggesting a deficit in adapting to social norms, possibly related to impairment in learning from negative outcomes (i.e., bad offers). In contrast, patients with VMF damage were intact in this respect, but were overall more likely to accept offers, even when they were very low. Notably, previous work has found the opposite pattern of behavior in VMF patients (i.e., an increased tendency to reject low offers) in another version of the same game, taken as evidence for a role for this region in regulating an emotional response to unfair deals [89]. However, these studies together are perhaps better explained by a more fundamental impairment in offer valuation, which might lead to different results depending on the details of the experimental task. A third study of ultimatum game performance after VMF damage supports this more general account [90].

Value and Attention

Several studies have reemphasized the importance of attentional mechanisms in decision-making. Early studies examining value-related neural signals considered attentional effects as a confound [91]. Subsequent efforts to separate the influences of task goals and (incidental) rewards on stimulus processing have shown that visual features previously paired with reward act as distractors even when the feature is nonsalient and task-irrelevant, a phenomenon termed "value-based attentional capture" [92]. Concurrently, there has been increased awareness that attentional factors (i.e., visual saliency and fixation times) can also influence value-based choice [93,94].

The brain mechanisms by which value and attention interact have mainly been studied in the visual system, where value effects have been demonstrated in primary visual cortex, the lateral intraparietal area, and more broadly within the ventral visual stream

[95–98]. The source of these value influences remains unclear, but the VMF is a plausible candidate region. It has abundant connections, directly and indirectly, with the ventral visual stream [99–101], and interactions between these regions have been shown with magnetoencephalography [102].

Whereas many visual and visual-attention processes, such as object recognition, hemispatial attention, and selective attention in the face of conflict, are unaffected by VMF damage [29], there is some evidence that this region is necessary for directing attention based on feedback, or emotionally relevant signals. Hodgson et al. [102a] investigated feedback-related changes in saccadic reaction time in a single patient with bilateral VMF damage. Whereas healthy subjects and other prefrontal-damaged subjects showed increased saccadic latencies to locations previously associated with negative feedback, the VMF patient did not. An electroencephalography study in a larger cohort of patients found that VMF damage reduced event-related potentials associated with attention to emotional stimuli [103]. We investigated the effects of VMF damage on directing attention to visual features that were predictive of rewards in the context of a visual search task [104]. Healthy controls, VMF-damaged patients, and patients with prefrontal damage sparing this region were asked to search for targets and ignore distractors that appeared in task-irrelevant colors. These colors could remain the same or switch between trials, resulting in a trial-by-trial priming effect whereby subjects were generally slower

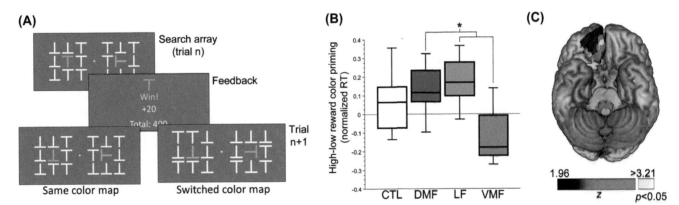

FIGURE 22.3   Effects of regional frontal damage on reward priming of visual search. (A) Task design. Subjects were presented with a search array that included a colored target (upright or upside-down T's) and a salient distractor (a T turned 90 degrees facing left or right) embedded in a field of white distractor T shapes. Subjects were asked to discriminate the orientation of the target, and received reward feedback that was incidentally associated with the color of the target. The salient distractor and target randomly appeared in orange or pink in each trial, so that the target and distractor colors would either remain the same or switch between trials. Subjects tended to take longer to respond after a color switch than when colors remained the same, an effect of color priming. (B) Reward modulation of the color priming effects. Distractors that appeared in a color associated with high reward exerted a larger influence on search reaction time (RT) than colors associated with low rewards in control subjects and patients with frontal damage sparing the VMF, but not VMF-damaged subjects. (C) Voxel-based lesion symptom mapping analysis for reward priming on attention. Two clusters of voxels in the VMF were significantly associated with reduced reward priming. Color bar shows Brunner—Munzel z scores with the statistical threshold adjusted for multiple comparisons. *$p < .05$, Tukey—Kramer post hoc tests. CTL, control; DMF, dorsomedial frontal damage; LF, lateral frontal damage; VMF, ventromedial frontal damage. *Data are from Vaidya AR, Fellows LK. Ventromedial frontal damage in humans reduces attentional priming of rewarded visual features. J Neurosci 2015;35:12813—23.*

to respond when these colors switched (Fig. 22.3A). Importantly, these colors were incidentally associated with different probabilities of a large reward over the course of the experiment. Colors previously associated with larger rewards exerted a greater distracting influence compared to colors associated with smaller rewards in healthy controls and prefrontal patients with intact VMF. In contrast, this effect of reward—color association on attention was significantly decreased in VMF-damaged patients, arguing for a regionally specific deficit in directing attention to reward-predictive visual features (Fig. 22.3B and C).

Deficits in directing attention to rewarded stimulus features could have consequences for associative learning. We frequently must deliberate between options with multiple features, e.g., potential lunch spots may be assessed on their location, price, and nutritional value. Devoting attention to features that are predictive of positive outcomes (e.g., nutrition) and ignoring less relevant (but perhaps more salient) information (e.g., the restaurant logo) can facilitate learning and promote adaptive future choices. In this same vein, learning theorists have speculated the existence of an attentional system that adaptively scales the learning rate for new information based on the predictive value of cues [105]. Failure to adaptively attend to reward-predictive cues could diminish the gain on relevant information, impairing learning in environments with dynamic value associations. Relatedly, learning deficits following VMF damage in macaques have been linked to the phenomenon of "spread of effect" [58,59]. This phrase refers to impairment in linking recent feedback to the contingent choice, resulting in a spread of feedback attribution to past (irrelevant) choices. This deficit arises from a failure to focus attention on relevant experiences, reducing the precise trial-by-trial tuning of associative learning. Consistent with this view, imaging work has pointed to a role for the lateral OFC in flexibly specifying relevant stimuli and stimulus attributes during decision-making [55,72]. These data collectively argue that this region has a role in filtering reward-predictive from nonpredictive features of the environment.

While this work has focused on how value associations can influence attention, a separate line of work has emphasized the influence of attention on value-based choice [93,106]. Imaging work has suggested that the value comparison process in the VMF is dependent on the locus of fixation, arguing that decision-related signals in this region are dynamically modulated by attention, biasing choice toward fixated options [107]. We investigated the necessary contributions of the VMF and other prefrontal subregions to this fixation bias phenomenon, tracking the eye movements of lesion patients and healthy controls while they chose between pairs of artwork [34]. In contrast to suggestions from imaging work, we found no effect of VMF damage on this fixation bias. However, DMF damage increased the influence of fixations on choice. A model-based analysis of this dataset indicated that DMF damage was associated with greater discounting of the value of the item not currently fixated. This functional explanation aligns with current theories of DMF function in foraging tasks, which have implicated this region in representing the value of switching away from a default option to explore alternative possibilities [38].

Decision Strategies

The work reviewed here shows an evolution in thinking about how value may be constructed through the interplay of several processes, relying on distinct neural mechanisms. It challenges a very simple view of a region within the VMF (i.e., the VMPFC) as uniquely critical in representing value. Instead, it highlights that even when examining a single decision option, many potential features could contribute to value estimation, and argues for contributions of the VMF to selecting, learning about, and perhaps integrating the currently relevant information. A quite different literature, arising from psychology studies of decision-making processes, is also pertinent to this general question. Just as the neuroscience findings reviewed earlier show that a "value" estimate is still produced despite, for example, VMF damage, but that it is produced from different inputs, so the decision process literature has identified multiple strategies that healthy people may follow when faced with explicitly multiple-attribute or multiple-option choice problems. At one extreme, the decider could engage a heuristic: a simplifying rule that will allow a quick decision in the face of complexity [108,109]. Rather than estimating and comparing the values of all 36 flavors of ice cream on offer, perhaps I choose my predetermined favorite, or perhaps I follow a more whimsical rule, like choosing whatever the person in line ahead of me chooses. Given that all ice cream is pretty tasty, chances are this will yield a good outcome. By saving processing time and effort, it may be considered more optimal than a more "rational" exhaustive comparison, particularly when options are numerous and very similar in value.

A second, more nuanced, strategic distinction has been drawn between maximizing and satisficing in making value-based decisions. Maximizing means seeking the most valuable available option, the best. In contrast, satisficing involves setting a minimum criterion and taking the first option that meets that threshold. This produces a choice that is "good enough," rather than necessarily "the best." Again, when there are many options that can be compared on many attributes, as is often the case in consumer choices, for example, this may be a far more effective strategy in the trade-off between optimizing the value of the outcome and the investment of time and energy in arriving at the decision. Interestingly, there is some evidence that tendencies toward maximizing versus satisficing can be conceived of as personality traits [110]. This literature also points out that different outcome features may matter more to different people. Maximizers are motivated to avoid regret (i.e., a postdecisional effect, when the outcome of the chosen option is compared to outcomes of the unchosen options), whereas satisficers focus principally on the value of the chosen option against some absolute benchmark.

These individual differences in decision strategy have not yet received much attention in the neuroscience literature, yet these may be important determinants of the factors contributing to value construction in the real world. Some years ago, we explored whether frontal lobe damage disrupted information acquisition in explicitly multiattribute choices. We found that those with VMF damage acquired information in a very different way compared to controls or those with other frontal damage, behaving as though they were satisficing rather than maximizing [111]. This strategic difference in behavior might relate to a failure to consider counterfactual options (i.e., a foregone ice cream flavor), as suggested by other work showing that such damage may disrupt the experience of regret [112]. However, the specific subregion within the VMF important in regret remains a matter of debate [113].

Decisions Over Time

The role of time in decision-making has been studied extensively in economics, in psychology, and, as mentioned above, in relation to foraging. Time is always a consideration in decisions, because choices are always, in some sense, about the future, i.e., they are made on the basis of the expectation of an outcome that has not yet occurred. Sometimes the delay to experiencing an outcome is explicitly considered in the choice. For example, classic intertemporal choice problems ask people to choose between an amount of money "now" and a larger amount that will be received only after a specified delay. These tasks demonstrate a phenomenon that is also evident in real life: the value of future reward is discounted sharply by delay. Decades of behavioral research have characterized delay discounting (also termed temporal discounting), showing that the relationship between subjective value and delay can be fit by a hyperbolic function and addressing how the shape

of that function varies across healthy individuals, how it differs in clinical populations, and the situational factors that can modify it [114,115]. When applied to money, discounting measured in this way is clearly "irrational" in the classical economic sense of that term. Even the most shallow temporal discounters downgrade value far more steeply than rational considerations of opportunity costs, interest rates, and inflation would support.

More often delay is implicit, one of many "costs" in ecological decision settings, i.e., when the choice involves acquiring the option by doing more than just pressing a button in an experimental paradigm. You might much prefer an after-dinner snack of chocolate cake to a piece of fruit, but if the cake requires driving to the grocery store and the fruit is on the counter in your kitchen, you may well select the fruit. In asking why this is, one could propose a temporal discounting account: the cake reward will be delayed. This can be restated as a cost: the time you must invest to obtain the cake is greater, time you might prefer to invest in other rewarding activities. The time window comes with the context—if you want a snack, you probably want it in the near future—and helps to establish the menu of options you spontaneously consider. Put another way, time is part of the decision-making "landscape," i.e., time considerations constrain the options that are spontaneously considered in any given decision context. Often, this makes ecological sense. If you are deciding on your food intake, you could consider an immediate snack against another, better meal in a few hours, but there is no point in weighing that against an extravagantly delicious banquet in several weeks, no matter how much more rewarding it might be, because you will die of starvation while you wait. However, optimal choices involving more abstract considerations, such as financial investments, may reasonably involve consideration of very long time windows. The "shortsighted" among us may not even undertake such decisions, because the outcomes are so distant that they do not qualify as plausible options.

Both temporal discounting in particular and future thinking (prospection) more generally have been related to regions within the frontal lobes. Two opposing views of temporal discounting have emerged from the fMRI literature. One proposes that separate neural systems representing "now" and "later" options compete for control of behavior. The immediate reward is proposed to engage the medial OFC, VMPFC, and ventral striatum, whereas the delayed reward is related to activity in the lateral prefrontal cortex (PFC) as well as parietal areas [116]. An alternative account is that both immediate and delayed subjective reward values are represented in the VMPFC and ventral striatum (as well as the posterior cingulate cortex), with the delayed reward value representation greater when the subjective value of that outcome is greater as reflected in choice behavior [117]. Both these models predict a critical role for the VMF, but the first also proposes a role for the lateral PFC. Can lesion evidence speak to these competing models? The two-system account makes clear predictions: VMF damage should lead to shallower temporal discounting, because presumably the intact lateral PFC—parietal system would dominate, leading to more choices of the larger reward—longer delay option. Lateral PFC damage would lead to the opposite effect, i.e., steeper discounting. The single-system view predicts no impact of lateral PFC damage on these choices, and that VMF damage should lead to more inconsistent intertemporal choices, due to a reduced ability to make fine-grained subjective value comparisons in general, rather than a tendency to become more impulsive or more patient.

We studied this question in patients with damage to the frontal lobes and patients with focal damage to non-frontal brain regions, as a control for nonspecific effects of brain injury or the experience of a serious illness. We found no systematic effect of regional frontal damage on temporal discounting rate, a finding that does not support the two-system model outlined earlier [45]. The published paper did not consider the question of whether choices were more inconsistent in any group, but an analysis we undertook after publication found that the VMF group was more likely to make individual choices that departed from the discounting rate predicted by their choices across the whole task (Fellows, unpublished observation), similar to their tendency to make inconsistent choices in pairwise preference tasks discussed above. Both these findings are consistent with the one-system model.

To our knowledge, only one other study has examined the effects of focal frontal lesions on temporal discounting, finding steeper discounting in the VMF group and no effect of the other frontal damage. It is not entirely clear how to reconcile the two lesion studies, which used similar, although not identical, tasks and similar lesion classification. The null finding comes from a larger sample. This could argue that the smaller sample is a false positive [118], particularly likely with this task, given the wide individual variability in discounting rates in healthy populations. Alternatively, the larger sample may have included patients with damage that, while falling within VMF boundaries, spared a putatively critical region within those boundaries, diluting the effect in the group [119], a potential limitation of ROI designs that can make it difficult to compare across studies [20]. In any case, neither finding supports the two-system model, which makes the opposite prediction.

There has also been lesion work on the neural basis of future thinking, defined in various ways. The same study described earlier included a measure of "future time perspective," the window of future time spontaneously considered when participants are asked to provide a list of events that come to mind when they think about their own futures. In contrast to the null results for the temporal discounting measure, the VMF group had a significantly shorter future time perspective compared to the other two patient groups [45]. Similar deficits in future thinking have been identified in patients with amnesia after medial temporal lobe damage. Interestingly, these patients showed a correspondence between the extent of autobiographical memory impairment and their diminished ability to imagine the future [120].

SUMMARY

Decision neuroscience is, in many ways, still at the beginning of its project of defining the brain basis of value-based choice. Here, we reviewed some of the early (and replicated) and more recent studies testing the necessary roles of brain regions, particularly within the frontal lobes, in various aspects of subjective value construction and learning and deciding based on value. This work provides useful converging evidence to constrain the interpretation of results obtained from correlational approaches, notably fMRI. Overall, component processes of decision-making and reward learning do seem to rely on definable frontal lobe substrates, and this framework has been particularly helpful in advancing our understanding of the VMF, until recently a region that was relatively little studied. The challenge in the coming years will be to link the ideas emerging from decision-making research to better understood processes underlying attention, action selection, and memory, to yield a more comprehensive view on the brain basis of flexible, motivated behavior.

Acknowledgments

The authors acknowledge operating support from the Canadian Institutes of Health Research (MOP-97821 and MOP-119291), salary support from the Fonds de Recherche en Santé du Québec (L.K.F.), and a Desjardins Outstanding Student Award (A.R.V.).

References

[1] Fellows LK. The cognitive neuroscience of human decision making: a review and conceptual framework. Behav Cogn Neurosci Rev 2004;3:159—72.

[2] Fellows LK. The role of orbitofrontal cortex in decision making: a component process account. Ann NY Acad Sci 2007;1121: 421—30.

[3] Fellows L. The neurology of value. In: Gottfried J, editor. Neurobiology of sensation and reward. Boca Raton (FL): CRC Press; 2011.

[4] Hastie R. Problems for judgment and decision making. Annu Rev Psychol 2001;52:653–83.

[5] Rorie AE, Newsome WT. A general mechanism for decision-making in the human brain? Trends Cogn Sci 2005:41–3.

[6] Sugrue LP, Corrado GS, Newsome WT. Choosing the greater of two goods: neural currencies for valuation and decision making. Nat Rev Neurosci 2005;6:363–75.

[7] Fellows LK. Current concepts in decision-making research from bench to bedside. J Int Neuropsychol Soc 2012;18:937–41.

[8] Rahman S, Sahakian BJ, Cardinal RN, Rogers RD, Robbins TW. Decision making and neuropsychiatry. Trends Cogn Sci 2001: 271–7.

[9] Sharp C, Monterosso J, Montague PR. Neuroeconomics: a bridge for translational research. Biol Psychiatry 2012:87–92.

[10] Bechara A, Damasio H, Tranel D, Damasio AR. Deciding advantageously before knowing the advantageous strategy. Science 1997;275:1293–5.

[11] Volkow ND, Fowler JS, Wang G-J. The addicted human brain: insights from imaging studies. J Clin Invest 2003;111:1444–51.

[12] Boyle PA, Wilson RS, Yu L, Buchman AS, Bennett DA. Poor decision making is associated with an increased risk of mortality among community-dwelling older persons without dementia. Neuroepidemiology 2013:247–52.

[13] http://www.nimh.nih.gov/research-priorities/rdoc/nimh-research-domain-criteria-rdoc.shtml.

[14] Vainik U, Dagher A, Dubé L, Fellows LK. Neurobehavioural correlates of body mass index and eating behaviours in adults: a systematic review. Neurosci Biobehav Rev 2013:279–99.

[15] Volkow ND, Wang GJ, Baler RD. Reward, dopamine and the control of food intake: implications for obesity. Trends Cogn Sci 2011: 37–46.

[16] Dubé L, Bechara A, Böckenholt U, Ansari A, Dagher A, Daniel M, et al. Towards a brain-to-society systems model of individual choice. Mark Lett 2009:105–6.

[17] Tang DW, Fellows LK, Small DM, Dagher A. Food and drug cues activate similar brain regions: a meta-analysis of functional MRI studies. Physiol Behav 2012;106:317–24.

[18] Dunn JC, Kirsner K. Editorial what can we infer from double dissociations?2; 2003. p. 1–7.

[19] Fellows LK, Heberlein AS, Morales DA, Shivde G, Waller S, Wu DH. Method matters: an empirical study of impact in cognitive neuroscience. J Cogn Neurosci 2005;17:850–4.

[20] Fellows LK. Group studies in experimental neuropsychology. In: Cooper H, Camic PM, Long DL, Panter AT, Rindskopf D, Sher KJ, editors. American Psychological Association handbook of research methods in psychology. Research designs: quantitative, qualitative, neuropsychological and biological, vol. 2; 2012. p. 647–59. Washington, DC.

[21] Kable JW. The cognitive neuroscience toolkit for the neuroeconomist: a functional overview. J Neurosci Psychol Econ 2011:63–84.

[22] Chatterjee A. A madness to the methods in cognitive neuroscience? J Cogn Neurosci 2005;17:847–9.

[23] Fellows LK, Stark M, Berg A, Chatterjee A. Patient registries in cognitive neuroscience research: advantages, challenges, and practical advice. J Cogn Neurosci 2008;20:1107–13.

[24] Bates E, Wilson SM, Saygin AP, Dick F, Sereno MI, Knight RT, et al. Voxel-based lesion-symptom mapping. Nat Neurosci 2003;6:448–50.

[25] Rorden C, Karnath H-O, Bonilha L. Improving lesion-symptom mapping. J Cogn Neurosci 2007;19:1081–8.

[26] Kimberg DY, Coslett HB, Schwartz MF. Power in voxel-based lesion-symptom mapping. J Cogn Neurosci 2007;19:1067–80.

[27] Milner B, Petrides M, Smith ML. Frontal lobes and the temporal organization of memory. Hum Neurobiol 1985;4: 137–42.

[28] Picton TW, Stuss DT, Alexander MP, Shallice T, Binns MA, Gillingham S. Effects of focal frontal lesions on response inhibition. Cereb Cortex 2007;17:826–38.

[29] Tsuchida A, Fellows LK. Are core component processes of executive function dissociable within the frontal lobes? Evidence from humans with focal prefrontal damage. Cortex 2013;49: 1790–800.

[30] Petrides M, Tomaiuolo F, Yeterian EH, Pandya DN. The prefrontal cortex: comparative architectonic organization in the human and the macaque monkey brains. Cortex 2012;48:46–57.

[31] Croxson PL, Johansen-Berg H, Behrens TEJ, Robson MD, Pinsk MA, Gross CG, et al. Quantitative investigation of connections of the prefrontal cortex in the human and macaque using probabilistic diffusion tractography. J Neurosci 2005;25: 8854–66.

[32] Koechlin E, Ody C, Kouneiher F. The architecture of cognitive control in the human prefrontal cortex. Science 2003;302:1181–5.

[33] Badre D, Hoffman J, Cooney JW, D'Esposito M. Hierarchical cognitive control deficits following damage to the human frontal lobe. Nat Neurosci 2009;12:515–22.

[34] Vaidya A, Fellows L. Critical prefrontal contributions to fixation-based value assessment and updating. Nat Commun 2015;6: 10120.

[35] Kahneman D, Tversky A. Prospect theory: an analysis of decision under risk. Econom J Econom Soc 1979;47:263–92.

[36] Monterosso J, Ainslie G. The behavioral economics of will in recovery from addiction. Drug Alcohol Depend 2007;90(Suppl. 1): S100–11.

[37] Camerer C. Three cheers – psychological, theoretical, empirical – for loss aversion. J Mark Res May 2005;43(2):129–33.

[38] Kolling N, Behrens TEJ, Mars RB, Rushworth MFS. Neural mechanisms of foraging. Science 2012;336:95–8.

[39] Frank MJ, Badre D. Mechanisms of hierarchical reinforcement learning in corticostriatal circuits 1: computational analysis. Cereb Cortex 2012;22:509–26.

[40] Berridge KC. From prediction error to incentive salience: mesolimbic computation of reward motivation. Eur J Neurosci 2012; 35:1124–43.

[41] Niv Y, Edlund J, Dayan P, O'Doherty J. Neural prediction errors reveal a risk-sensitive reinforcement-learning process in the human brain. J Neurosci 2012;32:551–62.

[42] Baldo JV, Shimamura AP, Delis DC, Kramer J, Kaplan E. Verbal and design fluency in patients with frontal lobe lesions. J Int Neuropsychol Soc 2001;7:586–96.

[43] Goel V, Grafman J. Are the frontal lobes implicated in "planning" functions? Interpreting data from the tower of Hanoi. Neuropsychologia 1995:623–42.

[44] Goel V, Grafman J, Tajik J, Gana S, Danto D. A study of the performance of patients with frontal lobe lesions in a financial planning task. Brain 1997;120(Pt 1):1805–22.

[45] Fellows LK, Farah MJ. Dissociable elements of human foresight: a role for the ventromedial frontal lobes in framing the future, but not in discounting future rewards. Neuropsychologia 2005; 43:1214–21.

[46] Robinson G, Shallice T, Bozzali M, Cipolotti L. The differing roles of the frontal cortex in fluency tests. Brain 2012;135(Pt 7): 2202–14.

[47] Goel V, Vartanian O, Bartolo A, Hakim L, Maria Ferraro A, Isella V, et al. Lesions to right prefrontal cortex impair real-world planning through premature commitments. Neuropsychologia 2013;51:713–24.

[48] Burgess PW, Alderman N, Forbes C, Costello A, Coates LM, Dawson DR, et al. The case for the development and use of "ecologically valid" measures of executive function in experimental and clinical neuropsychology. J Int Neuropsychol Soc 2006;12:194–209.

[49] Botvinick MM, Cohen JD. The computational and neural basis of cognitive control: charted territory and new frontiers. Cogn Sci 2014.

[50] Camille N, Griffiths CA, Vo K, Fellows LK, Kable JW. Ventromedial frontal lobe damage disrupts value maximization in humans. J Neurosci 2011;31:7527—32.

[51] Bartra O, McGuire JT, Kable JW. The valuation system: a coordinate-based meta-analysis of BOLD fMRI experiments examining neural correlates of subjective value. NeuroImage 2013;76:412—27.

[52] Levy DJ, Glimcher PW. The root of all value: a neural common currency for choice. Curr Opin Neurobiol 2012.

[53] Stalnaker TA, Cooch NK, Schoenbaum G. What the orbitofrontal cortex does not do. Nat Neurosci 2015;18:620—7.

[54] O'Doherty JP. The problem with value. Neurosci Biobehav Rev 2014;43:259—68.

[55] Walton ME, Chau BK, Kennerley SW. Prioritising the relevant information for learning and decision making within orbital and ventromedial prefrontal cortex. Curr Opin Behav Sci 2015;1:78—85.

[56] De Martino B, Kumaran D, Seymour B, Dolan RJ. Frames, biases, and rational decision-making in the human brain. Science 2006; 313:684—7.

[57] Tsuchida A, Doll BB, Fellows LK. Beyond reversal: a critical role for human orbitofrontal cortex in flexible learning from probabilistic feedback. J Neurosci 2010;30:16868—75.

[58] Noonan MP, Kolling N, Walton ME, Rushworth MFS. Re-evaluating the role of the orbitofrontal cortex in reward and reinforcement. Eur J Neurosci 2012;35:997—1010.

[59] Walton ME, Behrens TEJ, Noonan MP, Rushworth MFS. Giving credit where credit is due: orbitofrontal cortex and valuation in an uncertain world. Ann NY Acad Sci 2011;1239:14—24.

[60] Henri-Bhargava A, Simioni A, Fellows LK. Ventromedial frontal lobe damage disrupts the accuracy, but not the speed, of value-based preference judgments. Neuropsychologia 2012;50: 1536—42.

[61] Fellows LK, Farah MJ. The role of ventromedial prefrontal cortex in decision making: judgment under uncertainty or judgment per se? Cereb Cortex 2007;17:2669—74.

[62] Xia C, Stolle D, Gidengil E, Fellows LK. Lateral orbitofrontal cortex links social impressions to political choices. J Neurosci 2015; 35:8507—14.

[63] Brown S, Gao X, Tisdelle L, Eickhoff SB, Liotti M. Naturalizing aesthetics: brain areas for aesthetic appraisal across sensory modalities. NeuroImage 2011;58:250—8.

[64] Rudebeck PH, Murray EA. Review the orbitofrontal oracle: cortical mechanisms for the prediction and evaluation of specific behavioral outcomes. Neuron 2014;84:1143—56.

[65] Schultz W. Neural coding of basic reward terms of animal learning theory, game theory, microeconomics and behavioural ecology. Curr Opin Neurobiol 2004;14:139—47.

[66] Fellows LK, Farah MJ. Ventromedial frontal cortex mediates affective shifting in humans: evidence from a reversal learning paradigm. Brain 2003;126:1830—7.

[67] Rolls ET, Hornak J, Wade D, McGrath J. Emotion-related learning in patients with social and emotional changes associated with frontal lobe damage. J Neurol Neurosurg Psychiatry 1994;57:1518—24.

[68] Dias R, Robbins TW, Roberts AC. Dissociation in prefrontal cortex of affective and attentional shifts. Nature 1996;380:69—72.

[69] Butter C. Perseveration in extinction and in discrimination reversal tasks following selective frontal ablations in *Macaca mulatta*. Physiol Behav 1968;4:163—71.

[70] Jones B, Mishkin M. Limbic lesions and the problem of stimulus—reinforcement associations. Exp Neurol 1972;36:362—77.

[71] Rudebeck PH, Saunders RC, Prescott AT, Chau LS, Murray EA. Prefrontal mechanisms of behavioral flexibility, emotion regulation and value updating. Nat Neurosci 2013;16:1140—5.

[72] Chau B, Sallet J, Papageorgiou G, Noonan M, Bell A, Walton M, et al. Contrasting roles for orbitofrontal cortex and amygdala in credit assignment and learning in macaques. Neuron 2015;87: 1106—18.

[73] Hornak J, O'Doherty J, Bramham J, Rolls ET, Morris RG, Bullock PR, et al. Reward-related reversal learning after surgical excisions in orbito-frontal or dorsolateral prefrontal cortex in humans. J Cogn Neurosci 2004;16:463—78.

[74] Glascher J, Adolphs R, Damasio H, Bechara A, Rudrauf D, Calamia M, et al. Lesion mapping of cognitive control and value-based decision making in the prefrontal cortex. Proc Natl Acad Sci USA 2012:14681—6.

[75] Fellows LK, Farah MJ. Different underlying impairments in decision-making following ventromedial and dorsolateral frontal lobe damage in humans. Cereb Cortex 2005;15: 58—63.

[76] Rudebeck PH, Behrens TE, Kennerley SW, Baxter MG, Buckley MJ, Walton ME, et al. Frontal cortex subregions play distinct roles in choices between actions and stimuli. J Neurosci 2008;28:13775—85.

[77] Camille N, Tsuchida A, Fellows LK. Double dissociation of stimulus-value and action-value learning in humans with orbitofrontal or anterior cingulate cortex damage. J Neurosci 2011: 15048—52.

[78] Shenhav A, Botvinick MM, Cohen JD. The expected value of control: an integrative theory of anterior cingulate cortex function. Neuron 2013;79:217—40.

[79] Kolling N, Wittmann M, Rushworth MFS. Multiple neural mechanisms of decision making and their competition under changing risk pressure. Neuron 2014;81:1190—202.

[80] Cisek P. Making decisions through a distributed consensus. Curr Opin Neurobiol 2012.

[81] Silvetti M, Alexander W, Verguts T, Brown JW. From conflict management to reward-based decision making: actors and critics in primate medial frontal cortex. Neurosci Biobehav Rev 2013.

[82] Pearson JM, Watson KK, Platt ML. Decision making: the neuro-ethological turn. Neuron 2014:950—65.

[83] Daw ND, O'Doherty JP, Dayan P, Seymour B, Dolan RJ. Cortical substrates for exploratory decisions in humans. Nature 2006;441: 876—9.

[84] Boorman ED, Behrens TEJ, Woolrich MW, Rushworth MFS. How green is the grass on the other side? Frontopolar cortex and the evidence in favor of alternative courses of action. Neuron 2009; 62:733—43.

[85] Kovach CK, Daw ND, Rudrauf D, Tranel D, O'Doherty JP, Adolphs R. Anterior prefrontal cortex contributes to action selection through tracking of recent reward trends. J Neurosci 2012;32: 8434—42.

[86] Burgess PW, Veitch E, De Lacy Costello A, Shallice T. The cognitive and neuroanatomical correlates of multitasking. Neuropsychologia 2000;38:848—63.

[87] Volle E, Gonen-Yaacovi G, de Lacy Costello A, Gilbert SJ, Burgess PW. The role of rostral prefrontal cortex in prospective memory: a voxel-based lesion study. Neuropsychologia 2011; 49:2185—98.

[88] Gu X, Wang X, Hula A, Wang S, Xu S, Lohrenz TM, et al. Necessary, yet dissociable contributions of the insular and ventromedial prefrontal cortices to norm adaptation: computational and lesion evidence in humans. J Neurosci 2015;35: 467—73.

[89] Koenigs M, Tranel D. Irrational economic decision-making after ventromedial prefrontal damage: evidence from the ultimatum game. J Neurosci 2007;27:951—6.

[90] Moretti L, Dragone D, di Pellegrino G. Reward and social valuation deficits following ventromedial prefrontal damage. J Cogn Neurosci 2009;21:128—40.

[91] Maunsell JHR. Neuronal representations of cognitive state: reward or attention? Trends Cogn Sci 2004;8:261–5.

[92] Anderson BA, Laurent PA, Yantis S. Value-driven attentional capture. Proc Natl Acad Sci USA 2011;108:10367–71.

[93] Krajbich I, Armel C, Rangel A. Visual fixations and the computation and comparison of value in simple choice. Nat Neurosci 2010;13:1292–8.

[94] Navalpakkam V, Koch C, Rangel A, Perona P. Optimal reward harvesting in complex perceptual environments. Proc Natl Acad Sci USA 2010;107:5232–7.

[95] Van Der TC, Pennartz CMA, Roelfsema PR. A unified selection signal for attention and reward in primary visual cortex. 2013.

[96] Serences JT. Value-based modulations in human visual cortex. Neuron 2008;60:1169–81.

[97] Peck CJ, Jangraw DC, Suzuki M, Efem R, Gottlieb J. Reward modulates attention independently of action value in posterior parietal cortex. J Neurosci 2009;29:11182–91.

[98] Dorris MC, Glimcher PW. Activity in posterior parietal cortex is correlated with the relative subjective desirability of action. Neuron 2004;44:365–78.

[99] Cavada C, Compañy T, Tejedor J, Cruz-Rizzolo RJ, Reinoso-Suárez F. The anatomical connections of the macaque monkey orbitofrontal cortex. A review. Cereb Cortex 2000;10:220–42.

[100] Price JL. Definition of the orbital cortex in relation to specific connections with limbic and visceral structures and other cortical regions. Ann NY Acad Sci 2007;1121:54–71.

[101] Goodale MA, Milner AD. Separate visual pathways for perception and action. Trends Neurosci 1992:20–5.

[102] Bar M, Kassam KS, Ghuman AS, Boshyan J, Schmid AM, Dale AM, et al. Top-down facilitation of visual recognition. Proc Natl Acad Sci USA 2006;103:449–54.

[102a] Hodgson TL, Mort D, Chamberlain MM, Hutton SB, O'Neill KS, Kennard C. Orbitofrontal cortex mediates inhibition of return. Neuropsychologia 2002;40:1891–901.

[103] Hartikainen KM, Ogawa KH, Knight RT. Orbitofrontal cortex biases attention to emotional events. J Clin Exp Neuropsychol 2012;34:588–97.

[104] Vaidya AR, Fellows LK. Ventromedial frontal damage in humans reduces attentional priming of rewarded visual features. J Neurosci 2015;35:12813–23.

[105] Mackintosh NJ. A theory of attention: variations in the associability of stimuli with reinforcement. Psychol Rev 1975;82:276–98.

[106] Towal RB, Mormann M, Koch C. Simultaneous modeling of visual saliency and value computation improves predictions of economic choice. Proc Natl Acad Sci USA 2013;110: E3858–67.

[107] Lim S-L, O'Doherty JP, Rangel A. The decision value computations in the vmPFC and striatum use a relative value code that is guided by visual attention. J Neurosci 2011;31:13214–23.

[108] Marewski JN, Gaissmaier W, Gigerenzer G. Good judgments do not require complex cognition. Cogn Process 2010;11:103–21.

[109] Gigerenzer G, Gaissmaier W. Heuristic decision making. Annu Rev Psychol 2011;62:451–82.

[110] Schwartz B, Ward A, Monterosso J, Lyubomirsky S, White K, Lehman DR. Maximizing versus satisficing: happiness is a matter of choice. J Pers Soc Psychol 2002;83:1178–97.

[111] Fellows LK. Deciding how to decide: ventromedial frontal lobe damage affects information acquisition in multi-attribute decision making. Brain 2006;129:944–52.

[112] Camille N, Coricelli G, Sallet J, Pradat-Diehl P, Duhamel J-R, Sirigu A. The involvement of the orbitofrontal cortex in the experience of regret. Science 2004;304:1167–70.

[113] Levens SM, Larsen JT, Bruss J, Tranel D, Bechara A, Mellers BA. What might have been? The role of the ventromedial prefrontal cortex and lateral orbitofrontal cortex in counterfactual emotions and choice. Neuropsychologia 2014;54:77–86.

[114] Berns GS, Laibson D, Loewenstein G. Intertemporal choice—toward an integrative framework. Trends Cogn Sci 2007;11: 482–8.

[115] Monterosso J, Piray P, Luo S. Neuroeconomics and the study of addiction. Biol Psychiatry 2012:107–12.

[116] McClure SM, Laibson DI, Loewenstein G, Cohen JD. Separate neural systems value immediate and delayed monetary rewards. Science 2004;306:503–7.

[117] Kable JW, Glimcher PW. The neural correlates of subjective value during intertemporal choice. Nat Neurosci 2007;10: 1625–33.

[118] Button KS, Ioannidis JPA, Mokrysz C, Nosek BA, Flint J, Robinson ESJ, et al. Power failure: why small sample size undermines the reliability of neuroscience. Nat Rev Neurosci 2013;14: 365–76.

[119] Peters J. The role of the medial orbitofrontal cortex in intertemporal choice: prospection or valuation? J Neurosci 2011;31:5889–90.

[120] Race E, Keane MM, Verfaellie M. Medial temporal lobe damage causes deficits in episodic memory and episodic future thinking not attributable to deficits in narrative construction. J Neurosci 2011;31:10262–9.

23

Opponent Brain Systems for Reward and Punishment Learning: Causal Evidence From Drug and Lesion Studies in Humans

S. Palminteri[1,2], M. Pessiglione[3,4]

[1]University College London, London, United Kingdom; [2]Ecole Normale Supérieure, Paris, France; [3]Institut du Cerveau et de la Moelle (ICM), Inserm U1127, Paris, France; [4]Université Pierre et Marie Curie (UPMC-Paris 6), Paris, France

Abstract

Approaching rewards and avoiding punishments are core principles that govern the adaptation of behavior to the environment. The machine learning literature has proposed formal algorithms to account for how agents adapt their decisions to optimize outcomes. In principle, these reinforcement learning models could be equally applied to positive and negative outcomes, ie, rewards and punishments. Yet many neuroscience studies have suggested that reward and punishment learning might be underpinned by distinct brain systems. Reward learning has been shown to recruit midbrain dopaminergic nuclei and ventral prefrontostriatal circuits. The picture is less clear regarding the existence and anatomy of an opponent system: several hypotheses have been formulated for the neural implementation of punishment learning. In this chapter, we review the evidence for and against each hypothesis, focusing on human studies that compare the effects of neural perturbation, following drug administration and/or pathological conditions, on reward and punishment learning.

Good and evil, reward and punishment, are the only motives to a rational creature: these are the spur and reins whereby all mankind are set on work, and guided.

These famous words by John Locke suggest that rewards and punishments are not on a continuum from positive to negative: they pertain to distinct categories of events that we can imagine or experience. Indeed rewards and punishments trigger different kinds of subjective feelings (such as pleasure versus pain or desire versus dread) and elicit different types of behaviors (approach versus avoidance or invigoration versus inhibition). These considerations might suggest the idea that rewards and punishments are processed by different parts of the brain. In this chapter we examine this idea in the context of reinforcement learning, a computational process that could in principle apply equally to rewards and punishments. We start by summarizing the computational principles underlying reinforcement learning (Box 23.1 and Fig. 23.1) and by describing typical tasks that implement a comparison between reward and punishment learning (Box 23.2 and Fig. 23.2). Then we expose the current hypotheses about the possible implementation of reward and punishment learning systems in the brain (Fig. 23.3). Last,

---

BOX 23.1

COMPUTATIONAL MODELS OF REINFORCEMENT LEARNING

The first reinforcement learning (RL) models come from the behaviorist tradition, in the form of mathematical laws describing learning curves [82] or formal descriptions of associative conditioning [2]. Subsequently, in the 1980s, computational investigation of RL received a significant boost when it grabbed the attention of machine learning scholars, who were aiming at developing algorithms for goal-oriented artificial agents [1]. In the

Continued

BOX 23.1   (cont'd)

machine learning literature, the typical RL problem involves an agent navigating through a series of states (s) by performing actions (a), while collecting some numeric reward (r). The goal of the agent is to select actions that maximize the future cumulative reward (also referred to as "return"). The typical RL algorithm has therefore two main functions: the value function, which stores reward predictions, and the policy function, which determines which action has to be taken to maximize the reward.

A variety of solutions to the RL problem have been proposed. The most relevant RL models for psychologists and neurobiologists revolve around the notion of reward-prediction error, which is equivalent to temporal difference error in most choice tasks, in which there is only one transition step between stimuli and outcomes. Reward-prediction error (RPE) is the difference between obtained and expected outcomes. Thus, after receiving a reward r at trial t, an RPE δ is calculated based on the current estimate of the state value $V(s)$ as follows:

$$\delta_t = r_t - V(s)_t \tag{23.1}$$

This error term is subsequently used to update (improve) the reward prediction through a learning rule. The most commonly used is the delta rule, in which the impact of each RPE on future expectation is scaled by a learning rate α:

$$V(s)_{t+1} = V(s)_t + \alpha^* \delta_t \tag{23.2}$$

In RL models action selection can rely on "direct" or "indirect" policy functions. Direct policy implies that, instead of representing only state-based values [$V(s)$], the agent represents values that are both action- and state-dependent [$Q(s,a)$]. Whereas state value represents the reward expected in a given situation, action values represent the reward expected from taking a particular action in a given situation. Action values can also be learned via prediction errors and delta rules, and directly compared to make a choice, as implemented in the Q-learning model—a model very frequently used in human and animal studies [83] (Fig. 23.1A, left). Indirect policy involves two separate representations for value prediction and action propensity, as famously implemented in the actor–critic model (Fig. 23.1A, right). In this model, the actor makes choices by comparing state-dependent action propensities [$\pi(s,a)$]. The obtained reward r generates an RPE δ, relative to the state value stored by the critic [$V(s)$]. The RPE is then used to improve ("criticize") future reward expectations, as in Eq. (23.2), as well as action propensities in the following equation:

$$\pi(s,a)_{t+1} = \pi(s,a)_t + \alpha^* \delta_t \tag{23.3}$$

An important problem that all RL algorithms must address is the trade-off between exploiting current knowledge and exploring alternative options. This exploitation/exploration trade-off is particularly relevant in probabilistic and changing environments, in which sticking to first impressions may prove ruinous. The simplest way to address this issue is to allow some stochasticity in the decision process. Thus, instead of systematically picking the highest value action (hard maximization or greedy decision rule), a softmax decision rule has been proposed [84] in which the probability of choosing A over B is a sigmoid function of the value difference between A and B:

$$P(A) = 1/(1 + \exp((Q(B) - Q(A))/\beta)) \tag{23.4}$$

Here β is a temperature parameter that adjusts the steepness of the sigmoid function. The softmax function implies that exploration is maximal when $Q(A) \approx Q(B)$. Note that this is the same function that is used in logistic regression, in which the β weight on the value difference would be equivalent to an inverse temperature.

Computational models are useful for the experimental exploration of human RL abilities in many respects. First, they may be used to generate trial-by-trial estimates of value prediction and prediction errors, which can then be mapped to brain activity using fMRI, an approach termed model-based fMRI [85]. Second, computational models may help finesse the analysis of learning deficits induced by drugs and lesions, compared to aggregate behavioral measures (Fig. 23.1B). For example, consider a case in which two different treatments are shown to impair instrumental learning, as evidenced by a decrease in correct choice rate compared to placebo. We might be tempted to conclude that the two drugs have "similar effects." However, computational analysis may reveal that one drug affects the learning rate and the other the choice temperature parameter, thus dissociating their effects on the update versus selection process [86].

we review the evidence in favor or against the various hypotheses, focusing on studies in humans that compare reward and punishment learning and that employ approaches that assess causal implications, by observing the behavioral effects of pharmacological manipulations and brain lesions.

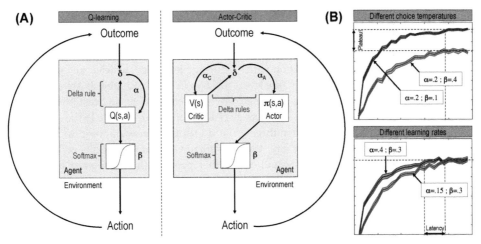

FIGURE 23.1 **Basic computational models of reinforcement learning.** (A) Computational architectures of standard models using direct and indirect policy rules (Q-learning and actor–critic models, respectively). In both architectures the decision process (policy) is modeled with a softmax function, whose exploration/exploitation trade-off is governed by parameter β (temperature). In the Q-learning model decisions are made by comparing Q-values, which are action-specific estimates of the expected future reward. When the outcome is received, a prediction error δ is calculated as the difference between the chosen action Q-value and the actual reward. The prediction error is then used to update the chosen action Q-value via a delta rule adjusted by parameter α (learning rate). In the actor–critic model decisions are made by comparing policy values π, which are action-specific estimates of action probabilities, stored by the actor component. When the outcome is received, a prediction error δ is calculated as the difference between the state value, stored by the critic component, and the actual reward. The prediction error is then used in a delta rule to update reward prediction (ie, the state value) as well as the action probability (ie, the π-value), via different learning rates. (B) Macroscopic effects of varying key Q-learning parameters (learning rate α and choice temperature β) on average learning curves. Crucially, the learning rate affects only the latency, ie, the number of trials required to reach a given performance level, whereas the temperature also affects the plateau, ie, the performance level after convergence.

BOX 23.2

BEHAVIORAL TASKS USED TO COMPARE REWARD AND PUNISHMENT LEARNING

The aim of such tasks is to dissociate valence-specific and valence-independent processes. It is important to implement the comparison within the same task to avoid confounds with details of the design and to avoid framing effects. Indeed subjects might reframe their expectations if they realize they are in a reward- or punishment-learning task, ie, they might change their reference point and, for instance, take an absence of reward as a punishment or an absence of punishment as a reward [76–78,87]. Contrasting reward and punishment in the same protocol entails the challenging problem of comparing stimuli that do not necessary share the same properties and whose values are not necessarily in the same range. A simple solution is to opt for secondary reinforcers such as money or even abstract "points." In this case, rewards and punishments share the same sensory properties and, being numeric in nature, they can be directly fed to model-based analyses. Yet the generalizability to other forms of reinforcers, perhaps more natural for the brain, of results obtained using money or points is not automatically granted.

In the following we focus on instrumental (or operant) learning tasks but note that Pavlovian (or classical) learning tasks could also be used. In Pavlovian tasks, the occurrence of reinforcers is not contingent on the behavior of the subject, who can remain entirely passive. In the reinforcement learning (RL) framework, Pavlovian tasks elicit only a state value function, which can be used to fit implicit measures of learning such as pupil diameter or skin conductance response. Yet these measures are noisy, and model fitting implies specifying a function that relates them to state values, which is not as straightforward as in the case of choices [8]. Also, for Pavlovian tasks that do not require any overt behavior, it may be harder to control the engagement of subjects, an issue that is particularly problematic with patients. However, Pavlovian tasks avoid the issue of possible confounds with motor responses, which must be carefully orthogonalized with respect to outcome valence in instrumental tasks. This is because reward obtainment and punishment avoidance might be more naturally associated with "go" and "no-go" responses, respectively [88,89].

Instrumental learning tasks in humans have most frequently taken the form of two-armed bandits. Subjects are repeatedly presented with a choice between two abstract stimuli (often fractal images or letters taken from exotic alphabets) representing the two available actions.

Continued

BOX 23.2   (cont'd)

Their task is to find out, by trial and error, the action with the highest expected value. One influential design (the Hiragana task, Fig. 23.2A) differentiates between a training and a test phase [38,45]. During the training phase subjects are asked to play several two-armed bandits (often three) until they reach a performance criterion. Crucially, the average value (ie, the policy-independent state value) of the bandits is zero because the two stimuli have reciprocal (and symmetrical) probabilities of winning/losing a point. In other terms, choices observed during the training phase do not discriminate between reward and punishment learning, because both outcomes can be experienced with the same bandit. After training, subjects are asked to recognize the most advantageous stimulus among new pairings of stimuli, in the absence of feedback. Choices in this phase are taken as indicators of learning that occurred during the training phase. The capacity to select the best possible stimulus is considered a measure of reward learning and the capacity to avoid the worst stimulus a measure of punishment learning.

Another influential design (the Agathodaimon task, Fig. 23.2B) directly considers instrumental choices made during learning, instead of postlearning preferences [25,66]. In this task, subjects are also asked to play several two-armed bandits (at least two, sometimes four). The crucial difference relative to the other task is that winning and losing are now opposed to neutral outcomes. So the average value of the bandits is different from zero, being either positive (in the reward maximization conditions) or negative (in the punishment minimization conditions). In other terms, this task directly discriminates between reward-seeking and punishment-avoidance performance, because both outcomes cannot be experienced in the same bandit. The correct response rates in the two conditions are thus considered as measures of reward and punishment learning, respectively. Importantly, this task allows fitting RL models on choice learning curves separately for reward and punishment conditions, and therefore characterizing valence-specific deficits in computational terms. Also, this task allows disentangling between two oppositions that were confounded in the previous task: reward versus punishment (outcome valence) and positive versus negative prediction errors (which are both present in the two conditions).

THE NEURAL CANDIDATES FOR REWARD AND PUNISHMENT LEARNING SYSTEMS

Reinforcement learning (RL) refers to a class of processes through which an agent builds associations between stimuli and actions under the influence of rewards or punishments, ie, events that possess a positive or negative value for the agent's well-being. RL processes have been formally captured by a variety of algorithms in the machine learning literature (see Box 23.1 and Fig. 23.1). The typical RL algorithm learns, by trial and error, to select between available actions in an environment characterized by a given set of possible states, so as to maximize some notion of cumulative reward [1]. A key principle, inherited from the animal learning literature, that is common to most RL algorithms is learning being driven by reward-prediction error (RPE), a sort of signed surprise defined as the difference between expected and obtained rewards [2].

Electrophysiology studies have consistently reported RPE encoding in the dopaminergic midbrain areas of both human and nonhuman primates [3,4]. More precisely, it has been reported that the firing rate of dopamine (DA) neurons positively and parametrically scales with RPE [5]. Neuroimaging studies in humans have confirmed and extended these results by showing functional magnetic resonance imaging (fMRI) correlates of RPE in the midbrain [6], as well as in the main DA projection sites, such as the striatum and the prefrontal cortex, especially in their ventral parts [7–9].

In RL algorithms, there is no reason why the reward term could not take negative values, and hence capture the notion of punishment. This implies that, at least in principle, the same computations could be used for reward and punishment learning. However, physiological constraints suggest that a single neural system might not be able to encode in an equally efficient manner both RPEs and punishment-prediction errors (PPEs). This is simply because firing rate cannot be negative, and therefore neurons that have low spontaneous activity, as those in the midbrain, do not have sufficient range to encode PPE precisely with decreasing firing rate [10]. To obviate this physiological constraint, one solution is to assume that an opponent system might positively respond to PPE, just as the dopaminergic midbrain does for RPE [11]. So far,

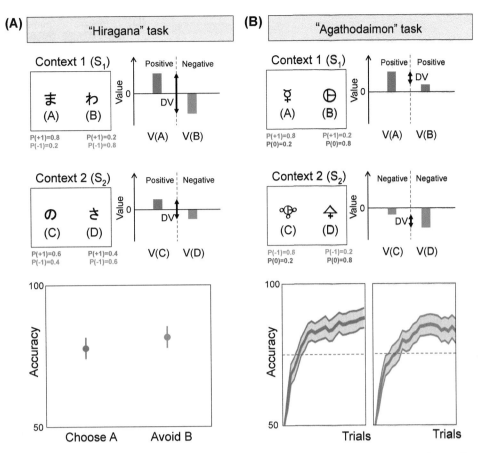

FIGURE 23.2 **Two classical tasks used to compare reward and punishment learning.** Presented for both tasks are decision screens in two possible contexts (pairs of symbols), the probabilistic contingencies associated with each symbol, the two option values (DV is the decision value, ie, the difference between the two options), and the main performance measure (choice accuracy) expected from a normal subject. Note that values are the actual values that subjects have to learn, before learning they are equal to zero. (A) The Hiragana task is designed so as to reveal in a test session the type of learning (reward seeking versus punishment avoidance) that was operant during the training session [38]. During the training session subjects are presented with fixed pairs of options (typically three pairs), materialized by Hiragana symbols and associated with different, reciprocal probabilities of winning or losing, P(+1) and P(−1). During the test session subjects are asked to identify the best option, among novel binary combinations, in the absence of feedback. The capacity to correctly identify the best option (choose A) and reject the worst (avoid B) is taken as a measure of the capacity to learn from positive and negative prediction errors. (B) The Agathodaimon task is designed to compare reward and punishment learning directly during the training session [25]. Subjects are also presented with fixed pairs of symbols (typically two pairs), now materialized by Agathodaimon symbols, with the crucial difference that rewards and punishments are never mixed within a pair. Some pairs of options are associated with reciprocal probabilities of winning or getting nothing, ie, P(+1) and P(0), and others with reciprocal probabilities of losing or getting nothing, ie, P(−1) and P(0). Typically, the rate of correct choice (ie, choosing the most rewarding or the least punishing option) is extracted on a trial-by-trial basis to assess the capacity to learn from rewards versus punishments.

several hypotheses have been formulated concerning the neural implementation of this tentative opponent system, but this remains extremely controversial [12–15] (see also Ref. [16] of the present volume). We have identified and describe here four main hypotheses (see also Fig. 23.3).

Hypothesis 1: No Opponent System

A first hypothesis is that there is actually no opponent system and that punishment avoidance is also resolved by the midbrain DA system within the basal ganglia circuits. It has been argued that, whereas phasic burst in DA activity encodes positive prediction errors, the duration of

pauses in DA activity might encode negative prediction errors [17]. In this framework, an important role may be played by the habenula, an epithalamic nucleus whose activity has been shown to provide inhibitory input to the midbrain DA neurons following reward omission in monkeys [18]. Consistently, the habenula has been shown, in humans, to encode aversive events such as electric shocks and to have an impact on striatal activity [19].

Hypothesis 2: Dopaminergic Opponent System

A second hypothesis also supposes that avoidance learning is driven by DA activity, but thanks to a

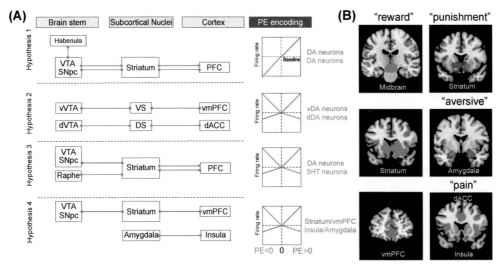

FIGURE 23.3   **Neural implementation of the opponency between reward and punishment learning.** (A) Various hypotheses concerning the neural implementation of punishment avoidance (in red), as opposed to reward seeking (in green). For each hypothesis, the key regions and connections of each opponent system are shown on the left, with their theoretical pattern of activity as a function of prediction error (PE) plotted on the right. (B) Maps resulting from automatized large-scale meta-analyses as implemented in Neurosynth [90]. Meta-analyses confirm the implication of the VTA—VS—vmPFC circuit in reward encoding, as is assumed in most hypotheses. Negative affects such as "punishment," "aversive," and "pain" involve both similar (striatum) and specific brain regions (notably dACC, insula, amygdala). Note that Neurosynth treats similarly positive and negative correlations between experimental factors and brain activity, leaving undecided whether the striatal representation of "punishment" is driven by activation with avoided punishment (hypothesis 1) or received punishment (hypothesis 2). *5HT*, serotonin; *DA*, dopamine; *dACC*, dorsal anterior cingulate cortex; *dDA*, dorsal dopamine; *DS*, dorsal striatum; *dVTA*, dorsal VTA; *PFC*, prefrontal cortex; *SNpc*, substantia nigra pars compacta; *vDA*, ventral dopamine; *vmPFC*, ventromedial PFC; *VS*, ventral striatum; *VTA*, ventral tegmental area; *vVTA*, ventral VTA.

different subset of midbrain neurons that positively encode punishments [20]. This hypothesis is based on electrophysiological observations that an anatomically segregated population of DA neurons in the nonhuman primate midbrain positively respond to aversive stimuli [21]. Consistently, human fMRI studies have shown positive encoding of punishment anticipation and PPE in the ventral tegmental area (VTA) and downstream in the striatum during aversive conditioning tasks [22–24]. fMRI data are generally consistent with the idea of a functional gradient, such that the ventral parts of the frontostriatal circuits would be preferentially concerned with reward seeking and dorsal parts with punishment avoidance [22,25,26].

Hypothesis 3: Serotoninergic Opponent System

A third hypothesis states that the neuromodulator serotonin (5-HT) could play the role of an opponent signal by encoding PPE [11]. A vast body of literature in rodents, linking the serotoninergic system (especially the dorsal raphe) to behavioral inhibition and "fight or flight" responses (generally induced by aversive events), originally motivated this hypothesis [27,28]. Further supporting this idea, at the electrophysiological level, 5-HT has been shown to antagonize DA function in the VTA and striatum [29,30].

Hypothesis 4: Other Opponent Systems

Finally, the fourth hypothesis postulates that punishment avoidance involves circuits outside the frontostriatal projections of the brain-stem neuromodulator systems. According to this hypothesis (which would be more appropriately considered as a collection of hypotheses), punishment learning is mediated by aversive signals encoded in other cortical and subcortical areas, such as the insula and amygdala (see also Ref. [31] in the present volume). A consistent body of electrophysiological, pharmacological, and lesion studies in animals supports the implication of these regions in punishment avoidance [32,33]. These results from animal studies align with some fMRI studies in humans, as well as with meta-analyses [34–37].

EVIDENCE FROM DRUG AND LESION STUDIES

For the sake of simplicity, we have assumed that the opponent systems encoding rewards and punishments both have a better precision (or gain) with increasing firing rates. This makes the prediction that damage to a subset of this system should preferentially degrade either reward or punishment learning and therefore produce effects on choice behavior that should interact with outcome valence. In the following we examine this prediction, under the four hypotheses regarding the implementation of the opponent system underlying punishment learning. We focus on tasks that were employed in humans to compare reward and punishment learning directly (see examples in Box 23.2 and Fig. 23.2).

The first hypothesis states that midbrain DA bursts and dips are necessary and sufficient for reward and punishment learning, respectively. Thus, according to this hypothesis, enhancing DA function should increase reward learning to the detriment of punishment learning and DA blocking should produce the opposite effects. These predictions have been verified in Parkinson's disease (PD) patients [38]. PD is characterized by DA neuronal loss and treated with DA enhancers, either metabolic precursors of DA or direct agonists of DA receptors [39]. A group of patients was tested twice (once ON and once OFF medication) with a probabilistic learning task (the Hiragana task in Fig. 23.2A). Results showed a significant medication by valence interaction, with ON-PD patients being better at reward learning than OFF-PD patients, and vice versa for punishment learning. This result has been interpreted within the framework of a neural network model of the basal ganglia that formalizes action selection as the result of a competition between a direct "go" and an indirect "no-go" pathway [40]. On one side, the go pathway expresses D1 DA receptors and is reinforced by DA bursts, leading to an increased probability of choosing options followed by reward. On the other side, the no-go pathway expresses D2 DA receptors and is reinforced by DA dips, leading to a reduced probability of choosing options followed by punishment. The interaction between medication status and outcome valence has been replicated several times in PD patients [41,42]. Another study further suggested that improvement in reward seeking observed under pro-DA modulation is specific to PD patients with DA dysregulation syndrome, whereas impairment of punishment avoidance stands only for nondysregulated patients [43]. This neurobiological model has received support from genetic studies, indicating specific roles for D1 and D2 polymorphisms in reward seeking and punishment avoidance behaviors, respectively [44,45].

Thus, investigations of RL abilities in PD with and without dopaminergic treatment were consistent with the idea that DA dips are necessary for punishment learning. The questions remained (1) whether these effects arose from the modulation of explicit learning processes or rather from implicit processes and (2) whether these effects were specific to PD patients and their medication or can be generalized to other conditions and treatments. To examine these questions, we adapted an instrumental learning task (the Agathodaimon task in

Fig. 23.2B) such that symbolic cues indicating the state of the environment were not consciously perceived, and we tested patients with Tourette syndrome (TS) in addition to PD patients [46,47]. TS is an interesting model for studying RL because it is characterized by hyper-DA symptoms and treated with neuroleptics, an anti-DA medication (Fig. 23.4A) [48]. Our results concerning PD replicated previous findings, indicating that the interaction between medication status and outcome valence holds for implicit learning processes (Fig. 23.4B). Interestingly, TS patients displayed the opposite double dissociation, with OFF-TS patients being better at reward seeking and ON-TS patients at punishment learning. Thus, untreated PD and treated TS might receive the same interpretation: DA levels are too low for RPE-related DA bursts to reinforce approach behavior (Fig. 23.4C). Reciprocally, in treated PD and untreated TS, DA levels might be too high for PPE-related DA dips to reinforce avoidance behavior.

However, despite this suggestive evidence for the implication of DA dips in punishment learning, other studies directly challenged these findings. In fact, while enhancing reward learning by increasing DA levels has been almost systematically observed, results regarding punishment learning have been less consistent, with several studies showing no effect of dopaminergic drugs on avoidance behavior, even with doses that were efficient on reward seeking [25,49–51]. Some of these studies fitted RL models to the observed choices to identify the computational parameter that would best capture drug effects. Interestingly, the positive effect of levodopa on reward learning was best accounted for by increasing the reward parameter, and not the learning rate that modulates RPE [25]. Thus, it could be that dopaminergic drugs just amplify reward representation, without affecting learning per se. By contrast, there was no significant effect on the punishment parameter.

To our knowledge, there is no pharmacological evidence that pro-DA drugs could improve punishment avoidance, as would be predicted by the second hypothesis, according to which a distinct (dorsal) population of DA neurons positively encodes aversive signals and underpins punishment learning. A corollary of this second hypothesis is the idea of a selective implication of the anterior caudate (commonly referred to as "dorsal striatum" in human fMRI literature) in punishment processing [22], whereas the nucleus accumbens (commonly referred to as "ventral striatum" in human fMRI literature) would be more involved in reward processing. To test this idea, we administered a probabilistic instrumental learning task (Agathodaimon task, Fig. 23.2B) to patients with Huntington disease (HD). This disease is a rare genetic condition characterized by choreic movements and caused by degeneration of

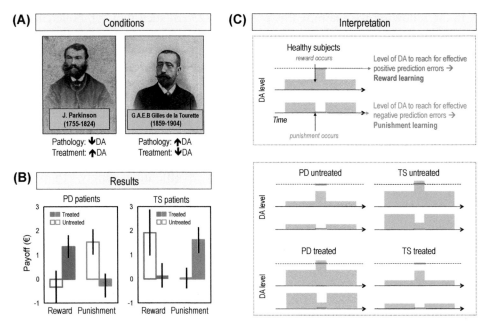

FIGURE 23.4   **Double dissociation of dopaminergic medication effects on reward and punishment learning.** (A) Parkinson's disease *(PD)* and Tourette syndrome *(TS)*, named after James Parkinson and Georges Albert Édouard Brutus Gilles de la Tourette, have proven to be useful models to study the role of dopamine (DA) in reinforcement learning. They represent opposite conditions in terms of symptoms (hypo- vs hyperkinesia), pathophysiology (hypo- vs hyperdopaminergia), and medication (pro- vs antidopaminergic) [91]. (B) Histograms in each graph show additional correct choices (in euros) in the reward-seeking (green) and punishment-avoidance (red) conditions relative to a neutral condition, observed in the Agathodaimon task. The results show an interaction between outcome valence (positive or negative) and medication status (treated or untreated) in both conditions, with opposite patterns [47]. (C) A schematic interpretation of DA activity (gray) following positive and negative outcomes, in healthy subjects as well as PD and TS patients. The *green and red lines*, respectively, represent the level to reach (either above or below the baseline) to express a signal strong enough to induce either positive or negative value update. The pathological conditions and pharmacological manipulations shift the baseline, either downward (untreated PD and treated TS), such that positive prediction errors fail to induce reward learning, or upward (treated PD and untreated TS), such that negative prediction errors fail to induce punishment learning [92]. This schematic is based on the results originally reported by Schultz and colleagues (1997) and largely inspired by the theory of Frank and colleagues (2004).

the striatum (Fig. 23.5A) [52]. The neural degeneration starts in the dorsal parts of the striatum, in the caudate head mostly, before the motor symptoms become apparent [53]. This makes presymptomatic HD a relevant lesion model to investigate dorsal striatal function. Our results were consistent with the idea that the dorsal striatum is specifically involved in punishment learning (Fig. 23.5B). However, our computational analyses revealed that the deficit observed in presymptomatic HD patients was best explained by increasing choice stochasticity, and not reducing the punishment parameter or punishment learning rate. Thus the dorsal striatum (anterior caudate) system might not be implicated in learning per se but in selecting between actions that lead to negative outcomes (the lesser of two evils). This could relate to the notion that dorsal prefrontostriatal circuits are responsible for response inhibition or avoidance behavior.

Despite its great theoretical appeal, the third hypothesis, which assigns to 5-HT the role of an opponent neuromodulator system, has received mixed evidence in human studies. While some studies did provide support for a specific role of 5-HT in punishment learning, other studies found nonspecific effects or even provided evidence for a specific role in reward learning [54–59]. Such inconsistent results have called for a revision of the original theory, which now incorporates a behavioral dimension—approach versus withdrawal—in addition to that of outcome valence—reward versus punishment [12,60]. In this theoretical framework, 5-HT would be needed to avoid punishment through response inhibition (no-go), but not when avoiding punishment implies response invigoration (go). Yet this addendum does not resolve every discrepancy reported in the literature. For instance, using the same subliminal learning task as in PD and TS patients [47], we found that 5-HT reuptake inhibitors, given as treatment for obsessive compulsive disorder, improved reward and punishment learning to similar extents and whether the response was implemented as a go or a no-go [58]. The complexity of the role of 5-HT in RL is further highlighted by its proven implication in the temporal discounting of reward [61,62]. It has even been argued that because the serotoninergic system is much more anatomically widespread

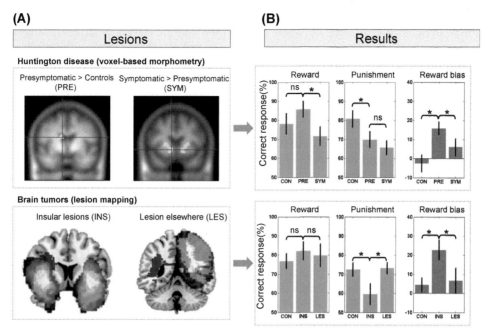

FIGURE 23.5 **The causal role of the anterior insula and dorsal striatum in punishment (but not reward) learning.** (A) Mapping of brain lesions. Striatal lesions were caused by neurodegenerative processes occurring in presymptomatic (*PRE*; dorsal striatum) and early symptomatic (*SYM*; both dorsal and ventral striatum) patients with Huntington disease. Insular lesions (INS) as well as control lesions (LES) outside the insula were due to low-grade gliomas. (B) Behavioral results from the Agathodaimon task. Both the PRE and the INS groups showed reduced punishment learning and consequently a positive reward bias (correct choice rate in reward minus punishment context). This demonstrates the critical and specific implication of both dorsal striatum and anterior insula in punishment learning.

and genetically complex compared to the dopaminergic system, it might be impossible to delineate a single functional domain for this neuromodulator [63]. Indeed the role of 5-HT might generalize to any sort of aversive signaling, including punishment (negative outcomes) but also effort (invigoration of action) and delay (opportunity cost).

Finally, the fourth hypothesis, which opposes structures such as the amygdala and anterior insula to the ventral striatum and prefrontal cortex, has been partially confirmed by investigations of patients with brain damage. Bilateral damage to the amygdala has been shown to impair implicit punishment learning, which was spared by bilateral damage to the hippocampus [64]. This classical observation has been more recently backed up by the finding that bilateral calcification in the amygdala abolishes loss aversion in economic decision-making [65]. To assess the implication of the insula in punishment avoidance, we administered to patients with brain tumors (Fig. 23.5A) the same task as that used with HD patients (Fig. 23.2B). Results (Fig. 23.5B) showed that insular lesions specifically impair punishment learning [66]. Computational analyses indicated that the deficit was best captured by decreasing the punishment parameter, which is consistent with fMRI studies reporting PPE encoding in the anterior insula [25,67,68]. Recent findings have shown the implication

of the insula in effort learning, suggesting that this region might represent aversive signals across different domains [69].

Prolonging our investigation of HD using our probabilistic instrumental learning task (Fig. 23.3B), we found that at a symptomatic but still early stage of the disease, when neural degeneration affects both dorsal and ventral striatum, patients exhibited deficits in both punishment and reward learning [66]. The deficit in reward learning was best explained by reducing the reward parameter in the RL model. This aligns well with the position of the ventral striatum as a main output of the mesolimbic DA pathway, because DA drug effects were also captured by adjusting the reward parameter [25]. Results regarding the ventromedial prefrontal cortex (VMPFC) are not that clear-cut. Because activity in this region has been repeatedly shown to encode value positively, across both appetitive and aversive items [70,71], one would expect VMPFC damage to impair reward learning. Yet patients with VMPFC lesions were found to have more difficulty with learning from negative feedback [72] in a probabilistic learning task (Hiragana task, Fig. 23.2A). This could mean that contrary to the assumption of hypothesis 4 (Fig. 23.3A, bottom), the precision of coding in some cortical regions might actually be better for decreasing firing rates, unlike what was seen in neuromodulatory systems. Note that in other studies from the same group, VMPFC patients also exhibited deficits in choosing between rewards [73].

CONCLUSIONS, LIMITATIONS, AND PERSPECTIVES

Whereas the implication of dopaminergic midbrain nuclei and ventral prefrontostriatal circuits in reward learning is quite well established, the delineation of an opponent system responsible for punishment learning is still a matter of debate. Our succinct review of the literature comparing reward and punishment learning in humans brings evidence to all four hypotheses regarding the neural implementation of a tentative punishment learning system. This could mean that various brain structures play a role in punishment learning: first those that were implicated in reward learning (DA, ventral striatum, VMPFC), second other neuromodulators such as 5-HT, and third other subcortical and cortical structures such as amygdala and anterior insula.

Yet this complicated picture might arise from some limitations in the approaches that were reviewed in this chapter. It is likely that the behavioral tasks do not purely target instrumental learning processes as implemented in RL models. Obviously a good fit of behavioral choices is no proof that the brain actually implements the computations operated in the models. Several learning systems might work in parallel to solve the choice problems, irrespective of outcome valence [74,75]. For instance, model-based and Pavlovian systems might interact with the model-free instrumental system that is formalized by the computational models commonly used to account for choice behavior. Another issue is that expected values are both subject- and context-dependent, which means that once state values are learned, some individuals might reframe their expectations (change their reference point), such that not winning can be perceived as a punishment, and not losing as a reward [76–78]. A related issue is that rewards and punishments are often confounded with positive and negative prediction errors. In fact, positive prediction errors can occur during punishment learning (when expected punishment is not received) and negative prediction errors during reward learning (when expected reward is not received). It is not clear at present whether the most relevant distinction for dividing brain systems is negative versus positive outcome (reward versus punishment) or negative versus positive prediction error. Finally, it is striking that most computationally characterized deficits were explained by changes in the outcome parameter (reward or punishment magnitude), or the choice temperature, but not the

learning rate [25,43,51,66,79]. This leaves open the possibility that the effects of drugs and lesions reported here were not affecting learning processes per se, but biased other representations that had an impact on the learning plateau.

There are other limitations that are common to any drug or lesion study. Notably, pharmacological manipulations have tonic effects that only indirectly influence the phasic signals assumed to drive learning. Also, given the multiplicity of receptors and the complex interactions with genotypes, it is naïve to expect similar effects across subjects and across drugs that target the same neuromodulatory system. Nonetheless, characterizing deficits in computational terms might provide insight into pathological conditions and help predict the effects of treatment. For instance, PD patients who exhibit a strong increase in reward sensitivity following administration of a DA agonist might be at risk of developing an impulse control disorder [43] (see also Ref. [80] in the present volume). Computational analyses of learning abilities might also help us understand psychiatric diseases such as schizophrenia, which has been shown to impair processing of prediction errors and consequently lead to distorted representations of the environment [81].

References

[1] Barto AG, Sutton RS. Reinforcement learning: an introduction. Cambridge: MIT Press; 1998. http://dx.doi.org/10.1109/TNN.1998.712192.

[2] Rescorla RA, Wagner AR. A theory of pavlovian conditioning: variations in the effectiveness of reinforcement and nonreinforcement. In: Black AH, Prokasy WF, editors. Classical conditioning II: current research and theory. New York: Appleton-Century-Crofts; 1972. p. 64—99.

[3] Schultz W, Dayan P, Montague PR. A neural substrate of prediction and reward. Science 1997;275:1593—9. http://dx.doi.org/10.1126/science.275.5306.1593.

[4] Zaghloul KA, Blanco JA, Weidemann CT, McGill K, Jaggi JL, Baltuch GH, et al. Human substantia nigra neurons encode unexpected financial rewards. Science 2009;323:1496—9. http://dx.doi.org/10.1126/science.1167342.

[5] Fiorillo CD, Tobler PN, Schultz W. Discrete coding of reward probability and uncertainty by dopamine neurons. Science 2003;299:1898—902. http://dx.doi.org/10.1126/science.1077349.

[6] D'Ardenne K, McClure SM, Nystrom LE, Cohen JD. BOLD responses reflecting dopaminergic signals in the human ventral tegmental area. Science 2008;319:1264—7. http://dx.doi.org/10.1126/science.1150605.

[7] Haruno M, Kawato M. Different neural correlates of reward expectation and reward expectation error in the putamen and caudate nucleus during stimulus-action-reward association learning. J Neurophysiol 2006;95:948—59. http://dx.doi.org/10.1152/jn.00382.2005.

[8] O'Doherty JP, Dayan P, Schultz J, Deichmann R, Friston K, Dolan RJ. Dissociable roles of ventral and dorsal striatum in instrumental conditioning. Science 2004;304:452—4. http://dx.doi.org/10.1126/science.1094285.

[9] Palminteri S, Boraud T, Lafargue G, Dubois B, Pessiglione M. Brain hemispheres selectively track the expected value of contralateral options. J Neurosci 2009;29:13465—72. http://dx.doi.org/10.1523/JNEUROSCI.1500-09.2009.

[10] Bayer HM, Glimcher PW. Midbrain dopamine neurons encode a quantitative reward prediction error signal. Neuron 2005;47:129—41. http://dx.doi.org/10.1016/j.neuron.2005.05.020.

[11] Daw ND, Kakade S, Dayan P. Opponent interactions between serotonin and dopamine. Neural Networks 2002;15:603—16. http://dx.doi.org/10.1016/S0893-6080(02)00052-7.

[12] Guitart-Masip M, Duzel E, Dolan R, Dayan P. Action versus valence in decision making. Trends Cogn Sci 2014;18:194—202. http://dx.doi.org/10.1016/j.tics.2014.01.003. Elsevier Ltd.

[13] Seymour B, Maruyama M, De Martino B. When is a loss a loss? Excitatory and inhibitory processes in loss-related decision-making. Curr Opin Behav Sci 2015;5:122—7. http://dx.doi.org/10.1016/j.cobeha.2015.09.003.

[14] Pessiglione M, Delgado MR. The good, the bad and the brain: neural correlates of appetitive and aversive values underlying decision making. Curr Opin Behav Sci 2015;5:78—84. http://dx.doi.org/10.1016/j.cobeha.2015.08.006.

[15] Knutson B, Katovich K, Suri G. Inferring affect from fMRI data. Trends Cogn Sci 2014:1—7. http://dx.doi.org/10.1016/j.tics.2014.04.006. Elsevier Ltd.

[16] Schultz W. Electrophysiological correlates of reward processing in dopamine neurons. In: Léon T, Dreher J-C, editors. Decision neuroscience.

[17] Maia TV, Frank MJ. From reinforcement learning models to psychiatric and neurological disorders. Nat Neurosci 2011;14:154—62. http://dx.doi.org/10.1038/nn.2723. Nature Publishing Group.

[18] Matsumoto M, Hikosaka O. Lateral habenula as a source of negative reward signals in dopamine neurons. Nature 2007;447:1111—5. http://dx.doi.org/10.1038/nature05860.

[19] Lawson RP, Seymour B, Loh E, Lutti A, Dolan RJ, Dayan P, et al. The habenula encodes negative motivational value associated with primary punishment in humans. Proc Natl Acad Sci 2014. http://dx.doi.org/10.1073/pnas.1323586111.

[20] Brooks AM, Berns GS. Aversive stimuli and loss in the mesocorticolimbic dopamine system. Trends Cogn Sci 2013:1—6. http://dx.doi.org/10.1016/j.tics.2013.04.001. Elsevier Ltd.

[21] Matsumoto M, Hikosaka O. Two types of dopamine neuron distinctly convey positive and negative motivational signals. Nature 2009;459:837—41. http://dx.doi.org/10.1038/nature08028. Macmillan Publishers Limited. All rights reserved.

[22] Seymour B, Daw N, Dayan P, Singer T, Dolan R. Differential encoding of losses and gains in the human striatum. J Neurosci 2007;27:4826—31. http://dx.doi.org/10.1523/JNEUROSCI.0400-07.2007.

[23] Delgado MR, Li J, Schiller D, Phelps EA. The role of the striatum in aversive learning and aversive prediction errors. Philos Trans R Soc Lond B Biol Sci 2008;363:3787—800. http://dx.doi.org/10.1098/rstb.2008.0161.

[24] Pauli WM, Larsen T, Collette S, Tyszka JM, Seymour B, O'Doherty JP. Distinct contributions of ventromedial and dorsolateral subregions of the human substantia nigra to appetitive and aversive learning. J Neurosci 2015;35:14220—33. http://dx.doi.org/10.1523/JNEUROSCI.2277-15.2015.

[25] Pessiglione M, Seymour B, Flandin G, Dolan RJ, Frith CD. Dopamine-dependent prediction errors underpin reward-seeking behaviour in humans. Nature 2006;442:1042—5. http://dx.doi.org/10.1038/nature05051.

[26] Shenhav A, Buckner RL. Neural correlates of dueling affective reactions to win-win choices. Proc Natl Acad Sci USA 2014;111:10978—83. http://dx.doi.org/10.1073/pnas.1405725111.

[27] Soubrié P. Reconciling the role of central serotonin neurons in human and animal behavior. Behav Brain Sci 1986;9:319. http://dx.doi.org/10.1017/S0140525X00022871.

[28] Deakin JF, Graeff FG. 5-HT and mechanisms of defence. J Psychopharmacol 1991;5:305—15. http://dx.doi.org/10.1177/026988119100500414.

[29] Kapur S, Remington G. Serotonin-dopamine interaction and its relevance to schizophrenia. Am J Psychiatry 1996;153:466—76. http://dx.doi.org/10.1176/ajp.153.4.466.

[30] Lorrain DS, Riolo JV, Matuszewich L, Hull EM. Lateral hypothalamic serotonin inhibits nucleus accumbens dopamine: implications for sexual satiety. J Neurosci 1999;19:7648—52. Available: http://www.ncbi.nlm.nih.gov/pubmed/10460270.

[31] SB, Salzman CD. Appetitive and aversive systems in the amygdala. In: Tremblay L, Dreher J-C, editors. Decision neuroscience.

[32] Hayes DJ, Duncan NW, Xu J, Northoff G. A comparison of neural responses to appetitive and aversive stimuli in humans and other mammals. Neurosci Biobehav Rev 2014;45:350—68. http://dx.doi.org/10.1016/j.neubiorev.2014.06.018.

[33] Namburi P, Al-Hasani R, Calhoon GG, Bruchas MR, Tye KM. Architectural representation of valence in the limbic system. Neuropsychopharmacology 2015. http://dx.doi.org/10.1038/npp.2015.358.

[34] Bartra O, McGuire JT, Kable JW. The valuation system: a coordinate-based meta-analysis of BOLD fMRI experiments examining neural correlates of subjective value. Neuroimage 2013;76:412—27. http://dx.doi.org/10.1016/j.neuroimage.2013.02.063. Elsevier Inc.

[35] Garrison J, Erdeniz B, Done J. Prediction error in reinforcement learning: a meta-analysis of neuroimaging studies. Neurosci Biobehav Rev 2013:1—14. http://dx.doi.org/10.1016/j.neubiorev.2013.03.023. Elsevier Ltd.

[36] Yacubian J, Gläscher J, Schroeder K, Sommer T, Braus DF, Büchel C. Dissociable systems for gain- and loss-related value predictions and errors of prediction in the human brain. J Neurosci 2006;26:9530—7. http://dx.doi.org/10.1523/JNEUROSCI.2915-06.2006.

[37] Büchel C, Dolan RJ. Classical fear conditioning in functional neuroimaging. Curr Opin Neurobiol 2000;10:219—23. http://dx.doi.org/10.1016/S0959-4388(00)00078-7.

[38] Frank MJ, Seeberger LC, Reilly RCO, O'Reilly RC. By carrot or by stick: cognitive reinforcement learning in parkinsonism. Science 2004;306:1940—3. http://dx.doi.org/10.1126/science.1102941.

[39] Samii A, Nutt JG, Ransom BR. Parkinson's disease. Lancet 2004;363:1783—93. http://dx.doi.org/10.1016/S0140-6736(04)16305-8.

[40] Frank MJ. Hold your horses: a dynamic computational role for the subthalamic nucleus in decision making. Neural Netw 2006;19:1120—36. http://dx.doi.org/10.1016/j.neunet.2006.03.006.

[41] Kéri S, Moustafa Aa, Myers CE, Benedek G, Gluck MA. α-Synuclein gene duplication impairs reward learning. Proc Natl Acad Sci USA 2010;107:15992—4. http://dx.doi.org/10.1073/pnas.1006068107. www.pnas.org/cgi/doi/10.1073/pnas.1006068107/-/DCSupplemental.

[42] Bódi N, Kéri S, Nagy H, Moustafa A, Myers CE, Daw N, et al. Reward-learning and the novelty-seeking personality: a between-and within-subjects study of the effects of dopamine agonists on young Parkinsons patients. Brain 2009;132:2385—95. http://dx.doi.org/10.1093/brain/awp094.

[43] Voon V, Pessiglione M, Brezing C, Gallea C, Fernandez HH, Dolan RJ, et al. Mechanisms underlying dopamine-mediated reward bias in compulsive behaviors. Neuron 2010;65:135—42. http://dx.doi.org/10.1016/j.neuron.2009.12.027. Elsevier Inc.

[44] Klein TA, Neumann J, Reuter M, Hennig J, von Cramon DY, Ullsperger M. Genetically determined differences in learning from errors. Science 2007;318:1642—5. http://dx.doi.org/10.1126/science.1145044. American Association for the Advancement of Science.

[45] Frank MJ, Moustafa AA, Haughey HM, Curran T, Hutchison KE. Genetic triple dissociation reveals multiple roles for dopamine in reinforcement learning. Proc Natl Acad Sci USA 2007;104: 16311—6. http://dx.doi.org/10.1073/pnas.0706111104.

[46] Pessiglione M, Petrovic P, Daunizeau J, Palminteri S, Dolan RJ, Frith CD. Subliminal instrumental conditioning demonstrated in the human brain. Neuron 2008;59:561—7. http://dx.doi.org/10.1016/j.neuron.2008.07.005.

[47] Palminteri S, Lebreton M, Worbe Y, Grabli D, Hartmann A, Pessiglione M. Pharmacological modulation of subliminal learning in Parkinson's and Tourette's syndromes. Proc Natl Acad Sci USA 2009;106:19179—84. http://dx.doi.org/10.1073/pnas.0904035106.

[48] Leckman JF. Tourette's syndrome. Lancet 2002;360:1577—86. http://dx.doi.org/10.1016/S0140-6736(02)11526-1.

[49] Rutledge RB, Lazzaro SC, Lau B, Myers CE, Gluck MA, Glimcher PW. Dopaminergic drugs modulate learning rates and perseveration in Parkinson's patients in a dynamic foraging task. J Neurosci 2009;29:15104—14. http://dx.doi.org/10.1523/JNEUROSCI.3524-09.2009.

[50] Jocham G, Klein Ta, Ullsperger M. Dopamine-mediated reinforcement learning signals in the striatum and ventromedial prefrontal cortex underlie value-based choices. J Neurosci 2011;31:1606—13. http://dx.doi.org/10.1523/JNEUROSCI.3904-10.2011.

[51] Eisenegger C, Naef M, Linssen A, Clark L, Gandamaneni PK, Mu U. Role of dopamine D2 receptors in human reinforcement learning. 2014. p. 1—10. http://dx.doi.org/10.1038/npp.2014.84.

[52] Walker FO. Huntington's disease. Semin Neurol 2007;27:143—50. http://dx.doi.org/10.1055/s-2007-971176.

[53] Douaud G, Gaura V, Ribeiro M-J, Lethimonnier F, Maroy R, Verny C, et al. Distribution of grey matter atrophy in Huntington's disease patients: a combined ROI-based and voxel-based morphometric study. Neuroimage 2006;32:1562—75. http://dx.doi.org/10.1016/j.neuroimage.2006.05.057.

[54] den Ouden HEM, Daw ND, Fernandez G, Elshout Ja, Rijpkema M, Hoogman M, et al. Dissociable effects of dopamine and serotonin on reversal learning. Neuron 2013;80:1090—100. http://dx.doi.org/10.1016/j.neuron.2013.08.030.

[55] Guitart-Masip M, Economides M, Huys QJM, Frank MJ, Chowdhury R, Duzel E, et al. Differential, but not opponent, effects of l-DOPA and citalopram on action learning with reward and punishment. Psychopharmacol (Berl) 2014;231:955—66. http://dx.doi.org/10.1007/s00213-013-3313-4.

[56] Cools R, Robinson OJ, Sahakian B. Acute tryptophan depletion in healthy volunteers enhances punishment prediction but does not affect reward prediction. Neuropsychopharmacology 2008;33: 2291—9. http://dx.doi.org/10.1038/sj.npp.1301598.

[57] Seymour B, Daw NDN, Roiser JPJ, Dayan P, Dolan R. Serotonin selectively modulates reward value in human decision-making. J Neurosci 2012;32:5833—42. http://dx.doi.org/10.1523/JNEUROSCI.0053-12.2012.

[58] Palminteri S, Clair A-HH, Mallet L, Pessiglione M. Similar improvement of reward and punishment learning by serotonin reuptake inhibitors in obsessive-compulsive disorder. Biol Psychiatry 2012;72:244—50. http://dx.doi.org/10.1016/j.biopsych.2011.12.028. Elsevier Inc.

[59] Crockett MJ, Clark L, Robbins TW. Reconciling the role of serotonin in behavioral inhibition and aversion: acute tryptophan depletion abolishes punishment-induced inhibition in humans. J Neurosci 2009;29:11993—9. http://dx.doi.org/10.1523/JNEUROSCI.2513-09.2009.

[60] Boureau Y-L, Dayan P. Opponency revisited: competition and cooperation between dopamine and serotonin. Neuropsychopharmacology 2011;36:74—97. http://dx.doi.org/10.1038/npp.2010.151. Nature Publishing Group.

[61] Tanaka SC, Schweighofer N, Asahi S, Shishida K, Okamoto Y, Yamawaki S, et al. Serotonin differentially regulates short- and long-term prediction of rewards in the ventral and dorsal striatum. PLoS One 2007;2:e1333. http://dx.doi.org/10.1371/journal.pone.0001333.

[62] Schweighofer N, Bertin M, Shishida K, Okamoto Y, Tanaka SC, Yamawaki S, et al. Low-serotonin levels increase delayed reward discounting in humans. J Neurosci 2008;28:4528—32. http://dx.doi.org/10.1523/JNEUROSCI.4982-07.2008.

[63] Spies M, Knudsen GM, Lanzenberger R, Kasper S. The serotonin transporter in psychiatric disorders: insights from PET imaging. The Lancet Psychiatry 2015;2:743—55. http://dx.doi.org/10.1016/S2215-0366(15)00232-1.

[64] Bechara A, Tranel D, Damasio H, Adolphs R, Rockland C, Damasio A. Double dissociation of conditioning and declarative knowledge relative to the amygdala and hippocampus in humans. Science 1995;269:1115—8. http://dx.doi.org/10.1126/science.7652558.

[65] De Martino B, Camerer CF, Adolphs R. Amygdala damage eliminates monetary loss aversion. Proc Natl Acad Sci USA 2010;107: 3788—92. http://dx.doi.org/10.1073/pnas.0910230107.

[66] Palminteri S, Justo D, Jauffret C, Pavlicek B, Dauta A, Delmaire C, et al. Critical roles for anterior insula and dorsal striatum in punishment-based avoidance learning. Neuron 2012;76: 998—1009. http://dx.doi.org/10.1016/j.neuron.2012.10.017.

[67] Kim H, Shimojo S, O'Doherty JP. Is avoiding an aversive outcome rewarding? Neural substrates of avoidance learning in the human brain. PLoS Biol 2006;4:e233. http://dx.doi.org/10.1371/journal.pbio.0040233.

[68] Seymour B, O'Doherty JP, Dayan P, Koltzenburg M, Jones AK, Dolan RJ, et al. Temporal difference models describe higher-order learning in humans. Nature 2004;429:664—7. http://dx.doi.org/10.1038/nature02636.1.

[69] Skvortsova V, Palminteri S, Pessiglione M. Learning to minimize efforts versus maximizing rewards: computational principles and neural correlates. J Neurosci 2014;34:15621—30. http://dx.doi.org/10.1523/JNEUROSCI.1350-14.2014.

[70] Tom SM, Fox CR, Trepel C, Poldrack RA. The neural basis of loss aversion in decision-making under risk. Science 2007;315:515—8. http://dx.doi.org/10.1126/science.1134239.

[71] Plassmann H, O'Doherty JP, Rangel A. Appetitive and aversive goal values are encoded in the medial orbitofrontal cortex at the time of decision making. J Neurosci 2010;30:10799—808. http://dx.doi.org/10.1523/JNEUROSCI.0788-10.2010.

[72] Wheeler EZ, Fellows LK. The human ventromedial frontal lobe is critical for learning from negative feedback. Brain 2008;131: 1323—31. http://dx.doi.org/10.1093/brain/awn041.

[73] Camille N, Griffiths Ca, Vo K, Fellows LK, Kable JW. Ventromedial frontal lobe damage disrupts value maximization in humans. J Neurosci 2011;31:7527—32. http://dx.doi.org/10.1523/JNEUROSCI.6527-10.2011.

[74] Daw ND. Advanced reinforcement learning. In: Glimcher PW, Fehr E, editors. Neuroeconomics decis mak brain. 2nd ed. London, UK: Neuroecono. Academic Press; 2013. p. 299—320. http://dx.doi.org/10.1016/B978-0-12-416008-8.00016-4.

[75] Dayan P. Twenty-five lessons from computational neuromodulation. Neuron 2012;76:240—56. http://dx.doi.org/10.1016/j.neuron.2012.09.027. Elsevier Inc.

[76] Vlaev I, Chater N, Stewart N, Brown GD. Does the brain calculate value? Trends Cogn Sci 2011;15:546—54. http://dx.doi.org/10.1016/j.tics.2011.09.008. Elsevier Ltd.

[77] Seymour B, McClure SM. Anchors, scales and the relative coding of value in the brain. Curr Opin Neurobiol 2008;18:173—8. http://dx.doi.org/10.1016/j.conb.2008.07.010.

[78] Rangel A, Clithero Ja. Value normalization in decision making: theory and evidence. Curr Opin Neurobiol 2012;22:970—81. http://dx.doi.org/10.1016/j.conb.2012.07.011. Elsevier Ltd.

[79] Shiner T, Seymour B, Wunderlich K, Hill C, Bhatia KP, Dayan P, et al. Dopamine and performance in a reinforcement learning task: evidence from Parkinson's disease. Brain 2012;135: 1871−83. http://dx.doi.org/10.1093/brain/aws083.

[80] Voon V. Decision making and impulse control disorders in Parkinson's disease. In: Tremblay L, Dreher J-C, editors. Decision neuroscience.

[81] Fletcher PC, Frith CD. Perceiving is believing: a Bayesian approach to explaining the positive symptoms of schizophrenia. Nat Rev Neurosci 2009;10:48−58. http://dx.doi.org/10.1038/nrn2536.

[82] Hull CL. Principles of behavior: an introduction to behavior theory [internet]. 1943. Available: https://books.google.fr/books/about/Principles_of_Behavior.html?id=6WB9AAAAMAAJ&pgis=1.

[83] Watkins CJCH, Dayan P. Q-learning. Mach Learn 1992;8:279−92. http://dx.doi.org/10.1007/BF00992698.

[84] Luce RD. Individual choice behavior: a theoretical analysis [internet]. Courier Corporation; 1959. Available: https://books.google.com/books?hl=en&lr=&id=D74qAwAAQBAJ&pgis=1.

[85] O'Doherty JP, Hampton A, Kim H. Model-based fMRI and its application to reward learning and decision making. Ann NY Acad Sci 2007;1104:35−53. http://dx.doi.org/10.1196/annals.1390.022.

[86] Montague PR, Dolan RJ, Friston KJ, Dayan P. Computational psychiatry. Trends Cogn Sci 2012;16:72−80. http://dx.doi.org/10.1016/j.tics.2011.11.018.

[87] Palminteri S, Khamassi M, Joffily M, Coricelli G. Contextual modulation of value signals in reward and punishment learning. Nat Commun 2015;6:8096. http://dx.doi.org/10.1038/ncomms9096. Nature Publishing Group.

[88] Guitart-Masip M, Fuentemilla L, Bach DR, Huys QJM, Dayan P, Dolan RJ, et al. Action dominates valence in anticipatory representations in the human striatum and dopaminergic midbrain. J Neurosci 2011;31:7867−75. http://dx.doi.org/10.1523/JNEUROSCI.6376-10.2011.

[89] Huys QJM, Cools R, Gölzer M, Friedel E, Heinz A, Dolan RJ, et al. Disentangling the roles of approach, activation and valence in instrumental and pavlovian responding. PLoS Comput Biol 2011;7:e1002028. http://dx.doi.org/10.1371/journal.pcbi.1002028.

[90] Yarkoni T, Poldrack Ra, Nichols TE, Van Essen DC, Wager TD. Large-scale automated synthesis of human functional neuroimaging data. Nat Methods 2011;8:665−70. http://dx.doi.org/10.1038/nmeth.1635.

[91] Singer HS. Tourette's syndrome: from behaviour to biology. Lancet Neurol 2005;4:149−59. http://dx.doi.org/10.1016/S1474-4422(05)01012-4.

[92] Palminteri S, Pessiglione M. Reinforcement learning and tourette syndrome [internet]. In: International review of neurobiology. 1st ed. Elsevier Inc.; 2013. http://dx.doi.org/10.1016/B978-0-12-411546-0.00005-6.

24

Decision-Making and Impulse Control Disorders in Parkinson's Disease

V. Voon[1,2]

[1]University of Cambridge, Cambridge, United Kingdom; [2]Cambridgeshire and Peterborough NHS Foundation Trust, Cambridge, United Kingdom

Abstract

Impulse control disorders associated with dopaminergic medications are common. Factors including elevated smoking and alcohol use, novelty seeking, impulsivity, depression, and anxiety suggest commonalities with other addictions. The parkinsonian lesion enhances the gain associated with levodopa, reinforcing properties of dopaminergic medications, and enhances delay discounting. Lower striatal dopamine transporter levels preceding medication exposure and decreased midbrain D2 autoreceptor sensitivity may underlie enhanced ventral striatal dopamine release and activity in response to salient reward cues, anticipated and unexpected rewards, and gambling tasks. Evidence supports enhanced learning from reward feedback, with a study highlighting reliance on a ventral striatal critic model of stimulus value with impaired learning from negative prediction error. Impairments in decisional impulsivity (delay discounting, reflection impulsivity, and risk taking) implicate the ventral striatum, orbitofrontal cortex, anterior insula, and dorsal cingulate. These findings provide insight into the role of dopamine on decision-making processes in addictions and potential therapeutic targets.

INTRODUCTION

Impulse control disorders (ICDs), or behavioral addictions, can emerge from chronic dopaminergic medications used in the treatment of Parkinson's disease (PD) [1,2]. The behaviors, including pathological gambling (PG), compulsive shopping or sexual behaviors, and binge eating, are common, reported in the multicenter DOMINION study at 17.1% of those on dopaminergic medications [2]. The behaviors are most commonly associated with D2/3-preferring dopamine (DA) agonists, which stimulate DA receptors and are independently associated with levodopa, a DA precursor [2], and amantadine, a dopaminergic and glutamatergic modulator [3]. Understanding mechanisms underlying these behaviors may shed light on the role of DA in behavioral and drug addictions in the general population and guide therapeutic targets. This chapter focuses on underlying cognitive mechanisms that might contribute to ICDs in the context of dopaminergic medications. Cognitive mechanisms represent an endophenotype mediating neural function and behavior. This chapter first considers broadly the role of dopaminergic medications, individual vulnerability and PD, and striatal dopaminergic activity. It then reviews the reinforcing role of dopaminergic medications, learning from feedback, risk taking, and decisional impulsivity.

THE ROLE OF DOPAMINERGIC MEDICATIONS AND INDIVIDUAL VULNERABILITY IN PARKINSON'S DISEASE

ICDs are hypothesized to be an interaction between dopaminergic medications and an underlying vulnerability, which may be an individual susceptibility to addiction or may be PD related [1] (Fig. 24.1). These behaviors occur only in a subset exposed to DA agonists (14% on DA agonists, odds ratio 2.72; 7.2% on levodopa, odds ratio 1.51) [2], suggesting a likely underlying vulnerability. This vulnerability includes an individual susceptibility to ICDs, which may include personal or family history of addictions, temperament, or cognitive, neural, genetic, or molecular differences.

Identified risk factors for ICDs in PD include current smoking, family history of gambling, young age, being unmarried, depression, anxiety, higher novelty seeking,

Decision Neuroscience
http://dx.doi.org/10.1016/B978-0-12-805308-9.00024-5

FIGURE 24.1   Impulse control disorders: interaction of dopaminergic medications with Parkinson's disease and individual vulnerability. *autoR*, autoreceptor; *DA*, dopamine agonists; *Dec.*, decreased; *FHx*, family history; *ICDs*, impulse control disorders.

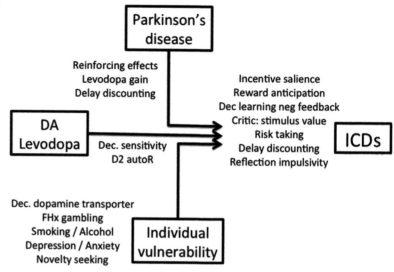

and impulsivity [2,4]. A small prospective study showed that ICDs were associated with greater caffeine use, cigarette smoking, and a trend toward higher alcohol use [5]. These factors have marked similarities to factors observed in disorders of addiction or PG, arguing for shared underlying neurobiological similarities [6]. Other investigated neurobiological factors (e.g., DA transporter striatal density, neural substrates, molecular signaling, or genetic factors) may also act as predictive factors.

The role of PD is also beginning to emerge. In PD, neurodegeneration of the substantia nigra pars compacta (SNpc) dopaminergic cells projecting to the dorsal striatum can affect up to 70% of dopaminergic cell bodies prior to the onset of parkinsonian motor symptoms, with mesial limbic neurodegeneration of the ventral tegmental area (VTA) projecting to the ventral striatum being much more variable. Thus, with dopaminergic therapy, greater preservation of dopaminergic projections to ventral limbic relative to dorsal motor regions may result in an "overdose" of otherwise intact limbic regions. The overdose hypothesis suggests that functioning of cognition or behavior follows a U-shaped curve with optimal functioning occurring at an optimal dopaminergic level with either higher or lower levels resulting in impairments [1,7–9]. Other considerations for PD include neurodegeneration affecting serotonergic and noradrenergic systems relevant to cognitive mechanisms [10], and also whether specific subtypes of PD might contribute to these symptoms.

Excessive ventral striatal DA release and activity has been demonstrated in PD patients with ICDs in specific contexts. Both PD subjects with single and those with multiple ICDs have enhanced ventral striatal DA release in response to reward-related visual stimuli [11,12] and

to a card gambling task [13]. Using an ecologically valid gambling task and [^{11}C]raclopride PET, there were no group differences in PD patients with PG, although enhanced DA release correlated with gambling symptom severity [14]. Unlike PD subjects with compulsive levodopa use, which show enhanced ventral striatal DA release to levodopa challenge [15], those with ICDs did not show a group difference in striatal DA release to a levodopa challenge [11]. PD patients with hypersexuality also show enhanced blood oxygen level-dependent (BOLD) activity specifically in response to sexual cues and not other rewards in the medial orbitofrontal cortex (OFC), ventral striatum, anterior cingulate, and hypothalamus [16]. PD + ICD subjects on DA agonists also have enhanced BOLD ventral striatal activity in response to expected and unexpected rewards [17]. Studies further implicate the OFC and cingulate: PD + ICD subjects show enhanced [^{18}F]fluorodopa uptake in the medial OFC, suggesting enhanced monoaminergic activity at baseline [18]. During a card gambling task and ^{15}H$_2$O PET, PD + PG subjects showed inhibition of activity with apomorphine challenge in the lateral OFC, rostral cingulate, amygdala, and external pallidum [19].

This enhanced striatal DA may be related to either decreased synaptic DA clearance or enhanced release. Emerging evidence suggests a role for lower striatal DA transporter (DAT) levels in PD patients with ICDs [20,21] that appear to predate and thus predict the onset of ICDs [22] prior to the onset of dopaminergic medications [23]. DA reuptake via the DAT is the primary mechanism by which striatal DA is removed from the synaptic cleft and DA neurotransmission regulated and terminated. Lower DAT levels would thus result in an enhancement of physiological phasic or tonic DA at the synaptic cleft due to impaired clearance. Alternatively, preliminary evidence in a small PD + ICD sample using [11C]FLB-457 PET showed decreased midbrain D2/D3 autoreceptor sensitivity, which may also enhance DA release [24]. Although one [11C]raclopride PET study suggested lower striatal D2/3 receptor levels during a motor control task [13], no evidence of lower striatal D2/3 receptor levels at baseline have been observed in subsequent studies [11,12]. These findings converge with the lack of difference in striatal D2/3 receptor levels in PG in the general population [14,25] and contrast with decreased D2/3 receptor levels observed in drug addiction in the general population.

DA signaling can act on either a tonic or a phasic level [26–29]. Converging evidence shows that phasic DA acts as prediction errors or a reinforcement signal to guide motivated behaviors [30], increasing with novel and unexpected rewards (positive prediction error). With associative learning, the phasic DA signal becomes associated with the cue predicting reward. The patterns of phasic DA on novelty and reward are dissociable: DA activity scales as a function of reward magnitude but is suggested to be coded in an absolute manner with novelty [31]. Unexpected loss or lack of a reward is associated with a pause or cessation of DA activity (negative prediction error). Tonic DA has been hypothesized to play several functions: in rodents, tonic DA increases with upcoming rewards, thus playing a role in reward anticipation and motivation [32], to represent an average reward signal over time relevant to opportunity cost and motivation [33] and to play a role during high uncertainty [27].

D1 and D2 receptors in the dorsal striatal system are segregated into the direct and indirect pathways involved in facilitating and inhibiting action selection, respectively. Phasic DA is proposed to promote learning from positive outcomes via D1 receptors of the "go" pathway and promote learning from negative outcomes via D2 receptors of the "no-go" pathway [34]. High-affinity D2 receptors may be more sensitive to low tonic activity and transient pauses associated with negative prediction errors, whereas both low-affinity D1 receptors and D2 receptors may be sensitive to phasic bursts with large increases in synaptic DA associated with positive prediction errors.

In rodent parkinsonian models, levodopa exposure results in aberrant expression of D3 receptors in the denervated dorsal striatum [35]. However, the role of D3 receptors in PD subjects with ICD behaviors is less clear. The role of the D3 receptor was initially raised given the observation of ICDs associated with pramipexole, a D3-preferring DA agonist [36–38], and has been highlighted in the recent FDA Adverse Event Reporting System reports emphasizing pramipexole and ropinirole and the D3 partial agonist aripiprazole

in ICDs [39]. A 2015 review described a trend showing that a proportion of reported ICD behaviors in PD is related to the relative D3 receptor selectivity of DA agonists [40]. The role of D3 receptors in PG in the general population has been emphasized. Although there were no differences in D2 or D3 levels in the striatum or substantia nigra in PG in the general population compared to matched controls, D3 receptor levels correlated with gambling severity symptoms [41]. In [11C]PHNO PET imaging, PG in the general population is associated with enhanced amphetamine-induced dorsal striatal DA release [42] with the extent of DA release predicted by substantia nigral D3 receptor levels and related to gambling severity. However, the role of D3 receptors in PD + ICD subjects is not yet clear, as a study showed lower [11C]PHNO ventral striatal binding, probably related to enhanced DA release.

The chronicity of dopaminergic medication exposure may also play an influential role. Whereas acute pramipexole decreases the dopaminergic mean firing rate in rodents by acting on D2 autoreceptors, subchronic pramipexole normalizes activity, although burst firing remains slightly decreased below baseline [43]. How variability in DA signaling might influence these findings in ICDs remains to be investigated. Similarly, chronic levodopa in 6-hydroxydopamine (6-OHDA)-depleted rodents influences the gain associated with DA activity by enhancing the proportion of spontaneously active DA neurons or those capable of phasic activity in response to a salient stimulus [44].

REINFORCING EFFECTS AND ASSOCIATIVE LEARNING

Both DA agonists and levodopa demonstrate reinforcing effects in rodent models. In rodents, DA agonists act similar to stimulants promoting conditioned place preference [45]. Rodents with 6-OHDA injected into the dorsolateral striatum required lower pramipexole doses to achieve the same conditioned place preference effect as sham, suggesting that the parkinsonian lesion may enhance the rewarding effects of pramipexole. The 6-OHDA dorsolateral striatal lesion results in selective loss of dorsolateral dopaminergic terminals and a significant decrease in SNpc dopaminergic cell bodies. However, with 6-OHDA injected into the rodent midbrain with loss of dopaminergic SNpc (−51%) and VTA (−31%) neurons, there were no differences as a function of the parkinsonian lesion in the reinforcing strength of pramipexole tested using a progressive ratio. The reinforcing strength of pramipexole across both lesioned and control rodents was moderate (5–22 lever presses for one infusion of pramipexole) relative to the breakpoints for cocaine (100) [46]. Pramipexole was associated with ΔFosB expression in ventro- and dorso-medial striatum, with a positive correlation observed between pramipexole motivation on the progressive ratio task and nucleus accumbens ΔFosB expression.

In a rodent model using viral-mediated α-synuclein accumulation in the SNpc, levodopa induced conditioned place preference in the lesioned rodents but not in the controls, hypothesized to reflect supersensitivity of postsynaptic DA receptors [47]. Levodopa also decreased sweetened water consumption in lesioned rodents, similar to the effect of psychostimulants on nondrug rewards. These findings in levodopa converge with observations of compulsive medication use in 3–4% of PD patients on levodopa and apomorphine [48] and that ICDs are associated with the presence of levodopa and levodopa dose [2].

These reports are also consistent with human studies of PD patients with ICDs and dopaminergic activity in response to salient cues. In [^{11}C]raclopride PET imaging, PD patients with mixed ICDs had heightened striatal DA release to heterogeneous reward-related visual cues [11]. Similarly, in fMRI, PD patients with hypersexuality had greater ventral striatal activity to sexual cues, an effect that correlated with subjective sexual desire of wanting but not liking [16], suggested to support an incentive salience process. Similarly, greater ventral striatal activity to gambling-related cues was demonstrated in a small fMRI study in PD patients with ICDs [49]. These studies are consistent with enhanced striatal DA release to drug cues in drug dependence studies [50].

A role for novelty has also been observed, with enhanced novelty seeking in PD patients particularly with compulsive shopping, with a trend for PG but not compulsive sexual activity or binge eating [4]. Similarly, PD + ICD subjects were shown to prefer novel stimuli on a probabilistic learning task [51].

LEARNING FROM FEEDBACK

A compelling hypothesis predicts that exogenous DA may enhance learning from reward feedback and impair learning from negative feedback [52]. Tonic stimulation of D2 receptors has been hypothesized to impair the detection of negative prediction errors or pauses in DA signaling [53]. Unmedicated PD subjects without ICDs show impaired learning from reward outcomes on a probabilistic classification task, which normalized after chronic DA agonists. In contrast, learning from loss feedback was intact in patients that were unmedicated and impaired with chronic DA agonist exposure [54].

The role of learning from reward and loss feedback in PD + ICD subjects is less clear. In a two-choice probabilistic discrimination task, PD + ICD subjects on levodopa were shown to be better at learning from loss feedback relative to reward, with the opposite observed when off medications relative to those without ICDs [55]. Using emotional facial stimuli in a similar task design, the authors replicated these findings [56]. Thus, these two studies suggest that in PD patients with ICDs, dopaminergic medications were associated with better learning from negative feedback relative to reward on a difficult ($p = .60$ to $.75$) probabilistic discrimination task.

In contrast, two separate studies have shown convergent behavioral findings. In a two-choice probabilistic discrimination task PD + ICD subjects tested on DA agonists showed better learning (optimal choices and calculated learning rate) from reward feedback compared to when off medications, whereas PD controls on DA agonists were slower to learn from loss feedback [17] (Fig. 24.2). Gain and loss feedback were separately assessed with easier probabilities of $p = .80$. Using a Q-learning algorithm based on action values, PD + ICD subjects on DA agonists showed greater ventral striatal activity in response to positive prediction error and expected reward compared to those off, with the opposite in PD controls. Greater posterior putaminal activity to expected rewards was also observed in PD + ICD subjects, possibly consistent with a habit-learning hypothesis. In PD controls on medication, greater striatal and anterior insular activity to negative prediction error and expected loss was observed. This study thus emphasized the influence of DA agonists on learning from reward feedback in PD subjects with ICDs while impairing learning from negative feedback in PD controls.

In a 2014 study, similar findings were reported in the reward domain with divergent findings in the loss domain. In a probabilistic classification task, PD + ICD subjects tested on medications were better at reward learning (more optimal choices) relative to those off or to PD controls and were also worse at punishment learning relative to healthy controls [57] (Fig. 24.2). PD controls off medications were impaired at reward learning and better at punishment learning relative to the other three groups. Subjects decided whether stimuli belonged to category A or B based on learning from probabilistic reward (+25 points) or punishment (−25 points) feedback ($p = .80$).

The authors then assessed model fits of reinforcement learning algorithms. The DA overdose hypothesis suggests that dopaminergic medications may "overdose" intact VTA dopaminergic projections [8]. This theory also has implications for actor–critic models of reinforcement learning, which differentially model ventral and dorsal striatal function, respectively [58–60]. The actor–critic model suggests that the ventral striatal critic uses prediction error to learn stimulus value to update

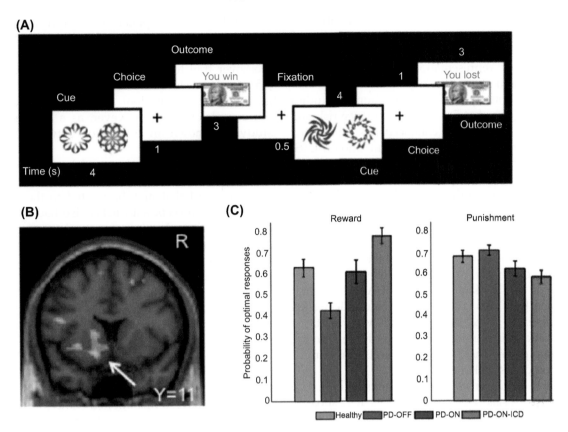

FIGURE 24.2 **Reinforcement learning**. (A) Probabilistic choice task and (B) enhanced ventral striatal activity in response to anticipated value in Parkinson patients (PD) with impulse control disorders (ICD) on dopamine agonists [17]. (C) Enhanced learning from reward outcomes in PD ICD patients on medications (PD-ON-ICD) relative to PD and healthy controls (left) and impaired learning from punishment outcomes in PD-ON-ICD relative to healthy controls (right) [57].

expected future rewards, whereas the dorsal striatal actor uses the prediction error signal to encode action valuation and selection leading to rewards. The study addressed model fits of the Q-learning framework, which computes prediction error based on the value of the action, or the actor–critic framework, in which the critic calculates prediction error based on the stimulus value independent of the selected action and the actor computes action values and selects the appropriate action based on the actor's Q values. The data showed a main effect of feedback and no main effect of optimality of action; the former is consistent with either learning strategy, whereas the latter is consistent with an actor–critic strategy. Thus the authors show that the actor–critic strategy had a better model fit than the Q-learning strategy. Using the actor–critic algorithm, PD + ICD subjects on medications showed greater reliance on a ventral striatal critic model based on stimulus value with particular impairments in learning from negative prediction errors. In contrast, PD controls on medication were more reliant on a dorsal striatal actor model based on action values with higher learning rates for positive prediction error.

In summary, the evidence for learning from reward and loss feedback is somewhat mixed. Two studies show that PD + ICD subjects on dopaminergic medications learn better from losses relative to rewards. In contrast, using differing and easier tasks (higher probability and either separate reward and loss conditions or classification), two studies showed better learning from positive feedback, with one also showing impaired learning from negative feedback. The latter study further extends the computational analyses to show that PD + ICD patients are reliant on a ventral striatal critic model with impaired learning from negative prediction error, whereas PD controls are reliant on a dorsal striatal actor model with improved learning from positive prediction error.

RISK AND UNCERTAINTY

Pathological behavioral choices are associated with decisions between anticipating a positive reward (the behavior) and the negative financial, social, and occupational consequences with either known (risk) or unknown probabilities (ambiguity). The evaluation of risk involves the representation of both anticipated reward and loss value, the integration of the two, and the calculation or representation of probability and, possibly, learning from feedback.

Both a rodent study and multiple human studies converge, suggesting that DA agonists enhance risk taking. In rodents, pramipexole enhanced risk taking [61] in a study using intracranial self-stimulation (ICSS) implanted to stimulate the medial forebrain bundle "reward" pathway as the positive reinforcer [61]. ICSS acts as a potent reinforcer that is rapidly acquired relative to food reinforcement with stable performance and without issues of satiety. Pramipexole enhanced risk taking with greater preference for the larger risky reinforcer (90% of effective current) at low probabilities compared to the small certain reinforcer (40% of effective current) regardless of the parkinsonian 6-OHDA dorsolateral striatal lesion [62]. This risk-seeking behavior decreased with pramipexole discontinuation and was reinstated with reinitiation of pramipexole. Unlike the reinforcing effect of pramipexole, there is no clear effect of the parkinsonian lesion on risk-taking behaviors.

In humans, DA agonists also increase risk taking. DA agonists enhance risk taking in both PD + ICD subjects and PD controls relative to healthy volunteers, with the PG subgroup showing greater risk taking than PD controls [55]. In a task selecting between sure and risky choices in separate gain and loss conditions, PD + ICD subjects on DA agonists showed increased risk taking relative to PD controls, particularly in response to gain but not loss anticipation. That this task did not involve feedback suggests an effect of anticipation of gain. Similarly PD + ICD subjects on medications compared to PD controls show greater risk taking in the Balloon Analogue Risk Task (BART) across both high- and low-risk conditions. The BART, in which subjects pump up a balloon, accumulating reward with an increasing likelihood of the balloon bursting, measures risk taking under ambiguity and with feedback [63]. Both groups similarly decreased risk taking following negative feedback without an effect of ICD status or DA agonist.

PD subjects on DA agonists compared to off had lower ventral striatal activity in the Gamble Risk (i.e., the difference between possible gain and loss outcomes controlled for expected value) [55] and in an arterial spin-labeling study using the BART [64]. The ventral striatum both encodes risk probability and in a bidirectional manner represents the anticipation of gain and loss values. These findings focusing on the Gamble Risk suggest that DA agonists might impair the representation of risk. PD + ICD subjects also had lower OFC and anterior insular activity or lower correlation of the Gamble Risk to the gain relative to the loss condition.

Risk taking also appears to be influenced by therapeutic interventions. Amantadine, a weak DA and glutamatergic modulator, was shown in a randomized crossover trial to be effective for PG symptoms in PD [65] and also to decrease risk taking in those with gambling symptoms [66]. The role of amantadine is not completely clear, however, as a large multicenter study has shown an association between amantadine and ICDs [3].

Put together, both rodent and human studies show that DA agonists enhance risk taking in PD + ICD subjects, an effect that may be related to the anticipation of gain. PD subjects with ICDs also had lower ventral striatal activity when on compared to off medications under both risk and ambiguous conditions. Amantadine appears to decrease risk taking in PD subjects with gambling behaviors.

IMPULSIVITY

Emerging evidence suggests that PD + ICD subjects have impairments in decisional impulsivity but not motor impulsivity. Impulsivity is the tendency toward rapid ill-considered disinhibited choices and is a heterogeneous construct with subtypes associated with distinct but overlapping neural substrates. Impulsivity can be broadly divided into decisional forms, including delay discounting (preference of a small immediate over a larger delayed reward), reflection impulsivity (rapid decision-making), and risk taking, and motor forms, including response inhibition (inhibition of a prepotent response) [10].

Delay discounting, also known as impulsive or intertemporal choice, has been consistently shown to be impaired in PD + ICD subjects compared to PD controls. Delay discounting involves the choice between an immediate small reward and a larger delayed reward. There are several processes that contribute to delay discounting, including the incentive value of the immediate choice, the process of waiting, delay aversion, marginal sensitivity, uncertainty avoidance, and timing deficits.

Both PD itself and DA agonists appear to have independent effects in enhancing delay discounting. In an ICSS model, rodents with 6-OHDA dorsolateral striatal lesions compared to sham showed a preference for the smaller (50 Hz) immediate reinforcer relative to the larger delayed (160 Hz delayed by 5–15 s) reinforcer [67]. There were no differences in reward sensitivity or evidence for habitual responding. These findings converge with human studies in which PD controls showed elevated delay discounting regardless of medication status [68]. Never-medicated PD patients also show higher discounting rates relative to healthy controls, which normalize with dopaminergic medications [69]. Pramipexole enhanced delay discounting in PD controls associated with greater medial prefrontal cortex and posterior cingulate activity and lower ventral striatal activity [70].

Converging studies show that PD + ICD subjects have enhanced delay discounting relative to PD controls while on, relative to off, medications [71] and while on medications compared to PD controls on medications [72]. Both tasks with hypothetical monetary rewards and long delays (days to weeks) and the Experiential Discounting Task with monetary feedback in real time and short delays (7–30 s) show similar findings [73]. In a large multi-center case–control study, PD patients with compulsive shopping and PG had elevated delay discounting relative to controls but neither those with compulsive shopping nor those with sexual behaviors demonstrated such differences [4]. Impulsive choice normally demonstrates a magnitude effect, whereby lower impulsive choices accompany increasing reward magnitude. This magnitude effect in delay discounting is more pronounced in PD patients with ICDs [4]. Whereas healthy controls normally experience diminishing marginal sensitivity or a decrease in subjective value with increasing objective value, PD patients with ICDs may show more of an effect. These observations suggest that DA agonists in those with ICDs may be associated with enhanced diminishing marginal sensitivity or greater subjective devaluation of the delayed, higher reward magnitude and hence greater impulsive choice for the immediate reward [4].

There are several potential mechanisms that might account for these delay discounting findings. Impaired delay discounting with intact reward incentive performance has been interpreted as evidence for a potential impairment in waiting for the delayed reward, rather than an enhanced incentive toward the immediate reward [72]. Converging evidence implicates a role for DA in delay discounting, particularly implicating the nucleus accumbens core and OFC [10]. In PD + ICD subjects, greater delay discounting is associated with greater dopaminergic terminal function in the anterior putamen as measured using [^{18}F]fluorodopa [74]. This contrasts with dopaminergic lesions of the dorsolateral striatum enhancing delay discounting in rodents [67] and with PD in humans [68,69]. Thus, either dissociable influences from different striatal regions or DA tone may influence delay discounting in a U-shaped manner as has been shown for prefrontal DA [75].

The medial and lateral OFC have also shown opposing effects on delay discounting with medial lesions enhancing and lateral lesions decreasing delay discounting [10,76]. These lesion studies converge with studies of single-neuron activity in the OFC showing higher activity with time-discounted rewards after a short delay and lower activity after a long delay,

independent of the encoding for absolute reward magnitude [77]. Similarly, stroke-induced lesions of the medial OFC increased delay discounting in humans [78]. Thus, the OFC appears to play a specific role in encoding time-discounted rewards beyond value encoding to guide choice behavior. PD + ICD subjects have shown greater dopaminergic terminal function in the medial OFC at baseline [18] with DA agonists inhibiting OFC activity [19], thus suggesting potential impairments of delay representation mediated via the OFC. Dopaminergic striatal mechanisms have also implicated perceptual timing accuracy particularly with interval timing deficits in the seconds to minutes range, which may related to impairments in waiting for the delayed reward [79].

Put together, both PD and DA agonists have independent effects on enhancing delay discounting. PD itself, and particularly dopaminergic lesions of the dorsolateral striatum, enhances delay discounting, thus suggesting that PD may play a role as a risk factor. DA agonists also increased delay discounting in PD patients along with decreasing ventral striatal activity. PD + ICD subjects have enhanced delay discounting in response to DA agonists, particularly in those with gambling and shopping behaviors, but not binge eating or compulsive sexual behaviors.

Greater reflection impulsivity or lower amount of evidence accumulated prior to a decision has also been shown to be affected in PD + ICD subjects. In a probabilistic decision-making task PD + ICD subjects accumulated less evidence prior to a decision compared to PD controls and pathological gamblers and healthy controls, with similar rates compared to patients with substance use disorders [80]. In PD subjects, DA agonists increased reflection impulsivity, but not levodopa or deep brain stimulation [81]. In a perceptual decision-making task in which subjects estimated whether a stimulus contained more red or blue pixels, PD + ICD subjects also made more incorrect decisions [82]. PD subjects on DA agonists similarly were faster than those on levodopa monotherapy or controls.

In contrast to impairments in decisional impulsivity, two studies have not shown any differences in motor response inhibition between PD subjects with and without ICDs [71], with one study demonstrating better response inhibition on the Stop Signal Task in those with ICDs irrespective of medication status [83]. In rodent studies, response inhibition is not influenced by DA and implicates more dorsomedial rather than ventral striatal regions [10,84].

SUMMARY

Converging evidence emphasizes a role for both DA agonists and levodopa interacting with an underlying individual vulnerability and PD resulting in ICD behaviors (reviewed in Fig. 24.1). Emerging evidence implicates an imbalance in ventral and dorsal striatal function. Several lines of evidence suggest a role for lower DAT levels particularly in the ventral striatum that might precede the exposure to DA agonists and the onset of ICDs. Lower DAT levels with intact neuronal integrity may be associated with prolongation of tonic and phasic physiological dopaminergic activity. Preliminary evidence suggests a role for decreased sensitivity of midbrain D2 autoreceptors with no differences in striatal D2/3 receptor density. These findings contrast with consistent decreases in striatal D2 receptor density in drug addictions and obesity, with intact levels observed in PG. Chronic levodopa in rodent PD models has been shown to increase the number of spontaneously active DA neurons capable of phasic firing in response to salient stimuli, hence enhancing gain. These mechanisms would be consistent with enhanced dopaminergic activity observed in the context of conditioned cues supporting incentive motivation theories, reward anticipation and unexpected rewards, gambling tasks, and novelty preference.

Rodent models of PD particularly targeting dopaminergic dorsolateral striatal projections enhance the reinforcing properties of DA agonists and levodopa. In rodents, DA agonists show moderate reinforcing properties along with enhanced ΔFosB accumulation in the nucleus accumbens correlating with motivation for DA agonists. Evidence from studies on learning from feedback show mixed results, which may be related to task-related differences. Some but not all studies show enhanced learning from reward feedback with mixed results for learning from loss feedback. An intriguing study highlights the dissociable roles of the ventral and dorsal striatum in learning, emphasizing the dopaminergic overdose theory, and supports the hypothesis of DA agonists impairing learning from negative feedback presumably acting on low-affinity D2 receptors. Here, PD patients with ICDs show reliance on a ventral striatal critic model of stimulus value with impaired learning from negative prediction error, whereas PD controls show reliance on a dorsal striatal actor model of action value with improved learning from positive prediction error. Impulsivity in the decisional (delay discounting, reflection impulsivity, and risk taking) rather than the motoric domain (response inhibition) appears to be impaired in PD subjects with ICDs, which may be in part related to the relative engagement of ventral versus dorsal striatal regions. Both rodent models and human studies demonstrate a role for DA agonists in enhancing risk taking with a particular emphasis on gain anticipation, with decreased activity in the ventral striatum and OFC in risk representation. Both rodent models and human studies emphasize a role for PD in

enhancing delay discounting, thus emphasizing a specific influence of PD on reinforcing properties of dopaminergic medications and delay discounting in the onset of ICD behaviors. DA agonists also enhance delay discounting in PD subjects, with converging studies demonstrating greater enhancement in those with ICDs and particularly those with gambling and shopping behaviors. Potential mechanisms may include impaired waiting, timing deficits, effects on diminishing marginal sensitivity, and dopaminergic influences on delay representation of the ventral striatum and OFC. A greater understanding of the cognitive mechanisms underlying dopaminergic medication-induced ICDs may provide a greater understanding of the role of DA on cognitive processes and behavioral and substance addictions and provide potential therapeutic targets.

References

[1] Voon V, Potenza MN, Thomsen T. Medication-related impulse control and repetitive behaviors in Parkinson's disease. Curr Opin Neurol 2007;20:484—92.

[2] Weintraub D, Koester J, Potenza MN, Siderowf AD, Stacy M, Voon V, et al. Impulse control disorders in Parkinson disease: a cross-sectional study of 3090 patients. Arch Neurol 2010;67: 589—95.

[3] Weintraub D, Sohr M, Potenza MN, Siderowf AD, Stacy M, Voon V, et al. Amantadine use associated with impulse control disorders in Parkinson disease in cross-sectional study. Ann Neurol 2010;68:963—8.

[4] Voon V, Sohr M, Lang AE, Potenza MN, Siderowf AD, Whetteckey J, et al. Impulse control disorders in Parkinson disease: a multicenter case-control study. Ann Neurol 2011;69(6): 986—96.

[5] Bastiaens J, Dorfman BJ, Christos PJ, Nirenberg MJ. Prospective cohort study of impulse control disorders in Parkinson's disease. Mov Disord 2013;28:327—33.

[6] Johansson A, Grant JE, Kim SW, Odlaug BL, Gotestam KG. Risk factors for problematic gambling: a critical literature review. J Gambl Stud 2009;25:67—92.

[7] Voon V, Fox SH. Medication-related impulse control and repetitive behaviors in Parkinson disease. Arch Neurol 2007;64: 1089—96.

[8] Cools R. Dopaminergic modulation of cognitive function-implications for L-DOPA treatment in Parkinson's disease. Neurosci Biobehav Rev 2006;30:1—23.

[9] Tremblay L, Worbe Y, Thobois S, Sgambato-Faure V, Feger J. Selective dysfunction of basal ganglia subterritories: from movement to behavioral disorders. Mov Disord 2015;30:1155—70.

[10] Robbins TW, Dalley JW. Impulsivity, risk choice and impulse control disorders: animal models. In: Dreher JC, Tremblay L, editors. Decision neuroscience. Elsevier; 2016.

[11] O'Sullivan SS, Wu K, Politis M, Lawrence AD, Evans AH, Bose SK, et al. Cue-induced striatal dopamine release in Parkinson's disease-associated impulsive-compulsive behaviours. Brain 2011;134(Pt 4).

[12] Wu K, Politis M, O'Sullivan SS, Lawrence AD, Warsi S, Bose S, et al. Single versus multiple impulse control disorders in Parkinson's disease: an (1)(1)C-raclopride positron emission tomography study of reward cue-evoked striatal dopamine release. J Neurol 2015;262:1504—14.

[13] Steeves TD, Miyasaki J, Zurowski M, Lang AE, Pellecchia G, Van Eimeren T, et al. Increased striatal dopamine release in Parkinsonian patients with pathological gambling: a [^{11}C] raclopride PET study. Brain 2009;132:1376–85.

[14] Joutsa J, Johansson J, Niemela S, Ollikainen A, Hirvonen MM, Piepponen P, et al. Mesolimbic dopamine release is linked to symptom severity in pathological gambling. Neuroimage 2012; 60:1992–9.

[15] Evans AH, Pavese N, Lawrence AD, Tai YF, Appel S, Doder M, et al. Compulsive drug use linked to sensitized ventral striatal dopamine transmission. Ann Neurol 2006;59:852–8.

[16] Politis M, Loane C, Wu K, O'Sullivan SS, Woodhead Z, Kiferle L, et al. Neural response to visual sexual cues in dopamine treatment-linked hypersexuality in Parkinson's disease. Brain 2013;136:400–11.

[17] Voon V, Pessiglione M, Brezing C, Gallea C, Fernandez HH, Dolan RJ, et al. Mechanisms underlying dopamine-mediated reward bias in compulsive behaviors. Neuron 2010;65:135–42.

[18] Joutsa J, Martikainen K, Niemela S, Johansson J, Forsback S, Rinne JO, et al. Increased medial orbitofrontal [^{18}F]fluorodopa uptake in Parkinsonian impulse control disorders. Mov Disord 2012; 27:778–82.

[19] van Eimeren T, Pellecchia G, Cilia R, Ballanger B, Steeves TD, Houle S, et al. Drug-induced deactivation of inhibitory networks predicts pathological gambling in PD. Neurology 2010;75:1711–6.

[20] Voon V, Rizos A, Chakravartty R, Mulholland N, Robinson S, Howell NA, et al. Impulse control disorders in Parkinson's disease: decreased striatal dopamine transporter levels. J Neurol Neurosurg Psychiatry 2014;85:148–52.

[21] Cilia R, Ko JH, Cho SS, van Eimeren T, Marotta G, Pellecchia G, et al. Reduced dopamine transporter density in the ventral striatum of patients with Parkinson's disease and pathological gambling. Neurobiol Dis 2010;39:98–104.

[22] Smith KM, Xie SX, Weintraub D. Incident impulse control disorder symptoms and dopamine transporter imaging in Parkinson disease. J Neurol Neurosurg Psychiatry 2015. http://dx.doi.org/10.1136/jnnp-2015-311827.

[23] Vriend C, Nordbeck AH, Booij J, van der Werf YD, Pattij T, Voorn P, et al. Reduced dopamine transporter binding predates impulse control disorders in Parkinson's disease. Mov Disord 2014;29:904–11.

[24] Ray NJ, Miyasaki JM, Zurowski M, Ko JH, Cho SS, Pellecchia G, et al. Extrastriatal dopaminergic abnormalities of DA homeostasis in Parkinson's patients with medication-induced pathological gambling: a [^{11}C] FLB-457 and PET study. Neurobiol Dis 2012; 48:519–25.

[25] Clark L, Stokes PR, Wu K, Michalczuk R, Benecke A, Watson BJ, et al. Striatal dopamine D(2)/D(3) receptor binding in pathological gambling is correlated with mood-related impulsivity. Neuroimage 2012;63:40–6.

[26] Grace AA. The tonic/phasic model of dopamine system regulation and its implications for understanding alcohol and psychostimulant craving. Addiction 2000;95(Suppl. 2):S119–28.

[27] Fiorillo CD, Tobler PN, Schultz W. Discrete coding of reward probability and uncertainty by dopamine neurons. Science 2003; 299:1898–902.

[28] Schultz W. Dopamine neurons and their role in reward mechanisms. Curr Opin Neurobiol 1997;7:191–7.

[29] Schultz W. Electrophysiological correlates of reward processing in dopamine neurons. In: Dreher JC, Tremblay L, editors. Decision neuroscience. Elsevier; 2016.

[30] Schultz W, Dayan P, Montague PR. A neural substrate of prediction and reward. Science 1997;275:1593–9.

[31] Bunzeck N, Duzel E. Absolute coding of stimulus novelty in the human substantia nigra/VTA. Neuron 2006;51:369–79.

[32] Howe MW, Tierney PL, Sandberg SG, Phillips PE, Graybiel AM. Prolonged dopamine signalling in striatum signals proximity and value of distant rewards. Nature 2013;500:575–9.

[33] Niv Y, Daw ND, Joel D, Dayan P. Tonic dopamine: opportunity costs and the control of response vigor. Psychopharmacology (Berl) 2007;191:507–20.

[34] Cohen MX, Frank MJ. Neurocomputational models of basal ganglia function in learning, memory and choice. Behav Brain Res 2009;199:141–56.

[35] Bordet R, Ridray S, Carboni S, Diaz J, Sokoloff P, Schwartz JC. Induction of dopamine D3 receptor expression as a mechanism of behavioral sensitization to levodopa. Proc Natl Acad Sci USA 1997;94:3363–7.

[36] Driver-Dunckley E, Samanta J, Stacy M. Pathological gambling associated with dopamine agonist therapy in Parkinson's disease. Neurology 2003;61:422–3.

[37] Dodd ML, Klos KJ, Bower JH, Geda YE, Josephs KA, Ahlskog JE. Pathological gambling caused by drugs used to treat Parkinson disease. Arch Neurol 2005;62:1377–81.

[38] Szarfman A, Doraiswamy PM, Tonning JM, Levine JG. Association between pathologic gambling and parkinsonian therapy as detected in the Food and Drug Administration Adverse Event database. Arch Neurol 2006;63:299–300. author reply 300.

[39] Moore TJ, Glenmullen J, Mattison DR. Reports of pathological gambling, hypersexuality, and compulsive shopping associated with dopamine receptor agonist drugs. JAMA Intern Med 2014; 174:1930–3.

[40] Seeman P. Parkinson's disease treatment may cause impulse-control disorder via dopamine D3 receptors. Synapse 2015;69: 183–9.

[41] Boileau I, Payer D, Chugani B, Lobo D, Behzadi A, Rusjan PM, et al. The D2/3 dopamine receptor in pathological gambling: a positron emission tomography study with [^{11}C]-(+)-propyl-hexahydro-naphtho-oxazin and [^{11}C]raclopride. Addiction 2013; 108:953–63.

[42] Boileau I, Payer D, Chugani B, Lobo DS, Houle S, Wilson AA, et al. In vivo evidence for greater amphetamine-induced dopamine release in pathological gambling: a positron emission tomography study with [(11)C]-(+)-PHNO. Mol Psychiatry 2014;19:1305–13.

[43] Chernoloz O, El Mansari M, Blier P. Sustained administration of pramipexole modifies the spontaneous firing of dopamine, norepinephrine, and serotonin neurons in the rat brain. Neuropsychopharmacology 2009;34:651–61.

[44] Harden DG, Grace AA. Activation of dopamine cell firing by repeated L-DOPA administration to dopamine-depleted rats: its potential role in mediating the therapeutic response to L-DOPA treatment. J Neurosci 1995;15:6157–66.

[45] Riddle JL, Rokosik SL, Napier TC. Pramipexole- and methamphetamine-induced reward-mediated behavior in a rodent model of Parkinson's disease and controls. Behav Brain Res 2012;233:15–23.

[46] Engeln M, Ahmed SH, Vouillac C, Tison F, Bezard E, Fernagut PO. Reinforcing properties of pramipexole in normal and parkinsonian rats. Neurobiol Dis 2013;49:79–86.

[47] Engeln M, Fasano S, Ahmed SH, Cador M, Baekelandt V, Bezard E, et al. Levodopa gains psychostimulant-like properties after nigral dopaminergic loss. Ann Neurol 2013;74:140–4.

[48] Evans AH, Lees AJ. Dopamine dysregulation syndrome in Parkinson's disease. Curr Opin Neurol 2004;17:393–8.

[49] Frosini D, Pesaresi I, Cosottini M, Belmonte G, Rossi C, Dell'Osso L, et al. Parkinson's disease and pathological gambling: results from a functional MRI study. Mov Disord 2010;25:2449–53.

[50] Volkow ND, Wang GJ, Telang F, Fowler JS, Logan J, Childress AR, et al. Cocaine cues and dopamine in dorsal striatum: mechanism of craving in cocaine addiction. J Neurosci 2006;26:6583–8.

[51] Djamshidian A, O'Sullivan SS, Wittmann BC, Lees AJ, Averbeck BB. Novelty seeking behaviour in Parkinson's disease. Neuropsychologia 2011;49:2483–8.

[52] Frank MJ, Seeberger LC, O'Reilly RC. By carrot or by stick: cognitive reinforcement learning in parkinsonism. Science 2004;306: 1940–3.

[53] Frank MJ. Dynamic dopamine modulation in the basal ganglia: a neurocomputational account of cognitive deficits in medicated and nonmedicated Parkinsonism. J Cogn Neurosci 2005;17:51–72.

[54] Bodi N, Keri S, Nagy H, Moustafa A, Myers CE, Daw N, et al. Reward-learning and the novelty-seeking personality: a between- and within-subjects study of the effects of dopamine agonists on young Parkinson's patients. Brain 2009;132:2385–95.

[55] Djamshidian A, Jha A, O'Sullivan SS, Silveira-Moriyama L, Jacobson C, Brown P, et al. Risk and learning in impulsive and nonimpulsive patients with Parkinson's disease. Mov Disord 2010;25:2203–10.

[56] Djamshidian A, O'Sullivan SS, Lees A, Averbeck BB. Effects of dopamine on sensitivity to social bias in Parkinson's disease. PLoS One 2012;7:e32889.

[57] Piray P, Zeighami Y, Bahrami F, Eissa AM, Hewedi DH, Moustafa AA. Impulse control disorders in Parkinson's disease are associated with dysfunction in stimulus valuation but not action valuation. J Neurosci 2014;34:7814–24.

[58] Dayan P, Balleine BW. Reward, motivation, and reinforcement learning. Neuron 2002;36:285–98.

[59] Sutton RS, Barto AG. Reinforcement learning: an introduction. , Cambridge, MA: MIT Press; 1998.

[60] O'Doherty JP. Reward representations and reward-related learning in the human brain: insights from neuroimaging. Curr Opin Neurobiol 2004;14:769–76.

[61] Rokosik SL, Napier TC. Intracranial self-stimulation as a positive reinforcer to study impulsivity in a probability discounting paradigm. J Neurosci Methods 2011;198:260–9.

[62] Rokosik SL, Napier TC. Pramipexole-induced increased probabilistic discounting: comparison between a rodent model of Parkinson's disease and controls. Neuropsychopharmacology 2012;37: 1397–408.

[63] Claassen DO, van den Wildenberg WP, Ridderinkhof KR, Jessup CK, Harrison MB, Wooten GF, et al. The risky business of dopamine agonists in Parkinson disease and impulse control disorders. Behav Neurosci 2011;125:492–500.

[64] Rao H, Mamikonyan E, Detre JA, Siderowf AD, Stern MB, Potenza MN, et al. Decreased ventral striatal activity with impulse control disorders in Parkinson's disease. Mov Disord 2010;25: 1660–9.

[65] Thomas A, Bonanni L, Gambi F, Di Iorio A, Onofrj M. Pathological gambling in Parkinson disease is reduced by amantadine. Ann Neurol 2010;68:400–4.

[66] Cera N, Bifolchetti S, Martinotti G, Gambi F, Sepede G, Onofrj M, et al. Amantadine and cognitive flexibility: decision making in Parkinson's patients with severe pathological gambling and other impulse control disorders. Neuropsychiatr Dis Treat 2014;10: 1093–101.

[67] Tedford SE, Persons AL, Napier TC. Dopaminergic lesions of the dorsolateral striatum in rats increase delay discounting in an impulsive choice task. PLoS One 2015;10:e0122063.

[68] Milenkova M, Mohammadi B, Kollewe K, Schrader C, Fellbrich A, Wittfoth M, et al. Intertemporal choice in Parkinson's disease. Mov Disord 2011;26(11).

[69] Al-Khaled M, Heldmann M, Bolstorff I, Hagenah J, Munte TF. Intertemporal choice in Parkinson's disease and restless legs syndrome. Parkinsonism Relat Disord 2015;21:1330–5.

[70] Antonelli F, Ko JH, Miyasaki J, Lang AE, Houle S, Valzania F, et al. Dopamine-agonists and impulsivity in Parkinson's disease: impulsive choices vs. impulsive actions. Hum Brain Mapp 2014; 35:2499–506.

[71] Leroi I, Barraclough M, McKie S, Hinvest N, Evans J, Elliott R, et al. Dopaminergic influences on executive function and impulsive behaviour in impulse control disorders in Parkinson's disease. J Neuropsychol 2013;7:306–25.

[72] Housden CR, O'Sullivan SS, Joyce EM, Lees AJ, Roiser JP. Intact reward learning but elevated delay discounting in Parkinson's disease patients with impulsive-compulsive spectrum behaviors. Neuropsychopharmacology 2010;35(11).

[73] Voon V, Reynolds B, Brezing C, Gallea C, Skaljic M, Ekanayake V, et al. Impulsive choice and response in dopamine agonist-related impulse control behaviors. Psychopharmacology (Berl) 2010;207: 645–59.

[74] Joutsa J, Voon V, Johansson J, Niemela S, Bergman J, Kaasinen V. Dopaminergic function and intertemporal choice. Transl Psychiatry 2015;5:e491.

[75] Kayser AS, Allen DC, Navarro-Cebrian A, Mitchell JM, Fields HL. Dopamine, corticostriatal connectivity, and intertemporal choice. J Neurosci 2012;32:9402–9.

[76] Mar AC, Walker AL, Theobald DE, Eagle DM, Robbins TW. Dissociable effects of lesions to orbitofrontal cortex subregions on impulsive choice in the rat. J Neurosci 2011;31:6398–404.

[77] Roesch MR, Taylor AR, Schoenbaum G. Encoding of time-discounted rewards in orbitofrontal cortex is independent of value representation. Neuron 2006;51:509–20.

[78] Sellitto M, Ciaramelli E, di Pellegrino G. Myopic discounting of future rewards after medial orbitofrontal damage in humans. J Neurosci 2010;30:16429–36.

[79] Meck WH. Frontal cortex lesions eliminate the clock speed effect of dopaminergic drugs on interval timing. Brain Res 2006;1108: 157–67.

[80] Djamshidian A, O'Sullivan SS, Sanotsky Y, Sharman S, Matviyenko Y, Foltynie T, et al. Decision making, impulsivity, and addictions: do Parkinson's disease patients jump to conclusions? Mov Disord 2012;27:1137–45.

[81] Djamshidian A, O'Sullivan SS, Foltynie T, Aviles-Olmos I, Limousin P, Noyce A, et al. Dopamine agonists rather than deep brain stimulation cause reflection impulsivity in Parkinson's disease. J Parkinson's Dis 2013;3:139–44.

[82] Djamshidian A, Sanotsky Y, Matviyenko Y, O'Sullivan SS, Sharman S, Selikhova M, et al. Increased reflection impulsivity in patients with ephedrone-induced Parkinsonism. Addiction 2013;108:771–9.

[83] Claassen DO, van den Wildenberg WP, Harrison MB, van Wouwe NC, Kanoff K, Neimat JS, et al. Proficient motor impulse control in Parkinson disease patients with impulsive and compulsive behaviors. Pharmacol Biochem Behav 2015;129: 19–25.

[84] Voon V, Dalley JW. Translatable and back-translatable measurement of impulsivity and compulsivity: convergent and divergent processes. In: Robbins TW, Sahakian BJ, editors. Current Topics in behavioural Neuroscience; 2015.

25

The Subthalamic Nucleus in Impulsivity

K. Witt

Christian Albrecht University, Kiel, Germany

Abstract

Motor control is the result of a perfect balance between activation and inhibition of movement patterns. The basal ganglia are connected to the motor cortex, including a direct pathway that facilitates the preparation of movement patterns and an indirect pathway suppressing movement preparation. Anatomically and functionally the subthalamic nucleus (STN) holds a central position within the indirect pathway, acting like a brake on the motor system. Using a hyperdirect pathway—a monosynaptic connection between the key nodes of the inhibitory network—the presupplementary motor area and the inferior frontal gyrus are connected to the STN. Within the process of decision-making, the inhibitory network employs the STN function to pause the motor system. This has been specifically shown in decision conflicts, i.e., whenever an automatic response must be suppressed to have more time to choose between alternative responses. The cortical inhibitory network and the STN dynamically regulate the decision threshold as a function of decision conflict. In this way cognitive processes of active inhibition influence the motor system via the STN to regulate behavior.

INTRODUCTION

The basal ganglia are thought to be important in the selection of wanted and the suppression of unwanted motor patterns according to explicit or implicit rules [1]. The subthalamic nucleus (STN; *corpus Luysi*) is involved in this basic mechanism of response selection and response inhibition [1]. This nucleus is a small component of the basal ganglia, measuring 5.9 mm in the anteroposterior, 3.7 mm in the mediolateral, and 5 mm in the dorsoventral dimension [2]. Anatomically it is located between the thalamus and the substantia nigra, lateral of the nucleus ruber. The STN has gained extensive research interest because its spontaneous firing rate is significantly increased in Parkinson's disease (PD) [3]. An STN lesion leads to an excessive increase in involuntary movements in the contralateral extremities (hemiballismus) [4,5]. Growing knowledge about functions within the corticobasal ganglia (cortico-BG) loops, the characterization of the direct and the indirect pathway, and the finding of increased neuronal activity in the STN in PD led to the breakthrough experiment by Bergman and colleagues demonstrating that an STN lesion reverses parkinsonian symptoms in an animal model of parkinsonism [3]. Shortly after this finding, Benazzouz et al. demonstrated a reversible, gradual, and controllable beneficial effect of STN deep brain stimulation (DBS) in MPTP-treated monkeys [6]. In humans Benabid and colleagues implanted DBS electrodes in the STN of PD patients. Although the exact mechanism of electrical stimulation of the STN is not completely understood [7], stimulation of above 100 Hz in STN improves motor function in PD [8]. Further randomized clinical trials demonstrated the beneficial effect of STN DBS on parkinsonian motor symptoms and quality of life in PD [9]. As a result, DBS has now become an evidence-based treatment for advanced PD. Some of the side effects seen after STN DBS illustrate the impact of the STN on cognitive and behavioral aspects specifically affecting features of impulsivity. These findings and a growing knowledge about the functional anatomy and physiology of the cortico-BG circuits has triggered a number of studies examining the impact of the STN on decision-making in more detail.

ANATOMY, PHYSIOLOGY, AND FUNCTION OF CORTICOBASAL GANGLIA CIRCUITS

Clinical and experimental research since 1985 has implicated neuroanatomic loops connecting the frontal cortex to the BG and thalamus as being involved in aspects such as motor control, cognition, and emotion [10]. The STN is a component of BG circuits that has

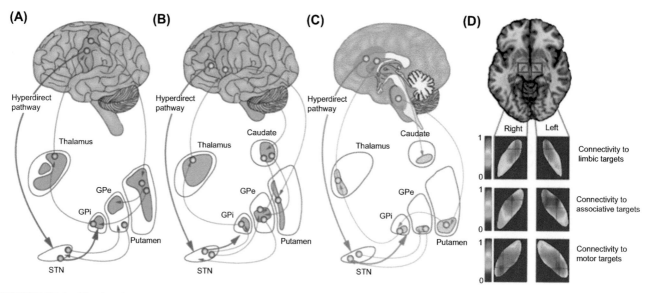

FIGURE 25.1   The basal ganglia (BG) form anatomically and functionally segregated neuronal circuits with thalamic nuclei and frontal cortical areas. (A) The motor circuit involves the motor and supplementary motor cortices, the posterolateral part of the putamen, the GPe and GPi, the dorsolateral STN, and the ventrolateral thalamus. (B) The associative loop and (C) the limbic loop connect medioprefrontal and laterofrontal cortices with distinct regions within the BG and the thalamus. In the STN, a functional gradient has been found, with a motor representation in the dorsolateral region of the nucleus, cognitive-associative functions in the intermediate zone, and limbic function in the ventromedial region via a "hyperdirect" pathway, the STN receives direct projections from the motor, prefrontal, and anterior cingulate cortices that detect and integrate response conflicts. In this position the STN is a second input node for information in a cortico-BG direction. (D) The probabilistic connectivity gradients for motor, associative, and limbic cortical areas (*hot colors* represent high connectivity probability). The overlapping areas might represent partly interlocked loops as an anatomical indication of an integrative role of the STN combining motor, cognitive, and limbic functions. *GPe*, external globus pallidus; *GPi*, internal globus pallidus; *STN*, subthalamic nucleus. *Permission obtained from John Wiley & Sons, Inc.*

traditionally been considered to be a relay station in the "indirect" pathway controlling thalamocortical excitability [11,12]. Now, the STN is also thought to be an important input nucleus of the BG [13]. Such nuclei receive input not only from large parts of the frontal cortex, but also from various thalamic, brain-stem structures and the amygdala [14]. The BG—via parallel cortico-BG loops—modulate the activity of distinct frontal cortical areas, including the motor and premotor cortices, the dorsolateral and inferior prefrontal cortex, the anterior cingulate cortex, and the frontal eye field (Fig. 25.1) [11,15]. In more detail, two pathways build the intrinsic BG organization. The *direct pathway* inhibits the thalamus, which has an excitatory influence on the cortex [15,16]. Thereby the cascade of signals within the direct pathway helps the cortex execute the intended movement or behavior. This feature has lent the direct pathway its alternative name, the "go pathway." The *indirect pathway* has a contrary effect on the cortex. Within the indirect pathway, the cascade of signals has an excitatory effect on the BG output nuclei. The internal globus pallidus inhibits thereby the cortex via the thalamus. It is assumed that the net effect of the indirect pathway enables the BG output nuclei to suppress inappropriate parts of competing motor programs, or to send a global "no-go signal" [17].

The STN, as part of the indirect pathway, receives monosynaptic projections (a "hyperdirect" pathway) from the cortex and, importantly, it also serves as an input nucleus [13].

As with other BG nuclei, the STN can be subdivided into functionally segregated territories (motor, associative, and limbic) [13,18,19]. Traditionally these functionally segregated territories have also been considered to be strictly separated anatomically [11,15,20]. The limbic territory of the STN is in an anterior position, whereas the motor territory lies in a posterior–lateral orientation (Fig. 25.1D) [15]. The associative territory receives projection from the dorsal, mid- and inferior prefrontal cortex and lies between the motor and the limbic territory of the STN. Further studies involving nonhuman primates showed that, in part, motor, associative, and limbic territories overlap within the STN [21,22]. Overlapping STN territories could also be shown in a recent in vivo study using probabilistic diffusion tractography to identify STN subregions with predominantly motor, associative, and limbic connectivity (Fig. 25.1D) [23]. The overlapping areas of motor, associative, and limbic maps might shift the concept of parallel segregated loops to partly interlocked loops and provide an anatomical indication of the integrative role of the STN in combining motor, cognitive, and limbic functions.

Behavioral experiments and neurocomputational models suggest that the cortico-BG circuits have two different roles in the decision-making process in a reinforcement context [24]. Phasic dopamine signals in the ventral striatum (so-called "bursts" and "dips") provide a reward-associated teaching signal [25] driving impulses of the type "go learning" (through D1 receptors and the direct BG pathway) to seek reward and those of the type "no-go learning," through D2 receptors and the indirect BG pathway, to avoid unpleasant events or actions. (See also Chapters 4 and 23.)

THE SUBTHALAMIC NUCLEUS AND DECISION-MAKING: EVIDENCE FROM ANIMAL STUDIES

One of the first studies investigating the effect of an STN lesion in monkeys reported hemiballismus of the contralateral body part [5,26]. Baunez et al. investigated the effects of STN lesioning but also the effects of bilateral STN muscimol injection on response accuracy in a simple reaction time task and a five-choice serial reaction time task (5CSRTT; Box 25.1) [27]. The latter task is especially useful to investigate visuospatial attention, perseveration tendencies, sustained attention, and motor impulsivity [27]. Bilateral lesion of the STN increased premature responding on this task, a finding that could be replicated in several studies [27,28]. Behavioral effects following excitotoxic lesions to different regions of the rat cortex and striatum on performance of the 5CSRTT showed impaired premature "impulsive" responding tendencies after dorsomedial and ventral lesion of the cortex or the striatum [29–31]. Dorsomedial lesioning of the prefrontal cortex or striatum also induces deficits in response accuracy, which can be interpreted as an attentional deficit. This finding also offers the possibility that attentional deficits contribute to the finding of premature responding in the following studies [32].

The go/no-go task and the stop signal task (SST; Box 25.1) are further paradigms to assess inhibitory control and thereby impulsivity in the motor domain. The behavioral measures of both tasks are positively correlated, highlighting the relations of both investigated behavioral characteristics [33]. In rats, an STN lesion prevented the animals from being able to stop an ongoing action [34]. Rats were also impaired in inhibiting a motor response in a modified go/no-go task [34], and electrophysiological recordings from the monkey STN identified different neuronal responses at no-go compared to go cues [35]. These studies showed direct involvement of the STN in response selection in animal models.

THE SUBTHALAMIC NUCLEUS AND IMPULSIVITY: EVIDENCE FROM BEHAVIORAL OBSERVATIONS

The inference of the specific role of the STN on complex behavior is challenging, because behavioral changes are often a creeping process influenced by more than one factor. In animal studies STN lesions increase the incentive motivation for food. Electrophysiological recordings have identified neurons within the STN that are specifically active when the animal is expecting its reward [36]. Furthermore, it has been shown that STN lesions or high-frequency stimulation of the STN reduces the motivation for consuming drugs of abuse such as cocaine [37], but increases the motivation for drinking alcohol in "high drinker" rats [38]. These results lead to the conclusion that the reward-prediction error as a teaching signal can be modified by the STN. Generally, patients with PD gain weight after STN DBS [39]. However, this could also have a metabolic or motor explanation rather than be a causal result of the motivational state after STN DBS surgery. There are no data available regarding traditional drugs of abuse in PD patients receiving STN DBS, but impulse control disorders and dopamine dysregulation syndrome (see Chapter 24)—associated with an addiction to dopaminergic treatment in the course of PD—improve after STN DBS [40]. These effects might be better explained by a desensitization of postsynaptic dopamine receptors in the ventral striatum due to a reduction in daily medication intake after STN DBS surgery rather than a specific effect of the STN DBS itself [41].

After STN DBS some affective and behavioral changes have been described: 4–15% of patients show postoperative euphoria and/or hypomania within the first 3 months after surgery. Manic psychosis is less frequent (0.9–1.7%). Both euphoria and manic states are typically seen in close association with the initiation of DBS; thus, such states seem to be a direct consequence of neuromodulation affecting the limbic BG pathways [10]. Apathy is a more frequently seen phenomenon after STN DBS [42]. It typically occurs months after the initiation of STN DBS electrodes. Apathy is the result of a complex interplay between individual vulnerability that depends on the dopaminergic signaling, especially in the ventral striatum, and the withdrawal of dopaminergic medication in the chronic STN DBS setting [43]. Therefore, apathy does not seem to be directly related to STN functions. A wide spectrum of abnormal behavior has been described in PD after STN DBS. Of note, the behavioral changes after STN DBS were not reported in randomized controlled studies but reported in single case reports or case series, which highlights the fact that abnormal behavior after STN DBS is very

BOX 25.1

THESE TASKS ASSESS ASPECTS OF DECISION-MAKING IN THE CONTEXT OF THE SUBTHALAMIC NUCLEUS FUNCTIONS, MOSTLY UNDER TIME PRESSURE AND IN A RESPONSE CONFLICT

Description of the Task	Tested Functions
Five-Choice Serial Reaction Time Task (5CSRTT) [27] Rats were trained within an operant chamber with at least five holes that could be illuminated. The 5CSRTT required the rat to correctly identify by means of a nosepoke which of the five holes was illuminated to receive a reward. Between every trial, there was a short interval (5 s) wherein the animal had to withhold all responses. Any response during this interval was recorded as failure to inhibit control.	5CSRTT has a separate measure for impulsivity and is a precursor of models for gambling and decision-making, visuospatial attention, perseveration tendencies, sustained attention, and motor impulsivity.
Go/No-Go Task Test-stimuli were presented in a continuous stream and participants made a binary decision based on each stimulus. One of the outcomes required participants to make a motor response ("go" signal), whereas the other required participants to withhold a response ("no-go" cue). The relevant stimulus was presented randomly to make the appearance unpredictable by the subject.	Accuracy data show the participant's ability to inhibit a response. Reaction times give more global information regarding motor readiness, attention, and motor execution.
Stop Signal Task (SST) [92] The SST consisted of a primary go task in which the animal (or human participant) had to respond to a (most often visual) stimulus by pressing a button (go cue). On a subset of trials, the go cue was followed shortly by a stop signal, which could be either visual or auditory. The task this time was to stop the planned or ongoing response to the go cue. Humans were instructed to try to respond as fast as possible to the go signal without slowing down their responses to anticipate the occurrence of a stop trial. The stop signals occurred after different time lapses and the probability of inhibition was high when the stop signal was presented close to the beginning of the go signal or shortly after. It became lower the longer the time interval between the go cue and the stop signal was. The goal of the task was to define the time required to present the stop signal after the go signal to successfully inhibit 50% of the signal responses.	In the SST participants are requested to stop an action that is already initiated. The amount of time between the go signal and the stop signal that inhibits 50% of all responses is a valid measure for inhibitory control and thereby impulsivity in the motor domain. One disadvantage is that the SST can only produce estimates by finding the specific delay at which accuracy begins to fall below 50%. Therefore, it is not possible to dissect accuracy from speed measures.
Stroop Test [93] Participants were requested to read the name of a color that was printed in a color not denoted by the name (e.g., "red" and "blue" are the answers for blue and red). Naming the color of the word takes longer and is more prone to cause errors than when the color of the ink matches the name of the color.	The Stroop effect is a demonstration of interference in the reaction time of a task. Selective attention, processing speed, automaticity, and cognitive control are measured.
Probabilistic Selection Task [24] Stimulus pairs included six stimuli. Two stimuli were presented simultaneously in a forced-choice task. One stimulus has a higher probability for positive feedback and the second stimulus holds the complementary probability for a positive outcome (pair 1, A 80%/B 20%; pair 2, C 70%/D 30%; pair 3, E 60%/F 40%). In a learning period the stimulus pairs were fixed and participants learned the probabilistic setting by trial and error. In the test phase the stimuli were newly arranged (e.g., A/C, B/D, or C/B). Again participants performed the forced-choice task, but no feedback was given.	Results of the test phase (presentation of novel test pairs) allows a differentiation among several learning strategies (having learned from positive feedback, or having learned from negative feedback, or both) [24]. Furthermore, participants' decision-making can be explored by analyzing trials with low conflict (e.g., choice between A or D) or in a high-conflict setting (e.g., choice between A or C) [54].
Sensory Integration Task (Moving Dots) [87] Participants were asked to decide if dots moved from right to left or vice versa. In nonconflict trials all dots slowly began to move in the same direction. To make the task harder the number of dots moving in one direction increased over time. In the conflict trials 10% of the dots moved in the opposite direction. These trials made it harder for the participant to decidvve which direction the majority of dots were moving.	This is a perceptual decision task. The advantage of this perceptual decision-making task is the gradual changes in conflict over time: the faster the participant is able to make a decision, the less conflict is presented to the participant.

rare. Impulse control disorders such as hypersexuality and gambling in close relation to the STN DBS initiation might be interpreted as a behavioral consequence of the stimulation of the limbic part of the STN resulting in a disinhibiting, more impulsive behavior. Suicide rates increase specifically in the first months after STN DBS surgery [44]. Postoperative suicide in PD might be the part of the spectrum of impulsive behavior associated with treatments that interfere with the normal inhibitory role of the STN in impulsivity. Nevertheless, impulsivity in the context of complex behavior such as suicide is surely facilitated and triggered by multiple factors rather than by a single neurobiological factor. Patients have been described as exhibiting difficulties in their relations with their spouses, their families, and their social and professional environment [45]. STN DBS significantly improves mobility and DBS surgery may open doors in patients' everyday lives that were often characterized by a fixed patient environment interaction before surgery.

THE SUBTHALAMIC NUCLEUS AND DECISION-MAKING: EVIDENCE FROM NEUROPSYCHOLOGICAL STUDIES

The most elucidative findings demonstrating the role of the STN on decision-making comes from studies assessing STN DBS, imaging studies, and computational models. Comparisons from pre to postsurgery, stimulation settings (on vs off stimulation), and right- or left-sided subthalamotomy have demonstrated behavioral changes in relation to STN functions. Electrophysiological recordings from the scalp, the STN, or both in combination, as well as studies using functional magnetic resonance imaging (fMRI), more closely clarify the interplay of specific brain areas such as frontal lobe areas and subcortical brain structures. Most of these studies were carried out in patients suffering from PD. The pathophysiology of PD directly involves the cortoco-BG circuits [46] but PD also affects neurons outside the BG, especially in advanced stages of the disease [47]. Therefore a generalization of these results should be made with caution. Basically five different tasks were used to get a deeper insight into the role of the STN in decision-making. These paradigms are listed in Box 25.1.

One early and consistent finding is impaired performance on the Stroop task after STN DBS [48–50]. Patients made more errors in the stimulation on setting, a finding that was interpreted as a behavioral correlate of premature responding seen in animal studies [50]. The specific impairment in Stroop task performance might reflect a selective impairment after STN DBS to overcoming the automatic mode of a reading habit rather than reflecting a generalized dysfunction in

attention. STN DBS induces decreased activation in the right anterior cingulate cortex and ventral striatum, which is associated with Stroop task performance [51]. Moreover, it is associated with ventral electrode localization [52]. Thus, these findings provide evidence that the STN modulates nonmotor cortico-BG circuitry particularly in a decision conflict. Macroelectrode recordings from STN electrodes in PD patients showed a desynchronization in the beta band (15–35 Hz) in a simple (non-conflict) reading condition [53], which is in line with previous results demonstrating that decreased beta activity is an electrophysiological prerequisite of the STN that allows the execution of a motor response. On incongruent trials of the Stroop task, the beta-band resynchronization was seen before the response. In error trials the resynchronization in the beta band occurred after response onset [53]. These results suggest that the beta-band resynchronization pauses the motor system until a conflict can be solved [53].

A computational model of decision-making predicts that the STN dynamically modulates decision thresholds in proportion to reinforcement and decision conflict [16]. In this model, the STN regulates the decision thresholds when dealing with multiple good options. In a decision conflict the STN provides a no-go signal ("hold your horses"), buying more time to settle on the assumed best option [16]. Indeed, compared to STN DBS in an off stimulation mode, turning on STN DBS significantly shortens responses in a win–win situation of a probabilistic selection task [54]. So the conflict-induced slowing effect on response reaction times (RTs) vanishes after switching on the STN DBS, whereas response RTs in low-conflict situations remain the same. These results were further confirmed by microelectrode recordings within the STN that showed increased single-unit STN activity when participants dealt with a decision and that there is a positive correlation between the level of spiking activity and the degree of decision conflict [55]. This "hold your horse" function of the STN is embedded in a medial prefrontal cortex (MPFC)–STN dialogue demonstrated by Cavanagh and colleagues [56]. EEG theta power (4–8 Hz) over the MPFC is associated with a low-frequency oscillation in the STN (2.5–5 Hz) and the theta power of the MPFC predicts a slowing in RT performance in high-conflict situations in healthy participants and PD patients without STN DBS. This MPFC–STN coupling is specific for high-conflict situations. STN DBS compromises this dynamic function and self-regulation in high-conflict situations leading to impulsive behavior [56]. These results lead to the conclusion that low-frequency communication of MPFC and STN reflects the neural substrate for cortico-BG communication during a conflict that is related to behavioral adjustments [56] and that a disruption of the normal STN function by DBS reverses the impact of frontal theta and induces a reduction in the decision threshold [56]. A further advance in the MPFC–STN model comes from a study that combined EEG and fMRI measurements using a probabilistic selection task. The results showed that the coactivity between MPFC and the STN is related to a dynamic adjustment of the decision threshold as a function of conflict in reinforcement values [57], meaning that their coupling dynamically changes in a trial-by-trial learning task. Moreover, the simultaneous fMRI and EEG measurements identified the presupplementary motor area (pre-SMA) as the neural source of the MPFC theta signal that had been attributed to the decision threshold adjustment [57]. The preSMA is part of the anterior cingulate cortex (ACC) known to be a cortical target of the cortico-BG loops [15]. STN DBS might also disrupt the formation of complex (probabilistically) learned reinforcement values when multiple pieces of acquired information must be combined and rated [58].

Whereas STN DBS impairs performance on the Stroop task and the probabilistic selection task, the results of the SST are heterogeneous, demonstrating that STN DBS improved stop signal RTs [59–61], sometimes deteriorated performance [62,63], or even showed no change [64]. These findings might be explained by various patient characteristics, testing different parts of the extremities (STN DBS had a more extensive effect on proximal parts of the extremity compared to distal parts), or variables in task settings such as unilateral or bilateral STN DBS or exact electrode localization. One disadvantage of the SST is that the "stop" performance can be estimated only by finding the specific delay at which accuracy begins to fall below 50%. When measured in this way, stop signal RTs are determined not only by the speed of inhibitory processes, but also by the accuracy with which these processes take place. Therefore, it is impossible to determine if DBS affects the accuracy or the speed of response inhibition or both. The improvement of stop signal performance has been explained by the argument that STN DBS restores cortico-BG signaling by suppressing STN overactivity. In line with this argumentation, Ray et al. reported beta-band activity in the STN in relation to the inhibition of motor action in response to stop signals [65]. For scalp EEG, there was greater beta power around the time of stopping for patients in an STN DBS on mode compared to STN DBS off mode [60]. Gamma-band activity within the STN is associated with (or may even trigger) voluntary movements [66]. Two studies examined the gamma activity within the STN during the SST and reported increased gamma activity when the voluntary movement was executed. If the stop signal means the prepared motor response is successfully stopped, there is a decrease in gamma activity in the STN; thus, there is direct electrophysiological evidence that the involvement of the STN controls the inhibition of motor response

[65,67]. Notably, a study examining microelectrode recordings during DBS surgery of the associative limbic part of the STN (in patients suffering from obsessive–compulsive disorder) reported selective neuronal activation before and during motor response (go signal), during successful withholding of a planned movement (inhibitory control in stop trials), and while error monitoring [68]. These results showed neuronal populations that respond differently to several components of executive control in nonparkinsonian patients.

Several fMRI studies showed activation in the STN area during SST performance. These studies clearly demonstrated the involvement of the STN in stopping motor responses [69–72]. These and other fMRI studies embedded STN activation in a (mostly right-sided) cortico-BG network including the preSMA, the inferior frontal gyrus (IFG), and the STN responsible for stop signal detection, error monitoring, and response inhibition. Various models assigned these functions to other areas and connections. First, the preSMA might translate the stop signal into the action system and the IFG plays an inhibitory control role in slowing down responses and also stopping preprogrammed responses via direct communication with the STN [73]. Second, the IFG is responsible for stop signal detection and the preSMA executes cortical inhibitory control via the STN [74]. Third, a nonhierarchical approach favors a network account including the preSMA, IFG, and STN [60,75]. Here the preSMA plays a preferential role in both executive processes of error monitoring and implementation of the stop response in the action system. A further model also highlights the anterior putamen as an important player via its connections to the preSMA. Swann and colleagues investigated the interplay of the preSMA and the IFG in a patient with an intracranial grid (implanted for a diagnostic procedure for epilepsy) using electrocorticography [76]. Low-density stimulation of the preSMA induced cortical evoked potentials within 30 ms in the IFG. Together with previous studies using diffusion-weighted MRI [69], these results demonstrate the structural and the functional connectivity between both cortical areas [69,76]. Preparing to stop showed high gamma amplitude increases with pre-SMA activity preceding that of the IFG [76]. For the stopping trials, again, a high gamma amplitude occurred earlier for the preSMA than for the IFG. This time shift of high gamma amplitude might favor the hypothesis that the preSMA was active, translating the stopping rule into the action system to prepare nodes in the inhibitory action control network [73], whereas the IFG was "putting the brakes" on behavior [76,77]. However, the exact dynamical interplay between the direct, the hyperdirect, and the indirect cortico-BG pathways and their associated cortical regions remains to be solved.

Further evidence of the inhibitory control mechanisms of the STN comes from subthalamotomy, a surgical lesion technique to disrupt STN function in PD. Using a conditional SST, Obeso and colleagues [78] tested PD patients treated medically with unilateral subthalamotomy as well as healthy controls. The conditional SST enabled them to investigate action inhibition and proactive inhibition, given the fact that participants were informed that in some trials the stop signal could be ignored. In this way the decision threshold could be adjusted, foreknowing that stop signals are behaviorally irrelevant. Unoperated PD patients showed the highest response threshold that was normalized after subthalamotomy, but operated patients failed to show a context-dependent strategic modulation of the response threshold, a possible physiologic function of the STN [78].

In most of the studies STN DBS impaired go/no-go task performance in such a way that RT was not altered by stimulation, but either response accuracy decreased after turning on the stimulator or faster RT was associated with less accuracy in task performance [79–81]. Specifically, a more ventral stimulation of limbic and associative territories of the STN affected go/no-go performance, decreasing hits and increasing false alarms [80]. A PET study of the go/no-go task in PD patients found that decreases in response inhibition ability correlated with increased activity in the ACC [82], resembling imaging results of the SST. The basic mechanisms of response inhibition mediated by the STN were investigated by Kühn and colleagues. They recorded local field potentials in the region of the STN in go and no-go trials in PD patients and demonstrated an event-related desynchronization in association with the movement execution. In contrast, an event-related synchronization is associated with the inhibition of a movement in no-go trials, pointing to the importance of the beta band and the braking function of the STN [83]. Moreover, a PET study examining seven PD patients in an STN DBS on and off stimulation setting reported reduced activation in the MPFC, preSMA, dorsal ACC, and IFG [79]. So this study showed that, in a go/no-go task, STN DBS altered the cortical network known to be important in response inhibition and error processing [84]. In addition, comparing brain metabolism in an STN DBS on and off setting, this study reported reduced activation in the precuneus, posterior cingulate cortex, and left inferior parietal cortex. Activation of the precuneus was negatively correlated with the number of omission errors. These results led to the hypothesis that STN activity also alters the activation of a posterior sensorimotor network [79] involved in the initiation of motor programs [85,86] and serving as a structure that supports proactive inhibitory control of movement-triggering mechanisms [85].

In comparison to the SST and the go/no-go task, the moving dots task holds two advantages. First, there is a gradual change in a set of visual stimuli, so this task includes an early period of conflict during which motor response might contaminate electrophysiological recordings less. Second, trials with different degrees of conflict can be compared. Electrophysiological recordings within the STN and simultaneous EEG recordings confirmed previous findings showing an increase in the theta and delta band coherence between the MPFC and the STN in high-conflict situations [87]. Moreover, a Granger causality method demonstrated that the elevated theta—delta power in the MPFC drives cortico-STN coherence [87]. Possibly the MPFC uses the STN as a brake to pause the motor system in high decision conflicts via a theta—delta rhythm.

A few studies have focused on the impact of STN DBS on risky decisions and reported no serious changes [88,89]. In fact, decision-making under risk seems to improve, which might be the consequence of medication reduction after surgery [89].

A MODEL OF THE IMPACT OF THE SUBTHALAMIC NUCLEUS ON DECISION-MAKING

Different terms have been used to describe the impact of STN DBS on the decision-making process. "Premature responding" [48,50], a shift in "accuracy—speed trade-off" [78], a "hold your horse" function [16,54,90], changes in "decision threshold" [56], all terms concerning aspects of (motor) impulsivity [1]. Impulsivity in the motor and the cognitive domain is a trade-off between spontaneously, maybe automatically, responding by using a set of limited responses often used and taking a pause to think before settling on a response that might be more advantageous for reaching a future goal. Indeed, many studies show the close relationship between the STN and the pause, giving more time to choose between response alternatives [16,56]. However, the qualitative behavioral consequence of decision-making does not seem to be significantly altered after STN DBS.

The STN has a central function in the motor system [10]. Anatomically part of the indirect pathway, it is able to suppress unwanted motor patterns and inhibit prepotent movements (Fig. 25.2) [1]. Consequently, the STN is able to stop and to modify planned movements. The cognitive domain employs the STN via an inhibitory network that includes the preSMA and the IFG [69]. The inhibitory cortical network introduces explicit rules and performs error monitoring and error detection, regulating the STN function to pause or even to stop an oncoming response. The cortical inhibitory network and the STN regulate the decision threshold dynamically as a function of decision conflict. This network

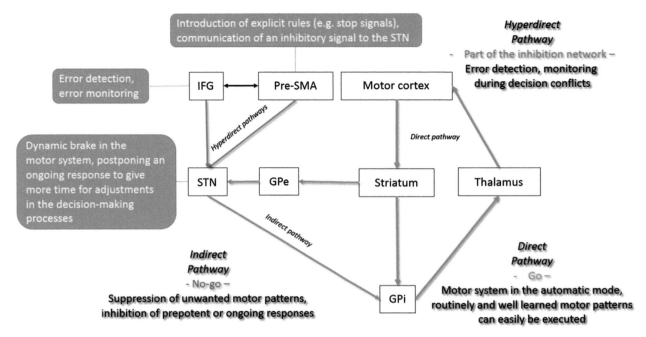

FIGURE 25.2   Anatomical and functional organization of the basal ganglia in decision-making. The impact of the STN on decision-making is explained in the text. *GPe*, external globus pallidus; *GPi*, internal globus pallidus; *IFG*, inferior frontal gyrus; *PreSMA*, supplemental motor area; *STN*, subthalamic nucleus.

operates by using different frequencies of neural oscillations. Slow oscillations in the range of the delta–theta band between preSMA and STN are used to increase decision thresholds [56,87]. Synchronization in the beta band within the STN serves as a brake to communicate a stop signal in the motor system [83]. In this way, cognitive processes of active and proactive inhibition influence the motor system to regulate behavior. This model is surely oversimplified and many questions remain unanswered. The exact role of the SMA and the IFG in the regulation of inhibition is still controversial and the influence of the posterior sensorimotor system in motor preparation and inhibition needs to be clarified. There is still a debate about the laterality of the inhibitory cortical network [69,79,91]. Furthermore, it remains to be seen whether the dynamic decision threshold mediated via the hyperdirect pathway is a universal mechanism for different kinds of conflicts (e.g., risky decisions, emotional conflicts, conflicts arising from information learned implicitly).

References

[1] Jahanshahi M, Obeso I, Baunez C, Alegre M, Krack P. Parkinson's disease, the subthalamic nucleus, inhibition, and impulsivity. Mov Disord 2015;30(2):128–40.

[2] Richter EO, Hoque T, Halliday W, Lozano AM, Saint-Cyr JA. Determining the position and size of the subthalamic nucleus based on magnetic resonance imaging results in patients with advanced Parkinson disease. J Neurosurg 2004;100(3):541–6.

[3] Bergman H, Wichmann T, DeLong MR. Reversal of experimental parkinsonism by lesions of the subthalamic nucleus. Science 1990; 249(4975):1436–8.

[4] Postuma RB, Lang AE. Hemiballism: revisiting a classic disorder. Lancet Neurol 2003;2(11):661–8.

[5] Whittier JR. Ballism and the subthalamic nucleus hypothalamicus; corpus luysi) review of the literature and study of 30 cases. Arch Neurol Psychiatry 1947;58(6):672–92.

[6] Benazzouz A, Gross C, Feger J, Boraud T, Bioulac B. Reversal of rigidity and improvement in motor performance by subthalamic high-frequency stimulation in MPTP-treated monkeys. Eur J Neurosci 1993;5(4):382–9.

[7] Agnesi F, Johnson MD, Vitek JL. Deep brain stimulation: how does it work? Handb Clin Neurol 2013;116:39–54.

[8] Limousin P, Pollak P, Benazzouz A, Hoffmann D, Le Bas JF, Broussolle E, et al. Effect of parkinsonian signs and symptoms of bilateral subthalamic nucleus stimulation. Lancet 1995; 345(8942):91–5.

[9] Deuschl G, Fogel W, Hahne M, Kupsch A, Muller D, Oechsner M, et al. Deep-brain stimulation for Parkinson's disease. J Neurol 2002;249(Suppl. 3):III/36–39.

[10] Volkmann J, Daniels C, Witt K. Neuropsychiatric effects of subthalamic neurostimulation in Parkinson disease. Nat Rev Neurol 2010;6(9):487–98.

[11] DeLong MR, Wichmann T. Circuits and circuit disorders of the basal ganglia. Arch Neurol 2007;64(1):20–4.

[12] Parent A, Hazrati LN. Functional anatomy of the basal ganglia. II. The place of subthalamic nucleus and external pallidum in basal ganglia circuitry. Brain Res Brain Res Rev 1995;20(1): 128–54.

[13] Nambu A, Tokuno H, Takada M. Functional significance of the cortico-subthalamo-pallidal "hyperdirect" pathway. Neurosci Res 2002;43(2):111–7.

[14] Parent A, Hazrati LN. Functional anatomy of the basal ganglia. I. The cortico-basal ganglia-thalamo-cortical loop. Brain Res Brain Res Rev 1995;20(1):91–127.

[15] Alexander GE, Crutcher MD. Functional architecture of basal ganglia circuits: neural substrates of parallel processing. Trends Neurosci 1990;13(7):266−71.

[16] Frank MJ. Hold your horses: a dynamic computational role for the subthalamic nucleus in decision making. Neural Netw 2006;19(8): 1120−36.

[17] Mink JW. The basal ganglia: focused selection and inhibition of competing motor programs. Prog Neurobiol 1996;50(4):381−425.

[18] Bevan MD, Atherton JF, Baufreton J. Cellular principles underlying normal and pathological activity in the subthalamic nucleus. Curr Opin Neurobiol 2006;16(6):621−8.

[19] Romanelli P, Heit G, Hill BC, Kraus A, Hastie T, Bronte-Stewart HM. Microelectrode recording revealing a somatotopic body map in the subthalamic nucleus in humans with Parkinson disease. J Neurosurg 2004;100(4):611−8.

[20] DeLong MR, Crutcher MD, Georgopoulos AP. Primate globus pallidus and subthalamic nucleus: functional organization. J Neurophysiol 1985;53(2):530−43.

[21] Alkemade A. Subdivisions and anatomical boundaries of the subthalamic nucleus. J Neurosci 2013;33(22):9233−4.

[22] Haynes WI, Haber SN. The organization of prefrontal-subthalamic inputs in primates provides an anatomical substrate for both functional specificity and integration: implications for basal ganglia models and deep brain stimulation. J Neurosci 2013;33(11):4804−14.

[23] Accolla EA, Dukart J, Helms G, Weiskopf N, Kherif F, Lutti A, et al. Brain tissue properties differentiate between motor and limbic basal ganglia circuits. Hum Brain Mapp 2014;35(10): 5083−92.

[24] Frank MJ, Seeberger LC, O'Reilly RC. By carrot or by stick: cognitive reinforcement learning in parkinsonism. Science 2004; 306(5703):1940−3.

[25] Schultz W. Behavioral dopamine signals. Trends Neurosci 2007; 30(5):203−10.

[26] Whittier JR, Mettler FA. Studies on the subthalamus of the rhesus monkey. II. Hyperkinesia and other physiologic effects of subthalamic lesions, with special reference to the subthalamic nucleus of Luys. J Comp Neurol 1949;90(3):319−72.

[27] Robbins TW. The 5-choice serial reaction time task: behavioural pharmacology and functional neurochemistry. Psychopharmacology (Berl) 2002;163(3−4):362−80.

[28] Baunez C, Nieoullon A, Amalric M. In a rat model of parkinsonism, lesions of the subthalamic nucleus reverse increases of reaction time but induce a dramatic premature responding deficit. J Neurosci 1995;15(10):6531−41.

[29] Christakou A, Robbins TW, Everitt BJ. Functional disconnection of a prefrontal cortical-dorsal striatal system disrupts choice reaction time performance: implications for attentional function. Behav Neurosci 2001;115(4):812−25.

[30] Chudasama Y, Passetti F, Rhodes SE, Lopian D, Desai A, Robbins TW. Dissociable aspects of performance on the 5-choice serial reaction time task following lesions of the dorsal anterior cingulate, infralimbic and orbitofrontal cortex in the rat: differential effects on selectivity, impulsivity and compulsivity. Behav Brain Res 2003;146(1−2):105−19.

[31] Rogers RD, Baunez C, Everitt BJ, Robbins TW. Lesions of the medial and lateral striatum in the rat produce differential deficits in attentional performance. Behav Neurosci 2001; 115(4):799−811.

[32] Chudasama Y, Robbins TW. Functions of frontostriatal systems in cognition: comparative neuropsychopharmacological studies in rats, monkeys and humans. Biol Psychol 2006;73(1):19−38.

[33] Bari A, Robbins TW. Inhibition and impulsivity: behavioral and neural basis of response control. Prog Neurobiol 2013;108:44−79.

[34] Eagle DM, Baunez C, Hutcheson DM, Lehmann O, Shah AP, Robbins TW. Stop-signal reaction-time task performance: role of

[35] Isoda M, Hikosaka O. Cortico-basal ganglia mechanisms for overcoming innate, habitual and motivational behaviors. Eur J Neurosci 2011;33(11):2058−69.

[36] Matsumura M, Kojima J, Gardiner TW, Hikosaka O. Visual and oculomotor functions of monkey subthalamic nucleus. J Neurophysiol 1992;67(6):1615−32.

[37] Baunez C, Dias C, Cador M, Amalric M. The subthalamic nucleus exerts opposite control on cocaine and "natural" rewards. Nat Neurosci 2005;8(4):484−9.

[38] Lardeux S, Paleressompoulle D, Pernaud R, Cador M, Baunez C. Different populations of subthalamic neurons encode cocaine vs. sucrose reward and predict future error. J Neurophysiol 2013; 110(7):1497−510.

[39] Montaurier C, Morio B, Bannier S, Derost P, Arnaud P, Brandolini-Bunlon M, et al. Mechanisms of body weight gain in patients with Parkinson's disease after subthalamic stimulation. Brain 2007; 130(Pt 7):1808−18.

[40] Ardouin C, Voon V, Worbe Y, Abouazar N, Czernecki V, Hosseini H, et al. Pathological gambling in Parkinson's disease improves on chronic subthalamic nucleus stimulation. Mov Disord 2006;21(11):1941−6.

[41] Lhommee E, Klinger H, Thobois S, Schmitt E, Ardouin C, Bichon A, et al. Subthalamic stimulation in Parkinson's disease: restoring the balance of motivated behaviours. Brain 2012;135(Pt 5):1463−77.

[42] Antonini A, Isaias IU, Rodolfi G, Landi A, Natuzzi F, Siri C, et al. A 5-year prospective assessment of advanced Parkinson disease patients treated with subcutaneous apomorphine infusion or deep brain stimulation. J Neurol 2010;258(4):579−85.

[43] Thobois S, Ardouin C, Lhommee E, Klinger H, Lagrange C, Xie J, et al. Non-motor dopamine withdrawal syndrome after surgery for Parkinson's disease: predictors and underlying mesolimbic denervation. Brain 2010;133(Pt 4):1111−27.

[44] Voon V, Krack P, Lang AE, Lozano AM, Dujardin K, Schupbach M, et al. A multicentre study on suicide outcomes following subthalamic stimulation for Parkinson's disease. Brain 2008;131(Pt 10): 2720−8.

[45] Schupbach M, Gargiulo M, Welter ML, Mallet L, Behar C, Houeto JL, et al. Neurosurgery in Parkinson disease: a distressed mind in a repaired body? Neurology 2006;66(12):1811−6.

[46] Brown P, Oliviero A, Mazzone P, Insola A, Tonali P, Di Lazzaro V. Dopamine dependency of oscillations between subthalamic nucleus and pallidum in Parkinson's disease. J Neurosci 2001; 21(3):1033−8.

[47] Braak H, Del Tredici K. Invited Article: nervous system pathology in sporadic Parkinson disease. Neurology 2008;70(20):1916−25.

[48] Jahanshahi M, Ardouin CM, Brown RG, Rothwell JC, Obeso J, Albanese A, et al. The impact of deep brain stimulation on executive function in Parkinson's disease. Brain 2000;123(Pt 6):1142−54.

[49] Witt K, Daniels C, Reiff J, Krack P, Volkmann J, Pinsker MO, et al. Neuropsychological and psychiatric changes after deep brain stimulation for Parkinson's disease: a randomised, multicentre study. Lancet Neurol 2008;7(7):605−14.

[50] Witt K, Pulkowski U, Herzog J, Lorenz D, Hamel W, Deuschl G, et al. Deep brain stimulation of the subthalamic nucleus improves cognitive flexibility but impairs response inhibition in Parkinson disease. Arch Neurol 2004;61(5):697−700.

[51] Schroeder U, Kuehler A, Haslinger B, Erhard P, Fogel W, Tronnier VM, et al. Subthalamic nucleus stimulation affects striato-anterior cingulate cortex circuit in a response conflict task: a PET study. Brain 2002;125(Pt 9):1995−2004.

[52] Witt K, Granert O, Daniels C, Volkmann J, Falk D, van Eimeren T, et al. Relation of lead trajectory and electrode position to neuropsychological outcomes of subthalamic neurostimulation in

prefrontal cortex and subthalamic nucleus. Cereb Cortex 2008; 18(1):178−88.

Parkinson's disease: results from a randomized trial. Brain 2013; 136(Pt 7):2109–19.

[53] Brittain JS, Watkins KE, Joundi RA, Ray NJ, Holland P, Green AL, et al. A role for the subthalamic nucleus in response inhibition during conflict. J Neurosci 2012;32(39):13396–401.

[54] Frank MJ, Samanta J, Moustafa AA, Sherman SJ. Hold your horses: impulsivity, deep brain stimulation, and medication in parkinsonism. Science 2007;318(5854):1309–12.

[55] Zaghloul KA, Weidemann CT, Lega BC, Jaggi JL, Baltuch GH, Kahana MJ. Neuronal activity in the human subthalamic nucleus encodes decision conflict during action selection. J Neurosci 2012; 32(7):2453–60.

[56] Cavanagh JF, Wiecki TV, Cohen MX, Figueroa CM, Samanta J, Sherman SJ, et al. Subthalamic nucleus stimulation reverses mediofrontal influence over decision threshold. Nat Neurosci 2011; 14(11):1462–7.

[57] Frank MJ, Gagne C, Nyhus E, Masters S, Wiecki TV, Cavanagh JF, et al. fMRI and EEG predictors of dynamic decision parameters during human reinforcement learning. J Neurosci 2015;35(2): 485–94.

[58] Coulthard EJ, Bogacz R, Javed S, Mooney LK, Murphy G, Keeley S, et al. Distinct roles of dopamine and subthalamic nucleus in learning and probabilistic decision making. Brain 2012; 135(Pt 12):3721–34.

[59] Mirabella G, Iaconelli S, Romanelli P, Modugno N, Lena F, Manfredi M, et al. Deep brain stimulation of subthalamic nuclei affects arm response inhibition in Parkinson's patients. Cereb Cortex 2012;22(5):1124–32.

[60] Swann N, Poizner H, Houser M, Gould S, Greenhouse I, Cai W, et al. Deep brain stimulation of the subthalamic nucleus alters the cortical profile of response inhibition in the beta frequency band: a scalp EEG study in Parkinson's disease. J Neurosci 2011; 31(15):5721–9.

[61] van den Wildenberg WP, van Boxtel GJ, van der Molen MW, Bosch DA, Speelman JD, Brunia CH. Stimulation of the subthalamic region facilitates the selection and inhibition of motor responses in Parkinson's disease. J Cogn Neurosci 2006;18(4):626–36.

[62] Obeso I, Wilkinson L, Rodriguez-Oroz MC, Obeso JA, Jahanshahi M. Bilateral stimulation of the subthalamic nucleus has differential effects on reactive and proactive inhibition and conflict-induced slowing in Parkinson's disease. Exp Brain Res 2013;226(3):451–62.

[63] Ray NJ, Jenkinson N, Brittain J, Holland P, Joint C, Nandi D, et al. The role of the subthalamic nucleus in response inhibition: evidence from deep brain stimulation for Parkinson's disease. Neuropsychologia 2009;47(13):2828–34.

[64] Greenhouse I, Gould S, Houser M, Hicks G, Gross J, Aron AR. Stimulation at dorsal and ventral electrode contacts targeted at the subthalamic nucleus has different effects on motor and emotion functions in Parkinson's disease. Neuropsychologia 2011;49(3):528–34.

[65] Ray NJ, Brittain JS, Holland P, Joundi RA, Stein JF, Aziz TZ, et al. The role of the subthalamic nucleus in response inhibition: evidence from local field potential recordings in the human subthalamic nucleus. NeuroImage 2012;60(1):271–8.

[66] Tan H, Pogosyan A, Anzak A, Ashkan K, Bogdanovic M, Green AL, et al. Complementary roles of different oscillatory activities in the subthalamic nucleus in coding motor effort in Parkinsonism. Exp Neurol 2013;248:187–95.

[67] Alegre M, Lopez-Azcarate J, Obeso I, Wilkinson L, Rodriguez-Oroz MC, Valencia M, et al. The subthalamic nucleus is involved in successful inhibition in the stop-signal task: a local field potential study in Parkinson's disease. Exp Neurol 2013;239:1–12.

[68] Bastin J, Polosan M, Benis D, Goetz L, Bhattacharjee M, Piallat B, et al. Inhibitory control and error monitoring by human subthalamic neurons. Transl Psychiatry 2014;4:e439.

[69] Aron AR, Behrens TE, Smith S, Frank MJ, Poldrack RA. Triangulating a cognitive control network using diffusion-weighted magnetic resonance imaging (MRI) and functional MRI. J Neurosci 2007;27(14):3743–52.

[70] Aron AR, Poldrack RA. Cortical and subcortical contributions to Stop signal response inhibition: role of the subthalamic nucleus. J Neurosci 2006;26(9):2424–33.

[71] Forstmann BU, Keuken MC, Jahfari S, Bazin PL, Neumann J, Schafer A, et al. Cortico-subthalamic white matter tract strength predicts interindividual efficacy in stopping a motor response. NeuroImage 2012;60(1):370–5.

[72] Li CS, Yan P, Sinha R, Lee TW. Subcortical processes of motor response inhibition during a stop signal task. NeuroImage 2008; 41(4):1352–63.

[73] Rushworth MF, Walton ME, Kennerley SW, Bannerman DM. Action sets and decisions in the medial frontal cortex. Trends Cogn Sci 2004;8(9):410–7.

[74] Duann JR, Ide JS, Luo X, Li CS. Functional connectivity delineates distinct roles of the inferior frontal cortex and presupplementary motor area in stop signal inhibition. J Neurosci 2009;29(32):10171–9.

[75] Swann N, Tandon N, Canolty R, Ellmore TM, McEvoy LK, Dreyer S, et al. Intracranial EEG reveals a time- and frequency-specific role for the right inferior frontal gyrus and primary motor cortex in stopping initiated responses. J Neurosci 2009;29(40): 12675–85.

[76] Swann NC, Cai W, Conner CR, Pieters TA, Claffey MP, George JS, et al. Roles for the pre-supplementary motor area and the right inferior frontal gyrus in stopping action: electrophysiological responses and functional and structural connectivity. NeuroImage 2012;59(3):2860–70.

[77] Jahfari S, Stinear CM, Claffey M, Verbruggen F, Aron AR. Responding with restraint: what are the neurocognitive mechanisms? J Cogn Neurosci 2010;22(7):1479–92.

[78] Obeso I, Wilkinson L, Casabona E, Speekenbrink M, Luisa Bringas M, Alvarez M, et al. The subthalamic nucleus and inhibitory control: impact of subthalamotomy in Parkinson's disease. Brain 2014;137(Pt 5):1470–80.

[79] Ballanger B, van Eimeren T, Moro E, Lozano AM, Hamani C, Boulinguez P, et al. Stimulation of the subthalamic nucleus and impulsivity: release your horses. Ann Neurol 2009;66(6):817–24.

[80] Hershey T, Campbell MC, Videen TO, Lugar HM, Weaver PM, Hartlein J, et al. Mapping Go-No-Go performance within the subthalamic nucleus region. Brain 2010;133(Pt 12):3625–34.

[81] Hershey T, Revilla FJ, Wernle A, Gibson PS, Dowling JL, Perlmutter JS. Stimulation of STN impairs aspects of cognitive control in PD. Neurology 2004;62(7):1110–4.

[82] Campbell MC, Karimi M, Weaver PM, Wu J, Perantie DC, Golchin NA, et al. Neural correlates of STN DBS-induced cognitive variability in Parkinson disease. Neuropsychologia 2008; 46(13):3162–9.

[83] Kuhn AA, Williams D, Kupsch A, Limousin P, Hariz M, Schneider GH, et al. Event-related beta desynchronization in human subthalamic nucleus correlates with motor performance. Brain 2004;127(Pt 4):735–46.

[84] Steele VR, Claus ED, Aharoni E, Harenski C, Calhoun VD, Pearlson G, et al. A large scale (N = 102) functional neuroimaging study of error processing in a Go/NoGo task. Behav Brain Res 2014;268:127–38.

[85] Jaffard M, Longcamp M, Velay JL, Anton JL, Roth M, Nazarian B, et al. Proactive inhibitory control of movement assessed by event-related fMRI. NeuroImage 2008;42(3):1196–206.

[86] Mattingley JB, Husain M, Rorden C, Kennard C, Driver J. Motor role of human inferior parietal lobe revealed in unilateral neglect patients. Nature 1998;392(6672):179–82.

[87] Zavala BA, Tan H, Little S, Ashkan K, Hariz M, Foltynie T, et al. Midline frontal cortex low-frequency activity drives subthalamic

nucleus oscillations during conflict. J Neurosci 2014;34(21): 7322—33.

[88] Boller JK, Barbe MT, Pauls KA, Reck C, Brand M, Maier F, et al. Decision-making under risk is improved by both dopaminergic medication and subthalamic stimulation in Parkinson's disease. Exp Neurol 2014;254:70—7.

[89] Castrioto A, Funkiewiez A, Debu B, Cools R, Lhommee E, Ardouin C, et al. Iowa gambling task impairment in Parkinson's disease can be normalised by reduction of dopaminergic medication after subthalamic stimulation. J Neurol Neurosurg Psychiatry 2015;86(2):186—90.

[90] Wiecki TV, Frank MJ. A computational model of inhibitory control in frontal cortex and basal ganglia. Psychol Rev 2013;120(2): 329—55.

[91] Aron AR, Fletcher PC, Bullmore ET, Sahakian BJ, Robbins TW. Stop-signal inhibition disrupted by damage to right inferior frontal gyrus in humans. Nat Neurosci 2003;6(2):115—6.

[92] Logan GD, Cowan WB, Davis KA. On the ability to inhibit simple and choice reaction time responses: a model and a method. J Exp Psychol Hum Percept Perform 1984;10(2):276—91.

[93] Lezak MD, Howieson DB, Loring DW. In: Neuropsychological assessment. 4th ed. New York: Oxford University Press; 2004.

26

Decision-Making in Anxiety and Its Disorders

D.W. Grupe

University of Wisconsin—Madison, Madison, WI, United States

Abstract

Anxiety disorders are characterized in part by heightened sensitivity to unpredictable future threat, which is manifest across diverse cognitive, affective, physiological, and behavioral symptoms of these disorders. Although less frequently discussed than processes such as altered fear learning and memory, emerging evidence suggests that chief among these behavioral symptoms are disruptions to value-based decision-making. In focusing a broad array of data through a decision-making lens, this chapter highlights alterations to five discrete processes in anxiety and its disorders: decision representation, valuation, action selection, outcome evaluation, and learning. Distinct anxious phenotypes may be characterized by differential alterations to these five processes and their associated neurobiological mechanisms. This framework suggests as-of-yet untested hypotheses related to each of these processes, which may lead to an increased understanding of the neurobiology of pathological anxiety while also refining the nosology of anxiety to include information regarding disrupted decision-making processes.

INTRODUCTION

We are all familiar with the experience of anxiety: a racing heart and jittery hands, increased attention to signs of threat, worried thoughts about meeting deadlines or pleasing our family. These physiological, behavioral, and cognitive features of anxiety attune the individual to potential negative events in the (current or future) environment, and the aversive subjective experience of anxiety serves as a motivating force to avoid these events or obtain greater certainty about them. In this way, anxiety serves an adaptive and functional role, so long as it is experienced in appropriate contexts and in a controllable, time-limited manner.

For many individuals, however, this state is triggered in nonthreatening contexts, is experienced excessively and uncontrollably, and interferes with daily function, work performance, and relationships with friends and family. Anxiety disorders are the most frequently experienced and costly class of mental disorders, with lifetime prevalence rates of almost 30% [1] and an estimated economic impact of $42 billion within the United States alone [2].

Although there are distinct diagnostic labels for different anxiety disorders, these conditions frequently co-occur with one another and other psychiatric disorders (particularly depression), and efforts to identify unique biological markers for specific diagnoses have been largely unsuccessful. Rather than categorizing these disorders on the basis of observable symptoms, there is a growing push to investigate biological mechanisms that may confer general risk for an array of psychiatric conditions, with specific manifestations of disorders influenced by unique environmental factors experienced at different developmental time points [3,4].

In the search for biological mechanisms of anxiety disorders, the research domain of decision-making has been largely ignored. Human neuroimaging studies typically induce fear or anxiety using passive procedures, such as fear conditioning or the presentation of threatening faces or aversive images. Although this approach helps link the experience of fear or anxiety to brain activation patterns, it fails to explain how this experience influences the choices we constantly consider and execute in our day-to-day lives. The ultimate impact of anxiety disorders is reflected not merely in phenomenological experience, but in the behavioral consequences of this experience, including the choices one makes between potentially rewarding and punishing outcomes.

This chapter explores neural mechanisms of anxiety and decision-making in the context of an influential model of value-based decision-making [5]. According to this perspective, the decision-making process can be decomposed into five stages that unfold sequentially to promote adaptive choice behavior (Fig. 26.1). These stages are (1) the *representation* of the decision, as

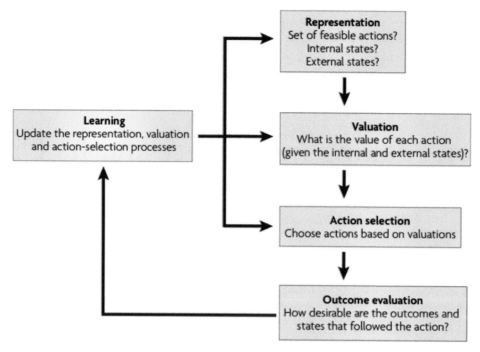

FIGURE 26.1   Five normative stages of adaptive decision-making that may be disrupted in anxiety disorders. *Figure reprinted with permission from Rangel A, Camerer CF, Montague PR. A framework for studying the neurobiology of value-based decision making. Nat Rev Neurosci 2008;9:545–56.*

influenced by one's internal state and external contextual factors; (2) the *valuation* of various options; (3) *action selection*; (4) *outcome evaluation*; and (5) *learning*, which feeds back to earlier stages to optimize future choice behavior.

Each section briefly discusses the normative role of each stage before reviewing extant empirical data and clinical observations related to their alterations in subclinical and clinical manifestations of anxiety.[1] Because anxiety and its disorders have not traditionally been considered in terms of aberrant decision-making, studies explicitly investigating specific facets of decision-making in anxiety are somewhat scarce. Nevertheless, a diverse array of neuroimaging, behavioral, and peripheral physiological data—when re-viewed through a decision-making lens—reveal pervasive alterations to decision-making processes in highly anxious individuals. Critically, altered or maladaptive

processing at any stage of the decision-making process can have deleterious consequences that cascade through the rest of the process, in addition to feeding back on earlier stages.

An array of neural and behavioral mechanisms is proposed to account for maladaptive responses to uncertainty that are observed across anxiety disorders [6]. As the general themes of uncertainty and decision-making are largely inextricable from one another, many of these mechanisms also account for decision-making alterations in anxiety (Fig. 26.2). In particular, a core feature of anxiety is elevated "intolerance of uncertainty," or difficulty tolerating the slightest possibility that some negative event may occur in the future. As such, anxious individuals are likely to forego greater gains associated with even small amounts of uncertainty to obtain smaller, more certain gains. Importantly, this tendency can have adaptive as well as maladaptive

---

[1] A common strategy when conducting research on anxiety disorders is to investigate individual differences, within a nonclinical sample, in self-reported anxiety characteristics using a measure such as the State–Trait Anxiety Inventory (STAI). It is important to recognize that critical differences exist between elevated "trait anxiety" (which is a relatively nonspecific construct more indicative of neuroticism or negative affect) and clinical anxiety disorders. That being said, these investigations in convenience samples provide important insight into disorders of anxiety at a considerable savings of cost and effort and, given the high rates of conversion from elevated trait anxiety to clinical disorders [91], allow for the identification of potential risk factors prior to the development of a disorder, particularly when using developmental and longitudinal research designs. Consistent with the growing acceptance of dimensional perspectives in psychiatry, data from nonclinical samples will be presented here, including measures indexing general negative affect/state anxiety (such as the STAI) as well as those targeting more specific dimensions of anxiety, including the Anxiety Sensitivity Index, Penn State Worry Questionnaire, and Intolerance of Uncertainty Scale.

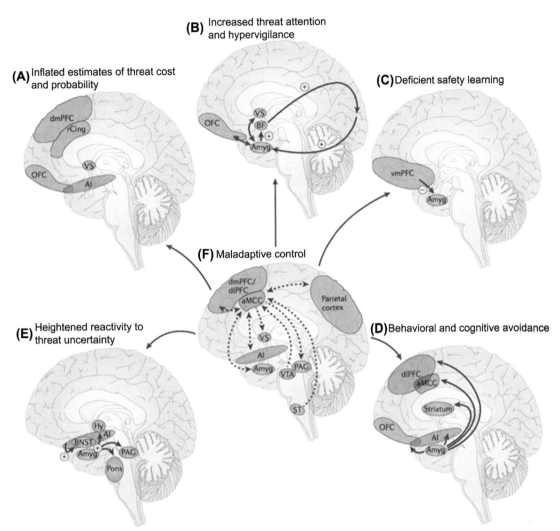

FIGURE 26.2 Proposed neural circuitry associated with alterations to maladaptive psychological responses to conditions of uncertainty in clinical anxiety disorders [6]. (A) Inflated estimates of the probability or cost of aversive outcomes are associated with disruptions to the dorsomedial prefrontal cortex (dmPFC), rostral cingulate (rCing), orbitofrontal cortex (OFC), ventral striatum (VS), and anterior insula (AI), and the overweighting of aversive events biases valuation and subsequent choice. (B) Elevated amygdala (Amyg) and basal forebrain (BF) activity modulates sensory input and contributes to hypervigilance, and interactions with the OFC and VS exacerbate attentional biases toward threat, thus boosting the salience of external sources of threat for subsequent choice behavior. (C) Deficient safety learning reflects disrupted inhibitory ventromedial PFC (vmPFC)—Amyg circuitry; it is unknown to what extent safety learning and valuation rely on similar versus distinct aspects of the vmPFC. (D) Behavioral and cognitive avoidance may reflect disruptions to circuitry critical for selecting the most adaptive choice when faced with approach—avoidance conflict, including the Amyg, OFC, dorsolateral PFC (dlPFC), striatum, anterior midcingulate cortex (AMCC), and AI. (E) Elevated activity in the bed nucleus of the stria terminalis (BNST) and Amyg in response to sustained, unpredictable threat modulates defensive responding, as mediated by the hypothalamus (Hy), pons, periaqueductal gray (PAG), and other midbrain and brain-stem structures. Greater sensitivity to punishing outcomes may contribute to loss-aversive behavior in anxiety. (F) Dysfunction of the AMCC, or disrupted structural connectivity between the AMCC and interconnected regions, prevents individuals from selecting and executing adaptive actions from competing options under conditions of uncertainty and contributes to each of the disruptions highlighted in (A—E). Lateral cortical regions are shown in *blue*, medial cortical regions in *pink*, and subcortical regions in *green*. Arrows in (A—E) depict functional pathways (*plus signs* indicate excitatory pathways and *minus signs* indicate inhibitory pathways). *Dashed arrows* in (F) depict known structural connections (directionality indicated by *arrowheads*). *ST*, spinothalamic tract; *VTA*, ventral tegmental area. *Figure and portions of legend reprinted with permission from Grupe DW, Nitschke JB. Uncertainty and anticipation in anxiety: an integrated neurobiological and psychological perspective. Nat Rev Neurosci 2013;14:488–501.*

consequences. Depending on the particular decision context, there may be physical, emotional, or survival benefits associated with minimizing risk or uncertainty and prioritizing harm minimization, even if this results in "suboptimal" decisions from a normative perspective.

The presence of anxiety and its behavioral consequences are not unequivocally negative, but should be considered relative to these contextual factors and the consequences of actions on the global well-being of the individual and the individual's ability to maintain healthy relationships and function effectively in daily life.

Finally, this chapter will not emphasize associations between specific anxiety disorders and particular decision-making disruptions or neural abnormalities. Consistent with the movement toward dimensional rather than categorical classification systems in psychiatry [3,4], the emphasis is on shared decision-making disruptions across different manifestations of anxiety. By adopting this process-oriented perspective, this review can inform future research that focuses on disruptions to specific neural circuits and psychological processes, rather than collections of observable symptoms. The identification of shared phenotypes characterized by impairments to specific decision-making processes may contribute to a refined conceptualization of anxiety disorders.

Representation

Before even approaching a particular decision scenario, myriad factors related to one's internal state and the external environment are silently setting the stage for the ultimate choice. For example, the decision to attend a social outing could depend both on one's internal state (Am I too tired? Do I feel a cold coming on?) and on external factors (What's the traffic like? Is it too cold to bike today?).

The representation of external factors in anxiety is influenced by attentional, cognitive, and perceptual biases that inflate perceptions of threat in the environment, thus coloring subsequent decision-making processes. Attentional bias to threat is a robust phenomenon in anxiety, with effects observed in different tasks and across different diagnoses and elevated trait anxiety [7,8]. Facilitated attention to threat, and difficulty disengaging this attention once captured, partially reflects increased activity in the amygdala, which is involved in the conscious and unconscious detection of biologically relevant stimuli and the deployment of attentional and perceptual resources toward external threat [9,10] (Fig. 26.2B). Increased amygdala sensitivity to potential danger appears to be a trait-like marker of susceptibility to anxiety disorders [11,12] that increases the salience of external sources of threat for subsequent choice behavior.

This construal of the external world as an overtly dangerous place is further facilitated by elevated "interpretation bias," or the tendency to interpret ambiguous stimuli as threatening. This cognitive bias is observed in generalized anxiety disorder (for spoken words with multiple meanings [13]), social anxiety disorder (for ambiguous social scenarios [14] and facial expressions [15]), and veterans with posttraumatic stress disorder (PTSD; for ambiguous sentences related to combat [16]).

Individual differences in sensory thresholds for threat also affect external representations, as illustrated by a study conducted in a group of individuals preparing for their first skydive [17]. Immediately before jumping, participants completed a threat perception task in which they indicated the moment when a neutral face morphed into an angry face. Greater sensation-seeking was associated with a higher threshold for perceiving this shift, and increased threat detection thresholds were in turn associated with lower cortisol responses during the jump and reduced anticipatory amygdala activation to threat in a separate fMRI session. This study of individuals at the phenotypic extreme of risk-avoidant anxious individuals allows insight into biological mechanisms of altered threat perception that, along with attentional and cognitive biases, can influence subsequent choice behavior.

Anxiety disorders have been associated with superior attention to and perception of the physiological state of the body, or interoception [18]. Emotion theorists have long proposed that the central processing of peripheral physiological signals influences cognition and behavior in domains including decision-making [19]. The somatic marker hypothesis (SMH) has provided an influential framework for how increased interoceptive sensitivity or accuracy influences decision-making [20] (although it has been heavily criticized for its lack of mechanistic specificity; see [21]). The SMH proposes that peripheral autonomic changes are represented as "somatic markers" in brain regions including the ventromedial prefrontal cortex (VMPFC), amygdala, and insula, and that reactivation of these central markers influences approach or avoidance behavior [20,22].

The SMH is classically tested using the Iowa Gambling Task (IGT), in which participants select cards from decks that are advantageous or disadvantageous (leading to greater long-term wins or losses, respectively). Recordings of peripheral autonomic activity show increased skin conductance responses preceding selection from "bad" relative to "good" decks [22], interpreted as support for the adaptive influence of somatic markers on decision-making.

Studies in nonanxious populations have demonstrated an association between improved interoceptive accuracy and both IGT performance [23] and the overweighting of large losses relative to large gains [24]. Consistent with the association between anxiety and greater interoceptive accuracy [18], several studies have reported greater IGT performance in participants with elevated trait anxiety [25,26] or generalized anxiety disorder [27]. A separate study reported *poorer* IGT performance in high trait anxiety [28], with discrepant results potentially attributable to interactions between trait and state anxiety [29].

Although there is tentative evidence for a link between interoceptive sensitivity and adaptive decision-making in anxiety, no experiment has been conducted as of this writing to directly test the hypothesis that increased interoceptive accuracy mediates the relationship between anxiety and adaptive decision-making. The idea that anxiety may have adaptive benefits for risky decision-making is intriguing, and further research on the adaptive behavioral advantages of anxiety could provide a refreshing counterpoint to the overwhelming emphasis on its negative consequences.

Valuation

The subjective value of different outcomes is influenced by the decision representation as well as additional choice-specific factors, including the temporal delay between the decision and the outcome and the uncertainty associated with various options. Greater delay and greater uncertainty contribute to less accurate predictions of decision outcomes, and each can affect decision-making in anxiety.

Exaggerated devaluation of future outcomes has been implicated in other psychiatric disorders, including drug addiction and pathological gambling [30]. Anxious individuals may show similar discounting behavior, given that distal events are associated with greater uncertainty, a particularly intolerable state for these individuals [31]. Indeed, steeper discounting of future rewards has been observed in participants with social anxiety symptoms [32] and elevated trait worry characteristics [33]. Individuals with greater intolerance of uncertainty also demonstrated a preference for immediately available outcomes with a lower expected value when the alternative was waiting in a state of uncertainty, even just a few seconds, to receive payouts with higher expected value [34].

Although sparse, these data suggest that elevated trait anxiety is associated with decreased valuation of future rewards, perhaps owing to the uncertainty associated with waiting. Each of these studies, however, compared decisions between immediate and delayed rewards. For situations involving punishment or loss, dispositional anxiety or worry may be associated with a distinct profile of discounting, as the more future-oriented perspective that is a signature of anxiety [27,35] could imbue future negative outcomes with relatively greater weight than proximal outcomes (Fig. 26.2A).

Temporal factors aside, the motivation to avoid uncertainty contributes to greater preferences for safe options versus risky alternatives in high trait anxiety [36]. Self-reported risk-taking across domains is associated with anxiety symptoms in healthy populations and in students with clinically significant anxiety symptoms [37]. Highlighting the functional significance of this tendency, reduced risk-taking in social anxiety and

generalized anxiety disorders is associated with reduced treatment-seeking [38].

The behavioral assessment of risk-taking provides a more objective indicator of risk aversion than studies relying on self-report. A commonly used behavioral task of risk aversion is the Balloon Analogue Risk Task (BART; [39]). On each trial, individuals decide to either take the safe option of cashing out points or pump up a virtual balloon to earn additional points, at the risk of exploding the balloon and losing all of their points. Risk-averse BART behavior is related to greater trait anxiety and social anxiety symptoms [37,40], even after controlling for differences in general negative affect [37]. Patients with panic disorder and generalized anxiety disorder also showed more conservative behavior when choosing between gambles of small- versus large-magnitude gains and losses [40]. In contrast to these studies, social anxiety symptoms were positively associated with risky decision-making in simulated social interactions, which the authors interpreted as a compensatory emotional regulation strategy to avoid feelings of anxiety arising from social interactions [41].

Studies utilizing behavioral and self-report measures thus provide support for a domain-general tendency to avoid risky outcomes in anxiety. Risk avoidance is not related to specific fears or objects of anxiety, but is instead, in the words of Maner [37], "relatively basic and 'content-free' ... [R]isk-avoidant decision-making can be viewed as the output of a motivational process—initiated by the experience of anxiety—that leads individuals to avoid threats associated with potentially risky courses of action."

Additional research is needed that relates anxiety to parametrically altered risk, to better quantify risk aversion and the value that different individuals place on certainty. Additionally, although anxious individuals interpret ambiguous scenarios as more negative than do nonanxious individuals [13,42], it is unclear how this tendency translates into behavioral choice under conditions of ambiguity. Implementing both types of tasks in the same individuals may reveal distinct anxious phenotypes characterized by risk and ambiguity aversion, with corresponding neurobiological differences.

In terms of neural mechanisms of valuation, activity in the posterior VMPFC reflects a "common currency" of valuation across different stimulus modalities, tasks, and valuation systems [43]. The VMPFC, ventral striatum, and posterior cingulate track the subjective value of choices under conditions of risk as well as ambiguity [44] and over different temporal delays [45]. The anterior insula may represent subjective value in a common currency in the opposite manner, such that increasing anterior insula activity is associated with decreasing subjective value [46]. Individual differences in this neural circuitry may contribute to devaluation of risky, ambiguous, or temporally delayed outcomes in highly anxious individuals.

Indeed, an fMRI study of choices between risky gambles and safe alternatives revealed striking alterations in neural valuation signals for decision-making under conditions of threat [47]. The neural representation of expected value in the VMPFC and ventral striatum was diminished when participants were exposed to threat of unpredictable shock, and during threat trials the anterior insula tracked the negative expected value of gambles. Furthermore, trial-by-trial activation in these regions differentially predicted choice behavior between safety (in the VMPFC and striatum) and threat (in the anterior insula). Extrapolating from this within-subject manipulation of threat to between-subject differences in anxiety, devaluation of risky outcomes may be driven by avoidance of potential losses, with altered valuation signals in the anterior insula, VMPFC, and ventral striatum mediating the relationship between anxiety and choice behavior (see Chapter 4).

Action Selection

Theoretically, action selection should directly reflect valuation, as an individual acting in his or her best interest should always choose the action with the greatest expected value. However, the translation of valuation into action selection is complicated by the presence of different neural systems—Pavlovian, habitual, and goal-directed—that can come into conflict with one another [5]. Selecting between options with competing costs and benefits encoded in distinct valuation systems is a tall order, and knowledge of how the brain implements this process is still quite limited. Rangel and colleagues [5] have proposed that executive dysfunction in psychopathology reflects difficulty mitigating conflicts between valuation systems and selecting the optimal course of action. For example, addiction involves a failure to adaptively resolve conflicts between appetitive Pavlovian responses to primary rewards and conflicting goal-directed responses for long-term outcomes.

Behavioral, cognitive, and emotional avoidance are transdiagnostic features of all anxiety disorders [48,49]. Behavioral avoidance can be thought of as maladaptive action selection in the face of approach/avoidance conflict, such that anxious individuals are excessively motivated to avoid potential harm or unpredictability. Although this avoidance may be adaptive in the short term (in that it ameliorates the sense of acute threat or panic), avoidance allows anxiety to be sustained and enhanced over the long run, as it prevents the disconfirmation of biased threat expectancies [37,50]. Exploring neural processes related to approach/avoidance conflict

is critical for understanding the nature of maladaptive avoidance in anxiety [51] (see Chapter 4 and Fig. 26.2D).

A critical clinical feature of avoidance is the corresponding sacrifice of potential gains: in the absence of some conflict between potential gains and losses, avoidance would lead to less distress and would not be considered "maladaptive" [51]. Researchers have adapted tasks initially developed in rodents to investigate approach/avoidance conflict (AAC) in humans. These tasks present choices between the delivery of both reward and punishment versus neither. The administration of anxiolytic drugs biases rodents' behavior toward approach rather than avoidance [52], suggesting that these tasks may have translational relevance for understanding human anxiety. As reviewed by Saga and Tremblay (Chapter 4), corticobasal ganglia loops involving the ventral striatum and ventral pallidum are critical for choosing an adaptive course of action from competing approach and avoidance outcomes in nonhuman primates.

In a translational adaptation of AAC tasks [53], human participants moved an avatar toward or away from an icon indicating increasing probability of viewing an unpleasant image. Critically, for trials on which participants viewed a "punishing" negative image, they also received a point reward, such that avoidance comes at the cost of reward consumption. Within a healthy control sample, correlations were observed between avoidance behavior and greater physical symptoms of anxiety (in males) and between approach behavior and self-reported approach tendencies (in females; [53]). This novel AAC task could be modified to parametrically alter outcomes on different trials (thus allowing for the identification of indifference points between relative amounts of reward versus punishment) or to include disorder-relevant punishments (e.g., threat of peer judgment in social anxiety disorder).

The negative consequences of avoidance on optimal decision-making were demonstrated in a modification of the traditional IGT in which advantageous and disadvantageous decks were paired with pictures of either spiders or butterflies, which were irrelevant to the task [54]. Individuals with spider phobia tended to avoid spider decks, resulting in reduced performance and fewer earnings than the control group on those trials when the spider decks were in fact associated with a higher long-term payout.

Competing options are associated with differences in not only expected value, but also the cost of the actions required to obtain these outcomes. The dorsal anterior cingulate cortex (DACC)/anterior midcingulate cortex (AMCC) are connected with cortical and subcortical motor circuitry and with brain regions involved in value representation, such as the ventral striatum and orbitofrontal cortex ([55], see also Chapter 4) (Fig. 26.2F).

Activity in the DACC and AMCC reflects the cost of executing various actions [56], and damage to these regions is associated with disrupted learning of the values associated with various actions [57]. Further, stimulation of the adjacent pregenual cingulate in macaques was shown to bias choice behavior toward avoidance in an AAC task, an effect that was blocked by delivery of an anxiolytic agent [58]. It has been proposed that the AMCC serves as a "hub, where information about pain, and other, more abstract kinds of punishment and negative feedback could be linked to motor centres responsible for ... coordinating aversively motivated instrumental behaviours" [55].

Given its proposed roles in representing the cost of actions and in selecting adaptive actions in the face of uncertainty, the AMCC is a prime candidate for maladaptive action selection in anxiety disorders. Relative to healthy controls, subjects with PTSD have shown increased activation of this region in response to cues formerly associated with punishment [59] and in response to cognitive interference tasks [60]. Altered functional activation and connectivity of this region in response to uncertainty has been reported in social anxiety disorder [61], pediatric anxiety disorders [62], and specific phobias [63]. These data highlight a general role for altered AMCC function across diverse manifestations of anxiety. Whether these general disruptions can be linked more specifically to maladaptive action selection—and increased avoidance behavior—remains to be explicitly tested in anxious populations.

Outcome Evaluation

Following action selection, one receives feedback about whether the chosen action led to reward maximization or harm mitigation. One manner in which feedback processing may be altered in highly anxious individuals is via elevated sensitivity to punishment, consistent with the drive to avoid potential punishment [64]. Although the anticipated cost of future negative events is elevated in highly anxious individuals [36], these expectancies could simply reflect a greater "affective forecasting bias," or the tendency to overestimate how bad imagined losses or challenges will make one feel, relative to the actual experience of loss [65].

Data reviewed by Shackman and colleagues [66], however, suggest that anxiety is indeed associated with elevated affective and physiological responses to punishment. Dispositionally anxious individuals report greater negative emotional responses to everyday stressors, interpersonal conflicts, and controlled laboratory stressors [66]. These exaggerated self-reported responses to stressors are paralleled by observations of increased behavioral, peripheral physiological, and neuroendocrine responses in anxious or behaviorally inhibited individuals to mild punishments or otherwise unpleasant laboratory stressors [67,68]. Neural and peripheral physiological responses to laboratory punishment (e.g., mild shock) under conditions of uncertainty are particularly elevated in anxious individuals, again highlighting the disruptive impact of uncertainty in anxiety [6] (Fig. 26.2E).

This enhanced reactivity to aversive outcomes under conditions of uncertainty could result in greater loss aversion. There is an asymmetric relationship between the hedonic impact of gains and losses, with the latter wielding relatively greater power over the former [69]. Loss aversion has been linked to increased physiological responses to losses [70] and increased interoceptive accuracy [24]. Avoidance of punishment may be driven either by increased physiological responses to punishment or by stronger subjective responses to equivalent physiological responses, owing to enhanced interoceptive function. These factors may also interact such that "sensitivity to physiological responses subsequently leads to an amplification of the bodily signal itself, which is then perceived even more intensely" [24]. Although it makes sense that increased loss aversion would be observed in highly anxious individuals—who display both increased sensitivity to punishment and elevated interoceptive ability—the only published study on this topic as of this writing, in clinically anxious adolescents, failed to find clear support for this hypothesis [71].

One neural mechanism for hypersensitivity to punishment in anxiety is heightened error- or surprise-related theta activity measured from frontomedial sites, or FMθ. FMθ, which is assessed using electrophysiological measurements of error-related negativity or feedback-related negativity, is thought to reflect conflict, surprise, or an unsigned prediction error signal. This signal drives adaptive adjustments to behavior by, for example, changing learning rates in more volatile environments or altering attentional allocation [72].

A 2015 metaanalysis provided evidence that individuals with elevated levels of dispositional anxiety show larger FMθ control signals, even in emotion-neutral cognitive control tasks [73]. This metaanalysis also showed that greater FMθ is consistently associated with more cautious behavioral responses following errors or punishment, in the form of both posterror slowing and response switching following errors. Similarly, an earlier fMRI study showed an association between elevated trait anxiety and increased DACC/medial prefrontal error signals during a forced-choice decision-making task [74]. Elevated FMθ control signals appear to reflect increased sensitivity to punishment in elevated trait anxiety and contribute to more cautious and avoidant behavior.

Evidence is also emerging for relative insensitivity to rewarding outcomes in PTSD and other anxiety disorders, which could further bias choices toward outcomes that minimize punishment rather than maximize gain. The high comorbidity of depression with PTSD, and the prominence of emotional numbing symptoms and reduced enjoyment of rewarding activities, hints at the possibility of reward-related abnormalities in this disorder [75]. Indeed, Vietnam veterans with PTSD reported less satisfaction from rewards obtained during a "wheel of fortune" task [76], and a follow-up fMRI study [77] demonstrated less striatal activation to gains versus losses in veterans with PTSD, particularly in veterans who reported a loss of interest in pleasurable activities. Similarly, trauma-exposed women with PTSD showed reduced ventral striatal feedback to monetary rewards relative to control subjects [78].

Decreased sensitivity to reward was also observed in behaviorally inhibited adolescents, who showed less dorsal striatum and VMPFC activation than their peers in response to monetary reward [79]. Reduced resting-state functional connectivity has been observed in social phobia between the nucleus accumbens and the VMPFC [80], which was interpreted by the authors as additional evidence for alterations to reward circuit function in anxiety disorders. Finally, in a pediatric community sample [81], dimensions of social anxiety and generalized anxiety showed opposing relationships with an event-related potential measure of reward sensitivity, interpreted as increased sensitivity to reward in social anxiety (see [82]) but decreased reward sensitivity in generalized anxiety.

Although it is relatively easy to evaluate decision outcomes in laboratory settings, these contrived choice scenarios may not capture the complexity of outcome evaluation "in the wild." Days, weeks, or years can pass before the consequences of real-world decisions are fully understood, and the outcomes of these decisions are rarely unambiguously positive or negative. For clinically anxious individuals, it may be particularly important to bring explicit awareness to the consequences (negative or positive) of real-world decisions, a strategy that is central to cognitive behavioral therapy. Explicitly identifying these consequences—and noting how avoidance of punishment or inadequate weighting of reward contributes to suboptimal outcomes—could serve as a "prosthetic" feedback system to help improve future choice behavior.

Learning

Even if each of the preceding stages is fully intact, an individual who fails to learn from suboptimal decisions will continue to make the same mistakes in the future.

Prediction error models posit that learning occurs only when predicted and actual outcomes conflict with one another. In anxiety disorders, excessive avoidance may preclude one from being exposed to a feared situation and thus from experiencing a negative prediction error (i.e., discovering that a horrible anticipated event has not occurred). By limiting exposure to surprising or unpredictable environments, anxious individuals may reduce the short-term experience of anxiety but allow for the long-term perpetuation of erroneous expectations about the nature of the world. Superstitious beliefs or actions (as reflected in the compulsions of an individual with obsessive–compulsive disorder, or persistent worry in generalized anxiety disorder) could also prevent learning from taking place. In these cases, anxious individuals learn to associate their actions or thoughts with the nonoccurrence of negative events, and fail to learn that those events would not have occurred even if they did not take those actions or think those thoughts.

In addition to avoidance and superstition, a failure to learn from prediction errors may help to sustain anxiety. This hypothesis was explored in a neuroimaging study using naturalistic social interactions in socially anxious adolescents [83]. Anxious adolescents showed heightened activation consistent with a *larger* prediction error signal in the striatum when high-value peers unexpectedly decided to chat with them (i.e., a violation of negative expectations). Functional connectivity between the striatum and the medial prefrontal cortex was reduced in response to this unexpected positive feedback, and reduced connectivity predicted poorer recollection of unexpected positive feedback in a postscan memory test. These results suggest that individuals with social anxiety have difficulty utilizing prediction error signals to learn from unexpectedly positive outcomes, which may contribute to persistent negative social expectations in this population.

In contrast to these alterations in positive prediction error signaling, a larger role could be hypothesized for deficient *negative* prediction error signaling in anxiety—that is, signals resulting from the nonoccurrence of expected events. The neural basis of this error signal is less well understood than that of positive prediction errors, but there is some evidence for involvement of the habenula, a perithalamic region that has strong inhibitory projections to midbrain dopaminergic cells [84]. During a difficult perceptual decision-making task, Ullsperger and von Cramon [85] observed comparable habenula activation for positive prediction errors for punishment *and* negative prediction errors for reward, two conditions that both reflect "worse-than-expected" outcomes.

It is not clear from the extant literature whether negative prediction errors for *punishment* share a neural basis with positive prediction errors for reward (as both reflect a "better-than-expected" outcome) or with negative prediction errors for reward (as both reflect the nonoccurrence of an expected outcome). A study in mice showed that positive and negative prediction errors for punishment were reflected in opposing patterns of basolateral amygdala responses [86], suggesting that this region is a prime candidate to explore with regard to negative prediction error signaling in anxiety.

Whereas unexpected outcomes that occur in a stable, unchanging environment might best be interpreted as fluke occurrences, in a more volatile environment one should update expectancies in response to unexpected occurrences. One study found that, whereas individuals with low levels of trait anxiety updated learning rates more quickly in volatile environments, matching the performance of an optimal Bayesian model, individuals with elevated trait anxiety failed to make this adjustment and had poorer task performance in volatile environments [87]. Trait anxiety was also associated with altered pupil dilation in volatile environments, suggesting a relationship between reduced noradrenergic function and deficient learning in dynamic environments. As mentioned above, one function of FMθ control signals—which show consistent alterations in elevated trait anxiety [73]—is to update learning rates in more volatile environments, suggesting a related neurobiological mechanism. These results are relevant for elevated intolerance of uncertainty in anxiety: if individuals fail to update learning rates appropriately in volatile environments, aversive events may seem to occur more unpredictably and uncontrollably.

SUMMARY AND CONCLUSIONS

Studies of decision-making in clinical anxiety disorders are still relatively uncommon, especially relative to phenomena such as fear learning and memory. However, the data reviewed here provide compelling evidence for alterations throughout the decision-making process that are ripe for further investigation. Several considerations are worth bearing with regard to this future research.

Much of the research reviewed here has been conducted in healthy populations varying in self-reported "trait anxiety" or related traits. Consistent with the growing movement toward dimensional models of psychopathology [3,4], these investigations clearly are relevant for understanding clinical disorders, but caution should be exercised in overgeneralizing results from these investigations to clinical populations. Furthermore, "trait anxiety" as it is typically assessed may index general negative affect, as opposed to constructs more specific to anxious pathology. Greater insight can be gained by looking at particular decision-making

alterations in relation to specific constructs that may reflect different courses of illness, or different "flavors" of anxiety, such as intolerance of uncertainty, anxiety sensitivity, or worry.

Additionally, many brain imaging studies in anxiety disorders are largely descriptive and correlational in nature, in that they primarily report brain regions that show more or less activation in a patient population. To identify the functional consequences of these differences, researchers must incorporate meaningful behavioral measures into experiments and assess relationships between brain activation and behavior. More sophisticated paradigms and manipulations from the world of neuroeconomics will help advance our knowledge of which decision-making processes are intact or disrupted across diverse anxious phenotypes (e.g., Refs. [24,47,53,54,83,87–89]).

The use of well-validated and targeted behavioral tasks allows one to draw inferences regarding underlying neural mechanisms that can be explicitly tested in subsequent studies. Following are several hypotheses that have been raised in preceding sections, many of which can be tested using purely behavioral assessments or peripheral physiology:

- More anxious individuals will demonstrate a reduced perceptual threshold for threat detection that contributes to more risk-averse decision-making (section Representation).
- Increased interoceptive accuracy will mediate the relationship between anxiety symptoms and improved performance on the IGT or other tasks involving risky outcomes (section Representation).
- In contrast to steeper discounting for future rewards, individuals with elevated anxiety symptoms will show less discounting for future punishments (section Valuation).
- Highly anxious individuals will show shifts in indifference points for decisions involving risk or ambiguity, and the quantifiable value that different individuals place on certainty will be reflected in neural valuation parameters during decision-making under uncertainty, in regions such as the ventral striatum, posterior VMPFC, and anterior insula (section Valuation).
- Extending findings from trait anxiety, individuals with clinical anxiety disorders will demonstrate behavioral shifts in AAC tasks, such that greater reward is needed to overcome the threat of punishment and shift behavior toward approach. The incorporation of disorder-specific punishments will differentially influence approach/avoidance behavior in heterogeneous anxious samples. Increased AMCC activation may mediate the relationship between anxiety and increased

avoidance behavior on these tasks (section Action Selection).

- The investigation of loss aversion in decisions absent of risk [90] will reveal relationships between anxiety symptoms and increased loss aversion, and this loss aversion will be related to increased interoceptive sensitivity and/or elevated physiological responses to loss, including greater skin conductance responses or FMθ signals (section Outcome Evaluation).
- Prediction error signaling will be disrupted in individuals with elevated anxiety symptoms. Specifically, there will be reductions in negative prediction errors for punishment or the ability to translate these negative prediction errors into reduced expectations of future punishment (section Learning).

Although it seems intuitively obvious that anxiety can dramatically influence one's day-to-day choice behavior, the extant data provide many more questions than definitive answers regarding the involvement of specific processes. By refocusing these data through an influential framework of decision-making [5], this chapter has attempted to provide an avenue for other researchers to tackle these questions head-on. More explicit and focused investigations of decision-making in anxiety disorders will allow for a reconceptualization of these disorders in terms of differential involvement of one or more of these processes, and as such will offer advancements to existing nosology, diagnosis, and treatment.

References

[1] Kessler RC, Berglund P, Demler O, Jin R, Merikangas KR, Walters EE. Lifetime prevalence and age-of-onset distributions of DSM—IV disorders in the National Comorbidity Survey Replication. Arch Gen Psychiatry 2005;62:593—602. http://dx.doi.org/10.1001/archpsyc.62.6.593.

[2] Greenberg PE, Sisitsky T, Kessler RC, Finkelstein SN, Berndt ER, Davidson JR, et al. The economic burden of anxiety disorders in the 1990s. J Clin Psychiatry 1999;60:427—35. http://dx.doi.org/10.4088/JCP.v60n0702.

[3] Brown TA, Barlow DH. A proposal for a dimensional classification system based on the shared features of the DSM—IV anxiety and mood disorders: implications for assessment and treatment. Psychol Assess 2009;21:256—71. http://dx.doi.org/10.1037/a0016608.

[4] Insel T, Cuthbert B, Garvey M, Heinssen R, Pine DS, Quinn K, et al. Research domain criteria (RDoC): toward a new classification framework for research on mental disorders. Am J Psychiatry 2010;167:748—51. http://dx.doi.org/10.1176/appi.ajp.2010.09091379.

[5] Rangel A, Camerer CF, Montague PR. A framework for studying the neurobiology of value-based decision making. Nat Rev Neurosci 2008;9:545—56. http://dx.doi.org/10.1038/nrn2357.

[6] Grupe DW, Nitschke JB. Uncertainty and anticipation in anxiety: an integrated neurobiological and psychological perspective. Nat Rev Neurosci 2013;14:488—501. http://dx.doi.org/10.1038/nrn3524.

[7] Bar-Haim Y, Lamy D, Pergamin L, Bakermans-Kranenburg MJ, van Ijzendoorn MH. Threat-related attentional bias in anxious and nonanxious individuals: a meta-analytic study. Psychol Bull 2007;133:1—24. http://dx.doi.org/10.1037/0033-2909.133.1.1.

[8] Cisler JM, Koster EH. Mechanisms of attentional biases towards threat in anxiety disorders: an integrative review. Clin Psychol Rev 2010;30:203—16. http://dx.doi.org/10.1016/j.cpr.2009.11.003.

[9] Davis M, Whalen PJ. The amygdala: vigilance and emotion. Mol Psychiatry 2001;6:13—34.

[10] Pessoa L. Emotion and cognition and the amygdala: from "What is it?" to "What's to be done?". Neuropsychologia 2010;48:3416—29. http://dx.doi.org/10.1016/j.neuropsychologia.2010.06.038.

[11] Schwartz CE, Wright CI, Shin LM, Kagan J, Rauch SL. Inhibited and uninhibited infants "grown up": adult amygdalar response novelty. Science 2003;300:1952—3. http://dx.doi.org/10.1126/science.1083703.

[12] Schwartz CE, Kunwar PS, Greve DN, Kagan J, Snidman NC, Bloch RB. A phenotype of early infancy predicts reactivity of the amygdala in male adults. Mol Psychiatry 2011;17:1042—50. http://dx.doi.org/10.1038/mp.2011.96.

[13] Mathews A, Richards A, Eysenck M. Interpretation of homophones related to threat in anxiety states. J Abnorm Psychol 1989;98:31—4.

[14] Stopa L, Clark DM. Social phobia and interpretation of social events. Behav Res Ther 2000;38:273—83.

[15] Yoon KL, Zinbarg RE. Threat is in the eye of the beholder: social anxiety and the interpretation of ambiguous facial expressions. Behav Res Ther 2007;45:839—47. http://dx.doi.org/10.1016/j.brat.2006.05.004.

[16] Kimble MO, Kaufman ML, Leonard LL, Nestor PG, Riggs DS, Kaloupek DG, et al. Sentence completion test in combat veterans with and without PTSD: preliminary findings. Psychiatry Res 2002;113:303—7.

[17] Mujica-Parodi LR, Carlson JM, Cha J, Rubin D. The fine line between "brave" and "reckless": amygdala reactivity and regulation predict recognition of risk. Neuroimage 2014. http://dx.doi.org/10.1016/j.neuroimage.2014.08.038.

[18] Domschke K, Stevens S, Pfleiderer B, Gerlach AL. Interoceptive sensitivity in anxiety and anxiety disorders: an overview and integration of neurobiological findings. Clin Psychol Rev 2010;30:1—11. http://dx.doi.org/10.1016/j.cpr.2009.08.008.

[19] James W, Lange CG. The emotions. Baltimore: Williams & Wilkins Co.; 1922.

[20] Damasio AR. Descartes' error: emotion, reason, and the human brain. New York: Penguin Books; 1994.

[21] Dunn BD, Dalgleish T, Lawrence AD. The somatic marker hypothesis: a critical evaluation. Neurosci Biobehav Rev 2006;30:239—71. http://dx.doi.org/10.1016/j.neubiorev.2005.07.001.

[22] Bechara A, Tranel D, Damasio H, Damasio AR. Failure to respond autonomically to anticipated future outcomes following damage to prefrontal cortex. Cereb Cortex 1996;6:215—25.

[23] Dunn BD, Galton HC, Morgan R, Evans D, Oliver C, Meyer M, et al. Listening to your heart. How interoception shapes emotion experience and intuitive decision making. Psychol Sci 2010;21:1835—44. http://dx.doi.org/10.1177/0956797610389191.

[24] Sokol-Hessner P, Hartley CA, Hamilton JR, Phelps EA. Interoceptive ability predicts aversion to losses. Cogn Emot 2014:1—7. http://dx.doi.org/10.1080/02699931.2014.925426.

[25] Werner NS, Duschek S, Schandry R. Relationships between affective states and decision-making. Int J Psychophysiol 2009;74:259—65. http://dx.doi.org/10.1016/j.ijpsycho.2009.09.010.

[26] Kirsch M, Windmann S. The role of anxiety in decision-making. Rev Psychol 2009;16:19—28.

[27] Mueller EM, Nguyen J, Ray WJ, Borkovec TD. Future-oriented decision-making in generalized anxiety disorder is evident across different versions of the Iowa gambling task. J Behav Ther Exp

Psychiatry 2010;41:165−71. http://dx.doi.org/10.1016/j.jbtep.2009.12.002.

[28] Miu AC, Heilman RM, Houser D. Anxiety impairs decision-making: psychophysiological evidence from an Iowa gambling task. Biol Psychol 2008;77:353−8. http://dx.doi.org/10.1016/j.biopsycho.2007.11.010.

[29] Robinson OJ, Bond RL, Roiser JP. The impact of stress on financial decision-making varies as a function of depression and anxiety symptoms. PeerJ 2015;3:e770. http://dx.doi.org/10.7717/peerj.770.

[30] Bickel WK, Jarmolowicz DP, Mueller ET, Koffarnus MN, Gatchalian KM. Excessive discounting of delayed reinforcers as a trans-disease process contributing to addiction and other disease-related vulnerabilities: emerging evidence. Pharmacol Ther 2012;134:287−97. http://dx.doi.org/10.1016/j.pharmthera.2012.02.004.

[31] Buhr K, Dugas MJ. The intolerance of uncertainty scale: psychometric properties of the English version. Behav Res Ther 2002;40:931−45.

[32] Rounds JS, Beck JG, Grant DM. Is the delay discounting paradigm useful in understanding social anxiety? Behav Res Ther 2007;45:729−35. http://dx.doi.org/10.1016/j.brat.2006.06.007.

[33] Worthy DA, Byrne KA, Fields S. Effects of emotion on prospection during decision-making. Front Psychol 2014;5:1−12. http://dx.doi.org/10.3389/fpsyg.2014.00591.

[34] Luhmann CC, Ishida K, Hajcak G. Intolerance of uncertainty and decisions about delayed, probabilistic rewards. Behav Ther 2011;42:378−86. http://dx.doi.org/10.1016/j.beth.2010.09.002.

[35] Borkovec TD. Life in the future versus life in the present. Clin Psychol Sci Pract 2002;9:76−80. http://dx.doi.org/10.1093/clipsy.9.1.76.

[36] Mitte K. Anxiety and risky decision-making: the role of subjective probability and subjective costs of negative events. Pers Individ Differ 2007;43:243−53. http://dx.doi.org/10.1016/j.paid.2006.11.028.

[37] Maner JK, Richey JA, Cromer K, Mallott M, Lejuez CW, Joiner TE, et al. Dispositional anxiety and risk-avoidant decision-making. Pers Individ Differ 2007;42:665−75. http://dx.doi.org/10.1016/j.paid.2006.08.016.

[38] Lorian CN, Grisham JR. Clinical implications of risk aversion: an online study of risk-avoidance and treatment utilization in pathological anxiety. J Anxiety Disord 2011;25:840−8. http://dx.doi.org/10.1016/j.janxdis.2011.04.008.

[39] Lejuez CW, Read JP, Kahler CW, Richards JB, Ramsey SE, Stuart GL, et al. Evaluation of a behavioral measure of risk taking: the balloon analogue risk task (BART). J Exp Psychol Appl 2002;8:75−84. http://dx.doi.org/10.1037//1076-898X.8.2.75.

[40] Giorgetta C, Grecucci A, Zuanon S, Perini L, Balestrieri M, Bonini N, et al. Reduced risk-taking behavior as a trait feature of anxiety. Emotion 2012;12:1373−83. http://dx.doi.org/10.1037/a0029119.

[41] Tang GS, van den Bos W, Andrade EB, McClure SM. Social anxiety modulates risk sensitivity through activity in the anterior insula. Front Neurosci 2011;5:142. http://dx.doi.org/10.3389/fnins.2011.00142.

[42] Hazlett-Stevens H, Borkovec TD. Interpretive cues and ambiguity in generalized anxiety disorder. Behav Res Ther 2004;42:881−92. http://dx.doi.org/10.1016/S0005-7967(03)00204-3.

[43] Levy DJ, Glimcher PW. The root of all value: a neural common currency for choice. Curr Opin Neurobiol 2012;22:1027−38. http://dx.doi.org/10.1016/j.conb.2012.06.001.

[44] Levy I, Snell J, Nelson AJ, Rustichini A, Glimcher PW. Neural representation of subjective value under risk and ambiguity. J Neurophysiol 2010;103:1036−47. http://dx.doi.org/10.1152/jn.00853.2009.

[45] Kable JW, Glimcher PW. The neural correlates of subjective value during intertemporal choice. Nat Neurosci 2007;10:1625−33. http://dx.doi.org/10.1038/nn2007.

[46] Grabenhorst F, Rolls ET. Different representations of relative and absolute subjective value in the human brain. Neuroimage 2009;48:258−68. http://dx.doi.org/10.1016/j.neuroimage.2009.06.045.

[47] Engelmann JB, Meyer F, Fehr E, Ruff CC. Anticipatory anxiety disrupts neural valuation during risky choice. J Neurosci 2015;35:3085−99. http://dx.doi.org/10.1523/JNEUROSCI.2880-14.2015.

[48] Craske MG, Barlow DH. A review of the relationship between panic and avoidance. Clin Psychol Rev 1988;8:667−85. http://dx.doi.org/10.1016/0272-7358(88)90086-4.

[49] Borkovec TD, Alcaine OM, Behar E. Avoidance theory of worry and generalized anxiety disorder. In: Heimberg RG, Turk CL, Mennin DS, editors. Gen. anxiety disord. adv. res. pract. 1st ed. New York: Guilford Press; 2004. p. 77−108.

[50] Lovibond PF, Mitchell CJ, Minard E, Brady A, Menzies RG. Safety behaviours preserve threat beliefs: protection from extinction of human fear conditioning by an avoidance response. Behav Res Ther 2009;47:716−20. http://dx.doi.org/10.1016/j.brat.2009.04.013.

[51] Aupperle RL, Paulus MP. Neural systems underlying approach and avoidance in anxiety disorders. Dialogues Clin Neurosci 2010;12:517−31.

[52] Millan MJ, Brocco M. The Vogel conflict test: procedural aspects, gamma-aminobutyric acid, glutamate and monoamines. Eur J Pharmacol 2003;463:67−96. http://dx.doi.org/10.1016/S0014-2999(03)01275-5.

[53] Aupperle RL, Sullivan S, Melrose AJ, Paulus MP, Stein MB. A reverse translational approach to quantify approach-avoidance conflict in humans. Behav Brain Res 2011;225:455−63. http://dx.doi.org/10.1016/j.bbr.2011.08.003.

[54] Pittig A, Brand M, Pawlikowski M, Alpers GW. The cost of fear: avoidant decision making in a spider gambling task. J Anxiety Disord 2014;28:326−34. http://dx.doi.org/10.1016/j.janxdis.2014.03.001.

[55] Shackman AJ, V Salomons T, Slagter HA, Fox AS, Winter JJ, Davidson RJ. The integration of negative affect, pain and cognitive control in the cingulate cortex. Nat Rev Neurosci 2011;12:154−67. http://dx.doi.org/10.1038/nrn2994.

[56] Croxson PL, Walton ME, O'Reilly JX, Behrens TEJ, Rushworth MFS. Effort-based cost-benefit valuation and the human brain. J Neurosci 2009;29:4531−41. http://dx.doi.org/10.1523/JNEUROSCI.4515-08.2009.

[57] Camille N, Tsuchida A, Fellows LK. Double dissociation of stimulus-value and action-value learning in humans with orbitofrontal or anterior cingulate cortex damage. J Neurosci 2011;31:15048−52. http://dx.doi.org/10.1523/JNEUROSCI.3164-11.2011.

[58] Amemori K, Graybiel AM. Localized microstimulation of primate pregenual cingulate cortex induces negative decision-making. Nat Neurosci 2012;15:776−85. http://dx.doi.org/10.1038/nn.3088.

[59] Milad MR, Pitman RK, Ellis CB, Gold AL, Shin LM, Lasko NB, et al. Neurobiological basis of failure to recall extinction memory in posttraumatic stress disorder. Biol Psychiatry 2009;66:1075−82. http://dx.doi.org/10.1016/j.biopsych.2009.06.026.

[60] Bryant RA, Felmingham KL, Kemp AH, Barton M, Peduto AS, Rennie C, et al. Neural networks of information processing in posttraumatic stress disorder: a functional magnetic resonance imaging study. Biol Psychiatry 2005;58:111−8. http://dx.doi.org/10.1016/j.biopsych.2005.03.021.

[61] Klumpp H, Angstadt M, Phan KL. Insula reactivity and connectivity to anterior cingulate cortex when processing threat in generalized social anxiety disorder. Biol Psychol 2012;89:273−6. http://dx.doi.org/10.1016/j.biopsycho.2011.10.010.

[62] Krain AL, Gotimer K, Hefton S, Ernst M, Castellanos FX, Pine DS, et al. A functional magnetic resonance imaging investigation of uncertainty in adolescents with anxiety disorders. Biol Psychiatry 2008;63:563−8. http://dx.doi.org/10.1016/j.biopsych.2007.06.011.

[63] Straube T, Mentzel HJ, Miltner WH. Waiting for spiders: brain activation during anticipatory anxiety in spider phobics. Neuroimage 2007;37:1427—36. http://dx.doi.org/10.1016/j.neuroimage.2007.06.023.

[64] Torrubia R, Ávila C, Moltó J, Caseras X. The sensitivity to punishment and sensitivity to reward questionnaire (SPSRQ) as a measure of Gray's anxiety and impulsivity dimensions. Pers Individ Differ 2001;31:837—62. http://dx.doi.org/10.1016/S0191-8869(00)00183-5.

[65] Kermer DA, Driver-Linn E, Wilson TD, Gilbert DT. Loss aversion is an affective forecasting error. Psychol Sci 2006;17:649—56. http://dx.doi.org/10.1111/j.1467-9280.2006.01760.x.

[66] Shackman AJ, Stockbridge MD, LeMay EP, Fox AS. Pathways linking emotional traits to emotional states. In: Nat. emot. fundam. quest. 2nd ed. Oxford: Oxford University Press; 2016.

[67] Kagan J, Snidman NC, Kahn V, Towsley S. The preservation of two infant temperaments into adolescence. Monogr Soc Res Child Dev 2007;72:1—75. http://dx.doi.org/10.1111/j.1540-5834.2007.00436.x.

[68] Fox AS, Kalin NH. A translational neuroscience approach to understanding the development of social anxiety disorder and its pathophysiology. Am J Psychiatry 2014;171:1162—73.

[69] Kahneman D, Tversky A. Prospect theory: an analysis of decision under risk. Econometrica 1979;47:263—91.

[70] Sokol-Hessner P, Hsu M, Curley NG, Delgado MR, Camerer CF, Phelps EA. Thinking like a trader selectively reduces individuals' loss aversion. Proc Natl Acad Sci USA 2009;106:5035—40. http://dx.doi.org/10.1073/pnas.0806761106.

[71] Ernst M, Plate RC, Carlisi CO, Gorodetsky E, Goldman D, Pine DS. Loss aversion and 5HTT gene variants in adolescent anxiety. Dev Cogn Neurosci 2014;8:77—85. http://dx.doi.org/10.1016/j.dcn.2013.10.002.

[72] Cavanagh JF, Frank MJ. Frontal theta as a mechanism for cognitive control. Trends Cogn Sci 2014;18:414—21. http://dx.doi.org/10.1016/j.tics.2014.04.012.

[73] Cavanagh JF, Shackman AJ. Frontal midline theta reflects anxiety and cognitive control: meta-analytic evidence. J Physiol Paris 2015;109:3—15.

[74] Paulus MP, Feinstein JS, Simmons A, Stein MB. Anterior cingulate activation in high trait anxious subjects is related to altered error processing during decision making. Biol Psychiatry 2004;55:1179—87. http://dx.doi.org/10.1016/j.biopsych.2004.02.023.

[75] Stein MB, Paulus MP. Imbalance of approach and avoidance: the yin and yang of anxiety disorders. Biol Psychiatry 2009;66:1072—4. http://dx.doi.org/10.1016/j.biopsych.2009.09.023.

[76] Hopper JW, Pitman RK, Su Z, Heyman GM, Lasko NB, Macklin ML, et al. Probing reward function in posttraumatic stress disorder: expectancy and satisfaction with monetary gains and losses. J Psychiatr Res 2008;42:802—7. http://dx.doi.org/10.1016/j.jpsychires.2007.10.008.

[77] Elman I, Lowen S, Frederick BB, Chi W, Becerra L, Pitman RK. Functional neuroimaging of reward circuitry responsivity to monetary gains and losses in posttraumatic stress disorder. Biol Psychiatry 2009;66:1083—90. http://dx.doi.org/10.1016/j.biopsych.2009.06.006.

[78] Sailer U, Robinson S, Fischmeister FP, König D, Oppenauer C, Lueger-Schuster B, et al. Altered reward processing in the nucleus accumbens and mesial prefrontal cortex of patients with posttraumatic stress disorder. Neuropsychologia 2008;46:2836—44. http://dx.doi.org/10.1016/j.neuropsychologia.2008.05.022.

[79] Helfinstein SM, Benson B, Perez-Edgar K, Bar-Haim Y, Detloff A, Pine DS, et al. Striatal responses to negative monetary outcomes differ between temperamentally inhibited and non-inhibited adolescents. Neuropsychologia 2011;49:479—85. http://dx.doi.org/10.1016/j.neuropsychologia.2010.12.015.

[80] Manning J, Reynolds G, Saygin ZM, Hofmann SG, Pollack M, Gabrieli JD, et al. Altered resting-state functional connectivity of the frontal-striatal reward system in social anxiety disorder. PLoS One 2015;10:e0125286. http://dx.doi.org/10.1371/journal.pone.0125286.

[81] Kessel EM, Kujawa A, Hajcak Proudfit G, Klein DN. Neural reactivity to monetary rewards and losses differentiates social from generalized anxiety in children. J Child Psychol Psychiatry 2014;56:792—800. http://dx.doi.org/10.1111/jcpp.12355.

[82] Guyer AE, Nelson EE, Perez-Edgar K, Hardin MG, Roberson-Nay R, Monk CS, et al. Striatal functional alteration in adolescents characterized by early childhood behavioral inhibition. J Neurosci 2006;26:6399—405. http://dx.doi.org/10.1523/JNEUROSCI.0666-06.2006.

[83] Jarcho JM, Romer AL, Shechner T, Galvan A, Guyer AE, Leibenluft E, et al. Forgetting the best when predicting the worst: preliminary observations on neural circuit function in adolescent social anxiety. Dev Cogn Neurosci 2015;13:21—31. http://dx.doi.org/10.1016/j.dcn.2015.03.002.

[84] Hikosaka O. The habenula: from stress evasion to value-based decision-making. Nat Rev Neurosci 2010;11:503—13. http://dx.doi.org/10.1038/nrn2866.

[85] Ullsperger M, von Cramon DY. Error monitoring using external feedback: specific roles of the habenular complex, the reward system, and the cingulate motor area revealed by functional magnetic resonance imaging. J Neurosci 2003;23:4308—14.

[86] McHugh SB, Barkus C, Huber A, Capitão L, Lima J, Lowry JP, et al. Aversive prediction error signals in the amygdala. J Neurosci 2014;34:9024—33. http://dx.doi.org/10.1523/JNEUROSCI.4465-13.2014.

[87] Browning M, Behrens TE, Jocham G, O'Reilly JX, Bishop SJ. Anxious individuals have difficulty learning the causal statistics of aversive environments. Nat Neurosci 2015;18:590—6. http://dx.doi.org/10.1038/nn.3961.

[88] van Duijvenvoorde AC, Huizenga HM, Somerville LH, Delgado MR, Powers A, Weeda WD, et al. Neural correlates of expected risks and returns in risky choice across development. J Neurosci 2015;35:1549—60. http://dx.doi.org/10.1523/JNEUROSCI.1924-14.2015.

[89] McGuire JT, Kable JW. Medial prefrontal cortical activity reflects dynamic re-evaluation during voluntary persistence. Nat Neurosci 2015;18:1—10. http://dx.doi.org/10.1038/nn.3994.

[90] Gächter S, Johnson EJ, Whitehead K. Individual-level loss aversion in riskless and risky choices. 2007.

[91] Kendler KS, Gardner CO. A longitudinal etiologic model for symptoms of anxiety and depression in women. Psychol Med 2011;41:2035—45. http://dx.doi.org/10.1017/S0033291711000225.

27

Decision-Making in Gambling Disorder: Understanding Behavioral Addictions

L. Clark

University of British Columbia, Vancouver, BC, Canada

Abstract

Examples from real-world gambling behavior have been a frequent source of inspiration to the fields of behavioral economics and decision neuroscience, but it is surprising how little is known about decisional processes in regular gamblers and especially individuals with gambling disorder. This chapter presents a conceptualization of disordered gambling as a behavioral addiction driven by an exaggeration of multiple psychological distortions that are characteristic of human decision-making and underpinned by neural circuitry subserving appetitive behavior, reinforcement learning, and choice selection. Focusing on examples of loss aversion, probability weighting, perceptions of randomness, and the illusion of control, I consider the evidence that these phenomena are related to actual gambling behavior, their relevance to gambling disorder, and insights from neuroscience experiments as to their brain basis.

INTRODUCTION: GAMBLING AND DISORDERED GAMBLING

Gambling is a widespread and ancient form of entertainment in human societies. In the modern, commercialized gambling sector, this activity takes many diverse forms; for example, state lotteries are typically the most popular games across many jurisdictions [1,2], and within casino environments, much of the casino floor space is allocated to "electronic gaming machines," an umbrella term that includes modern slot machines. Most readers will be broadly familiar with the other dominant varieties of gambling: scratch cards (or "instant lotteries"), casino table games such as roulette and blackjack, and various forms of sports betting. For a broad psychological definition that captures this array of activities, gambling is the wagering of money on the uncertain prospect of a larger monetary prize [3,4]. Inevitably, such definitions suffer from a substantial "gray area" of risk-taking activities that include private investment in the stock market and entrepreneurship [4].

Gambling is a fertile area for psychologists and neuroscientists with interests in decision-making. First and foremost, it is an extremely widespread behavior. Prevalence data from a North American household survey found that 78% of adults reported lifetime gambling [5], and past-year gambling involvement hovers around 70% across many jurisdictions [2]. There is evidence that gambling has existed in at least some human societies for thousands of years [6], and it is certainly a global phenomenon in the 21st century. From a decision-making perspective, how can we understand the appeal of gambling? Here, it is critical to acknowledge that modern commercial forms of gambling are offered in a way that ensures a steady profit for the casino, bookmaker, or slot machine—this is generally termed the "house edge." For example, a slot machine is hard-programmed with a "return-to-player" value set between 85% and 95%, which means that 5–15% of the amount bet will be held by the machine, and the remainder returned to the gambler as wins [7].

Given the traditional debate in economics and philosophy as to whether human choice is fundamentally rational (*Homo economicus*), it is provocative to question why gambling exists at all. Here we should recognize that house edge is not an inherent aspect of gambling: historical forms of gambling (and some modern forms) were zero-sum games that may have served useful prosocial purposes of redistributing valued items (e.g., tools) among the members of a community [6]. The attraction to such games may reflect a human fascination with chance itself, which may be expressed neurobiologically in the psychological conditions that best stimulate the ascending dopamine projection [8,9].

In addition to considering the general appeal of gambling, it is increasingly recognized that gambling causes significant negative consequences in a subset of players [10]. In its most severe form, gambling disorder is a psychiatric syndrome in the *Diagnostic and Statistical Manual of Mental Disorders*, fifth edition (DSM-5), in which it is grouped alongside the substance use disorders as the only recognized behavioral addiction [11]. This condition is diagnosed if an individual meets at least four of nine criteria, which include some symptoms reminiscent of drug addiction, namely withdrawal and tolerance, as well as some of the negative consequences of excessive gambling, for instance, a disruption of interpersonal or occupational functioning. The prevalence of the full DSM-diagnosed disorder is around 1% in most jurisdictions [12], but it is recognized that gambling harms increase in direct proportion to gambling involvement [13], such that 3—5% of the population displays subclinical features, which have been labeled "problem gambling" or "at-risk gambling" [10].

As we have seen so far, gambling research is a multidisciplinary field with critical contributions from sociology, public health, and psychiatry (to name just a few), as well as psychology and neuroscience. Yet ultimately, all forms of gambling can be broken down to a series of decisions made by the player. Single gambling events can be decomposed to a selection (e.g., which slot machine to play), an anticipatory period (the reel spin), and the gain or loss feedback. Understanding the psychological and neuroscience mechanisms that underlie these three stages of decision-making will inform the research priorities described above [14]. Furthermore, many cardinal examples of psychological illusions and decision-making biases can be observed in gambling behavior, to the extent that gambling can be fruitfully viewed as a paradigm for studying risky decision-making itself [15]. The following sections are organized as follows. I will focus on four phenomena that were originally conceptualized within decision-making research. Taking each effect in turn, I will consider three questions. To what extent are these phenomena evident in real gambling behavior? To what extent are these phenomena altered (and hypothetically exacerbated) in individuals with disordered gambling? And how have decision neuroscience experiments contributed to our understanding of these biases and the neural circuitry that underpins them?

LOSS AVERSION

The axiom that "losses loom larger than gains" is one of the best-known effects in behavioral economics. If people are offered a gamble on a coin toss, on which they would win $10 if a tail is thrown, but lose $10 if a head is thrown, most participants decline this gamble [16]. As the expected value of the coin toss is zero (it is risk neutral), we might expect participants to be indifferent regarding these two options, but they are not. The win amount can be gradually increased in a series of steps while keeping the loss constant. Participants typically require the win to be approximately twice as large as the loss (i.e., win $20, lose $10) before reliably accepting the gamble. This is parameterized by λ values of around 2, and these values are prone to substantial individual differences [17]. Loss aversion is explained within Prospect Theory by the relative steepness of the value function in the loss domain relative to the gain domain [18].

Prima facie, this phenomenon is obviously relevant to gambling behavior. On one hand, loss aversion may be a deterrent for many people to engage in gambling in the first place; the risk of losing one's bet overrides any potential gain. From this, we might expect problem gamblers to display reduced loss aversion relative to healthy participants; that is to say, their value functions should be *similarly* steep in the gain and loss domains. Such effects have been described in some neurological studies of patients with brain injury. In a coin-toss task similar to the aforementioned scenario, a mixed group of brain injury patients with damage to the ventromedial prefrontal cortex, amygdala, or insula were more likely to accept a positive expected value coin toss (win $2.50, lose $1) than healthy controls, with the healthy participants declining a substantial proportion of such bets owing to loss aversion [19]. These findings were later confirmed in two cases with selective amygdala damage as a result of Urbach—Wiethe disease [20].

However, reduced loss aversion is not the only prediction for real-world gamblers. Loss aversion is conceptually related to loss chasing, a key symptom of gambling disorder in which the gambler persists with betting in a desperate attempt to recoup mounting debts. Loss chasing is regarded by some as the defining feature of the pathological gambler [21], and empirically, it is the most frequently endorsed item on the DSM checklist [13]. In a compelling field demonstration in online gamblers who closed their accounts because of gambling-related problems, Xuan and Shaffer [22] observed that these gamblers (relative to a control group who did not close their accounts) responded to mounting losses by increasing their bet size, rather than placing more bets or taking bets at longer odds.

Could loss chasing arise from *increased* loss aversion? Such an explanation requires an additional feature of Prospect Theory: the reference point [18]. Most decisions are made from the origin of the value function, such that between any two choices, people "re-reference" so that the options are evaluated independently. In a gambling episode, players may fail to re-reference between successive bets, so that a player on a losing streak progressively shifts leftward on the value function, getting steadily farther from the origin. As the loss curve plateaus, the subjective value of a win becomes dramatic, whereas the subjective value of sustaining a further loss becomes negligible.

A 2014 experiment sought to test this hypothesis, quantifying loss aversion in patients with gambling disorder [23] using a straightforward behavioral economics procedure to titrate a λ coefficient for each participant. Notably, the gamblers did not differ in the overall group comparison of loss aversion values. Nevertheless, individual differences in loss aversion were related to the length of time in treatment, such that a subgroup tested later in the course of treatment was more loss averse. It would be worthwhile to explore further any bimodality in the loss aversion scores, whereby gamblers may polarize toward extreme high or low values, with the former group hypothetically disposed to loss chasing. It is also worth noting that loss chasing could be driven by several distinct characteristics of the value function, such as the relative steepness of the gain and loss domains, the distance to plateau, or the tendency to re-reference between gambles.

Characterizing loss aversion in gambling disorder may also benefit from a greater emphasis on the underlying neuroscience. Psychophysiological measures like skin conductance provide objective markers of the relative impact of losses and gains: the skin conductance response to a losing outcome is greater than that to an equivalently sized gain [17]. As participants select bets during a gambling task, skin conductance levels scale closely with bet size and are further correlated with the sensitivity to losing outcomes [24,25] (see Fig. 27.1). Skin conductance levels also change following near losses on a "wheel of fortune" task, and these near losses similarly "loom larger" than equivalent near wins [24]. Using fMRI, a seminal experiment in neuroeconomics by Tom et al. [26] showed widespread neural sensitivity to losses over gains in brain regions that included the striatum and medial prefrontal cortex. Converging with neuropsychological studies, the amygdala response to losing outcomes also scales with behavioral estimates of loss aversion [27].

PROBABILITY WEIGHTING

The probability weighting function is a second core component of Prospect Theory. This function describes the relationship between the objective probability of an event and the subjective perception of that probability [28]. Classically, low-probability events such as the chances of dying in a freak accident are overestimated, and high-probability events (such as mortality rates for heart disease or cancer) are underestimated [18]. This is expressed as a nonlinear (sigmoidal) function that can be estimated for a participant using an adjustment procedure similar to that described earlier for loss aversion [29].

Individual differences in the shape of the probability weighting function may be highly relevant to gambling

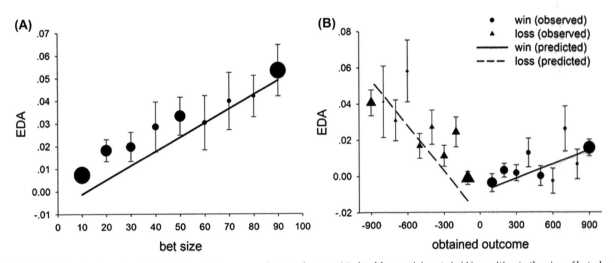

FIGURE 27.1 Electrodermal activity (*EDA;* also known as skin conductance) in healthy participants is (A) sensitive to the size of bet placed on the spin of a wheel of fortune and (B) disproportionately sensitive to losing outcomes relative to gain outcomes. The available bets were from 10 to 90 points (in 10-point increments). As bet size was selected by the participant, the number of trials contributing to each data point varied and is indicated by the size of each data point in proportion to the number of observations. *Reprinted from Wu Y, van Dijk E, Aitken MRF, Clark L. Missed losses loom larger than missed gains: electrodermal reactivity to decision choices and outcomes in a gambling task. Cogn Affect Behav Neurosci 2016;16: 353–61. http://dx.doi.org/10.3758/s13415-015-0395-y.*

behavior. Lotteries serve as a clear example of this. The chances of winning the jackpot on the Canadian Lotto 649 are 1 in 13,983,816. Humans do not have a rigorous statistical concept of 1 in 14 million, and even a weak tendency to overweight this probability may prompt people to buy a ticket in the hope of a jackpot win. In analyzing lottery ticket sales in two US states, Lyon and Ghezzi [30] found no sensitivity to successive shifts in the odds of winning, but sales increased with the introduction of a minimum jackpot over $1 million. Of course, the probabilities of winning in other forms of gambling are often less extreme; for example, guessing red or black in roulette. But an increased degree of distortion in the probability weighting function may also diminish the gambler's sensitivity to small changes in odds in the intermediate range [29]. In a behavioral study comparing experienced (but nonpathological) poker players against pathological poker players, the pathological gamblers showed greater degree of error in their probability estimations for different hands and were more likely to play on low-probability hands, compared to the expert group [31].

In relating the probability weighting function to disordered gambling, an exemplary study by Ligneul et al. [29] tested pathological gamblers and healthy controls using the certainty equivalents procedure. Rather than displaying a greater degree of distortion (i.e., an exaggerated S shape), the pathological gamblers showed an overall elevation in subjective probability across the full range of odds from 0 to 1, an effect that was further correlated with gambling severity in the pathological gamblers and "gambling affinity" in the healthy group. These data converge with those from a neuropsychological study in which problem gamblers placed higher bets on the Cambridge Gambling Task, across the range of probabilities from high (60%) to low (90%) uncertainty [32].

Some further insights into probability assessment in disordered gambling have come from studies of probabilistic discounting, referring to the tendency to select certain rewards over larger uncertain rewards. Whereas delay discounting has been studied extensively in the context of both substance addictions and disordered gambling [33], probabilistic discounting has received less attention but is phenomenologically the more relevant construct (see also Chapter 7). In their consistent tendency to select risky uncertain gambles, we may expect problem gamblers to show less steep probabilistic discounting, and this has been confirmed in one study [34]. Miedl et al. [35] scanned individuals with gambling disorder as they made both delay and probabilistic discounting decisions. The gambling group showed steeper delay discounting, and a trend effect toward shallower probabilistic discounting (which was probably underpowered given the small sample size

for neuroimaging). In modeling how the neural response to reward varied as a function of its delay and uncertainty, the gamblers showed stronger value representations in ventral striatum during delay discounting (consistent with their preference for immediate rewards), combined with weaker value representations in the same region during probabilistic discounting. In seeking to characterize the probability weighting function more formally, a PET study using a dopamine D1 receptor ligand in healthy participants found that lower D1 binding in the striatum predicted the strongest tendency to overweight low probabilities and underweight high probabilities [36]. In contrast, using fMRI, Tobler et al. [37] highlighted prefrontal cortical contributions to the distorted processing of probabilities: probability coding was approximately linear within the striatum, whereas the dorsolateral prefrontal cortex (PFC) overweighted low-probability events and underweighted high-probability events, reflecting the typical human bias.

PERCEPTIONS OF RANDOMNESS

Some of the more salient psychological biases seen in gambling pertain to predictions made in long sequences of events, such as guessing heads or tails in a series of coin tosses or choosing red or black in a series of roulette spins. The classic gambler's fallacy refers to the tendency to predict against the recent run of outcomes; for example, after observing four consecutive heads in coin flips, participants are more likely to guess "tails" on the next prediction. This phenomenon of "negative recency" is robust both in laboratory studies and in the field. For example, using security video footage of casino roulette wheels, Croson and Sundali [38] were able to show a stepwise decrease in the choice of red or black as a function of the prior run length of that color.

The dominant psychological account of the gambler's fallacy is as an instance of the representativeness heuristic [39], such that people expect short sequences of events to be representative of the full population from which they are drawn. In reality, such runs and streaks occur in truly random data more often than we expect [40]. The gambler's fallacy in successive predictions exists alongside a number of other misperceptions of randomness. For example, when healthy participants are asked to generate random sequences of binary outcomes, typical sequences display excessive switch rates (i.e., probability of alternation >0.5) [41]. Similarly, if participants are presented with binary sequences, and asked to select how those sequences were created (e.g., a basketball player or a coin toss), the most "random" sequences [i.e., $p(\text{Alt}) = .5$] are attributed to the basketball player, and highly "switchy" sequences [$p(\text{Alt}) = .8$] are attributed to the coin toss [42].

The psychological factors underlying the gambler's fallacy may also contribute to loss chasing: if a gambler experiences a long series of losses, thoughts about negative recency may bolster a belief that a win must be "due." In describing the "end of the day effect" in horse-race bettors, McGlothlin [43] observed that wagers at the racetrack increased throughout the day and peaked on the final race and, more specifically, that the fewer bettors who won on any given race, the greater the amount wagered on the subsequent race was. However, in binary choice settings, some nuances arise from the overlapping sequences generated by the outcome runs (e.g., reds or blacks) and the feedback streaks (whether the participant's guesses are correct or incorrect): we expect winning and losing streaks to continue, in line with positive recency. This is often termed the "hot hand belief" [44], a term from basketball referring to the player who has successfully shot with the last few shots, and is thus expected to be "hot" or "in the zone." Indeed, although the gambler's fallacy and hot hand belief are sometimes treated as opposites, the two beliefs may coexist based upon our assumptions about whether the causal agent that generates the sequence is a mechanical device or an intentional human agent [38,42].

To what extent are these biases in sequential predictions relevant to gambling disorder? Marmurek et al. [45] recruited university students with some past-year gambling, and examined predictors of color choices on a laboratory roulette task: the gambler's fallacy was stronger in male participants and correlated with scores on the Gambling-Related Cognitions Scale. Other work has examined gamblers' preferences between two simulated slot machines that differed in their alternation rates [46,47]. Whereas nongamblers were largely indifferent between a more "streaky" machine [set to $p(\text{Alt}) = .5$] and a more switchy machine [set to $p(\text{Alt}) = .7$], regular casino patrons preferred the more streaky machine, and this preference increased with problem gambling scores.

How are these sequential biases expressed in the underlying brain circuitry? The picture that emerges here is similar to the story from probability weighting, that PFC regions may detect these (false) associations between consecutive events in a sequence. An early study passively presented a binary sequence of abstract images: significant signal change was observed in dorsolateral PFC when a pattern was violated, in instances in which the pattern was either repeating (XXXXXXXO) or alternating (XOXOXOXX) [48]. Using a rewarded guessing task, Akitsuki et al. [49] saw that the striatal responses to winning outcomes did not differ between the first and the fourth successive win, but that the medial

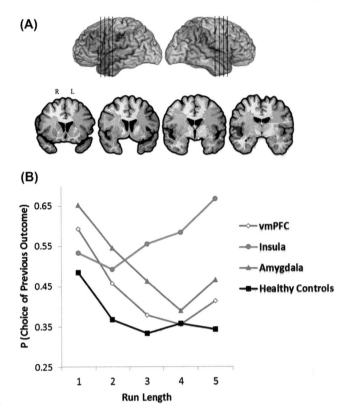

FIGURE 27.2 (A) Neuropsychological data on the effects of insula damage (displayed in coronal sections, for seven of eight patients for whom MRI scans were available) on sequential binary choice. (B) On the roulette task, involving sequential red/black color predictions, the gambler's fallacy is seen as a decrease in the likelihood of choosing either color as a function of the preceding run length of that color. VMPFC, ventromedial prefrontal cortex. *Data reproduced from Clark L, Studer B, Bruss J, Tranel D, Bechara A. Damage to insula abolishes cognitive distortions during simulated gambling. Proc Natl Acad Sci USA 2014;111: 6098−103. http://dx.doi.org/10.1073/pnas.1322295111.*

PFC was selectively recruited during longer streaks of both winning and losing predictions.

More recent work has sought to extend this hypothesis using neuropsychological designs. Using both fMRI and brain stimulation, Xue et al. [50] found convergent evidence for left lateral PFC involvement in predicting the break of a run of outcomes (i.e., gambler's fallacy). A study in patients with focal brain injury also highlighted a role for the insula in this effect [51]. Compared to lesion groups with damage to the amygdala or ventromedial PFC, a group of cases with insula damage showed a complete abolition of the gambler's fallacy on a roulette task involving red/black predictions (see Fig. 27.2). The insula group was also unresponsive to "near miss" outcomes on a second task based upon a simulated slot machine. The absence of these two gambling distortions following insula damage logically predicts hyperactivity in the insula region in pathological gamblers, who appear to be more sensitive to these erroneous thinking styles.

ILLUSORY CONTROL

The "illusion of control" is a further bias that is pervasive across many forms of gambling. In its original conceptualization by Langer [52], the illusion of control refers to an "expectancy of personal success probability [that is] inappropriately higher than the objective probability would warrant" (p. 311). This heightened expectancy of winning is created when features of skill-based situations are present in games in which the outcomes are determined by chance. As such, Langer's thesis is often summarized as that of "skill—chance confusion" [53]. One classic means of creating illusory control is via an "irrelevant choice." For example, your chances of winning a lottery are equivalent whether you select your favorite numbers or take a "lucky dip," but many people prefer to choose their own numbers. Langer [52] reported that participants who personally chose their lottery tickets demanded a higher price to sell back those tickets (mean $8.67, for a ticket that originally cost $1) than a control group whose ticket was selected for them (mean $1.96).

Skill-based features can be introduced into games of chance in a wide variety of other ways, making this a broad class of psychological distortion. Many studies (and real-world examples) make use of an instrumental action, such as throwing a dice or ball. In a field study of craps players, players who were personally "shooting the dice" were more likely to bet, placed higher bets, and placed bets with longer odds, compared to decisions made on other players' throws [54]. Familiarity with the stimulus materials and practice with the game itself are important factors [52,55], and trial-by-trial feedback can be manipulated to imbue a sense that one's performance is improving [56]. Finally, an element of competition may create a sense of skill, as highlighted in work showing that the tendency to anthropomorphize slot machines (to personify as if a human opponent) is associated with more persistent gambling [57].

While these examples are often regarded as compelling demonstrations of human bias, the underlying psychological processes are less well specified. Gamblers may refuse to exchange lottery tickets for various reasons, including beliefs about tempting fate [58] or an endowment effect. It is difficult to distinguish the subjective probability of winning (implied in Langer's original definition) from a heightened emotional response to outcomes deriving from one's own actions [59]. Much of the experimental literature relies on confidence ratings or arbitrary decisions of no consequence to the participant; these constitute weak demonstrations of an illusion [53,60]. By a stronger definition, the participant should be prepared to actively disadvantage himself or herself to exert irrelevant control over a chance outcome. This was evident in Langer's original study [52, experiment 3], in which the participants who chose their lottery ticket later rejected an opportunity to swap their ticket for one with a higher objective chance of winning, but in later studies, there were marked boundary conditions to this effect [61,62].

There is reasonable evidence to support the hypothesis that individuals with gambling disorder are more susceptible to the illusion of control. Questionnaire measures of gambling distortions like the Gambling-Related Cognitions Scale [63] unanimously include at least one subscale probing beliefs about superstitions, rituals, and the tendency to interpret wins as evidence of skill acquisition. Groups with gambling disorder reliably show elevated scores on such scales [64,65]. In a 2013 behavioral experiment, pathological gamblers in treatment reported higher control ratings on a zero-contingency task akin to procedures used to elicit "depressive realism" [66]. One common question is how this distortion might operate in games like poker, in which skill and experience play genuine roles. Clearly, this bias should not be regarded as an illusion in such games, but it is nevertheless observed that skill-based gamblers overestimate their personal ability and degree of skill [67], and in the specific case of poker, this is particularly true in online players [68].

A number of functional neuroimaging experiments have begun to examine brain responses associated with the illusion of control. While the distortions reviewed in the previous sections were manifested in higher-level cortical activity rather than at the level of the striatum, striatal responses are clearly sensitive to skill-related psychological variables. For example, outcome-related striatal responses to monetary wins were greater when the participant chose between two gambles, compared to when the gamble was selected by the computer, after equating outcome magnitude ([69], see also [70]). Studies have sought to extend this story to consider the perception of control. One experiment involving repeated choice between three gambles manipulated whether the participant's selection was either approved or vetoed by the computer: in this study, it was the medial PFC and posterior cingulate that scaled with (illusory) control, whereas the striatum did not [71]. However, two subsequent studies have refuted this conclusion. Leotti and Delgado [59] found that the ventral striatum was sensitive to the opportunity to make a choice (over a forced choice), even though any choice was randomly associated with gain or loss. In a large sample ($n = 79$) of adolescents playing a simulated slot machine, a subgroup who experienced illusory control during the game ($n = 19$, e.g., timing their button presses) showed greater activity in both striatum and

PFC compared to the subgroup who perceived the game more objectively [72]. The sensitivity of these regions to skill and control is highly congruent with the established role of frontostriatal circuitry in reinforcement learning and detection of agency.

CONCLUSION

In this chapter, I have conceptualized gambling disorder in terms of dysregulated decision-making, effectively of "pathological choice" [15]. I have highlighted four aspects of decision-making that are putatively relevant to why some individuals develop disordered gambling and to the maintenance of persistent betting in individuals with gambling disorder. Notably, two of these examples (loss aversion and probability weighting) are constructs from behavioral economics, and while these constructs are intuitively pertinent to gambling, the lack of empirical attention that they have received within gambling research is remarkable. The other two examples (perceptions of randomness and the illusion of control) serve as better illustrations of how basic research from cognitive psychology and neuroscience has at least begun to inspire clinically oriented questions concerning problem gambling.

Each of the four constructs that I have considered has been investigated from a neuroscience perspective, with functional imaging data (in addition to some convergent techniques like examination of patients with brain injury) implicating shared brain circuitry across each of the distortions, which includes striatum, insula, and PFC. These areas are robustly implicated in decision-making and outcome processing, and I have argued that the distortions of decision-making (which are the focus of the current chapter) are expressed in the anomalous recruitment of these same regions [73]. Clinical neuroimaging studies also implicate pathophysiology within these areas in groups with gambling disorder, as well as in patients with substance use disorders [74]. Such findings strengthen the hypothesis that faulty decision processes lie at the core of addictive disorders more broadly [75], and support the DSM-5 decision to reclassify gambling disorder (previously pathological gambling) alongside the substance use disorders, as the first recognized behavioral addiction.

Nonetheless, a few caveats should be noted. First, while psychological distortions such as loss aversion, overweighting low probabilities, and illusory control have intuitive value in understanding gambling disorder, it is unclear to what extent (if any) these distortions contribute to substance use disorders. Is it conceivable that these decisional constructs may be specifically relevant to gambling? A second issue that is clear from neuroscience data on experimental animals is that drugs of abuse are capable of stimulating the brain circuitry of interest (which relies upon the mesolimbic dopamine projection) more potently than any natural reinforcers like food or sex [76]. In my opinion, this presents an unresolved challenge for behavioral addictions: how do natural reinforcers (or conditioned reinforcers like money) come to "hijack" this neural system in a manner comparable to that of drugs of abuse? I have suggested elsewhere [77] that the multitude of psychological distortions that occur in gambling constitute an added ingredient that can persistently stimulate motivational circuitry in the brain and is capable of elevating this behavior to pathological levels in vulnerable individuals. Comparable mechanisms may operate for online video gaming, in which overlapping schedules of reinforcement combined with recent developments, like the avatar and the continuation of the game when the player is absent, may create a "perfect storm" for excessive behavior. But it is less clear what the analogous decisional mechanisms might be in other putative behavioral addictions like compulsive shopping or food addiction, and here we should recognize concerns about the "overpathologization" of everyday behaviors [78] and the null hypothesis that an addictions lens may not be appropriate for all excessive behaviors.

Disclosures

Dr. Clark is the Director of the Centre for Gambling Research at UBC, which is supported by funding from the Province of British Columbia and the British Columbia Lottery Corporation. He holds research funding in the UK from the Medical Research Council (G1100554). Dr. Clark has provided paid consultancy to Cambridge Cognition Ltd. on issues relating to neurocognitive assessment and has received a speaker honorarium from Svenska Spel (Sweden). He has not received any further direct or indirect payments from the gambling industry, or any other groups substantially funded by gambling, to conduct research or to speak at conferences or events.

References

[1] Ariyabuddhiphongs V. Lottery gambling: a review. J Gambl Stud 2011;27:15–33. http://dx.doi.org/10.1007/s10899-010-9194-0.

[2] Wardle H, Moody A, Spence S, Orford J, Volberg R, Jotangia D, et al. British gambling prevalence survey. London (UK): National Centre for Social Research; 2011.

[3] Bolen DW, Boyd WH. Gambling and the gambler: a review and preliminary findings. Arch Gen Psychiatry 1968;18:617–30.

[4] Reber AS. The EVF model: a novel framework for understanding gambling and, by extension, poker. Gaming Res Rev J 2012;16: 59–76.

[5] Kessler RC, Hwang I, LaBrie R, Petukhova M, Sampson NA, Winters KC, et al. DSM-IV pathological gambling in the national comorbidity survey replication. Psychol Med 2008;38:1351–60. http://dx.doi.org/10.1017/S0033291708002900.

[6] Binde P. Gambling across cultures: mapping worldwide occurrence and learning from ethnographic comparison. Int Gambl Stud 2005; 5:1–27. http://dx.doi.org/10.1080/14459790500097913.

[7] Turner NE, Horbay R. How do slot machines and other electronic gambling machines actually work? J Gambl Issues 2004:11.

[8] Fiorillo CD, Tobler PN, Schultz W. Discrete coding of reward probability and uncertainty by dopamine neurons. Science 2003; 299:1898–902.

[9] Zald DH, Boileau I, El-Dearedy W, Gunn R, McGlone F, Dichter GS, et al. Dopamine transmission in the human striatum during monetary reward tasks. J Neurosci 2004;24:4105–12.

[10] Hodgins DC, Stea JN, Grant JE. Gambling disorders. Lancet 2011; 378:1874–84. http://dx.doi.org/10.1016/S0140-6736(10)62185-X.

[11] Petry NM, Blanco C, Auriacombe M, Borges G, Bucholz K, Crowley TJ, et al. An overview of and rationale for changes proposed for pathological gambling in DSM-5. J Gambl Stud 2013. http://dx.doi.org/10.1007/s10899-013-9370-0.

[12] Shaffer HJ, Hall MN, Vander Bilt J. Estimating the prevalence of disordered gambling behavior in the United States and Canada: a research synthesis. Am J Public Health 1997;89:1369–76.

[13] Toce-Gerstein M, Gerstein DR, Volberg RA. A hierarchy of gambling disorders in the community. Addiction 2003;98: 1661–72. http://dx.doi.org/10.1111/j.1360-0443.2003.00545.x.

[14] Murch WS, Clark L. Games in the brain: neural substrates of gambling addiction. Neuroscientist 2015. http://dx.doi.org/10.1177/1073858415591474.

[15] Clark L, Averbeck BB, Payer D, Sescousse G, Winstanley CA, Xue G. Pathological choice: the neuroscience of gambling and gambling addiction. J Neurosci 2013;33:17617–23. http://dx.doi.org/10.1523/JNEUROSCI.3231-13.2013.

[16] Kahneman D, Tversky A. Prospect theory: an analysis of decision under risk. Econometrica 1979;47:264–91. From: http://www.jstor.org/stable/1914185.

[17] Sokol-Hessner P, Hsu M, Curley NG, Delgado MR, Camerer CF, Phelps EA. Thinking like a trader selectively reduces individuals' loss aversion. Proc Natl Acad Sci USA 2009;106:5035–40. http://dx.doi.org/10.1073/pnas.0806761106.

[18] Fox CR, Poldrack RA. Prospect theory and the brain. In: Neuroeconomics Decis. Mak. Brain. Elsevier; 2009. p. 533–67. http://dx.doi.org/10.1016/B978-0-12-416008-8.00042-5.

[19] Shiv B, Loewenstein G, Bechara A, Damasio H, Damasio AR. Investment behavior and the negative side of emotion. Psychol Sci 2005;16:435–9.

[20] De Martino B, Camerer CF, Adolphs R. Amygdala damage eliminates monetary loss aversion. Proc Natl Acad Sci USA 2010;107: 3788–92. http://dx.doi.org/10.1073/pnas.0910230107.

[21] Lesieur HR. The chase: the compulsive gambler. Rochester (VT): Schenkman; 1984.

[22] Xuan Z, Shaffer H. How do gamblers end gambling: longitudinal analysis of Internet gambling behaviors prior to account closure due to gambling related problems. J Gambl Stud 2009;25: 239–52. http://dx.doi.org/10.1007/s10899-009-9118-z.

[23] Giorgetta C, Grecucci A, Rattin A, Guerreschi C, Sanfey AG, Bonini N. To play or not to play: a personal dilemma in pathological gambling. Psychiatry Res 2014;219:562–9. http://dx.doi.org/10.1016/j.psychres.2014.06.042.

[24] Wu Y, van Dijk E, Aitken MRF, Clark L. Missed losses loom larger than missed gains: electrodermal reactivity to decision choices and outcomes in a gambling task. Cogn Affect Behav Neurosci 2016;16:353–61. http://dx.doi.org/10.3758/s13415-015-0395-y.

[25] Studer B, Clark L. Place your bets: psychophysiological correlates of decision-making under risk. Cogn Affect Behav Neurosci 2011; 11:144–58. http://dx.doi.org/10.3758/s13415-011-0025-2.

[26] Tom SM, Fox CR, Trepel C, Poldrack RA. The neural basis of loss aversion in decision-making under risk. Science 2007;315:515–8.

[27] Sokol-Hessner P, Camerer CF, Phelps EA. Emotion regulation reduces loss aversion and decreases amygdala responses to losses. Soc Cogn Affect Neurosci 2013;8:341–50. http://dx.doi.org/10.1093/scan/nss002.

[28] Gonzalez R, Wu G. On the shape of the probability weighting function. Cogn Psychol 1999;166:129–66.

[29] Ligneul R, Sescousse G, Barbalat G, Domenech P, Dreher J-C. Shifted risk preferences in pathological gambling. Psychol Med 2013;43:1059–68. http://dx.doi.org/10.1017/S0033291712001900.

[30] Lyons CA, Ghezzi PM. Wagering on a large scale: relationships between public gambling and game manipulations in two state lotteries. J Appl Behav Anal 1995;28:127–37. http://dx.doi.org/10.1901/jaba.1995.28-127.

[31] Linnet J, Frøslev M, Ramsgaard S, Gebauer L, Mouridsen K, Wohlert V. Impaired probability estimation and decision-making in pathological gambling poker players. J Gambl Stud 2012;28: 113–22. http://dx.doi.org/10.1007/s10899-011-9244-2.

[32] Lawrence AJ, Luty J, Bogdan NA, Sahakian BJ, Clark L. Problem gamblers share deficits in impulsive decision-making with alcohol-dependent individuals. Addiction 2009;104:1006–15. http://dx.doi.org/10.1111/j.1360-0443.2009.02533.x. ADD2533.

[33] Reynolds B. A review of delay-discounting research with humans: relations to drug use and gambling. Behav Pharmacol 2006;17: 651–67.

[34] Madden GJ, Petry NM, Johnson PS. Pathological gamblers discount probabilistic rewards less steeply than matched controls. Exp Clin Psychopharmacol 2009;17:283–90. http://dx.doi.org/10.1037/a0016806.

[35] Miedl SF, Peters J, Buchel C. Altered neural reward representations in pathological gamblers revealed by delay and probability discounting. Arch Gen Psychiatry 2012;69:177–86. http://dx.doi.org/10.1001/archgenpsychiatry.2011.1552.

[36] Takahashi H, Matsui H, Camerer C, Takano H, Kodaka F, Ideno T, et al. Dopamine D(1) receptors and nonlinear probability weighting in risky choice. J Neurosci 2010;30:16567–72. http://dx.doi.org/10.1523/JNEUROSCI.3933-10.2010.

[37] Tobler PN, Christopoulos GI, O'Doherty JP, Dolan RJ, Schultz W. Neuronal distortions of reward probability without choice. J Neurosci 2008;28:11703–11. http://dx.doi.org/10.1523/JNEUROSCI.2870-08.2008.

[38] Croson R, Sundali J. The gambler's fallacy and the hot hand: empirical data from casinos. J Risk Uncertain 2005;30:195–209. http://dx.doi.org/10.1007/s11166-005-1153-2.

[39] Kahneman D, Tversky A. Subjective probability: a judgment of representativeness. Cogn Psychol 1972;3:430–54. http://dx.doi.org/10.1016/0010-0285(72)90016-3.

[40] Oskarsson AT, Van Boven L, McClelland GH, Hastie R. What's next? Judging sequences of binary events. Psychol Bull 2009;135: 262–85. http://dx.doi.org/10.1037/a0014821.

[41] Wagenaar WA. Generation of random sequences by human subjects: a critical survey of literature. Psychol Bull 1972;77:65–72.

[42] Ayton P, Fischer I. The hot hand fallacy and the gambler's fallacy: two faces of subjective randomness? Mem Cogn 2004;32:1369–78. http://dx.doi.org/10.3758/BF03206327.

[43] McGlothlin WH. Stability of choices among uncertain alternatives. Am J Psychol 1956;69:604–15.

[44] Gilovich T, Vallone R, Tversky A. The hot hand in basketball: on the misperception of random sequences. Cogn Psychol 1985;17: 295–314.

[45] Marmurek HHC, Switzer J, D'Alvise J. Impulsivity, gambling cognitions, and the gambler's fallacy in university students. J Gambl Stud 2013. http://dx.doi.org/10.1007/s10899-013-9421-6.

[46] Scheibehenne B, Wilke A, Todd PM. Expectations of clumpy resources influence predictions of sequential events. Evol Hum Behav 2011;32:326–33. http://dx.doi.org/10.1016/j.evolhumbehav.2010.11.003.

[47] Wilke A, Scheibehenne B, Gaissmaier W, McCanney P, Barrett HC. Illusionary pattern detection in habitual gamblers. Evol Hum Behav 2014. http://dx.doi.org/10.1016/j.evolhumbehav.2014.02.010.

[48] Huettel SA, Mack PB, McCarthy G. Perceiving patterns in random series: dynamic processing of sequence in prefrontal cortex. Nat Neurosci 2002;5:485–90. http://dx.doi.org/10.1038/nn841nn841.

[49] Akitsuki Y, Sugiura M, Watanabe J, Yamashita K, Sassa Y, Awata S, et al. Context-dependent cortical activation in response to financial reward and penalty: an event-related fMRI study. NeuroImage 2003;19:1674–85. http://dx.doi.org/10.1016/S1053-8119(03)00250-7.

[50] Xue G, Juan CH, Chang CF, Lu ZL, Dong Q. Lateral prefrontal cortex contributes to maladaptive decisions. Proc Natl Acad Sci USA 2012;109:4401–6. http://dx.doi.org/10.1073/pnas.1111927109.

[51] Clark L, Studer B, Bruss J, Tranel D, Bechara A. Damage to insula abolishes cognitive distortions during simulated gambling. Proc Natl Acad Sci USA 2014;111:6098–103. http://dx.doi.org/10.1073/pnas.1322295111.

[52] Langer EJ. The illusion of control. J Pers Soc Psychol 1975;32:311–28. http://dx.doi.org/10.1037/0022-3514.32.2.311.

[53] Stefan S, David D. Recent developments in the experimental investigation of the illusion of control. A meta-analytic review. J Appl Soc Psychol 2013;43:377–86.

[54] Davis D, Sundahl I, Lesbo M. Illusory personal control as a determinant of bet size and type in casino craps games. J Appl Soc Psychol 2000;30:1224–42.

[55] Ladouceur R, Walker M. A cognitive perspective on gambling. In: Salkovskis PM, editor. Trends in cognitive and behavioural therapies. Chichester (UK): Wiley & Sons; 1996. p. 89–120.

[56] Langer EJ, Roth J. Heads I win, tails it's chance: the illusion of control as a function of the sequence of outcomes in a purely chance task. J Pers Soc Psychol 1975;32:951–5. http://dx.doi.org/10.1037//0022-3514.32.6.951.

[57] Riva P, Sacchi S, Brambilla M. Humanizing machines: anthropomorphization of slot machines increases gambling. J Exp Psychol Appl 2015.

[58] Risen JL, Gilovich T. Another look at why people are reluctant to exchange lottery tickets. J Pers Soc Psychol 2007;93:12–22. http://dx.doi.org/10.1037/0022-3514.93.1.12.

[59] Leotti LA, Delgado MR. The value of exercising control over monetary gains and losses. Psychol Sci 2014. http://dx.doi.org/10.1177/0956797613514589.

[60] Koehler JJ, Gibbs BJ, Hogarth RM. Shattering the illusion of control: multi-shot versus Single-shot gambles. J Behav Decis Mak 1994;7:183–91.

[61] Dunn DS, Wilson TD. When the stakes are high: a limit to the illusion-of-control effect. Soc Cogn 1990;8:305–23. http://dx.doi.org/10.1521/soco.1990.8.3.305.

[62] Grou B, Tabak BM. Ambiguity aversion and illusion of control: experimental evidence in an emerging market. J Behav Financ 2008;9:22–9. http://dx.doi.org/10.1080/15427560801897162.

[63] Raylu N, Oei TPS. The Gambling Related Cognitions Scale (GRCS): development, confirmatory factor validation and psychometric properties. Addiction 2004;99:757–69. http://dx.doi.org/10.1111/j.1360-0443.2004.00753.x.

[64] Michalczuk R, Bowden-Jones H, Verdejo-Garcia A, Clark L. Impulsivity and cognitive distortions in pathological gamblers attending the UK National Problem Gambling Clinic: a preliminary report. Psychol Med 2011;41:2625–35.

[65] Goodie AS, Fortune EE. Measuring cognitive distortions in pathological gambling: review and meta-analyses. Psychol Addict Behav 2013;27:730–43. http://dx.doi.org/10.1037/a0031892.

[66] Orgaz C, Estevez A, Matute H. Pathological gamblers are more vulnerable to the illusion of control in a standard associative learning task. Front Psychol 2013;4:306. http://dx.doi.org/10.3389/fpsyg.2013.00306.

[67] Myrseth H, Brunborg GS, Eidem M. Differences in cognitive distortions between pathological and non-pathological gamblers with preferences for chance or skill games. J Gambl Stud 2010;26:561–9. http://dx.doi.org/10.1007/s10899-010-9180-6.

[68] Mackay T-L, Bard N, Bowling M, Hodgins DC. Do pokers players know how good they are? Accuracy of poker skill estimation in online and offline players. Comput Hum Behav 2014;31:419–24.

[69] Coricelli G, Critchley HD, Joffily M, O'Doherty JP, Sirigu A, Dolan RJ. Regret and its avoidance: a neuroimaging study of choice behavior. Nat Neurosci 2005;8:1255–62. http://dx.doi.org/10.1038/nn1514.

[70] Tricomi EM, Delgado MR, Fiez JA. Modulation of caudate activity by action contingency. Neuron 2004;41:281–92.

[71] Kool W, Getz SJ, Botvinick MM. Neural representation of reward probability: evidence from the illusion of control. J Cogn Neurosci 2013;25:852–61. http://dx.doi.org/10.1162/jocn.

[72] Lorenz RC, Gleich T, Kühn S, Pöhland L, Pelz P, Wüstenberg T, et al. Subjective illusion of control modulates striatal reward anticipation in adolescence. NeuroImage 2015;117:250–7. http://dx.doi.org/10.1016/j.neuroimage.2015.05.024.

[73] Clark L. Decision-making during gambling: an integration of cognitive and psychobiological approaches. Philos Trans R Soc Lond B Biol Sci 2010;365:319–30. http://dx.doi.org/10.1098/rstb.2009.0147.

[74] Limbrick-Oldfield EH, Van Holst RJ, Clark L. Fronto-striatal dysregulation in drug addiction and pathological gambling: consistent inconsistencies? NeuroImage Clin 2013;2:385–93. http://dx.doi.org/10.1016/j.nicl.2013.02.005.

[75] Monterosso J, Piray P, Luo S. Neuroeconomics and the study of addiction. Biol Psychiatry 2012;72:107–12. http://dx.doi.org/10.1016/j.biopsych.2012.03.012.

[76] Wise RA. Dopamine, learning and motivation. Nat Rev Neurosci 2004;5:483–94.

[77] Clark L. Disordered gambling: the evolving concept of behavioral addiction. Ann NY Acad Sci 2014;1327:46–61. http://dx.doi.org/10.1111/nyas.12558.

[78] Billieux J, Schimmenti A, Khazaal Y, Maurage P, Heeren A. Are we overpathologizing everyday life? a tenable blueprint for behavioral addiction research. J Behav Addict 2015;4:119–23. http://dx.doi.org/10.1556/2006.4.2015.009.

GENETIC AND HORMONAL INFLUENCES ON MOTIVATION AND SOCIAL BEHAVIOR

28

Decision-Making in Fish: Genetics and Social Behavior

R.D. Fernald

Stanford University, Stanford, CA, United States

Abstract

In evolution, social behavior has been a potent selective force in shaping brains to control action. Social animals interact with others and their environment to survive and reproduce. To be successful, animals must acquire, evaluate, and translate information about their social and physical environments to decide what to do next. The resulting decisions can profoundly alter an animal's behavior and physiology, producing a diverse array of cellular and molecular responses. Clearly the brain controls behavior but how does an animal's behavior sculpt its brain? Using a particularly suitable fish model system with complex social interactions, we report how the social context of behavior shapes the brain and alters the behavior and neural circuitry of interacting animals. We have shown that social opportunities produce rapid changes in gene expression in key nuclei in the brain and these genomic responses prepare the individual to modify its behavior moving into a different social niche. Both social success and failure produce changes in neuronal cell size and connectivity in key brain nuclei. Understanding mechanisms through which social information is transduced into cellular and molecular changes provides a deeper understanding of the brain systems responsible for animal decisions.

During evolution, social behavior has been a potent selective force for generating brains that control essential actions, and especially for deciding what to do next. Social animals that are engaged in finding mates and food must modify their behavior continuously to survive and be successful in producing offspring. Social interactions clearly influence the brain and circulating hormonal levels, but just how does the social environment regulate the physiological, cellular, and molecular processes of the brain, particularly related to making choices? How do animals decide what to do next? Elucidating this connection between neural and behavioral processes is critically important. Ultimately, social interactions, especially those related to reproduction, are essential for the evolutionary success of all animals.

Decision-making in the regulation of vertebrate reproduction offers a unique opportunity for understanding how the brain controls behavior and, in turn, how behavior influences the brain for two reasons. First, reproductive behaviors are often genetically encoded and hence stereotypic, making them relatively easy to observe and quantify. Second, the central control of reproduction is hierarchically organized in a regulatory cascade known as the brain—pituitary—gonadal axis. In all vertebrates, at the top of this axis are neurons in the hypothalamus containing gonadotropin-releasing hormone 1 (GnRH1), a decapeptide that is delivered to the pituitary where it regulates release of additional hormones that ultimately control reproduction. This means that the GnRH1 neurons integrate the external and internal factors that control reproduction. Upstream from these neurons are the circuits that deliver the information that determines the reproductive outcome. This chapter will describe how we can establish a mechanistic understanding of an animal's behavior from its ecosystem to its social brain [1].

Understanding behavioral regulation of the brain requires an animal model in which: (1) social interchange is essential for reproductive success; (2) animals can be studied in a seminatural context; (3) key molecular, cellular, and physiological processes are accessible; and (4) behavior and physiology of both individuals and groups of animals can be readily analyzed.

SOCIAL SYSTEM OF THE AFRICAN CICHLID FISH, *ASTATOTILAPIA BURTONI*

A. burtoni is endemic to Lake Tanganyika in the Rift Valley system in east Africa where it lives in shallow shore pools and river estuaries [2]. *A. burtoni* males exist as one of two reversible, socially controlled phenotypes: reproductively competent dominant (D) males and reproductively incompetent nondominant (ND) males (see Fig. 28.1). These two male phenotypes and their hypothalamic—pituitary—gonadal (HPG) axis activity—and hence reproductive capacity—are tightly coupled to their social status [3,4]. In their natural habitat, D or territorial males represent a small percentage of the population (10—30%) and are brightly colored (blue or yellow) with a black stripe through the eye (eye-bar), an opercular black spot at the caudal tip of the gill cover, prominent egg-spots on the anal fin, and a red humeral patch on the side of the body (Fig. 28.1). D males defend territories vigorously against encroachment by rival males and they spend the majority of their time actively courting and spawning with females [5,6]. In contrast, subordinate (also called nonterritorial) males make up the majority of the male population (70—90%), are duller in coloration (lacking eye-bar and humeral patch), do not hold territories or reproduce, school with females and other subordinate males, and flee from the aggressive attacks of D males. *A. burtoni* lives in a lek-like social system (Fig. 28.2) in which D males defend territories over a food source, which they guard from rival males.

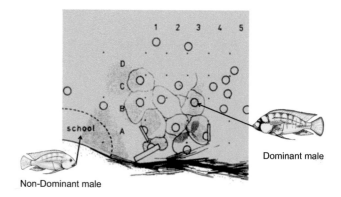

Non-Dominant male

Dominant male

FIGURE 28.2   Sketch of an observation area in Lake Tanganyika, Burundi, Africa. *Solid dots* are grid stakes spaced ∼50 cm apart and labeled (1—4; A—D) for identification. *Circles* represent spawning pit locations of dominant males. *Lighter colored outlines* circumscribe the territories of individuals. Nondominant males and females school near the territorial area. *Based on Fernald RD, Hirata NR. Field study of* Haplochromis burtoni: *quantitative behavioral observations. Anim Behav 1977;25:964—75.*

D males perform 19 distinct behavioral patterns during social interactions that are associated with territoriality and reproduction [6]. For example, D males establish a spawning area by digging a pit in their territory, engaging in agonistic threat displays and border

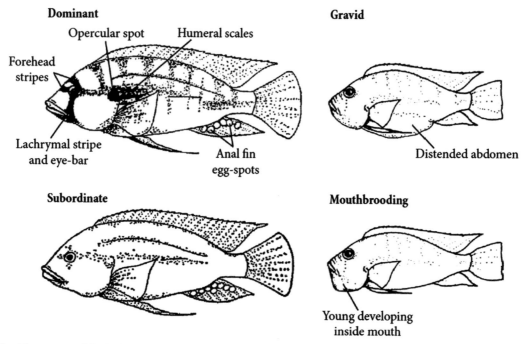

FIGURE 28.1   Illustrations of the body patterns for typical dominant (territorial) and subordinate (nonterritorial) *Astatotilapia burtoni* males, and sexually receptive gravid and mouthbrooding parental females. Dominant males are brightly colored (yellow or blue) and have distinct yellow-orange egg-spots on their anal fins, dark forehead stripes, a dark opercular spot on the caudal edge of the gill cover, a dark lachrymal stripe or eye-bar extending from the eye to the lower jaw, and a bright orange-red patch on the humeral scales. Subordinate males lack the robust markings of their dominant counterparts and are more similar in coloration to females. Females cycle between a gravid receptive phase in which they develop distended abdomens from growing oocytes as they get closer to spawning and a mouthbrooding phase in which their jaws protrude forward to accommodate the developing young inside the mouth. *Modified in part from Fernald RD. Quantitative behavioural observations of* Haplochromis burtoni *under semi-natural conditions. Anim Behav 1977;25:643—53.*

disputes with neighboring D males, chasing subordinate males away from their territory, and performing courtship quivering behavior toward passing females in an attempt to lead them into their territorial pit to spawn. D status is transient and depends on the male's ability to maintain his territory against attack and, importantly, to elude predators. When a D male loses his territory by losing a fight or being eaten, multiple ND males will compete for the territory and the successful male will rapidly (seconds) turn on colors characteristic of dominance (see later). Because defensible territory substrate for spawning and feeding is often limited, and females are less likely to mate outside the protection of a spawning shelter, there is fierce competition for this resource. Consequently, at any given time, only a minority of males defend a territory and mate.

Once a receptive (gravid, "ripe with eggs") female follows a D male into his territory and is appropriately stimulated, she will deposit eggs on the substrate and then immediately collect them in her mouth. The male then displays his anal fin egg-spots on the substrate in front of her, and while she attempts to collect the egg-spots on his fin, the male releases sperm near her mouth to fertilize the eggs. There are typically several bouts of egg-laying and fertilization that may be briefly interrupted as the D male chases away intruders or interacts with neighboring males. When spawning and fertilization are complete, the female leaves the territory to brood the young in her mouth (mouthbrooding) for approximately 2 weeks until releasing them as fully developed fry. Meanwhile the D male resumes his territorial defense and continues to court other receptive females [5].

Social System of *Astatotilapia burtoni*: Consequences of Status

The distinctive external differences between *A. burtoni* male phenotypes reflect major physiological and neural responses to differences in social status. When a male switches phenotypes, changes including expression of the black bar through the eye, brightening of the body color, and changed behavioral repertoires, are expressed in minutes, whereas concomitant physiological and neural changes happen over hours to days.

ND males attend closely to activity in the social scene, assessing when they might be able to gain a territory by defeating a resident male. In deciding whether to take on a higher ranking male, the ND male must assess the costs and benefits of this choice. He must weigh the cost of potential physical damage against the benefit of being able to spawn and thus produce offspring. When a male chooses to try and take over a territory, he typically initiates a dramatic fight in which the two

males engage in mouth-to-mouth biting, hitting each other with their bodies and nipping at each other's fins. If the ND male successfully defeats the resident, he rapidly turns on his bright body colors [5,7] and will quickly begin performing the 17 distinct behaviors characteristic of D males. Beginning immediately and continuing over the next few days, the reproductive system of the ascending male is remodeled, rendering him reproductively competent. Changes are evident at several levels along the HPG axis [8].

As noted earlier, in *A. burtoni*, as in all vertebrates, reproduction is controlled by GnRH1-containing neurons in the hypothalamus that deliver the eponymously named GnRH decapeptide to the pituitary. When a male ascends (ND → D), delivery of this molecule sets in motion a cascade of actions ultimately resulting in reproductive competence and release of sex steroids from the gonads. This social change rapidly increases production of hypothalamic GnRH1 mRNA [7] and GnRH1 peptide [9], and in a few days, the GnRH1 neurons increase in volume by eightfold [3] and extend their secondary dendrites [4]. We have shown that in D males, the GnRH1 neurons are connected to one another via gap junctions, thought to promote synchronous firing required for reproductive competence [10].

In striking contrast, when a D male is moved into a social system with larger D males (>5% greater in standard length), it abruptly loses its color (<1 min) and joins other ND males and females in a school. Its GnRH-containing neurons in the preoptic area shrink to one-eighth their volume and produce less GnRH mRNA and peptide, causing hypogonadism and loss of reproductive competence (~2 weeks) [3,11]. Similarly, androgen, estrogen, and GnRH receptor mRNA expression levels depend on social status [12−14] as do electrical properties of the GnRH neurons themselves [15].

DOMAINS OF *ASTATOTILAPIA BURTONI* SOCIAL DECISIONS

Can Males Be Deceptive?

Could or would a male *A. burtoni* deceive another male, and under what circumstances? We asked this question experimentally using a novel paradigm in which two differently sized males share a tank, divided in half by a clear, watertight divider and a black removable divider, which allowed us to control when the animals could see one another [16]. Half of a terra cotta pot was halved again lengthwise and placed so that a shelter of ¼ pot was on each side of the pair of dividers, allowing each male to occupy this "shared" shelter. Importantly, with the black divider in place, neither animal

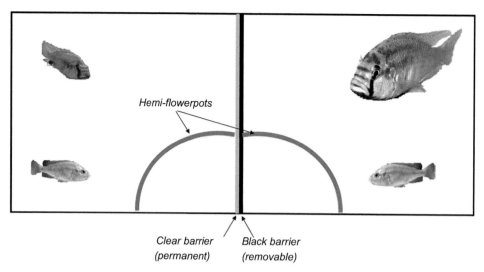

FIGURE 28.3   Front view of the aquarium (45 L) divided in half with a watertight, clear divider (*gray midline*) and a removable opaque barrier (*black midline*). There are a small male fish (left compartment) and a large male fish (~4× larger; right compartment). The half terra cotta pot was cut so that both fish "shared" the same shelter, although they were not aware of each other's presence. This "shared" shelter was hemisected by both center dividers. A layer of gravel covered the bottom of the tank. *Modified from Chen C-C, Fernald RD. Visual information alone changes behavior and physiology during social interactions in a cichlid fish* (Astatotilapia burtoni). *PLoS One 2011;6:e20313.*

could know that the other animal was present (Fig. 28.3). One male was about four times larger than the other and each male was paired with an appropriately sized female in his half of the tank.

Both the small and the large fish were habituated to this starting condition with the opaque barrier in place for 2 days, during which time each behaved like a normal D male in his territory: excavating gravel from his hemi-pot, courting the female in his half of the tank, leading the female back to the shelter, and performing typical courtship and territorial male behaviors, all of which were quantified. On the third day, the opaque barrier was lifted, and although there was no physical or chemical contact possible, the larger male made several "attacks" toward the small male, who quickly lost his coloration, including the eye-bar. This is typical behavior for a male losing his territorial status as was confirmed by quantifying the male's behavior. Indeed, the smaller males essentially abandoned their part of the shelter, digging a new pit remote from that corner. Interestingly, this suppression of dominance, which is based entirely on visual signals, as reflected by the behavioral quantification, also reduced the expression of androgen hormones, but only for the first 3 days [16]. Seven days after the black barrier was lifted, the smaller animals recovered their hormone expression levels and other brain markers of dominance while maintaining the coloration of an ND male. Moreover, they could be seen courting their females when out of view of the D male.

Thus, the effects of the visual suppression resulted in changes in the expression of aggressive, territorial behavioral responses by the smaller male but did not result in sustained physiological changes. This suggests that the smaller males uncoupled changes in circulating hormones from their effects on outward appearance, seemingly presenting a false outward appearance not consonant with internal changes. This appears to be an example of deceptive behavior on the part of the male, allowing him to continue his courtship but not be influenced by the larger male. We assume that the smaller animal learned that the clear barrier prevented the large male from actual physical attacks and this recognition led the smaller animals to a novel strategy. The small male attended carefully to the D male when carrying out his behavior. The attention that the ND male paid to the D male is a more general attribute that we find in an attention hierarchy these animals have.

Attention Hierarchy Among Male Fish

Individual *A. burtoni* monitor the behavior of other individuals within their group. In particular, subordinate animals attend to the behavior of D animals in what is called in primates an "attention hierarchy" [17]. Attention hierarchies have been identified in humans, particularly in groups of children [18], and describe the conditions under which individuals modulate their behavior depending on their own status relative to that of others (see also Chapter 20). Within a hierarchy, when a high-ranking individual attacks a lower-ranking individual, the lower-ranking individual often

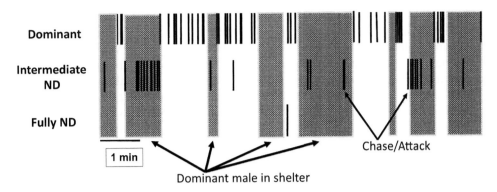

FIGURE 28.4 Schematic illustration of typical dominant male behavior in the presence of an intermediate nondominant (ND) male trying to ascend to dominance. *Large rectangles* represent the dominant male in his shelter and *small black bars* show when an individual attacks. Note that intermediate males attack other males only when the dominant male cannot see them because he is in his shelter.

then subsequently attacks an individual lower in rank [19]. In addition to humans, attention hierarchies have been described in baboons and mandrills [20] as well as in reptiles [21] and fish [22].

In *A. burtoni*, we assembled groups of animals that we marked to allow identification of individuals. We videotaped these groups ($n = 20$/group; four replicates) and quantified interactions between D and ND males [23]. We discovered several facts about how these animals interact. First, D and ND males never behaved aggressively toward another animal at the same time. Specifically, when D animals behaved, ND animals did not. Even more interesting was that when the D male was out of view in a shelter, ND males that were larger and attempting to ascend behaved aggressively and even courted females, behaviors that never occurred when the D male could see them (Fig. 28.4). These animals were termed "intermediate males" because they were the individuals most likely to ascend at the next opportunity.

As shown (Fig. 28.4), each time a D animal is out of view, the intermediate male attacks an ND male until the D male reappears. When the D male returns to the scene, he attacks within a few seconds but does not specifically target fish that have been aggressive to others in his absence [23], which makes sense because he cannot know who has been doing what in his absence.

These data show that ND males are attending to D males and altering their behavior by acting aggressively, which is not possible when the D male is present. In addition, these males on occasion will approach and court females when the D male cannot see them, another behavior that does not occur when a D male is present. These behaviors reflect a sophisticated social calculus in which ND males are doing the most they can to increase their chances of becoming D. It is possible that they are also learning about D

behavior through watching, a skill we have reported elsewhere [24].

When Being Observed, Males Change Their Behavior

During their lifetimes, like all animals, *A. burtoni* must continually make important decisions regarding reproduction and survival [25]. Such decisions are essential for their reproductive success because, as described above, territorial ownership and defense is the key to successful reproduction. Both physical and social changes can drastically reorganize social hierarchies by changing the number and location of available territories as animals are preyed upon, exposing fish to new social environments. Storms, intruders, and other disruptions of the substrate contribute to the instability of the social system, so continual maintenance of territory is a critical social behavior [5]. Based on the attention hierarchy described earlier, we reasoned that animals might alter their behavior when being observed by conspecifics, consistent with the notion that they could deceive other individuals that are watching them.

To test this notion, we designed experiments in which two D male fish were in a tank, separated by a clear, watertight barrier, and both could be observed from a third compartment [26]. When two size-matched D males are placed in these compartments, they will fight through the transparent barrier for the 20-min duration of the experiment while they are observed by a noninteracting audience. We asked how the behavior of the fighting animals would vary as a function of the composition of the audience and found that exactly what they did depended on the composition of the audience. Aggression is energetically costly and also increases visibility and hence danger from predation, so animals may assess opponents or likely opponents to evaluate competitive ability before engaging in a fight [27–29].

How might audience members interpret such differences in fighting intensity? Fish decreasing their aggressive intensity when being watched by a larger male may allow the audience male to infer that the displaying males are a threat to him in his future attempts to reproduce. Similarly, increasing aggression in the presence of gravid females might allow females to infer that the fighting males are more dominant than they actually are, increasing the displaying male's chances for reproduction. This change in aggressiveness could be seen as social manipulation [30], which has been reported in primates [31,32], children [33], and fish [34]. These data extend the view that bystanders are gathering information that may be useful in future encounters.

In addition to the behavioral observations, we measured brain responses to animals being observed while fighting. To do this, we used the expression of immediate-early genes to identify which brain areas were activated [26]. We found that depending on the nature of the audience, immediate-early gene expression in key brain nuclei was differentially influenced. Both when an audience of larger males watched fighting males and when they were watching larger males fighting, nuclei in the brain considered homologous with mammalian nuclei known to be associated with anxiety showed increased activity. When males were in the presence of any audience or when males saw any other males fighting, nuclei in the brain known to be involved in reproduction and aggression were differentially activated relative to control animals. In all cases, there was a close relationship between patterns of brain gene expression between fighters and observers. This suggests that the network of brain regions known as the social behavior network, common across vertebrates, is activated not only in association with the expression of social behavior but also by the reception of social information.

Transitive Inference by Males

An ongoing goal of ND *A. burtoni* males is to ascend in the social hierarchy to become a D male, be reproductively active, find mates, and reproduce. In Lake Tanganyika, colonies of *A. burtoni* range in size from a few dozen animals to over 100, depending on the number of fish the local substrate can support. More generally, males are faced with the challenge of deciding whom to fight because it is through such contests that dominance is decided. How do males decide whom to fight? Clearly, fighting with many animals to find a D male weak enough to beat would require energetic and physical demands that are prohibitive. Because our previous studies revealed that animals attended to one another during social encounters, we wondered whether their observational skills might allow them to predict the outcome of male–male encounters. Could they choose

to engage in fights they had a reasonable chance of winning? That is, could males infer their chances of winning a fight simply from watching other animals fight?

The logic of this process, known as transitive inference, is that if you know that A is taller than B and B is taller than C, you can infer that A is taller than C by constructing a virtual cognitive hierarchy without needing to see A, B, and C lined up for comparison. This ability was one of the developmental milestones first described by Piaget [35] and has since been identified in humans older than ~3 years of age as well as in nonhuman primates [36–38], rats [39,40], and birds [41–44].

To discover whether *A. burtoni* had the ability to infer fighting abilities of other fish, we tested whether bystander males could synthesize information from observing pairwise fights into an implied hierarchy of male fighting abilities. We tested bystander fish by having them observe staged fights between five size-matched males (A–E) in which fighting prowess was A > B, B > C, C > D, and D > E, which has the implied hierarchy of A > B > C > D > E [45]. In a specially built aquarium, fights were staged by moving one rival into another rival's territory, which resulted in the intruder animal losing (Fig. 28.5). For the control animals, there was no implied hierarchy (e.g., A = B = C = D = E).

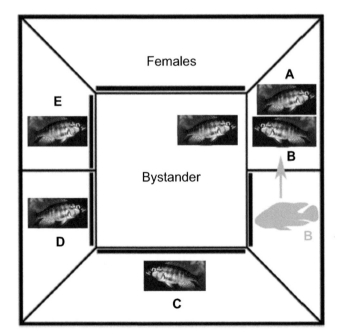

FIGURE 28.5 Tank arrangement and bystander training. Five rival males (A, B, C, D, and E) were arranged in visually, chemically, and physically isolated compartments around the central bystander unit. To train a bystander on a particular fight, the male scheduled to be the "loser" was removed from its unit and placed in the territory of the scheduled "winner." The opaque barrier separating the bystander from the rivals was then removed to allow the bystander to view the fight. The fight A1 versus B2 (A wins, B loses) is shown here in diagrammatic form. *Modified from Grosenick L, Clement TS, Fernald RD. Fish can infer social rank by observation alone. Nature 2007;445:429–32.*

The bystander males were trained by watching pairwise fights, and we tested their preference between rivals they had never seen fight, that is, pairs that were not in the testing sequence. After the bystander had seen five animals in four pairwise fights, we tested whether it could identify that B would beat D. Observers consistently chose D as the weaker animal by swimming to be closer to B than D, because fish will move toward the rival they perceive to be weaker [46,47].

The fact that *A. burtoni* can perform transitive inference in this important situation, choosing which male to attack, is consistent with the behavioral needs and ecological context in their natural habitat. In the temporary shore pools and estuaries of their native habitat, there is regular disruption of established territories by movement of hippopotamuses, the wind, and predation [5]. So being able to judge their rivals based on featural representations, independent of context, would be invaluable for them to increase their chances of reproductive success. We have shown that social ascent upon gaining a territory is swift and activates many behavioral, physiological, and molecular processes, allowing the ascending male a chance at reproductive success [7,8]. Since our original demonstration, transitive inference has been shown in another teleost, brook trout (*Salvelinus fontinalis*) [48]. It seems likely that transitive inference could be found in other colony-living animals that face similar constraints on reproduction that can be tested using methods that exploit a natural context and behavioral elements related to behavioral acts in the animal's natural life.

Genomic Consequences of Female Mate Choice

In females of many species, information about potential mates can change reproductive physiology, including gene expression. Because female mate choice is a large and active field of research in behavior, it is important to understand how a female brain responds to social information about a prospective mate. This specific experimental question depends on deciding both how and where to look for a signal in the brain that reflects female mate choice. We chose to look for changes in brain activity marked by gene expression using an immediate-early gene (IEG), *egr-1*, following a behavioral mate choice paradigm. IEGs are inducible transcription factors that make up the first wave of gene expression induced in neurons. Prior work showed that IEG expression is induced by a range of natural experiences including sensory stimuli and, consequently, it has been used extensively in mammals and birds ([49,50]). More recently, we showed that *egr-1* is highly conserved in *A. burtoni* and that functionally it responds robustly within 30 min of stimulation [51]. As for the brain location, we hypothesized that the conserved vertebrate social behavior

network (SBN) would be a logical place to look for responses to mate choice. The SBN was originally described by Newman [52] as a collection of brain nuclei implicated in a variety of social behaviors including male mating behavior, female sexual behavior, parental behavior, and aggressive behavior. Anatomical homologs to this network have subsequently been identified in birds and fish [53,54] and served to guide a variety of experiments. It was unknown, however, whether IEGs might respond to social information as well as to behavioral actions. Our previous experiments on female choice had demonstrated that reproductively ready (i.e., gravid) females prefer to associate with D, reproductively active males, whereas nongravid females prefer ND, nonreproductive males [47].

Using a paradigm similar to that described in Clement et al. [47], we placed females in an aquarium with one male at each end, behind a clear Plexiglas barrier that was watertight to eliminate chemical cues. Thus the female received only visual information. We observed the female, recording behavior and position relative to the males, for 20 min and could score her preference for one of the two males based on these data. Following this, we staged a fight between the two males and at random allowed the female's chosen male to win or lose that fight. Our control condition was for the female to choose between two males and not see a subsequent fight. We hypothesized that the female's choice of a mate would produce distinct patterns of brain gene expression. Subsequently, we measured and compared the mRNA levels of *egr-1* and another IEG, *cfos*, in six brain nuclei constituting the SBN. Surprisingly, females seeing their selected male win or lose a fight produced dramatically different brain IEG expression patterns. Females who saw their preferred male win a fight activated brain nuclei associated with reproduction and reproductive behaviors, specifically, the anterior hypothalamus, ventromedial hypothalamus, preoptic area, and periaqueductal gray. In striking contrast, females who saw their preferred male lose a fight had a much higher expression of IEGs in a part of the brain associated with anxiety, the lateral septum [55]. A remarkable aspect of these results is that the females were responding to visual information alone. It is important to remember that the IEG expression we measured is surely only a very small part of the total brain responses and hence is just a glimpse of the presumed complete genomic response to social information, but its differential effect on specific brain areas shows that females are activating their brains based on visual information and may use this to guide rapid decisions about what to do.

One additional question is how this information might inform the female's choice of a mate. In a separate experiment, we performed the same protocol, but after the female had chosen and seen the staged fight, she

TABLE 28.1   Tabulation of Female *Astatotilapia burtoni* Choices After Seeing Their Preferred Male Win or Lose a Fight Compared to Their Choices After Not Seeing a Fight

	Number of Switches/Total Choices	
	Preferred Male Wins	Preferred Male Loses
Female saw a fight	2/13	11/12
Female did not see a fight	0/10	1/9

After seeing her preferred male win a fight, the female rarely switched her choice (Fisher's exact test: $p = .0002$), but after seeing her preferred male lose, she nearly always switched her choice (Fisher's exact test: $p = .0004$; Ref. [56]).

had to choose again. In this second choice, if she had seen her preferred male lose, she almost always switched her choice, and if he won, she rarely switched her choice (Table 28.1).

SUMMARY

This review of research on social behavior in *A. burtoni* has highlighted the important role of social status in regulating the social life in this species. Because *A. burtoni* is a resource-guarding species with a hierarchy of males, this feature has been the focus of much of the research. Our observations in the field and experiments in the laboratory reveal that these animals collect information on conspecifics and use that information to decide what to do in particular circumstances. Both males and females use information gathered through observation to guide their behavior and make decisions. We have followed this information into the brain to identify where it acts using IEGs to mark active areas. As expected, males can change quickly what they decide to do depending on what they perceive is happening in their surroundings, and these changes cause corresponding changes in the brain in specific cells, receptors, and circuits, preparing the brain of the animal for a phase of life in a new status. How social information is transduced into cell and molecular changes in the brain, however, remains a mystery.

References

[1] Robinson GE, Fernald RD, Clayton DF. Genes and social behavior. Science 2008;322:896–900.

[2] Fernald RD, Hirata NR. Field study of *Haplochromis burtoni*: habitats and co-habitants. Environ Biol Fishes 1977;2:299–308.

[3] Davis MR, Fernald RD. Social control of neuronal soma size. J Neurobiol 1990;21:1180–8.

[4] Fernald RD. Social control of the brain. Annu Rev Neurosci 2012;35:133–51.

[5] Fernald RD, Hirata NR. Field study of *Haplochromis burtoni*: quantitative behavioral observations. Anim Behav 1977;25:964–75.

[6] Fernald RD. Quantitative behavioural observations of *Haplochromis burtoni* under semi-natural conditions. Anim Behav 1977;25:643–53.

[7] Burmeister SS, Jarvis ED, Fernald RD. Rapid behavioral and genomic responses to social opportunity. PLoS Biol 2005;3:e363.

[8] Maruska KP, Fernald RD. Social regulation of gene expression in the African cichlid fish *Astatotilapia burtoni*. In: Oxford handbook of molecular psychology. Oxford University Press; 2014. p. 1–22.

[9] White SA, Nguyen T, Fernald RD. Social regulation of gonadotropin-releasing hormone. J Exp Biol 2002;205:2567–81.

[10] Ma Y, Juntti SA, Hu CK, Huguenard JR, Fernald RD. Electrical synapses connect a network of gonadotropin releasing hormone neurons in a cichlid fish. Proc Natl Acad Sci USA 2015;112:3805–10.

[11] Francis RC, Soma K, Fernald RD. Social regulation of the brain-pituitary-gonadal axis. Proc Natl Acad Sci USA 1993;90:7794–8.

[12] Au TM, Greenwood AK, Fernald RD. Differential social regulation of two pituitary gonadotropin-releasing hormone receptors. Behav Brain Res 2006;170:342–6.

[13] Burmeister SS, Kailasanath V, Fernald RD. Social dominance regulates androgen and estrogen receptor gene expression. Horm Behav 2007;51:164–70.

[14] Harbott LK, Burmeister SS, White RB, Vagell M, Fernald RD. Androgen receptors in a cichlid fish, *Astatotilapia burtoni*: structure, localization, and expression levels. J Comp Neurol 2007;504:57–73.

[15] Greenwood AK, Fernald RD. Social regulation of the electrical properties of gonadotropin-releasing hormone neurons in a cichlid fish (*Astatotilapia burtoni*). Biol Reprod 2004;71:909–18.

[16] Chen C-C, Fernald RD. Visual information alone changes behavior and physiology during social interactions in a cichlid fish (*Astatotilapia burtoni*). PLoS One 2011;6:e20313.

[17] Chance MRA. Social structure of attention. John Wiley & Sons Ltd; 1976.

[18] Boulton M, Smith PK. Affective bias in children's perceptions of dominance relationships. Child Dev 1990;61:221–9.

[19] Vaughn BE, Waters WE. Attention structure, sociometric status and dominance: interrelations, behavioral correlates and relationships to social competence. Dev Psy 1981;17:275–88.

[20] Emory GR. Aspects of attention, orientation, and status hierarchy in mandrills (*Mandrillus sphinx*) and gelada baboons (*Theropithecus gelada*). Behaviour 1976;59:70–87.

[21] Summers CH, Forster GL, Korzan WJ, Watt MJ, Larson ET, et al. Dynamics and mechanics of social rank reversal. J Comp Phys A 2005;191:241–52.

[22] Øverli Ø KW, Höglund E, Winberg S, Bollig H, et al. Stress coping style predicts aggression and social dominance in rainbow trout. Horm Behav 2004;45:235–41.

[23] Desjardins JK, Hofmann HA, Fernald RD. Social context influences aggressive and courtship behavior in a cichlid fish. PLoS One 2012;7:e32781.

[24] Alcazar RM, Hilliard AT, Becker L, Bernaba M, Fernald RD. Brains over brawn: experience overcomes a size disadvantage in fish social hierarchies. J Exp Biol 2014;217:1462–8.

[25] Darwin C. On the origin of species by means of natural selection, or the preservation of favoured races in the struggle for life. London (UK): Murray; 1859.

[26] Desjardins JK, Becker L, Fernald RD. The effect of observers on behavior and the brain during aggressive encounters. Behav Brain Res 2015;292:174–83.

[27] Enquist M, Leimar O. Evolution of fighting behaviour: decision rules and assessment strategies. J Theor Biol 1983;102:387.

[28] Maynard-Smith J, Harper DGC. Animal signals: models and terminology. J Theor Biol 1995;177:305–11.

[29] Parker GA. Assessment strategy and the evolution of fighting behavior. J Theor Biol 1974;47:223e243.

[30] Getty T. Deception: the correct path to enlightenment? Trends Ecol Evol 1999;12:159—61.

[31] Whiten A, Byrne RW. Machiavellian intelligence: social expertise and the evolution of intellect. Oxford University Press; 1985.

[32] Whiten A, Byrne RW. Tactical deception by primates. Behav Brain Sci 2010;11:233—44.

[33] Wimmer H, Perner J. Beliefs about beliefs: representation and constraining function of wrong beliefs in young children. Cognition 1983;1:103—28.

[34] Plath MRS, Tiedemann R, Schlupp I. Male fish deceive competitors about mating preferences. Curr Biol 2008;18:1138—41.

[35] Piaget J. Judgement and reasoning in the child. London: Kegan, Paul, Trench & Trubner; 1928.

[36] Gillian DJ. Reasoning in the chimpanzee: II. Transitive inference. J Exp Psychol Anim Behav Process 1981;7:87—108.

[37] McGonigle BO, Chalmers M. Are monkeys logical? Nature 1977; 267:694—6.

[38] Rapp PR, Kansky MT, Eichenbaum H. Learning and memory for hierarchical relationships in the monkey: effects of aging. Behav Neurosci 1996;110:887—97.

[39] Davis H. Transitive inference in rats (*Rattus norvegicus*). J Comp Psychol 1992;106:342—9.

[40] Roberts WA, Phelps MT. Transitive inference in rats-a test of the spatial coding hypothesis. Psychol Sci 1994;5:368—74.

[41] Bond AB, Kamil AC, Balda RP. Social complexity and transitive inference in corvids. Anim Behav 2003;65:479—87.

[42] Steirn JN, Weaver JE, Zentall TR. Transitive inference in pigeons: simplified procedures and a test of value transfer theory. Anim Learn Behav 1995;23:76—82.

[43] von Fersen L, Wynne CDL, Delius JD, Staddon JER. Transitive inference formation in pigeons. J Exp Psychol Anim Behav Process 1991;17:334—41.

[44] Kumaran D, Melo HL, Duzel E. The emergence and representation of knowledge about social and nonsocial hierarchies. Neuron 2012;76:653—66.

[45] Grosenick L, Clement TS, Fernald RD. Fish can infer social rank by observation alone. Nature 2007;445:429—32.

[46] Oliveira RF, McGregor PK, Burford FRL, CustOdio MR, Latruffe C. Functions of mudballing behaviour in the European fiddler crab *Uca tangeri*. Anim Behav 1998;55:1299—309.

[47] Clement TS, Grens KE, Fernald RD. Female affiliative preference depends on reproductive state in the African cichlid fish, *Astatotilapia burtoni*. Behav Ecol 2005;16:83—8.

[48] White SL, Gowan C. Brook trout use individual recognition and transitive inference to determine social rank. Behav Ecol 2012. http://dx.doi.org/10.1093/beheco/ars136.

[49] Mello CV, Vicario DS, Clayton DF. Song presentation induces gene expression in the songbird forebrain. Proc Natl Acad Sci USA 1992;89:6818—22.

[50] Rusak B, Robertson HA, Wisden W, Hunt SP. Light pulses that shift rhythms induce gene expression in the suprachiasmatic nucleus. Science 1990;248:1237—40.

[51] Burmeister SS, Fernald RD. Evolutionary conservation of the egr-1 immediate-early gene response in a teleost. J Comp Neurol 2005; 481:220—32.

[52] Newman SW. The medial extended amygdala in male reproductive behavior. A node in the mammalian social behavior network. Ann NY Acad Sci 1999;877:242—57.

[53] Goodson JL. The vertebrate social behavior network: evolutionary themes and variations. Horm Behav 2005;48:11—22.

[54] Goodson JL, Bass AH. Vocal-acoustic circuitry and descending vocal pathways in teleost fish: convergence with terrestrial vertebrates reveals conserved traits. J Comp Neurol 2002;448:298—322.

[55] Desjardins JK, Klausner JQ, Fernald RD. Female genomic response to mate information. Proc Natl Acad Sci USA 2010;107: 21176—80.

[56] Klausner JQ. The neurological and physiological effects of receiving social information in female fish [Honors thesis]. Stanford University 2009.

29

Imaging Genetics in Humans: Major Depressive Disorder and Decision-Making

U. Rabl, N. Ortner, L. Pezawas

Medical University of Vienna, Vienna, Austria

Abstract

Whereas the global health burden of major depressive disorder (MDD) is increasing, treatments for MDD remain only partly effective. Despite an abundance of promising preclinical reports investigating genes, neurotransmitter systems, and brain circuits, none of the potential biological mechanisms has so far consistently been proven to be useful in the diagnosis or treatment of MDD. Recently, the cognitive symptoms of MDD have gained more attention from researchers, because they provide a promising new avenue given the increasing understanding of cognition and decision-making in healthy subjects. This chapter examines the relationship between MDD and decision-making and the role of imaging genetics in its exploration.

INTRODUCTION

Major depressive disorder (MDD) is among the most prevalent diseases and a leading contributor to disability and mortality in young adults. Available antidepressant treatments show only modest response rates on average, and attempts to improve their efficacy by delivering treatments based on a patient's individual biosignature have failed so far [1]. Given the tremendous socioeconomic damage caused by MDD, a better understanding of its underlying neurobiology is urgently needed to open roads to innovative diagnostic and therapeutic strategies [1]. Impairments of cognition and decision-making are core symptoms as well as diagnostic criteria of MDD [2]. Intriguingly, there are obvious similarities between conclusions reached from studies investigating brain circuits, genes, and neurotransmitter systems implicated in depression and decision-making. This chapter provides a comprehensive review of these findings and a synthesis of the commonalities of these fields of research.

MAJOR DEPRESSIVE DISORDER AS A DISORDER OF DECISION-MAKING

Indecisiveness is a diagnostic psychopathological symptom of MDD that can be found within its operationalization in the *Diagnostic and Statistical Manual of Mental Disorders*, fifth edition (DSM-5), in which MDD is defined as requiring the presence of at least five of the following symptoms for a minimum of 2 weeks: (1) depressed mood or (2) a lack of interest or pleasure in daily activities; (3) weight or appetite alterations; (4) insomnia or hypersomnia; (5) psychomotor agitation or retardation; (6) fatigue or loss of energy; (7) feelings of worthlessness or guilt; (8) diminished ability to think or concentrate, or indecisiveness; and (9) recurrent thoughts of death, suicidal ideation, or suicide plans or attempts. Symptoms need to cause significant distress and to include at least one of the first two criteria [2]. Given the lack of any obligatory symptom within its definition, the clinical face of MDD is highly variable [3]. Importantly, its operationalization is not based on any disease mechanism, which is key to future development of modern diagnostic schemes and therapies [4]. Research on indecisiveness in MDD patients, a rather neglected area, appears to be promising due to the increasing neurobiological understanding of decision-making in healthy subjects and its tight association with clinically relevant disease features compared to, for example, neurovegetative symptoms of MDD [5,6]. The clinical importance of cognitive symptoms is further highlighted by psychological theories of MDD such as learned helplessness or Beck's cognitive model of depression, which postulate a causal relationship between cognitive processes and resulting depressive symptoms [7,8]. In contrast, pharmacological theories such as the "monoamine hypothesis" propose that lacking monoamines is causal for MDD. While they are mutually exclusive, research has demonstrated more common ground between these two positions than originally anticipated [8,9].

Frequently, MDD patients exhibit deficits on neuropsychological tests of cognitive function such as psychomotor speed, attention, visual learning and memory, attentional switching, verbal fluency, and cognitive flexibility [10]. On a psychopathological level, these deficits might be reflected as indecisiveness and difficulties in thinking or concentration in general. MDD has been found to result in a shift from a reflective and model-driven mode of decision-making that employs a rich representation of the environment to a more habitual and reflexive strategy that utilizes a minimal task representation and results in more exploratory and stochastic choices [11]. This decrease in model-based action is related to alterations in the use of reward and punishment information [12]. Depressed individuals exhibit reward hyposensitivity, characterized by reduced emotional reactions in the anticipation of reward and decreased likelihood to modulate behavior in the presence of reward information. Reward hyposensitivity is closely related to anhedonia, which is an obligatory criterion (lack of interest or pleasure in daily activities) in the diagnostic definition of MDD [13]. MDD patients also exhibit maladaptive responses to punishment such as an increased error likelihood after the occurrence of an error. These results have been interpreted as punishment hypersensitivity or "catastrophic response to perceived failure," but may alternatively be viewed as failure to use negative feedback to improve future performance. Accordingly, MDD patients exhibit a general tendency to disregard rewarding and adverse environmental information to guide behavior [13]. This diminished capacity to integrate environmental feedback is also reflected in autobiographical memory deficits. MDD patients report more general, superficial, and negative autobiographical memories and tend to recollect fewer recent events than controls [14]. Related to these findings, acute and remitted patients exhibit a tendency toward rumination, a mode of mind wandering in response to adversity that involves repetitively and passively focusing on the possible causes and consequences of distress [15,16]. Notably, mind wandering dominated by unpleasant topics also occurs frequently in the general population, in which it has been found to be directly related to subsequent feelings of unhappiness [17]. However, in contrast to healthy subjects, depressed individuals are unable to push aside negatively valenced information from the attention focus and to terminate ruminations [15]. While MDD patients seem to be insensitive to reward and punishment, they are nevertheless highly responsive to emotional cues. Performance differences in cognitive tasks between MDD patients and healthy controls seem to be more pronounced in emotionally "hot" tasks than in "cold" tasks that use neutral stimuli [18]. Moreover, MDD patients even tend to interpret ostensibly neutral cues in an emotional manner that is typical for "hot" cognition, resulting in intrusion effects by irrelevant emotional material and negatively skewed interpretation of neutral stimuli [18]. This "negativity bias" is found across the full range of cognitive processes, including perception, attention, memory, motivation, and reward processing [9].

Apart from the aforementioned detrimental effects, MDD patients may also exhibit beneficial responses under certain circumstances. MDD patients are more reluctant to accept options and may make better choices in complex decision-making tasks that require a slow, deliberate mode of thinking [19]. Correspondingly, MDD patients are less prone to overestimating their degree of control over outcomes, a finding that resulted in the theory of "depressive realism" [20]. Accordingly, the depressive mode may result in maladaptive responses only under gain-maximization conditions, but could even be beneficial under loss-minimization conditions [21].

Interesting theoretical approaches to explain these behavioral alterations and their relationship to environmental adversity have been derived from Bayesian reinforcement learning models. These theories argue that depression results from a mismatch between the present environment and its internal representation. This framework resonates well with the concept of learned helplessness, which proposes that MDD patients do not avoid punishments because they initially learned to have no control over the outcome of an unpleasant situation. While the adapted behavioral mode may be advantageous in a truly adverse environment, it is maladaptive in an environment in which the exploitation of reinforcement strategies is beneficial [22]. Evidence suggests that humans employ a simple Pavlovian pruning strategy during goal-directed choices, in which mental evaluation of a sequence of choices is reflexively curtailed as soon as the agent encounters a large loss. Interestingly, the tendency toward this pruning strategy correlates with subclinical mood disturbance, suggesting a role in mood disorders and a link between goal-directed behavior and Pavlovian behavioral inhibition, which has also been related to serotonin (5-HT) signaling [23]. These theories provide interesting links between the neurobiology and the phenomenology of MDD as well as theoretical accounts of decision-making [22,23].

One of the earliest neuroimaging findings in MDD has been an increase in amygdala activation in response to negatively valenced cues [24]. Metaanalytic evidence extends this finding to the salience network (SN), including the insula and the dorsal anterior cingulate cortex (dACC) [24]. The SN induces attention shifts in response to internal or external salient cues to mediate error monitoring, task initiation, adjustment, and maintenance [25]. The SN has been proposed to be an important regulator of brain states that interacts with the locus coeruleus—noradrenergic (LC—NA) system mediating global NA-driven network resets that facilitate reorientation to salient cues [26—28]. Interestingly, increased baseline activity of the thalamic pulvinar nucleus has also been found [24]. The pulvinar acts as an information transmission gateway that regulates amygdala input via monosynaptic projections based on attentional demands [29]. Attentional gating by the pulvinar provides a neural mechanism for balancing between bottom-up and top-down controls and is primarily controlled by cortical inputs, including strong bidirectional connections with the SN [24]. Exaggerated tonic pulvinar activation in MDD may initiate an attentional bias toward negative information that propagates to the SN, resulting in increased SN response to negative cues, a potential mechanism underlying the depressive "negativity bias" [9,24]. The SN also controls the balance between the extrinsic mode network (EMN) and the default mode network (DMN). These large-scale networks are reciprocally activated during externally versus internally directed processes [30,31]. The DMN consistently deactivates during goal-directed, attention-demanding, non-self-referential behaviors, but activates at rest, accounting for the major part of the brain's energy consumption [32]. In contrast, the EMN is recruited by externally focused tasks [30]. MDD has been associated with an increased dominance of the DMN over the EMN, probably related to biased SN engagement [33]. The involvement of the DMN in MDD is probably also related to the hippocampus, which is a critical—though sporadic—contributor to the DMN [34]. Hippocampal volume is reduced in acutely ill and at-risk subjects and increases during successful treatment [35]. The hippocampus provides a relational "allocentric" space and time map of the experienced external world [36]. This prior allocentric information is collated to current egocentric sensory information via a pathway involving the posterior DMN [37]. The hippocampus also interacts intensively with the medial prefrontal cortex (mPFC), forming a network that is important for fear and autobiographical memory processing and memory-guided decision-making [38—40]. This network is active during both retrieval of past memories and imagination of future experiences. Interestingly, hippocampal lesion patients show marked impairments in their predictive and imaginative potential, which has led to the hypothesis that the hippocampus is required for mental scene construction. However, evidence shows that this impairment is constrained to tasks that require memory consolidation or reconsolidation, whereas retrieval of remote memories remains possible [41]. The aforementioned deficits in memory-dependent tasks in MDD subjects are surprisingly reminiscent of hippocampal lesion patients, though much less pronounced. The formation of distinct mnemonic representations from similar experiences—called pattern separation—requires hippocampal neurogenesis and plasticity, which is critically dependent on brain-derived neurotrophic factor (BDNF) signaling [42,43]. Prolonged stress is associated with BDNF downregulation, resulting in decreased neurogenesis, probably contributing to alterations in hypothalamus—pituitary—adrenal axis signaling [35,44,45]. Most antidepressant treatments seem to counteract this effect by increasing BDNF and hippocampal neuroplasticity [46]. Rumination and autobiographical memory deficits in acute and remitted patients have also been related to increased activation of the mPFC [16,33,47]. Further, MDD is characterized by mPFC—hippocampal network alterations, which may be related to characteristic shifts in the activation and connectivity between the anterior and the posterior DMN [48—50]. Importantly, the mPFC exhibits a remarkable dichotomy into dorsal—caudal "cognitive" and ventral—rostral "affective" subdivisions [51]. The ventral—rostral subdivision, termed subgenual or subcallosal anterior cingulate cortex (sACC), shows metabolic and histological changes in MDD. The sACC is implicated in top-down control of the amygdala and a primary target for deep brain stimulation [52,53]. Further, it is a key target of 5-HT that is activated by distant threats and implicated in fear extinction [54]. The integrity of the sACC—amygdala network also correlates with anxious traits and is reduced in MDD [55,56]. In contrast, the dorsal—caudal subdivision of the mPFC blends into the dACC of the SN and is implicated in active avoidance of threats, including fight or flight responses [54]. Notably, this suggests a spectrum from safety to threat across the mPFC ranging from sACC to dACC, in which the former is mostly controlled by 5-HT mediating behavioral inhibition, and the latter relates to NA-driven reorienting and active avoidance. Given that the SN has also been implicated in cognitive task shifts, this function is not necessarily confined to acute threats, but may rather serve as a general mechanism for resetting thought and subsequent behavior. This dichotomy resonates well with accumulating evidence for antagonistic actions of 5-HT and NA in the control of the cortical information flow within the cortical—striatal—pallidal—thalamic—cortical circuit [57]. Interestingly, both impairments of 5-HT and NA signaling have been linked to MDD and alterations in their balance may give rise to specific symptoms and brain level changes [58]. Prolonged uncontrollable stress

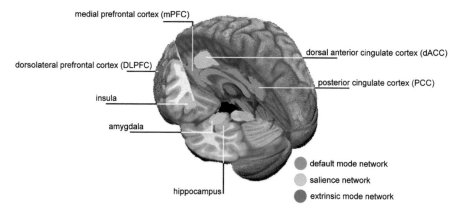

FIGURE 29.1   Brain regions implicated in major depressive disorder (MDD) and their respective large-scale networks as indicated by blood oxygen level-dependent signal coactivation or functional connectivity. Exemplary clusters have been extracted by searching related terms in Neurosynth and displayed on the ICBM 2009b nonlinear asymmetric template [119,120].

has been shown to result in alterations of the density of key regulatory proteins such as 5-HT$_{1A}$ and 5-HT$_{2A}$ receptors, which seem to be inversely affected in MDD [59−61]. Putatively, these regulatory mechanisms may fine-tune brain networks toward a negatively or positively skewed mode [9]. Interestingly, 5-HT$_{1A}$ and 5-HT$_{2A}$ receptors have also been implicated in differential regulation of the anterior versus posterior DMN and the amygdala [62−64]. In contrast to these regional effects, an indirect measure of 5-HT transporter reuptake function correlated negatively with global DMN activity and positively with a cluster in the EMN [65]. Beyond 5-HT and NA, there is evidence for alterations in dopaminergic signaling [24]. MDD patients exhibit an attenuated ventral striatal response to unexpected reward as well as lower activation in the dorsal striatum and the dorsolateral prefrontal cortex (DLPFC) in response to negative stimuli [24,66]. These findings may suggest impaired amygdala and hippocampal input to the ventral striatum as well as diminished DLPFC−dorsal striatum interactions, which have been related to novelty seeking and reward dependence, respectively [67]. Of note, the subthalamic nucleus and the habenula, two nuclei that have been related to 5-HT signaling and depression, also seem to be important for the control of information flow in the basal ganglia by integrating signals from the mPFC and limbic areas [68,69]. Accordingly, pathologic brain function in MDD may lead to a ruminative vicious circle that prevents the proper processing of environmental information via limbic areas and its propagation up the ascending cortical−striatal−pallidal−thalamic−cortical loop. As a result, novel environmental information processed in the limbic circuitries may be kept from entering executive cortical−striatal−pallidal−thalamic−cortical circuits, culminating in a habitual, reflexive mode that cannot exploit the reinforcing properties of the environment. Fig. 29.1 presents a diagram of the brain regions implicated in MDD and their respective large-scale networks.

IMAGING GENETICS OF MAJOR DEPRESSIVE DISORDER

Despite the modest heritability of MDD (30−40%), the genetic makeup of a patient is crucial because it moderates the relationship between stressors and clinical symptoms [70]. Moreover, the discovery of disease-related genetic mechanisms may open the door for targeted drug development and biological diagnostics, which are still not available in psychiatry in contrast to all other medical fields [71]. Because initial attempts to relate genes to DSM categories by linkage analysis or candidate gene association studies were nonreplicable [72], geneticists have met this challenge with ever-increasing resolution of genetic mapping techniques and growing sample sizes [73]. Another promising approach was taken by the introduction of intermediate measures with a clear genetic linkage, so-called endophenotypes. Because no measure has been discovered that fulfills all of the strict criteria for endophenotypes, the term intermediate phenotype has been coined, referring to a simpler and evident phenotype with stronger genetic association compared to its complex behavioral counterpart. While some behavioral intermediate phenotypes have been reported, neuroimaging has revealed a vast number of potential intermediate phenotypes across all psychiatric disorders. Today many of those discovered neuroimaging-based intermediate phenotypes are used for drug discovery among other techniques because this approach is capable of mapping genetics onto a brain systems level in the living human [74].

Initially, imaging genetics research has been focused on a small set of candidate genes. Most of the investigated polymorphisms play marginal roles as risk factors for MDD when considered alone. As an example, Val[66]Met (rs6265), the most intensively investigated polymorphism in the *BDNF* gene, did not associate with MDD in metaanalyses, though evidence for a small, but clinically insignificant, association with antidepressant drug response exists [75,76]. However, the effects of these polymorphisms have been shown more consistently on the brain level, providing a window into the mechanisms linking genes with MDD.

The S allele of 5-HTTLPR, a polymorphism in the promoter region of the 5-HT transporter gene (*SLC6A4*), has been associated with risk for depression in interaction with environmental adversity [77]. This variant also exacerbates the effects of tryptophan depletion, although it reduces 5-HT reuptake, similar to selective serotonin reuptake inhibitors [78]. These paradoxical effects have been explained by increased levels of 5-HT (and behavioral inhibition) throughout development in S carriers that may result in larger differences in functional 5-HT levels following acute depletion [78]. In a hallmark study, the first investigating genotype effects with fMRI, amygdala activation was found to be increased in healthy carriers of the S allele [79]. This effect has been supported by subsequent metaanalysis, though substantial heterogeneity between studies exists. These inconsistencies may partly arise from technical issues, because the amygdala signal is especially prone to confounders [80]. Further, there is evidence for moderation by task characteristics indicating specifically increased responses to negative stimuli, reminiscent of the "negativity bias" in MDD [24,81,82]. Correspondingly, the S allele drives alterations in function and structure of the amygdala–cingulate network that are associated with harm avoidance and susceptibility to framing effects in decision-making [56,83,84]. Similar to MDD, it has been shown that 5-HTTLPR effects in healthy subjects extend to the SN [85]. Further, the S allele of 5-HTTLPR has been associated with increased volume and activation of the pulvinar, corresponding to heightened baseline activity in MDD [86,87]. These results suggest a general bias toward SN and amygdala activation in response to neutral or negative stimuli in healthy S carriers that corresponds well to alterations reported for acute MDD [24].

Val[66]Met BDNF impairs activity-dependent BDNF signaling in Met carriers and has been investigated with similar intensity compared to 5-HTTLPR. Mirroring the important role of BDNF in hippocampal neuroplasticity, the Met allele of Val[66]Met BDNF has been associated with decreased volume of the hippocampus as well as alterations in hippocampal activation. These effects on hippocampal function and anatomy have been supported in a recent metaanalysis and shown to translate to differences in declarative memory performance [88]. Moreover, Val[66]Met BDNF seems to have an influence on fear extinction learning, with Met carriers exhibiting increased fear generalization and reduced performance in conditioned fear response extinction, accompanied by decreased activity of the mPFC and increased activity of the amygdala in Met carriers [89,90]. These findings are putatively related to hippocampal BDNF effects, given the strong interaction between the hippocampus and the mPFC [38,39].

Another intensively investigated polymorphism is Val[158]Met in the catechol-*O*-methyltransferase gene (*COMT*), which encodes an enzyme especially critical for prefrontal dopaminergic signaling [91]. This polymorphism has been associated with prefrontal activation as well as limbic responses to unpleasant stimuli, suggesting a trade-off between favorable responses in executive cognition in Met carriers, whereas the Val allele may be beneficial in emotional contexts [92]. This mechanism exemplifies the importance of environmental contexts as modulators of genotype effects. Further complexity is added by epistatic gene–gene interactions, which can change the effect of a specific allele entirely in the context of other variants. Increasingly, these complex interaction effects are studied with imaging methods [93,94]. With this regard, it has been shown that the BDNF Met allele protects against the deleterious effects of the 5-HTTLPR S allele on mPFC–amygdala circuitry structure, whereas both alleles have disadvantageous main effects [93]. Further, these *SLC6A4*, *BDNF*, and *COMT* polymorphisms have been shown to additively determine the relationship between stressful life events and hippocampal volume, suggesting a combined downstream effect that resonates with the role of hippocampal integrity in MDD risk and treatment [35]. More recently, direct genome-wide analyses of MDD-relevant intermediate phenotypes are under way that have the potential to unravel previously unknown genetic variants that are inaccessible for standard genome-wide association approaches [95].

CONCLUSION

Because of the efficacy of monoaminergic drugs in clinical trials, the "monoamine deficiency" hypothesis has been favored despite the lack of direct molecular evidence in MDD patients [96]. Meanwhile, alternative hypotheses have been developed, including neurotrophic mechanisms, stress axis hyperactivity, and inflammation increases, but so far causality has not been established because of the lack of effective antidepressants specifically targeting the proposed molecular pathways [97]. Stronger emphasis on cognitive symptoms in MDD research, however, may provide new insights for disease models of MDD given its limited exploration in clinical studies [13]. As reviewed here, increasing evidence highlights the importance of dysfunctional cognitions with respect to environmental control and their association with negativity bias in MDD. Notably, the notion of loss of control does not necessarily have to be detrimental. In fact, rodents, and potentially humans, can even actively switch to a stochastic behavioral mode when predictive approaches are inefficient, a behavior that is mediated by LC–NA-driven deactivation of the mPFC [98]. However, most decisions are influenced by previous experience, which sets a reference point that biases future decision-making toward loss aversion or risk seeking. This notion of an individual reference point also is the key distinction between the traditional expected utility theory and modern theoretical accounts of economic decision-making such as Prospect Theory [99]. MDD can be conceptualized as an extreme manifestation of this bias, in which the reference point is severely at odds with reality because of an extreme shift to the negative side (note that mania exhibits the opposite pattern of inappropriate overconfidence). Interestingly, neural circuits tracking environmental variability to bias decision-making are even present in *Caenorhabditis elegans*, where a dopaminergic mechanism mediates between sensory, learning, and behavioral circuits [100]. In the same model animal, 5-HT and dopamine (DA) have been characterized as universal switches between fundamental behaviors, raising the possibility that a similar role has been retained throughout evolution [101]. Indeed, it has been suggested that the decision reference point is related to tonic 5-HT and DA, which may reflect average punishment and reward [99]. However, given that 5-HT and DA both have been implicated in reward as well as punishment it is highly unlikely that these monoamines code absolute state value signals [102]. Hence, it is tempting to speculate that monoamines may in the first place control the brain's information flow by phasic signaling, similar to traffic lights controlling road transportation, whereas their long-term, tonic signal may set the reference point for future decision-making [57,102]. Interestingly, whereas there is increasing evidence arguing against opposing roles of DA and 5-HT, a potential antagonism between 5-HT and NA in controlling cortical information flow gains momentum, whereby 5-HT seems to keep computations between intratelencephalic neurons within the cortical–striatal–thalamic–cortical loop, whereas NA is required to force action via pyramidal tract neurons. Consequently, 5-HT and NA might control the amount of preaction computation in terms of a speed–accuracy trade-off, whereas DA might be instrumental for action selection [57]. In this regard, 5-HT may signal retreat from the environment, which resonates well with its associations with delayed reward, patience, behavioral inhibition, response to uncontrollable stress,

punishment, pain, its potential involvement in DMN control, and its negative association with impulsivity [62–65]. These findings may help to explain observations that are difficult to reconcile with the "monoamine deficiency" hypothesis alone [103,104], which can hardly be verified directly in living humans [105]. It may be insightful to consider that an increase in the amount of a signaling molecule does not necessarily equate to a gain in the downstream signal, as known in type 2 diabetes. This metabolic disorder is substantially comorbid with MDD at least in elderly and more severely ill patients and shares several hallmarks such as the critical role of early environmental stress and genetic risk as predisposing factors and the involvement of homeo- or allostatic imbalance in disease progression [106–108]. Interestingly, hyperinsulinemia is a key pathogenic mechanism in the progress from prediabetes to manifest illness and also a congenital risk factor [108,109]. Putatively, chronically elevated 5-HT levels due to genetic variation or chronic stress may be accompanied by compensatory resistance in 5-HT's downstream targets, similar to insulin resistance in metabolic syndrome. In diabetes, the combination of hyperinsulinemia and insulin resistance can progress to β-cell insufficiency, finally resulting in insulin deficiency [108]. A similar mechanism may explain the distribution of monoamine metabolic markers in MDD, which has been found to be bimodal with indices of low 5-HT present only in severely suicidal patients, who also exhibit marked alterations of dorsal raphe structure [110,111]. Whereas momentary retreat from the environment is probably crucial to make sense of it and may literally occur in the blink of an eye [112], overstraining of this mechanism may be detrimental and promote progression from neuroticism to manifest MDD and suicidality. The finding that healthy 5-HTTLPR S carriers, who are thought to have increased 5-HT levels, exhibit increased activity in circuits related to fear conditioning and MDD fits well into this model, suggesting the utility of imaging genetics to link brain level alterations and psychopathology with the underlying molecular pathways [24,85,113].

Potentially, these and other approaches will resolve the open question if MDD patients share a common etiology or in fact exhibit distinct etiological pathways that are indiscernible at the level of psychopathology [97]. In fact, the latter was already postulated 100 years ago and is clinically observable during treatment with specific drugs (e.g., interferon) or as a prodromal or comorbid condition in somatic diseases such as Parkinson's disease and many more [114–117]. Therefore, it is not surprising that the abundance of promising neuroscientific reports on differences between MDD patients and healthy subjects have so far not been reliably verified in clinical trials and even overwhelming preclinical evidence such as the importance of the amygdala in fear conditioning has not been convincingly shown in humans [81,113]. Moreover, none of the potential pathological molecular pathways of MDD has been identified beyond reasonable doubt [97]. In fact, most evidence of the underlying neurochemical alterations in MDD still stems from clinical antidepressant trials, which are able to make causal inferences between treatments and outcomes. In contrast, most neuroscientific studies (including imaging studies) in humans are cross-sectional and therefore unsuited to infer causality. To bridge the gap between laboratory and bedside, it will be necessary to integrate clinical and neuroscience research, e.g., by including placebo-controlled pharmacological manipulations in imaging genetics studies. With increasing sophistication of imaging, molecular, and statistical methods, this endeavor may finally culminate in a functional approach that could open up new vistas for diagnosis and treatment of psychiatric illness, in which neuroscientific research on decision-making may eventually play an important role [118].

References

[1] Kupfer DJ, Frank E, Phillips ML. Major depressive disorder: new clinical, neurobiological, and treatment perspectives. Lancet 2012;379(9820):1045−55.

[2] American Psychiatric Association. Diagnostic and statistical manual of mental disorders. 5th ed. 2013.

[3] Fried EI, Nesse RM. Depression is not a consistent syndrome: an investigation of unique symptom patterns in the STAR*D study. J Affect Disord 2014;172:96−102.

[4] Stephan KE, et al. Charting the landscape of priority problems in psychiatry, part 1: classification and diagnosis. Lancet Psychiatry 2015;3(1):77−83.

[5] Fried EI, Nesse RM. The impact of individual depressive symptoms on impairment of psychosocial functioning. PLoS One 2014; 9(2):e90311.

[6] Lux V, Kendler KS. Deconstructing major depression: a validation study of the DSM-IV symptomatic criteria. Psychol Med 2010;40(10):1679−90.

[7] Henkel V, et al. Cognitive-behavioural theories of helplessness/hopelessness: valid models of depression? Eur Arch Psychiatry Clin Neurosci 2002;252(5):240−9.

[8] Disner SG, et al. Neural mechanisms of the cognitive model of depression. Nat Rev Neurosci 2011;12(8):467−77.

[9] Roiser JP, Elliott R, Sahakian BJ. Cognitive mechanisms of treatment in depression. Neuropsychopharmacology 2012;37(1): 117−36.

[10] Lee RS, et al. A meta-analysis of cognitive deficits in first-episode major depressive disorder. J Affect Disord 2012;140(2): 113−24.

[11] Blanco NJ, et al. The influence of depression symptoms on exploratory decision-making. Cognition 2013;129(3):563−8.

[12] Kunisato Y, et al. Effects of depression on reward-based decision making and variability of action in probabilistic learning. J Behav Ther Exp Psychiatry 2012;43(4):1088−94.

[13] Eshel N, Roiser JP. Reward and punishment processing in depression. Biol Psychiatry 2010;68(2):118−24.

[14] Young KD, et al. Functional neuroimaging correlates of autobiographical memory deficits in subjects at risk for depression. Brain Sci. Multidisciplinary Digital Publishing Institute; 2015. p. 144−64.

[15] Berman MG, et al. Depression, rumination and the default network. Soc Cogn Affect Neurosci 2011;6(5):548−55.

[16] Bartova L, et al. Reduced default mode network suppression during a working memory task in remitted major depression. J Psychiatr Res 2015;64:9−18.

[17] Killingsworth MA, Gilbert DT. A wandering mind is an unhappy mind. Science 2010;330(6006):932.

[18] Roiser JP, Sahakian BJ. Hot and cold cognition in depression. CNS Spectr. Cambridge University Press; 2013. p. 139−49.

[19] von Helversen B, et al. Performance benefits of depression: sequential decision making in a healthy sample and a clinically depressed sample. J Abnorm Psychol 2011;120(4):962−8.

[20] Moore MT, Fresco DM. Depressive realism: a meta-analytic review. Clin Psychol Rev 2012;32(6):496−509.

[21] Maddox WT, et al. Depressive symptoms enhance loss-minimization, but attenuate gain-maximization in history-dependent decision-making. Cognition 2012:118−24.

[22] Huys QJ, Dayan P. A Bayesian formulation of behavioral control. Cognition 2009;113(3):314−28.

[23] Huys QJ, et al. Bonsai trees in your head: how the pavlovian system sculpts goal-directed choices by pruning decision trees. PLoS Comput Biol 2012;8(3):e1002410.

[24] Hamilton JP, et al. Functional neuroimaging of major depressive disorder: a meta-analysis and new integration of base line activation and neural response data. Am J Psychiatry. Arlington (VA): American Psychiatric Publishing; 2012. p. 693−703.

[25] Uddin LQ. Salience processing and insular cortical function and dysfunction. Nat Rev Neurosci. Nature Publishing Group; 2015. p. 55−61.

[26] Hermans EJ, et al. Stress-related noradrenergic activity prompts large-scale neural network reconfiguration. Science. New York (NY): American Association for the Advancement of Science; 2011. p. 1151−3.

[27] Menon V. Large-scale brain networks and psychopathology: a unifying triple network model. Trends Cogn Sci 2011:483−506.

[28] Corbetta M, Patel G, Shulman GL. The reorienting system of the human brain: from environment to theory of mind. Neuron 2008: 306−24.

[29] Pessoa L, Adolphs R. Emotion processing and the amygdala: from a low road to many roads of evaluating biological significance. Nat Rev Neurosci. Nature Publishing Group; 2010. p. 773−83.

[30] Hugdahl K, et al. On the existence of a generalized non-specific task-dependent network. Front Hum Neurosci. Frontiers; 2015. p. 430.

[31] Menon V. Salience network. Brain Mapp. Elsevier; 2015. p. 29−35.

[32] Raichle ME. The brain's default mode network. Annu Rev Neurosci 2015;38:433−47.

[33] Hamilton JP, et al. Default-mode and task-positive network activity in major depressive disorder: implications for adaptive and maladaptive rumination. Biol Psychiatry 2011;70(4):327−33.

[34] Buckner RL, Andrews-Hanna JR, Schacter DL. The brain's default network: anatomy, function, and relevance to disease. Ann N Y Acad Sci. Blackwell Publishing Inc.; 2008. p. 1−38.

[35] Rabl U, et al. Additive gene-environment effects on hippocampal structure in healthy humans. J Neurosci. Society for Neuroscience; 2014. p. 9917−26.

[36] Nielson DM, et al. Human hippocampus represents space and time during retrieval of real-world memories. Proc Natl Acad Sci USA. National Acad Sciences; 2015. p. 11078−83.

[37] Kravitz DJ, et al. A new neural framework for visuospatial processing. Nat Rev Neurosci. Nature Publishing Group; 2011. p. 217−30.

[38] Xu W, Sudhof TC. A neural circuit for memory specificity and generalization. Science 2013;339(6125):1290–5.

[39] Gluth S, et al. Effective connectivity between hippocampus and ventromedial prefrontal cortex controls preferential choices from memory. Neuron 2015;86(4):1078–90.

[40] Bonnici HM, et al. Detecting representations of recent and remote autobiographical memories in vmPFC and hippocampus. J Neurosci 2012;32(47):16982–91.

[41] Kim S, et al. Memory, scene construction, and the human hippocampus. Proc Natl Acad Sci USA 2015;112(15):4767–72.

[42] Clelland CD, et al. A functional role for adult hippocampal neurogenesis in spatial pattern separation. Science. New York (NY): American Association for the Advancement of Science; 2009. p. 210–3.

[43] Sahay A, et al. Increasing adult hippocampal neurogenesis is sufficient to improve pattern separation. Nature 2011:466–70. England.

[44] Autry AE, Monteggia LM. Brain-derived neurotrophic factor and neuropsychiatric disorders. Pharmacol Rev 2012;64(2):238–58.

[45] Besnard A, Sahay A. Adult hippocampal neurogenesis, fear generalization, and stress. Neuropsychopharmacology. Nature Publishing Group; 2015.

[46] Patrício P, et al. Differential and converging molecular mechanisms of antidepressants' action in the hippocampal dentate gyrus. Neuropsychopharmacology. Nature Publishing Group; 2015. p. 338–49.

[47] Young KD, et al. Behavioral and neurophysiological correlates of autobiographical memory deficits in patients with depression and individuals at high risk for depression. JAMA Psychiatry. American Medical Association; 2013. p. 698–708.

[48] Zhu X, et al. Evidence of a dissociation pattern in resting-state default mode network connectivity in first-episode, treatment-naive major depression patients. Biol Psychiatry 2012;71(7):611–7.

[49] Diener C, et al. A meta-analysis of neurofunctional imaging studies of emotion and cognition in major depression. Neuroimage 2012;61(3):677–85.

[50] de Kwaasteniet B, et al. Relation between structural and functional connectivity in major depressive disorder. Biol Psychiatry 2013;74(1):40–7.

[51] Etkin A, Egner T, Kalisch R. Emotional processing in anterior cingulate and medial prefrontal cortex. Trends Cogn Sci 2011:85–93.

[52] Hamani C, et al. The subcallosal cingulate gyrus in the context of major depression. Biol Psychiatry 2011;69(4):301–8.

[53] Johnstone T, et al. Failure to regulate: counterproductive recruitment of top-down prefrontal-subcortical circuitry in major depression. J Neurosci 2007;27(33):8877–84.

[54] Mobbs D, et al. From threat to fear: the neural organization of defensive fear systems in humans. J Neurosci 2009;29(39):12236–43.

[55] Kim MJ, Whalen PJ. The structural integrity of an amygdala-prefrontal pathway predicts trait anxiety. J Neurosci 2009;29(37):11614–8.

[56] Pezawas L, et al. 5-HTTLPR polymorphism impacts human cingulate-amygdala interactions: a genetic susceptibility mechanism for depression. Nat Neurosci 2005;8(6):828–34.

[57] Shepherd GMG. Corticostriatal connectivity and its role in disease. Nat Rev Neurosci. Nature Publishing Group; 2013. p. 278–91.

[58] Homan P, et al. Serotonin versus catecholamine deficiency: behavioral and neural effects of experimental depletion in remitted depression. Transl Psychiatry 2015;5:e532.

[59] Dean B, et al. Lower cortical serotonin 2A receptors in major depressive disorder, suicide and in rats after administration of imipramine. Int J Neuropsychopharmacol 2014;17(6):895–906.

[60] Savitz JB, Drevets WC. Neuroreceptor imaging in depression. Neurobiol Dis 2013:49–65.

[61] Jovanovic H, et al. Chronic stress is linked to 5-HT(1A) receptor changes and functional disintegration of the limbic networks. Neuroimage 2011;55(3):1178–88.

[62] Hahn A, et al. Differential modulation of the default mode network via serotonin-1A receptors. Proc Natl Acad Sci USA 2012;109(7):2619–24.

[63] Carhart-Harris RL, et al. Neural correlates of the psychedelic state as determined by fMRI studies with psilocybin. Proc Natl Acad Sci USA. National Acad Sciences; 2012. p. 2138–43.

[64] Fisher PM, et al. Medial prefrontal cortex 5-HT(2A) density is correlated with amygdala reactivity, response habituation, and functional coupling. Cereb Cortex 2009;19(11):2499–507.

[65] Scharinger C, et al. Platelet serotonin transporter function predicts default-mode network activity. PLoS One. Public Library of Science; 2014:e92543.

[66] Robinson OJ, et al. Ventral striatum response during reward and punishment reversal learning in unmedicated major depressive disorder. Am J Psychiatry 2012;169(2):152–9.

[67] Cohen MX, et al. Connectivity-based segregation of the human striatum predicts personality characteristics. Nat Neurosci 2009:32–4.

[68] Cavanagh JF, et al. Subthalamic nucleus stimulation reverses mediofrontal influence over decision threshold. Nat Neurosci 2011;14(11):1462–7.

[69] Proulx CD, Hikosaka O, Malinow R. Reward processing by the lateral habenula in normal and depressive behaviors. Nat Neurosci. Nature Publishing Group; 2014. p. 1146–52.

[70] Caspi A, et al. Genetic sensitivity to the environment: the case of the serotonin transporter gene and its implications for studying complex diseases and traits. Am J Psychiatry 2010;167(5):509–27.

[71] Hyman SE. Can neuroscience be integrated into the DSM-V? Nat Rev Neurosci. Nature Publishing Group; 2007. p. 725–32.

[72] Sullivan PF. Spurious genetic associations. Biol Psychiatry 2007:1121–6.

[73] Abbott A. Psychiatric genetics: the brains of the family. Nature 2008;454(7201):154–7.

[74] Meyer-Lindenberg A, Weinberger DR. Intermediate phenotypes and genetic mechanisms of psychiatric disorders. Nat Rev Neurosci. Nature Publishing Group; 2006. p. 818–27.

[75] Niitsu T, et al. Pharmacogenetics in major depression: a comprehensive meta-analysis. Prog Neuropsychopharmacol Biol Psychiatry 2013;45:183–94.

[76] Gyekis JP, et al. No association of genetic variants in BDNF with major depression: a meta- and gene-based analysis. Am J Med Genet B Neuropsychiatr Genet 2013;162(1):61–70.

[77] Rocha TB, et al. Gene-environment interaction in youth depression: replication of the 5-HTTLPR moderation in a diverse setting. Am J Psychiatry 2015;172(10):978–85.

[78] Dayan P, Huys QJ. Serotonin, inhibition, and negative mood. PLoS Comput Biol 2008;4(2):e4.

[79] Hariri AR, et al. Serotonin transporter genetic variation and the response of the human amygdala. Science 2002;297(5580):400–3.

[80] Boubela RN, et al. fMRI measurements of amygdala activation are confounded by stimulus correlated signal fluctuation in nearby veins draining distant brain regions. Sci Rep 2015;5:10499.

[81] Pezawas L. Serotonin transporter linked polymorphic region: from behavior to neural mechanisms. Biol Psychiatry 2015;78(8):522–4.

[82] Murphy SE, et al. The effect of the serotonin transporter polymorphism (5-HTTLPR) on amygdala function: a meta-analysis. Mol Psychiatry 2013;18(4):512–20.

[83] Pacheco J, et al. Frontal-limbic white matter pathway associations with the serotonin transporter gene promoter region (5-HTTLPR) polymorphism. J Neurosci 2009;29(19):6229—33.

[84] Roiser JP, et al. A genetically mediated bias in decision making driven by failure of amygdala control. J Neurosci 2009;29(18):5985—91.

[85] Klumpers F, et al. Dorsomedial prefrontal cortex mediates the impact of serotonin transporter linked polymorphic region genotype on anticipatory threat reactions. Biol Psychiatry 2015;78(8):582—9.

[86] Young KA, et al. 5HTTLPR polymorphism and enlargement of the pulvinar: unlocking the backdoor to the limbic system. Biol Psychiatry 2007;61(6):813—8.

[87] Drabant EM, et al. Neural mechanisms underlying 5-HTTLPR-related sensitivity to acute stress. Am J Psychiatry 2012;169(4):397—405.

[88] Kambeitz JP, et al. Effect of BDNF val(66)met polymorphism on declarative memory and its neural substrate: a meta-analysis. Neurosci Biobehav Rev 2012;36(9):2165—77.

[89] Muhlberger A, et al. The BDNF Val66Met polymorphism modulates the generalization of cued fear responses to a novel context. Neuropsychopharmacology 2014;39(5):1187—95.

[90] Soliman F, et al. A genetic variant BDNF polymorphism alters extinction learning in both mouse and human. Science 2010;327(5967):863—6.

[91] Bilder RM, et al. The catechol-O-methyltransferase polymorphism: relations to the tonic-phasic dopamine hypothesis and neuropsychiatric phenotypes. Neuropsychopharmacology 2004:1943—61. United States.

[92] Mier D, Kirsch P, Meyer-Lindenberg A. Neural substrates of pleiotropic action of genetic variation in COMT: a meta-analysis. Mol Psychiatry 2010;15(9):918—27.

[93] Pezawas L, et al. Evidence of biologic epistasis between BDNF and SLC6A4 and implications for depression. Mol Psychiatry 2008;13(7):709—16.

[94] Bogdan R, et al. Genetic moderation of stress effects on cortico-limbic circuitry. Neuropsychopharmacology 2015;41(1):275—96.

[95] Geschwind DH, Flint J. Genetics and genomics of psychiatric disease. Science. New York (NY): American Association for the Advancement of Science; 2015. p. 1489—94.

[96] Lacasse JR, Leo J. Serotonin and depression: a disconnect between the advertisements and the scientific literature. PLoS Med 2005;2(12):e392.

[97] Hasler G. Pathophysiology of depression: do we have any solid evidence of interest to clinicians? World Psychiatry 2010;9(3):155—61.

[98] Tervo DGR, et al. Behavioral variability through stochastic choice and its gating by anterior cingulate cortex. Cell 2014:21—32.

[99] Cools R, Nakamura K, Daw ND. Serotonin and dopamine: unifying affective, activational, and decision functions. Neuropsychopharmacology 2011;36(1):98—113.

[100] Calhoun AJ, et al. Neural mechanisms for evaluating environmental variability in *Caenorhabditis elegans*. Neuron 2015;86(2):428—41.

[101] Vidal-Gadea A, et al. *Caenorhabditis elegans* selects distinct crawling and swimming gaits via dopamine and serotonin. Proc Natl Acad Sci USA. National Acad Sciences; 2011. p. 17504—9.

[102] Cohen JY, Amoroso MW, Uchida N. Serotonergic neurons signal reward and punishment on multiple timescales. eLife 2015;4.

[103] Barton DA, et al. Elevated brain serotonin turnover in patients with depression: effect of genotype and therapy. Arch Gen Psychiatry 2008;65(1):38—46.

[104] Frick A, et al. Serotonin synthesis and reuptake in social anxiety disorder: a positron emission tomography study. JAMA Psychiatry 2015;72(8):794—802.

[105] Andrews PW, et al. Is serotonin an upper or a downer? The evolution of the serotonergic system and its role in depression and the antidepressant response. Neurosci Biobehav Rev 2015;51:164—88.

[106] Thurner S, et al. Quantification of excess risk for diabetes for those born in times of hunger, in an entire population of a nation, across a century. Proc Natl Acad Sci USA 2013;110(12):4703—7.

[107] McEwen BS, et al. Mechanisms of stress in the brain. Nat Neurosci 2015:1353—63.

[108] Donath MY, Shoelson SE. Type 2 diabetes as an inflammatory disease. Nat Rev Immunol 2011;11(2):98—107.

[109] Kapoor RR, et al. Persistent hyperinsulinemic hypoglycemia and maturity-onset diabetes of the young due to heterozygous HNF4A mutations. Diabetes 2008;57(6):1659—63.

[110] Mann JJ. The serotonergic system in mood disorders and suicidal behaviour. Philos Trans R Soc Lond B Biol Sci 2013;368(1615):20120537.

[111] Asberg M, et al. "Serotonin depression" — a biochemical subgroup within the affective disorders? Science 1976;191(4226):478—80.

[112] Nakano T, et al. Blink-related momentary activation of the default mode network while viewing videos. Proc Natl Acad Sci USA. National Acad Sciences; 2013. p. 702—6.

[113] Fullana MA, et al. Neural signatures of human fear conditioning: an updated and extended meta-analysis of fMRI studies. Mol Psychiatry 2015;21(4):500—8.

[114] Udina M, et al. Interferon-induced depression in chronic hepatitis C: a systematic review and meta-analysis. J Clin Psychiatry 2012;73(8):1128—38.

[115] Miller JM, et al. Anhedonia after a selective bilateral lesion of the globus pallidus. Am J Psychiatry 2006:786—8.

[116] Gustafsson H, Nordstrom A, Nordstrom P. Depression and subsequent risk of Parkinson disease: a nationwide cohort study. Neurology 2015;84(24):2422—9.

[117] Bonhoeffer K. Die exogenen Reaktionstypen. Archiv für Psychiatrie und Nervenkrankheiten 1917;58(1):58—70.

[118] van Praag HM, et al. Denosologization of biological psychiatry or the specificity of 5-HT disturbances in psychiatric disorders. J Affect Disord 1987;13(1):1—8.

[119] Fonov V, et al. Unbiased average age-appropriate atlases for pediatric studies. Neuroimage 2011;54(1):313—27.

[120] Yarkoni T, et al. Large-scale automated synthesis of human functional neuroimaging data. Nat Methods 2011;8(8):665—70.

30

Time-Dependent Shifts in Neural Systems Supporting Decision-Making Under Stress

E.J. Hermans[1], M.J.A.G. Henckens[1], M. Joëls[2,a], G. Fernández[1,a]

[1]Radboud University Medical Centre, Nijmegen, The Netherlands;
[2]University Medical Center Utrecht, Utrecht, The Netherlands

Abstract

Acute stress profoundly affects cognitive functions such as decision-making. This promotes rapid decision-making in threatening situations, but impairs flexible and elaborate decisions. Dual-systems models of decision-making propose that these two types of decisions rely on distinct neural systems, but their regulation under stress remains unclear. Here, we integrate existing knowledge of effects of stress at the neuroendocrine, cellular, brain systems, and behavioral levels to describe how stress-related neuromodulators trigger time-dependent shifts in balance between specific large-scale neural systems. We argue that stress triggers a reallocation of neural resources toward a "salience" network, which supports rapid but rigid decisions, at the expense of an "executive control" network, which supports flexible, elaborate decisions. This resource reallocation actively reverses during recovery from stress, presumably to support adjustment of long-term goals. We conclude by showing how this biphasic-reciprocal model elucidates paradoxical findings reported in human studies on stress and cognition.

INTRODUCTION

We make many of our most important decisions under stress. These include complex rational decisions, such as whether to take a mortgage and buy a house, but also split-second decisions, for instance, to dodge a looming object. As most of us have experienced, these two types of decision-making are not equally affected by stress. Acute stress appears to hamper complex types of judgment, but at the same time, it makes us alert and ready to decide promptly and reflexively in acutely threatening situations. A widely held view is therefore that stress triggers a breakdown of neural systems supporting higher-order, goal-directed types of cognition, such that as a consequence, decision-making is relinquished to phylogenetically older neural systems that control reflexive or habitual behavior [1,2]. This apparent shift from a strategic to a tactical mode of decision-making allows us to make rapid, life-saving decisions, but does so without cognitive flexibility or regard for long-term consequences.

The notion of a shutdown of higher-order cognition during acute stress is rooted in dual-systems models of decision-making. Such models distinguish cognitive or neural systems characterized, for instance, as slow versus fast [3], controlled versus automatic [4], reflective versus reflexive [5], or model-based versus model-free [6]. While these distinctions have a long history in psychology and cognitive neuroscience, exactly which "dual systems" are involved in these two types of decision-making, and how these are differentially affected by stress, has remained underspecified and poorly understood.

In this chapter, we describe a model of how acute stress affects two distinct neural systems supporting these different types of decision-making [7]. This model integrates existing data across multiple levels of analysis. We will explain how, at the neuroendocrine level, stressors trigger a precisely timed concatenation of events throughout the body [8,9]. At the cellular level, these reactions can modulate neural excitability and plasticity in a manner that is both regionally and temporally specific [8,10,11]. By considering an emerging

---

[a] Authors contributed equally.

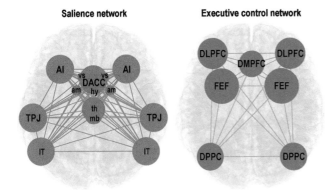

FIGURE 30.1 Anatomical overview of salience network (SN) and executive control network (ECN) nodes. Sphere sizes illustrate relative sizes of the clusters that exhibit functional connectivity with the respective networks. SN nodes: *AI*, anterior insula; *am*, amygdala; *DACC*, dorsal anterior cingulate cortex; *hy*, hypothalamus; *IT*, inferotemporal cortex; *mb*, midbrain; *th*, thalamus; *TPJ*, temporoparietal junction; *vs*, ventral striatum. ECN nodes: *DLPFC*, dorsolateral prefrontal cortex; *DMPFC*, dorsomedial prefrontal cortex; *DPPC*, dorsal posterior parietal cortex; *FEF*, frontal eye fields (precentral/superior frontal sulci). *Adapted from Hermans EJ, Henckens MJAG, Joëls M, Fernández G. Dynamic adaptation of large-scale brain networks in response to acute stressors. Trends Neurosci 2014;37:304—14; Hermans EJ, van Marle HJF, Ossewaarde L, Henckens MJAG, Qin S, van Kesteren MTR, et al. Stress-related noradrenergic activity prompts large-scale neural network reconfiguration. Science 2011;334:1151—3.*

literature on large-scale neurocognitive systems in the brain, we will next explore how such effects can generate coordinated, brain-wide shifts in neural functioning at the brain systems level [12,13]. Finally, we will show how an understanding of the spatiotemporal dynamics of effects of stress on these large-scale neural systems can shed new light on the vast and sometimes paradoxical literature on the effects of acute stress on decision-making at the behavioral level. In the following section, we start by introducing the concept of large-scale neurocognitive systems.

LARGE-SCALE NEUROCOGNITIVE SYSTEMS AND SHIFTS IN RESOURCE ALLOCATION

Recent years have seen a rapid proliferation of research into large-scale neurocognitive systems in the human brain. Whereas human neuroimaging initially focused on mapping specific cognitive functions onto individual brain regions, it turned out that activation, and also deactivation, of certain sets of brain regions is consistently associated with broader cognitive domains [14]. Furthermore, such sets of brain regions not only respond coherently during task execution, but also maintain coherent activity even when not performing any specific task [15,16]. Regions that exhibit such

intrinsic functional connectivity also show stronger structural connectivity [17]. These observations have led to the recognition of large-scale neurocognitive systems as an important level of organization within the architectural hierarchy of the human brain [15,18—20], and have caused a paradigm shift in the field of human neuroimaging toward mapping functional connectivity of the human brain [21].

A growing number of such large-scale neurocognitive systems have been discovered and described since 2005 [19,22]. The first was a set of midline structures that activates whenever an individual does not direct attention to the outside world. This "default mode," or task-negative, network has been shown to play a role in memory processes and in forms of internally directed cognition such as self-referential processing and mind wandering [14].

Later work revealed two dissociable "task-positive" networks, which activate during externally directed cognition (Fig. 30.1). One of these is the salience network (SN), a network that integrates a variety of neurocognitive functions associated with processing of, and responding to, salient stimuli. These functions include regulation of reflexive and habitual behaviors, exogenous (bottom-up) attention, and regulation of autonomic and neuroendocrine systems [22]. The other task-positive network is referred to as the executive control network (ECN), a large-scale network that regulates high-order cognitive functions such as complex decision-making, endogenous (top-down) attention, and working memory [22—24]. Notably, the alleged functions of the ECN and SN map remarkably well onto the slow and fast, or reflective and reflexive, systems for decision-making, respectively, as mentioned above [3,5].

One explanation for the observed effects of stress on cognitive functioning is therefore that neurotransmitters and hormones released during acute stress somehow trigger a shift of resources from the ECN to the SN. This notion of a "reallocation" of neural resources is consistent with bioenergetic studies, which have shown that the net increase in the brain's energy consumption in response to mental challenges is remarkably small. For instance, glucose consumption increases only a few percent in response to moderately demanding cognitive tasks, and up to 12% during acute stress [25]. One reason for this may be that even at rest, the brain already uses an excessive amount of energy (20%) that is disproportionate to its weight (2%). Furthermore, during acute stress, the brain must compete with other organs for energy resources [26]. The ability to reallocate the brain's energy resources within the boundaries of energy supply may therefore yield a more efficient way of dealing with sudden changes in cognitive demands during challenging situations.

Studies into the interactions between large-scale connectivity networks are beginning to shed light on how such shifts in resource allocation may be accomplished. For instance, task-positive and task-negative networks activate reciprocally, that is, their activity at rest is negatively correlated [23]. Such observations suggest that neurocognitive functions subserved by one network can be suspended in favor of another, and that different large-scale networks may thus compete for limited resources [27]. It remains unclear, however, how such "network switches" are established. It is possible that one system simply uses all its available resources and thereby "steals" resources from another system. Another possibility is that resources are reallocated between large-scale systems partly by active suppression of neural systems that are not currently needed [28].

One way in which an active shift in resource allocation may be achieved is through activation of stress-sensitive neuromodulatory systems, which target neural tissues throughout the brain. Research in animals has demonstrated how acute stressors trigger multiple waves of neurochemical changes. Among the most rapid of these are neuropeptidergic (e.g., corticotropin-releasing factor) and catecholaminergic (e.g., norepinephrine and dopamine) responses [8], which initiate almost immediately and generally subside not long after termination of the stressor. Stressors furthermore increase the release of corticosteroids through activation of the hypothalamic—pituitary—adrenal (HPA) axis. Corticosteroids easily cross the blood—brain barrier, but peak concentrations in the brain are not reached within 20 min after onset of a stressor, and remain elevated longer than catecholaminergic changes, at least in rodents [29]. In addition to temporally specific effects, catecholaminergic and corticosteroid actions can also be spatially specific, owing to, for instance, regional variation in receptor distribution, affinity, and downstream signaling cascades. Thus, as we will detail below, these different classes of stress-sensitive neuromodulators (in particular catecholamines and corticosteroids) may allow for well-timed shifts between distinct neural circuits.

In the remainder of this chapter, we will explain how spatially and temporally specific effects of stress-sensitive hormones and neurotransmitters at the cellular level may dynamically alter the balance between large-scale neurocognitive systems involved in distinct types of decision-making. We will first describe how the SN and ECN are rapidly affected during the immediate phase of acute stress, before turning to slower effects during the phase of recovery from stress, return to homeostasis, and preparation for future challenges.

SALIENCE NETWORK AND ACUTE STRESS

Responding adaptively in threatening circumstances requires the ability to make prompt decisions to change one's current course of action, to reorient attention toward the currently most relevant sensory information, and to free necessary energy reserves to take appropriate action. The SN (see Fig. 30.1) has been proposed as a neurocognitive system that integrates such functions [19,22,30]. It includes regions associated with sensory vigilance and attentional reorienting (amygdala, thalamus, and inferotemporal/temporoparietal regions) [24,31,32], habitual behavior (striatum) [33], autonomic-neuroendocrine control (dorsal anterior cingulate cortex, hypothalamus) [34], visceral perception (anterior insula), and catecholaminergic signaling (brain-stem/midbrain nuclei) [35]. Metaanalyses of human functional neuroimaging studies show that these regions respond consistently to a variety of salient stimuli, including aversive affective material, conditioned stimuli [36], and pain [37]. Thus, SN activity is closely associated with characteristics of stressor-induced arousal, which suggests that neurotransmitters and hormones released during acute stress and emotional arousal may modulate activity within this network.

Salience Network: Neuroendocrine and Cellular Effects of Acute Stress

Stress-induced arousal is primarily mediated by rapid changes in catecholaminergic signaling. For instance, acute stress almost instantly alters functioning of the locus coeruleus—norepinephrine (LC—NE) system, the primary source of NE (also known as noradrenaline) in the central nervous system [38]. Single-cell recording studies in monkeys have demonstrated that this alteration is best characterized as a shift from phasic activity (i.e., selective responsiveness to sensory stimuli) to tonic activity (increased nonselective background activity) [39]. This centrally mediated activation of the LC—NE system is furthermore accompanied by peripheral activation of the sympathoadrenomedullary system, which triggers the release of epinephrine from the adrenal medulla. In turn, although epinephrine cannot cross the blood—brain barrier, peripheral epinephrine release further increases activity of the LC—NE system through ascending vagal projections to the nucleus of the solitary tract (NTS) [40]. The LC—NE system has wide projections throughout the brain, including the entire cerebral cortex, hypothalamus, thalamus, and amygdala [41]. However, effects of NE can also be regionally specific because of local differences in the expression of different receptor types, which in turn have different affinities for NE. For instance, whereas α2A-adrenoceptors in the prefrontal cortex (PFC) are occupied at moderate levels of NE [42], the lower-affinity α1-adrenoceptors in the PFC and the β1-adrenoceptors in the amygdala are engaged at stress levels only [43]. Stress levels of NE may therefore have opposing effects on neural functioning in the PFC and the amygdala [12,44], which are critical regions within the ECN and the SN, respectively.

Stressors are also known to potently activate the dopaminergic system [45,46]. This leads to increases in dopamine levels in the PFC, but also in ventral (nucleus accumbens) and dorsal striatal regions [47]. Similar to the LC–NE system, stress mainly increases tonic, rather than phasic, firing in dopaminergic neurons [48,49]. Although dopaminergic activity is often associated mainly with reward processing and appetitive motivation, it has been shown that distinct subpopulations of dopaminergic neurons respond selectively to salient aversive stimuli [50]. For instance, in monkeys, neurons that respond to aversive stimuli have been identified in dorsolateral parts of the midbrain [51]. Furthermore, aversive stimuli trigger dopamine release in the dorsal striatum and nucleus accumbens in rats [52]. Effects of dopaminergic projections are also regionally specific. Whereas moderate stimulation of D1 receptors in the PFC reduced neuronal firing in response to noise stimulation in both rats and monkeys, excessive levels of D1 stimulation reached during acute stress suppress all neuronal firing [53], which explains the known detrimental effects of supranormal D1 stimulation on PFC-dependent functions [54]. This pattern again appears to be different in the amygdala, where the expression of conditioned fear, for instance, has been shown to depend on D1 receptor availability [55].

These findings suggest that both the LC–NE system and the dopamine system contribute to opposing regulation of the SN and the ECN during acute stress, but they may play separate roles. For instance, noradrenergic projections are more widespread and include regions involved in attentional and sensory processes, whereas dopaminergic innervation is more pronounced in the striatum, a region implicated in rigid and habitual behavior [56]. The distinct roles of these two neuromodulatory systems in mediating the central stress response, however, remain to be explored, in particular in humans.

Stress-induced corticosteroid release may also play an important role during the early phase of the stress response by activating rapid, nongenomic pathways. In vitro studies in rodents have shown that corticosteroids exert rapid effects through low-affinity membrane-bound glucocorticoid receptors (GRs) and mineralocorticoid receptors (MRs). Such effects can initiate as soon as corticosteroids reach target tissues [57]. In the paraventricular nucleus of the hypothalamus, corticosteroids have been shown to rapidly and reversibly decrease neuronal excitability through GRs, which may contribute to rapid negative feedback of the HPA axis [58]. On the other hand, MR activation has been shown to increase neuronal excitability in the basolateral amygdala [59] and hippocampus [57]. Thus, although these nongenomic pathways are just beginning to be understood and have been examined in only a small number of brain regions, such effects may also be regionally specific and thereby contribute to the regulation of broader neural systems [9].

Corticosteroids may furthermore interact with catecholaminergic activity [40]. For instance, noradrenergic activity has been shown to be amplified by GR activation at the level of the NTS and the LC [60], and NE levels in the amygdala are increased after systemic injections of corticosteroids [61]. Corticosteroids furthermore enhance the effect of acute stress on dopamine release [62], and blockade of GRs within the rat PFC reduces stress-induced dopamine release from the ventral tegmental area [63]. Complex interactions between β-adrenoceptor activation and glucocorticoids also exist regarding synaptic plasticity in the basolateral amygdala [64]. Behaviorally, corticosteroids postsynaptically augment the effects of NE on amygdala function by enhancing the β-adrenoceptor–cAMP system [65–67]. Similar effects have been described in the nucleus accumbens [68]. On the other hand, the negative effects of supraoptimal levels of catecholaminergic activity on PFC function can also be exacerbated by corticosteroids [69] through membrane-bound GRs [70].

Salience Network: Behavioral Effects of Acute Stress

Notably, there is substantial overlap between the functions attributed to catecholaminergic, in particular noradrenergic, signaling, on the one hand, and those ascribed to the SN on the other hand. Both systems are strongly implicated in attentional (re)orienting [24,71–73]. Whereas phasic LC–NE signaling is thought to subserve selective processing of stimuli relevant to current task sets, tonically elevated LC activity would lead to enhanced environmental scanning and facilitated reorienting of attention toward unexpected and potentially threatening stimuli [24,39,72,74]. Noradrenergic projections from the LC are further thought to regulate sensory gain to optimize the signal-to-noise ratio (SNR) for sensory gating and sustained attention [39,75,76]. This implies that stress-induced changes in noradrenergic level or activity may amplify sensory information to a supraoptimal level and thereby cause a decrease in SNR and sensory gating ability due to amplification of sensory noise. Functionally, such a shift in sensory function would benefit attentional vigilance, but at the expense of increased distractibility. In agreement, stimulation of the LC resulted in a suppression of phasic discharge of thalamic and barrel cortex neurons in response to sensory stimuli, which was accompanied by an increase in spontaneous activity [77]. Both stress induction [78] and pharmacological stimulation of the LC [79] in rats have furthermore been shown to impair prepulse inhibition, the phenomenon that a weak preceding stimulus causes an inhibition of the startle reflex in response to a subsequent stronger stimulus, which is seen as an index of sensory gating.

Similar findings have been reported in humans. For instance, oral administration of hydrocortisone rapidly impairs prepulse inhibition [80]. Sensory gating in humans, however, is more commonly investigated using the attentional blink paradigm. The attentional blink refers to the inability to detect the second of two targets when presented with a 200- to 500-ms stimulus onset asynchrony within a rapid serial visual presentation. Computational models of LC–NE function postulate that this attentional blink occurs as a consequence of LC–NE firing in response to the first target. This phasic response would exert a local inhibitory effect on the LC through α2-autoreceptors, thus creating a brief refractory period of suppression of LC responses to subsequent stimuli [81]. If stress shifts the LC from a phasic to a tonic mode of activity [39], one would expect this refractory period to disappear. This "rescue" of the detection of the second target is indeed what is observed, both following administration of the selective NE reuptake inhibitor reboxetine [82] and after stress induction [83], although in the latter study it was not certain that this finding (20 min after stress) was obtained within the time window of elevated catecholaminergic availability. Stress induction has also been shown to selectively modulate distinct components of event-related potentials evoked by sensory stimulation in humans [84–86]. Together, these findings indicate that acute stress induces a hypervigilant state, which is also reflected in the style of decision-making [87], most likely involving tonically elevated LC–NE activity.

As noted above, the SN also comprises striatal regions that are implicated in rigid stimulus-response learning and habitual behaviors [33]. The shift from goal-directed to rigid stimulus-response behavior is a well-documented effect of acute stress in both animals [88,89] and humans [90]. For instance, briefly following corticosterone delivery, rodents exhibit increased persistence in making disadvantageous, high-risk decisions in a rodent model of the Iowa Gambling Task [91], behavior that resembles anxious behavior exhibited by rats in the elevated-plus maze [92]. Such findings concur with clinical observations of stress-related relapse in addiction as well as exacerbation of symptoms in various psychiatric disorders including obsessive–compulsive disorder [2,93]. In humans, there is furthermore experimental evidence that stress enhances simple stimulus-response learning such as classical conditioning using a negative unconditioned stimulus [94].

A number of studies furthermore suggest that there may be sex differences in the rapid effect of stress on decision-making. Stress induction has been shown to increase persistence in making disadvantageous choices in the Iowa Gambling Task in men, but decrease it in women [95–97]. Thus, it appears that, whereas men tend to shift toward a habitual style of decision-making under stress, women may have a stronger tendency to become more inhibited and vigilant [98]. However, what causes this putative sex difference remains unclear [99].

Salience Network: Brain Systems-Level Effects of Acute Stress

Several human neuroimaging studies have reported increased responsiveness of the amygdala, one of the core regions of the SN, during the immediate phase of acute stress following experimental stress induction [100–102]. A number of studies have furthermore shown hyperactivation in limbic and subcortical regions, including the amygdala, after symptom provocation in anxiety disorder patients [103–109]. In line with the notion that such effects are driven by increased noradrenergic activity, pharmacologically increasing central NE levels, through administration of the α2-adrenoreceptor antagonist yohimbine [110] or the selective NE reuptake inhibitor reboxetine [111], mimics these effects, in particular when combined with additional administration of hydrocortisone [112]. Contrariwise, administration of the β-adrenergic receptor blocker propranolol diminishes the amygdala response [113]. It was furthermore shown that carriers of a common functional deletion in the gene coding for the α2B-adrenoreceptor (ADRA2B), which reduces the negative feedback function of the noradrenergic system, exhibit stronger stress-induced increases in the phasic amygdala response [102]. Thus, human neuroimaging studies consistently show that NE plays an important role in regulating amygdala responsiveness.

There is also substantial evidence that activity in other regions of the SN increases during acute stress. For instance, activity in the dorsal anterior cingulate cortex and the anterior insula correlates with physiological markers of stress, such as increased heart rate [114], elevated blood pressure [115], and increased cortisol levels [116], and also with reduced heart rate variability

[117], a parasympathetic index of stress [118]. A number of studies furthermore used network-based methods to examine functional connectivity within the entire SN. One early study showed that spontaneous activity within the SN is associated with electroencephalographic signatures of alertness [119]. Furthermore, it was shown that connectivity of the amygdala with a number of SN regions was elevated in the context of stress induction [120]. These regions included a cluster of activation that corresponded with the anatomical location of the LC, although this finding should be interpreted with some caution given the small size of this region [121]. Another study showed that multiple physiological and psychological measures of acute stress, including the cortisol response, changes in α-amylase (a peripheral marker of noradrenergic activity), and subjective mood changes are all associated with increased connectivity within the SN during exposure to a stressor consisting of aversive cinematographic material [122]. Increased interconnectivity within the SN was also found in response to threat of mild electrical shock [123,124]. SN function was furthermore causally linked with noradrenergic activity in a study that showed that administration of the β-adrenergic blocker propranolol diminished connectivity within the SN, whereas blocking endogenous corticosteroid production through administration of metyrapone had no effect [122]. Interconnectivity of the ventral striatum within the SN was finally shown to be increased after administration of L-dopa, which increases dopamine availability [125]. In sum, both the activity of and the interconnectivity of the SN appear to be boosted by catecholaminergic activity during the early phase of the response to acute stressors.

EXECUTIVE CONTROL NETWORK AND ACUTE STRESS

If a shift toward a hypervigilant and rigid style of decision-making is adaptive in situations of acute stress, it is reasonable to assume that suspending slower and more elaborate types of decision-making also benefits short-term survival [44]. Such goal-directed types of decision-making are thought to be supported by a frontoparietal network involving dorsal frontal (dorsolateral PFC; precentral/superior frontal sulci, or frontal eye fields; and dorsomedial PFC) and dorsal posterior parietal areas [23], which is commonly referred to as the ECN [19] (see Fig. 30.1).

As mentioned earlier, rodent and nonhuman primate work has demonstrated that excessive catecholaminergic activity during acute stress impairs PFC function through occupation of the lower-affinity α1- and β1-adrenoceptors, as well as through supraoptimal dopamine D1 receptor

stimulation [2,11]. Inhibitory effects of NE on PFC function are furthermore exacerbated by corticosteroids [70]. A consequence of this supraoptimal catecholaminergic activity is that PFC neurons lose the capacity to maintain persistent patterns of spiking activity, which is seen as an important neurophysiological substrate of working memory maintenance [126].

Executive Control Network: Behavioral Effects of Acute Stress

In humans, a vast majority of studies investigating the effects of acute stress on working memory, one of the cognitive functions supported by the ECN, have found impairments, particularly when testing was performed while catecholaminergic activity was still elevated [94,127–131] (for exceptions see Refs. [132,133]). Consistent with the notion that the ECN supports flexible goal-directed action, stress induction has also been shown to negatively affect task switching [134,135] and cognitive flexibility [136]. In the realm of decision-making, studies have found that acute stress reduces elaboration of alternative problem-solving strategies [137] and more generally impairs strategic decision-making [138–140]. Furthermore, stress increases the subjective valuation of immediate rewards despite negative consequences, as shown, for instance, in increased risky decisions in the Game of Dice Task [141] (but see Refs. [142,143]) and increased rates of delay discounting (i.e., the degree to which one discounts the subjective value of future rewards) [144] (but see Ref. [145]). Finally, also suggesting impaired ability to flexibly adjust one's behavior given changing circumstances, stress has been shown to reduce the sensitivity of instrumental responding to goal devaluation [90,146,147].

Pharmacological studies have provided insight into the mechanisms underlying these effects. One study showed an insensitivity to outcome devaluation similar to what is seen under stress after combined administration of yohimbine and hydrocortisone [148]. Furthermore, stress-induced impairments in cognitive flexibility tasks [149], as well as insensitivity to outcome devaluation [150], are blocked by the β-adrenergic blocker propranolol. Thus, the impairments in goal-directed decision-making seen during acute stress appear to be mediated by rapid catecholaminergic activation, possibly in interaction with corticosteroids.

Executive Control Network: Brain Systems-Level Effects of Acute Stress

Several neuroimaging studies in humans have investigated the effects of stress induction and pharmacological

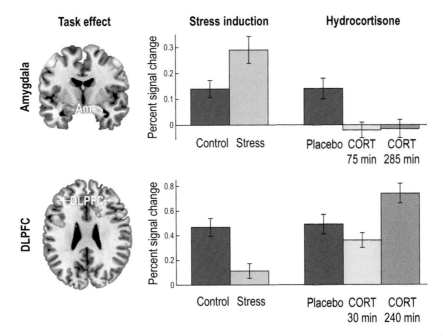

FIGURE 30.2 Opposite effects of stress induction (at short time intervals) and hydrocortisone administration (at different time intervals) on amygdala versus DLPFC. Data from four studies probing amygdala (major node of the salience network) and DLPFC (major node of the executive control network) function are shown. Shortly following stress induction, the amygdala response to salient stimuli increases [100], whereas DLPFC activity related to an executive control task decreases [151]. Contrariwise, administration of hydrocortisone reduces amygdala activity at both 75- and 285-min delays [168] and enhances DLPFC activity only at a 240-min delay [171], which is sufficient to allow genomic effects. For amygdala, bar graphs indicate percentage signal change for emotional facial expressions versus baseline. For DLPFC, bar graphs indicate percentage signal change for a two-back working memory task condition versus a zero-back control. *Am*, amygdala; *CORT*, hydrocortisone; *DLPFC*, dorsolateral prefrontal cortex. *Adapted from Hermans EJ, Henckens MJAG, Joëls M, Fernández G. Dynamic adaptation of large-scale brain networks in response to acute stressors. Trends Neurosci 2014;37:304−14.*

manipulations on neural activity in brain regions within the ECN. One study revealed reduced working memory-related activation in the dorsolateral PFC (DLPFC) immediately after experimental induction of stress [151] (see Fig. 30.2). This negative effect of stress on working memory-related activation of the DLPFC was furthermore shown to be stronger in Met homozygotes for the gene coding catechol-*O*-methyltransferase, who are thought to have higher baseline levels of catecholamines [130]. A similar suppressive effect on PFC activity was found in a task involving reward-related value representation [152], which may explain stress-induced impairments in reward value-based decision-making described in the previous paragraphs. During stress, working memory-related activation of the ECN is further accompanied by a failure to suppress activity in regions that are part of other large-scale networks (e.g., default mode network [151,153]), which is in line with the notion that acute stress impairs selective allocation of processing resources to the ECN. Finally, pharmacological MRI work has shown that combined administration of yohimbine and hydrocortisone was again more effective in reducing PFC activity than either drug alone [110,154]. Thus, human neuroimaging work is in line with animal work [70] and demonstrates that

the ECN is suppressed when catecholaminergic effects dominate or coincide with corticosteroid elevation.

SALIENCE NETWORK AND EXECUTIVE CONTROL NETWORK AND RECOVERY FROM STRESS

Excessive and prolonged activation of hormonal responses to stressors exhausts our energy reserves and in the long term results in an increased allostatic load: a failure to maintain homeostatic balance [155]. Therefore, a timely limitation and termination of the stress response is a critical aspect of healthy and adaptive functioning under stress. There is now considerable evidence that corticosteroids play a critical role in this process.

As described before, corticosteroid receptor binding leads to altered gene transcription, a relatively slow process that results in changes in levels of multiple proteins that affect neuronal function [9]. These genomic effects take at least an hour to initiate but can continue for at least several hours. Notably, in many brain regions, such slow genomic effects may oppose the rapid nongenomic effects described above. GR-mediated effects 4 h and more after stress, for instance, have been shown to enhance PFC neuronal function and facilitate working memory in rats [156], presumably by increasing GR/serum- and glucocorticoid-inducible kinase-induced glutamate receptor trafficking [157]. Similar effects were found in the dorsal hippocampus [158], but these effects are opposite those observed in the ventral hippocampus [159] and (basolateral) amygdala [160]. Thus, like other mechanisms of action, these genomic actions have also been shown to exhibit regional specificity. Genomic actions of corticosteroids may thereby provide a mechanism that actively complements—in an opposite direction—the rapid effects of various stress mediators described above.

Early studies in rodents have indeed shown that administration of corticosteroids can have anxiolytic rather than anxiogenic effects [161]. In agreement with such findings, human studies have demonstrated that hydrocortisone administration at various time intervals before testing reduces emotional interference in cognitive tasks [162,163] and has a protective effect on self-reported mood in stress-induction paradigms [164]. Pharmacological inhibition of cortisol production by means of administration of metyrapone had the opposite effect of increasing sympathetic arousal in response to stressors [165]. Furthermore, hydrocortisone administration in anxiety disorder patients enhances extinction-based psychotherapy [166] and reduces phobic symptoms [167]. Thus, elevation of corticosteroids, in particular at longer delays and when not accompanied by catecholaminergic activation, appears to lead to a suppression of neurocognitive processes supported by the SN.

Recent neuroimaging work in humans has tested whether such slow vigilance-reducing effects of hydrocortisone may indeed be explained by a reduction in responsiveness in SN regions such as the amygdala. Approximately 4–5 h after administration of hydrocortisone, amygdala responsiveness to negative or positive information was reduced [168] (Fig. 30.2). Importantly, salivary cortisol levels at this time point had already returned to baseline, strongly suggesting that these effects specifically involve slow genomic actions. Furthermore, hydrocortisone administration ±2 h before testing reduced functional connectivity between the amygdala and a brain-stem cluster corresponding to the anatomical location of the LC [169], a finding opposite that of the aforementioned study assessing amygdala–LC coupling directly after stress [120]. Effects of hydrocortisone administration on amygdala activity at shorter delays are somewhat less clear, with a reduction at 75 min in a task involving passive viewing of emotional facial expressions [168] and an increased response at 60 min in a task involving emotional distractors [170]. This

discrepancy may be explained by different levels of coinciding catecholaminergic activity elicited by these tasks, but also by the critical delay necessary to induce genomic actions, which is approximately 60 min. Nonetheless, hydrocortisone administration studies support the notion that corticosteroids actively contribute to the gradual downregulation of SN regions and thus help in normalizing this aspect of the stress response.

There is also considerable evidence of an effect in the opposite direction on ECN functioning several hours after stress. As mentioned earlier, rodent findings show that stress potentiates excitatory neurotransmission through a slow GR-dependent mechanism in the PFC. This effect was accompanied by improved working memory performance 4 h after stress [156], a delay that is long enough to allow genomic effects to develop [9]. A human study administered hydrocortisone at different time intervals to subjects before they performed a working memory task [171] (Fig. 30.2). Hydrocortisone administered 4 h before testing positively affected working memory performance and DLPFC activation, whereas hydrocortisone administered 30 min before testing did not have such an effect. These findings thus suggest that several hours after stress, genomic actions of corticosteroids can complement—in an opposite direction—the stress-induced impairment in executive control functioning seen directly after stress, which may serve to support flexible decision-making and the adjustment of long-term goals. Of note, these slow and presumably gene-mediated effects after stress have as of this writing not specifically been tested with regard to decision-making.

SUMMARY AND CONCLUSION

Dual-systems models of decision-making have long assumed that acute stress induces a shift from slow to fast [3], or reflective to reflexive [5], systems for decision-making. In the previous sections, we attempted to provide a neural basis for these notions by integrating the relevant existing knowledge at multiple levels of analysis. We have summarized rodent data showing that distinct waves of stress-related hormones and neurotransmitters exert specific effects in widely distributed brain regions. A recurrent finding within this literature is that effects in regions such as the PFC are accompanied by opposing effects in limbic and subcortical structures such as amygdala and striatum. This suggests that such brain regions are differentially regulated as part of multiple large-scale functional networks [2,12].

A comprehensive account of the architecture and interactions within and between such networks can be obtained only by investigating the resulting changes at the brain systems level. We therefore attempted to integrate these findings with an emerging literature in the field of functional connectivity network modeling in humans, which is beginning to reveal how regions shown to be differentially affected by stress play a role in distinct large-scale neural systems. We have argued that the SN and ECN [19] are the most critical neurocognitive systems targeted by such effects. However, as the network architecture of the human brain is parsed in more and more detail [16], it appears likely that a more complex picture will emerge in the future. Nonetheless, taking these networks and the putative roles they play in different types of decision-making as a starting point, we argue that a classification into broad cognitive domains supported by the SN and the ECN is a critical factor in understanding how stress may affect decision-making.

A second critical factor is timing: as we explained earlier, the effects of stress on the SN and the ECN appear to be opposite during the acute phase compared to the slow phase. In this chapter, we have shown how these two phases can be linked to dominating catecholaminergic and corticosteroid actions, respectively. However, future research will probably reveal more complexity also in this domain, for instance, by revealing how neurotransmitters and hormones act outside of their typical temporal domain [8], or by elaborating the roles of, for instance, serotonin

and neuropeptides in the central stress response [11], which have received relatively little attention in human research.

Nonetheless, an initial classification across the two factors of network and timing creates order in an otherwise confusing and paradoxical body of empirical research on the effects of stress on human cognition. Fig. 30.3 visualizes the findings from 38 empirical research articles cited in this chapter. All these studies involved either controlled induction of stress (mainly through psychological manipulations) or administration of hydrocortisone. As can be seen in Fig. 30.3A, when all cognitive tasks are taken together, the effect sizes (Cohen's d; standardized mean differences) associated with stress induction or hydrocortisone administration reveal no systematic bias toward better or worse performance. Similarly, when data points are split for tasks tapping into ECN functions and SN functions, no clear bias is visible. Adding the timing factor (Fig. 30.3B), it becomes apparent that the two different systems are modulated in a reciprocal manner over time, with trend lines that cross after approximately 1 h, when genomic effects of corticosteroids start to develop.

In Fig. 30.4, we summarize all the literature reviewed in this chapter within a biphasic-reciprocal model of stress-induced reallocation of neural resources. While this heuristic model integrates all the relevant data to the best of our knowledge, many open questions remain.

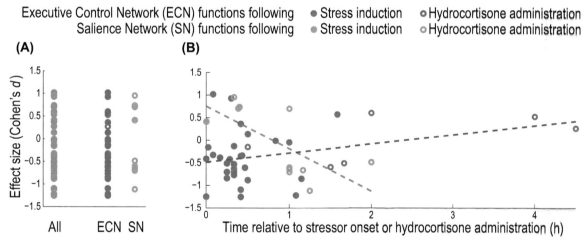

FIGURE 30.3 Effect sizes (Cohen's d) of stress induction or hydrocortisone administration (A) overall and sorted by cognitive functions supported by the executive control network (ECN) versus the salience network (SN) and (B) plotted as a function of the time point at which measurement of cognitive performance started relative to stressor onset or administration. Timing information was obtained from descriptions of the experimental procedures in the respective papers. *Dashed lines* indicate trends over time for each neurocognitive function. Data points taken from studies tapping into ECN-dependent functions are plotted in blue (digit span [127,131–133]; reading span [94]; n-back [128,130,151,171]; Sternberg [129] with emotional distractors [163]; dual task performance, task shifting, selective attention, and cognitive flexibility [134–136,149]; delay discounting [144]; rational and strategic decision-making [137–140]; risky decision-making [141–143,174]; and instrumental learning and extinction [146–148,150,154]). In red are shown SN-dependent functions (attentional blink [83], prepulse inhibition [80], emotional interference [101,162,170], subjective measures of negative mood [164] and fear [166,167], and hypervigilant style of decision-making [87]). *Adapted and extended from Hermans EJ, Henckens MJAG, Joëls M, Fernández G. Dynamic adaptation of large-scale brain networks in response to acute stressors. Trends Neurosci 2014;37:304–14.*

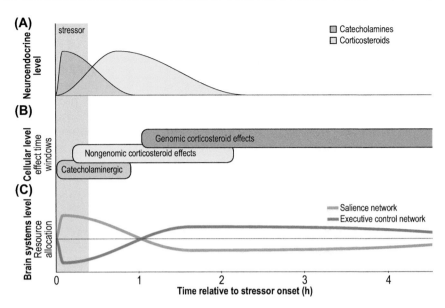

FIGURE 30.4   Biphasic-reciprocal model of reallocation of neural resources in response to stress. The links between effects of stress at (A) neuroendocrine, (B) cellular, and (C) brain systems levels are illustrated. (A) Neuroendocrine level: following exposure to a stressor, central levels of catecholamines (norepinephrine, dopamine) increase promptly and normalize not long after stressor offset. Corticosteroid levels in the brain rise more slowly and remain elevated for a longer period. (B) Cellular level: cellular effects occur within distinct effect time windows. Catecholamines primarily exert immediate effects through G-protein-coupled receptors. Corticosteroids have rapid nongenomic effects that may overlap and interact with catecholaminergic effects in an early time window, but also exert slower genomic effects. (C) Brain systems level: owing to local differences in receptor distribution and signaling cascades, opposite effects occur within different neurocognitive systems. We propose that this causes a dynamic reallocation of neural resources to systems responsible for attentional vigilance (*SN*, salience network) and executive control (*ECN*, executive control network; see Fig. 30.1 for an anatomical overview of both networks). Critically, our model proposes that stress-related hormones and neurotransmitters strengthen SN activity during the acute stress phase at the cost of ECN function, but subsequently contribute actively to the return to homeostasis by reversing this balance (see "Salience Network and Executive Control Network and Recovery From Stress" section on slow effects of corticosteroids for an explanation of underlying cellular mechanisms, and see Fig. 30.2 for empirical evidence at the neural level). By introducing time dependency and cognitive domain dependency as critical factors, our model explains substantial variability observed in the effects of stress on cognition at the behavioral level. *Adapted from Hermans EJ, Henckens MJAG, Joëls M, Fernández G. Dynamic adaptation of large-scale brain networks in response to acute stressors. Trends Neurosci 2014;37:304—14.*

Most importantly, how and why do individuals differ in their sensitivity to stressors, and why does stress have pathological consequences in some individuals? A suggestion that follows from our model is that individual differences, and ultimately maladaptation, may result from an inability to constrain sympathetic activation by subsequently released corticosteroids, a problem that may occur in posttraumatic stress disorder [172,173]. Our model would predict that in such conditions the SN is strongly activated, whereas later complementary action of the ECN is inadequate, potentially impairing an individual's ability to exert cognitive control over the emotional aspects of the stressful event. Our framework thus aligns with an ongoing paradigm shift within neuropsychiatry from identifying foci of abnormality toward developing a global understanding of aberrations at the level of large-scale networks [19].

In conclusion, in this chapter we integrated neuroendocrine, cellular, brain systems, and behavioral levels of analysis to propose a framework describing global and dynamic shifts in network resource allocation in response to acute stressors. We believe that this

framework provides an initial neural foundation for understanding shifts between strategic and tactical modes of decision-making observed under stress.

References

[1] Ohman A, Mineka S. Fears, phobias, and preparedness: toward an evolved module of fear and fear learning. Psychol Rev 2001; 108:483—522.

[2] Arnsten AFT. Stress signalling pathways that impair prefrontal cortex structure and function. Nat Rev Neurosci 2009;10:410—22.

[3] Kahneman D. Thinking, fast and slow. New York: Farrar, Straus and Giroux; 2011.

[4] McNally RJ. Automaticity and the anxiety disorders. Behav Res Ther 1995;33:747—54.

[5] Lieberman MD. The X- and C-systems: the neural basis of reflexive and reflective social cognition. In: Harmon-Jones E, Winkielman P, editors. Fundamentals of social neuroscience; 2007. p. 353—75. New York.

[6] Daw ND, Niv Y, Dayan P. Uncertainty-based competition between prefrontal and dorsolateral striatal systems for behavioral control. Nat Neurosci 2005;8:1704—11.

[7] Hermans EJ, Henckens MJAG, Joëls M, Fernández G. Dynamic adaptation of large-scale brain networks in response to acute stressors. Trends Neurosci 2014;37:304—14.

[8] Joëls M, Baram TZ. The neuro-symphony of stress. Nat Rev Neurosci 2009;10:459–66.

[9] Joëls M, Sarabdjitsingh RA, Karst H. Unraveling the time domains of corticosteroid hormone influences on brain activity: rapid, slow, and chronic modes. Pharmacol Rev 2012;64:901–38.

[10] Joëls M. Corticosteroid effects in the brain: U-shape it. Trends Pharmacol Sci 2006;27:244–50.

[11] Arnsten AF, Wang MJ, Paspalas CD. Neuromodulation of thought: flexibilities and vulnerabilities in prefrontal cortical network synapses. Neuron 2012;76:223–39.

[12] Arnsten AF. The biology of being frazzled. Science 1998;280: 1711–2.

[13] Packard MG, Cahill L. Affective modulation of multiple memory systems. Curr Opin Neurobiol 2001;11:752–6.

[14] Raichle ME, MacLeod AM, Snyder AZ, Powers WJ, Gusnard DA, Shulman GL. A default mode of brain function. Proc Natl Acad Sci USA 2001;98:676–82.

[15] Smith SM, Fox PT, Miller KL, Glahn DC, Fox PM, Mackay CE, et al. Correspondence of the brain's functional architecture during activation and rest. Proc Natl Acad Sci USA 2009;106: 13040–5.

[16] Laird AR, Fox PM, Eickhoff SB, Turner JA, Ray KL, McKay DR, et al. Behavioral interpretations of intrinsic connectivity networks. J Cogn Neurosci 2011;23:4022–37.

[17] Greicius MD, Supekar K, Menon V, Dougherty RF. Resting-state functional connectivity reflects structural connectivity in the default mode network. Cereb Cortex 2008;19:72–8.

[18] Raichle ME. Neuroscience. The brain's dark energy. Science 2006; 314:1249–50.

[19] Menon V. Large-scale brain networks and psychopathology: a unifying triple network model. Trends Cogn Sci 2011;15: 483–506.

[20] Park HJ, Friston K. Structural and functional brain networks: from connections to cognition. Science 2013;342:1238411.

[21] Raichle ME. A brief history of human brain mapping. Trends Neurosci 2009;32:118–26.

[22] Seeley WW, Menon V, Schatzberg AF, Keller J, Glover GH, Kenna H, et al. Dissociable intrinsic connectivity networks for salience processing and executive control. J Neurosci 2007;27: 2349–56.

[23] Vincent JL, Kahn I, Snyder AZ, Raichle ME, Buckner RL. Evidence for a frontoparietal control system revealed by intrinsic functional connectivity. J Neurophysiol 2008;100:3328–42.

[24] Corbetta M, Patel G, Shulman GL. The reorienting system of the human brain: from environment to theory of mind. Neuron 2008; 58:306–24.

[25] Madsen PL, Hasselbalch SG, Hagemann LP, Olsen KS, Bülow J, Holm S, et al. Persistent resetting of the cerebral oxygen/glucose uptake ratio by brain activation: evidence obtained with the Kety-Schmidt technique. J Cereb Blood Flow Metab 1995;15: 485–91.

[26] Peters A. The selfish brain: competition for energy resources. Am J Hum Biol 2011;23:29–34.

[27] Fox MD, Zhang D, Snyder AZ, Raichle ME. The global signal and observed anticorrelated resting state brain networks. J Neurophysiol 2009;101:3270–83.

[28] Singh KD, Fawcett IP. Transient and linearly graded deactivation of the human default-mode network by a visual detection task. NeuroImage 2008;41:100–12.

[29] Droste SK, de Groote L, Atkinson HC, Lightman SL, Reul JMHM, Linthorst ACE. Corticosterone levels in the brain show a distinct ultradian rhythm but a delayed response to forced swim stress. Endocrinology 2008;149:3244–53.

[30] Dosenbach NUF, Fair DA, Cohen AL, Schlaggar BL, Petersen SE. A dual-networks architecture of top-down control. Trends Cogn Sci 2008;12:99–105.

[31] Davis M, Whalen PJ. The amygdala: vigilance and emotion. Mol Psychiatry 2000;6:13–34.

[32] Nieuwenhuis S, Aston-Jones G, Cohen JD. Decision making, the P3, and the locus coeruleus–norepinephrine system. Psychol Bull 2005;131:510–32.

[33] Ashby FG, Turner BO, Horvitz JC. Cortical and basal ganglia contributions to habit learning and automaticity. Trends Cogn Sci 2010;14:208–15.

[34] Critchley HD. Neural mechanisms of autonomic, affective, and cognitive integration. J Comp Neurol 2005;493:154–66.

[35] Ulrich-Lai YM, Herman JP. Neural regulation of endocrine and autonomic stress responses. Nat Rev Neurosci 2009;10: 397–409.

[36] Fullana MA, Harrison BJ, Soriano-Mas C, Vervliet B, Cardoner N, Àvila-Parcet A, et al. Neural signatures of human fear conditioning: an updated and extended meta-analysis of fMRI studies. Mol Psychiatry 2016;21:500–8.

[37] Kober H, Barrett LF, Joseph J, Bliss-Moreau E, Lindquist K, Wager TD. Functional grouping and cortical-subcortical interactions in emotion: a meta-analysis of neuroimaging studies. NeuroImage 2008;42:998–1031.

[38] Valentino RJ, van Bockstaele E. Convergent regulation of locus coeruleus activity as an adaptive response to stress. Eur J Pharmacol 2008;583:194–203.

[39] Aston-Jones G, Cohen JD. Adaptive gain and the role of the locus coeruleus-norepinephrine system in optimal performance. J Comp Neurol 2005;493:99–110.

[40] Roozendaal B, McEwen BS, Chattarji S. Stress, memory and the amygdala. Nat Rev Neurosci 2009;10:423–33.

[41] Foote SL, Morrison JH. Extrathalamic modulation of cortical function. Annu Rev Neurosci 1987;10:67–95.

[42] Wang M, Ramos BP, Paspalas CD, Shu Y, Simen A, Duque A, et al. Alpha2A-adrenoceptors strengthen working memory networks by inhibiting cAMP-HCN channel signaling in prefrontal cortex. Cell 2007;129:397–410.

[43] Birnbaum S, Gobeske KT, Auerbach J, Taylor JR, Arnsten AF. A role for norepinephrine in stress-induced cognitive deficits: alpha-1-adrenoceptor mediation in the prefrontal cortex. Biol Psychiatry 1999;46:1266–74.

[44] Arnsten AF. Through the looking glass: differential noradenergic modulation of prefrontal cortical function. Neural Plast 2000;7: 133–46.

[45] Thierry AM, Tassin JP, Blanc G, Glowinski J. Selective activation of mesocortical DA system by stress. Nature 1976;263:242–4.

[46] Reinhard JF, Bannon MJ, Roth RH. Acceleration by stress of dopamine synthesis and metabolism in prefrontal cortex: antagonism by diazepam. Naunyn Schmiedebergs Arch Pharmacol 1982;318:374–7.

[47] Abercrombie ED, Keefe KA, DiFrischia DS, Zigmond MJ. Differential effect of stress on in vivo dopamine release in striatum, nucleus accumbens, and medial frontal cortex. J Neurochem 1989; 52:1655–8.

[48] Anstrom KK, Woodward DJ. Restraint increases dopaminergic burst firing in awake rats. Neuropsychopharmacology 2005;30: 1832–40.

[49] Ungless MA, Argilli E, Bonci A. Effects of stress and aversion on dopamine neurons: implications for addiction. Neurosci Biobehav Rev 2010;35:151–6.

[50] Bromberg-Martin ES, Matsumoto M, Hikosaka O. Dopamine in motivational control: rewarding, aversive, and alerting. Neuron 2010;68:815–34.

[51] Matsumoto M, Hikosaka O. Two types of dopamine neuron distinctly convey positive and negative motivational signals. Nature 2009;459:837–41.

[52] Budygin EA, Park J, Bass CE, Grinevich VP, Bonin KD, Wightman RM. Aversive stimulus differentially triggers

subsecond dopamine release in reward regions. Neuroscience 2012;201:331—7.

[53] Vijayraghavan S, Wang M, Birnbaum SG, Williams GV, Arnsten AFT. Inverted-U dopamine D1 receptor actions on prefrontal neurons engaged in working memory. Nat Neurosci 2007; 10:376—84.

[54] Zahrt J, Taylor JR, Mathew RG, Arnsten AFT. Supranormal stimulation of D1 dopamine receptors in the rodent prefrontal cortex impairs spatial working memory performance. J Neurosci 1997; 17:8528—35.

[55] Lamont EW, Kokkinidis L. Infusion of the dopamine D1 receptor antagonist SCH 23390 into the amygdala blocks fear expression in a potentiated startle paradigm. Brain Res 1998;795:128—36.

[56] Martel MM. Sexual selection and sex differences in the prevalence of childhood externalizing and adolescent internalizing disorders. Psychol Bull 2013;139:1221—59.

[57] Karst H, Berger S, Turiault M, Tronche F, Schütz G, Joëls M. Mineralocorticoid receptors are indispensable for nongenomic modulation of hippocampal glutamate transmission by corticosterone. Proc Natl Acad Sci USA 2005;102:19204—7.

[58] Tasker JG, Di S, Malcher-Lopes R. Minireview: rapid glucocorticoid signaling via membrane-associated receptors. Endocrinology 2006;147:5549—56.

[59] Karst H, Berger S, Erdmann G, Schütz G, Joëls M. Metaplasticity of amygdalar responses to the stress hormone corticosterone. Proc Natl Acad Sci USA 2010;107:14449—54.

[60] Roozendaal B, Williams CL, McGaugh JL. Glucocorticoid receptor activation in the rat nucleus of the solitary tract facilitates memory consolidation: involvement of the basolateral amygdala. Eur J Neurosci 1999;11:1317—23.

[61] McReynolds JR, Donowho K, Abdi A, McGaugh JL, Roozendaal B, McIntyre CK. Memory-enhancing corticosterone treatment increases amygdala norepinephrine and Arc protein expression in hippocampal synaptic fractions. Neurobiol Learn Mem 2010;93:312—21.

[62] Saal D, Dong Y, Bonci A, Malenka RC. Drugs of abuse and stress trigger a common synaptic adaptation in dopamine neurons. Neuron 2003;37:577—82.

[63] Butts KA, Weinberg J, Young AH, Phillips AG. Glucocorticoid receptors in the prefrontal cortex regulate stress-evoked dopamine efflux and aspects of executive function. Proc Natl Acad Sci USA 2011;108:18459—64.

[64] Sarabdjitsingh RA, Joëls M, de Kloet ER. Glucocorticoid pulsatility and rapid corticosteroid actions in the central stress response. Physiol Behav 2012;106:73—80.

[65] Roozendaal B, Quirarte GL, McGaugh JL. Glucocorticoids interact with the basolateral amygdala beta-adrenoceptor—cAMP/ cAMP/PKA system in influencing memory consolidation. Eur J Neurosci 2002;15:553—60.

[66] Roozendaal B, Okuda S, van der Zee EA, McGaugh JL. Glucocorticoid enhancement of memory requires arousal-induced noradrenergic activation in the basolateral amygdala. Proc Natl Acad Sci USA 2006;103:6741—6.

[67] Roozendaal B, Schelling G, McGaugh JL. Corticotropin-releasing factor in the basolateral amygdala enhances memory consolidation via an interaction with the beta-adrenoceptor-cAMP pathway: dependence on glucocorticoid receptor activation. J Neurosci 2008;28:6642—51.

[68] Wichmann R, Fornari RV, Roozendaal B. Glucocorticoids interact with the noradrenergic arousal system in the nucleus accumbens shell to enhance memory consolidation of both appetitive and aversive taste learning. Neurobiol Learn Mem 2012;98:197—205.

[69] Roozendaal B, McReynolds JR, McGaugh JL. The basolateral amygdala interacts with the medial prefrontal cortex in regulating glucocorticoid effects on working memory impairment. J Neurosci 2004;24:1385—92.

[70] Barsegyan A, Mackenzie SM, Kurose BD, McGaugh JL, Roozendaal B. Glucocorticoids in the prefrontal cortex enhance memory consolidation and impair working memory by a common neural mechanism. Proc Natl Acad Sci USA 2010;107: 16655—60.

[71] Corbetta M, Shulman GL. Control of goal-directed and stimulus-driven attention in the brain. Nat Rev Neurosci 2002;3:201—15.

[72] Sara SJ, Bouret S. Orienting and reorienting: the locus coeruleus mediates cognition through arousal. Neuron 2012;76:130—41.

[73] Weissman DH, Prado J. Heightened activity in a key region of the ventral attention network is linked to reduced activity in a key region of the dorsal attention network during unexpected shifts of covert visual spatial attention. NeuroImage 2012;61:798—804.

[74] Bouret S, Sara SJ. Network reset: a simplified overarching theory of locus coeruleus noradrenaline function. Trends Neurosci 2005; 28:574—82.

[75] Berridge CW, Waterhouse BD. The locus coeruleus-noradrenergic system: modulation of behavioral state and state-dependent cognitive processes. Brain Res Rev 2003;42: 33—84.

[76] Eldar E, Cohen JD, Niv Y. The effects of neural gain on attention and learning. Nat Neurosci 2013;16:1146—53.

[77] Devilbiss DM, Waterhouse BD, Berridge CW, Valentino R. Corticotropin-releasing factor acting at the locus coeruleus disrupts thalamic and cortical sensory-evoked responses. Neuropsychopharmacology 2012;37:2020—30.

[78] Ellenbroek BA, van den Kroonenberg PT, Cools AR. The effects of an early stressful life event on sensorimotor gating in adult rats. Schizophr Res 1998;30:251—60.

[79] Alsene KM, Bakshi VP. Pharmacological stimulation of locus coeruleus reveals a new antipsychotic-responsive pathway for deficient sensorimotor gating. Neuropsychopharmacology 2011;36:1656—67.

[80] Richter S, Schulz A, Zech CM, Oitzl MS, Daskalakis NP, Blumenthal TD, et al. Cortisol rapidly disrupts prepulse inhibition in healthy men. Psychoneuroendocrinology 2011;36:109—14.

[81] Nieuwenhuis S, Gilzenrat MS, Holmes BD, Cohen JD. The role of the locus coeruleus in mediating the attentional blink: a neuro-computational theory. J Exp Psychol Gen 2005;134:291—307.

[82] de Martino B, Strange BA, Dolan RJ. Noradrenergic neuromodulation of human attention for emotional and neutral stimuli. Psychopharmacology 2008;197:127—36.

[83] Schwabe L, Wolf OT. Emotional modulation of the attentional blink: is there an effect of stress? Emotion 2010;10:283—8.

[84] Weymar M, Schwabe L, Löw A, Hamm AO. Stress sensitizes the brain: increased processing of unpleasant pictures after exposure to acute stress. J Cogn Neurosci 2012;24:1511—8.

[85] Elling L, Steinberg C, Bröckelmann A-K, Dobel C, Bölte J, Junghofer M. Acute stress alters auditory selective attention in humans independent of HPA: a study of evoked potentials. PLoS One 2011;6:e18009.

[86] Shackman AJ, Maxwell JS, McMenamin BW, Greischar LL, Davidson RJ. Stress potentiates early and attenuates late stages of visual processing. J Neurosci 2011;31:1156—61.

[87] Johnston JH, Driskell JE, Salas E. Vigilant and hypervigilant decision making. J Appl Psychol 1997;82:614—22.

[88] Packard MG, Teather LA. Amygdala modulation of multiple memory systems: hippocampus and caudate-putamen. Neurobiol Learn Mem 1998;69:163—203.

[89] Elliott AE, Packard MG. Intra-amygdala anxiogenic drug infusion prior to retrieval biases rats towards the use of habit memory. Neurobiol Learn Mem 2008;90:616—23.

[90] Schwabe L, Wolf OT. Stress prompts habit behavior in humans. J Neurosci 2009;29:7191—8.

[91] Koot S, Baars A, Hesseling P, van den Bos R, Joëls M. Time-dependent effects of corticosterone on reward-based decision-

making in a rodent model of the Iowa Gambling Task. Neuropharmacology 2013;70:306−15.

[92] de Visser L, Baars AM, Lavrijsen M, van der Weerd CMM, van den Bos R. Decision-making performance is related to levels of anxiety and differential recruitment of frontostriatal areas in male rats. Neuroscience 2011;184:97−106.

[93] Sinha R, Jastreboff AM. Stress as a common risk factor for obesity and addiction. Biol Psychiatry 2013;73:827−35.

[94] Luethi M, Meier B, Sandi C. Stress effects on working memory, explicit memory, and implicit memory for neutral and emotional stimuli in healthy men. Front Behav Neurosci 2008;2:5.

[95] Lighthall NR, Mather M, Gorlick MA. Acute stress increases sex differences in risk seeking in the balloon analogue risk task. PLoS One 2008;4:e6002.

[96] van den Bos R, Harteveld M, Stoop H. Stress and decision-making in humans: performance is related to cortisol reactivity, albeit differently in men and women. Psychoneuroendocrinology 2009;34:1449−58.

[97] van den Bos R. Sex matters, as do individual differences.... Trends Neurosci 2015;38:401−2.

[98] Mather M, Lighthall NR. Both risk and reward are processed differently in decisions made under stress. Curr Dir Psychol Sci 2012;21:36−41.

[99] Hermans EJ, Henckens MJAG, Joëls M, Fernández G. Toward a mechanistic understanding of interindividual differences in cognitive changes after stress: reply to van den Bos. Trends Neurosci 2015;38:403−4.

[100] van Marle HJF, Hermans EJ, Qin S, Fernández G. From specificity to sensitivity: how acute stress affects amygdala processing of biologically salient stimuli. Biol Psychiatry 2009;66:649−55.

[101] Oei NYL, Veer IM, Wolf OT, Spinhoven P, Rombouts SARB, Elzinga BM. Stress shifts brain activation towards ventral "affective" areas during emotional distraction. Soc Cogn Affect Neurosci 2012;7:403−12.

[102] Cousijn H, Rijpkema M, Qin S, van Marle HJF, Franke B, Hermans EJ, et al. Acute stress modulates genotype effects on amygdala processing in humans. Proc Natl Acad Sci USA 2010; 107:9867−72.

[103] Tillfors M, Furmark T, Marteinsdottir I, Fredrikson M. Cerebral blood flow during anticipation of public speaking in social phobia: a PET study. Biol Psychiatry 2002;52:1113−9.

[104] Tillfors M, Furmark T, Marteinsdottir I, Fischer H, Pissiota A, Långström B, et al. Cerebral blood flow in subjects with social phobia during stressful speaking tasks: a PET study. Am J Psychiatry 2001;158:1220−6.

[105] Pissiota A, Frans O, Fernandez M, von Knorring L, Fischer H, Fredrikson M. Neurofunctional correlates of posttraumatic stress disorder: a PET symptom provocation study. Eur Arch Psychiatry Clin Neurosci 2002;252:68−75.

[106] Fredrikson M, Furmark T. Amygdaloid regional cerebral blood flow and subjective fear during symptom provocation in anxiety disorders. Ann NY Acad Sci 2003;985:341−7.

[107] Pissiota A, Frans O, Michelgård A, Appel L, Långström B, Flaten MA, et al. Amygdala and anterior cingulate cortex activation during affective startle modulation: a PET study of fear. Eur J Neurosci 2003;18:1325−31.

[108] Ahs F, Furmark T, Michelgård A, Långström B, Appel L, Wolf OT, et al. Hypothalamic blood flow correlates positively with stress-induced cortisol levels in subjects with social anxiety disorder. Psychosom Med 2006;68:859−62.

[109] Osuch EA, Willis MW, Bluhm R, CSTS Neuroimaging Study Group, Ursano RJ, Drevets WC. Neurophysiological responses to traumatic reminders in the acute aftermath of serious motor vehicle collisions using [^{15}O]-H$_2$O positron emission tomography. Biol Psychiatry 2008;64:327−35.

[110] van Stegeren AH, Roozendaal B, Kindt M, Wolf OT, Joëls M. Interacting noradrenergic and corticosteroid systems shift human brain activation patterns during encoding. Neurobiol Learn Mem 2010;93:56−65.

[111] Onur OA, Walter H, Schlaepfer TE, Rehme AK, Schmidt C, Keysers C, et al. Noradrenergic enhancement of amygdala responses to fear. Soc Cogn Affect Neurosci 2009;4:119−26.

[112] Kukolja J, Schläpfer TE, Keysers C, Klingmüller D, Maier W, Fink GR, et al. Modeling a negative response bias in the human amygdala by noradrenergic-glucocorticoid interactions. J Neurosci 2008;28:12868−76.

[113] Hurlemann R, Walter H, Rehme AK, Kukolja J, Santoro SC, Schmidt C, et al. Human amygdala reactivity is diminished by the β-noradrenergic antagonist propranolol. Psychol Med 2010; 40:1839−48.

[114] Wager TD, Waugh CE, Lindquist M, Noll DC, Fredrickson BL, Taylor SF. Brain mediators of cardiovascular responses to social threat: part I: reciprocal dorsal and ventral sub-regions of the medial prefrontal cortex and heart-rate reactivity. NeuroImage 2009;47:821−35.

[115] Gianaros PJ, Sheu LK, Matthews KA, Jennings JR, Manuck SB, Hariri AR. Individual differences in stressor-evoked blood pressure reactivity vary with activation, volume, and functional connectivity of the amygdala. J Neurosci 2008;28:990−9.

[116] Pruessner J, Dedovic K, Khalilimahani N, Engert V, Pruessner M, Buss C, et al. Deactivation of the limbic system during acute psychosocial stress: evidence from positron emission tomography and functional magnetic resonance imaging studies. Biol Psychiatry 2008;63:234−40.

[117] Ahs F, Sollers III JJ, Furmark T, Fredrikson M, Thayer JF. High-frequency heart rate variability and cortico-striatal activity in men and women with social phobia. NeuroImage 2009;47: 815−20.

[118] Porges SW. Cardiac vagal tone: a physiological index of stress. Neurosci Biobehav Rev 1995;19:225−33.

[119] Sadaghiani S, Scheeringa R, Lehongre K, Morillon B, Giraud A-L, Kleinschmidt A. Intrinsic connectivity networks, alpha oscillations, and tonic alertness: a simultaneous electroencephalography/functional magnetic resonance imaging study. J Neurosci 2010;30:10243−50.

[120] van Marle HJF, Hermans EJ, Qin S, Fernández G. Enhanced resting-state connectivity of amygdala in the immediate aftermath of acute psychological stress. NeuroImage 2010;53:348−54.

[121] Astafiev SV, Snyder AZ, Shulman GL, Corbetta M. Comment on "Modafinil shifts human locus coeruleus to low-tonic, high-phasic activity during functional MRI" and "Homeostatic sleep pressure and responses to sustained attention in the suprachiasmatic area." Science 2010;328:309 [author reply 309].

[122] Hermans EJ, van Marle HJF, Ossewaarde L, Henckens MJAG, Qin S, van Kesteren MTR, et al. Stress-related noradrenergic activity prompts large-scale neural network reconfiguration. Science 2011;334:1151−3.

[123] McMenamin BW, Langeslag SJE, Sirbu M, Padmala S, Pessoa L. Network organization unfolds over time during periods of anxious anticipation. J Neurosci 2014;34:11261−73.

[124] McMenamin BW, Pessoa L. Discovering networks altered by potential threat ("anxiety") using quadratic discriminant analysis. NeuroImage 2015;116:1−9.

[125] Cole DM, Oei NYL, Soeter RP, Both S, van Gerven JMA, Rombouts SARB, et al. Dopamine-dependent architecture of cortico-subcortical network connectivity. Cereb Cortex 2013;23: 1509−16.

[126] Devilbiss DM, Jenison RL, Berridge CW. Stress-induced impairment of a working memory task: role of spiking rate and spiking history predicted discharge. PLoS Comput Biol 2012;8:e1002681.

[127] Elzinga BM, Roelofs K. Cortisol-induced impairments of working memory require acute sympathetic activation. Behav Neurosci 2005;119:98–103.

[128] Schoofs D, Preuss D, Wolf OT. Psychosocial stress induces working memory impairments in an n-back paradigm. Psychoneuroendocrinology 2008;33:643–53.

[129] Oei NYL, Everaerd WTAM, Elzinga BM, van Well S, Bermond B. Psychosocial stress impairs working memory at high loads: an association with cortisol levels and memory retrieval. Stress 2006;9:133–41.

[130] Qin S, Cousijn H, Rijpkema M, Luo J, Franke B, Hermans EJ, et al. The effect of moderate acute psychological stress on working memory-related neural activity is modulated by a genetic variation in catecholaminergic function in humans. Front Integr Neurosci 2012;6:16.

[131] Schoofs D, Wolf OT, Smeets T. Cold pressor stress impairs performance on working memory tasks requiring executive functions in healthy young men. Behav Neurosci 2009;123:1066–75.

[132] Hoffman R, Al'Absi M. The effect of acute stress on subsequent neuropsychological test performance (2003). Arch Clin Neuropsychol 2004;19:497–506.

[133] Kuhlmann S, Piel M, Wolf OT. Impaired memory retrieval after psychosocial stress in healthy young men. J Neurosci 2005;25:2977–82.

[134] Plessow F, Kiesel A, Kirschbaum C. The stressed prefrontal cortex and goal-directed behaviour: acute psychosocial stress impairs the flexible implementation of task goals. Exp Brain Res 2012;216:397–408.

[135] Steinhauser M, Maier M, Hübner R. Cognitive control under stress: how stress affects strategies of task-set reconfiguration. Psych Sci 2007;18:540–5.

[136] Plessow F, Fischer R, Kirschbaum C, Goschke T. Inflexibly focused under stress: acute psychosocial stress increases shielding of action goals at the expense of reduced cognitive flexibility with increasing time lag to the stressor. J Cogn Neurosci 2011;23:3218–27.

[137] Keinan G. Decision making under stress: scanning of alternatives under controllable and uncontrollable threats. J Pers Soc Psychol 1987;52:639–44.

[138] Vinkers CH, Zorn JV, Cornelisse S, Koot S, Houtepen LC, Olivier B, et al. Time-dependent changes in altruistic punishment following stress. Psychoneuroendocrinology 2013;38:1467–75.

[139] Leder J, Häusser JA, Mojzisch A. Stress and strategic decision-making in the beauty contest game. Psychoneuroendocrinology 2013;38:1503–11.

[140] Youssef FF, Dookeeram K, Basdeo V, Francis E, Doman M, Mamed D, et al. Stress alters personal moral decision making. Psychoneuroendocrinology 2012;37:491–8.

[141] Starcke K, Wolf OT, Markowitsch HJ, Brand M. Anticipatory stress influences decision making under explicit risk conditions. Behav Neurosci 2008;122:1352–60.

[142] Pabst S, Brand M, Wolf OT. Stress effects on framed decisions: there are differences for gains and losses. Front Behav Neurosci 2013;7:142.

[143] Pabst S, Brand M, Wolf OT. Stress and decision making: a few minutes make all the difference. Behav Brain Res 2013;250:39–45.

[144] Kimura K, Izawa S, Sugaya N, Ogawa N, Yamada KC, Shirotsuki K, et al. The biological effects of acute psychosocial stress on delay discounting. Psychoneuroendocrinology 2013;38:2300–8.

[145] Haushofer J, Cornelisse S, Seinstra M, Fehr E, JoÄ ls M, Kalenscher T. No effects of psychosocial stress on intertemporal choice. PLoS One 2013;8:e78597.

[146] Schwabe L, Wolf OT. Stress increases behavioral resistance to extinction. Psychoneuroendocrinology 2011;36:1287–93.

[147] Schwabe L, Wolf OT. Socially evaluated cold pressor stress after instrumental learning favors habits over goal-directed action. Psychoneuroendocrinology 2010;35:977–86.

[148] Schwabe L, Tegenthoff M, Hoffken O, Wolf OT. Concurrent glucocorticoid and noradrenergic activity shifts instrumental behavior from goal-directed to habitual control. J Neurosci 2010;30:8190–6.

[149] Alexander JK, Hillier A, Smith RM, Tivarus ME, Beversdorf DQ. Beta-adrenergic modulation of cognitive flexibility during stress. J Cogn Neurosci 2007;19:468–78.

[150] Schwabe L, Hoffken O, Tegenthoff M, Wolf OT. Preventing the stress-induced shift from goal-directed to habit action with a β-adrenergic antagonist. J Neurosci 2011;31:17317–25.

[151] Qin S, Hermans EJ, van Marle HJF, Luo J, Fernández G. Acute psychological stress reduces working memory-related activity in the dorsolateral prefrontal cortex. Biol Psychiatry 2009;66:25–32.

[152] Ossewaarde L, Qin S, van Marle HJF, van Wingen GA, Fernández G, Hermans EJ. Stress-induced reduction in reward-related prefrontal cortex function. NeuroImage 2011;55:345–52.

[153] Cousijn H, Rijpkema M, Qin S, van Wingen GA, Fernández G. Phasic deactivation of the medial temporal lobe enables working memory processing under stress. NeuroImage 2012;59:1161–7.

[154] Schwabe L, Tegenthoff M, Hoffken O, Wolf OT. Simultaneous glucocorticoid and noradrenergic activity disrupts the neural basis of goal-directed action in the human brain. J Neurosci 2012;32:10146–55.

[155] McEwen BS, Gianaros PJ. Stress- and allostasis-induced brain plasticity. Annu Rev Med 2011;62:431–45.

[156] Yuen EY, Liu W, Karatsoreos IN, Feng J, McEwen BS, Yan Z. Acute stress enhances glutamatergic transmission in prefrontal cortex and facilitates working memory. Proc Natl Acad Sci USA 2009;106:14075–9.

[157] Yuen EY, Liu W, Karatsoreos IN, Ren Y, Feng J, McEwen BS, et al. Mechanisms for acute stress-induced enhancement of glutamatergic transmission and working memory. Mol Psychiatry 2011;16:156–70.

[158] Karst H, Joëls M. Corticosterone slowly enhances miniature excitatory postsynaptic current amplitude in mice CA1 hippocampal cells. J Neurophysiol 2005;94:3479–86.

[159] Maggio N, Segal M. Differential corticosteroid modulation of inhibitory synaptic currents in the dorsal and ventral hippocampus. J Neurosci 2009;29:2857–66.

[160] Duvarci S, Paré D. Glucocorticoids enhance the excitability of principal basolateral amygdala neurons. J Neurosci 2007;27:4482–91.

[161] File SE, Vellucci SV, Wendlandt S. Corticosterone – an anxiogenic or an anxiolytic agent? J Pharm Pharmacol 1979;31:300–5.

[162] Putman P, Hermans EJ, Koppeschaar H, van Schijndel A, van Honk J. A single administration of cortisol acutely reduces preconscious attention for fear in anxious young men. Psychoneuroendocrinology 2007;32:793–802.

[163] Oei NYL, Tollenaar MS, Spinhoven P, Elzinga BM. Hydrocortisone reduces emotional distracter interference in working memory. Psychoneuroendocrinology 2009;34:1284–93.

[164] Het S, Wolf OT. Mood changes in response to psychosocial stress in healthy young women: effects of pretreatment with cortisol. Behav Neurosci 2007;121:11–20.

[165] Maheu FS, Joober R, Lupien SJ. Declarative memory after stress in humans: differential involvement of the beta-adrenergic and corticosteroid systems. J Clin Endocrinol Metab 2005;90:1697–704.

[166] de Quervain DJ-F, Bentz D, Michael T, Bolt OC, Wiederhold BK, Margraf J, et al. Glucocorticoids enhance extinction-based psychotherapy. Proc Natl Acad Sci USA 2011;108:6621–5.

[167] Soravia LM, Heinrichs M, Aerni A, Maroni C, Schelling G, Ehlert U, et al. Glucocorticoids reduce phobic fear in humans. Proc Natl Acad Sci USA 2006;103:5585–90.

[168] Henckens MJAG, van Wingen GA, Joëls M, Fernández G. Time-dependent effects of corticosteroids on human amygdala processing. J Neurosci 2010;30:12725–32.

[169] Henckens MJAG, van Wingen GA, Joëls M, Fernández G. Corticosteroid induced decoupling of the amygdala in men. Cereb Cortex 2012;22:2336–45.

[170] Henckens MJAG, van Wingen GA, Joëls M, Fernández G. Time-dependent effects of cortisol on selective attention and emotional interference: a functional MRI study. Front Integr Neurosci 2012; 6:1–14.

[171] Henckens MJAG, van Wingen GA, Joëls M, Fernández G. Time-dependent corticosteroid modulation of prefrontal working memory processing. Proc Natl Acad Sci USA 2011;108:5801–6.

[172] Schelling G. Efficacy of hydrocortisone in preventing posttraumatic stress disorder following critical illness and major surgery. Ann NY Acad Sci 2006;1071:46–53.

[173] Lupien SJ, McEwen BS, Gunnar MR, Heim C. Effects of stress throughout the lifespan on the brain, behaviour and cognition. Nat Rev Neurosci 2009;10:434–45.

[174] Preston SD, Buchanan TW, Stansfield RB, Bechara A. Effects of anticipatory stress on decision making in a gambling task. Behav Neurosci 2007;121:257–63.

31

Oxytocin's Influence on Social Decision-Making

A. Lefevre[1,2], A. Sirigu[1,2]

[1]Institut des Sciences Cognitives Marc Jeannerod, UMR 5229, CNRS, Bron, France; [2]Université Claude Bernard Lyon 1, Lyon, France

Abstract

The aim of this chapter is to review evidence for a role of oxytocin in each step of decision-making, i.e., from perception to decision and learning. Several findings suggest that oxytocin modulates decision processes via many pathways. Here we will describe oxytocin's action within these networks and the selective influence of this hormone on social decisions. Furthermore, we will discuss the physiological origins of oxytocin's effects in both humans and animals. Finally, we will discuss the evolutionary origin of oxytocin and the synergistic expansion of social brain mechanisms and of the oxytocinergic system.

INTRODUCTION

The role of oxytocin in social behavior is now firmly established. This peptide has attracted a lot of public attention and research interest given its potential as a therapeutic target in a wide range of developmental and psychiatric disorders such as autism, schizophrenia, or Williams syndrome. In this chapter we review studies demonstrating the role played by oxytocin during social decisions. We present evidence showing oxytocin intervention at every step of decision-making, from perception, choice, and reward reinforcement and learning. We discuss fMRI and neurotransmission studies demonstrating oxytocin's modulatory effect on brain regions important for reward perception and social evaluation. We argue that studying the neurobiology of oxytocin is promising if we wish to identify new therapies to alleviate patients' social disturbances.

The Social Brain

The so-called social brain circuit has been divided into two main networks [1]: the first, centered in the amygdala, is thought to process the emotional significance of social stimuli [2]; the second, centered in the nucleus accumbens, is known for coding the rewarding nature of objects and events [3]. Both networks are present in most vertebrates and their neurochemical properties are very similar [1]. Among several neurotransmitters, oxytocin (OT), a neurohormone important for social affiliation and attachment, has receptors in many of these regions. It has been suggested that social decisions involving approach/avoidance or partner/in-group affiliation are controlled by a twofold neurochemical action of OT. According to this view, by lowering the amount of stress generated by social contact, OT heightens the rewarding value of social contexts to bias perception. As social expertise differs among species (e.g., species-specific courtship displays [4]), OT activity is adapted to species-specific social interaction modes [5].

Oxytocin's Dual Mode of Release in the Brain

A fascinating property of OT is the way it is released in the brain. It can be diffused from dendrites of magnocellular neurons (volume transmission, see Ref. [6]) or liberated through the classical axonal route. Axonal transmission is run by two distinct neuronal populations, the parvocellular and the magnocellular neurons, which project axons or axon collaterals into the central nervous system. Additionally, magnocellular neurons' main projections are directed toward the posterior pituitary region where OT is released in blood vessels [7] (see Fig. 31.1). Time-course and regional activity of dendritic and axonal paths are different and this allow OT to apply a broad range of actions, ranging from local neurotransmission to large-scale neuromodulation. This anatomical organization of the OT system has evolved in parallel with the increasing complexity of social behaviors [8].

Another interesting property of the oxytocinergic system is the variety of G proteins that can potentially bind to OT receptors [9]. This represents an opportunity for

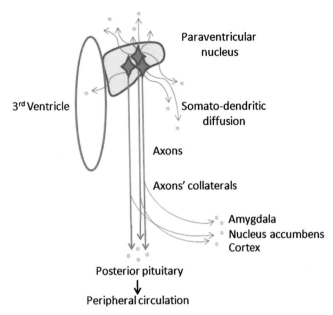

FIGURE 31.1 **Oxytocin (OT) modes of release.** Magnocellular neurons from the paraventricular nucleus of the hypothalamus diffuse OT from dendrites, into the extracellular fluid and the cerebrospinal fluid. OT is also released via axons into the blood and via axon collaterals into various distant brain regions. Note that parvocellular neurons are not shown here, but although they are known to project axons at the central level, few studies have tried to characterize their pathways. *Green dots* represent OT.

OT to apply an umbrella of agonist and antagonist actions on cells expressing this receptor. Because the configuration of the receptor varies depending on which G protein it is bound to, many synthetic agonists and antagonists can potentially activate only a subpopulation of OT receptors [10]. Hence, to understand OT effects on behavior it seems important to characterize the neural processes linked to the variety of subtypes of OT receptors.

OXYTOCIN AND PERCEPTION OF SOCIAL STIMULI

Sensory Perception

In many species, OT receptors are distributed in many sensory regions [11,12]. Studies have revealed the molecular mechanisms mediated by OT during sensory processing. For instance, in mice, after birth OT stimulates sensory plasticity in tactile, auditory, and visual areas [13], a process taking place via OT diffusion from dendritic release given that there are no axon projections from the paraventricular nucleus to sensory regions. In adult mice, these neural events are crucial because OT orients maternal behavior in response to pup calls by regulating inhibitory/excitatory activity in auditory cortex [14]. It seems that the presence of OT receptors in

sensory cortex has an evolutionary meaning. The amount of OT receptors in primary sensory areas is, in a given species, function of the importance of these regions in social perception. In other words, OT receptors are preferentially distributed according to the species' preferred social modality, i.e., olfaction for rodents, auditory for birds, and visual for primates [7]. Data showing improvements in eye contact after OT administration in primates [15–17] or increasing time of olfactory search in rodents [18] support this idea. Of course, OT's effects are broader and not restricted to a main sensory modality. Nevertheless, OT action on primary sensory regions can be seen as a winning strategy to rapidly process social cues, objects, and events. Hence, a first route for OT to influence decision-making is by primarily modulating incoming information via sensory mechanisms.

Emotion Perception

A large number of studies have demonstrated that OT is important for emotion perception. Intranasal administration of OT in humans leads to a better recognition of facial emotional expressions [19] and emotional valence [20]. Neurally this is explained by the large influence of OT on amygdala activity, a limbic region involved in emotional control and in the processing of social fear. fMRI investigations in humans have shown that OT reduces amygdala response to arousing emotional stimuli [21,22]. Research in rodents has provided a detailed description of OT neurotransmission for the control of emotional behavior. When OT is liberated in the centrolateral amygdala, GABAergic neurons become active in the centromedial amygdala where they suppress output neural signals from this region. The consequence of these chemical events on the animal's behavior is a decreasing fear reaction (freezing) to various social and nonsocial stressors [23–25].

OT actions on socioemotional behavior have been embedded in the context of a general framework proposed by Bethlehem and colleagues [26], who suggested that OT reduction of anxiety has the purpose of increasing the saliency of social stimuli. These independent but complementary effects are probably controlled by distinct neural processes. For instance, whereas anxiety and stress can be regulated by fast OT axonal release in the amygdala, the modulation of sensory areas can be achieved via slow OT hormonal-like dendritic diffusion. Interestingly, OT dendritic diffusion can happen while OT release from terminals in the neurohypophysis is temporarily inhibited, suggesting that the two ways of OT liberation are partially independent [27–29]. Further studies should investigate the potential correlation between the time course of OT action in cortical areas and in limbic regions (see Fig. 31.2).

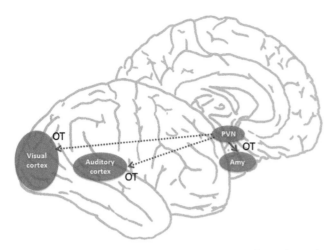

FIGURE 31.2 Oxytocin and social perception. Oxytocin modulates GABA interneurons in sensory cortices and amygdala, thereby modulating perception of social and emotional stimuli. *Red line* indicates direct axonal projection, *dotted red line* indicates unknown mode of transport (axonal way or volume transmission). *Amy*, amygdala; *OT*, oxytocin; *PVN*, paraventricular nucleus of the hypothalamus.

OXYTOCIN AND SOCIAL DECISIONS

In addition to sensory areas, OT receptors are expressed in reward-related regions like the nucleus accumbens [5], the ventral tegmental area, and the substantia nigra [30]. Moreover, they are also present in regions important for decision-making like the medial prefrontal cortex (MPFC) [31] or the anterior cingulate cortex (ACC) [9]. However, the role of OT in these regions at the cellular level remains unclear.

Choices and social decisions can be made at three different levels: for self, for others, and in compliance with social norms [3]. Hence, various experimental approaches have been used to test OT effects in tasks involving (1) trust, (2) empathy, and (3) moral judgments.

Trust Games and Cooperative Behaviors

The finding that OT can bias social decision has attracted wide media coverage. To understand the relation between OT and neuroeconomic contexts, many studies have designed tasks with the purpose of mimicking real-life human economic transactions.

Kosfeld et al.'s study is probably the first to show OT effects on trust behavior [32]. Since then, a large number of reports have been published. Although some have successfully replicated the findings of Kosfeld et al., many have failed to show any effect of OT on trust (for a metaanalysis see Refs. [33,34]). A tentative conclusion from this body of research is that OT has a weak to moderate effect on trust that seems to be context-dependent.

Two fMRI studies using trust games have found that OT has a main action on the amygdala, insula, and

prefrontal cortex, three regions belonging to the emotional/social brain network [35,36]. Enhanced activity in these areas can be linked to increasing expectation of a future (social) reward ("if I trust this person, he will trust me back"), although the exact meaning of OT's action is unknown because the blood oxygen level-dependent signal does not allow to distinguish excitatory from inhibitory activity. In fact, as already shown by animal research, whether OT is inhibiting the amygdala while facilitating activity in other cortical regions is unclear. Additionally, these fMRI studies also found increased activity in the caudate and the putamen, two regions where no anatomical or biological evidence suggest a possible action of OT. Indeed, PET-scan investigations have shown that OT does not seem to modulate activity in the basal ganglia [37,38].

Another way to study OT's effects in the brain is by looking at how it modifies functional connectivity between associative and reward areas. Studies have reported increased connectivity between the amygdala and both the MPFC [39] and the ACC [40] or between the hypothalamus and the dorsolateral prefrontal cortex [41]. Again, the meaning of these results for understanding regional and large-scale OT action in the brain remains limited. To this end, experiments in nonhuman primates are needed to elucidate the excitatory/inhibitory steps orchestrated by OT across large scale brain regions (see Fig. 31.3).

Finally, it must be noted that most OT experiments in humans have been conducted using intranasal administration. This method is currently highly debated because

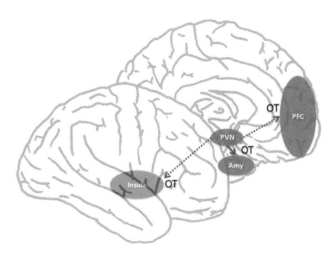

FIGURE 31.3 Oxytocin and social decisions. Oxytocin modulates GABA interneurons in the prefrontal cortex, insula, and amygdala, thereby regulating approach/avoidance balance important for trust, cooperation, and empathy. *Red line* indicates direct axonal projection, *dotted red line* indicates unknown mode of transport (axonal way or volume transmission). *Amy*, amygdala; *OT*, oxytocin; *PFC*, prefrontal cortex; *PVN*, paraventricular nucleus of the hypothalamus.

of various methodological and physiological issues (see Refs. [42,43]) regarding whether OT inhalation reaches the brain and, if such is the case, what are the modes of its release (dendritic, axonal, or both).

Empathy and Prosociality

Few human experiments have explored the problem of prosocial behavior and empathy after OT administration. In this kind of experiments, typically, individuals are prone to give money or to help others without receiving back any compensation or utility. Human studies have shown increasing prosociality toward in-group but not out-group members after OT administration [34,44]. Concerning empathy, again fMRI experiments have mostly focused on empathy for others' pain [48], but no findings are available on the neural correlates of empathic choices in humans after OT intake. OT prosocial effects have been documented also in nonhuman primates [45,46] although these were not observed during neurophysiological recordings. Prosocial paradigms developed in rodents [47] are promising, given the large array of molecular tools like optogenetics and DREADD (designed receptor exclusively activated by designed drug) that are currently available.

Social Norms

Other prosocial choices humans can make are those for the benefit of the society. These include, for instance, giving money to charity, respecting norms and laws, punishing free riders, etc. Of course this field has been investigated mostly in humans. In males, OT was found to increase the amount of money donated to a charity although it did not increase the number of participants who gave money, suggesting that OT enhances donation behavior in individuals already keen to donate [49]. Using the same paradigm in females, another study found that OT's effect was similar to that found in males but limited to participants who experienced low parental love-withdrawal [50]. Finally, in a social dilemma task in which subjects could keep or distribute money to members of their group or to all participants, OT significantly increased decisions to send money to all players [51] thus suggesting, that OT facilitates prosocial choices regardless of group membership. The neural basis of this facilitation remains, however, unknown.

A related issue is how OT acts on cooperative behaviors. For this topic we refer the reader to a 2015 review [34].

In sum, in humans OT has been convincingly associated with various types of social decisions. Nevertheless, the most promising developments will probably come from nonhuman primate experiments using tasks

involving prosocial choices (animals have the option to give or not give a reward to a conspecific independent of any personal utility) associated with invasive neural recording.

OXYTOCIN AND SOCIAL REWARD

As stated earlier, OT receptors are present in the reward system [9], and OT is known to modulate reward in various ways. A key point is that OT action is expected when reward has a salient social dimension.

Social Reward

Several types of social attachment, like maternal, pair-bonding, and consociation (friendship), are processed in the brain [52]. Note that although maternal attachment is present in most mammals, only a minority of species (~5%) display monogamous behavior [53].

In voles, a monogamous species, OT modifies pair bonding via dopamine modulation in the nucleus accumbens (for a review see Ref. [54]). Similar mechanisms seem active in humans as well. OT increases activity in reward areas when subjects see their own partner, whereas the same activity decreases when they are seeing unfamiliar faces (opposite sex). This modulation seems mediated by the dopamine D1/D2 receptors balance [55]. OT and dopamine interaction also modulates sexual behavior (control of penile erection in rats) in nonmonogamous species without provoking pair bonding [56]. Common OT and dopamine mechanisms subserve attachment/parenting behavior in both rats [57,58] and primates [59,60]. The nucleus accumbens, the hypothalamus (paraventricular nucleus), and the MPFC constitute the neural circuit in which both neuromodulators promote basic form of social behavior (see Ref. [61] for review). In the nucleus accumbens OT action is also interfaced with opioid receptor activity [62], another chemical coupling highly relevant for social rewards [63]. The triadic relation formed by OT, dopamine, and opioids certainly deserves further investigation [64,65].

A second type of social reward occurs during interaction with unfamiliar conspecifics (neither the partner nor the offspring). Little work has been conducted so far, but the few available data show that OT modulates this type of interaction in a context-dependent manner. In women, activity in the ventral tegmental area, a brain region synthesizing dopamine, was enhanced by OT during the presentation of cues signaling a friendly face [66]. OT seems also to strengthen functional connectivity between the amygdala and the caudate nucleus during a social learning task [67].

Data also suggest that serotonin, a neurotransmitter important for approach/avoidance behavior, interacts with OT [68]. Results in mice show that OT stimulates the release of serotonin from raphe nuclei projections, which in turn modulates activity in the nucleus accumbens [68] (see Fig. 31.4). This OT/serotonin interaction has been found in humans as well in one study performed by our group with PET-scan imaging. Using a radiotracer specific for the serotonin 1A receptor, we found that intranasal OT significantly modifies the activity of this system in various brain regions important for social behavior [38] (see Fig. 31.4).

OT also reinforces social bonds among individuals of different species. For instance, human/animal interaction can produce release of OT. In dogs, administration of intranasal OT increases affiliative behavior toward the owner [69]. Moreover, dog/human interaction triggers peripheral OT (measured in the urine) [70]. Finally, in lambs OT mediates the stress reaction experienced by the animal after the departure of the human caregiver [71], and in rats, gentle stroking activates OT neurons in the hypothalamus [72].

Nonsocial Reward

A review of how OT modulates reward cannot disregard findings showing OT effects on addiction. Briefly, OT has been found to reduce drug seeking and consumption [73,74], especially when it is directly administered in the nucleus accumbens [75,76]. This raises the question of an unsuspected role of exogenous OT on suppression of addiction through stimulation of the reward pathway.

OT has also been linked to feeding behavior. When centrally released, OT induces satiety feelings via the activation of the hypothalamus and the nucleus of the tractus solitarius [77]. Of interest here, OT-induced satiety may be linked once again to the activation of reward mechanisms [78–80]. Because of these potential therapeutic properties, OT has been also under investigation as a potential treatment for several eating-related disorders such as obesity and anorexia [81–84].

In sum, OT has been shown to signal reward and it is regarded as a potential therapy for various pathologies associated with reward dysfunctions (Fig. 31.5).

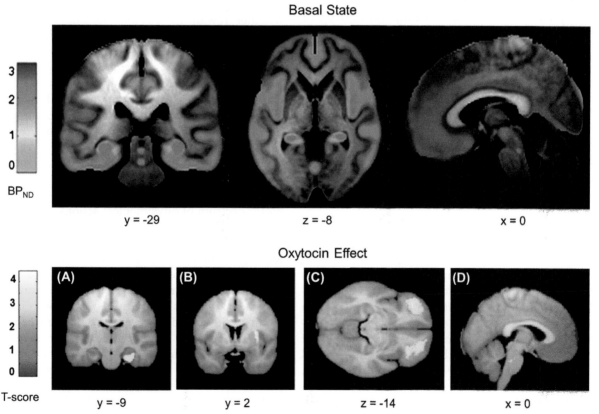

Basal State

BP_{ND}

y = -29   z = -8   x = 0

Oxytocin Effect

T-score

(A) y = -9   (B) y = 2   (C) z = -14   (D) x = 0

FIGURE 31.4 **MPPF binding on serotonin 5-HT1A receptors in the basal state and OT effects.** (Top) Brain mapping of the 2'-methoxyphenyl-(N-2'-pyridinyl)-p-[18F]fluoro-benzamidoethylpiperazine (MPPF) binding potential (BP_{ND}) in the basal state ($n = 24$). 5-HT1A binding is localized in the amygdala, hippocampus and parahippocampus, insula, dorsalis raphe nucleus, orbitofrontal cortex, and anterior cingulate cortex. (Bottom) T-map from Statistical Parametric Mapping analysis ($p < .01$) showing the effect of oxytocin administration on MPPF BP_{ND} in the oxytocin group ($n = 12$) compared to the basal state: (A) Right amygdala/hippocampus/parahippocampus complex, (B) right anterior insula, (C) right and left orbitofrontal cortex, and (D) dorsalis raphe nucleus. No significant effect in the placebo group was found ($n = 12$; not shown). *Reprinted with permission from Mottolese R, Redouté J, Costes N, Le Bars D, Sirigu A, Switching brain serotonin with oxytocin. Proc Natl Acad Sci USA 2014;111:8637–42. http://dx.doi.org/10.1073/pnas.1319810111.*

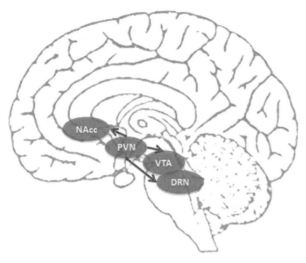

FIGURE 31.5 **Oxytocin and reward.** Oxytocin activates dopaminergic and serotoninergic systems in the reward network. *Red line* indicates direct axonal projections. *DRN*, dorsal raphe nucleus; *NAcc*, nucleus accumbens; *PVN*, paraventricular nucleus of the hypothalamus; *VTA*, ventral tegmental area.

It remains unclear, however, how endogenous OT modulates reward-seeking behavior that is not social (i.e., compulsory search and consumption of drug or food). The chemical interactions with dopamine, serotonin, and opioids are probably a key to understanding how OT controls behaviors in a variety of contexts, for a variety of rewards and their by-products.

A consensus on how dendritic, axon collateral, or parvocellular axonal pathways are associated with social reward processing has not been reached yet [5,7,52]. Answering these questions is important to refine therapeutic strategies, like combining drugs to improve neurotransmitters' action or targeting specific receptor subtypes using partial agonists, etc.

OXYTOCIN, LEARNING, AND MEMORY

Another way for OT to influence decisions is through its action on learning and memory, two processes important for decisions. Since the 1960s, studies have shown a role for OT in memory (see Ref. [85] for a complete review). For instance, this hormone has a negative effect on recall of nonsocial stimuli but it facilitates the social or the emotional ones [86–89]. At the neuronal level, OT can improve the efficiency of hippocampal neurotransmission by increasing the firing rate of fast-spiking interneurons [90]. This in turn lowers the spontaneous activity of hippocampal pyramidal cells and enhances signal-to-noise ratio. This demonstrates that OT modulates hippocampal excitatory/inhibitory balance. It must be noted, however, that OT receptors were not found in the hippocampus of humans and nonhuman

primates. For a deeper discussion of these findings, we refer the reader to a 2013 review of OT effects on learning and memory [91].

PERSPECTIVES

We hope this chapter made it clear that OT influences decision-making processes via two key mechanisms: (1) first by modulating activity within the amygdala and sensory cortices to modify social perception in order to produce socially oriented behaviors (such as pup retrieval in rodents); (2) second by modulating dopaminergic and serotoninergic inputs in the nucleus accumbens to enhance social reward. Of course, OT also has strong effects on the activity of many other regions involved in decision-making processes such as the insula and the frontal cortex, although its action in these areas remains unknown.

To conclude, we would like to suggest future research directions. A central interrogation is how socially rewarding stimuli activate OT release (dendrites, magnocellular collaterals, or parvocellular axons) and whether rewarding but not socially relevant stimuli (drugs, food) borrow the same circuits. Regarding reward, depicting the interactions of OT with dopamine, serotonin, and opioids is critical. Although neurotransmission studies are difficult in humans, PET-scan methods may offer a way to understand how the oxytocinergic system interfaces with other neurotransmitters to produce adaptive response to reward and social contexts [38].

Finally we would like to point on the evolutionary role of OT. Beyond the known activity in parturition and lactation, this hormone has been linked to a very wide spectrum of physiological actions, such as skeletal homeostasis [92], muscle maintenance [93], temperature regulation [94], cardiovascular regulation [95,96], feeding behavior [77], pain reduction [97–99], and even cancer proliferation [100,101]. Such a variety of fundamental actions is phylogenetically ancient and not totally surprising because OT is associated with feeding and reproduction even in *Caenorhabditis elegans* [102,103]. But why does a single peptide have so many functions and what is its evolutionary advantage for social behavior in mammals? Across species studies may tentatively provide some answers to this question.

To summarize, OT directly and indirectly modulates decisions, and preferentially decisions in social contexts. However, OT's action is not necessarily always prosocial. Accounts have challenged the prosocial specificity of this hormone [104], or argue that it is a system whose action has consequences mainly on broad homeostatic/interoceptive body regulation [105].

Last but not least, OT remains a great target for psychiatric disorders such as depression and autism in which social behavior and reward processing are specifically impaired. This is a sufficiently important goal to foster research for understanding the functioning of the oxytocinergic system.

Acknowledgments

We thank CNRS for funding this research. This work was also supported by the LABEX CORTEX (ANR-11-LABX-0042) of the Université de Lyon, within the program "Investissements d'Avenir" (ANR-11-IDEX-0007) operated by the French National Research Agency (ANR).

References

[1] O'Connell LA, Hofmann HA. Evolution of a vertebrate social decision-making network. Science 2012;336:1154—7. http://dx.doi.org/10.1126/science.1218889.

[2] Dalgleish T. The emotional brain. Nat Rev Neurosci 2004;5:583—9. http://dx.doi.org/10.1038/nrn1432.

[3] Ruff CC, Fehr E. The neurobiology of rewards and values in social decision making. Nat Rev Neurosci 2014;15:549—62. http://dx.doi.org/10.1038/nrn3776.

[4] West-Eberhard MJ. Darwin's forgotten idea: the social essence of sexual selection. Neurosci Biobehav Rev 2014;46(Pt 4):501—8. http://dx.doi.org/10.1016/j.neubiorev.2014.06.015.

[5] Ross HE, Freeman SM, Spiegel LL, Ren X, Terwilliger EF, Young LJ. Variation in oxytocin receptor density in the nucleus accumbens has differential effects on affiliative behaviors in monogamous and polygamous voles. J Neurosci 2009;29:1312—8. http://dx.doi.org/10.1523/JNEUROSCI.5039-08.2009.

[6] Veening JG, de Jong T, Barendregt HP. Oxytocin-messages via the cerebrospinal fluid: behavioral effects; a review. Physiol Behav 2010;101:193—210. http://dx.doi.org/10.1016/j.physbeh.2010.05.004.

[7] Grinevich V, Knobloch-Bollmann HS, Eliava M, Busnelli M, Chini B. Assembling the puzzle: pathways of oxytocin signaling in the brain. Biol Psychiatry 2015. http://dx.doi.org/10.1016/j.biopsych.2015.04.013.

[8] Knobloch HS, Grinevich V. Evolution of oxytocin pathways in the brain of vertebrates. Front Behav Neurosci 2014;8:31. http://dx.doi.org/10.3389/fnbeh.2014.00031.

[9] Stoop R. Neuromodulation by oxytocin and vasopressin. Neuron 2012;76:142—59. http://dx.doi.org/10.1016/j.neuron.2012.09.025.

[10] Busnelli M, Saulière A, Manning M, Bouvier M, Galés C, Chini B. Functional selective oxytocin-derived agonists discriminate between individual G protein family subtypes. J Biol Chem 2012;287:3617—29. http://dx.doi.org/10.1074/jbc.M111.277178.

[11] Boccia ML, Petrusz P, Suzuki K, Marson L, Pedersen CA. Immunohistochemical localization of oxytocin receptors in human brain. Neuroscience 2013. http://dx.doi.org/10.1016/j.neuroscience.2013.08.048.

[12] Gimpl G, Fahrenholz F. The oxytocin receptor system: structure, function, and regulation. Physiol Rev 2001;81:629—83.

[13] Zheng J-J, Li S-J, Zhang X-D, Miao W-Y, Zhang D, Yao H, et al. Oxytocin mediates early experience-dependent cross-modal plasticity in the sensory cortices. Nat Neurosci 2014;17:391—9. http://dx.doi.org/10.1038/nn.3634.

[14] Marlin BJ, Mitre M, D'amour JA, Chao MV, Froemke RC. Oxytocin enables maternal behaviour by balancing cortical inhibition. Nature 2015. http://dx.doi.org/10.1038/nature14402.

[15] Ebitz RB, Watson KK, Platt ML. Oxytocin blunts social vigilance in the rhesus macaque. Proc Natl Acad Sci USA 2013;110:11630—5. http://dx.doi.org/10.1073/pnas.1305230110.

[16] Guastella AJ, Mitchell PB, Dadds MR. Oxytocin increases gaze to the eye region of human faces. Biol Psychiatry 2008;63:3—5. http://dx.doi.org/10.1016/j.biopsych.2007.06.026.

[17] Dal Monte O, Noble PL, Costa VD, Averbeck BB. Oxytocin enhances attention to the eye region in rhesus monkeys. Front Neurosci 2014;8:41. http://dx.doi.org/10.3389/fnins.2014.00041.

[18] Witt DM, Winslow JT, Insel TR. Enhanced social interactions in rats following chronic, centrally infused oxytocin. Pharmacol Biochem Behav 1992;43:855—61.

[19] Shahrestani S, Kemp AH, Guastella AJ. The impact of a single administration of intranasal oxytocin on the recognition of basic emotions in humans: a meta-analysis. Neuropsychopharmacology 2013;38:1929—36. http://dx.doi.org/10.1038/npp.2013.86.

[20] Cardoso C, Ellenbogen MA, Linnen A-M. The effect of intranasal oxytocin on perceiving and understanding emotion on the Mayer-Salovey-Caruso emotional intelligence test (MSCEIT). Emotion (Wash DC) 2013. http://dx.doi.org/10.1037/a0034314.

[21] Kirsch P, Esslinger C, Chen Q, Mier D, Lis S, Siddhanti S, et al. Oxytocin modulates neural circuitry for social cognition and fear in humans. J Neurosci 2005;25:11489—93. http://dx.doi.org/10.1523/JNEUROSCI.3984-05.2005.

[22] Wigton R, Radua J, Allen P, Averbeck B, Meyer-Lindenberg A, McGuire P, et al. Neurophysiological effects of acute oxytocin administration: systematic review and meta-analysis of placebo-controlled imaging studies. J Psychiatry Neurosci 2015;40:E1—22.

[23] Knobloch HS, Charlet A, Hoffmann LC, Eliava M, Khrulev S, Cetin AH, et al. Evoked axonal oxytocin release in the central amygdala attenuates fear response. Neuron 2012;73:553—66. http://dx.doi.org/10.1016/j.neuron.2011.11.030.

[24] Viviani D, Charlet A, van den Burg E, Robinet C, Hurni N, Abatis M, et al. Oxytocin selectively gates fear responses through distinct outputs from the central amygdala. Science 2011;333:104—7. http://dx.doi.org/10.1126/science.1201043.

[25] Peters S, Slattery DA, Uschold-Schmidt N, Reber SO, Neumann ID. Dose-dependent effects of chronic central infusion of oxytocin on anxiety, oxytocin receptor binding and stress-related parameters in mice. Psychoneuroendocrinology 2014;42:225—36.

[26] Bethlehem RAI, Baron-Cohen S, van Honk J, Auyeung B, Bos PA. The oxytocin paradox. Front Behav Neurosci 2014;8:48. http://dx.doi.org/10.3389/fnbeh.2014.00048.

[27] Sabatier N, Caquineau C, Dayanithi G, Bull P, Douglas AJ, Guan XMM, et al. α-Melanocyte-stimulating hormone stimulates oxytocin release from the dendrites of hypothalamic neurons while inhibiting oxytocin release from their terminals in the neurohypophysis. J Neurosci 2003;23:10351—8.

[28] Ludwig M, Leng G. Dendritic peptide release and peptide-dependent behaviours. Nat Rev Neurosci 2006;7:126—36. http://dx.doi.org/10.1038/nrn1845.

[29] Wotjak CT, Ganster J, Kohl G, Holsboer F, Landgraf R, Engelmann M. Dissociated central and peripheral release of vasopressin, but not oxytocin, in response to repeated swim stress: new insights into the secretory capacities of peptidergic neurons. Neuroscience 1998;85:1209—22. http://dx.doi.org/10.1016/S0306-4522(97)00683-0.

[30] Vaccari C, Lolait SJ, Ostrowski NL. Comparative distribution of vasopressin V1b and oxytocin receptor messenger ribonucleic acids in brain. Endocrinology 1998;139:5015—33. http://dx.doi.org/10.1210/en.139.12.5015.

[31] Nakajima M, Görlich A, Heintz N. Oxytocin modulates female sociosexual behavior through a specific class of prefrontal

cortical interneurons. Cell 2014;159:295—305. http://dx.doi.org/10.1016/j.cell.2014.09.020.

[32] Kosfeld M, Heinrichs M, Zak PJ, Fischbacher U, Fehr E. Oxytocin increases trust in humans. Nature 2005;435:673—6. http://dx.doi.org/10.1038/nature03701.

[33] Bakermans-Kranenburg MJ, van IJzendoorn MH. Sniffing around oxytocin: review and meta-analyses of trials in healthy and clinical groups with implications for pharmacotherapy. Transl Psychiatry 2013;3:e258. http://dx.doi.org/10.1038/tp.2013.34.

[34] De Dreu CKW, Kret ME. Oxytocin conditions intergroup relations through upregulated in-group empathy, cooperation, conformity, and defense. Biol Psychiatry 2015. http://dx.doi.org/10.1016/j.biopsych.2015.03.020.

[35] Baumgartner T, Heinrichs M, Vonlanthen A, Fischbacher U, Fehr E. Oxytocin shapes the neural circuitry of trust and trust adaptation in humans. Neuron 2008;58:639—50. http://dx.doi.org/10.1016/j.neuron.2008.04.009.

[36] Rilling JK, DeMarco AC, Hackett PD, Thompson R, Ditzen B, Patel R, et al. Effects of intranasal oxytocin and vasopressin on cooperative behavior and associated brain activity in men. Psychoneuroendocrinology 2012;37:447—61. http://dx.doi.org/10.1016/j.psyneuen.2011.07.013.

[37] Striepens N, Matusch A, Kendrick KM, Mihov Y, Elmenhorst D, Becker B, et al. Oxytocin enhances attractiveness of unfamiliar female faces independent of the dopamine reward system. Psychoneuroendocrinology 2014;39:74—87. http://dx.doi.org/10.1016/j.psyneuen.2013.09.026.

[38] Mottolese R, Redouté J, Costes N, Le Bars D, Sirigu A. Switching brain serotonin with oxytocin. Proc Natl Acad Sci USA 2014;111:8637—42. http://dx.doi.org/10.1073/pnas.1319810111.

[39] Sripada CS, Phan KL, Labuschagne I, Welsh R, Nathan PJ, Wood AG. Oxytocin enhances resting-state connectivity between amygdala and medial frontal cortex. Int J Neuropsychopharmacol 2012:1—6. http://dx.doi.org/10.1017/S1461145712000533. FirstView.

[40] Kovács B, Kéri S. Off-label intranasal oxytocin use in adults is associated with increased amygdala-cingulate resting-state connectivity. Eur Psychiatry 2015. http://dx.doi.org/10.1016/j.eurpsy.2015.02.010.

[41] Wang J, Qin W, Liu B, Wang D, Zhang Y, Jiang T, et al. Variant in OXTR gene and functional connectivity of the hypothalamus in normal subjects. NeuroImage 2013;81:199—204. http://dx.doi.org/10.1016/j.neuroimage.2013.05.029.

[42] Leng G, Ludwig M. Intranasal oxytocin: myths and delusions. Biol Psychiatry 2015. http://dx.doi.org/10.1016/j.biopsych.2015.05.003.

[43] Quintana DS, Alvares GA, Hickie IB, Guastella AJ. Do delivery routes of intranasally administered oxytocin account for observed effects on social cognition and behavior? A two-level model. Neurosci Biobehav Rev 2014. http://dx.doi.org/10.1016/j.neubiorev.2014.12.011.

[44] Bartz JA, Zaki J, Bolger N, Ochsner KN. Social effects of oxytocin in humans: context and person matter. Trends Cogn Sci 2011;15:301—9. http://dx.doi.org/10.1016/j.tics.2011.05.002.

[45] Singer T, Snozzi R, Bird G, Petrovic P, Silani G, Heinrichs M, et al. Effects of oxytocin and prosocial behavior on brain responses to direct and vicariously experienced pain. Emotion (Wash DC) 2008;8:781—91. http://dx.doi.org/10.1037/a0014195.

[46] Chang SWC, Barter JW, Ebitz RB, Watson KK, Platt ML. Inhaled oxytocin amplifies both vicarious reinforcement and self reinforcement in rhesus macaques (Macaca mulatta). Proc Natl Acad Sci USA 2012;109:959—64. http://dx.doi.org/10.1073/pnas.1114621109.

[47] Mustoe AC, Cavanaugh J, Harnisch AM, Thompson BE, French JA. Do marmosets care to share? Oxytocin treatment reduces prosocial behavior toward strangers. Horm Behav 2015. http://dx.doi.org/10.1016/j.yhbeh.2015.04.015.

[48] Hernandez-Lallement J, van Wingerden M, Marx C, Srejic M, Kalenscher T. Rats prefer mutual rewards in a prosocial choice task. Front Neurosci 2014;8:443. http://dx.doi.org/10.3389/fnins.2014.00443.

[49] Barraza JA, McCullough ME, Ahmadi S, Zak PJ. Oxytocin infusion increases charitable donations regardless of monetary resources. Horm Behav 2011;60:148—51. http://dx.doi.org/10.1016/j.yhbeh.2011.04.008.

[50] van Ijzendoorn MH, Huffmeijer R, Alink LRA, Bakermans-Kranenburg MJ, Tops M. The impact of oxytocin administration on charitable donating is moderated by experiences of parental love-withdrawal. Front Psychol 2011;2:258. http://dx.doi.org/10.3389/fpsyg.2011.00258.

[51] Israel S, Weisel O, Ebstein RP, Bornstein G. Oxytocin, but not vasopressin, increases both parochial and universal altruism. Psychoneuroendocrinology 2012;37:1341—4. http://dx.doi.org/10.1016/j.psyneuen.2012.02.001.

[52] Dölen G. Oxytocin: parallel processing in the social brain? J Neuroendocrinol 2015. http://dx.doi.org/10.1111/jne.12284.

[53] Numan M, Young LJ. Neural mechanisms of mother-infant bonding and pair bonding: similarities, differences, and broader implications. Horm Behav 2015. http://dx.doi.org/10.1016/j.yhbeh.2015.05.015.

[54] Young LJ, Wang Z. The neurobiology of pair bonding. Nat Neurosci 2004;7:1048—54. http://dx.doi.org/10.1038/nn1327.

[55] Scheele D, Wille A, Kendrick KM, Stoffel-Wagner B, Becker B, Güntürkün O, et al. Oxytocin enhances brain reward system responses in men viewing the face of their female partner. Proc Natl Acad Sci USA 2013;110(50):20308—13. http://dx.doi.org/10.1073/pnas.1314190110.

[56] Melis MR, Argiolas A. Central control of penile erection: a revisitation of the role of oxytocin and its interaction with dopamine and glutamic acid in male rats. Neurosci Biobehav Rev 2011;35:939—55. http://dx.doi.org/10.1016/j.neubiorev.2010.10.014.

[57] D'Cunha TM, King SJ, Fleming AS, Lévy F. Oxytocin receptors in the nucleus accumbens shell are involved in the consolidation of maternal memory in postpartum rats. Horm Behav 2011;59:14—21. http://dx.doi.org/10.1016/j.yhbeh.2010.09.007.

[58] Shahrokh DK, Zhang T-Y, Diorio J, Gratton A, Meaney MJ. Oxytocin-dopamine interactions mediate variations in maternal behavior in the rat. Endocrinology 2010;151:2276—86. http://dx.doi.org/10.1210/en.2009-1271.

[59] Damiano CR, Aloi J, Dunlap K, Burrus CJ, Mosner MG, Kozink RV, et al. Association between the oxytocin receptor (OXTR) gene and mesolimbic responses to rewards. Mol Autism 2014;5:7. http://dx.doi.org/10.1186/2040-2392-5-7.

[60] Strathearn L. Maternal neglect: oxytocin, dopamine and the neurobiology of attachment. J Neuroendocrinol 2011;23:1054—65. http://dx.doi.org/10.1111/j.1365-2826.2011.02228.x.

[61] Love TM. Oxytocin, motivation and the role of dopamine. Pharmacol Biochem Behav 2013. http://dx.doi.org/10.1016/j.pbb.2013.06.011.

[62] Gu X-L, Yu L-C. Involvement of opioid receptors in oxytocin-induced antinociception in the nucleus accumbens of rats. J Pain 2007;8:85—90. http://dx.doi.org/10.1016/j.jpain.2006.07.001.

[63] Resendez SL, Dome M, Gormley G, Franco D, Nevárez N, Hamid AA, et al. μ-Opioid receptors within subregions of the striatum mediate pair bond formation through parallel yet distinct reward mechanisms. J Neurosci 2013;33:9140—9. http://dx.doi.org/10.1523/JNEUROSCI.4123-12.2013.

[64] Brown CH, Russell JA, Leng G. Opioid modulation of magnocellular neurosecretory cell activity. Neurosci Res 2000;36:97—120.

[65] Csiffáry A, Ruttner Z, Tóth Z, Palkovits M. Oxytocin nerve fibers innervate beta-endorphin neurons in the arcuate nucleus of the rat hypothalamus. Neuroendocrinology 1992; 56:429–35.

[66] Groppe SE, Gossen A, Rademacher L, Hahn A, Westphal L, Gründer G, et al. Oxytocin influences processing of socially relevant cues in the ventral tegmental area of the human brain. Biol Psychiatry 2013;74:172–9. http://dx.doi.org/10.1016/j.biopsych.2012.12.023.

[67] Hu J, Qi S, Becker B, Luo L, Gao S, Gong Q, et al. Oxytocin selectively facilitates learning with social feedback and increases activity and functional connectivity in emotional memory and reward processing regions. Hum Brain Mapp 2015;36:2132–46. http://dx.doi.org/10.1002/hbm.22760.

[68] Dölen G, Darvishzadeh A, Huang KW, Malenka RC. Social reward requires coordinated activity of nucleus accumbens oxytocin and serotonin. Nature 2013;501:179–84. http://dx.doi.org/10.1038/nature12518.

[69] Kis A, Hernádi A, Kanizsár O, Gácsi M, Topál J. Oxytocin induces positive expectations about ambivalent stimuli (cognitive bias) in dogs. Horm Behav 2015;69:1–7.

[70] Romero T, Nagasawa M, Mogi K, Hasegawa T, Kikusui T. Oxytocin promotes social bonding in dogs. Proc Natl Acad Sci USA 2014;111: 9085–90. http://dx.doi.org/10.1073/pnas.1322868111.

[71] Coulon M, Nowak R, Andanson S, Ravel C, Marnet PG, Boissy A, et al. Human-lamb bonding: oxytocin, cortisol and behavioural responses of lambs to human contacts and social separation. Psychoneuroendocrinology 2013;38:499–508. http://dx.doi.org/10.1016/j.psyneuen.2012.07.008.

[72] Okabe S, Yoshida M, Takayanagi Y, Onaka T. Activation of hypothalamic oxytocin neurons following tactile stimuli in rats. Neurosci Lett 2015. http://dx.doi.org/10.1016/j.neulet.2015.05.055.

[73] Walker SC, McGlone FP. The social brain: neurobiological basis of affiliative behaviours and psychological well-being. Neuropeptides 2013. http://dx.doi.org/10.1016/j.npep.2013.10.008.

[74] Sarnyai Z, Kovács GL. Oxytocin in learning and addiction: from early discoveries to the present. Pharmacol Biochem Behav 2014; 119:3–9. http://dx.doi.org/10.1016/j.pbb.2013.11.019.

[75] McGregor IS, Bowen MT. Breaking the loop: oxytocin as a potential treatment for drug addiction. Horm Behav 2012;61:331–9. http://dx.doi.org/10.1016/j.yhbeh.2011.12.001.

[76] Baracz SJ, Everett NA, McGregor IS, Cornish JL. Oxytocin in the nucleus accumbens core reduces reinstatement of methamphetamine-seeking behaviour in rats. Addict Biol 2014. http://dx.doi.org/10.1111/adb.12198.

[77] Sabatier N, Leng G, Menzies J. Oxytocin, feeding, and satiety. Front Endocrinol 2013;4:35. http://dx.doi.org/10.3389/fendo.2013.00035.

[78] Klockars A, Levine AS, Olszewski PK. Central oxytocin and food intake: focus on macronutrient-driven reward. Front Endocrinol 2015;6:65. http://dx.doi.org/10.3389/fendo.2015.00065.

[79] Mullis K, Kay K, Williams DL. Oxytocin action in the ventral tegmental area affects sucrose intake. Brain Res 2013;1513: 85–91. http://dx.doi.org/10.1016/j.brainres.2013.03.026.

[80] Ott V, Finlayson G, Lehnert H, Heitmann B, Heinrichs M, Born J, et al. Oxytocin reduces reward-driven food intake in humans. Diabetes 2013. http://dx.doi.org/10.2337/db13-0663.

[81] Cai D, Purkayastha S. A new horizon: oxytocin as a novel therapeutic option for obesity and diabetes. Drug Discov Today Dis Mech 2013;10:e63–8. http://dx.doi.org/10.1016/j.ddmec.2013.05.006.

[82] Kim Y-R, Kim C-H, Cardi V, Eom J-S, Seong Y, Treasure J. Intranasal oxytocin attenuates attentional bias for eating and fat shape stimuli in patients with anorexia nervosa. Psychoneuroendocrinology 2014. http://dx.doi.org/10.1016/j.psyneuen.2014.02.019.

[83] Maguire S, O'Dell A, Touyz L, Russell J. Oxytocin and anorexia nervosa: a review of the emerging literature. Eur Eat Disord Rev 2013. http://dx.doi.org/10.1002/erv.2252.

[84] Odent M. Autism and anorexia nervosa: two facets of the same disease? Med Hypotheses 2010;75:79–81. http://dx.doi.org/10.1016/j.mehy.2010.01.039.

[85] McEwen BB. The roles of vasopressin and oxytocin in memory processing. In: Advances in pharmacology. Academic Press; 2004. p. 1–740. http://www.sciencedirect.com/science/article/pii/S1054358904500182.

[86] Dantzer R, Bluthe RM, Koob GF, Le Moal M. Modulation of social memory in male rats by neurohypophyseal peptides. Psychopharmacology (Berl) 1987;91:363–8.

[87] Gur R, Tendler A, Wagner S. Long-term social recognition memory is mediated by oxytocin-dependent synaptic plasticity in the medial amygdala. Biol Psychiatry 2014. http://dx.doi.org/10.1016/j.biopsych.2014.03.022.

[88] Hurlemann R, Patin A, Onur OA, Cohen MX, Baumgartner T, Metzler S, et al. Oxytocin enhances amygdala-dependent, socially reinforced learning and emotional empathy in humans. J Neurosci 2010;30:4999–5007. http://dx.doi.org/10.1523/JNEUROSCI.5538-09.2010.

[89] Savaskan E, Ehrhardt R, Schulz A, Walter M, Schächinger H. Post-learning intranasal oxytocin modulates human memory for facial identity. Psychoneuroendocrinology 2008;33:368–74. http://dx.doi.org/10.1016/j.psyneuen.2007.12.004.

[90] Owen SF, Tuncdemir SN, Bader PL, Tirko NN, Fishell G, Tsien RW. Oxytocin enhances hippocampal spike transmission by modulating fast-spiking interneurons. Nature 2013;500: 458–62. http://dx.doi.org/10.1038/nature12330.

[91] Chini B, Leonzino M, Braida D, Sala M. Learning about oxytocin: pharmacologic and behavioral issues. Biol Psychiatry 2013. http://dx.doi.org/10.1016/j.biopsych.2013.08.029.

[92] Colaianni G, Tamma R, Di Benedetto A, Yuen T, Sun L, Zaidi M, et al. The oxytocin-bone axis. J Neuroendocrinol 2013. http://dx.doi.org/10.1111/jne.12120.

[93] Elabd C, Cousin W, Upadhyayula P, Chen RY, Chooljian MS, Li J, et al. Oxytocin is an age-specific circulating hormone that is necessary for muscle maintenance and regeneration. Nat Commun 2014;5:4082. http://dx.doi.org/10.1038/ncomms5082.

[94] Kasahara Y, Sato K, Takayanagi Y, Mizukami H, Ozawa K, Hidema S, et al. Oxytocin receptor in the hypothalamus is sufficient to rescue normal thermoregulatory function in male oxytocin receptor knockout mice. Endocrinology 2013. http://dx.doi.org/10.1210/en.2012-2206.

[95] Gutkowska J, Jankowski M. Oxytocin revisited: its role in cardiovascular regulation. J Neuroendocrinol 2012;24:599–608. http://dx.doi.org/10.1111/j.1365-2826.2011.02235.x.

[96] Kemp AH, Quintana DS, Kuhnert R-L, Griffiths K, Hickie IB, Guastella AJ. Oxytocin increases heart rate variability in humans at rest: implications for social approach-related motivation and capacity for social engagement. PLoS One 2012;7:e44014. http://dx.doi.org/10.1371/journal.pone.0044014.

[97] Juif P-E, Poisbeau P. Neurohormonal effects of oxytocin and vasopressin receptor agonists on spinal pain processing in male rats. Pain 2013. http://dx.doi.org/10.1016/j.pain.2013.05.003.

[98] Martínez-Lorenzana G, Espinosa-López L, Carranza M, Aramburo C, Paz-Tres C, Rojas-Piloni G, et al. PVN electrical stimulation prolongs withdrawal latencies and releases oxytocin in cerebrospinal fluid, plasma, and spinal cord tissue in intact and neuropathic rats. Pain 2008;140:265–73. http://dx.doi.org/10.1016/j.pain.2008.08.015.

[99] Rash JA, Aguirre-Camacho A, Campbell TS. Oxytocin and pain: a systematic review and synthesis of findings. Clin J Pain 2013. http://dx.doi.org/10.1097/AJP.0b013e31829f57df.

V. GENETIC AND HORMONAL INFLUENCES ON MOTIVATION AND SOCIAL BEHAVIOR

[100] Imanieh MH, Bagheri F, Alizadeh AM, Ashkani-Esfahani S. Oxytocin has therapeutic effects on cancer, a hypothesis. Eur J Pharmacol 2014;741C:112–23. http://dx.doi.org/10.1016/j.ejphar.2014.07.053.

[101] Péqueux C, Breton C, Hendrick J-C, Hagelstein M-T, Martens H, Winkler R, et al. Oxytocin synthesis and oxytocin receptor expression by cell lines of human small cell carcinoma of the lung stimulate tumor growth through autocrine/paracrine signaling. Cancer Res 2002;62:4623–9.

[102] Beets I, Janssen T, Meelkop E, Temmerman L, Suetens N, Rademakers S, et al. Vasopressin/oxytocin-related signaling regulates gustatory associative learning in *C. elegans*. Science 2012; 338:543–5. http://dx.doi.org/10.1126/science.1226860.

[103] Garrison JL, Macosko EZ, Bernstein S, Pokala N, Albrecht DR, Bargmann CI. Oxytocin/vasopressin-related peptides have an ancient role in reproductive behavior. Science 2012;338:540–3. http://dx.doi.org/10.1126/science.1226201.

[104] Harari-Dahan O, Bernstein A. A general approach — avoidance hypothesis of oxytocin: accounting for social and non-social effects of oxytocin. Neurosci Biobehav Rev 2014;47C:506–19. http://dx.doi.org/10.1016/j.neubiorev.2014.10.007.

[105] Porges SW. The polyvagal theory: neurophysiological foundations of emotions, attachment, communication. Self-Regulation 2011. http://www.ncbi.nlm.nih.gov/pmc/articles/PMC3490536/.

32

Appetite as Motivated Choice: Hormonal and Environmental Influences

A. Dagher, S. Neseliler, J.-E. Han

McGill University, Montreal, QC, Canada

Abstract

The alarming increase in the incidence of obesity over the past half-century has led to a vigorous research endeavor to understand the neurobehavioral factors that determine food choices and food intake. Evidence from personality and neurocognitive research consistently demonstrates that body weight is related to poorer self-control and increased impulsivity. Obesity can thus be understood as an example of consistently maladaptive choices, placing it in the realm of decision neuroscience. On the other hand, feeding is also under the control of homeostatic signals from the periphery that convey both immediate and long-term information on energy balance to the brain. Here we review the literature on the interaction between human brain systems that mediate eating behavior and the gut and adipose peptides that signal the current state of energy balance.

INTRODUCTION

The brain controls energy homeostasis by adapting behavior to guarantee a supply of fuels to maintain tissue function and long-term energy storage [1]. It receives multiple signals from the body informing it of current energy balance. These signals consist of levels of circulating fuels such as glucose and lipids, information on gut contents transferred via the vagus nerve, and peptide hormones, notably insulin, leptin, ghrelin, cholecystokinin, glucagon-like peptide 1 (GLP-1), and peptide YY (PYY), among others, that relay information about short- and long-term energy balance. Although these signals reach the brain via different routes, they are integrated in the hypothalamus. The hypothalamus in turn conveys this information to brain centers implicated in learning, motivation, and decision-making. Woods has argued that the immediate aspects of feeding (i.e., quantity and content of meal, initiation, and termination) may be influenced by signals that reflect energy intake such as circulating glucose, but that there are advantages to regulating feeding by using so-called distal cues such as the sight, taste, and smell of food. These cues have acquired predictive value through past experience. He argues that in a stable food environment distal cues are good predictors of the nutrient content of foods and can be used to reliably guide eating choices, allowing for stable eating patterns [1,2]. In an unstable environment the relationship between distal cues and nutrient content is lessened and signals such as glucose levels are slow to reflect the energy content of ingested foods. Therefore "having to rely upon more proximal cues comes at a cost of risking consuming too many calories at once or else adopting a pattern of eating smaller and more frequent meals, thus perhaps interfering with other behaviors" [1]. The ability to guide eating behavior based on conditioned distal cues allows the planning of regular meals, freeing up the individual for other important activities. It also places eating behavior within the domain of decision-making.

The decision to eat is often dichotomized into homeostatic versus hedonic. According to this framework, there are two types of eating: one that occurs in response to energy balance signals from the periphery, and another that occurs in response to appetitive cues, mostly from the environment. The flaws in this view are apparent when one considers that feeding requires animals or humans to expend effort, take risks, or spend money to eat. Eating requires motivation. And indeed, research shows that homeostatic signals such as ghrelin, leptin, and insulin, act on incentive systems in the brain. Conversely, food cues, which increase appetite and potentiate feeding, regulate the secretion of these peripheral peptides. The homeostatic/hedonic dichotomy

FIGURE 32.1 (A) The homeostatic/hedonic dichotomy views eating as occurring under the control of two different systems, with different inputs. (B) A model based on the allostatic model of Sterling and Eyer [3] that views feeding behavior from the standpoint of decision neuroscience. The weighing of needs and opportunities relates feeding to behavioral economics. *GLP-1*, glucagon-like peptide 1; *PYY*, peptide YY.

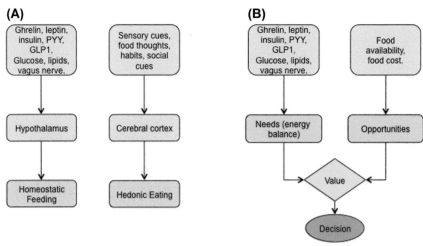

APPETITIVE BRAIN SYSTEMS PROMOTE FOOD INTAKE

also fails to explain the behavior of a shopper at the supermarket, who is selecting foods that will not be consumed for many days. Moreover, human eating behavior is influenced by social factors, habits, environmental cues, stress, physical activity, and the cost of food. The allostatic model of Sterling and Eyer [3] may better account for the regulation of eating behavior (Fig. 32.1).

Nonetheless, the brain regulates energy intake in response to peripheral signals of energy balance. Energy stores (body adiposity) are under regulatory control. For example, voluntary weight loss from dieting leads to compensatory increases in appetite and food intake. There are two types of peripheral hormones that regulate feeding: satiation signals such as cholecystokinin and GLP-1, which relay information about gut nutrient contents, and longer-term energy balance signals such as insulin, leptin, and ghrelin. All of these peptides cross the blood—brain barrier and act on numerous targets within the central nervous system to affect food intake. Here we review the role of decision-making in feeding behavior and obesity and the interplay between hormonal signals and self-regulation and decision-making.

Food intake can be understood as an interplay between homeostatic signals and environmental signals of food availability [4]. In this context, the brain networks that regulate eating behavior can be divided into three interrelated brain systems (Fig. 32.2): (1) the hypothalamus, which maintains energy homeostasis [5]; (2) the appetitive network, that is, the limbic and cortical systems involved in the coding of food reward value and in engaging appropriate motor responses

[6]; and (3) the ventromedial and dorsolateral prefrontal cortical systems implicated in goal-oriented and self-regulated behavior [4].

The main portal of entry of energy balance information into the central nervous system (CNS) is the hypothalamus. The metabolic state of the body is communicated to the rest of the brain via the neurons in the arcuate nucleus of the hypothalamus [5]. These neurons integrate the input from signals of peripheral energy stores (e.g., leptin and insulin), the gut (e.g., ghrelin), and short-term meal-related signals (e.g., macronutrients and gut-derived satiety signals). This integrated signal from the hypothalamus modulates the activity of the higher-order networks that mediate incentive salience and food choice. Functional connectivity analysis of the hypothalamus at rest has shown that it is indeed connected with the appetitive and goal-directed systems [7].

An appetitive network is formed by four interconnected regions: the hippocampus and the amygdala, the insula, the striatum, and the orbitofrontal cortex (OFC) [6]. These regions receive dopaminergic input from the midbrain and are implicated in emotion and reward processing. Increased fMRI activation in structures in this network while viewing food cues has been correlated with preference for highly palatable foods, increased caloric intake [8,9], and weight gain over one year [10,11]. We will review the role of each of these structures in detail next.

Hippocampus and Amygdala

The role of the hippocampus (HC) in food intake is often linked to learning and memory. The blood oxygen level-dependent (BOLD) signal in the HC measured with fMRI is increased selectively by food pictures compared to nonfood pictures and modulated by

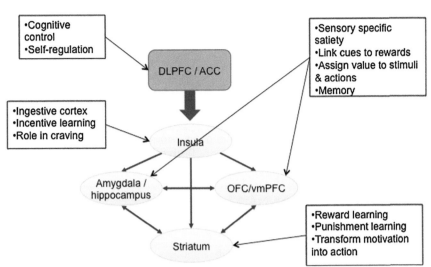

FIGURE 32.2 The appetitive network (see text for details). Interactions among these brain regions are important for learning about foods (e.g., their nutrient content), regulating the incentive salience of food cues, and setting the motivation to eat. These regions are all under the influence of energy balance signals from the body. *ACC*, anterior cingulate cortex; *DLPFC*, dorsolateral prefrontal cortex; *OFC*, orbitofrontal cortex; *VMPFC*, ventromedial prefrontal cortex.

hunger, subjective craving, and metabolic signals [12–14]. Obese individuals compared to lean controls show increased activation in the HC in response to both visual and olfactory food cues [15,16] and reduced cerebral blood flow in the HC in response to liquid taste after a satiating amount [17]. These results suggest that the role of the HC in food intake might exceed learning and memory and include modulation of food cue reactivity by internal states (energy balance).

The amygdala is interconnected with the forebrain structures implicated in reward as well as the hypothalamic centers important for homeostatic regulation. It plays an important role in emotional learning and assigning incentive value to sensory stimuli such as food cues [18–21]. In humans, metaanalyses reveal amygdala responses consistently to both gustatory and visual food cues [22,23]. An intact amygdala is required for palatability-related coding in other parts of the brain such as the taste cortex [24].

The activity in the amygdala appears to reflect the incentive value of foods. In human fMRI studies, its response to food cues is modulated by factors that also affect motivational salience such outcome devaluation, hunger, and stress [12,25–28], as well as nutritional labels [29]. Relative insensitivity of the amygdala to satiation, as reflected by activation in response to milk shake in the absence of hunger, is predictive of weight gain [30].

Insula

The midinsula and the frontal operculum make up the taste cortex (TC), which encodes the features of pure taste stimuli such as quality and intensity [31,32]. In addition to gustatory stimuli, TC activity is modulated by other aspects of food intake, such as somatosensory (e.g., oral texture) and olfactory properties of food [33]. Information on the various sensory properties of foods (gustatory, olfactory, visual, somatosensory) converges in the TC to form the flavor percept that guides food selection [34]. A metaanalysis identified the left insula as the only region that is activated to food cues in all sensory modalities [35].

In addition to being a multimodal sensory area, the TC plays an important role in flavor–nutrient learning: associating each flavor with the postingestive effects of the food, for example, caloric content. Activity in the TC at rest correlates with the changes in plasma concentrations of several gut hormones [28]. In response to food images, BOLD signal in the TC is modulated by the metabolic state as measured by peripheral glucose levels [36]. Animals without an insular cortex cannot assign proper incentive value to ingested foods based on their caloric content [37].

The TC also appears to be sensitive to a number of functions that reflect value coding, including variations in perceived pleasantness [30,38], internal state [39,40], and expectancy and beliefs [41]. Indeed, craving for favorite foods activates the TC [14] and its activity is modulated by higher-order cognitive functions such as attention. In one fMRI study, subjects who tried to detect a taste in a tasteless solution showed increased insular activation compared to passively sampling it [42]. Rodent experiments further confirm the role of the insula in higher-order processing of food cues. For example, the expectancy of sucrose solution produced sucrose-like activation in the insular cortex even when the sucrose reward was omitted [43] and silencing of insular neurons inhibits food-approach behavior [44]. In sum, the insula is more than a TC; it is central to reward and expectation processing of food cues that drive feeding behavior.

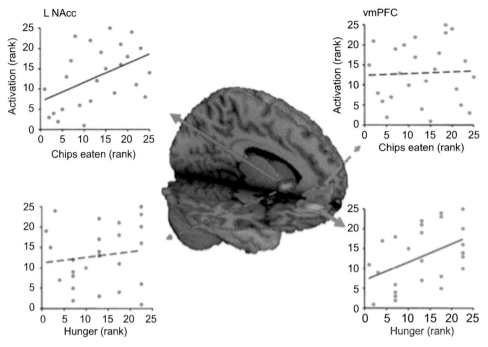

FIGURE 32.3   Nucleus accumbens response to food cues predicts snack consumption. Correlation between potato chips eaten after the scanning session (ad lib) and blood oxygen level-dependent response in left nucleus accumbens (*L NAcc*) and ventromedial prefrontal cortex (*vmPFC*). Bold lines/arrows indicate significant correlations. *Reprinted with permission from Lawrence NS, Hinton EC, Parkinson JA, Lawrence AD. Nucleus accumbens response to food cues predicts subsequent snack consumption in women and increased body mass index in those with reduced self-control. NeuroImage 2012;63:415—22. http://dx.doi.org/10.1016/j.neuroimage.2012.06.070.*

Orbitofrontal Cortex

As for the insula, inputs from all sensory modalities converge in the OFC and probably contribute to the flavor percept of the food [45]. OFC activity correlates with pleasantness ratings of a liquid taste [46] and tracks the current reward value of food cues [25]. The OFC response to food cues is modulated by internal states such as hunger [26]. It is also higher for high-calorie food items [16], and BOLD signal in the medial OFC predicts subsequent ad lib choosing of foods with high fat content [8]. This suggests that the OFC might play a key role in food choice in favor of the food cues that have "gained" incentive salience [47]. Consistent with this idea, obesity is marked by an inability to downregulate the OFC response to high-calorie food cues despite satiation [48]. OFC activation to food cues also predicts weight gain in a prospective manner [49].

Striatum

The striatum is interconnected with the amygdala, HC, OFC, ventromedial prefrontal cortex (VMPFC), insula, and dorsolateral prefrontal cortex (DLPFC). This position is suited to integrating and evaluating food stimuli and to transforming the food-related information into an action or motivated behavior, as well as learning from the consequences of these actions to reshape future behavior [50]. The striatum is consistently activated by food cues [23,26]. fMRI activation of the striatum in response to food cues correlates with subjective craving [14] and is higher for subjectively tasty foods [51], in the obese, and in the fasted state [52]. Additionally, higher striatal activity to food cues predicts chocolate consumption ad lib [53], weight gain [10,54], and less successful weight-loss maintenance [55]. Moreover, obesity has been associated with increased striatal reactivity to food cues (especially highly palatable foods) [16,17,56,57].

Striatal activity in response to food cues reflects their incentive salience. This response may interact with poor self-control to ultimately lead to increased food consumption and weight gain. Indeed, striatal cue reactivity correlates with increased snack consumption through an interaction with reduced self-control [11] (see Fig. 32.3), and cognitive self-control may manifest as reduced striatal activation mediated by the lateral PFC [58].

Prefrontal Regions

Food intake depends on more than appetitive drive. Cultural and social factors, habits (e.g., time of day), and a desire to be healthy also contribute to feeding behavior. A neuroeconomics approach sees food choice as based on computing both the basic attributes of food cues (i.e., palatability) and an individual's long-term goals (i.e., health, diet). Brain systems would therefore compute what, when, and how much we eat based on our cognitive goals, attributes of the foods, and current physiological state (e.g., hunger and energy balance) [4,59].

One way to measure the current value of foods is to calculate the willingness to pay (WTP) using an auction paradigm [60]. Several lines of research suggest that the current value of food cues is represented in the VMPFC and in the adjacent OFC. In fMRI metaanalysis studies, the VMPFC is the area that most consistently tracks the subjective value of various types of stimuli, including foods [61,62].

The VMPFC has connections with all of the appetitive regions, as well as with the hypothalamus and the prefrontal regions associated with self-control. This allows the value signal in the VMPFC to be computed based on several attributes of food stimuli and the current state. For example, the fMRI signal in the VMPFC during exposure to food cues is modulated by the caloric density of the items [63] and by hunger [64]. The VMPFC also exhibits functional connectivity with the appetitive network structures at the time of decision, especially the striatum [63,64]. Interestingly, increased connectivity between the striatum and the VMPFC has been observed during resting-state fMRI in obese individuals versus lean controls [65]. Note, however, that activity in numerous brain regions in addition to the VMPFC appears to track the subjective value of foods, as determined by WTP (Fig. 32.4).

The VMPFC value signal can also be modulated by the lateral PFC, and this interaction may be a neural correlate of self-control [66]. In nondieting individuals, cognitive strategies that require a participant to consider the healthiness of or regulate his or her craving for food items result in increased activity in the DLPFC and modulate the activity of the VMPFC [51,58,59].

Increased reactivity in the DLPFC in response to highly palatable foods predicts reduced subsequent chocolate intake [53] and better weight loss due to dieting [55] or bariatric surgery [67].

SELF-CONTROL AND LATERAL PREFRONTAL CORTEX: ROLE IN APPETITE REGULATION AND OBESITY

Paradoxically, obesity is rapidly becoming a leading cause of preventable death worldwide [68] despite the enormous amount of information about food and health available to us [69]. Individuals with higher body mass index (BMI) tend to make maladaptive decisions [70] and a comprehensive review identified a consistent correlation between BMI and personality and cognitive tests of executive function [71].

Of the myriad of food-related decisions that we make everyday, several require weighing the value of immediate, tempting rewards against potential long-term consequences that may conflict with goals such as losing weight and leading a healthy lifestyle [72]. In such conflicting situations, self-control describes the mental processes that allow us to select an action that is in line with our long-term goals [73]. Dietary self-control has been identified as a factor associated with eating behavior and obesity. For example, individuals with weak or depleted self-control appear to endorse a greater desire for food, consume more unhealthy foods, be more likely to fail at dieting, and gain more weight [74–81]. On the other hand, successful self-controllers tend to eat less, are more likely to succeed in making dietary changes, and lose more weight [82–84].

The lateral PFC is most frequently implicated in dietary self-control. Damage to the frontal cortex has been linked to binge eating and strong cravings for foods high in refined sugar [85,86]. Volitionally suppressing desire for

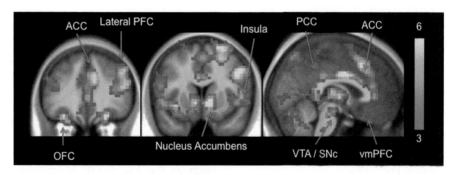

FIGURE 32.4 Brain regions that track subjective value. Blood oxygen level-dependent signal correlating with willingness to pay for food items. *ACC*, anterior cingulate cortex; *OFC*, orbitofrontal cortex; *PCC*, posterior cingulate cortex; *PFC*, prefrontal cortex; *SNc*, substantia nigra pars compacta; *vmPFC*, ventromedial prefrontal cortex; *VTA*, ventral tegmental area. *Reprinted with permission from Tang DW, Fellows LK, Dagher A. Behavioral and neural valuation of foods is driven by implicit knowledge of caloric content. Psychol Sci 2014;25:2168–76. http://dx.doi.org/10.1177/0956797614552081.*

tasty or craved foods using different reappraisal strategies induces increased activity in the lateral PFC, including the inferior frontal gyrus (IFG) and DLPFC [51,87]. Inhibiting prepotent responses to appetizing foods is associated with increased activity in the DLPFC, IFG, and superior frontal gyrus [88,89]. In support of these findings, a 2014 study with a more ecologically valid paradigm showed that people with greater IFG activity related to response inhibition tend to experience food desire to a lesser extent, be more capable of resisting food temptations, and be able to consume less food in their daily life [90]. Furthermore, greater left DLPFC activation in response to food pictures was detected in individuals who placed higher value on diet importance [91] and in those who exhibited successful weight loss [67,92]; it also predicted less subsequent food intake [53]. On the other hand, obese individuals compared to the nonobese and the previously obese show reduced activation in the left DLPFC postprandially, which is negatively correlated with their percentage body fat [93]. Researchers have also used food-related decision-making paradigms to probe dietary self-control. Across three separate studies, Rangel et al. showed that self-controllers avoid selecting unhealthy-but-tasty foods while recruiting the left DLPFC to downmodulate value-related activity in the VMPFC [59,66]. Using electroencephalography, they further demonstrated the involvement of the DLPFC in attentional filtering [94]. In support of these findings, Weygandt and colleagues showed that stronger functional connectivity between the VMPFC and the DLPFC predicted dietary success [95]. Another study observed that connectivity between these regions could be reduced by cognitive resource depletion affecting self-regulatory capacity [96].

A personality trait strongly linked to eating behaviors and body weight is conscientiousness [71,97,98], as measured by the Five-Factor Model of personality. It assesses the degree to which an individual is organized, strong-willed, self-disciplined, and persistent in the pursuit of long-term goals [99]. Conscientiousness robustly overlaps with self-control at both the behavioral and the neural levels. In addition to being highly correlated with various questionnaire measures of self-control [100], conscientiousness is positively associated with gray matter density in the lateral PFC [101]. Individuals who score high on conscientiousness tend to consume more vegetables and fruits and exhibit less emotional, external, and binge eating behaviors and greater restraint in eating [98,102—104]. Furthermore, conscientiousness may serve as a protective factor against stress-induced overeating [105,106].

The obesogenic environment, replete with cheap, highly palatable, calorically dense foods and their cues (e.g., advertisements), is thought to be an important contributor to the increased incidence of obesity [107—109].

Food cues act as conditioned stimuli to trigger hunger and food craving and intake [110]. When confronted with appealing food odors or images, overeaters compared to normal eaters show greater and more sustained cephalic phase responding (e.g., salivation and insulin release), reflecting greater appetite, followed by increased food consumption and, eventually, weight gain [2,111—113]. There is considerable neuroimaging evidence that brain areas involved in self-control can modulate cue-induced food intake and weight gain. In normal-weight individuals, passive viewing of food images is associated with increased fMRI activity in prefrontal brain regions including the left IFG [26], and abnormalities in the lateral PFC have been reported in obese individuals (see Refs. [114,115] for reviews). Also, greater left DLPFC activation in response to food images predicts weight-loss success [67,92]. Intrinsic differences in the neural systems that control appetite may predispose individuals to obesity, forming a set of endophenotypes (Fig. 32.5).

INTERACTION BETWEEN ENERGY BALANCE SIGNALS AND DECISION-MAKING

Hormonal signals of satiety or energy balance can mediate hunger/satiety and affect food intake. They act on appetitive circuitry directly and indirectly via the hypothalamus and ascending dopaminergic projections [116].

Insulin

Insulin is secreted from the islet cells of the pancreas in response to anticipated and ingested calories to maintain energy homeostasis [117]. Insulin levels after a meal signal satiety and assist in meal termination by reducing the rewarding effects of food [117—119]. These effects are partly mediated through the depression of dopamine neuron activity [120]. Fasting plasma insulin levels correlate with increased BOLD response to high-calorie food images in the HC and TC in lean and obese individuals [121]. In addition, fasting insulin levels correlate positively, and insulin sensitivity index negatively, with the strength of functional connectivity of the striatum and the OFC at rest [118].

Parenteral insulin administration leads to attenuated BOLD signal in the visual cortex, hypothalamus, and PFC at rest [122] as well as in the appetitive brain centers such as the HC [123]. Manipulation of insulin levels either by feeding or by intranasal insulin administration also affects the hypothalamic and appetitive network fMRI responsivity to food pictures [122]. For example, the BOLD responses to food cues in the visual cortex,

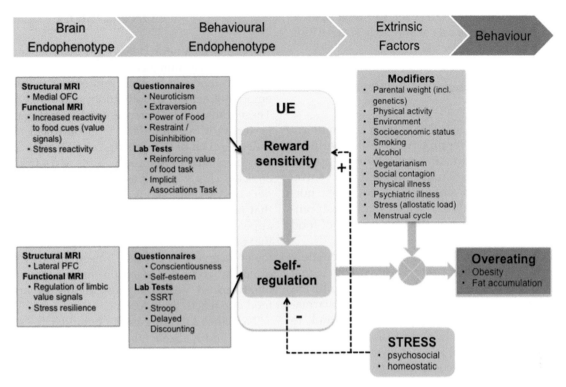

FIGURE 32.5 An endophenotype approach to obesity vulnerability. Uncontrolled eating (*UE*) results from an interaction of heightened reward sensitivity and poor self-regulation. These can be assessed using neurocognitive tests, personality questionnaires, and brain imaging [71,153]. *OFC*, orbitofrontal cortex; *PFC*, prefrontal cortex, *SSRT*, stop signal reaction time task.

insula, striatum, and OFC all decrease as a function of insulin levels [28,119,123,124].

Leptin

Leptin, secreted by adipose tissue, signals the size of fat energy stores. It can reduce food intake both by lessening the hedonic and incentive value of food cues and by increasing cognitive control and inhibition [125–128]. Like insulin, leptin appears to exert these effects partly by reducing dopamine neuron activity [129]. fMRI studies in patients with a rare genetic defect that leads to leptin deficiency find that leptin treatment reduces the ventral striatal activation to food cues [125]. Three patients with congenital leptin deficiency upon discontinuing their leptin treatment showed increased activation to high-calorie food pictures in the insula and decreased activation in the PFC (including VMPFC). Leptin administration reduced BOLD response to food pictures in patients with lipodystrophy in the amygdala, HC, and striatum [130]. In obese subjects, both BMI (positively) and leptin levels (negatively) correlate with ventral striatal BOLD response to food cues [126]. In another study leptin was administered to obese participants ($n = 6$) undergoing fMRI after they lost 10% of their body weight. Leptin reversed the weight-loss-induced changes in neural reactivity to food cues [131].

Ghrelin

In contrast to leptin, ghrelin, a hunger hormone secreted by the stomach, increases food intake in lean and obese subjects [132]. Its actions are mediated by activation of agouti-related peptide neurons in the hypothalamus, as well as dopamine neurons directly [133]. Its levels increase before meal initiation, and drop upon eating [134], but it also signals long-term energy balance [135]. Postprandial ghrelin levels correlate with activation in the OFC, amygdala, and insula at rest [28]. Fasting ghrelin levels correlate with BOLD response to food pictures in areas linked to visual processing, and in the appetitive centers (i.e., amygdala, striatum, TC) as well the midbrain and the hypothalamus to food pictures [136]. Ghrelin administration to lean subjects at pharmacological levels increases BOLD responses to food cues in all of the appetitive regions, along with the dopaminergic midbrain and visual areas such as the pulvinar and the fusiform gyrus [12]. In this study, ghrelin also increased self-reported hunger proportionally to the increased BOLD activation in the amygdala, the OFC, and the pulvinar. Ghrelin administration to recently fed individuals increased the hedonic and OFC BOLD responses to food cues back to fasted levels [137]. It is also argued based on the animal literature that stress-induced increases in ghrelin levels may mediate stress eating [138]. However, in humans, as of this writing, studies on the role of ghrelin in stress eating have been inconsistent [139–142].

Satiety Hormones: Glucagon-Like Peptide 1 and Peptide YY

The results cited suggest that signals of long-term energy balance, insulin, leptin, and ghrelin, can influence food intake by modulating the activity of appetitive centers (i.e., mesolimbic dopamine, ventral striatum, OFC, HC, and amygdala) [143]. Other hormones such as PYY and GLP-1, which signal postprandial satiety [1,116], also act on the CNS control of food intake.

GLP-1 is secreted from the jejunum both immediately prior to and during a meal and acts as a satiety signal that reduces food intake [144]. GLP-1 agonist infusion reduces hunger ratings in humans. In addition, pharmacological GLP-1 agonists are consistently found to reduce ad libitum food intake and result in weight loss [145–147]. At physiological levels, however, the effects of GLP-1 infusions are less consistent or not statistically significant [148].

Similarly, human brain-imaging data suggest that GLP-1 actions on the brain mediate satiety in humans. Postprandial GLP-1 levels negatively correlate with BOLD activity in the OFC and TC at rest [28] and GLP-1 agonist administration attenuated the BOLD response to food pictures in the entire appetitive network, especially in the insula [149].

The brain response to GLP-1 varies depending on the current metabolic status of the individual. Studies compared the effects of GLP-1 agonist administration in lean, obese, and obese with type 2 diabetes individuals. Anticipation of a palatable food (chocolate milk shake) led to activation in the appetitive network, including striatum and insula, proportional to BMI. This activation was reversed by the GLP-1 agonist exenatide, along with activation in the OFC and amygdala [150]. Similar results were reported in a paradigm using food pictures [151].

Combination of GLP-1 with other satiety hormones might be a more realistic representation of the postmeal scenario, as GLP-1 is usually coreleased with PYY in response to food intake. Indeed, similar effects on fMRI activation have been demonstrated, with a general blunting of activity in appetitive areas following PYY infusion [149,152]. Ad lib food intake following either PYY or GLP-1 injection alone does not typically decrease. However, coadministration of PYY and GLP-1 significantly reduces food intake coupled with a synergistic effect on reducing the BOLD signal in all the appetitive regions (especially the insula) [149].

CONCLUSION

Obesity as a neurobehavioral disorder has been tied to various psychological traits such as increased attention to food, increased reinforcing value of food, emotional eating, disinhibition, and binge eating [71]. These traits appear to represent different manifestations (severity) of a single factor, uncontrolled eating [153]. Indeed, much of the variation in BMI in the population may be related to reduced voluntary control over eating choices. This places obesity within the realm of disorders of decision-making. A second piece of evidence is that much of the recent rise in the incidence of obesity appears to be attributable to the lower cost of food and that obesity due to unhealthy energy-dense diets targets lower income groups [154]. Indeed, brain systems implicated in economic decision-making are consistently identified in neuroimaging studies of appetite and food cue reactivity [4,6,66].

Nonetheless, eating has to be under the strong influence of bodily signals that signal both short- and long-term measures of energy storage and balance. Here we have shown that the brain systems implicated in motivation and decision-making are consistently found to be under the influence of blood-borne peptides secreted by the gut and adipose tissues. This demonstrates that homeostatic signals and cognitive and motivational processes are completely intertwined in matching feeding behavior to bodily needs and environmental opportunities.

References

[1] Woods SC. The control of food intake: behavioral versus molecular perspectives. Cell Metab 2009;9:489–98.

[2] Woods SC. The eating paradox: how we tolerate food. Psychol Rev 1991;98:488–505.

[3] Sterling P. Allostasis: a model of predictive regulation. Psychol Rev 2012;106:5–15. http://dx.doi.org/10.1016/j.physbeh.2011.06.004.

[4] Rangel A. Regulation of dietary choice by the decision-making circuitry. Nat Neurosci 2013;16:1717–24. http://dx.doi.org/10.1038/nn.3561.

[5] Morton GJ, Cummings DE, Baskin DG, Barsh GS, Schwartz MW. Central nervous system control of food intake and body weight. Nature 2006;443:289–95. http://dx.doi.org/10.1038/nature05026.

[6] Dagher A. Functional brain imaging of appetite. Trends Endocrinol Metab 2012;23:250–60. http://dx.doi.org/10.1016/j.tem.2012.02.009.

[7] Kullmann S, Heni M, Linder K, Zipfel S, Häring H-U, Veit R, et al. Resting-state functional connectivity of the human hypothalamus. Hum Brain Mapp 2014;35:6088–96. http://dx.doi.org/10.1002/hbm.22607.

[8] Mehta S, Melhorn SJ, Smeraglio A, Tyagi V, Grabowski T, Schwartz MW, et al. Regional brain response to visual food cues is a marker of satiety that predicts food choice. Am J Clin Nutr 2012;96:989–99. http://dx.doi.org/10.3945/ajcn.112.042341.

[9] Tryon MS, Carter CS, Decant R, Laugero KD. Chronic stress exposure may affect the brain's response to high calorie food cues and

predispose to obesogenic eating habits. Physiol Behav 2013;120: 233–42. http://dx.doi.org/10.1016/j.physbeh.2013.08.010.

[10] Demos KE, Heatherton TF, Kelley WM. Individual differences in nucleus accumbens activity to food and sexual images predict weight gain and sexual behavior. J Neurosci 2012;32:5549–52. http://dx.doi.org/10.1523/JNEUROSCI.5958-11.2012.Individual.

[11] Lawrence NS, Hinton EC, Parkinson JA, Lawrence AD. Nucleus accumbens response to food cues predicts subsequent snack consumption in women and increased body mass index in those with reduced self-control. NeuroImage 2012;63:415–22. http://dx.doi.org/10.1016/j.neuroimage.2012.06.070.

[12] Malik S, McGlone F, Bedrossian D, Dagher A. Ghrelin modulates brain activity in areas that control appetitive behavior. Cell Metab 2008;7:400–9. http://dx.doi.org/10.1016/j.cmet.2008.03.007.

[13] Davidson TL, Kanoski SE, Schier LA, Clegg DJ, Benoit SC. A potential role for the hippocampus in energy intake and body weight regulation. Curr Opin Pharmacol 2007;7:613–6. http://dx.doi.org/10.1016/j.coph.2007.10.008.

[14] Pelchat ML, Johnson A, Chan R, Valdez J, Ragland JD. Images of desire: food-craving activation during fMRI. NeuroImage 2004; 23:1486–93. http://dx.doi.org/10.1016/j.neuroimage.2004.08.023.

[15] Bragulat V, Dzemidzic M, Bruno C, Cox CA, Talavage T, Considine RV, et al. Food-related odor probes of brain reward circuits during hunger: a pilot FMRI study. Obesity 2010;18: 1566–71. http://dx.doi.org/10.1038/oby.2010.57.

[16] Stoeckel LE, Weller RE, Cook EW, Twieg DB, Knowlton RC, Cox JE. Widespread reward-system activation in obese women in response to pictures of high-calorie foods. NeuroImage 2008; 41:636–47. http://dx.doi.org/10.1016/j.neuroimage.2008.02.031.

[17] DelParigi A, Chen K, Salbe AD, Hill JO, Wing RR, Reiman EM, et al. Persistence of abnormal neural responses to a meal in post-obese individuals. Int J Obes 2004;28:370–7. http://dx.doi.org/10.1038/sj.ijo.0802558.

[18] Zhang Q, Li H, Guo F. Amygdala, an important regulator for food intake. Front Biol 2011;6:82–5. http://dx.doi.org/10.1007/s11515-011-0950-z.

[19] Mahler SV, Berridge KC. What and when to "want"? Amygdala-based focusing of incentive salience upon sugar and sex. Psychopharmacol (Berl) 2011;221:407–26. http://dx.doi.org/10.1007/s00213-011-2588-6.

[20] Yasoshima Y, Yoshizawa H, Shimura T, Miyamoto T. The basolateral nucleus of the amygdala mediates caloric sugar preference over a non-caloric sweetener in mice. Neuroscience 2015;291: 203–15. http://dx.doi.org/10.1016/j.neuroscience.2015.02.009.

[21] Baxter MG, Murray EA. The amygdala and reward. Nat Rev Neurosci 2002;3:563–73. http://dx.doi.org/10.1038/nrn875.

[22] Jastreboff AM, Sinha R, Lacadie C, Small DM, Sherwin RS, Potenza MN. Neural correlates of stress- and food cue-induced food craving in obesity: association with insulin levels. Diabetes Care 2013;36:394–402. http://dx.doi.org/10.2337/dc12-1112.

[23] Tang DW, Fellows LK, Small DM, Dagher A. Food and drug cues activate similar brain regions: a meta-analysis of functional MRI studies. Physiol Behav 2012;106:317–24. http://dx.doi.org/10.1016/j.physbeh.2012.03.009.

[24] Piette CE, Baez-Santiago MA, Reid EE, Katz DB, Moran A. Inactivation of basolateral amygdala specifically eliminates palatability-related information in cortical sensory responses. J Neurosci 2012;32:9981–91. http://dx.doi.org/10.1523/JNEUR-OSCI.0669-12.2012.

[25] Gottfried JA, O'Doherty J, Dolan RJ. Encoding predictive reward value in human amygdala and orbitofrontal cortex. Science 2003; 301:1104–7. http://dx.doi.org/10.1126/science.1087919.

[26] van der Laan LN, de Ridder DT, Viergever MA, Smeets PA. The first taste is always with the eyes: a meta-analysis on the neural correlates of processing visual food cues. NeuroImage 2011;55: 296–303. http://dx.doi.org/10.1016/j.neuroimage.2010.11.055.

[27] Rudenga KJ, Sinha R, Small DM. Acute stress potentiates brain response to milkshake as a function of body weight and chronic stress. Int J Obes 2012;37:309–16. http://dx.doi.org/10.1038/ijo.2012.39.

[28] Li J, An R, Zhang Y, Li X, Wang S. Correlations of macronutrient-induced functional magnetic resonance imaging signal changes in human brain and gut hormone responses. Am J Clin Nutr 2012;96:275–82. http://dx.doi.org/10.3945/ajcn.112.037440.

[29] Grabenhorst F, Schulte FP, Maderwald S, Brand M. Food labels promote healthy choices by a decision bias in the amygdala. NeuroImage 2013;74:152–63. http://dx.doi.org/10.1016/j.neuroimage.2013.02.012.

[30] Sun X, Kroemer NB, Veldhuizen MG, Babbs AE, de Araujo IE, Gitelman DR, et al. Basolateral amygdala response to food cues in the absence of hunger is associated with weight gain susceptibility. J Neurosci 2015;35:7964–76. http://dx.doi.org/10.1523/JNEUROSCI.3884-14.2015.

[31] Small DM. Taste representation in the human insula. Brain Struct Funct 2010;214:551–61. http://dx.doi.org/10.1007/s00429-010-0266-9.

[32] Bender G, Veldhuizen MG, Meltzer JA, Gitelman DR, Small DM. Neural correlates of evaluative compared with passive tasting. Eur J Neurosci 2009;30:327–38. http://dx.doi.org/10.1111/j.1460-9568.2009.06819.x.

[33] De Araujo IE, Rolls ET. Representation in the human brain of food texture and oral fat. J Neurosci 2004;24:3086–93. http://dx.doi.org/10.1523/JNEUROSCI.0130-04.2004.

[34] de Araujo IE, Geha P, Small DM. Orosensory and homeostatic functions of the insular taste cortex. Chemosens Percept 2012;5: 64–79. http://dx.doi.org/10.1007/s12078-012-9117-9.

[35] Huerta CI, Sarkar PR, Duong TQ, Laird AR, Fox PT. Neural bases of food perception: coordinate-based meta-analyses of neuroimaging studies in multiple modalities. Obesity 2014;22: 1439–46. http://dx.doi.org/10.1002/oby.20659.

[36] Simmons WK, Rapuano KM, Kallman SJ, Ingeholm JE, Miller B, Gotts SJ, et al. Category-specific integration of homeostatic signals in caudal but not rostral human insula. Nat Neurosci 2013; 16:1551–2. http://dx.doi.org/10.1038/nn.3535.

[37] Balleine BW, Dickinson A. The effect of lesions of the insular cortex on instrumental conditioning: evidence for a role in incentive memory. J Neurosci 2000;20:8954–64.

[38] Small DM, Gregory MD, Mak YE, Gitelman D, Mesulam MM, Parrish T. Dissociation of neural representation of intensity and affective valuation in human gustation. Neuron 2003;39:701–11.

[39] Small DM, Zatorre RJ, Dagher A, Evans AC, Jones-Gotman M. Changes in brain activity related to eating chocolate: from pleasure to aversion. Brain J Neurol 2001;124:1720–33.

[40] Smeets PAM, de Graaf C, Stafleu A, van Osch MJP, Nievelstein RAJ, van der Grond J. Effect of satiety on brain activation during chocolate tasting in men and women 1'2'3. Am J Clin Nutr 2006;83:1297–305.

[41] Nitschke JB, Dixon GE, Sarinopoulos I, Short SJ, Cohen JD, Smith EE, et al. Altering expectancy dampens neural response to aversive taste in primary taste cortex. Nat Neurosci 2006;9: 435–42. http://dx.doi.org/10.1038/nn1645.

[42] Veldhuizen MG, Bender G, Constable RT, Small DM. Trying to detect taste in a tasteless solution: modulation of early gustatory cortex by attention to taste. Chem Senses 2007;32:569–81. http://dx.doi.org/10.1093/chemse/bjm025.

[43] Gardner MPH, Fontanini A. Encoding and tracking of outcome-specific expectancy in the gustatory cortex of alert rats. J Neurosci 2014;34:13000–17. http://dx.doi.org/10.1523/JNEUROSCI.1820-14.2014.

V. GENETIC AND HORMONAL INFLUENCES ON MOTIVATION AND SOCIAL BEHAVIOR

[44] Kusumoto-Yoshida I, Liu H, Chen BT, Fontanini A, Bonci A. Central role for the insular cortex in mediating conditioned responses to anticipatory cues. Proc Natl Acad Sci USA 2015;112: 1190–5. http://dx.doi.org/10.1073/pnas.1416573112.

[45] Small DM, Prescott J. Odor/taste integration and the perception of flavor. Exp Brain Res 2005;166:345–57. http://dx.doi.org/10.1007/s00221-005-2376-9.

[46] Kringelbach ML, O'Doherty J, Rolls ET, Andrews C. Activation of the human orbitofrontal cortex to a liquid food stimulus is correlated with its subjective pleasantness. Cereb Cortex 2003; 1991(13):1064–71.

[47] Berridge KC, Ho CY, Richard JM, Difeliceantonio AG. The tempted brain eats: pleasure and desire circuits in obesity and eating disorders. Brain Res 2010:43–64.

[48] Dimitropoulos A, Tkach J, Ho A, Kennedy J. Greater corticolimbic activation to high-calorie food cues after eating in obese vs. normal-weight adults. Appetite 2012;58:303–12. http://dx.doi.org/10.1016/j.appet.2011.10.014.

[49] Yokum S, Ng J, Stice E. Attentional bias to food images associated with elevated weight and future weight gain: an fMRI study. Obesity 2011;19:1775–83. http://dx.doi.org/10.1038/oby.2011.168.

[50] Haber SN, Behrens TEJ. The neural network underlying incentive-based learning: implications for interpreting circuit disruptions in psychiatric disorders. Neuron 2014;83:1019–39. http://dx.doi.org/10.1016/j.neuron.2014.08.031.

[51] Hollmann M, Hellrung L, Pleger B, Schlögl H, Kabisch S, Stumvoll M, et al. Neural correlates of the volitional regulation of the desire for food. Int J Obes 2012;2005(36):648–55. http://dx.doi.org/10.1038/ijo.2011.125.

[52] Goldstone AP, Prechtl de Hernandez CG, Beaver JD, Muhammed K, Croese C, Bell G, et al. Fasting biases brain reward systems towards high-calorie foods. Eur J Neurosci 2009;30: 1625–35. http://dx.doi.org/10.1111/j.1460-9568.2009.06949.x.

[53] Frankort A, Roefs A, Siep N, Roebroeck A, Havermans R, Jansen A. Neural predictors of chocolate intake following chocolate exposure. Appetite 2015;87:98–107. http://dx.doi.org/10.1016/j.appet.2014.12.204.

[54] Stice E, Yokum S, Burger KS, Epstein LH, Small DM. Youth at risk for obesity show greater activation of striatal and somatosensory regions to food. J Neurosci 2011;31:4360–6. http://dx.doi.org/10.1523/JNEUROSCI.6604-10.2011.

[55] Murdaugh DL, Cox JE, Cook EW, Weller RE. fMRI reactivity to high-calorie food pictures predicts short- and long-term outcome in a weight-loss program. NeuroImage 2012;59:2709–21. http://dx.doi.org/10.1016/j.neuroimage.2011.10.071.

[56] Carnell S, Benson L, Pantazatos SP, Hirsch J, Geliebter A. Amodal brain activation and functional connectivity in response to high-energy-density food cues in obesity. Obesity 2014;22. http://dx.doi.org/10.1002/oby.20859.

[57] Rothemund Y, Preuschhof C, Bohner G, Bauknecht H-C, Klingebiel R, Flor H, et al. Differential activation of the dorsal striatum by high-calorie visual food stimuli in obese individuals. NeuroImage 2007;37:410–21. http://dx.doi.org/10.1016/j.neuroimage.2007.05.008.

[58] Siep N, Roefs A, Roebroeck A, Havermans R, Bonte M, Jansen A. Fighting food temptations: the modulating effects of short-term cognitive reappraisal, suppression and up-regulation on mesocorticolimbic activity related to appetitive motivation. NeuroImage 2012;60:213–20. http://dx.doi.org/10.1016/j.neuroimage. 2011.12.067.

[59] Hare TA, Malmaud J, Rangel A. Focusing attention on the health aspects of foods changes value signals in vmPFC and improves dietary choice. J Neurosci 2011;31:11077–87. http://dx.doi.org/10.1523/JNEUROSCI.6383-10.2011.

[60] Plassmann H, O'Doherty J, Rangel A. Orbitofrontal cortex encodes willingness to pay in everyday economic transactions.

J Neurosci 2007;27:9984–8. http://dx.doi.org/10.1523/JNEUROSCI.2131-07.2007.

[61] Bartra O, McGuire JT, Kable JW. The valuation system: a coordinate-based meta-analysis of BOLD fMRI experiments examining neural correlates of subjective value. NeuroImage 2013;76: 412–27. http://dx.doi.org/10.1016/j.neuroimage.2013.02.063.

[62] Clithero JA, Rangel A. Informatic parcellation of the network involved in the computation of subjective value. Soc Cogn Affect Neurosci 2014;9:1289–302. http://dx.doi.org/10.1093/scan/nst106.

[63] Tang DW, Fellows LK, Dagher A. Behavioral and neural valuation of foods is driven by implicit knowledge of caloric content. Psychol Sci 2014;25:2168–76. http://dx.doi.org/10.1177/0956797614552081.

[64] Thomas JM, Higgs S, Dourish CT, Hansen PC, Harmer CJ, McCabe C. Satiation attenuates BOLD activity in brain regions involved in reward and increases activity in dorsolateral prefrontal cortex: an fMRI study in healthy volunteers. Am J Clin Nutr 2015;101:697–704. http://dx.doi.org/10.3945/ajcn.114.097543.

[65] Coveleskie K, Gupta A, Kilpatrick LA, Mayer ED, Ashe-McNalley C, Stains J, et al. Altered functional connectivity within the central reward network in overweight and obese women. Nutr Diabetes 2015;5:e148. http://dx.doi.org/10.1038/nutd.2014.45.

[66] Hare TA, Camerer CF, Rangel A. Self-control in decision-making involves modulation of the vmPFC valuation system. Science 2009;324:646–8. http://dx.doi.org/10.1126/science.1168450.

[67] Goldman RL, Canterberry M, Borckardt JJ, Madan A, Byrne TK, George MS, et al. Executive control circuitry differentiates degree of success in weight loss following gastric-bypass surgery. Obesity 2013;21:2189–96. http://dx.doi.org/10.1002/oby.20575.

[68] Hennekens CH, Andreotti F. Leading avoidable cause of premature deaths worldwide: case for obesity. Am J Med 2013;126: 97–8. http://dx.doi.org/10.1016/j.amjmed.2012.06.018.

[69] Malhotra A, Noakes T, Phinney S. It is time to bust the myth of physical inactivity and obesity: you cannot outrun a bad diet. Br J Sports Med 2015. http://dx.doi.org/10.1136/bjsports-2015-094911. bjsports–2015–094911.

[70] Fitzpatrick S, Gilbert S, Serpell L. Systematic review: are overweight and obese individuals impaired on behavioural tasks of executive functioning? Neuropsychol Rev 2013;23:138–56. http://dx.doi.org/10.1007/s11065-013-9224-7.

[71] Vainik U, Dagher A, Dubé L, Fellows LK. Neurobehavioural correlates of body mass index and eating behaviours in adults: a systematic review. Neurosci Biobehav Rev 2013;37:279–99. http://dx.doi.org/10.1016/j.neubiorev.2012.11.008.

[72] Wansink B, Sobal J. Mindless eating the 200 daily food decisions we overlook. Environ Behav 2007;39:106–23. http://dx.doi.org/10.1177/0013916506295573.

[73] Hofmann W, Dillen LV. Desire the new hot spot in self-control research. Curr Dir Psychol Sci 2012;21:317–22. http://dx.doi.org/10.1177/0963721412453587.

[74] Allan JL, Johnston M, Campbell N. Unintentional eating. What determines goal-incongruent chocolate consumption? Appetite 2010;54:422–5. http://dx.doi.org/10.1016/j.appet.2010.01.009.

[75] Allom V, Mullan B. Individual differences in executive function predict distinct eating behaviours. Appetite 2014;80:123–30. http://dx.doi.org/10.1016/j.appet.2014.05.007.

[76] Guerrieri R, Nederkoorn C, Jansen A. How impulsiveness and variety influence food intake in a sample of healthy women. Appetite 2007;48:119–22. http://dx.doi.org/10.1016/j.appet.2006.06.004.

[77] Guerrieri R, Nederkoorn C, Jansen A. The interaction between impulsivity and a varied food environment: its influence on food intake and overweight. Int J Obes 2008;2005(32):708–14. http://dx.doi.org/10.1038/sj.ijo.0803770.

[78] Hofmann W, Rauch W, Gawronski B. And deplete us not into temptation: Automatic attitudes, dietary restraint, and self-

regulatory resources as determinants of eating behavior. J Exp Soc Psychol 2007;43:497—504. http://dx.doi.org/10.1016/j.jesp.2006.05.004.

[79] Jasinska AJ, Yasuda M, Burant CF, Gregor N, Khatri S, Sweet M, et al. Impulsivity and inhibitory control deficits are associated with unhealthy eating in young adults. Appetite 2012;59:738—47. http://dx.doi.org/10.1016/j.appet.2012.08.001.

[80] Nederkoorn C, Houben K, Hofmann W, Roefs A, Jansen A. Control yourself or just eat what you like? Weight gain over a year is predicted by an interactive effect of response inhibition and implicit preference for snack foods. Health Psychol 2010;29:389—93. http://dx.doi.org/10.1037/a0019921.

[81] Vohs KD, Heatherton TF. Self-regulatory failure: a resource-depletion approach. Psychol Sci 2000;11:249—54.

[82] Allan JL, Johnston M, Campbell N. Missed by an inch or a mile? Predicting the size of intention-behaviour gap from measures of executive control. Psychol Health 2011;26:635—50. http://dx.doi.org/10.1080/08870441003681307.

[83] Hall PA, Fong GT, Epp LJ, Elias LJ. Executive function moderates the intention-behavior link for physical activity and dietary behavior. Psychol Health 2008;23:309—26. http://dx.doi.org/10.1080/14768320701212099.

[84] Johnson F, Pratt M, Wardle J. Dietary restraint and self-regulation in eating behavior. Int J Obes 2012;2005(36):665—74. http://dx.doi.org/10.1038/ijo.2011.156.

[85] Mendez MF, Licht EA, Shapira JS. Changes in dietary or eating behavior in frontotemporal dementia versus Alzheimer's disease. Am J Alzheimers Dis Other Demen 2008;23:280—5. http://dx.doi.org/10.1177/1533317507313140.

[86] Piguet O. Eating disturbance in behavioural-variant frontotemporal dementia. J Mol Neurosci 2011;45:589—93. http://dx.doi.org/10.1007/s12031-011-9547-x.

[87] Giuliani NR, Mann T, Tomiyama AJ, Berkman ET. Neural systems underlying the reappraisal of personally craved foods. J Cogn Neurosci 2014;26:1390—402. http://dx.doi.org/10.1162/jocn_a_00563.

[88] Batterink L, Yokum S, Stice E. Body mass correlates inversely with inhibitory control in response to food among adolescent girls: an fMRI study. NeuroImage 2010;52:1696—703. http://dx.doi.org/10.1016/j.neuroimage.2010.05.059.

[89] He Q, Xiao L, Xue G, Wong S, Ames SL, Schembre SM, et al. Poor ability to resist tempting calorie rich food is linked to altered balance between neural systems involved in urge and self-control. Nutr J 2014;13:92. http://dx.doi.org/10.1186/1475-2891-13-92.

[90] Lopez RB, Hofmann W, Wagner DD, Kelley WM, Heatherton TF. Neural predictors of giving in to temptation in daily life. Psychol Sci 2014;25:1337—44. http://dx.doi.org/10.1177/0956797614531492.

[91] Smeets PAM, Kroese FM, Evers C, de Ridder DTD. Allured or alarmed: counteractive control responses to food temptations in the brain. Behav Brain Res 2013;248:41—5. http://dx.doi.org/10.1016/j.bbr.2013.03.041.

[92] Jensen CD, Kirwan CB. Functional brain response to food images in successful adolescent weight losers compared with normal-weight and overweight controls. Obesity 2015;23:630—6. http://dx.doi.org/10.1002/oby.21004.

[93] Le DS, Chen K, Pannacciulli N, Gluck M, Reiman EM, Krakoff J. Reanalysis of the obesity-related attenuation in the left dorsolateral prefrontal cortex response to a satiating meal using gyral regions-of-interest. J Am Coll Nutr 2009;28:667—73.

[94] Harris A, Hare T, Rangel A. Temporally dissociable mechanisms of self-control: early attentional filtering versus late value modulation. J Neurosci 2013;33:18917—31. http://dx.doi.org/10.1523/JNEUROSCI.5816-12.2013.

[95] Weygandt M, Mai K, Dommes E, Leupelt V, Hackmack K, Kahnt T, et al. The role of neural impulse control mechanisms

for dietary success in obesity. NeuroImage 2013;83:669—78. http://dx.doi.org/10.1016/j.neuroimage.2013.07.028.

[96] Wagner DD, Altman M, Boswell RG, Kelley WM, Heatherton TF. Self-regulatory depletion enhances neural responses to rewards and impairs top-down control. Psychol Sci 2013;24:2262—71. http://dx.doi.org/10.1177/0956797613492985.

[97] Sutin AR, Terracciano A. Personality traits and body mass index: modifiers and mechanisms. Psychol Health 2015:1—31. http://dx.doi.org/10.1080/08870446.2015.1082561.

[98] Keller C, Siegrist M. Does personality influence eating styles and food choices? Direct and indirect effects. Appetite 2015;84:128—38. http://dx.doi.org/10.1016/j.appet.2014.10.003.

[99] Costa PT, McCrae RR. Revised NEO personality inventory (NEO PI-R) and NEO five-factor inventory (NEO-FFI). Psychological Assessment Resources; 1992.

[100] McCrae RR, Löckenhoff CE. Self-regulation and the five-factor model of personality traits. In: Hoyle RH, editor. Handbook of personality and self-regulation. Wiley-Blackwell; 2010. pp. 145—68. http://onlinelibrary.wiley.com/doi/10.1002/9781444318111.ch7/summary.

[101] DeYoung CG, Hirsh JB, Shane MS, Papademetris X, Rajeevan N, Gray JR. Testing Predictions from personality neuroscience. Brain structure and the big five. Psychol Sci 2010;21:820—8. http://dx.doi.org/10.1177/0956797610370159.

[102] Heaven PCL, Mulligan K, Merrilees R, Woods T, Fairooz Y. Neuroticism and conscientiousness as predictors of emotional, external, and restrained eating behaviors. Int J Eat Disord 2001;30:161—6. http://dx.doi.org/10.1002/eat.1068.

[103] O'Connor DB, Conner M, Jones F, McMillan B, Ferguson E. Exploring the benefits of conscientiousness: an investigation of the role of daily stressors and health behaviors. Ann Behav Med 2009;37:184—96. http://dx.doi.org/10.1007/s12160-009-9087-6.

[104] Raynor DA, Levine H. Associations between the five-factor model of personality and health behaviors among college students. J Am Coll Health 2009;58:73—82. http://dx.doi.org/10.3200/JACH.58.1.73-82.

[105] Gartland N, O'Connor DB, Lawton R, Ferguson E. Investigating the effects of conscientiousness on daily stress, affect and physical symptom processes: a daily diary study. Br J Health Psychol 2014;19:311—28. http://dx.doi.org/10.1111/bjhp.12077.

[106] O'Connor DB, O'Connor RC. Perceived changes in food intake in response to stress: the role of conscientiousness. Stress Health 2004;20:279—91.

[107] Wansink B, Payne CR. Eating behavior and obesity at Chinese buffets. Obesity 2008;16:1957—60. http://dx.doi.org/10.1038/oby.2008.286.

[108] Wansink B, van Ittersum K, Painter JE. Ice cream illusions bowls, spoons, and self-served portion sizes. Am J Prev Med 2006;31:240—3. http://dx.doi.org/10.1016/j.amepre.2006.04.003.

[109] Levitsky DA, Youn T. The more food young adults are served, the more they overeat. J Nutr 2004;134:2546—9.

[110] Dagher A. The neurobiology of appetite: hunger as addiction. Int J Obes 2009;33:S30—3. http://dx.doi.org/10.1038/ijo.2009.69.

[111] Hays NP, Roberts SB. Aspects of eating behaviors "disinhibition" and "restraint" are related to weight gain and BMI in women. Obesity 2008;16:52—8. http://dx.doi.org/10.1038/oby.2007.12.

[112] Jansen A, Theunissen N, Slechten K, Nederkoorn C, Boon B, Mulkens S, et al. Overweight children overeat after exposure to food cues. Eat Behav 2003;4:197—209. http://dx.doi.org/10.1016/S1471-0153(03)00011-4.

[113] Wardle J. Conditioning processes and cue exposure in the modification of excessive eating. Addict Behav 1990;15:387—93.

[114] Brooks SJ, Cedernaes J, Schiöth HB. Increased prefrontal and parahippocampal activation with reduced dorsolateral prefrontal and insular cortex activation to food images in obesity: a meta-

analysis of fMRI studies. PLoS One 2013;8:e60393. http://dx.doi.org/10.1371/journal.pone.0060393.

[115] Pursey KM, Stanwell P, Callister RJ, Brain K, Collins CE, Burrows TL. Neural responses to visual food cues according to weight status: a systematic review of functional magnetic resonance imaging studies. Front Nutr 2014;1:7. http://dx.doi.org/10.3389/fnut.2014.00007.

[116] Salem V, Dhillo WS. IMAGING IN ENDOCRINOLOGY: The use of functional MRI to study the endocrinology of appetite. Eur J Endocrinol 2015;173:R59−68. http://dx.doi.org/10.1530/EJE-14-0716.

[117] Davis JF, Choi DL, Benoit SC. Insulin, leptin and reward. Trends Endocrinol Metab 2010;21:68−74. http://dx.doi.org/10.1016/j.tem.2009.08.004.

[118] Kullmann S, Heni M, Veit R, Ketterer C, Schick F, Häring H-U, et al. The obese brain: association of body mass index and insulin sensitivity with resting state network functional connectivity. Hum Brain Mapp 2012;33:1052−61. http://dx.doi.org/10.1002/hbm.21268.

[119] Kroemer NB, Krebs L, Kobiella A, Grimm O, Vollstädt-Klein S, Wolfensteller U, et al. (Still) longing for food: insulin reactivity modulates response to food pictures. Hum Brain Mapp 2013; 34:2367−80. http://dx.doi.org/10.1002/hbm.22071.

[120] Labouebe G, Liu S, Dias C, Zou H, Wong JCY, Karunakaran S, et al. Insulin induces long-term depression of ventral tegmental area dopamine neurons via endocannabinoids. Nat Neurosci 2013;16:300−8. http://dx.doi.org/10.1038/nn.3321.

[121] Wallner-Liebmann S, Koschutnig K, Reishofer G, Sorantin E, Blaschitz B, Kruschitz R, et al. Insulin and hippocampus activation in response to images of high-calorie food in normal weight and obese adolescents. Obesity 2010;18:1552−7. http://dx.doi.org/10.1038/oby.2010.26.

[122] Kullmann S, Heni M, Fritsche A, Preissl H. Insulin action in the human brain: evidence from neuroimaging studies. J Neuroendocrinol 2015. http://dx.doi.org/10.1111/jne.12254. n/a−n/a.

[123] Guthoff M, Grichisch Y, Canova C, Tschritter O, Veit R, Hallschmid M, et al. Insulin modulates food-related activity in the central nervous system. J Clin Endocrinol Metab 2010;95: 748−55. http://dx.doi.org/10.1210/jc.2009-1677.

[124] Luo S, Monterosso JR, Sarpelleh K, Page KA. Differential effects of fructose versus glucose on brain and appetitive responses to food cues and decisions for food rewards. Proc Natl Acad Sci USA 2015; 112:6509−14. http://dx.doi.org/10.1073/pnas.1503358112.

[125] Farooqi IS, Bullmore E, Keogh J, Gillard J, O'Rahilly S, Fletcher PC. Leptin regulates striatal regions and human eating behavior. Science 2007;317:1355. http://dx.doi.org/10.1126/science.1144599.

[126] Grosshans M, Vollmert C, Vollstädt-Klein S, Tost H, Leber S, Bach P, et al. Association of leptin with food cue-induced activation in human reward pathways. Arch Gen Psychiatry 2012;69: 529−37. http://dx.doi.org/10.1001/archgenpsychiatry.2011.1586.

[127] Baicy K, London ED, Monterosso J, Wong M-L, Delibasi T, Sharma A, et al. Leptin replacement alters brain response to food cues in genetically leptin-deficient adults. Proc Natl Acad Sci USA 2007;104:18276−9. http://dx.doi.org/10.1073/pnas.0706481104.

[128] Farr OM, Fiorenza C, Papageorgiou P, Brinkoetter M, Ziemke F, Koo B-B, et al. Leptin therapy alters appetite and neural responses to food stimuli in brain areas of leptin sensitive subjects without altering brain structure. J Clin Endocrinol Metab 2014. http://dx.doi.org/10.1210/jc.2014-2774. jc20142774.

[129] Fulton S, Pissios P, Manchon RP, Stiles L, Frank L, Pothos EN, et al. Leptin regulation of the mesoaccumbens dopamine pathway. Neuron 2006;51:811−22. http://dx.doi.org/10.1016/j.neuron.2006.09.006.

[130] Aotani D, Ebihara K, Sawamoto N, Kusakabe T, Aizawa-Abe M, Kataoka S, et al. Functional magnetic resonance imaging analysis of food-related brain activity in patients with lipodystrophy undergoing leptin replacement therapy. J Clin Endocrinol Metab 2012;97:3663−71. http://dx.doi.org/10.1210/jc.2012-1872.

[131] Rosenbaum M, Sy M, Pavlovich K, Leibel RL, Hirsch J. Leptin reverses weight loss-induced changes in regional neural activity responses to visual food stimuli. J Clin Invest 2008;118: 2583−91. http://dx.doi.org/10.1172/JCI35055.

[132] Müller TD, Nogueiras R, Andermann ML, Andrews ZB, Anker SD, Argente J, et al. Ghrelin. Mol Metab 2015;4:437−60. http://dx.doi.org/10.1016/j.molmet.2015.03.005.

[133] Abizaid A, Liu ZW, Andrews ZB, Shanabrough M, Borok E, Elsworth JD, et al. Ghrelin modulates the activity and synaptic input organization of midbrain dopamine neurons while promoting appetite. J Clin Invest 2006;116:3229−39. http://dx.doi.org/10.1172/JCI29867.

[134] Cummings DE, Purnell JQ, Frayo RS, Schmidova K, Wisse BE, Weigle DS. A preprandial rise in plasma ghrelin levels suggests a role in meal initiation in humans. Diabetes 2001;50:1714−9. http://dx.doi.org/10.2337/diabetes.50.8.1714.

[135] Cummings DE, Weigle DS, Frayo RS, Breen PA, Ma MK, Dellinger EP, et al. Plasma ghrelin levels after diet-induced weight loss or gastric bypass surgery. N Engl J Med 2002;346: 1623−30. http://dx.doi.org/10.1056/NEJMoa012908.

[136] Kroemer NB, Krebs L, Kobiella A, Grimm O, Pilhatsch M, Bidlingmaier M, et al. Fasting levels of ghrelin covary with the brain response to food pictures. Addict Biol 2013;18:855−62. http://dx.doi.org/10.1111/j.1369-1600.2012.00489.x.

[137] Goldstone AP, Prechtl CG, Scholtz S, Miras AD, Chhina N, Durighel G, et al. Ghrelin mimics fasting to enhance human hedonic, orbitofrontal cortex, and hippocampal responses to food. Am J Clin Nutr 2014;99:1319−30. http://dx.doi.org/10.3945/ajcn.113.075291.

[138] Spencer SJ, Xu L, Clarke MA, Lemus M, Reichenbach A, Geenen B, et al. Ghrelin regulates the hypothalamic-pituitary-adrenal axis and restricts anxiety after acute stress. Biol Psychiatry 2012;72: 457−65. http://dx.doi.org/10.1016/j.biopsych.2012.03.010.

[139] Rouach V, Bloch M, Rosenberg N, Gilad S, Limor R, Stern N, et al. The acute ghrelin response to a psychological stress challenge does not predict the post-stress urge to eat. Psychoneuroendocrinology 2007;32:693−702. http://dx.doi.org/10.1016/j.psyneuen.2007.04.010.

[140] Geliebter A, Carnell S, Gluck ME. Cortisol and ghrelin concentrations following a cold pressor stress test in overweight individuals with and without night eating. Int J Obes 2013;2005(37): 1104−8. http://dx.doi.org/10.1038/ijo.2012.166.

[141] Monteleone P, Tortorella A, Scognamiglio P, Serino I, Monteleone AM, Maj M. The acute salivary ghrelin response to a psychosocial stress is enhanced in symptomatic patients with bulimia nervosa: a pilot study. Neuropsychobiology 2012;66: 230−6. http://dx.doi.org/10.1159/000341877.

[142] Saegusa Y, Takeda H, Muto S, Nakagawa K, Ohnishi S, Sadakane C, et al. Decreased plasma ghrelin contributes to anorexia following novelty stress. Am J Physiol Endocrinol Metab 2011;301:E685−96. http://dx.doi.org/10.1152/ajpendo.00121.2011.

[143] Morton GJ, Meek TH, Schwartz MW. Neurobiology of food intake in health and disease. Nat Rev Neurosci 2014;15:367−78. http://dx.doi.org/10.1038/nrn3745.

[144] Begg DP, Woods SC. The endocrinology of food intake. Nat Rev Endocrinol 2013;9:584−97.

[145] Flint A, Raben A, Ersbøll AK, Holst JJ, Astrup A. The effect of physiological levels of glucagon-like peptide-1 on appetite, gastric emptying, energy and substrate metabolism in obesity. Int J Obes Relat Metab Disord 2001;25:781−92. http://dx.doi.org/10.1038/sj.ijo.0801627.

[146] Verdich C, Flint A, Gutzwiller J-P, Näslund E, Beglinger C, Hellström PM, et al. A meta-analysis of the effect of glucagon-like peptide-1 (7−36) amide on ad libitum energy intake in humans. J Clin Endocrinol Metab 2001;86:4382−9. http://dx.doi.org/10.1210/jcem.86.9.7877.

[147] Vilsbøll T, Christensen M, Junker AE, Knop FK, Gluud LL. Effects of glucagon-like peptide-1 receptor agonists on weight loss: systematic review and meta-analyses of randomised controlled trials. BMJ 2012;344:d7771. http://dx.doi.org/10.1136/bmj.d7771.

[148] van Bloemendaal L, ten Kulve JS, La Fleur SE, Ijzerman RG, Diamant M. Effects of glucagon-like peptide 1 on appetite and body weight: focus on the CNS. J Endocrinol 2014;221. http://dx.doi.org/10.1530/JOE-13-0414.

[149] De Silva A, Salem V, Long CJ, Makwana A, Newbould RD, Rabiner EA, et al. The gut hormones PYY 3-36 and GLP-1 7-36 amide reduce food intake and modulate brain activity in appetite centers in humans. Cell Metab 2011;14:700−6. http://dx.doi.org/10.1016/j.cmet.2011.09.010.

[150] van Bloemendaal L, Veltman DJ, Ten Kulve JS, Groot PFC, Ruhé HG, Barkhof F, et al. Brain reward-system activation in response to anticipation and consumption of palatable food is altered by glucagon-like peptide-1 receptor activation in humans. Diabetes Obes Metab 2015. http://dx.doi.org/10.1111/dom.12506.

[151] van Bloemendaal L, IJzerman RG, Ten Kulve JS, Barkhof F, Konrad RJ, Drent ML, et al. GLP-1 receptor activation modulates appetite- and reward-related brain areas in humans. Diabetes 2014;63:4186−96. http://dx.doi.org/10.2337/db14-0849.

[152] Batterham RL, Ffytche DH, Rosenthal JM, Zelaya FO, Barker GJ, Withers DJ, et al. PYY modulation of cortical and hypothalamic brain areas predicts feeding behaviour in humans. Nature 2007;450:106−9. http://dx.doi.org/10.1038/nature06212.

[153] Vainik U, Neseliler S, Konstabel K, Fellows LK, Dagher A. Eating traits questionnaires as a continuum of a single concept. Uncontrolled eating. Appetite 2015;90:229−39. http://dx.doi.org/10.1016/j.appet.2015.03.004.

[154] Rehm CD, Monsivais P, Drewnowski A. Relation between diet cost and healthy eating index 2010 scores among adults in the United States 2007−2010. Prev Med 2015;73:70−5. http://dx.doi.org/10.1016/j.ypmed.2015.01.019.

33

Perspectives

J.-C. Dreher[1], L. Tremblay[1], W. Schultz[2]

[1]Institute of Cognitive Science (CNRS), Lyon, France; [2]University of Cambridge, Cambridge, United Kingdom

Abstract

Discovering how the brain makes decisions is one of the most exciting challenges of neurosciences that has emerged in recent years. The evolution of the field of decision neuroscience has benefited from the advance of novel technological capabilities in neurosciences, and the pace at which these capabilities have been developed has accelerated dramatically in the past decade.

Discovering how the brain makes decisions is one of the most exciting challenges of neurosciences that has emerged in recent years. The evolution of the field of decision neuroscience has benefited from the advance of novel technological capabilities in neurosciences, and the pace at which these capabilities have been developed has accelerated dramatically since 2005.

It is certainly difficult to predict what will be the most exciting developments in decision neuroscience in the future and somewhat arbitrary to organize potential perspectives along a coherent line. We have asked the contributors to this book to give us their respective perspectives for developments in their research domains. We have taken the liberty to build these perspectives based on these views and along lines outlined in a neuroscience report from the Brain Research through Advancing Innovative Neurotechnologies (BRAIN) Initiative [1]. This BRAIN Initiative should help us develop and apply new tools and technologies to understand the brain at multiple levels. In parallel, the Human Brain Project from the EU, the Brain/MINDS Project from Japan (Brain Mapping by Integrated Neurotechnologies for Disease Studies), CanadaBrain, and a national brain project under way in China should also foster technological innovation for great discoveries that should lead to a revolution in our understanding of how the brain makes decisions, from a multilevel perspective.

IDENTIFYING FUNDAMENTAL COMPUTATIONAL PRINCIPLES: PRODUCE CONCEPTUAL FOUNDATIONS FOR UNDERSTANDING THE BIOLOGICAL BASIS OF MENTAL PROCESSES THROUGH DEVELOPMENT OF NEW THEORETICAL AND DATA ANALYSIS TOOLS

Theory and mathematical modeling are advancing our understanding of complex, nonlinear brain functions where human intuition fails. New kinds of data are accruing at increasing rates, mandating new methods of data analysis and interpretation. To enable progress in theory and data analysis, decision neuroscience will need to foster collaborations between experimentalists and researchers from physics, mathematics, engineering, and computer science.

One general approach widely used in the field of fMRI research and more recently applied to monkey electrophysiology is to use a so-called model-based approach, allowing us to identify the computations performed by a given brain region. This approach selects the best model fitting behavior among a set of models and allows us to regress brain activity with output parameters from these models. One classical model-based fMRI approach concerning learning of basic stimulus—reinforcer associations used prediction errors as regressors. Similar model-based fMRI approaches have been used to study social learning, such as learning social hierarchies based on victories and defeats in a competitive game, or modeling of strategic reasoning (see Ligneul and Dreher, Chapter 17; Palminteri and Pessiglione, Chapter 23; and Lee, Chapter 18).

One fundamental theoretical view about the brain, put forward by leading researchers such as Karl Friston and Rajesh Rao, is that the brain performs Bayesian computations in general, and particularly when making decisions. According to this view, decision-making and action selection are treated as an inference problem solving the problem of selecting behavioral sequences or policies. Choices are based upon beliefs about alternative policies, whereby the most likely policy minimizes the difference between attainable and desired outcomes. Policies are then selected under the prior belief that they minimize the difference (relative entropy) between a probability distribution over states that can be reached and states that agents believe they should occupy. Future developments in the field of decision neuroscience will be to test this Bayesian view of the brain in various contexts, not only for perceptual and value-based decisions but also for social decision-making. Perspectives for research include the articulation of neurocomputational definitions of value coding with a general Bayesian brain perspective. The pioneering work of Karl Friston has been developed along this line but will need to be extended to find predictable experimental validations.

Yet another perspective for future research is to extend classical reinforcement learning approaches to social decision-making (see Ligneul and Dreher, Chapter 17) and strategic reasoning (see Lee, Chapter 18). Social interactions are often repeated in a particular setting, making it possible for decision-makers to improve their strategies through experience. Therefore, the exact nature of learning algorithms utilized during iterative social decision-making and the corresponding neural substrates are important topics for psychological and neurobiological research. Previous research has shown that humans and nonhuman primates rely on a dynamic mixture of multiple learning algorithms for both social and nonsocial decision-making. For simple, model-free reinforcement learning, strategies are revised exclusively based on the observed outcomes from previously chosen actions, whereas for model-based reinforcement learning and belief learning, observed behaviors of other decision-makers and inferences about them also influence future choices. These different types of learning algorithms might be implemented in different regions of the association cortex and basal ganglia, but how the neuronal activity in each of these brain areas contributes to the specific types of computations for learning remains poorly understood. For example, how the brain can update the values for multiple actions through mental simulation is difficult to study, because the activity related to such simulation may not be tightly linked to any observable sensory and motor events. In addition, although the brain must continuously support multiple learning algorithms,

how the outputs of various learning algorithms are combined and how potential conflicts between them get resolved need to be investigated further.

UNDERSTANDING THE FUNCTIONAL ORGANIZATION OF THE PREFRONTAL CORTEX AND THE NATURE OF THE COMPUTATIONS PERFORMED IN VARIOUS SUBREGIONS: VALUE-CODING COMPUTATIONS

The subdivisions of the prefrontal cortex and the computations performed by these subregions will be key to providing a mechanistic understanding of decision-making. For example, the roles of subdivisions of the medial and lateral orbitofrontal cortex will need to be specified, together with their participation in multiple modulatory loops with other important structures such as the nucleus accumbens, ventral pallidum, amygdala, and hypothalamus, as well as modulation with autonomic input from the gut (see Dagher, Chapter 32).

Similarly, functional divisions of the dorsolateral prefrontal cortex will need to be further characterized. This brain region, considered the highest level of the executive hierarchy, temporally coordinates the perception—action cycle by means of its cognitive executive functions upon the posterior cortex (see Fuster, Chapter 8). In addition, the dorsolateral prefrontal cortex has the capability of anticipating (predicting) perception, action, and outcome; this confers to that cortex the functions of planning and preadapting that are critical for effective decision-making. The ventromedial prefrontal cortex closes the perception—action cycle by collecting neural feedback from reward, monitoring of outcome, and risk assessment; it also has predictive capability of anticipated reward. All our decisions are to some degree Bayesian, based on the updating of prior hypotheses of perception, action, or outcome, whether their "database" is conscious, unconscious, or intuitive; therefore, any reasonable computational neuroscience of decision-making should include probability as an essential variable.

One key organizing concept in the field of decision neuroscience is the concept of value. Understanding the computational principles of value coding in the brain has received considerable attention from researchers to understand the neurobiological basis of decision-making. This progress has illuminated both where decision processing occurs in the brain and what information is represented in relevant neural activity. For example, neurophysiological and neuroimaging studies have identified specific brain areas involved in option valuation and selection, including the frontal and parietal cortices, amygdala, and basal ganglia including midbrain dopamine neurons, ventral striatum, and pallidum. Neural activity in these areas has been shown to correlate with diverse decision variables relevant to choice behavior, such as reward magnitude, risk, ambiguity, and delay to reinforcement. A central principle derived from this research is that information about the idiosyncratic subjective value of choice options is represented in the neural activity of decision-related brain areas and that it is this idiosyncratic representation that appears to drive actual choice behavior. However, a critical aspect of decision-making processes (and one oddly relevant to economics and psychology) remains largely unexplored: how neural circuits represent value information. In an information-processing system, the form of information representation is a key intermediate level mediating the link between low-level implementation and high-level goals. Information coding is a particularly significant issue for biological systems, which face inherent constraints such as energetic costs and biophysical limitations. Because such constraints limit the information-coding capacity of neural systems, they require a transformation between the input (the variable to be encoded) and the output (the neural activity representing that variable) of a neural circuit. For example, the representation of the vast range of potential rewarding outcomes with the finite dynamic range of neural activity necessitates a compressive input–output computation that can have significant implications for what we choose and when we choose it.

Thus, to understand decision-related input–output functions (i.e., value coding), it will be critical for studies not simply to demonstrate correlation between neural activity and value but to quantify the precise relationships between the two. Experimental studies have begun to quantify these neural value-coding computations in brain regions such as the orbitofrontal and posterior parietal cortices. A notable finding of this initial work is that these value input–output functions are flexible and dynamic, changing in very specific ways in response to contextual influences such as the architecture of the choice set a decision-maker faces and the history of past rewards encountered by that decision-maker. Importantly, this contextual value coding is, at least in part, mediated by well-described computations such as divisive normalization that are prominent in sensory processing, an observation arguing for a general mechanism for information coding in the brain.

In addition to identifying, quantifying, and modeling these value-coding computations, two specific directions are important targets for future research. First,

how are value-coding computations related to the structure and connectivity of the underlying biological circuits? One important approach to answering this question will be the examination of various circuit components, including cells with different functional roles (i.e., excitation versus inhibition), laminar locations, and connectivity patterns. New devices and techniques, such as large electrode arrays and optogenetics, will be crucial to this process. Another promising approach in this direction is a dynamical analysis of neural activity, focusing on fast-timescale (i.e., millisecond level) changes in firing rates rather than activity averaged over long windows; such dynamics can reveal key details about the functional connectivity of neural circuits and the resulting patterns of information flow. Second, how do value-coding computations affect choice behavior? Given the inherent constraints of information processing in neural circuits, biological decision-making can never reach the optimality predicted by normative models that have no real biological constraints. Quantifying the relationship between value-coding computations and choice behavior will illuminate both the constraints faced by biological choice systems and how neural computational algorithms compensate for those constraints.

DEMONSTRATING CAUSALITY: LINKING BRAIN ACTIVITY TO BEHAVIOR BY DEVELOPING AND APPLYING PRECISE INTERVENTIONAL TOOLS THAT CHANGE NEURAL CIRCUIT DYNAMICS

To enable the immense potential of circuit manipulation, a new generation of tools for optogenetics, chemogenetics, and biochemical and electromagnetic modulation should be developed for use in animals and eventually in human patients.

Since the pioneering work by Wolfram Schultz and colleagues, we have learned a great deal about the nature of dopamine responses during learning. In particular, we know that dopamine neurons signal reward-prediction error, or the difference between the reward that an animal expects and the reward it actually receives. This signal is thought to reinforce rewarding actions and suppress alternative actions, potentially through corticobasal ganglia loops defined by expression of different dopamine receptors. However, we are only at the beginning stages of understanding how dopamine neurons calculate these responses. Given the number of different possible sources of input, how do dopamine neurons converge on such similar prediction error responses? In what ways are dopamine neurons homogeneous versus heterogeneous? Are they involved in learning from punishments as well as rewards? Furthermore, there are many unanswered questions about how dopamine release affects downstream circuits in vivo. What are the differential roles of phasic versus tonic dopamine firing in motivating learning and behavior? What is the effect of dopamine release on striatal and cortical neurons in vivo? How do striatal D1 and D2 neurons interact during behavior? What types of learning require dopamine, and what types are dopamine independent? To address these fundamental questions, newly developed molecular, genetic, and recording techniques will be critical.

By directly activating and inhibiting populations of neurons in a behavioral context, neuroscience is progressing from correlative measures to understanding of causal brain regions [transcranial magnetic stimulation (TMS), neuropsychology]. Methods such as TMS or transcranial direct current stimulation (tDCS) are thus likely (see Ruff, Chapter 19) to establish causal mechanisms for a given brain region, complementing classical neuropsychological approaches in patients with focal brain lesions (see Fellows, Chapter 22). A central challenge for a neuropsychological perspective on the role of the prefrontal cortex in value-based decision-making is to continue to dissect decision processes at the level of brain mechanisms. We have general guides to this now: clearly there are specific regions within the frontal lobes, for example, contributing in specific ways to value-based choice. However, the mechanisms that are engaged remain unclear, in part because the component processes of decision-making remain ill-defined, with likely multiple routes to decision-making in any given situation. We need to take advantage of converging methods to provide robust tests of well-specified, mechanistic models of decision-making. This of course is true for cognitive neuroscience in general, but it seems particularly true for decision neuroscience, in which, for the most part, models remain very general. Progress in the neuropsychological study of decision-making requires good behavioral measures of the constructs of interest. Although the past several years of work now better equip us in this regard, there is still much to be done. Creative approaches that go beyond button-press choices and reaction times, such as eye tracking, autonomic measures, and assessments of physical and cognitive effort, hold promise for uncovering the "microbehaviors" underlying value assessment and choice. It is also increasingly clear that we cannot take an isolationist perspective on decision-making. Decision behaviors do not emerge fully formed from some specialized "economic" module of the brain, but rather are interlinked with attention, memory,

social—emotional, and action-selection processes. A better understanding of these interactions will accelerate advances, particularly as many of these related processes are much more thoroughly studied. Finally, decision neuroscience must aim to understand value-based choice broadly construed: in economic, but also political, social, and esthetic contexts. Testing the generality of explanatory models across the whole gamut of motivated behavior will, in the end, yield the most powerful insights.

Finally, causality can also be assessed using computational models, which allow researchers to assess probabilistic causality in humans. Building on theories of nonlinear dynamical systems, whole-brain computational models have been used to efficiently characterize network-level communication across distributed sets of brain areas (i.e., functional connectivity) to investigate the spatiotemporal dynamics of brain organization and complex cognitive architectures [2]. This dynamic characterization can incorporate time-dependent activity operating on varying timescales, which may capture a more complete picture of the spatiotemporal properties inherent to decision-making.

MAPS AT MULTIPLE SCALES: GENERATE CIRCUIT DIAGRAMS THAT VARY IN RESOLUTION FROM SYNAPSES TO THE WHOLE BRAIN

It is increasingly possible to map connected neurons in local circuits and distributed brain systems, enabling an understanding of the relationship between neuronal structure and function. It is now possible to envision improved technologies—faster, less expensive, scalable—for anatomic reconstruction of neural circuits at all scales, from noninvasive whole human brain imaging to dense reconstruction of synaptic inputs and outputs at the subcellular level.

For example, understanding of the circuit diagrams that underlie impulsivity, risky choice, and impulse control disorders is now possible to attain based on animal models. Impulsivity has emerged as a major dimensional construct in psychiatry with relevance to a range of disorders from addiction to attention deficit hyperactivity disorder (ADHD) and from Parkinson's disease to depression, mania, and dementia. As a heritable, disorder-associated trait, impulsivity is broadly acknowledged to affect the quality of decision-making through effects on risk sensitivity, subjective value-based judgments (e.g., temporal discounting of delayed rewards), and cognitive control mechanisms responsible for the inhibition of ongoing behavior. Several decades of research in humans and experimental animals have revealed divergent but often interacting neural circuitry that underlies various impulsivity phenotypes, including the inability to await rewards, inability to terminate initiated behavior, preference for risky choice, or tendency to incompletely process information prior to decision-making. Yet formidable challenges lie ahead. For example, at present we lack a detailed understanding of the biological origins and neural circuitry of trait impulsivity, including environmental interactions, and how these collectively contribute to poor impulse control. Addressing this shortfall requires preclinical scientists to study predictive biomarkers and neurodevelopmental trajectories for impulsivity in much younger animals. By continuing to explore the behavioral diversity of impulsivity and adopting translational neural imaging, genomic, and objective behavioral approaches, we expect to see further advances in our understanding of trait impulsivity. This work requires a detailed dimensional analysis of impulsivity, characterized in aggregate by variation in genes, molecules, and circuits, in addition to a therapeutic focus away from brain monoaminergic systems (e.g., in the form of medication with Ritalin for ADHD) toward novel brain mechanisms and hence new neuropharmacological targets.

THE BRAIN IN ACTION: PRODUCE A DYNAMIC PICTURE OF THE FUNCTIONING BRAIN BY DEVELOPING AND APPLYING IMPROVED METHODS FOR LARGE-SCALE MONITORING OF NEURAL ACTIVITY

One important challenge in the future will be to record dynamic neuronal activity from densely sampled—and in some test cases complete—neural networks, over long periods of time, in all areas of the brain, in both mammalian systems and diverse model organisms, while making various types of decisions. There are promising opportunities both for improving existing technologies and for developing entirely new technologies for neuronal recording, including methods based on electrodes, optics, molecular genetics, and nanoscience and encompassing various facets of brain activity.

The combination of existing techniques using multimodal neuroimaging approaches in both nonhuman and human primates is also likely to bring insights into how the brain makes decisions. For example, the combination of intracranial EEG (iEEG) recordings in patients with epilepsy (whether with single cells or macroelectrodes) and fMRI, or single/multiple-cell recordings combined simultaneously with fMRI in monkeys, should bring a better understanding of the precise temporal dynamics at the systems level. Similarly, the new PET–fMRI scanners, which allow us to map

simultaneously both radiotracers and to acquire blood oxygen level-dependent (BOLD) responses during decision-making tasks, should bring exciting new findings to the community. Converging approaches using the same paradigms with different imaging modalities (e.g., EEG or MEG) and fMRI, together with physiological measures (e.g., pupil dilation, heart beat, etc.) should allow us to specify the dynamics of decisions together with a broader view at the neurophysiological level.

Another interesting perspective from our field comes from the observation that the social environment shapes neural structures and processes, and vice versa. In Chapter 28, Fernald gave a few examples of these interrelationships using genetics and social behavior in animals. Social animals interact with others and their environment to survive and reproduce if possible. To do this, animals acquire, evaluate, and translate information about their social and physical situation into decisions about what to do next. The information gathered and the resulting decisions can profoundly alter both the behavior and the physiology of an animal. These choices in the brain are both produced by and result in a diverse array of cellular and molecular actions. The challenge is to discover where decisions are made and, in particular, what information is used to guide specific choices. With new genetic techniques, animal studies directed at understanding how the brain decides are not restricted to a limited number of "model organisms" but any animal with an interesting decision-making behavior.

THE ANALYSIS OF CIRCUITS OF INTERACTING NEURONS

The circuits of interacting neurons are particularly rich in research opportunities, with potential for revolutionary advances. This area of research represents a real knowledge gap. We can now study the brain at very high resolution by examining individual genes, molecules, synapses and neurons, or we can study large brain areas at low resolution with whole-brain imaging. The challenge remaining is what lies in between—the thousands and millions of neurons that constitute functional circuits.

One example is to understand the essential circuitry that mediates the neural bases of goal-directed action. Bradfield and Balleine point that current research in neuroscience is predominantly technique driven and, as a consequence, it can be a challenge to maintain the balance between doing what is expedient and asking questions that are worth answering. Not all recently developed techniques are equally useful in studying complex psychological capacities, something that is particularly true of studies investigating goal-directed action in animals. In such experiments the events to which the nervous system is exposed are predominantly under the animal's control rather than the experimenters', meaning, therefore, that, because the initiating and terminating conditions for actions are fluid, the dynamics of the neural processes that mediate both acquisition and subsequent performance can be very complex. The challenge for the future is to bring this complexity under control. To the extent that is achieved it may become possible to address one of the most important open questions: it is still not known with any precision what learning rules mediate the acquisition of goal-directed actions. Establishing the essential circuitry supporting this learning process should help in that regard but there are important behavioral constraints to bear in mind. For example, different learning processes appear to be engaged at different rates by different schedules of reward: ratio schedules generate more consistent goal-directed learning and higher rates of performance than interval schedules even when parameters are selected that match rates of reward delivery or interresponse times. Whether such distinctions can be captured in associative or computational terms is still an open question. A number of researchers have recently claimed that goal-directed learning is best captured, computationally, by model-based reinforcement learning, using which a model of the environment is constructed to ensure that action selection maximizes long-run future reward. However, the performance of goal-directed actions respects the causal value of an action with respect to its specific outcome, and causal value does not necessarily coincide with reward maximization. Indeed, considerable evidence suggests that animals prefer causal actions to both equally rewarding noncausal actions and to performing no actions at all. Establishing the essential circuitry that mediates goal-directed action and the computational processes implemented in that circuit that make such actions possible is one of the most important research problems and most difficult challenges for future research.

DEVELOP INNOVATIVE TECHNOLOGIES AND SIMULTANEOUS MEASURES TO UNDERSTAND HOW THE BRAIN MAKES DECISIONS

Consenting humans who are undergoing diagnostic brain monitoring or receiving neurotechnology for clinical applications provide an extraordinary opportunity for scientific research. This setting enables research on human brain function, the mechanisms of human brain disorders, the effect of therapy, and the value of diagnostics. Seizing this opportunity requires closely integrated research teams performing according to the highest ethical standards of clinical care and research. New mechanisms are needed to maximize the collection of this priceless information and ensure that it benefits both patients and science.

Examples include linking hormones and BOLD response during behavioral tasks (see Hermans and Fernandez, Chapter 30, and Lefbvre and Sirigu, Chapter 31). Another related example concerns the effects of acute stress on decision-making, which are just beginning to be understood. Such stress-induced shift from "reflective" to "reflexive" behavior may map two distinct large-scale neural systems. This mapping is based on a vast body of animal findings of effects of stress-related neuromodulators within individual brain regions. It is essential that this cross-species inference is corroborated in humans. There is, however, a paucity of human pharmacological work detailing region-specific effects and time-dependent effects of catecholamines such as dopamine and norepinephrine. In particular, we highlight the lack of human work on stress-induced dopamine release, which to our knowledge is limited to one seminal paper showing increased dopamine release using PET. Understanding the specific roles of dopamine and norepinephrine in the central response to stressors will be critical to developing an understanding not only of immediate effects on decision-making processes, but also of the specific vulnerabilities that occur in response to acute stress in the realm of psychopathology. Regarding corticosteroids, a fruitful road for further exploration will be to specify the role of corticosteroids in limiting or terminating the acute response to stressors and promoting "reflective" types of decision-making to enhance long-term adaptation. In particular, this role of corticosteroids has not been explored fully in relation to stress-related psychopathology. In investigating this, it will be important to distinguish the roles of baseline shifts and phasic responses to stressors. One particularly promising avenue is to further explore the potential of corticosteroids in enhancing various forms of extinction-based therapy. Another large gap in our knowledge is how rapid and comprehensive shifts in neural activity are generated across large-scale neural systems. We highlighted the potential contribution of stress-related neuromodulators to this process, but these probably have downstream effects on the balance between excitatory and inhibitory neurotransmitters, which remain poorly understood. Finally, the combination of basic neuroscience work with network-level analyses using functional neuroimaging in humans has yielded important new insights about the architecture of human cognition and its regulation at various levels of stress and arousal. One important future challenge is to translate these network-level findings back to basic neuroscience, in which these network-level effects can be studied in much more spatiotemporal detail using, for instance, in vivo electrophysiological recordings and optogenetic manipulations.

ADVANCING HUMAN DECISION NEUROSCIENCE: UNDERSTANDING NEUROLOGICAL/PSYCHIATRIC DISORDERS AND TREATING BRAIN DISEASES

Clinical developments coming from the field of decision neuroscience and reward processing are vast and likely to bring new promises. For example, in Parkinson's disease, the main current treatment is the dopamine precursor drug, L-dopa, but its efficacy decreases over time while severe side effects increase. Understanding the brain's motor circuits and decisional system with deep brain stimulation, which can restore motor circuit function in patients with Parkinson's disease for up to several years, may also help to understand how we form a decision. Which factors specifically involve the inhibitory cortical network interacting with subthalamic nucleus (STN) function in the decision-making process? Is it the decision conflict per se or other factors such as choice difficulty, appetitive/aversive valence of the choices, or information integration that influence STN activity and adjustment of response thresholds? Changing dynamically the response threshold might be a universal function in decision conflict or might be task specific. Therefore it has to be shown if different neuronal circuits/mechanisms are involved, for example, adopting risk-taking strategies or acting under time pressure along the line of an accuracy—speed trade-off. In a clinical perspective the exact electrode position in relation to changes in inhibitory control should give us further insights into the exact fiber tracts that are involved in the adjustment of response threshold. High-frequency stimulation has a negative impact on decision threshold. In analogy it should be clarified if low-frequency stimulation improves the decision-making process, reflecting the other side. Similar research concerning deep brain stimulation of various areas into brain circuits for mood and emotion have the potential to advance psychiatry in similar ways.

As noted previously, reinforcement learning combined with model-based fMRI has proven a valuable tool to reveal the brain regions computing prediction errors during learning stimulus—reward/punishment associations. It is now possible to use such tool to understand various neurological and psychiatric diseases, such as schizophrenia. Critically, this perspective links clinical observations to a vibrant and rapidly developing cognitive neuroscience field. More complex and sophisticated models of reinforcement learning are beginning to demonstrate the importance of adaptations in key parameters such as prediction error and learning rate. By explicitly studying this adaptivity and how it may be perturbed in mental illness, we are likely to develop an ever-richer explanatory link between key symptoms of mental illness and alterations in brain, behavior, and cognition. Progress in refining our understanding in this regard could ultimately pave the way for the introduction of precision medicine (scientifically based, individually tailored treatment) interventions in psychiatry.

Similarly, understanding the neuronal bases of negative motivational behavior including avoidance will be crucial points to elucidate aversive behavior related to psychiatric disorders such as anxiety. Nonhuman primate models would be essential for preclinical study. It would be required to find a neuronal circuit for aversive behavior and observation of its abnormal state. These processes would pave the way to understanding psychiatric disorders and developing treatments.

One example comes from the field of anxiety disorders. Research on psychiatric disorders has increasingly focused on broad biological and psychological mechanisms that can confer risk for psychopathology generally speaking, with specific manifestations of disorders influenced by environmental factors experienced at different developmental time points. A huge challenge currently faced by the field is delineating what these key domains of functioning are that may confer such broad risk when disrupted, and how these disruptions are neurobiologically characterized, all to better understand who may develop these conditions and treat or ideally prevent clinical anxiety. In the search for these broad underlying mechanisms of anxiety disorders, the research domain of decision-making has been largely ignored, with most human neuroimaging studies focusing instead on the passive elicitation of fear or anxiety. While phenomenologically valid, this approach falls short in demonstrating the adaptive or maladaptive behavioral consequences of anxiety, including the choices one makes between potentially rewarding and punishing outcomes. Along with emerging investigations of value-based decision-making in anxiety and its disorders, extant data that do not explicitly probe decision-making processes provide evidence for disruptions to neurobiological mechanisms throughout the decision-making process. Future research that systematically explores alterations to specific aspects of the decision-making process and associated changes in brain function or structure, and links these changes with symptoms of anxiety and associated psychopathology, has the potential to advance our ability to diagnose, treat, and prevent the emergence of anxiety disorders.

Our understanding of brain mechanisms underlying decision-making is also likely to bring new knowledge to the understanding of drug and behavioral addictions. For example, gambling serves as a real-world example of risky decision-making and an activity that becomes excessive for some people. Chapter 27 by Clark explores what we currently know about decision-making and its underlying brain basis in gambling, with a focus on gambling disorder, the first recognized behavioral addiction in the *Diagnostic and Statistical Manual of Mental Disorders*, fifth edition. Despite long-standing discussion in behavioral economics as to why people play such games, given their negative expected value, it is only recently that researchers have begun to investigate phenomena like loss aversion and the illusion of control in groups of participants separated in terms of gambling involvement.

CONCLUSIONS

Collectively, the chapters from this book, *Decision Neuroscience*, illustrate that: (1) theories and experiments in neuroscience are helping to illuminate the mechanisms underlying decisions; (2) much remains to be done regarding complex decisions; (3) social decision neuroscience offers a special challenge of addressing more complex problems that depend on predicting the intentions of others; (4) the social environment shapes neural structures and processes, and vice versa; and (5) new experimental methods (optogenetics) or noninvasive causal methods (e.g., TMS, tDCS) will help researchers to decipher the necessary brain regions engaged in specific processes when making different types of decisions.

To conclude, this book opens up three main perspectives:

1. *Pursue human studies and nonhuman models in parallel.* The goal is to understand the human brain, but many methods and ideas are developed first in animal models, both vertebrate and invertebrate. Experiments should take advantage of the unique strengths of diverse species and experimental systems. The research on animals has been and will remain crucial to determining the neural basis of the underlying mechanisms of decision-making.
2. *Cross boundaries in interdisciplinary collaborations.* No single researcher or discovery will solve the brain's mysteries. The most exciting approaches will bridge fields, linking experiments to theories, biology to engineering, tool development to experimental application, human neuroscience to nonhuman models in innovative ways.
3. *Integrate spatial and temporal scales.* A unified view of the brain will cross spatial and temporal levels, recognizing that the nervous system consists of interacting molecules, cells, and circuits across the entire body, and important functions can occur in milliseconds or minutes, or take a lifetime.

The most important perspective of the field of decision neuroscience will be a comprehensive, mechanistic understanding of how the brain makes decisions that emerges from synergistic applications of new technologies and conceptual structures.

Reference

[1] Jorgenson LA, et al. The BRAIN Initiative: developing technology to catalyse neuroscience discovery. Philos Trans R Soc 2015; 370(1668).
[2] Deco G, Tononi G, Boly M, Kringelbach ML. Rethinking segregation and integration: contributions of whole-brain modelling. Nat Rev Neurosci July 2015;16(7):430—9.

Index

'Note: Page numbers followed by "f" indicate figures, "t" indicate tables and "b" indicate boxes.'